Trends and Challenges in Robotic Applications

Trends and Challenges in Robotic Applications

Editors

Luis Gracia
Carlos Perez-Vidal

MDPI

Basel • Beijing • Wuhan • Barcelona • Belgrade • Novi Sad • Cluj • Manchester

Editors
Luis Gracia
Universitat Politecnica de Valencia
Valencia
Spain

Carlos Perez-Vidal
Universidad Miguel Hernández de Elche
Elche
Spain

Editorial Office
MDPI
St. Alban-Anlage 66
4052 Basel, Switzerland

This is a reprint of articles from the Special Issue published online in the open access journal *Applied Sciences* (ISSN 2076-3417) (available at: https://www.mdpi.com/journal/applsci/special_issues/robotic_applications).

For citation purposes, cite each article independently as indicated on the article page online and as indicated below:

Lastname, A.A.; Lastname, B.B. Article Title. *Journal Name* **Year**, *Volume Number*, Page Range.

ISBN 978-3-0365-8634-2 (Hbk)
ISBN 978-3-0365-8635-9 (PDF)
doi.org/10.3390/books978-3-0365-8635-9

Contents

About the Editors

Luis Gracia

Luis Gracia (Prof. Dr.) received his B.Sc. degree in electronic engineering, his M.Sc. degree in control systems engineering, and his Ph.D. degree in automation and industrial computer science from the Technical University of Valencia (UPV), Spain, in 1998, 2000, and 2006, respectively. He is currently Professor at the Department of Systems Engineering and Control (DISA) of the UPV, where he has worked since 2001. His research interests include mobile robots, robotic manipulators, sliding-mode control, collaborative robots and system modeling and control.

Carlos Perez-Vidal

Carlos Perez-Vidal (Dr.) received his B.Sc. degree in industrial engineering in 1998, his M.Sc. degree in control engineering in 2000, and his Ph.D. degree in industrial technologies in 2008, from the Technical University of Valencia and the Universidad Miguel Hernández (UMH), both in Spain. He is currently Associate Professor of control and systems engineering at the UMH, where he has worked since 2001. His research interests include robotics, direct visual servoing, automation, control engineering and medical applications.

applied sciences

MDPI

Editorial

Special Issue on Trends and Challenges in Robotic Applications

Luis Gracia [1,*] and Carlos Perez-Vidal [2]

1 Instituto de Diseño y Fabricación (IDF), Universitat Politècnica de València, 46022 Valencia, Spain
2 Instituto de Investigación en Ingeniería I3E, Universidad Miguel Hernández de Elche, 03202 Elche, Spain; carlos.perez@umh.es
* Correspondence: luigraca@isa.upv.es

1. Introduction

The world of robotics has evolved rapidly in recent years, with groundbreaking advancements and innovative applications becoming increasingly prevalent. Robots are no longer limited to traditional industrial environments, and are now being utilized across different sectors, revolutionizing processes and transforming how humans perceive robotics. This Special Issue explores the latest trends and challenges in robotic applications, shedding light on how these technological advances are shaping the future of automation, with emphasis on robotics in particular.

This compendium of papers focuses on the current research fields, trends and challenges in robotic applications and includes a set of review papers that allow us to frame this Special Issue in a general context. Four of the presented articles describe areas of robotic application and their associated trends and challenges, including service robots [1], advanced applications in industry [2], multiple object tracking [3] and drone control and localization [4].

2. Fields of Application

Robots are continuously being introduced to new environments, and are their applications are becoming more diverse. This Special Issue contains articles from the following research fields: collaborative robots; service robotics; computer vision; mobile robots; and other advanced tools in robotics.

2.1. Collaborative Robots

Collaborative robots, or cobots, are designed to work alongside humans, creating safer and more efficient work environments and enabling the development of new strategic approaches to problems [5] and alternative technical solutions. Unlike their predecessors, cobots are equipped with advanced sensors and sophisticated algorithms that enable them to perceive and respond to human movements. As a result, they can assist in tasks that require human dexterity and decision making while reducing the risk of workplace accidents. From manufacturing assembly lines and assembly techniques [6] to healthcare settings, cobots are being adopted across multiple industries to enhance human capabilities.

The authors of [7] developed a human–robot collaboration system with good path-tracking accuracy. This enabled the implementation of an industrial robot with enhanced capabilities to improve system behavior.

2.2. Service Robotics

Service robots [1] are gaining popularity and significantly impacting how businesses operate and humans interact with technology. In the hospitality industry, robots are now employed to carry out room service, reception duties and concierge services. In healthcare, robots assist in patient care, including as nanoelectromechanical devices for medical applications [8], in rehabilitation support [9,10] and in surgical robotics [11]. Moreover, service

Citation: Gracia, L.; Perez-Vidal, C. Special Issue on Trends and Challenges in Robotic Applications. *Appl. Sci.* **2023**, *13*, 9131. https://doi.org/10.3390/app13169131

Received: 4 August 2023
Accepted: 9 August 2023
Published: 10 August 2023

robots are also making inroads in agriculture [12], retail and public spaces, presenting new opportunities for automation in areas that previously relied solely on human labor.

Robotic rehabilitation is a challenging field that involves a transversal flow of knowledge with other robotic technologies. For example, the similarity between lower limb rehabilitation robotics and humanoid auto-balanced walking robotics [13] lies in their focus on stability and balance. Both lower limb rehabilitation robotics and humanoid auto-balanced walking robotics prioritize maintaining stability and balance during movement. In lower limb rehabilitation robotics, ensuring the patient's safety and preventing falls are crucial aspects of the rehabilitation process. The robot's control algorithms and mechanical design are geared towards providing stable and controlled movements during therapy sessions. On the other hand, humanoid auto-balanced walking robotics aims to replicate human-like walking patterns while autonomously maintaining balance. These robots often have advanced sensors, such as inertial measurement units (IMUs), cameras and force/torque sensors, to continually assess their orientation and stability during walking. Advanced control algorithms are employed to adjust the robot's posture and foot placement in real-time, ensuring balance and stability, even on uneven or challenging terrain. Technologies and insights developed in one field can potentially inform and improve the other, fostering cross-disciplinary advancements in robotics research [14].

2.3. Computer Vision and Other Advanced Techniques

Computer vision focuses on enabling computers to interpret and understand visual information from the world, including objects and humans [15]. It involves the development of algorithms and techniques that allow machines to process, analyze and extract meaningful insights from images and videos. In recent years, computer vision has undergone significant advancements, leading to the emergence of new technologies that allow for new interaction applications; for example, in [16], a moving object is manipulated by means of batting primitive to play table tennis with a human player.

Pose estimation is a fundamental concept in computer vision that involves determining the position and orientation of an object or a camera relative to a specific coordinate system. It plays a crucial role in various applications, such as augmented reality [17], robotics, 3D scene reconstruction and human–computer interaction. In simple terms, pose estimation aims to answer the question: "Where is the object or camera located, and how is it oriented in 3D space?" To achieve this, computer vision algorithms analyze visual data, typically in the form of images or videos, and extract relevant features or keypoints from the objects of interest. Similarly, [18] presents an interesting approach to pose computation using a 3D reconstruction of data obtained using an RGB-D multi-camera. A groundbreaking piece of technology that is currently attracting attention due to its relatively low cost is Time-of-Flight (ToF) sensing, and the authors of [19] present a multi-perspective ToF laser ranging system using prisms and mirrors.

The authors of [20] discuss the significance of 6D pose estimation in industry and its application in functions like bin picking and autopilot. They highlight the evolution of various approaches, both learning-based and non-learning-based, and aim to provide an up-to-date, thorough review of methods for 6D pose estimation, as well as the challenges and future trends in this field. Their paper compares the performance of different methods and categorizes them into two types of approach: non-learning-based and learning-based.

Using data from external sensors that measure 3D/6D locations is crucial to determining robots' kinematic parameters and the transformation between the world coordinate frame and the robot base. The study presented in [21] shows that full pose measurements result in significantly smaller robot orientation errors compared to using 3D data alone, while the robot position errors remain similar in both cases.

2.4. Mobile Robots

Mobile robots, including autonomous vehicles, are an exciting and rapidly evolving technology that has the potential to revolutionize robotic transportation. As research and development in this field progress, several new topics are emerging that address the challenges and enhance the capabilities of autonomous mobile robots, including swarm robotics and its tools [22]. Examples of this flourishing field of research include [23], which is focused on automatically allocating space for parking autonomous and human-operated vehicles, and [24], which presents a control system developed to achieve mobile robot formations based on the leader–follower method. In [25], a LiDAR sensor is proposed to navigate featured environments.

A special case of this technology is robotic navigation inside tunnels and mines. The authors of [26] present a localization and navigation system for autonomous dump vehicles in tunnels. This work also presents an autonomous load system for this kind of vehicle [27].

2.5. Other Advanced Tools in Robotics

Advanced tools for robotics applications encompass a wide range of software and hardware technologies that enhance the capabilities and efficiency of robotic systems. These tools are designed to streamline development, improve control, and enable robots to carry out complex tasks with higher precision. For instance, the authors of [28] use robot audition methods in order to estimate the elevation and azimuth angles of birds' vocalizations.

In [29], the authors propose a new stochastic method that efficiently tracks the desired end-effector task-space motions for mechanisms with redundant actuation and is applicable to industrial and collaborative robots. It utilizes manipulability measures and null-space configurations to achieve better manipulability, together with a collision-free trajectory in the task-space, providing a computationally tractable alternative to optimal motion planning, and demonstrated promising results in simulations and real robot scenarios.

Kinematics is a fundamental aspect of robotics because it deals with the study of robots' mechanical link motions. The work presented in [30] discusses the challenges in selecting structural parameters for artificial neural networks, which often relies on trial-and-error procedures. It presents a design method based on neural networks and Genichi Taguchi's approach, which is applied to solving inverse kinematics in a robotic manipulator, leading to improved result accuracy with a prediction percentage above 90% and a margin of error under 5%.

3. Conclusions

The future of robotics is exciting and promising, with the discussed trends in robotic applications continuing to unfold. From collaborative robots enhancing workplace safety to AI-powered robots driving intelligent automation and service robots revolutionizing industries, the impact of robotics is becoming increasingly pervasive. As technology advances and becomes more accessible, we can expect robots to play an even more significant role in transforming various aspects of human life, contributing to a more efficient, productive, and innovative world. However, with these advancements also come ethical considerations and the need for thoughtful regulations to ensure the responsible and beneficial integration of robots into society.

4. Statistical Data

This Special Issue on "Trends and Challenges in Robotic Applications" focuses on a wide range of robotics applications. In total, 39 papers were received between 20 August 2020 and 16 February 2023. After the reviewing process, 30 papers were accepted and 9 were not accepted, indicating an acceptance ratio of 76.9%. The accepted papers originated from research centers based mainly in Latin America, Europe and China.

Author Contributions: Conceptualization, L.G. and C.P.-V.; investigation, L.G. and C.P.-V.; writing—original draft preparation, L.G. and C.P.-V.; writing—review and editing, L.G. and C.P.-V.; visualization, L.G. and C.P.-V.; supervision, L.G. and C.P.-V.; funding acquisition, L.G. and C.P.-V.; All authors have read and agreed to the published version of the manuscript.

Funding: This research was partially funded by the Spanish Government (Grant PID2020-117421RB-C21, funded by MCIN/AEI/10.13039/501100011033).

Acknowledgments: The Guest Editors would like to thank all the authors and reviewers who contributed to this Special Issue for their valuable work in disseminating and curating the information contained within it. The Guest Editors also thank the *Applied Sciences* journal for trusting them to manage this publication.

Conflicts of Interest: The authors declare no conflict of interest.

References

1. Gonzalez-Aguirre, J.A.; Osorio-Oliveros, R.; Rodríguez-Hernández, K.L.; Lizárraga-Iturralde, J.; Morales Menendez, R.; Ramírez-Mendoza, R.A.; Ramírez-Moreno, M.A.; Lozoya-Santos, J.d.J. Service Robots: Trends and Technology. *Appl. Sci.* **2021**, *11*, 10702. [CrossRef]
2. Dzedzickis, A.; Subaciute-Žemaitiene, J.; Šutinys, E.; Samukaite-Bubniene, U.; Bucinskas, V. Advanced Applications of Industrial Robotics: New Trends and Possibilities. *Appl. Sci.* **2022**, *12*, 135. [CrossRef]
3. Gad, A.; Basmaji, T.; Yaghi, M.; Alheeh, H.; Alkhedher, M.; Ghazal, M. Multiple Object Tracking in Robotic Applications: Trends and Challenges. *Appl. Sci.* **2022**, *12*, 9408. [CrossRef]
4. Yousaf, J.; Zia, H.; Alhalabi, M.; Yaghi, M.; Basmaji, T.; Shehhi, E.A.; Gad, A.; Alkhedher, M.; Ghazal, M. Drone and Controller Detection and Localization: Trends and Challenges. *Appl. Sci.* **2022**, *12*, 12612. [CrossRef]
5. Gutman, D.; Olatunji, S.; Edan, Y. Evaluating Levels of Automation in Human—Robot Collaboration at Different Workload Levels. *Appl. Sci.* **2021**, *11*, 7340. [CrossRef]
6. Ortega-Aranda, D.; Jimenez-Vielma, J.F.; Saha, B.N.; Lopez-Juarez, I. Dual-Arm Peg-in-Hole Assembly Using DNN with Double Force/Torque Sensor. *Appl. Sci.* **2021**, *11*, 6970. [CrossRef]
7. Reyes-Uquillas, D.; Hsiao, T. Compliant Human—Robot Collaboration with Accurate Path-Tracking Ability for a Robot Manipulator. *Appl. Sci.* **2021**, *11*, 5914. [CrossRef]
8. Šajic, J.L.; Langthaler, S.; Baumgartner, C. Creating a Novel Mathematical Model of the Kv10.1 Ion Channel and Controlling Channel Activity with Nanoelectromechanical Systems. *Appl. Sci.* **2022**, *12*, 3836. [CrossRef]
9. Menga, G. The Spherical Inverted Pendulum: Exact Solutions of Gait and Foot Placement Estimation Based on Symbolic Computation. *Appl. Sci.* **2021**, *11*, 1588. [CrossRef]
10. Bressi, F.; Santacaterina, F.; Cricenti, L.; Campagnola, B.; Nasto, F.; Assenza, C.; Morelli, D.; Cordella, F.; Lapresa, M.; Zollo, L.; et al. Robotic-Assisted Hand Therapy with Gloreha Sinfonia for the Improvement of Hand Function after Pediatric Stroke: A Case Report. *Appl. Sci.* **2022**, *12*, 4206. [CrossRef]
11. Rosa, M.; Liu, R.; Pitruzzello, G.; Tortora, G. A Smart Modular IoT Sensing Device for Enhancing Sensory Feedbacks in Surgical Robotics. *Appl. Sci.* **2022**, *12*, 8083. [CrossRef]
12. Mao, W.; Liu, Z.; Liu, H.; Yang, F.; Wang, M. Research Progress on Synergistic Technologies of Agricultural Multi-Robots. *Appl. Sci.* **2021**, *11*, 1448. [CrossRef]
13. Polakovic, D.; Juhás, M.; Juhásová, B.; Cervenanská, Z. Bio-Inspired Model-Based Design and Control of Bipedal Robot. *Appl. Sci.* **2022**, *12*, 10058. [CrossRef]
14. Nakamura, K.; Saga, N. A Symmetry Evaluation Method, Using Elevation Angle, for Lower Limb Movement Patterns during Sitting-to-Standing. *Appl. Sci.* **2022**, *12*, 9454. [CrossRef]
15. Khalifa, A.; Abdelrahman, A.A.; Strazdas, D.; Hintz, J.; Hempel, T.; Al-Hamadi, A. Face Recognition and Tracking Framework for Human–Robot Interaction. *Appl. Sci.* **2022**, *12*, 5568. [CrossRef]
16. Joe, H.-M.; Lee, J.; Oh, J.-H. Dynamic Nonprehensile Manipulation of a Moving Object Using a Batting Primitive. *Appl. Sci.* **2021**, *11*, 3920. [CrossRef]
17. García, A.; Solanes, J.E.; Muñoz, A.; Gracia, L.; Tornero, J. Augmented Reality-Based Interface for Bimanual Robot Teleoperation. *Appl. Sci.* **2022**, *12*, 4379. [CrossRef]
18. de Medeiros Esper, I.; Smolkin, O.; Manko, M.; Popov, A.; From, P.J.; Mason, A. Evaluation of RGB-D Multi-Camera Pose Estimation for 3D Reconstruction. *Appl. Sci.* **2022**, *12*, 4134. [CrossRef]
19. Pogacnik, L.; Munih, M. Towards a Multi-Perspective Time of Flight Laser Ranging Device Based on Mirrors and Prisms. *Appl. Sci.* **2022**, *12*, 7121. [CrossRef]
20. He, Z.; Feng, W.; Zhao, X.; Lv, Y. 6D Pose Estimation of Objects: Recent Technologies and Challenges. *Appl. Sci.* **2021**, *11*, 228. [CrossRef]
21. Franaszek, M.; Marvel, J.A. Using Full Pose Measurement for Serial Robot Calibration. *Appl. Sci.* **2022**, *12*, 3680. [CrossRef] [PubMed]
22. Aurecianus, S.; Ha, G.-H.; Park, H.-C.; Kang, T.-S. Longitudinal Mode System Identification of an Insect-like Tailless Flapping-Wing Micro Air Vehicle Using Onboard Sensors. *Appl. Sci.* **2022**, *12*, 2486. [CrossRef]
23. Wu, M.; Jiang, H.; Tan, C.-A. Automated Parking Space Allocation during Transition with both Human-Operated and Autonomous Vehicles. *Appl. Sci.* **2021**, *11*, 855. [CrossRef]

24. Hirata-Acosta, J.; Pliego-Jiménez, J.; Cruz-Hernández, C.; Martínez-Clark, R. Leader-Follower Formation Control of Wheeled Mobile Robots without Attitude Measurements. *Appl. Sci.* **2021**, *11*, 5639. [CrossRef]
25. Nguyen, P.T.-T.; Yan, S.-W.; Liao, J.-F.; Kuo, C.-H. Autonomous Mobile Robot Navigation in Sparse LiDAR Feature Environments. *Appl. Sci.* **2021**, *11*, 5963. [CrossRef]
26. Mascaró, M.; Parra-Tsunekawa, I.; Tampier, C.; Ruiz-del-Solar, J. Topological Navigation and Localization in Tunnels—Application to Autonomous Load-Haul-Dump Vehicles Operating in Underground Mines. *Appl. Sci.* **2021**, *11*, 6547. [CrossRef]
27. Tampier, C.; Mascaró, M.; Ruiz-del-Solar, J. Autonomous Loading System for Load-Haul-Dump (LHD) Machines Used in Underground Mining. *Appl. Sci.* **2021**, *11*, 8718. [CrossRef]
28. Suzuki, R.; Hayashi, K.; Osaka, H.; Matsubayashi, S.; Arita, T.; Nakadai, K.; Okuno, H.G. Estimating the Soundscape Structure and Dynamics of Forest Bird Vocalizations in an Azimuth- Elevation Space Using a Microphone Array. *Appl. Sci.* **2023**, *13*, 3607. [CrossRef]
29. Gil Aparicio, A.; Valls Miro, J. An Efficient Stochastic Constrained Path Planner for Redundant Manipulators. *Appl. Sci.* **2021**, *11*, 10636. [CrossRef]
30. Ibarra-Pérez, T.; Ortiz-Rodríguez, J.M.; Olivera-Domingo, F.; Guerrero-Osuna, H.A.; Gamboa-Rosales, H.; Martínez-Blanco, M.d.R. A Novel Inverse Kinematic Solution of a Six-DOF Robot Using Neural Networks Based on the Taguchi Optimization Technique. *Appl. Sci.* **2022**, *12*, 9512. [CrossRef]

applied sciences

MDPI

Review

Service Robots: Trends and Technology

Juan Angel Gonzalez-Aguirre †, Ricardo Osorio-Oliveros, Karen L. Rodríguez-Hernández, Javier Lizárraga-Iturralde, Rubén Morales Menendez, Ricardo A. Ramírez-Mendoza, Mauricio Adolfo Ramírez-Moreno and Jorge de Jesús Lozoya-Santos *

School of Engineering and Sciences, Tecnologico de Monterrey, Eugenio Garza Sada 2501 Sur, Monterrey 64849, Mexico; a01382394@itesm.mx (J.A.G.-A.); a00823223@itesm.mx (R.O.-O.); a01197734@itesm.mx (K.L.R.-H.); a01731990@itesm.mx (J.L.-I.); mm@tec.mx (R.M.M.); ricardo.ramirez@tec.mx (R.A.R.-M.); mauricio.ramirezm@tec.mx (M.A.R.-M.)

* Correspondence: jorge.lozoya@tec.mx

† Current address: Mechatronics Department, School of Engineering and Sciences, Tecnologico de Monterrey, Eugenio Garza Sada 2501 Sur, Monterrey 64849, Mexico.

Abstract: The 2021 sales volume in the market of service robots is attractive. Expert reports from the International Federation of Robotics confirm 27 billion USD in total market share. Moreover, the number of new startups with the denomination of service robots nowadays constitutes 29% of the total amount of robotic companies recorded in the United States. Those data, among other similar figures, remark the need for formal development in the service robots area, including knowledge transfer and literature reviews. Furthermore, the COVID-19 spread accelerated business units and some research groups to invest time and effort into the field of service robotics. Therefore, this research work intends to contribute to the formalization of service robots as an area of robotics, presenting a systematic review of scientific literature. First, a definition of service robots according to fundamental ontology is provided, followed by a detailed review covering technological applications; state-of-the-art, commercial technology; and application cases indexed on the consulted databases.

Keywords: robotics; service robots; human–robot interaction; healthcare robots; robot-as-a-service; smart cities; AGV; AMR

Citation: Gonzalez-Aguirre, J.A.; Osorio-Oliveros, R.; Rodríguez-Hernández, K.L.; Lizárraga-Iturralde, J.; Morales Menendez, R.; Ramírez-Mendoza, R.A.; Ramírez-Moreno, M.A.; Lozoya-Santos, J.d.J. Service Robots: Trends and Technology. *Appl. Sci.* **2021**, *11*, 10702. https://doi.org/10.3390/app112210702

Academic Editor: Luis Gracia and Carlos Perez-Vidal

Received: 1 October 2021
Accepted: 26 October 2021
Published: 12 November 2021

1. Introduction

The objectives of the present review focus on a specific area of robotics: service robots. Based on information gathering and database analysis, this work highlights definitions, main product research directions, and commercial technology currently used in the mentioned field. This study can provide valuable information for researchers and product developers to estimate the future value of investments in service robots research and development.

This work addresses the following research questions:

- What is a service robot?
- What are the main technological trends in the area of service robots?
- What are the most common applications of service robots?
- What are the main commercial technologies on-board of the most recent service robots?

Problem Formulation

The problem formulation for the present work is based on a systems-thinking perspective to avoid poor problem formulation.

- Undesirable situation: There exists a lack of specialization in trend predictions for service robots due to the nature of emerging technology.
- Assumptions: The development trends in this research area can be predicted through a literature review and will be useful among researchers.

- The Feasible Conceptual Future Desirable Situation: A compendium of actual service robots and literature trends will encourage technology development.
- The problem: To formulate a methodology that effectively helps to identify relevant information to provide significant figures and data.
- The solution: The development of a systematic literature review based on existing methodologies and its effective dissemination of results.

Based on the previous statements, this research problem is classified by its objective complexity as a Wicked Problem. It has no clear stopping rules since there is no definitive "problem" and "solution," and the problem ends with the consumption of the resources (time, energy, others).

2. Materials and Methods

This study used Scopus (Elsevier's abstract and citation database, Amsterdam, Netherlands), IEEE Xplore (Copyright 2021 IEEE - All rights reserved. A not-for-profit organization IEEE digital library, New York, NY, USA) and Google Scholar (Bibliographic database owned by Google LLC Mountain View, CA, USA) to conduct a systematic search of relevant scientific publications. The literature search presents six sectors: definitions and classifications, robot technology, state of the art, commercial technology, trends, and applications.

This study is a systematic review that searches for topics of service robots in scientific literature. To select and revise relevant papers from each section, the PRISMA approach was used [1], as well as the following criteria:

1. Recent (2010–2020) literature review ensures a revision of the current state of the art of the technologies applied to service robotics, prioritizing papers published during the 2015–2020 period. Figure 1a shows the distribution of the years of publications of the revised papers for this review.
2. Among the selected literature for each section, the most cited works were revised extensively. Figure 1b presents the distribution of the number of references with an increasing number of citations of the revised literature.
3. The search of papers prioritizes journals with high impact factors; most revised papers were from journals with an impact factor higher than 1.0. Figure 1c shows journal impact factor of the revised papers in this review.
4. Additionally, Figure 1d shows the type of references (journal, conference proceedings, books, technical reports, and others) selected and their percentage.

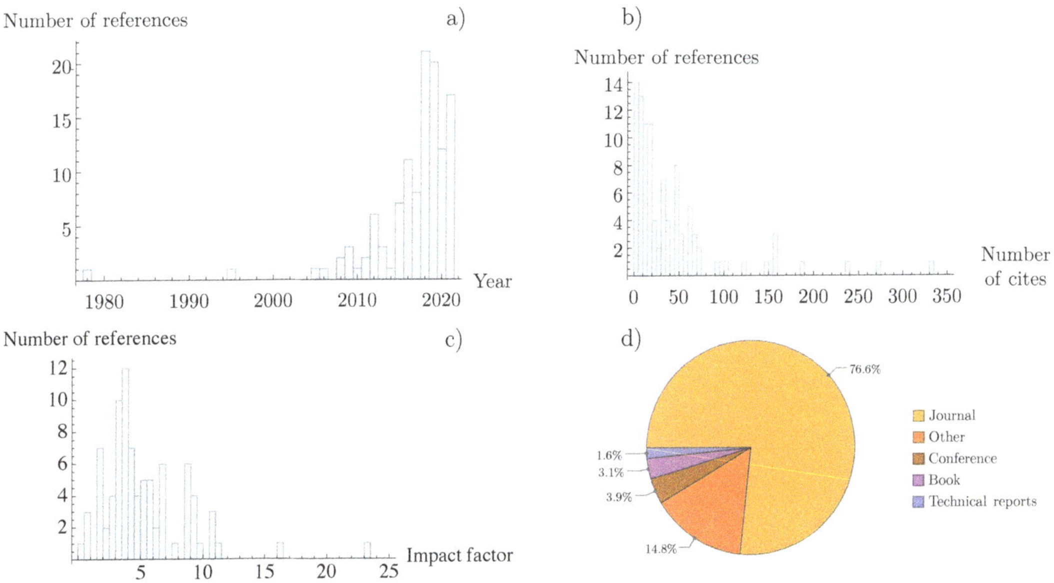

Figure 1. Statistics of the revised literature. A histogram shows the results of several reviewed references: (**a**) number of citations, (**b**) year of publication, and (**c**) impact factor. (**d**) Percentage of literature revised divided into journal papers, conferences, books, technical reports, and others.

3. Results

3.1. Definitions and Types

As part of robot categorization and definitions, the whole universe of robots can be divided into conventional and advanced robotics. Conventional robotics is also divided according to the fixation of the base; thus, one can identify mobile robots and robot manipulators, and conventional robotics also includes industrial robotics [2]. Following this idea, service robots belong to the field of advanced robotics. This general definition is valid but very broad, mainly because the field of robotics is a very vast research area, and proposing a general and universal definition is not a trivial task [3].

Towards the search of a better standardization, some committees have worked to develop an ontology that unifies entities in the field [4]. The ISO (International Organization for Standardization) organized an expert committee in 2007 to accomplish this task and the efforts materialized into the ISO 8373:2012 standard that defines a service robot as a robot that performs useful tasks for humans or equipment excluding industrial automation applications [5]. The International Federation of Robotics (IFR) agrees with such definition and appends that the robot needs to have a semi or fully autonomous behavior [6]. Other authors consider the incorporation of a natural communication system and the presence of artificial intelligence (AI) as a must in the category of service robots [7,8] to allow such robots to adapt to an uncontrollable environment. Moreover, the IEEE RAS (Robotics and Automation Society) has dedicated efforts in the task of unifying terms using advanced applied ontological methodologies [9] that aim to validate existing ontologies for consistency. By the time this work is written (2021), some advances exist on the generalization of definitions, entities, classes, and agents [10,11], but there is still work to be undertaken in this sense. Some benefits of standardization and generalization are the facilitation of design, production, knowledge, and technology transfer processes between groups and among the Research and Development areas to properly determine a paradigm in terminology and formalism.

The ISO standard, as well as the IFR [12], classifies service robots in two main classes, as can be observed in Figure 2. The first class mainly includes robots for personal and domestic use, such as robots that perform domestic tasks, entertainment robots, elderly and handicap assistance, personal transportation, home security and surveillance, and other types of domestic robots. The typical applications of these robots include tasks of non-commercial nature.

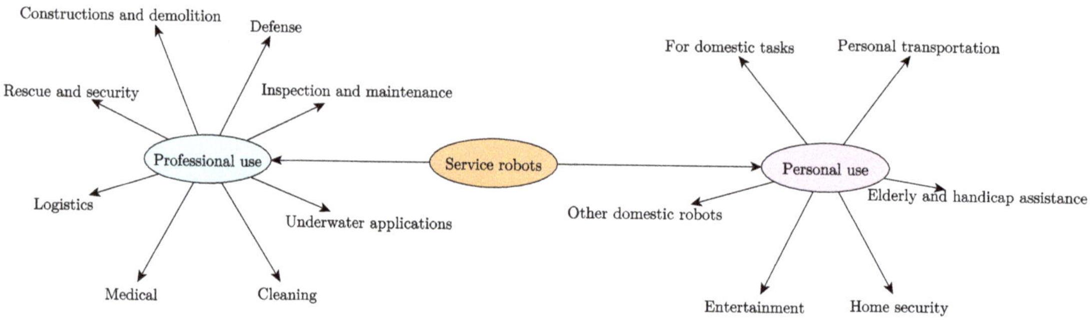

Figure 2. Taxonomy of service robots as proposed by the ISO 8373:2012.

Continuing with the classification, the second class is dedicated to service robots intended for professional use. The main taxonomy threads are field robotics, professional cleaning, inspections and maintenance systems, constructions and demolition, logistic systems, medical robotics, rescue and security applications, defense applications, underwater systems, and other professional service robots not specified above. The typical user for this kind of robot is an operator that has relevent former education or training in manual work. Contrary to personal robots, this specification involves commercial activities. The above classification can be better understood in Figure 3.

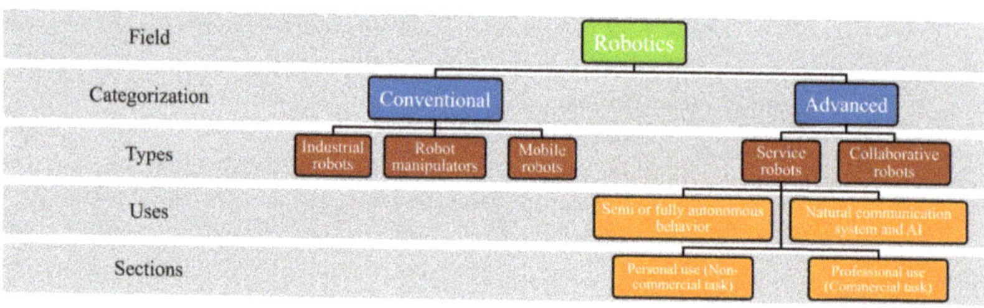

Figure 3. Robot categorization.

A complementary and more human-centric taxonomy is proposed by [9,13,14], where the main robotics definitions fall intro three classes depending on the human–robot interaction (HRI) level:

Class 1 The robot totally replaces the human worker in an environment that can be either hazardous or dirty and the task is usually of a tedious nature.

Class 2 The robot operates closely in cooperation with a human in order to increase comfort or minimize discomfort.

Class 3 The robot operates on the human body.

From the previous definitions arise some categorical problems. For instance, as pointed out by [9], an exoskeleton can be a medical robot when used in rehabilitation but can be a non-medical robot when used for assistance tasks, and can also be used in military tasks, contradicting the intended use of the robot. Further work is required to avoid ambiguous definitions, and ontology engineering provides the required framework for this disambiguation, as proposed by [15,16].

The technological stack of a service robot must address an uncontrolled environment and choose a clever combination of sensors in an integration effort to fulfill the required robot task to ensure a deep integration with humans [2]. Three main technical groups enable a service robot. These are software layers, contextualization, and human–robot interfaces. Software layers are primarily responsible for integrating the robotic device, connecting, and establishing a standard communication system for every component. Artificial intelligence is an important component of these robotic systems; tools such as TensorFlow or PyTorch execute machine learning algorithms and related tasks. Robotic systems need to know where they are and react properly based on their location, that is, to be aware of the spatial context of the environment [17]. This function is achieved via localization and sensorization. Finally, some human–robot interfaces implement its integration to the human workflow generally. Web dashboards, speech recognition, touchscreens, or even mobile device applications enable these interfaces. This technological stack is presented in a graphical form in Figure 4.

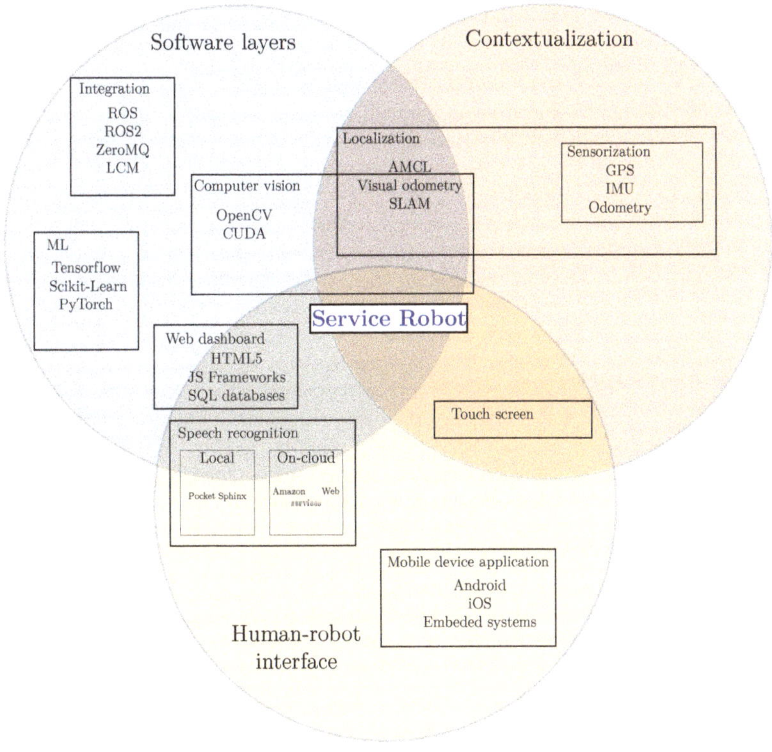

Figure 4. Common components of a service robot with examples of the technological stack available as in 2021.

Background

The technological evolution of service robotics is very vast and extensive. Its technological development will boost economic interest in different areas of growth and future market niches. There are some emerging visions and opportunities to develop new technologies and services, such as the Internet of Robotic Things [18] as a merge of pervasive sensors with robotic and autonomous systems. Another example is Robot Process Automation (RPA) as an emergent technology that mimics the steps taken by a human to complete a task [19]. RPA implies the automation of repetitive processes that involve routine tasks, structured data, and deterministic outcomes [20]. The optimization of RPA routines uses different techniques grouped in Robotic Process Mining (RPM) [21]. Its function is to determine the importance and prioritization of a routine or task, converting the RPM in a functional tool to RPA routines. Despite the progress made so far, it is important to state that to keep researching and developing new service robots and automating tasks, there must be at least one economic reason to do so. Complementing the above, the banking and financial industry is a fundamental pillar and one of the main drivers of digital disruption [22]. In the near future, new technologies, trends, and applications of service robots will emerge. However, is there a way to forecast a service robot's impact and the future trends that service robotics will have? To answer this question, we need to look into technological evolution to predict disruptive innovations and identify which service robotics technologies are likely to experiment with fast growth and development. Due to this reason, we are going to look into several theories related to technology evolution.

The theory of technological parasitism for the measurement of the evolution of technology and technological forecasting [23] aims to measure the evolution of technology. It takes the ecology of parasites and their evolvement as a reference to estimate technological growth and its dynamics. This approach foretells which innovations and developments are likely to have fast progress and an easy society acceptance. This research, to achieve the measurement of the evolution of technology, takes several approaches (hedonic, RAND, functional and structural, wholistic and holistic approaches) [23]. It concludes it is possible to have a measure on the technological evolution, but is challenging and complex to foretell which technological innovations are going to have fast growth.

A theory of classification and evolution of technologies within a generalized Darwinism [24]. Following the previous theory, the synergy between humans and technologies propagates and generates a parasitic ecosystem. Thus, this idea implies that all the agents participating in the ecosystem are supposed to benefit. In this theory, a taxonomy is presented to differentiate the possible human–technologies interactions (parasitism, commensalism, mutualism, symbiosis, amensalism, or competition) [24]. This classification between humans and different technologies proposes an explanation of how the technology evolves, how complex systems are going to be socially implemented, and the impacts the different interactions are going to have on the economics of innovation.

A theory of the evolution of technology: Technological parasitism and the implications for innovation management [25]. The adaptive behavior derived by high competition between firms and nations impels the technology evolution. In this theory, it is stated that host technologies that have a high number of technological parasites are more likely to have an accelerated evolution, rather than the ones who have low technological parasites. This condition results since having more "parasites" involves having more complex systems, with more interactions between technologies and more benefits driven by those interactions. It also considers that because a specific technology has more "parasites," more humans are focused on developing new operations and uses (could be performed through a host technology or a parasite technology).

The above can be a very general view of technological development. However, it allows us to lay the foundations to estimate future disruptive inventions and the impacts of innovations on social dynamics. Moreover, the change generated by disruptive technologies highly tends to change competitive advantages that a firm could have in a determined market. Some examples of firms that implement disruptive technologies are Apple Inc. (introducing wireless headphones to the market) and AstraZeneca (generating innovative treatments treat lung cancer) [26]. However, these new technologies imply a change in industrial behavior, leading companies to destroy (directly or indirectly) older products, goods, or technologies, to keep their leadership on their market, maximize profits, and/or protect competitive advantages [27]. Nevertheless, constantly generating innovations can be a problem for companies if they are not planned incrementally. To solve this, Coccia presented a conceptual framework of problem-driven innovation to explain industrial and technological change and the importance of solving problems by researching and developing radical innovations, either to maintain competitive advantages, maximize profits, or stay leaders in their sector [28]. Moreover, firms must consider the behavior that technologies have in the market since there is an asymmetry in the technological cycle of disruptive innovations (having a down phase shorter than the up phase) [29]. Empirically speaking, this behavior depends on the offer and demand on the markets, the grade of acceptance of a particular product that uses disruptive technologies, and by what firms want to sell.

Given all the above, it is crucial to consider the impact that service robots will have on the economics of innovation, its life-cycle, and that they could be a radical innovation in some fields that destroys or replaces the operation worth of preliminarily established technologies applied and used in markets [30]. There is a variety of tasks that researchers must develop for the use and implementation of service robotics, such as object detection, task/motion planning, activity recognition, navigation and localization, knowledge representation and retrieval, and the intertwining of perception/vision and machine learning techniques [31].

There are many areas of opportunity to apply service robots; however, there may be countries that have very incipient markets for the use of these new technologies, so one option may be to resort to foreign markets [32]. Besides, giving robots with cognitive and affective faculties, by working out infrastructures that allow them to establish compassionate connections with people, is a priority task [33]. Definitively, due to the constantly changing social mindset and current status quo of humans, the discussion on social, ethical implications, and concerns of using service robots is open, but the economic impact and social changes that service robots produce will likely accelerate their technological evolution.

3.2. Robots Technology

An under-studied field of research is the economics of technical change and technology management. A theory on the classification and evolution of technology considers the taxonomic characteristics of the interaction between technologies. The proposed classification makes an analogy with the evolution of parasites considering generalized Darwinism: parasitic technologies, commensal technologies, mutualistic technologies, and symbiotic technologies [24]. The classification of parasitic technologies is based on parasite–host relationships, and it has been shown that technologies with a high number of parasites have a high evolution. This theory provides a new perspective to explain and generalize the evolution of technology to sustain the competitive advantage of companies and nations [23].

The rapid development of service robots is mainly due to the fourth industrial revolution. In our current era, a person can obtain information and technology from the internet. However, due to the inherent speed of technological changes enjoyed by today's society, it is often overlooked and tends to be forgotten. All this technological research and development would change humankind and how it is and, according to Schwab [34], there are three main reasons why the ongoing fourth industrial revolution is changing our daily lives: The velocity in which current technology is evolving (exponential growth rather than linear growth), the breadth and depth the information has reached in today's society, and its impact on entire systems that are changing their paradigms from a micro to a macro level [7].

Before even starting to explain the new technologies that service robots are bringing and implementing, we need to go into depth into the levels that society will be affected, besides the tasks that will be executed by service robots. Taking up what was written by Wirtz et al. [7], alongside the micro-level, using service robots in different areas would bring advantages such as personalized service for each person/client, homogeneous quality service, accelerated learning, interconnection, to mention a few. Along the meso-level, service robots will become a solution to a market necessity, a commodity instead of a critical source, and reduce payroll expenses. Along the macro-level, service robots will reduce the number of unattractive, time-consuming jobs that imply task repetition and the need to be present in one place due to the nature of the job (as a receptionist for the hotel), which would directly lead to a reduction in expenditures in general.

Across the literature reviewed in this paper, and taking into account the classification presented by Rubio et al. [35], we can resume the current uses, developments, and applications of service robots in the different operations areas as shown in Figure 5. This search used the ScienceDirect and Scopus databases during September 2021. It should be noted that the research carried out to establish a state of the art (SoA) for this work could be biased due to the large fields of application that service robots have. We include the leading research and developments that other researchers and peers have been undertaking; however, there may be more information about other applications not considered in this paper.

Delving into each area reviewed in the literature, the leading technologies applied to service robots can be identified. In the area of health, especially in the area of mental health, chatbots and virtual embodied artificial intelligence (AI)-supported psychotherapeutic devices are being tested to deal with anxiety and depression [36]. In addition to this, some of the disruptive technologies applied for the analysis of COVID-19 are the Internet of Medical Things (IoMT), data science and big data, blockchain, virtual reality (VR), telemedicine, 5G, AI, drones, and autonomous robots [37]. Nowadays, the creation of a mobile healthcare robot is possible thanks to AI, machine learning, facial recognition, and teleportation technologies [38]. There are arising openings for the operation of robotics to endorse ubiquitous healthcare that may reflect in cheapening medical expenses and adding the amenity of cases and people in general [39]. Alongside the education field, educational robots are implementing different learning models to enhance learning student performance. There is a wide variety of models such as adaptive learning, agent-based learning, and smart learning [40]. Other implementations are teleoperated, autonomous, and convertible robots to assist elementary school teachers during classes [41,42].

In the technology and kinetics area, unmanned aerial vehicles (UAV) require a ground control station (GCS), batteries, fuel cells, or hybrid power sources to work, as well as power management strategies for real-time monitoring of power consumption (rule-based and fuzzy logic strategies) [18]. Unmanned ground vehicles (UGV) require the development of robotic frameworks and platforms. Some of them are the Robot Operating System (ROS), Middleware for Robotic Application (MIRA), Yet Another Robot Platform (YARP), Lightweight Communications and Marshalling (LCM), Mission Oriented Operating Suite (MOOS), and Universal Robotic (Urbi), to mention a few [43]. Moreover, rescue robotics as autonomous robots should use field-deployable technologies and work in real-world environments [44]. Some service robots and robotic platforms tested in farms and factories use IoT, edge, and cloud computing through virtualization and AI technologies, pushing its commercial adoption [45,46].

There are also great opportunities and growth areas in the field of leisure and recreation, starting with the tourism sector; technologies such as information-centric networking, cloud computing, big data, blockchain, AI systems, and IoT are essential in the development of robotics in tourism [47]. Traditional hotels will have to transform into smart hotels and implement interconnectivity and interoperability to support business partners' applications, use big data to forecast revenues more precisely, and use instant translation devices to avoid miscommunication [48]. Not only will resorts' experiences change, but the shopping customer experience will change due to service robots. The implementation of neuroscience, business process automation, blockchain, digital twins, VR, AI, mobile robots, location-based wearables, and machine-to-machine interaction through IoT by organizations are going to provide immersive and personalized environments to consumers [49]. Moreover, the inception of humanoid service robots (HSRs) by companies will generate competitive advantages against their competitors and trigger compensatory consumer behavior [50]. However, the use of service robots is not only limited to experiences and buildings; direct robot–human interaction will go further due to the use of wearable affective robots that will imitate cognitive competencies. Examples of these service robots are social robots that recognize emotions, affective robots, and intelligent brain wearables that recognize electroencephalography (EEG) data. They use natural language processing, pattern recognition, data mining, and other machine-learning techniques to achieve a human brain working mode simulation [51]. Lastly, service robots related to sexuality are no longer a fantasy due to the interest in human–robot interaction focused on sexual robots programmed with AI [52].

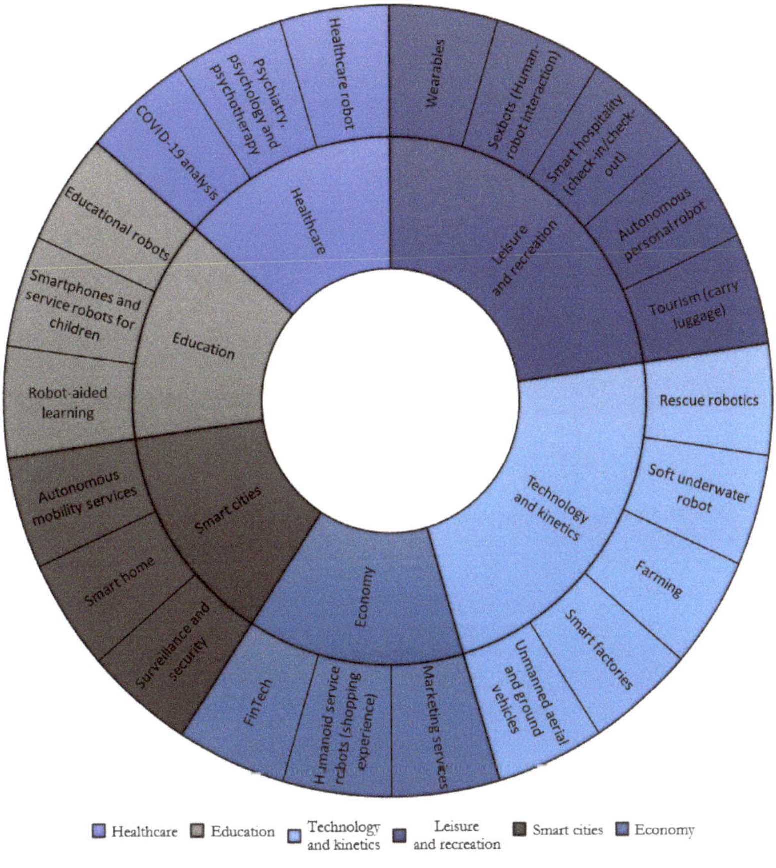

Figure 5. Uses and applications of service robots in different operation areas.

Thinking about smart urban environments, autonomous vehicles (AVs) is a concept that comes to mind. Technologies related to AVs are vehicles automation, automation, and electrification of public transportation, and electric propulsion [53]. Moreover, the term smart home is becoming more relevant, implying the use of cloud servers, cloud learning services, and machine-learning algorithms. Besides, a home service robot must be capable of recognizing human body activity, tracking a human position, sound-based human activity monitoring, and fall detection and rescue [54]. Assistive robots can also be used as caregivers in smart homes for elderly people [55]. Context awareness is an important topic related to surveillance. To achieve a context-aware model applied to an intelligent surveillance robot, techniques such as data mining, Bayesian network, collaborative filtering, and machine learning are applied [56]. People's economy is an important topic to consider; therefore, the development of financial technology (FinTech) supported by AI is vital for the world's economy [57]. Self-service technologies (SSTs) such as automated teller machines, self-checkouts, and self-service kiosks are likely to use and implement service robots with human-like characteristics such as memory, gaze, and gestures [58].

As shown below, the principal technologies related to service robots are artificial intelligence, the Internet of Things, human recognition, machine learning, blockchain, and big data. Table 1 states and summarizes some keywords used by authors.

Table 1. Operation areas, applications, and keywords related to service robots.

Operation Areas	Applications	Keywords
Healthcare	Interacts directly with humans, handling routine logistical tasks, disinfecting rooms, helping transport patients, moving heavy machinery.	Artificial intelligence, ethics, medicine, robotics, COVID-19, blockchain, Internet of Medical Things (iomt), Industry 4.0, healthcare, 5G.
Education	Serve as a tutor or a peer in a student's home. Teaches and quizzes a student on the topics they are having trouble with in the classroom, be controlled by a teacher.	Children's learning, N-screen, remote control, robot-based learning, streaming, hardware design, sign language, social child–robot interaction, service agent.
Technology and kinetics	Uses in military, scientific, agricultural, commercial, policing, surveillance, product deliveries, distribution and logistics, aerial photography fields, calculation, and decision making through artificial intelligence algorithm.	MATLAB, multibody dynamics, robotics, wheeled mobile robot, extreme environments, IoT, edge-computing, artificial intelligence, power supply, energy management, locomotion, navigation, perception, sensoring.
Leisure and recreation	Diversified booking method, improved pre-arrival experience, increased personalized data collection, new costumer experiences, shopping assistant, chatbots-as-a-service, exhibitions, and events.	Artificial intelligence, automated tourism, intelligent automation, service robots, customer experience challenges, physical and social realms, emotion cognition, social robot.
Smart cities	Establish the digital model of physical space and social space, collect data of the environment and its own operation, respond to various needs in real time.	Assistive technology, elderly care, home service robot, smart home, IoT, surveillance robot, intelligent service, context awareness, future service scenarios, value networks.
Economy	Increase in productivity in service organizations and their ability to generate insights. Robots can open spreadsheets and databases, copy data between programs, compare entries, and perform other routine tasks.	Artificial intelligence, finance, robo-advisors, robots, technology adoption, anthropomorphism, humanoid service robots, human–robot interaction, public service, trust, turn-taking.

3.3. Commercial Technology

Currently, the technology level in service robots is emerging; that is, the technology started to be commercialized by some vendors. Industry leaders have pilots and deployments in commercial service robots such as SoftBank Robotics, Furhat Robotics, Smart Robotics, and Temi. From a consumer perspective point of view, this level of maturity implies the very first generation of products, a very high price, and customization. Accordingly, a few firms dominate personal service robots, mainly taken by vacuum cleaners such as iRobot, approximating the market as an oligopoly.

Within the global market, there are different items of service robots, with some sample commercial robots presented in Table 2. There are applications on logistic, defense, public environmental, medical, field, exoskeletons, construction, inspection and maintenance, professional cleaning, and other uses. It is essential to mention that, for the most part, from 2018 to 2020, the sales of each item doubled. Moreover, the sales value of professional services robots has increased by 32%, which means 11.2 billion USD from 2018 to 2019 (data taken from EMIS). Furthermore, the COVID-19 pandemic can potentiate this growth. Robotic disinfection solutions, robotic logistics solutions in factories and warehouses, or robots for home delivery are examples of this trend, according to the World Robotics 2020—Service Robots report, presented by the International Federation of Robotics (IFR).

A significant sector that is currently adopting service robots is the hotel sector. As presented in [59], there exists a correlation in positive online reviews of hotel services to the use of service robots, and it also happens to increase the motivation of guests to write a review providing evidence of the service given by the robot. Another example of early commercial adoption of this technology is the catering and delivery business. According to [60], malls and university campus cafeterias are adopting delivery robots in order to reduce queue lines, thus reducing the mean delivery time. Such a study demonstrates an increase in business profit up to 95.4% when implementing the so-called Contactless Meal Order and Takeout Service (MOTS).

However, as of today, in 2021, market researchers, [61], have studied the relationship between the perception of value in consumers and users of service robots. A categorization presents the relative value of the robot as hedonic or utilitarian (hedonic refers to the value that enters via emotion or feelings, utilitarian value refers to value selected via rational behavior or monetary value) in different aspects such as hotels, hospitals, airports, and other tourism activities. The cited study throws two important conclusions: the utilitarian value is essential to obtain customers, and the hedonic value will attract more clients and *catch the eye* of the new possible users. Such values prompt a design guideline in the future of service robots. The following main conclusion suggests that, at the current stage, users are unlikely to pay attention to the utilitarian value of service robots. However, the more engaged society is with this kind of robot, the more likely it will increase actual utilitarian value.

The perceived responsibility in case of malfunction of a service robot triggers another pitfall when developing commercial technology. The work by [62] calculates the *degree of responsibility* in the errors that happen on the robot end and on the user end. The results happen to be inconsistent with the self-serving bias [63], which states that people attribute their successes but not the failures. In the study context, the adverse outcomes (errors and undesirable situations) are attributed to the service customer/user, and positive outcomes are attributed to the service robot.

Table 2. Sample commercial service robots.

Figure	Commercial Name	Developer	Classification	Application
	UVD Robot [64]	Blue Ocean Robotics	Professional use—Healthcare—Class 1	Cleaning surfaces using UV on hospitals.
	MyAppCafe—Street barista [65]	My App Cafe GmbH	Personal use—Food—Class 1	A robotic manipulator is installed in a cell and serves coffee with no human help and using a mobile application interface.
	EksoNR [66]	Ekso Bionics	Professional use—Healthcare—Class 3	Exoskeleton that aids and accelerates therapy among users that suffer spinal injury.
	Nuro R2 [67]	Nuro Inc.	Personal use—Delivery—Class 1	Autonomous mobile robot that drives through the city and delivers goods requested using a mobile application.
	Roomba [68]	iRobot	Personal use—Cleaning—Class 1	Floor vacuum floor cleaner.

Table 2. *Cont.*

Figure	Commercial Name	Developer	Classification	Application
	Pepper [69]	SoftBank Robotics	Personal use—Multiple uses—Class 1	Humanoid that recognizes faces and emotions is currently used in airports and schools to provide assistance.
	Turtlebot 2 [70]	OSRF	Professional use—Education—Class 2	Low-cost robot kit for prototyping and learning.

3.4. Scientific Literature

Due to the rapid advances in robot technology combined with AI, the creation and implementation of service robots in different industrial sectors have increased dramatically. Service robots can be in different forms; they can be virtual, chatbots, humanoids, and non-humanoids [61]. Thanks to the advances in robotics and the implementation of AI, machines can perform even more complex and repetitive tasks [59].

Some AI and natural language processing applications emerged as a COVID-19 response to protect and prevent further damage due to the health crisis that emerged in 2020. An example of this technology implementation is the Intelligent Voice Assistant for Coronavirus Disease Self-Assessment, a deployment that successfully merges natural language processing and cloud computing to create a virtual service robot that helps to diagnose symptoms related to COVID-19 [71].

The demand for service robots grew for the attention of social distancing and health-monitoring protocols due to the COVID-19 outbreak. Therefore, many industries have opted to include service robots as part of their staff to improve customer experience, service quality, and efficiency, as well as to reduce labor costs [72]. For example, service robots can provide more accurate services reducing mistakes and becoming more reliable and consistent than human employees. More importantly, robots can perform tasks without stopping, at a faster pace than humans, as well as carry multiple, repetitive, and mundane tasks without protesting [72].

Human–robot interaction is in constant development; the acceptance of service robots has flourished. To illustrate this idea, Figure 6 shows a prediction of the near future on the potential development of service robots. However, since it still is an early stage of development of the service robots, some groups of persons are open to their use, while others express concerns related to the negative consequences [61].

The recent and fast development of robotic technologies has inspired tourist corporations to adopt service robots. An excellent example of this is the service robot "Pepper". The benefits of this adoption were a noticeable increase in the customer's satisfaction, as well as creating a positive word-of-mouth [73]. For instance, medical robot assistants are being used to monitor patients and alert the medical staff when needed. Nowadays, the implementation of robotic medical assistants has increased due to the COVID-19 outbreak, resulting in a valuable and efficient way to monitor and control highly contagious diseases patients [74].

An exploratory study reveals that service robots are becoming a popular and more recurrent feature in tourism. This study suggests the tourism market perceives implementation of service robots as valuable, not only for interested technological visitors but also by an increasing number of customers [59]. Other studies have implemented service robots in restaurant companies to reduce work hours and improve service quality. The restaurant industry suffers from product losses constantly when the total demand exceeds the service production capacity. Results demonstrate the robots have reduced 20 work hours of the service staff, also improving labor productivity (sales per hour) and reducing the losses dramatically due to lack of production; with the implementation of service robots, the service production sustained a higher and more efficient pace [75]. Current robotics applications will be covered in the next section. Moreover, we wondered about the terms associated with our search, so we constructed two different word clouds. One including single keywords used by the authors in the reviewed papers (see Figure 7) and one including composed keywords (see Figure 8).

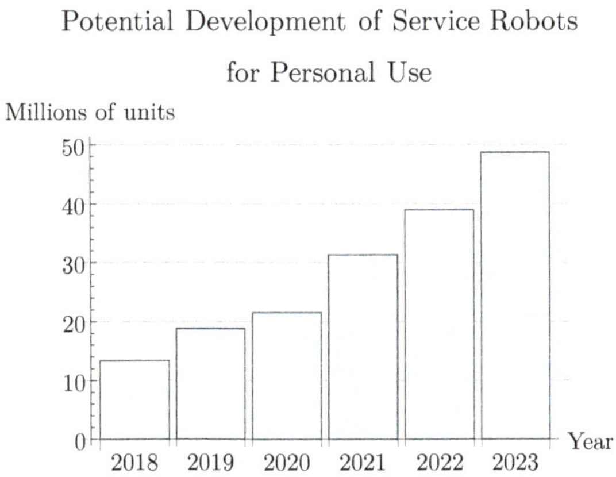

Figure 6. Service robots potential development for personal/domestic use.

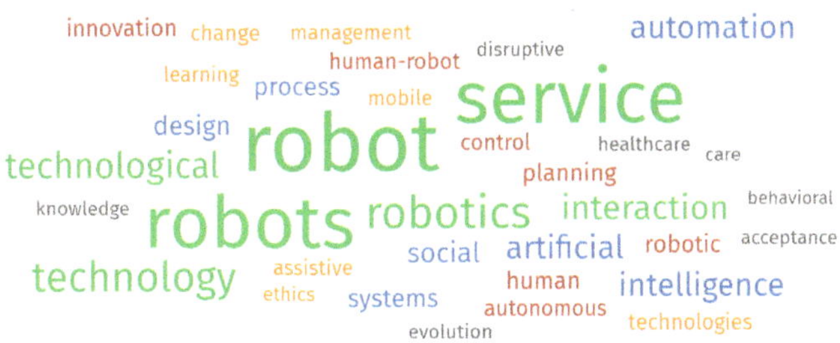

Figure 7. Word cloud of simple keywords used by authors in the reviewed literature.

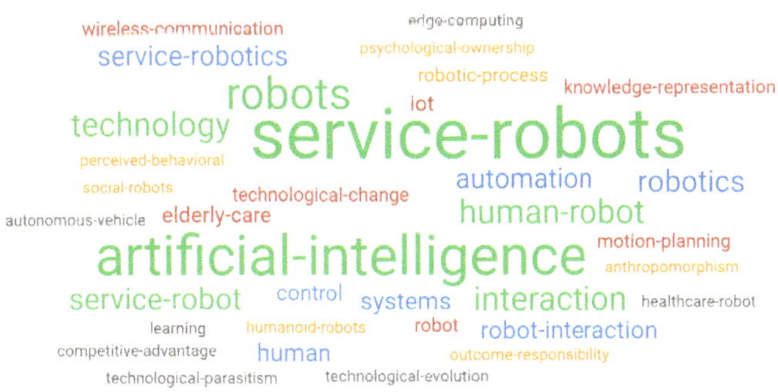

Figure 8. Word cloud of composite keywords used by authors in the reviewed literature.

3.5. Applications

There are several applications where service robots can add value; due to their versatility, the different needs of each industry will lead the design process.

A new frontier entering the market in industrial robotics are new, easy-to-use collaborative robotics solutions, where the robot advantages such as precision, speed, and repeatability come together with the flexibility and cognitive skills of human workers [76]. In the past years, industrial robots were robust, used to perform a specific task, and placed into a cage for the safety of others. However, collaborative robots are designed and prepared to interact and work alongside humans [77].

Recently, the development of collaborative industrial robots for the manufacturing process has increased. There are automatic guided vehicles (AGVs), as observed in Figure 9, that are a type of service robots whose primary functionality is to help in the realization of internal transport processes [78]. Another application is the human–robot interaction (HRI), which is becoming a new trend thanks to technology and advances in perception, cognition, and control algorithms. As interest in these robots increases, so do the benefits of their implementation, such as productivity and production line flexibility [79], resulting in increased production and demand from industries to integrate them [80].

A significant application for service robots is daily-life assistance. It is very complex because, for comfortable assistance, the robot must recognize its surroundings, including the motion of humans, the position of the objects, and obstacles such as stairs [81]. The principal objective in this application is to reach effective communication between the robot, its real-world environment, and the people in it [82]. In such a situation, the robot must have a manipulator that can grasp, transport, and place objects, as these are fundamental capabilities for this type of service robot [83].

Figure 9. Example of AGV robots [78].

A robot-integrated smart home (RiSH) refers to a house that contains at least one service robot, a sensor network, a mobile device, cloud servers, and remote caregivers [54], so the service robot controls everything inside the house from afar. A telepresence robot system performs assistive functions to improve the well-being of elderly persons. It can assist them to do daily activities independently, to encourage social interactions to combat the sense of isolation or loneliness, and to help the professional caregivers in routine care [84]. However, in this scope, it is crucial to consider the acceptance of service robots by elderly people considering the psychological variables for proper interaction between people and technology [85].

Another application where service robots are being incorporated and are considered the workforce of the future is in operation and management, including the hotel industry [86]. In this case, a bellboy robot performs hotel-related functions such as walking alongside guests and providing information about the city and the hotel [87]. Depending on these functionalities and the total interaction with hotel guests, the overall experience of the visitors will change [88].

In recent years, the primary purpose of developing more robots has been to improve productivity. However, with the current COVID-19 pandemic, a more urgent purpose has arrived [89] where robots present significant advantages. We have been involved in the revolutionary development of mobile healthcare robots as there is a need for people to avoid physical interaction [38]. It enables the closer analysis of this need in order to ensure the regulation of social distancing rules [90].

Hand gestures will be the usual method to manipulate human–computer interfaces (HCIs); however, to assist people with motor disabilities, an HCI must be designed especially for them. The help of service robot platforms in communication with three-dimensional (3D) imaging sensors and a wearable armband explores this solution [91].

As technology advances, society must adapt to new trends. Therefore, students and teachers have incorporated some of the latest technologies, overcoming many obstacles in the process. One of these technologies is the NAO humanoid robot that is currently being used in computer and science classes from elementary schools to university classes in many countries around the world [92].

Regarding business and financial institutions, there has already been considerable progress automating specific tasks implementing RPA and RPM concepts. As an example, Vodafone combines RPA and RPM to identify non-standard orders that require a high level of human interaction, which helps to reduce the time invested into checking complex orders before delivering to a supplier [93]. In the banking industry, some cases using RPA are: automatic report generation, opening an account, audit and compliance, chatbots, anti-money laundering, among others [94]. The combination of RPA and IA can lead to improved operational efficiency and increase the impact on the economics of innovation. However, RPA implementation is limited to business and banking; industries such as insurance, manufacturing, logistics, government, and public security can also take advantage of this technology [95], thus, it is of utmost importance to determine the current degree of automation in an organization's business processes to correctly identify tasks and processes that can be automated or improved [96]. RPA is an emerging technology that will have many applications, but one of the critical challenges to fully take advantage of this technology is to transfer digital tasks (performed in an environment with a virtual desktop interface) to cyber-physical tasks and processes [97].

Service robots can navigate through inaccessible or unsafe environments for the Urban Search and Rescue (USAR), where human teams cannot enter. The principal features that these types of robots must have are speed, weight, robustness, reliability, affordability, adaptability to different environments or tasks, and provide excellent two-way audio and/or video communication [98]. As well as the shown applications in the previous section, service robots have become a significant worldwide trend. The following section covers some robotics trends.

3.6. Trends

The revision of the literature on service robots identifies diverse trends in the fields of healthcare, industry, home service, and multi-purpose indoor environments. Regarding the type of robots, trends in autonomous navigation [99,100], mobile robots [35], unmanned autonomous systems [101,102], and imitation learning systems [103,104] were identified.

Healthcare: The current global population trend has shown an increase in elderly persons, as well as an increase in more populated cities; therefore, there is a higher demand for healthcare services such as medicine and nursing [105]. According to Archibald and Barnard [105], three types of service robots oriented to health exist: doctor, nurse, and home health care robots. Regarding nursing robots, applications have been implemented for feeding assistance, automated soaping and showering for the elderly, robotic therapeutic companions, pharmaceutical transporters, pick-and-place patients in bed, and ambulation assistants [105]. A human action recognition algorithm using depth cameras, object detection, and human joint identification techniques supports patients with mild cognitive impairment (MCI) at their homes. Assistive living robots could be of use to MCI patients, as they can identify potential risks due to errors and avoid dangerous accidents [106].

Other studies have implemented physical exercise programs in elderly health centers with success. In this study case, 41 volunteers underwent a training program with a humanoid Aldebaran NAO robot and with a human coach [107]. The study results showed a good response towards the humanoid coach, at least for the training program. When analyzing a factual information task between robot and human, the human coach was more efficient. A biomechanical rehabilitation, one degree of freedom (DOF) robotic handle for post-stroke patients applies an adaptive reinforcement learning (RL) algorithm [108]. By constantly adapting the difficulty level of a virtual "nut-catching" game, depending on the skill level of volunteers, patients can learn at a faster pace while not losing their motivation.

A telemedicine robot (see Figure 10) was developed in [109], composed of a four-wheel base, a robotic arm, and a tablet, acting as the head of the robot. The robot also includes an array of ultrasound sensors and cameras. The robot allows automated navigation combining its sensors and actuators, including obstacle avoidance and object manipulation tasks (e.g., the floor and shelves). It implements routines such as TakeMedicine, WallFollowing, and Doorpassing. Other applications are also included, such as fetching, providing reminders, calendar, and interpersonal communication [109].

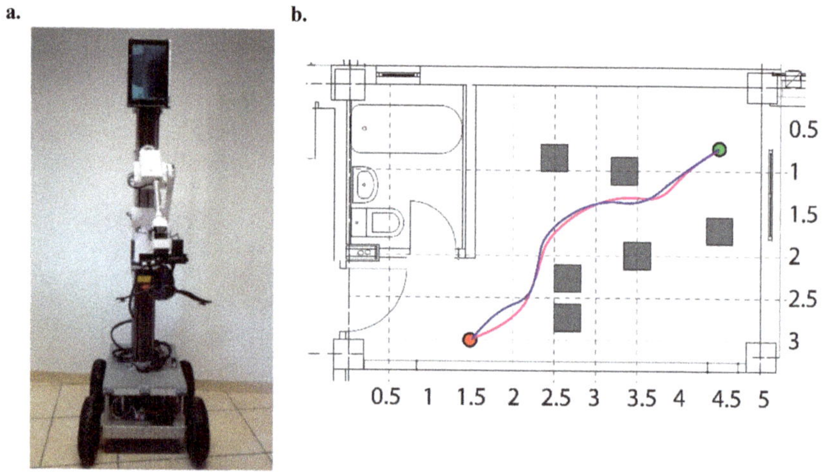

Figure 10. (**a**) Autonomous telemedicine robot system for assisted and independent living. (**b**) Simulated and experimental paths of an obstacle avoidance trial in a real-world scenario (center for the elderly), executed by the telemedicine robot developed in [109].

Industry: Different applications of service robots have been implemented for the industrial context. In [104], a machine learning technology enables a chatbot to provide support to customers in financial-product sales. Using robots to work continuously, text information from FAQs, call center response manuals, and office documents were used as input to a machine learning model to generate artificial conversation about bank services to interested customers. The use of unmanned aerial vehicles (UAV) has been reported in high risk tasks, such as transmission line inspection in China, Japan, Spain, and Britain [102]. By implementing UAV control and combining image processing and artificial intelligence, UAVs have performed autonomous inspection. These types of robots use a combination of visible light and thermal infrared sensing, as well as LiDAR technology [102]. Although UAVs are useful for safety purposes, manual inspection still outperforms in some scenarios, and it can also perform repairs, while UAV cannot.

Collaborative robots, usually robotic arms or semi-humanoid robots, represent an intermediate automation level between manual and fully automated manufacturing. In this approach, the robot acts as an assistant to a human during specific tasks. Vision-based collision prediction systems, capacitive sensing (skin detection), and safety design parameters, and routine instrument these types of robots [110]. Another typical application for industrial robots is object manipulation. An exciting study case is that of the first Amazon Picking Challenge [111], challenging 26 teams (primarily academics). The challenge involves designing autonomous robots to pick objects from a warehouse shelf. Objects of different shapes and sizes were used (Oreo cookies, an outlet protector, and a softcover book are some examples), and the teams used different approaches. Among the most common solutions were suction actuators, 3D imaging sensing, and a geometrical and/or color recognition approach for feature selection.

Home service: Some application trends in home service are in education, entertainment, household, social interactions, gaming, security, and rehabilitation [112]. A home voice-activated semi-autonomous vehicle robot was implemented in [113]. It consisted of a modified lawnmower and an IoT control module through voice-activated Alexa commands. Humanoid robots can help in house chores, grasping and carrying objects, opening doors, and entertainment [114].

Multi-purpose indoor environments: Positioning systems using sensor fusion for indoor positioning tasks mainly uses ultrasonic sensors and information from radar and odometry [115]. A mobile robot using a robust convolutional neural network (CNN) algorithm for person identification, tracking, and locking followed the identified persons through different rooms, with great accuracy [116]. The development of a robotic waiter system integrates different autonomous navigation algorithms and sensing approaches, such as IMUs, odometry, SLAM, and adaptive Monte Carlo simulation [100]. Artificial vision methods identify tables in a restaurant as well as persons. Another approach of indoor positioning, Steady Delivery, makes use of sensor fusion, involving radar, ultrasonic sensors, and odometry [117]. Table 3 presents the current trends of service robots.

Table 3. Service robots' current trends.

Field	Figure	Commercial Name	Robot Type	Application
Healthcare		Aldebaran NAO robot [107]	Humanoid robot	Physical exercise programs in elderly health centers.
		Care-o-Bot [109]	Mobile robot assistant	Implements routines such as taking medicine, wall following, and door passing.
Industry	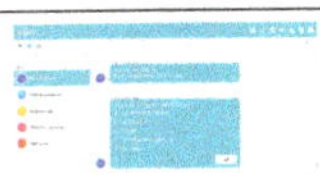	Financial Services Solution Finplex Robot Agent Platform (FRAP) [104]	Chatbot	Provide support to customers in financial-product sales.
		Yaskawa Motoman [111]	Autonomous robot	Autonomous robot to pick objects from a warehouse shelf.
Home service		R1 [114]	Humanoid robot	Help in house chores, grasping and carrying objects, opening doors, and entertainment.
Multi-purpose indoor environments		Festo Robotino [114]	Mobile robot	Indoor positioning system for service robot applications.

Table 3. *Cont.*

Field	Figure	Commercial Name	Robot Type	Application
		Pioneer P3-DX [115]	Mobile robot	Wall detection and obstacle avoidance, autonomous navigation.
		Beta-G [115]	Mobile robot	Waiter robot (identify tables in a restaurant, go to the target table to serve the food).

4. Conclusions

From the presented review, the emerging status of service robots technology in the world and as a research area becomes more evident. This carries essential opportunities for early research, development, and investment in commercial technology as a strategic decision for long-term profit [118]. Despite the fact of the good will of robotics, some challenges are still present. Some of them are the lack of generalization and formalism in classifications and taxonomy [10], the current perceived utilitarian value [61], battery and autonomy modelling and estimation [119], ethics [36], and even design problems related to gender biases based on the occupation of the robot [120]. These challenges are opportunities for future research questions or different research groups. Moreover, a solid field of service robots is healthcare and cleaning robots. Consequently, with the COVID-19 pandemic, a push in these technologies was observed [37,90] and must be taken into account when performing research in this area. After the *accelerated development* in these technologies (caused by the COVID-19 spread) reaches a plateau, it may be an interesting research question to study the degradation or improvement in service robot-related areas. As a final remark, despite the fact of current opportunities and observations in the field, estimations [50] point that, one way or another, service robots will undoubtedly be part of our daily life in the near future. This study will help researchers as it provides valuable information on recent developments of service robots. This work can serve as a starting point for researchers when studying this field. Moreover, we consider it helps to estimate the future value of the investment in service robot research and development. Consequently, the readers will contribute to this work by producing more studies and expanding the research area.

Author Contributions: Conceptualization, J.A.G.-A. and J.d.J.L.-S.; methodology, J.A.G.-A., R.O.-O., J.L.-I., K.L.R.-H. and M.A.R.-M.; investigation, J.A.G.-A., R.O.-O., J.L.-I., K.L.R.-H., and M.A.R.-M.; writing—original draft preparation, J.A.G.-A., R.O.-O., J.L.-I., K.L.R.-H. and M.A.R.-M.; writing—review and editing, J.d.J.L.-S., M.A.R.-M., and R.A.R.-M.; visualization, J.A.G.-A., R.O.-O., J.L.-I., K.L.R.-H., and M.A.R.-M.; validation, J.d.J.L.-S., R.M.M. and R.A.R.-M.; supervision, J.d.J.L.-S., R.M.M. and R.A.R.-M.; project administration, J.d.J.L.-S., R.M.M. and R.A.R.-M.; funding acquisition, J.d.J.L.-S. and R.A.R.-M. All authors have read and agreed to the published version of the manuscript.

Funding: This project is funded by the Campus City initiative from Tecnologico de Monterrey. The APC was funded by Tecnologico de Monterrey.

Institutional Review Board Statement: Not applicable.

Informed Consent Statement: Not applicable.

Data Availability Statement: Not applicable.

Acknowledgments: Authors would like to thank the support of Tecnologico de Monterrey through the Campus City initiative.

Conflicts of Interest: The authors declare no conflict of interest. The founders had no role in the writing of the manuscript.

References

1. Moher, D.; Liberati, A.; Tetzlaff, J.; Altman, D.G.; Group, T.P. Preferred Reporting Items for Systematic Reviews and Meta-Analyses: The PRISMA Statement. *PLoS Med.* **2009**, *6*, e1000097. [CrossRef]
2. Siciliano, B.; Sciavicco, L.; Villani, L.; Oriolo, G. *Robotics*; Advanced Textbooks in Control and Signal Processing; Springer: London, UK, 2009; [CrossRef]
3. Lin, P. Robot ethics: The ethical and social implications of robotics. *Choice Rev. Online* **2012**, *50*, 50-0844. [CrossRef]
4. Virk, G.S.; Moon, S.; Gelin, R. ISO standards for service robots. In *Advances in Mobile Robotics, Proceedings of the 11th International Conference on Climbing and Walking Robots and the Support Technologies for Mobile Machines, CLAWAR 2008, Coimbra, Portugal, 8–10 September 2008*; World Scientific: Singapore, 2008; pp. 133–138. [CrossRef]
5. ISO. Robots and Robotic Devices—Vocabulary ISO 8373:2012; Technical Report; ISO: Geneve, Switzerland, 2012.
6. International Federation of Robotics. Service Robots. Available online: https://ifr.org/service-robots (accessed on 27 October 2021).
7. Wirtz, J.; Patterson, P.G.; Kunz, W.H.; Gruber, T.; Lu, V.N.; Paluch, S.; Martins, A. Brave new world: Service robots in the frontline. *J. Serv. Manag.* **2018**, *29*, 907–931. [CrossRef]
8. Kopacek, P. Development Trends in Robotics. *IFAC-PapersOnLine* **2016**, *49*, 36–41. [CrossRef]
9. Haidegger, T.; Barreto, M.; Gonçalves, P.; Habib, M.K.; Ragavan, S.K.V.; Li, H.; Vaccarella, A.; Perrone, R.; Prestes, E. Applied ontologies and standards for service robots. *Robot. Auton. Syst.* **2013**, *61*, 1215–1223. [CrossRef]
10. Bayat, B.; Bermejo-Alonso, J.; Carbonera, J.; Facchinetti, T.; Fiorini, S.; Goncalves, P.; Jorge, V.A.; Habib, M.; Khamis, A.; Melo, K.; et al. Requirements for building an ontology for autonomous robots. *Ind. Robot. Int. J.* **2016**, *43*, 469–480. [CrossRef]
11. Schlenoff, C.; Prestes, E.; Madhavan, R.; Goncalves, P.; Li, H.; Balakirsky, S.; Kramer, T.; Miguelanez, E. An IEEE standard Ontology for Robotics and Automation. In Proceedings of the IEEE International Conference on Intelligent Robots and Systems, Vilamoura-Algarve, Portugal, 7–12 October 2012; pp. 1337–1342. [CrossRef]
12. Zielinska, T.T. *History of Service Robots and New Trends*; IGI Global: Hershey, PA, USA, 2019; pp. 158–187. [CrossRef]
13. Jorge, V.A.; Rey, V.F.; Maffei, R.; Rama Fiorini, S.; Carbonera, J.L.; Branchi, F.; Meireles, J.P.; Franco, G.S.; Farina, F.; Da Silva, T.S.; et al. Exploring the IEEE ontology for robotics and automation for heterogeneous agent interaction. *Robot. Comput.-Integr. Manuf.* **2015**, *33*, 12–20. [CrossRef]
14. Fiorini, S.R.; Carbonera, J.L.; Goncalves, P.; Jorge, V.A.; Rey, V.F.; Haidegger, T.; Abel, M.; Redfield, S.A.; Balakirsky, S.; Ragavan, V.; et al. Extensions to the core ontology for robotics and automation. *Robot. Comput.-Integr. Manuf.* **2015**, *33*, 3–11. [CrossRef]
15. Gasevic, D.; Devedzic, V.; Djuric, D. *Model Driven Engineering and Ontology Development*; Springer: Berlin/Heidelberg, Germany, 2009; [CrossRef]
16. Gruber, T.R. Toward Principles for the Design of Ontologies. *Int. J. Hum.-Comput. Stud.* **1995**, *43*, 907–928. [CrossRef]
17. Iocchi, L.; Holz, D.; Ruiz-Del-Solar, J.; Sugiura, K.; Van Der Zant, T. RoboCup@Home: Analysis and results of evolving competitions for domestic and service robots. *Artif. Intell.* **2015**, *229*, 258–281. [CrossRef]
18. Simoens, P.; Dragone, M.; Saffiotti, A. The Internet of Robotic Things: A review of concept, added value and applications. *Int. J. Adv. Robot. Syst.* **2018**, *15*, 1–11. [CrossRef]
19. Kommera, V. Robotic process automation. *Am. J. Intell. Syst.* **2019**, *9*, 49–53. [CrossRef]
20. Aguirre, S.; Rodriguez, A. Automation of a business process using robotic process automation (RPA): A case study. In *Workshop on Engineering Applications*; Springer: Berlin/Heidelberg, Germany, 2017; pp. 65–71.
21. Leno, V.; Polyvyanyy, A.; Dumas, M.; La Rosa, M.; Maggi, F.M. Robotic Process Mining: Vision and Challenges. *Bus. Inf. Syst. Eng.* **2021**, *63*, 301–314. [CrossRef]
22. Pokharkar, A.P. Robotic Process Automation: Concept, Benefits, Challenges in Banking Industry. *IIBM'S J. Manag. Res.* **2019**, *4*, 17–25. [CrossRef]
23. Coccia, M. The theory of technological parasitism for the measurement of the evolution of technology and technological forecasting. *Technol. Forecast. Soc. Chang.* **2019**, *141*, 289–304. [CrossRef]
24. Coccia, M. A theory of classification and evolution of technologies within a Generalised Darwinism. *Technol. Anal. Strateg. Manag.* **2019**, *31*, 517–531. [CrossRef]
25. Coccia, M.; Watts, J. A theory of the evolution of technology: Technological parasitism and the implications for innovation magement. *J. Eng. Technol. Manag. JET-M* **2020**, *55*, 101552. [CrossRef]
26. Coccia, M. Disruptive firms and technological change. *Quad. IRCrES-CNR* **2018**, *3*, 3–18. [CrossRef]
27. Coccia, M. Disruptive firms and industrial change. *J. Econ. Soc. Thought* **2018**, *4*, 437–450.
28. Coccia, M. Sources of technological innovation: Radical and incremental innovation problem-driven to support competitive advantage of firms. *Technol. Anal. Strateg. Manag.* **2017**, *29*, 1048–1061. [CrossRef]
29. Coccia, M. Asymmetry of the technological cycle of disruptive innovations. *Technol. Anal. Strateg. Manag.* **2020**, *32*, 1462–1477. [CrossRef]
30. Coccia, M. *Global Encyclopedia of Public Administration, Public Policy, and Governance*; Springer: Cham, Switzerland, 2020; [CrossRef]
31. Paulius, D.; Sun, Y. A Survey of Knowledge Representation in Service Robotics. *Robot. Auton. Syst.* **2019**, *118*, 13–30. [CrossRef]

32. Nezhnikova, E.; Pshinshev, K. The prospect of development of robotics in Russia. In Proceedings of the E3S Web of Conferences, Rostovon-Don, Russia, 26–28 February 2020; Volume 175, p. 05043.
33. Nocentini, O.; Fiorini, L.; Acerbi, G.; Sorrentino, A.; Mancioppi, G.; Cavallo, F. A survey of behavioral models for social robots. *Robotics* **2019**, *8*, 54. [CrossRef]
34. Schwab, K. *The Fourth Industrial Revolution*; Currency: New York, NY, USA, 2017.
35. Rubio, F.; Valero, F.; Llopis-Albert, C. A review of mobile robots: Concepts, methods, theoretical framework, and applications. *Int. J. Adv. Robot. Syst.* **2019**, *16*. [CrossRef]
36. Fiske, A.; Henningsen, P.; Buyx, A. Your robot therapist will see you now: Ethical implications of embodied artificial intelligence in psychiatry, psychology, and psychotherapy. *J. Med. Internet Res.* **2019**, *21*, 1–12. [CrossRef] [PubMed]
37. Abdel-Basset, M.; Chang, V.; Nabeeh, N.A. An intelligent framework using disruptive technologies for COVID-19 analysis. *Technol. Forecast. Soc. Chang.* **2021**, *163*, 120431. [CrossRef]
38. Wan, S.; Gu, Z.; Ni, Q. Cognitive computing and wireless communications on the edge for healthcare service robots. *Comput. Commun.* **2020**, *149*, 99–106. [CrossRef]
39. Santos, N.B.; Bavaresco, R.S.; Tavares, J.E.; Ramos, G.d.O.; Barbosa, J.L. A systematic mapping study of robotics in human care. *Robot. Auton. Syst.* **2021**, *144*, 103833. [CrossRef]
40. Jeong, G.M.; Park, C.W.; You, S.; Ji, S.H. A study on the education assistant system using smartphones and service robots for children. *Int. J. Adv. Robot. Syst.* **2014**, *11*, 71. [CrossRef]
41. Han, J. Robot-aided learning and r-learning services. In *Human-Robot Interaction*; InTech: Rijeka, Croatia, 2010; p. 288.
42. Meghdari, A.; Alemi, M.; Zakipour, M.; Kashanian, S.A. Design and Realization of a Sign Language Educational Humanoid Robot. *JOurnal Intell. Robot. Syst. Theory Appl.* **2019**, *95*, 3–17. [CrossRef]
43. Rivera, Z.B.; De Simone, M.C.; Guida, D. Unmanned ground vehicle modelling in Gazebo/ROS-based environments. *Machines* **2019**, *7*, 42. [CrossRef]
44. Delmerico, J.; Mintchev, S.; Giusti, A.; Gromov, B.; Melo, K.; Horvat, T.; Cadena, C.; Hutter, M.; Ijspeert, A.; Floreano, D.; et al. The current state and future outlook of rescue robotics. *J. Field Robot.* **2019**, *36*, 1171–1191. [CrossRef]
45. Hu, L.; Miao, Y.; Wu, G.; Hassan, M.M.; Humar, I. iRobot-Factory: An intelligent robot factory based on cognitive manufacturing and edge computing. *Future Gener. Comput. Syst.* **2019**, *90*, 569–577. [CrossRef]
46. Zamora-Izquierdo, M.A.; Santa, J.; Martínez, J.A.; Martínez, V.; Skarmeta, A.F. Smart farming IoT platform based on edge and cloud computing. *Biosyst. Eng.* **2019**, *177*, 4–17. [CrossRef]
47. Tussyadiah, I. A review of research into automation in tourism: Launching the Annals of Tourism Research Curated Collection on Artificial Intelligence and Robotics in Tourism. *Ann. Tour. Res.* **2020**, *81*, 102883. [CrossRef]
48. Leung, R. Smart hospitality: Taiwan hotel stakeholder perspectives. *Tour. Rev.* **2019**, *74*, 50–62. [CrossRef]
49. Bolton, R.N.; McColl-Kennedy, J.R.; Cheung, L.; Gallan, A.; Orsingher, C.; Witell, L.; Zaki, M. Customer experience challenges. *J. Serv. Manag.* **2018**, *29*, 776–808. [CrossRef]
50. Mende, M.; Scott, M.L.; van Doorn, J.; Grewal, D.; Shanks, I. Service Robots Rising: How Humanoid Robots Influence Service Experiences and Elicit Compensatory Consumer Responses. *J. Mark. Res.* **2019**, *56*, 535–556. [CrossRef]
51. Chen, M.; Zhou, J.; Tao, G.; Yang, J.; Hu, L. Wearable affective robot. *IEEE Access* **2018**, *6*, 64766–64776. [CrossRef]
52. González-González, C.S.; Gil-Iranzo, R.M.; Paderewski-Rodríguez, P. Human–robot interaction and sexbots: A systematic literature review. *Sensors* **2021**, *21*, 216.
53. Bösch, P.M.; Becker, F.; Becker, H.; Axhausen, K.W. Cost-based analysis of autonomous mobility services. *Transp. Policy* **2018**, *64*, 76–91. [CrossRef]
54. Do, H.M.; Pham, M.; Sheng, W.; Yang, D.; Liu, M. RiSH: A robot-integrated smart home for elderly care. *Robot. Auton. Syst.* **2018**, *101*, 74–92. [CrossRef]
55. Čaić, M.; Odekerken-Schröder, G.; Mahr, D. Service robots: Value co-creation and co-destruction in elderly care networks. *J. Serv. Manag.* **2018**, *29*, 178–205. [CrossRef]
56. Shin, M.S.; Kim, B.C.; Hwang, S.M.; Ko, M.C. Design and implementation of IoT-based intelligent surveillance robot. *Stud. Inform. Control.* **2016**, *25*, 422. [CrossRef]
57. Belanche, D.; Casaló, L.V.; Flavián, C. Artificial Intelligence in FinTech: Understanding robo-advisors adoption among customers. *Ind. Manag. Data Syst.* **2019**, *119*, 1411–1430. [CrossRef]
58. van Pinxteren, M.M.; Wetzels, R.W.; Rüger, J.; Pluymaekers, M.; Wetzels, M. Trust in humanoid robots: Implications for services marketing. *J. Serv. Mark.* **2019**, *33*, 507–518. [CrossRef]
59. Borghi, M.; Mariani, M.M. Service robots in online reviews: Online robotic discourse. *Ann. Tour. Res.* **2021**, *87*, 103036. [CrossRef]
60. Lin, T.Y.; Wu, K.R.; Chen, Y.S.; Huang, W.H.; Chen, Y.T. Takeout Service Automation with Trained Robots in the Pandemic-Transformed Catering Business. *IEEE Robot. Autom. Lett.* **2021**, *6*, 903–910. [CrossRef]
61. Hu, Y. An improvement or a gimmick? The importance of user perceived values, previous experience, and industry context in human–robot service interaction. *J. Destin. Mark. Manag.* **2021**, *21*, 100645. [CrossRef]
62. Jörling, M.; Böhm, R.; Paluch, S. Service Robots: Drivers of Perceived Responsibility for Service Outcomes. *J. Serv. Res.* **2019**, *22*, 404–420. [CrossRef]
63. Bradley, G.W. Self-serving biases in the attribution process: A reexamination of the fact or fiction question. *J. Personal. Soc. Psychol.* **1978**, *36*, 56–71. [CrossRef]

64. Blue Ocean Robotics. UVD Robot Web Page. Available online: http://uvd.blue-ocean-robotics.com (accessed on 20 September 2021).
65. My App Cafe GmbH. My App Cafe—Street Barista. Available online: https://www.my-app-cafe.com (accessed on 20 September 2021).
66. Ekso Bionics. Ekso Homepage. Available online: https://eksobionics.com (accessed on 20 September 2021).
67. Nuro Inc. Nuro Homepage. Available online: https://www.nuro.ai (accessed on 20 September 2021).
68. iRobot. Roomba Robot. Available online: https://global.irobot.com/en/Robot-Resources?Country=MX (accessed on 20 September 2021).
69. SoftBank Robotics. Pepper Robot Homepage. Available online: https://www.softbankrobotics.com/emea/en/pepper (accessed on 20 September 2021).
70. Open Source Robotics Foundation. TurtleBot Homepage. Available online: https://www.turtlebot.com (accessed on 20 September 2021).
71. Dhakal, P.; Damacharla, P.; Javaid, A.Y.; Vege, H.K.; Devabhaktuni, V.K. IVACS: I ntelligent v oice A ssistant for C oronavirus Disease (COVID-19) S elf-Assessment. In Proceedings of the 2020 International Conference on Artificial Intelligence and Modern Assistive Technology, ICAIMAT 2020, Riyadh, Saudi Arabia, 24–26 November 2020; [CrossRef]
72. Chuah, S.H.W.; Aw, E.C.X.; Yee, D. Unveiling the complexity of consumers' intention to use service robots: An fsQCA approach. *Comput. Hum. Behav.* **2021**, *123*, 106870. [CrossRef]
73. Hou, Y.; Zhang, K.; Li, G. Service robots or human staff: How social crowding shapes tourist preferences. *Tour. Manag.* **2021**, *83*, 104242. [CrossRef]
74. Prentice, C.; Nguyen, M. Robotic service quality—Scale development and validation. *J. Retail. Consum. Serv.* **2021**, *62*, 102661. [CrossRef]
75. Shimmura, T.; Ichikari, R.; Okuma, T.; Ito, H.; Okada, K.; Nonaka, T. *Service Robot Introduction to a Restaurant Enhances Both Labor Productivity and Service Quality*; Elsevier B.V.: Amsterdam, The Netherlands, 2020; Volume 88, pp. 589–594. [CrossRef]
76. Villani, V.; Pini, F.; Leali, F.; Secchi, C. Survey on human–robot collaboration in industrial settings: Safety, intuitive interfaces and applications. *Mechatronics* **2018**, *55*, 248–266. [CrossRef]
77. Mihelj, M.; Bajd, T.; Ude, A.; Lenarcic, J.; Stanovnik, A.; Munih, M.; Rejc, J.; Šlajpah, S. *Robotics*, 2nd ed.; Springer: Cham, Switzerland, 2018; pp. 1–251. [CrossRef]
78. Karabegović, I.; Karabegović, E.; Mahmić, M.; Husak, E. The application of service robots for logistics in manufacturing processes. *Adv. Prod. Eng. Manag.* **2015**, *10*, 185–194. [CrossRef]
79. Human–robot interaction review and challenges on task planning and programming. *Int. J. Comput. Integr. Manuf.* **2016**, *29*, 916–931. [CrossRef]
80. Dekle, R. Robots and industrial labor: Evidence from Japan. *J. Jpn. Int. Econ.* **2020**, *58*, 101108. [CrossRef]
81. Pyo, Y.; Nakashima, K.; Kuwahata, S.; Kurazume, R.; Tsuji, T.; Morooka, K.; Hasegawa, T. Service robot system with an informationally structured environment. *Robot. Auton. Syst.* **2015**, *74*, 148–165. [CrossRef]
82. Coronado, E.; Venture, G.; Yamanobe, N. Applying Kansei/Affective Engineering Methodologies in the Design of Social and Service Robots: A Systematic Review. *Int. J. Soc. Robot.* **2021**, *13*, 1161–1171. [CrossRef]
83. Zhang, B.; Xie, Y.; Zhou, J.; Wang, K.; Zhang, Z. State-of-the-art robotic grippers, grasping and control strategies, as well as their applications in agricultural robots: A review. *Comput. Electron. Agric.* **2020**, *177*, 105694. [CrossRef]
84. Koceski, S.; Koceska, N. Evaluation of an Assistive Telepresence Robot for Elderly Healthcare. *J. Med. Syst.* **2016**, *40*, 1–7. [CrossRef] [PubMed]
85. Baisch, S.; Kolling, T.; Schall, A.; Rühl, S.; Selic, S.; Kim, Z.; Rossberg, H.; Klein, B.; Pantel, J.; Oswald, F.; Knopf, M. Acceptance of Social Robots by Elder People: Does Psychosocial Functioning Matter? *Int. J. Soc. Robot.* **2017**, *9*, 293–307. [CrossRef]
86. Choi, Y.; Choi, M.; Oh, M.; Kim, S. Service robots in hotels: Understanding the service quality perceptions of human-robot interaction. *J. Hosp. Mark. Manag.* **2020**, *29*, 613–635. [CrossRef]
87. Pinillos, R.; Marcos, S.; Feliz, R.; Zalama, E.; Gómez-García-Bermejo, J. Long-term assessment of a service robot in a hotel environment. *Robot. Auton. Syst.* **2016**, *79*, 40–57. [CrossRef]
88. Fuentes-Moraleda, L.; Díaz-Pérez, P.; Orea-Giner, A.; Muñoz- Mazón, A.; Villacé-Molinero, T. Interaction between hotel service robots and humans: A hotel-specific Service Robot Acceptance Model (sRAM). *Tour. Manag. Perspect.* **2020**, *36*, 100751. [CrossRef]
89. Chiang, A.H.; Trimi, S. Impacts of service robots on service quality. *Serv. Bus.* **2020**, *14*, 439–459. [CrossRef]
90. Seyitoğlu, F.; Ivanov, S. Service robots as a tool for physical distancing in tourism. *Curr. Issues Tour.* **2021**, *24*, 1631–1634. [CrossRef]
91. Ding, I.J.; Lin, R.Z.; Lin, Z.Y. Service robot system with integration of wearable Myo armband for specialized hand gesture human–computer interfaces for people with disabilities with mobility problems. *Comput. Electr. Eng.* **2018**, *69*, 815–827. [CrossRef]
92. Reich-Stiebert, N.; Eyssel, F. Learning with Educational Companion Robots? Toward Attitudes on Education Robots, Predictors of Attitudes, and Application Potentials for Education Robots. *Int. J. Soc. Robot.* **2015**, *7*, 875–888. [CrossRef]
93. Geyer-Klingeberg, J.; Nakladal, J.; Baldauf, F.; Veit, F. Process Mining and Robotic Process Automation: A Perfect Match Process Mining as Enabler for RPA Implementation. In Proceedings of the 16th International Conference on Business Process Management, Sydney, Australia, 9–14 September 2018; Volume i.

94. Choubey, A.; Sharma, M. Implementation of robotics and its impact on sustainable banking: A futuristic study. *J. Phys. Conf. Ser.* **2021**, *1911*, 012013. [CrossRef]
95. Zhang, X.; Wen, Z. Thoughts on the development of artificial intelligence combined with RPA. *J. Phys. Conf. Ser.* **2021**, *1883*, 012151. [CrossRef]
96. Leopold, H.; van der Aa, H.; Reijers, H.A. Identifying candidate tasks for robotic process automation in textual process descriptions. *Lect. Notes Bus. Inf. Process.* **2018**, *318*, 67–81. [CrossRef]
97. Gavrilovskaya, N.V.; Kuvaldin, V.P.; Zlobina, I.S.; Lomakin, D.E.; Suchkova, E.E. Developing a robot with computer vision for automating business processes of the industrial complex. *J. Phys. Conf. Ser.* **2021**, *1889*, 022024. [CrossRef]
98. Stepanova, E.R.; von der Heyde, M.; Kitson, A.; Schiphorst, T.; Riecke, B.E. Gathering and applying guidelines for mobile robot design for urban search and rescue application. In Proceedings of the International Conference on Human-Computer Interaction, Vancouver, BC, Canada, 9–14 July 2017; pp. 562–581. [CrossRef]
99. Ravankar, A.; Ravankar, A.; Kobayashi, Y.; Hoshino, Y.; Peng, C.C. Path smoothing techniques in robot navigation: State-of-the-art, current and future challenges. *Sensors* **2018**, *18*, 3170. [CrossRef]
100. Cheong, A.; Lau, M.; Foo, E.; Hedley, J.; Bo, J. *Development of a Robotic Waiter System*; Elsevier: Amsterdam, The Netherlands, 2016; Volume 49, pp. 681–686. [CrossRef]
101. Zhang, T.; Li, Q.; Zhang, C.S.; Liang, H.W.; Li, P.; Wang, T.M.; Li, S.; Zhu, Y.L.; Wu, C. Current trends in the development of intelligent unmanned autonomous systems. *Front. Inf. Technol. Electron. Eng.* **2017**, *18*, 68–85. [CrossRef]
102. Li, X.; Li, Z.; Wang, H.; Li, W. Unmanned Aerial Vehicle for Transmission Line Inspection: Status, Standardization, and Perspectives. *Front. Energy Res.* **2021**, *9*, 1–13. [CrossRef]
103. Fang, B.; Jia, S.; Guo, D.; Xu, M.; Wen, S.; Sun, F. Survey of imitation learning for robotic manipulation. *Int. J. Intell. Robot. Appl.* **2019**, *3*, 362–369. [CrossRef]
104. Okuda, T.; Shoda, S. AI-based chatbot service for financial industry. *Fujitsu Sci. Tech. J.* **2018**, *54*, 4–8.
105. Archibald, M.; Barnard, A. Futurism in nursing: Technology, robotics and the fundamentals of care. *J. Clin. Nurs.* **2018**, *27*, 2473–2480. [CrossRef] [PubMed]
106. Stavropoulos, G.; Giakoumis, D.; Moustakas, K.; Tzovaras, D. Automatic action recognition for assistive robots to support MCI patients at home. In Proceedings of the 10th international conference on pervasive technologies related to assistive environments, Rhodes, Greece, 21–23 June 2017; pp. 366–371. [CrossRef]
107. Shen, Z.; Wu, Y. Investigation of practical use of humanoid robots in elderly care centres. In Proceedings of the Fourth International Conference on Human Agent Interaction, Singapore, 4–7 October 2016; pp. 63–66. [CrossRef]
108. Andrade, K.; Fernandes, G.; Caurin, G.; Siqueira, A.; Romero, R.; Pereira, R. Dynamic player modelling in serious games applied to rehabilitation robotics. In Proceedings of the 2014 Joint Conference on Robotics: SBR-LARS Robotics Symposium and Robocontrol, Sao Carlos, Brazil, 18–23 October 2014; pp. 211–216. [CrossRef]
109. Koceska, N.; Koceski, S.; Zobel, P.; Trajkovik, V.; Garcia, N. A telemedicine robot system for assisted and independent living. *Sensors* **2019**, *19*, 834. [CrossRef] [PubMed]
110. Vysocky, A.; Novak, P. Human—Robot collaboration in industry. *MM Sci. J.* **2016**, *9*, 903–906. [CrossRef]
111. Correll, N.; Bekris, K.; Berenson, D.; Brock, O.; Causo, A.; Hauser, K.; Okada, K.; Rodriguez, A.; Romano, J.; Wurman, P. Analysis and observations from the first Amazon picking challenge. *IEEE Trans. Autom. Sci. Eng.* **2018**, *15*, 172–188. [CrossRef]
112. Zachiotis, G.; Andrikopoulos, G.; Gornez, R.; Nakamura, K.; Nikolakopoulos, G. A Survey on the Application Trends of Home Service Robotics. In Proceedings of the 2018 IEEE International Conference on Robotics and Biomimetics (ROBIO), Kuala Lumpur, Malaysia, 12–15 December 2018; pp. 1999–2006. [CrossRef]
113. Solorio, J.; Garcia-Bravo, J.; Newell, B. Voice Activated Semi-Autonomous Vehicle Using off the Shelf Home Automation Hardware. *IEEE Internet Things J.* **2018**, *5*, 5046–5054. [CrossRef]
114. Parmiggiani, A.; Fiorio, L.; Scalzo, A.; Sureshbabu, A.; Randazzo, M.; Maggiali, M.; Pattacini, U.; Lehmann, H.; Tikhanoff, V.; Domenichelli, D.; et al. The design and validation of the R1 personal humanoid. In Proceedings of the 2017 IEEE/RSJ International Conference on Intelligent Robots and Systems (IROS), Vancouver, BC, Canada, 24–28 September 2017; pp. 674–680. [CrossRef]
115. Dobrev, Y.; Gulden, P.; Vossiek, M. An Indoor Positioning System Based on Wireless Range and Angle Measurements Assisted by Multi-Modal Sensor Fusion for Service Robot Applications. *IEEE Access* **2018**, *6*, 69036–69052. [CrossRef]
116. Jiang, S.; Yao, W.; Hong, Z.; Li, L.; Su, C.; Kuc, T.Y. A classification-lock tracking strategy allowing a person-following robot to operate in a complicated indoor environment. *Sensors* **2018**, *18*, 3903. [CrossRef] [PubMed]
117. Dobrev, Y.; Vossiek, M.; Christmann, M.; Bilous, I.; Gulden, P. Steady Delivery: Wireless Local Positioning Systems for Tracking and Autonomous Navigation of Transport Vehicles and Mobile Robots. *IEEE Microw. Mag.* **2017**, *18*, 26–37. [CrossRef]
118. Lu, L.; Cai, R.; Gursoy, D. Developing and validating a service robot integration willingness scale. *Int. J. Hosp. Manag.* **2019**, *80*, 36–51. [CrossRef]
119. Chang, K.; Rammos, P.; Wilkerson, S.A.; Bundy, M.; Gadsden, S.A. LiPo battery energy studies for improved flight performance of unmanned aerial systems. *Unmanned Syst. Technol. XVIII* **2016**, *9837*, 98370W. [CrossRef]
120. Wang, Z.; Huang, J.; Fiammetta, C. Analysis of Gender Stereotypes for the Design of Service Robots: Case Study on the Chinese Catering Market. In Proceedings of the DIS 2021—2021 ACM Designing Interactive Systems Conference: Nowhere and Everywhere, Online, 28 June–2 July 2021; pp. 1336–1344. [CrossRef]

applied sciences

MDPI

Review

Advanced Applications of Industrial Robotics: New Trends and Possibilities

Andrius Dzedzickis *, Jurga Subačiūtė-Žemaitienė, Ernestas Šutinys, Urtė Samukaitė-Bubnienė * and Vytautas Bučinskas

Department of Mechatronics, Robotics, and Digital Manufacturing, Vilnius Gediminas Technical University, J. Basanaviciaus Str. 28, LT-03224 Vilnius, Lithuania; jurga.subaciute-zemaitiene@vilniustech.lt (J.S.-Ž.); ernestas.sutinys@vilniustech.lt (E.Š.); vytautas.bucinskas@vilniustech.lt (V.B.)
* Correspondence: andrius.dzedzickis@vilniustech.lt (A.D.); urte.samukaite-bubniene@vilniustech.lt (U.S.-B.)

Abstract: This review is dedicated to the advanced applications of robotic technologies in the industrial field. Robotic solutions in areas with non-intensive applications are presented, and their implementations are analysed. We also provide an overview of survey publications and technical reports, classified by application criteria, and the development of the structure of existing solutions, and identify recent research gaps. The analysis results reveal the background to the existing obstacles and problems. These issues relate to the areas of psychology, human nature, special artificial intelligence (AI) implementation, and the robot-oriented object design paradigm. Analysis of robot applications shows that the existing emerging applications in robotics face technical and psychological obstacles. The results of this review revealed four directions of required advancement in robotics: development of intelligent companions; improved implementation of AI-based solutions; robot-oriented design of objects; and psychological solutions for robot–human collaboration.

Keywords: industrial robots; collaborative robots; machine learning in robotics; computer vision

Citation: Dzedzickis, A.; Subačiūtė-Žemaitienė, J.; Šutinys, E.; Samukaitė-Bubnienė, U.; Bučinskas, V. Advanced Applications of Industrial Robotics: New Trends and Possibilities. *Appl. Sci.* **2022**, *12*, 135. https://doi.org/10.3390/app12010135

Academic Editor: Luis Gracia

Received: 9 November 2021
Accepted: 19 December 2021
Published: 23 December 2021

1. Introduction

The industrial robotics sector is one of the most quickly growing industrial divisions, providing standardised technologies suitable for various automation processes. In ISO 8373:2012 standard [1], an industrial robot is defined as an automatically controlled, reprogrammable, multipurpose manipulator, programmable in three or more axes, which can be stationary or mobile for use in industrial automation applications. However, the same standard creates an exception for wider implementation. It states that the robot's classification into industrial, service, or other types is undertaken according to its intended application.

According to the International Federation of Robotics (IRF) [2], 373,000 industrial robots were sold globally in 2019. In 2020 the total number of industrial robots operating in factories globally reached 2.7 million. Successful application of industrial robots, their reliability and availability, and the active implementation of the *Industry 4.0* concept have stimulated growing interest in robots' optimisation and the research of new implementations in various areas, especially in non-manufacturing and non-typical applications. According to one of the biggest scientific databases, ScienceDirect [3], more than 4500 scientific papers were published in 2019 using the term "Industrial robot" as a keyword and, in 2020, the number of papers with a similar interest and research direction increased to 5300. Figure 1 shows the annual ratio of new robot installations vs. the number of scientific publications in the ScienceDirect database. Scientific interest in this field is based on a steady increase in the number of publications, independent of the political, economic, and social factors affecting the market for new robots.

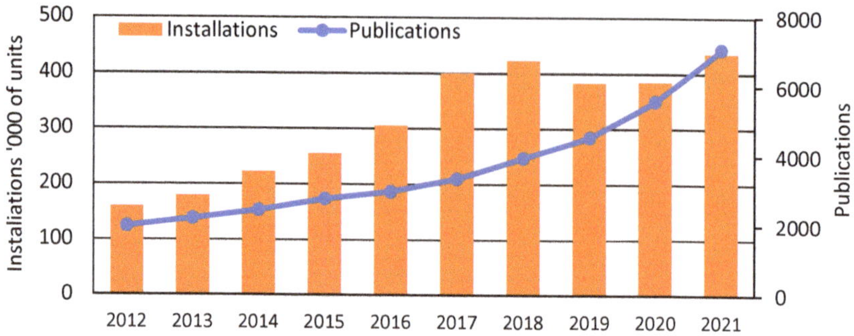

Figure 1. The annual ratio of publications to newly installed industrial robots.

This review assesses the recent development trends in robotics, and identifies some of the most relevant ethical, technological and, scientific uncertainties limiting wider implementation possibilities. This literature review is focused mainly on the 2018–2021 applications of industrial robots in fields in which endorsement of robotisation has traditionally been weak (i.e., medical applications, the food industry, agricultural applications, and the civil engineering industry). It also includes fundamental issues such as human–machine interaction, object recognition, path planning, and optimisation.

For this review, main keywords, such as industrial robots, collaborative robots, and robotics, were used to survey published papers over a four-year period. Because this is a widely researched and dynamic area, the review focused on a relatively short time period and encompassed the most recent sources to ensure the analysis conducted was novel.

According to the search request, Google Scholar returned 79,500 results, from which 115 publications were selected. The surveyed articles were selected according to the direction of the literature review and the indicated criteria (application area, novelty and significance of achievements, reliability, and feasibility of results).

Despite the ever-growing field of automation in daily life and society's accustomed use of smart devices, non-typical applications of robotics are still often viewed with considerable scepticism. The most common myth about robots is that they will occupy human workplaces, leaving human workers without a source of livelihood. Nevertheless, the research provided in [4], which aimed to evaluate the public outcry about robots taking over jobs in electronics and textiles industries in Japan, proved that such a point of view is incorrect. Evaluation of the use of the robots based on their number and real implementation price determined that implementation of robots positively affects productivity, which results in a positive impact for the most vulnerable workers in society, i.e., women, part-time workers, high-school graduates, and aged persons.

Technological and scientific uncertainties also require a special approach. Each robotisation task is unique in its own way. These tasks often require the use of individual tools, the creation of a corresponding working environment, the use of additional sensors or measurement systems, and the implementation of complex control algorithms to expand the functionalities or improve the characteristics of standard robots. In most applications, industrial robots form bigger units as robotic cells or automated/autonomous manufacturing lines. As a result, the robotisation of even a relatively simple task becomes a complex solution requiring a systemic approach.

Moreover, the issue of implementing an industrial robot remains complicated by its interdisciplinary nature: proper organisation of the work cycle is the object of manufacturing management sciences; the design of grippers and related equipment lies within the field of mechanical engineering; and the integration of all devices into a united system, sensor data analysis and whole system control are the objects of mechatronics.

This review focuses on the hardware and software methods used to implement industrial robots in various applications. The aim was to systematically classify the newest achievements in industrial robotics according to application fields without strong robotisation traditions. The analysis of this study was also undertaken from a multidisciplinary perspective, and considers the implementation of computer vision and machine learning for robotic applications.

2. Main Robotisation Strategies

According to the human–robot cooperation type, a review of the most recent trends in industrial robotics applications indicates two main robotisation strategies: classical and modern. In industrial robotics, five typical levels of human–robot cooperation are defined (Figure 2): (i) no collaboration; (ii) coexistence; (iii) synchronisation; (iv) cooperation; (v) collaboration [5].

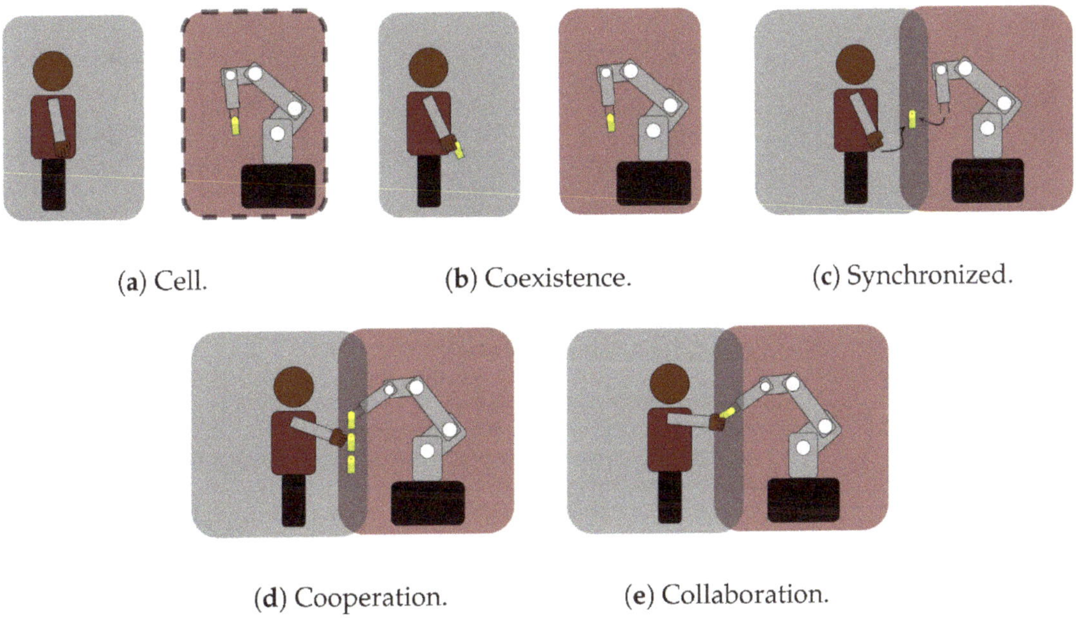

Figure 2. Human–robot cooperation levels [5]: (**a**) no collaboration, the robot remains inside a closed work cell; (**b**) coexistence, removed cells, but separate workspaces; (**c**) synchronisation, sharing of the workspace, but never at the same time; (**d**) cooperation, shared task and workspace, no physical interaction; (**e**) collaboration, operators and robots exchange forces.

The classical strategy encompasses the first cooperation level (Figure 2a). It is based on the approach that robots must limit humans in their workplace by creating closed robot cells in which human activity is unacceptable; if a human must enter the robot's workspace, the robot must be stopped. This approach uses various safety systems to detect and prevent human access to the robot's workspace. The modern strategy includes the remaining four cooperation levels (Figure 2b–e). This is based on an opposing approach, and states that robots and humans can work in one workplace and collaborate. Such an approach creates additional requirements for robot's design, control, and sensing systems. Robots adapted to operate in conjunction with human workers are usually defined as collaborative robots or cobots.

2.1. Classical Robotisation Strategy

Following the issuing of the patent for the first industrial robot to George Devol in 1954, the classical robotisation strategy has indicated that robots should replace human workers in routine tasks and unhealthy workplaces. This strategy suggests that humans should be removed from the robot's workspace (Figure 3a). Direct cooperation between the robot and humans is forbidden due to the potential danger for human health and safety. This approach was later expanded to encompass accuracy, reliability, productivity, and economic factors.

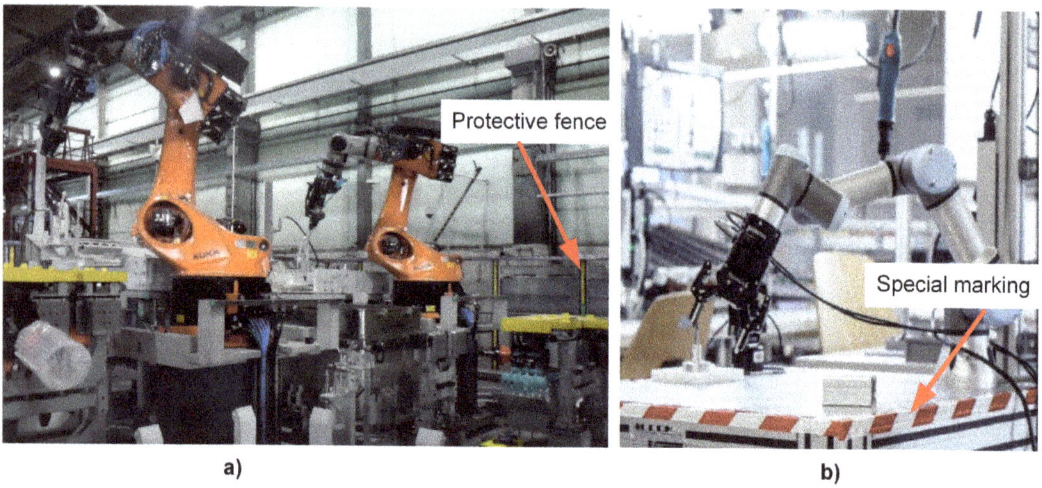

Figure 3. Comparison of the operating environment: (**a**) industrial robots (adapted from [6]); (**b**) collaborative robots (adapted from [7]).

Research provided in [8] analyses the possibilities of implementing service robots in hotels from social, economic, and technical perspectives. The authors indicated the need to evaluate hotel managers' perceptions regarding the advantages and disadvantages of service robots, compared to human workers, as the primary goal of their research, whereas determining tasks suitable for robotisation was of secondary importance. This approach confirms the assumption that the implementation of robotics in non-traditional applications is often limited not by technological issues, but by the company managers' attitudes. Analysing questionnaires completed by 79 hotel managers, it was concluded that robots have an advantage over human employees due to better data processing capabilities, work speed, protection of personal data, and fewer mistakes. The main disadvantages of robots were listed as: lack of capability to provide personalised service; inability to handle complaints; lack of friendliness and politeness; inability to implement a special request that goes beyond their programming; and the lack of understanding of emotions.

Despite the common doubts, implementing automation and robotic solutions has a positively impact in many cases. The study provided in [9] analysed the general impact of robot implementation in workplaces for packing furniture parts. The analysis focused on the ergonomic perspective, and found that implementation of robotics eliminates the risk of work-related musculoskeletal disorders. A similar study [8] analysed the design, engineering, and testing of adaptive automation assembly systems to increase automation levels, and to complement human workers' skills and capabilities in assembling industrial refrigerators. This study showed that automated assembly process productivity could be increased by more than 79%. Implementing an industrial robot instead of partial automation would likely result in an even more significant increase in productivity. Research comparing human capabilities with automated systems is also described in [10]. The authors compared human and automated vision recognition system capabilities to recognise and evaluate forest or mountain trails from a single monocular image acquired from the viewpoint of a robot travelling on the trail. The obtained results showed that a deep neural network-based system, trained on a large dataset, performs better than humans.

Neural network-based algorithms can also be used to control industrial robots to address imperfections in their mechanical systems, which typically behave as non-linear dynamic systems due to a large number of uncertainties. The research presented in [11,12] provides neural network-based methods for advanced control of robot movements. In [11], a perspective non-linear model-based predictive control method for robotic manipulators, which minimises the settling time and position overshoot of each joint, is provided.

The classic strategy is well suited to robotisation of mass production processes in various fields, and its main advantages are clear requirements for work process organisation, robotic cell design, and installation; the availability of a large variety of standard equipment and typical partial solutions; and higher productivity and reliability compared to the cases where human workers perform the same tasks. The main disadvantages are insufficient flexibility, unsuitability for unique production, and high economic costs when it is necessary to adapt the existing robotic cell to a new product or process. Applying a modern robotisation strategy can avoid some of these disadvantages (or at least minimise their impact).

2.2. Modern Robotisation Strategy

The modern robotisation strategy is based on implementing collaborative robots (cobots). According to [13], the definition of cobot was first used in a 1999 US patent [14] and was intended for "an apparatus and method for direct physical integration between a person and a general-purpose manipulator controlled by a computer." It was the result of the efforts of General Motors to implement robotics in the automotive sector to help humans in assembly operations. The first lightweight cobot, LBR3, designed by a German robotics company, was introduced in 2004 [13]. This has led to the broader development of a modern robotics strategy and new manufacturers in the market. In 2008, the Danish manufacturer Universal Robots released the UR5, a cobot that could safely operate alongside the employees, eliminating the need for safety caging or fencing (Figure 3b). This launched a new era of flexible, user-friendly, and cost-efficient collaborative robots [13], and resulted in the current situation, in which all of the major robot manufacturers have at least a few cobot models in their product range.

The fourth industrial revolution—*Industry 4.0*—significantly fostered the development of cobot's technologies, because the concept fitted well with *Industry 4.0* content, allowing human–robot collaboration to be realised and being suitable for flexible manufacturing systems. Contrary to typical industrial robots, next-generation robotics uses artificial intelligence (AI) to collaboratively perform tasks and is suitable for uncontrolled/unpredictable environments [15]. Moreover, due to favourable conditions (advances in AI, sensing technologies, and computer vision), collaborative industrial robots have become significantly smarter, showing the potential of reliable and secure cooperation, and increasing the productivity and efficiency of the involved processes [15]. However, it should be noted that *Industry 4.0* fostered not only the widespread of robotics, but also posed new challenges. When developing highly automated systems, most of the equipment is related through the Internet of Things (IoT) or other communication technologies. Therefore, cybersecurity and privacy protection of processes used to monitor and control data [16,17] must be considered. The issue of data protection is also becoming more critical due to the latest communication technologies, such as 5G and 6G [18]. These technologies allow the development of standardised wireless communication networks for various control levels (single-cell, production line, factory, network of factories) and, at the same time, makes systems more sensitive to external influences. The main impact of *Industry 4.0* and new communication technologies on industrial robots is that their controllers have an increasing number of connections, functions, and protocols to communicate with other "smart" devices.

The study presented in [19] analyses the possibilities of human–robot collaboration in aircraft assembly operations. The benefits of human–robot cooperation were examined in terms of the productivity increase and the levels of satisfaction of the human workers. The obtained results showed that humans and robots could simultaneously work safely in a common area without any physical separation, and significantly reduce time and costs compared with manual operations. Moreover, assessment of employee opinions showed that most employees positively evaluated the implementation of collaborative robots. Nevertheless, employee attitudes depend on their practical experience: it was noticed that experts felt more confident than beginners. This can be explained by the fact that experts better understand the overall manufacturing process and are more accustomed to operating with various equipment.

Compared to traditional industrial robots, cobots have more user-friendly control features and wider teaching options. A new assembly strategy was described in a previous study [20], in which a cobot learnt skills from manual teaching to perform peg-in-hole automatic assembly when the geometric profile and material elastic parameters of parts were inaccurate. The results showed that the manual assembly process could be analysed mathematically, splitting it into a few stages and implementing it as a model in robot control. Using an Elite EC75 manipulator (Elite Robot, Suzhou, CN, an assembly time of less than 20 s was achieved, ensuring a 100% success rate from 30 attempts when the relative error between the peg and hole was ±4.5 mm, and the clearance between the peg and the hole was 0.18 mm.

As a result of the development of sensor and imaging technologies, new applications in robotics are emerging, especially in human–robot collaboration. In [21], detailed research focused on identifying the main strengths and weaknesses of augmented reality (AR) in industrial robots applications. The analysis shows that AR is mainly used to control and program robotic arms, visualise general tasks or robot information, and visualise the industrial robot workspace. Results of the analysis indicate that AR systems are faster than traditional approaches; users have greater appreciation for AR systems in terms of likeability and usability; and AR seems to reduce physical workload, whereas the impact on mental workload depends on the interaction interface [16]. Nevertheless, industrial implementation of AR is still limited by insufficient accuracy, occlusion problems, and the limited field of view of wearable AR devices.

A summary of the analysed robotisation strategies indicates that they both have their specific implementation fields. The classical strategy is well suited to strictly controlled environments. The modern strategy ensures more flexible operation and is suitable for non-predictable environments. Nevertheless, it is necessary to note that the strict line between these strategies has gradually disappeared due to advances in sensing technologies, artificial intelligence, and computer vision. A typical industrial robot equipped with modern sensing and control systems can operate similarly to a cobot. According to [22], collaborative regimes can be realised using industrial robots, laser sensors, and vision systems, or controller alteration if compliance with the ISO/TS 15066 standard—which specifies parameters and materials adapted to safe activities with and near humans—is ensured [23]. This standard defines four main classes of safety requirements for collaborative robots: safety-rated monitored stop; hand-guiding; speed and separation monitoring; and power and force limiting.

In addition, it is essential to mention that all improvements and advances in robotics can be classified into two main types: universal and application dependent. The remaining part of this article reviews and classifies the latest advances in robotics according to the areas of their implementation.

3. Recent Achievements in Industrial Robotics Classified according to Implementation Area

3.1. *Human–Machine Interaction*

To date, manual human work has been often replaced by robotic systems in industry. However, within complex systems, the interaction between humans and machines/robots (HMI) still needs to occur. HMI is an area of research related to the development of robotic systems based on understanding, evaluation, and analysis, and this system combines various forms of cooperation or interaction with humans. Interaction requires communication between robots and humans, and human communication and collaboration with the robot system can take many forms. However, these forms are greatly influenced by whether the human is close to the robot and the context being used: (i) human–computer context—

keyboard, buttons, etc.; (ii) real procedures context—haptics, sensors; and (iii) close and exact interaction. Therefore, both human and robot communication or interaction can be divided into two main categories: remote interaction and exact interaction. Remote interaction takes place by remote operation or supervised control. Close interaction takes place by operation with an assistant or companion. Close interaction may include physical interaction. Because close interactions are the most difficult, it is crucial to consider a number of aspects to ensure a successful collaboration, i.e., a real-time algorithm, "touch" detection and analysis, autonomy, semantic understanding capabilities, and AI-aided anticipation skills. A summary of the relevant research focused on improving and developing HMI methods is provided in Table 1.

Table 1. Research focused on human–machine interaction.

Objective	Technology	Approach	Improvement	Ref.
To improve flexibility, productivity and quality of a multi-pass gas tungsten arc welding (GTAW) process performed by a collaborative robot.	A haptic interface. 6-axis robotic arm (Mitsubishi MELFA RV-13FM-D). The end effector with GTAW torch. A monitoring camera (Xiris XVC-1000). A Load Cell (ATI Industrial Automation Mini45-E) to evaluate tool force interactions with work pieces.	A haptic-based approach is designed and tested in a manufacturing scenario proposing light and low-cost real-time algorithms for "touch" detection.	Two main criteria were analysed to assess the performance: the 3-Sigma rule and the Hampel identifier. Experimental results showed better performance of the 3-Sigma rule in terms of precision percentage (mean value of 99.9%) and miss rate (mean value of 10%) concerning the Hampel identifier. Results confirmed the influence of the contamination level related to the dataset. This algorithm adds significant advances to enable the use of light and simple machine learning approaches in real-time applications.	[24,25]
To produce more advanced or complex forms of interaction by enabling cobots with semantic understanding capabilities or AI-aided anticipation skills.	Collaborative robots	Artificial intelligence.	The overview provides hints of future cobot developments and identifies future research frontiers related to economic, social, and technological dimensions.	[26]
To strike a balance in order to find a suitable level of autonomy for human operators.	Model of Remotely Instructed Robots (RIRs.)	Modelling method.	Developed model in which the robot is autonomous in task execution, but also aids the operator's ultimate decision-making process about what to do next. Presentation of the robot's own model of the work scene enables corrections to be made by the robot, as well as it can enhance the operator's confidence in the robot's work.	[27,28]

The interaction between humans and robots or mechatronic systems encompasses many interdisciplinary fields, including physical sciences, social sciences, psychology, artificial intelligence, computer science, robotics, and engineering. This interaction examines all possible situations in which a human and a robot can systematically collaborate or complement each other. Thus, the main goal is to provide robots with various competencies to facilitate their interaction with humans. To implement such competencies, modelling of real-life situations and predictions is necessary, applying models in interaction with robots, and trying to make this interaction as efficient as possible, i.e., inherently intuitive, based on human experience and artificial intelligence algorithms.

The role of various interfering aspects (Table 2.) in human–robot interaction may lead to different future perspectives.

Table 2. Interfering aspects in human–robot interaction.

Objective	Interaction	Approach	Solution	Ref.
Frustration	Close cooperative work	Controlled coordination	Sense of control of frustration, affective computing.	[29]
Emotion recognition	By collecting different kinds of data.	Discrete models describing emotions used, facial expression analysis, camera positioning.	Affective computing. Empowering robots to observe, interpret and express emotions. Endow robots with emotional intelligence.	[30]
Decoding of action observation	Elucidating the neural mechanisms of action observation and intention understanding.	Decoding the underlying neural processes.	The dynamic involvement of the mirror neuron systems (MNS) and the theory of mind ToM/mentalising network during action observation.	[31]
Verbal and non-verbal communication	Interactive communication.	Symbol grounding	Composition of grounded semantics, online negotiation of meaning, affective interaction and closed-loop affective dialogue, mixed speech-motor planning, massive acquisition of data-driven models for human–robot communication through crowd-sourced online games, real-time exploitation of online information and services for enhanced human–robot communication.	[32]

We can summarise that the growing widespread use of robots and the lack of highly skilled professionals in the market form clear guidelines for future development in the HMI area. The main aspirations are an intuitive, human-friendly interface, faster and simpler programming methods, advanced communication features, and robot reactions to human movements, mood, and even psychological state. Methods to monitor human actions and emotions [33], fusion of sensors' data, and machine learning are key technologies for further improvement in the HMI area.

3.2. Object Recognition

Object recognition is a typical issue in industrial robotics applications, such as sorting, packaging, grouping, pick and place, and assembling (Table 3). The appropriate recognition method and equipment selection mainly depends on the given task, object type, and the number of recognisable parameters. If there are a small number of parameters, simpler sensing technologies based on typical approaches (geometry measuring, weighing, material properties' evaluation) can be implemented. Alternatively, if there are a significant number of recognisable parameters, photo or video analysis is preferred. Required information in two- or three-dimensional form from image or video can be extracted using computer vision techniques such as object localisation and recognition. Various techniques of vision-based object recognition have been developed, such as appearance-, model-, template-, and region-based approaches. Most vision recognition methods are based on deep learning [34] and other machine learning methods.

In a previous study [35], a lightweight Franka Emika Panda, cobot with seven degrees of freedom and a Realsense D435 RGB-D camera, mounted on an end effector, was used to extend the default robots' function. Instead of using a large dataset-based machine learning technique, the authors proposed a method to program the robot from a single demonstration. This robotic system can detect various objects, regardless of their position and orientation, achieving an average success rate of more than 90% in less than 5 min of training time, using an Ubuntu 16.04 server running on an Intel(R) Core(TM) i5-2400 CPU (3.10 GHz) and an NVIDIA Titan X GPU.

Another approach for grasping randomly placed objects was presented in [36]. The authors proposed a set of performance metrics and compared four robotic systems for bin picking, and took first place in the Amazon Robotics Challenge 2017. The survey results show that the most promising solutions for such a task are RGB-D sensors and CNN-based algorithms for object recognition, and a combination of suction-based and typical two-finger grippers for grasping different objects (vacuum grippers for a stiff object with large and smooth surface areas, and two-finger grippers for air-permanent items).

Similar localisation and sorting tasks appear in the food and automotive industries, and in almost every production unit. In [37], an experimental method was proposed using a pneumatic robot arm for separation of objects from a set according to their colour. If the colour of the workpiece is recognisable, it is selected with the help of a robotic arm. If the workpiece colour does not meet the requirements, it is rejected. The described sorting system works according to an image processing algorithm in MATLAB software. More advanced object recognition methods based on simultaneous colour and height detection are presented in [38]. A robotic arm with six degrees of freedom (DoF) and a camera with computer vision software ensure a sorting efficiency of about 99%.

A Five DoF robot arm, "OWI Robotic Arm Edge", proposed by Pengchang Chen et al., was used to validate the practicality and feasibility of a faster region-based convolutional neural network (faster R-CNN) model using a dataset containing images of symmetric objects [39]. Objects were divided into classes based on colour, and defective and non-defective objects.

Despite significant progress in existing technologies, randomly placed unpredictable objects remain a challenge in robotics. The success of a sorting task often depends on the accuracy with which recognisable parameters can be defined. Yan Yu et al. [40] proposed an RGB-D-based method for solid waste object detection. The waste sorting system consists of a server, vision sensors, industrial robots, and rotational speedometer. Experiments performed on solid waste image analysis resulted in a mean average precision value of 49.1%.

Furthermore, Wen Xiao et al. designed an automatic sorting robot that uses height maps and near-infrared (NIR) hyperspectral images to locate the region of interest (ROI) of objects, and to perform online statistic pixel-based classification in contours [41]. This automatic sorting robot can automatically sort construction and demolition waste ranging in size from 0.05 to 0.5 m. The online recognition accuracy of the developed sorting system reaches almost 100% and ensures operation speed up to 2028 picks/h.

Another challenging issue in object recognition and manipulation is objects having an undefined shaped and contaminated by dust or smaller particles, such as minerals or coal. Quite often, such a task requires not only recognising the object but also determining the position of the centre of mass of the object. Man Li et al. [42] proposed an image processing-based coal and gangue sorting method. Particle analysis of coal and gangue samples is performed using morphological corrosion and expansion methods to obtain a complete, clean target sample. The object's mass centre is obtained using the centre of the mass method, consisting of particle removal and filling, image binarization, and separation of overlapping samples, reconstruction, and particle analysis. The presented method achieved identification accuracy of coal and gangue samples of 88.3% and 90.0%, and the average object mass centre coordinate errors in the x and y directions were 2.73% and 2.72%, respectively [42].

Intelligent autonomous robots for picking different kinds of objects were studied as a possible means to overcome the current limitations of existing robotic solutions for picking objects in cluttered environments [43]. This autonomous robot, which can also be used for commercial purposes, has an integrated two-finger gripper and a soft robot end effector to grab objects of various shapes. A special algorithm solves 3D perception problems caused by messy environments and selects the right grabbing point. When using lines, the time required depends significantly on the configuration of the objects, and ranges from 0.02 s when the objects have almost the same depth, to 0.06 s in the worst case when the depth of the tactile objects is greater than the lowest depth but not perceived [43].

In robotics, the task of object recognition often includes not only recognition and the determinaton of coordinates, but it also plays an essential role in the creation of a robot control program. Based on the ABB IRB 140 robot and a digital camera, a low-cost shapes identification system was developed and implemented, which is particularly important due to the high variability of welded products [44]. The authors developed an algorithm that recognises the required toolpath from a taken image. The algorithm defines a path as a complex polynomial. It later approximates it by simpler shapes with a lower number of coordinates (line, arc, spline) to realise the tool movement using standard robot programming language features.

Moreover, object recognition can be used for robot machine learning to analyse humans' behaviour. Such an approach was presented by Hiroaki et al. [45], where the authors studied the behaviour of a human crowd, and formulated a new forecasting task, called crowd density forecasting, using a fixed surveillance camera. The main goal of this experiment was to predict how the density of the crowd would change in unseen future frames. To address this issue, patch-based density forecasting networks (PDFNs) were developed. PDFNs project a variety of complex dynamics of crowd density throughout the scene, based on a set of spatially or spatially overlapping patches, thus adapting the receptive fields of fully convolutional networks. Such a solution could be used to train robotic swarms because they behave similarly to humans in crowded areas.

Table 3. Research focused on object recognition in robotics.

Objective	Technology	Approach	Improvement	Ref.
Extended default "program from demonstration" feature of collaborative robots to adapt them to environments with moving objects.	Franka Emika Panda cobot with 7 degrees of freedom, with a Realsense D435 RGB-D camera mounted on the end-effector.	Grasping method to fine-tune using reinforcement learning techniques.	The system can grasp various objects from a demonstration, regardless of their position and orientation, in less than 5 min of training time.	[35,46]
Introduction of a set of metrics for primary comparison of robotic systems' detailed functionality and performance.	Robot with different grippers.	Recognition method and the grasping method.	Developed original robot performance metrics and tested on four robot systems used in the Amazon Robotics Challenge competition. Results of analysis showed the difference between the systems and promising solutions for further improvements.	[36,45,47]
To build a low-cost system for identifying shapes to program industrial robots for the 2D welding process.	Robot ABB IRB 140 with a digital camera, which detects contours on a 2D surface.	A binarisation and contour recognition method.	A low-cost system based on an industrial vision was developed and implemented for the simple programming of the movement path.	[48,49]
The patch-based density forecasting networks (PDFNs) directly forecast crowd density maps of future frames instead of trajectories of each moving person in the crowd.	Fixed surveillance camera	Density Forecasting in Image Space. Density Forecasting in Latent Space. PDFNs. Spatio-Temporal Patch-Based Gaussian filter.	Proposed patch-based models, PDFN-S and PDFN-ST, outperformed baselines on all the datasets. PDFN-ST successfully forecasted dynamics of individuals, a small group, and a crowd. The approach cannot always forecast sudden changes in walking directions, especially when they happened in the later frames.	[45]
To separate the objects from a set according to their colour.	Pneumatic Robot arm	Force in response to applied pressure.	The proposed robotic arm may be considered for sorting. Servo motors and image processing cameras can be used to achieve higher repeatability and accuracy.	[37,50]
An image processing-based method for coal and gangue sorting. Development of a positioning and identification system.	Coal and gangue sorting robot	Threshold segmentation methods. Clustering method. Morphological corrosion and expansion methods. The centre of mass method.	Efficiency is evaluated using the images of coal and gangue, which are randomly picked from the production environment. The average coordinate errors in the x and y directions are 2.73% and 2.72%, and the identification accuracy of coal and gangue samples is 88.3% and 90.0%, respectively, and the sum of the time for identification, positioning, and opening the camera for a single sample averaged 0.130 s.	[41,51,52]
A computer vision-based robotic sorter is capable of simultaneously detecting and sorting objects by their colours and heights. Vision-based process encompasses identification, manipulation, selection, and sorting objects depending on colour and geometry.	A 5 or 6 DOF robotic arm and a camera with the computer vision software detecting various colours and heights and geometries.	Computer Vision methods with the Haar Cascade algorithm. The Canny edge detection algorithm is used for shape identification.	A robotic arm is used for picking and placing objects based on colour and height. In the proposed system, colour and height sorting efficiency is around 99%. Effectiveness, high accuracy and low cost of computer vision with a robotic arm in the sorting process according to color and shape are revealed.	[38,53,54]
A novel multimodal convolutional neural network for RGB-D object detection.	A base solid waste sorting system consisting of a server, vision sensors, industrial robot, and rotational speedometer.	Comparison with single modal methods. Washington RGB-D object recognition benchmark evaluated.	Meeting the real-time requirements and ensuring high precision. Achieved 49.1% mean average precision, processing images in real-time at 35.3 FPS on one single Nvidia GTX1080 GPU. Novel dataset.	[40,55]

Table 3. *Cont.*

Objective	Technology	Approach	Improvement	Ref.
Practicality and feasibility of a faster R-CNN model using a dataset containing images of symmetric objects.	Five DoF robot arm "OWI Robotic Arm Edge."	CNN learning algorithm that processes images with multiple layers (filters) and classifies objects in images. Regional Proposal Network (RPN)	The accuracy and precision rate are steadily enhanced. The accuracy rate of detecting defective and non-defective objects is successfully improved, increasing the training dataset to up to 400 images of defective and non-defective objects.	[39,56,57]
An automatic sorting robot with height maps and near-infrared (NIR) hyperspectral images to locate objects' ROI and conduct online statistic pixel-based classification in contours. 24/7 monitoring.	The robotic system with four modules: (1) the main conveyor, (2) a detection module, (3) a light source module, and (4) a manipulator. Mask-RCNN and YOLOv3 algorithms.	Method for an automatic sorting robot. Identification include pixel, sub-pixel, object-based methods.	The prototype machine can automatically sort construction and demolition waste with a size range of 0.05–0.5 m. The sorting efficiency can reach 2028 picks/h, and the online recognition accuracy nearly reaches 100%. Can be applied in technology for land monitoring.	[41,58,59]
Overcoming current limitations on the existing robotic solutions for picking objects in cluttered environments.	Intelligent autonomous robots for picking different kinds of objects. Universal jamming gripper.	A comparative study of the algorithmic performance of the proposed method.	When a corner is detected, it takes just 0.003 s to output the target point. With lines, the required time depends on the object's configuration, ranging from 0.02 s, when objects have almost the same depth, to 0.06 s in the worst-case scenario.	[43,60–62]

A few main trends can be highlighted from the research analysis related to object recognition in robotics. These can be defined as object recognition for localisation and further manipulation; object recognition for shape evaluation and automatic generation of the robot program code for the corresponding robot movement; and object recognition for behaviour analysis to use as initial data for machine learning algorithms. A large number of reliable solutions have been tested in the industrial environment for the first trend, in contrast to the second and third cases, which are currently being developed.

3.3. Medical Application

The da Vinci Surgical System is the best-known robotic manipulator used in surgery applications. Florian Richter et al. [63] presented a Patient Side Manipulator (PSM) arm technology to implement reinforcement learning algorithms for the surgical da Vinci robots. The authors presented the first open-source reinforcement learning environment for surgical robots, called dVRL [63]. This environment allows fast training of da Vinci robots for autonomous assistance, and collaborative or repetitive tasks, during surgery. During the experiments, the dVRL control policy was effectively learned, and it was found that it could be transferred to a realrobot- with minimal efforts. Although the proposed environment resulted in the simple and primitive actions of reaching and picking, it was useful for suction and debris removal in a real surgical setting.

Meanwhile, in their work, Yohannes Kassahun et al. reviewed the role of machine learning techniques in surgery, focusing on surgical robotics [64]. They found that currently, the research community faces many challenges in applying machine learning in surgery and robotic surgery. The main issues are a lack of high-quality medical and surgical data, a lack of reliable metrics that adequately reflect learning characteristics, and a lack of a structured approach to the effective transfer of surgical skills for automated execution [64]. Nevertheless, the application of deep learning in robotics is a very widely studied field. The article by Harry A. Pierson et al. in 2017 provides a recent review emphasising the benefits and challenges vis-à-vis robotics [65]. Similarly to [64], they found that the main limitations preventing deep learning in medical robotics are the huge volume of training data required and a relatively long training time.

Surgery is not the only field in medicine in which robotic manipulators can be used. Another autonomous robotic grasping system, described by John E. Downey et al., introduces shared control of a robotic arm based on the interaction of a brain–machine interface (BMI) and a vision guiding system [66]. A BMI is used to define a user's intent to grasp or transfer an object. Visual guidance is used for low-level control tasks, short-range movements, definition of the optimal grasping position, alignment of the robot end-effector, and grasping. Experiments proved that shared control movements were more accurate, efficient, and less complicated than transfer tasks using BMI alone.

Another case that requires fast robot programming methods and is implemented in medicine is the assessment of functional abilities in functional capacity evaluations (FCEs) [67]. Currently, there is no single rational solution that simulates all or many of the standard work tasks that can be used to improve the assessment and rehabilitation of injured workers. Therefore, the authors proposed that, with the use of the robotic system and machine learning algorithms, it is possible to simulate workplace tasks. Such a system can improve the assessment of functional abilities in FCEs and functional rehabilitation by performing reaching manoeuvres or more complex tasks learned from an experienced therapist. Although this type of research is still in its infancy, robotics with integrated machine learning algorithms can improve the assessment of functional abilities [67].

Although the main task of robotic manipulators is the direct manipulation of objects or tools in medicine, these manipulators can also be used for therapeutic purposes for people with mental or physical disorders. Such applications are often limited by the ability to automatically perceive and respond as needed to maintain an engaging interaction. Ognjen Rudovic et al. presented a personalised deep learning framework that can adapt robot perception [68]. The researchers in the experiment focused on robot perception, for which they developed an individualised deep learning system that could automatically assess a patient's emotional states and level of engagement. This makes it easier to monitor treatment progress and optimise the interaction between the patient and the robot.

Robotic technologies can also be applied in dentistry. To date, there has been a lack of implementation of fundamental ideas. In a comprehensive review of robotics and the application of artificial intelligence, Jasmin Grischke et al. present numerous approaches to apply these technologies [69]. Robotic technologies in dentistry can be used for maxillofacial surgery [70], tooth preparation [71], testing of toothbrushes [72], root canal treatment and plaque removal [73], orthodontics and jaw movement [74], tooth arrangement for full dentures [75], X-ray imaging radiography [76], swab sampling [77], etc.

A summary of research focused on robotics in medical applications is provided in Table 4. It can be seen that robots are still not very popular in this area, and technological and phycological/ethical factors can explain this. From the technical point of view, more active implementation is limited by the lack of fast and reliable robot program preparation methods. Regarding psychological and ethical factors, robots are still unreliable for a large portion of society. Therefore, they are only accepted with significant hesitation.

Table 4. Robotic solutions in medical applications.

Objective	Technology	Approach	Improvement	Ref.
Create bridge between reinforcement learning and the surgical robotics communities by presenting the first open-sourced reinforcement learning environments for surgical da Vinci robots.	Patient Side Manipulator (PSM) arm. Da VinciR©Surgical Robot. Large Needle Driver (LND), with a jaw gripper to grab objects such as suturing needle.	Reinforced learning, OpenAI Gym DDPG (Deep Deterministic Policy Gradients) and HER (Hindsight Experience Replay) V-REP physics simulator	Developed new reinforced learning environment for fast and effective training of surgical da Vinci robots for autonomous operations.	[63]
A method of shared control where the user controls a prosthetic arm using a brain–machine interface and receives assistance with positioning the hand when it approaches an object.	Brain–machine interface system. Robotic arm. RGB-D camera mounted above the arm base.	Shared control system. An autonomous robotic grasping system	Shared control system for a robotic manipulator, making control more accurate, more efficient, and less difficult than an alone control system.	[66]
A personalised deep learning framework can adapt robot perception of children's affective states and engagement to different cultures and individuals.	Unobtrusive audiovisual sensors and wearable sensors, providing the child's heart-rate, skin-conductance (EDA), body temperature, and accelerometer data.	Feed-forward multilayer neural networks. GPA-net	Achieved an average agreement of ~60% with human experts to estimate effect and engagement.	[68]

Table 4. *Cont.*

Objective	Technology	Approach	Improvement	Ref.
An overview of existing applications and concepts of robotic systems and artificial intelligence in dentistry, for functional capacity evaluations, of the role of ML in surgery using surgical robotics, of deep learning vis-à-vis physical robotic systems, focused on contemporary research.	An overview	An overview	An overview	[64,65,67,69]
Transoral robot towards COVID-19 swab sampling.	Flexible manipulator, an endoscope with a monitor, a master device.	Teleoperated configuration for swab sampling	A flexible transoral robot with a teleoperated configuration is proposed to address the surgeons' risks during the face-to-face COVID-19 swab sampling.	[77]

3.4. Path Planning, Path Optimisation

The process known as robotic navigation aims to achieve accurate positioning and avoiding obstacles in the pathway. It is essential to satisfy constraints such as limited operating space, distance, energy, and time [78]. The path trajectory formation process consists of these four separate modules: perception, when the robot receives the necessary information from the sensors; localisation, when the robot aims to control its position in the environment; path planning; and motion control [79]. The development of autonomous robot path planning and path optimisation algorithms is one of the most challenging current research areas. Nevertheless, any kind of path planning requires information about the initial robot position. In the stationary robot's case, such information is usually easily accessible, contrary to industrial manipulators mounted on mobile platforms. In mobile robots and automatically guided vehicles (AGV), accurate self-localisation in various environments [80,81] is a basis for further trajectory planning and optimisation.

According to the amount of available information, robot path planning can be categorised into two categories, namely, local and global path planning. Through a local path planning strategy, the robot has rather limited knowledge of the navigation environment. The robot has in-depth knowledge of the navigation environment when planning the global path to reach its destination by following a predetermined path. The robotic path planning method has been applied in many fields, such as reconstructive surgery, ocean and space exploration, and vehicle control. In the case of pure industrial robots, path planning refers to finding the best trajectory to transfer a tool or object to the destination in the robot workspace. It is essential to note that typical industrial robots are not feasible for real-time path planning. Usually, trajectories are prepared in advance using online or offline programming methods. One of the possible techniques is the implementation of specialised commercial computer-aided manufacturing (CAM) software such as Mastercam/Robotmaster or Sprutcam. However, the functionality of such software is relatively constrained and does not go beyond the framework of classical tasks, such as welding or milling. The use of CAM software also requires highly qualified professionals. As a result, the application of this software to individual installations is economically disadvantageous. As an alternative to CAM software, methods based on the copying movements of highly skilled specialists using commercially available equipment, such as MIMIC from Nordbo Robotics (Antvorskov, Denmark), may be used. This platform allows using demonstrations to teach robots smooth, complex paths by recording required movements that are smoothed and optimised. To overcome the limitations caused by the lack of real-time path planning features in robot controllers, additional external controllers and real-time communication with the manipulator is required. In the area of path planning and optimisation, experiments have been conducted for automatic object and 3D position detection [82] quasi-static path optimisation [83], image analysis [84], path smoothing [85], BIM [86], and accurate self-localisation in harsh industrial environments [80,81]. More information about methods and approaches proposed by researchers is listed in Table 5.

Table 5. Research focused on path planning and optimisation.

Objective	Technology	Approach	Improvement	Ref.
The position of the objects—possible trajectory to an object in real-time.	A robotic system consisting of an ABB IRB120 robot equipped with a gripper and a 3D Kinect sensor.	Detection of the workpieces. Object recognition techniques are applied using available algorithms in MATLAB's Computer Vision and Image Acquisition Toolbox.	The algorithm for finding 3D object position according to colour segmentation in real-time. The main focus was on finding the depth of an object from the Kinect sensor. Kinect could distinguish colour correctly, and the robot could accurately navigate to the detected object.	[82]
The combination of eye-tracking and computer vision automate the approach of a robot to its targeted point by acquiring its 3D location.	Eye-tracking device, webcam.	Image analysis and geometrical reconstruction.	The computed coordinates of the target 3D localisation have an average error of 5.5 cm, which is 92% more accurate than eye-tracking only for the point of gaze calculation, with an estimated error of 72 cm.	[87]
Computer vision technology for real-time seam tracking in robotic gas tungsten arc welding (GTAW).	Welding robot GTAW—the robot arm, the robot controller, the vision system, isolation unit, the weld power supply, and the host computer. Passive vision system.	Passive vision system image processing.	The developed method is feasible and sufficient to meet the specific precision requirements of some applications in robotic seam tracking.	[88]
A higher fidelity model for predicting the entire pose-dependent FRF of an industrial robot by combining the advantages of Experimental Modal Analysis (EMA) with Operational Modal Analysis for milling processes.	KUKA KR500-3 6 DOF industrial robot	Hybrid statistical modelling: Frequency Response Function (FRF) modelling method.	A Bayesian inference and hyperparameter updating approach for updating the EMA-calibrated GPR models of the robot FRF with OMA-based FRF data improved the model's compliance RMSE by 26% and 27% in the x and y direction tool paths, respectively, compared to only EMA-based calibration. The methodology reduced the average number of iterations and calibration times required to determine the optimal GPR model hyperparameters by 50.3% and 31.3%, respectively.	[84]
Safe trajectories without neglecting cognitive ergonomics and production efficiency aspects.	UR3 lightweight robot	Experimental tasks	The task's execution time was reduced by 13.1% regarding the robot's default planner and 19.6% concerning the minimum jerk smooth collaboration planner. This new approach is highly relevant for manufacturers of collaborative robots (e.g., for integration as a path option in the robot pendant software) and for users (e.g., an online service for calculating the optimal path and subsequent transfer to the robot).	[89]
An industrial robot moving between stud welding operations in a stud welding station.	Industrial robot	Quasi-static path optimisation for an industrial robot	The method was successfully applied to a stud welding station for an industrial robot moving between two stud welding operations. Even for a difficult case, the optimised path reduced the internal force in the dress pack. It kept the dressed robot from the surrounding geometry with a prescribed safety clearance during the entire robot motion.	[83]
An industrial assembly task for learning and optimisation, considering uncertainties.	A Franka EMIKA Panda manipulator	Task trajectory learning approach. Task optimisation approach.	The proposed approach made the robot learn the task execution and compensate for the task uncertainties. The HMM + BO methodology and the HMM algorithm without optimisation were compared. This comparison shows the capabilities of the optimisation stage to compensate for task uncertainties. In particular, the HMM + BO methodology shows an assembly task success rate of 93%, while the HMM algorithm shows a success rate of only 19%.	[90]
The postprocessing and path optimisation based on the non-linear errors to improve the accuracy of multi-joint industrial robot-based 3D printing.	Multi-joint industrial robot for 3D printing	Path smoothing method	Multi-joint industrial robot-based 3D printing can be used for the high-precision printing of complex freeform surfaces. An industrial robot with only three joints is used, and the solutions of joint angles for the tool orientations are not proposed, which is essential for printing the freeform surface.	[91]

Table 5. *Cont.*

Objective	Technology	Approach	Improvement	Ref.
A comparative study of robot pose optimisation using static and dynamic stiffness models for different cutting scenarios.	KUKA KR 500–3 industrial robot, aluminium 6061	Complete pose (CP) and the decoupled partial pose (DPP) methods. Effect of optimisation method on machining accuracy	A dynamic model-based robot pose optimisation yields significant improvement over a static model-based optimisation for cutting conditions where the time-varying cutting forces approach the robot's natural frequencies. A static model-based optimisation is sufficient when the frequency content of the cutting forces is not close to the robot's natural frequencies.	[92]
The feasibility and validity of proposed stiffness identification and configuration optimisation methods.	KUKA KR500 industrial robot	Robot stiffness characteristics and optimisation methods. Point selection method	The smooth processing strategy improves optimisation efficiency, ensuring minimal stiffness loss. According to the machining results of a cylinder head of a vehicle engine, the milling quality was improved obviously after the configuration optimisation, and the validity of these methods are verified.	[85]
Real-time compensation setups.	A standard KUKA KR120R2500 PRO industrial robot with a spindle end-effector	Real-time Closed Loop Compensation method	Real-time metrology feedback cannot fully compensate for the sudden error spikes caused by the backlash. The mitigation strategy of automatically reducing feed rate (ASC) was demonstrated to reduce backlash error significantly. However, ASC considerably increases the cycle time for a toolpath that involves many direction reversals and leads to uneven cutter chip load and variation in surface finish. Backlash, therefore, remains the largest source of residual error for a robot under real-time metrology compensation.	[93]
Building Information Model (BIM)-based robotic assembly model that contains all the required information for planning.	ABB IRB6700-235 robot (6 DOF), a construction plane (approximately 1.5 m × 0.9 m), a scene modelling camera (Sony a5100), and a modelling computer (Dell Precise).	Image-based 3D modelling method. Experimental method	A general IFC model for robotic assembly contains all the information needed for task-level planning; BIM and image-based modelling are used to calibrate robot pose for the unification of the robot coordinate system, construction area, and assembly task; a simple conversion process is presented to convert the 3D placement point coordinates of each brick into the robotic control instructions. In the process of experimental verification, task-level planning can maintain the same accuracy as that of the traditional method but saves time when facing more complex tasks.	[86]
A model of reversibly controlled industrial robots based on abstract semantics.	Robotic assembly	Error recovery using reverse execution	A programming model which enables robot assembly programs to be executed in reverse. Temporarily switching the direction of program execution can be an efficient error recovery mechanism. Additional benefits arise from supporting reversibility in robotic assembly language, namely, increased code reuse and automatically derived disassembly sequences.	[94]
The control strategies for robotic PiH assemblies and the limitations of the current robotic assembly technologies.	Robotic PiH assembly	Typical peg-in-hole (PiH) assembly methods	The system outperforms the operator performing the same task with magnified visual feedback regarding both completion time and the number of successful insertions. The proposed strategies can correctly diagnose the assembly process's position errors and effectively realise error recovery.	[95]
An overview of computer vision for preoperative, intraoperative, and postoperative surgical stages to assist with planning, tool detection, identification, pose tracking, and augmented reality, for surgical skill assessment and retrospective analysis of the procedure.	An overview	An overview	An overview	[96]

3.5. Food Industry

As the world's population grows, the demand for food also continues to grow. Food suppliers are under pressure to work more efficiently, and consumers want more convenient and sustainable food. Robotics and automation are a key part of the solution to this goal. The food production sector has been relatively slowly robotised compared to other industries [97]. Robotics is applied in food manufacture, packaging, delivery, and cookery (cake decoration) [98]. Although the food industry is ranked fourth in terms of the most-automated sectors, robotic devices capable of processing nutrients of different shapes and materials are in high demand. In addition, these devices help to avoid consequences such as food-borne illness caused directly by the contamination of nutrients by nutrient handlers [99]. For this purpose, a dual-mode soft gripper was developed that can grasp and suck various objects having a weight of up to 1 kg. Soft grippers prevent damage to food [100].

Artificial intelligence-enabled robotic applications are entering the restaurant industry in the food processing and guest service operations. In a review assessing the potential for process innovation in the restaurant sector, an information process for the use of new technologies for process innovation was developed [101]. However, the past year, particularly due to the circumstances of COVID-19, has been a breakthrough year in robotisation in the food industry. A more detailed overview of researches focused on robotising the food industry is provided in Table 6.

Table 6. Research focused on the food industry.

Objective	Technology	Approach	Improvement	Ref.
The applications of industrial robots in the food industry and their automation prospects. A 4-step Food Industrial Robot Methodology for selecting industrial robots for food processing operations.	Articulated robot, parallel robot, Cartesian robot	The four steps within the Food Industrial Robot Methodology (FIRM).	The FIRM presented in this paper outlined the ability to classify industrial robot capabilities and match them to specific characteristics of foodstuffs and requirements for their processing based on four steps that navigate eight tasks. This work also identified many factors that should lay the groundwork for future research in the application of industrial robots within food manufacturing.	[102]
Identification, analysis, and understanding robotics in one of the largest sectors, the food chain.	Robots in the food chain	Case study of a Delivery Bot	The emergence of robotics in business is widely seen across the world. However, the trust in human–robotic interaction appears to be underdeveloped. Reducing the number of repetitive jobs by replacing them with robots is not replacing jobs but paving the way for more intelligent jobs.	[98]
Maximise performance by utilising fewer resources	Dual model soft gripper for food packaging	Grasp and suck process for various types of objects having a weight up to 1 kg	The proposed dual-mode gripper can perform grasp and suck functions for multiple types of nutrients. Additional improvements may be automatic switching of the gripper finger configuration and distance adjustment.	[100]
Challenges in the application of industrial robots in the food industry	An overview	An overview	An overview	[97]
Path planning optimisation technique in the food industry	The proposed optimisation technique is based on the use of an off-axis tool	EPSON T6 SCARA robot	This path optimisation technique shortens the cycle time and reduces energy consumption.	[103]

3.6. Agricultural Applications

Agricultural robots are a specialised type of technology capable of assisting farmers with a wide range of operations. Their primary role is to tackle labour intensive, repetitive, and physically demanding tasks. Robots are used in planting, seedling identification, and sorting. Autonomous tractors perform the function of weeding and harvesting. Drones and autonomous ground vehicles are used for crop monitoring and condition assessment. In animal husbandry, robots are used for feeding cattle, milking, collecting and sorting eggs, and autonomous cleaning of pens. Cobots are also used in agriculture. These robots possess mechanical arms and make harvesting much easier for farmers. The agriculture robot market size is expected to reach USD 16,640.4 billion by 2026; however, specific robots, rather than industrial robots, will occupy the majority of the market. A detailed overview of research focused on implementing industrial robots in agricultural applications is provided in Table 7.

Table 7. Research focused on agricultural applications.

Objective	Technology	Approach	Improvement	Ref.
The potential applications in agriculture by presenting a variety of manipulators and various forms of sensors.	Parallel grippers, angular grippers, and biologically inspired grippers manufactured by Festo. Various sensors	Application methods.	State-of-the-art robotic grippers, grasping and control strategies, and their applications in agricultural robots. Applications of robotic grippers in food, agricultural, and bio-system engineering were summarised in detail.	[104]
A scheme that combines computer vision and multi-tasking processes to develop a small-scale smart agricultural machine that can automatically weed and perform variable rate irrigation within a cultivated field.	The frames of the machine, the weeding and watering mechanism, the image and soil moisture sensor, the actuator, and the graphical user interface (GUI)	Image processing methods such as HSV (hue (H), saturation (S), value (V)) colour conversion, estimation of thresholds during the image binary segmentation process, and morphology operator procedures. Fuzzy logic, multi-tasking processes	The system can classify plants and weeds in real time with an average classification rate of 90% or higher. This allows the machine to perform weeding and watering while maintaining the moisture content of the deep soil at 80 ± 10% and an average weeding rate of 90%.	[105]
A systematic overview aiming to identify the applicability of computer vision in precision agriculture to produce the five most-produced grains in the world: maise, rice, wheat, soybean, and barley. Different approaches to treat disease detection, grain quality, and phenotyping.	An overview	An overview	An overview	[106]

3.7. Civil Engineering Industry

In general, the construction industry is relatively inefficient from the perspective of automation. Robotics are seldom applied [107]. The main identified challenges for higher adoption of robotics in the construction industry were grouped into four categories: contractor-side economic factors; client-side economic factors; technical and work-culture factors; and weak business case factors. Technical and work-culture factors include an untrained workforce; unproven effectiveness and immature technology; and the current work culture and aversion to change [108].

The perspective of robotics in civil engineering is significantly better. Here, robotics provides considerable opportunities to increase productivity, efficiency, and flexibility, from automated modular house production to robotic welding, material handling on construction sites, and 3D printing of houses or certain structures. Robots make the industry safer and more economical, increase sustainability, and reduce its environmental impact, while improving quality and reducing waste. The total global value of the construction industry is forecast to grow by 85% to USD 15.5 trillion by 2030 [109]. Robots can make construction safer by handling large and heavy loads, working in hazardous locations, and enabling new, safer construction methods. Transferring repetitive and dangerous tasks that humans are increasingly reluctant to perform to robots means that automation can help address the labour and skills crisis, and make the construction industry more attractive [110,111]. Few classic robots are used in the construction process due to the dynamic and inaccurately described environment; however, work on 3D buildings and their environmental models reduces this limitation. A detailed overview of related references is provided in Table 8.

Table 8. Research focused on implementing robots in the construction and civil engineering industry.

Objective	Technology	Approach	Improvement	Ref.
A novel fabrication process for the assembly of full-scale masonry vaults without falsework.	Two industrial robotic arms (ABB 4600 2.55). The prototype of the robotically assembled brick vault.	The fabrication method is based on a cooperative assembly approach in which two robots alternate between placement and support first to build a stable central arch.	Cooperative robotic assembly methods can be applied to constructing a spanning structure built without a temporary falsework. Where traditional manufacturing techniques require geometric guides, this project shows how it can instead leverage the robots' precision to accurately place bricks in bespoke orientations.	[112]
A computer vision for real-time extrusion quality monitoring during robotic building construction.	Laboratory-scale concrete printer. Logitech 720p camera to capture extrusion videos. The extrusion videos are processed in real-time by a Raspberry Pi 3B.	OpenCV library, adopted, shape-based approach. Gaussian filter.	The developed system can print up to ten 120 cm long concrete layers. It uses an extrusion mechanism similar to the Contour Crafting machine to print layers having a height of 3.81 cm and a width of 2.54 cm, from concrete and mortar at different linear speeds (up to 10 cm/s) and deposition rates. The vision system detected all designed variation levels (± 5 to ± 15 L/m^3 change in the water in the mixture). In terms of accuracy and responsiveness to material variations, the obtained experimental results imply the excellent potential for using computer vision for automated quality monitoring of construction-scale 3D printing.	[113]
Presents the possibilities of applying lightweight cobots to individual tasks in the construction sector.	Presenting of light robotics together with 3D printing technology provides the rapid advantage of prototyping to test ideas and applications.	The simplest visual system was used to follow a simplified approach, which can be controlled directly by a robot controller.	Future research on increasing the dynamics of torsional tasks using a mobile robot with a scissor lift could result in the cobot and mobile platform covering the entire construction area.	[114]
To determine if improved robotic technologies have also been used in the building industry.	An overview	An overview	An overview	[115]
To determine how robotic automation can help in the construction industry.	A common framework for current technological innovation in this field and a development plan were outlined.	The projected impacts on traditional processes, construction sites, emerging technologies, and related professions are summarised to identify future implications and future directions toward self-sufficiency.	Artificial intelligence must be a successful factor in the involvement of robotic devices in the construction industry.	[110]
Provide a systematic overview of human-robot interactions concerning various types of robots	Human–robot interaction, human–robot cooperation (HRC).	An overview	Further investigation of multi-function robots, human–robot interaction in robotic fabrication, and multipurpose robots.	[116]
The main goal is to fully describe feedback based on sensor informed programs for process monitoring and fabrication data collection and analysis.	Additive manufacturing.	An overview	Effective robotic production still requires the communication and management of progressively improving materials and building systems.	[117]
Application of a Building Information Modelling (BIM) method for efficient and simple deployment of robot systems for building construction and operation	BIM integrative, collaborative robotics.	The robot is provided with a priori geometric and semantic information about the environment with the help of the BIM system.	Future improvements consist of the assessment of the actual applicability of the system on the construction site and closing the gap between robotic systems and the construction site.	[111]

4. Discussion

Implementing an industrial robot in practice is a complex procedure that requires answering many questions about the possibilities of using the robot and the process itself. The situation varies slightly depending on the industry area. Robots have been used in some areas for 30 or more years, whereas, in other areas, the implementation of robots is only beginning. In industrial sectors with a long tradition of robotics, new solutions are relatively more straightforward. These solutions are typically limited to implementing new tools, control algorithms, and robotic action quality control systems. Therefore, our article focuses on areas where traditions of implementing robots do not exist yet, and such solutions are just beginning to be implemented.

Despite the different application areas, some achievements in robotics can be successfully transferred from one industry to another. Furthermore, bypassing limitations in one area often ensures advances in robotics in other sectors. For example, the implementation of computer vision to localise and manipulate randomly placed mechanical parts on a conveyor fostered the robotisation of sorting processes in all industry fields.

This article provided an overview of the main areas where robots are beginning to be implemented, and identified the main challenges and limitations they face (Figure 4).

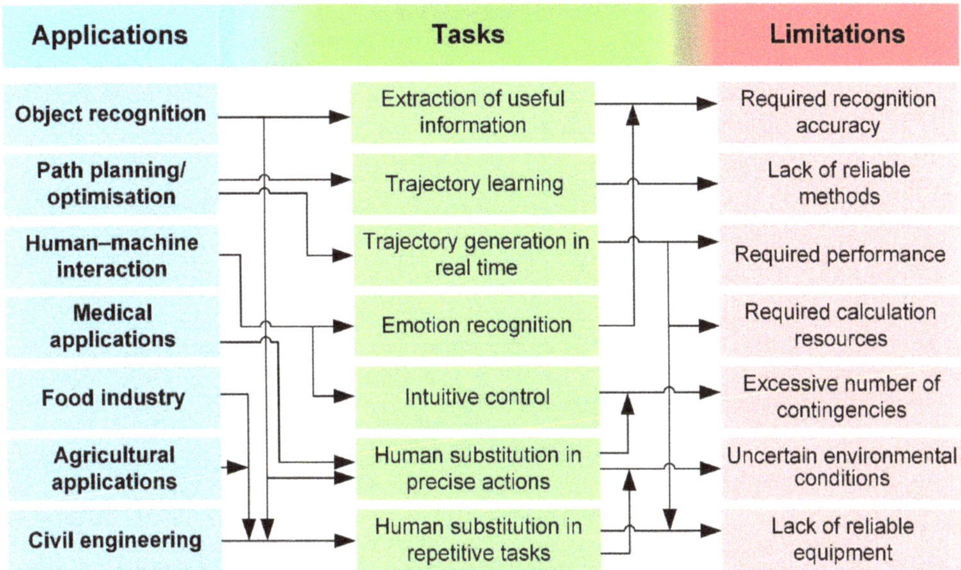

Figure 4. Relations between robot implementation areas, typical tasks and limitations.

The conclusion is that tasks performed by the robots and actual limitations are closely related to each other regardless of the implementation field. In this paper, the tasks for which robots are most preferred rather than humans were identified. Typically, these tasks are repetitive and extremely precise operations that require evaluating a considerable amount of data. For example, the implementation of robots for object recognition has three main functions in which robots replace humans: (1) extraction of useful information from massive data flow; (2) accurate movements to manipulate with an object or tool; and (3) repetitive action (sorting). In addition, the food, agriculture, and civil engineering industries aim to replace humans involved in repetitive actions. In contrast, medical applications are mainly related to accurate manipulation and hazardous environments.

Preparation of robots for an operation, particularly in dynamic, varying situations, is a time- and resource-consuming activity. Therefore, a large amount of research focuses on enhancing human–robot interaction and path planning/optimisation issues. The goal is to develop faster and more comfortable methods to operate robots in real time, and to create a possibility for the robot to react to the operator's emotional state.

Many different factors limit the implementation of industrial robots in typical tasks. The seven main limitations in the reviewed application fields were identified. In summary, the main limitations are the lack of suitable methods, high recognition accuracy, and performance requirements; varying environmental conditions; an excessive number of possible situations; and lack of reliable equipment (tools). Notably, these limitations are unrelated to the robot's mechanical systems (except the tools). Therefore, most modern robotic solutions are fostered by the development of additional equipment or control algorithms. Computer vision, sensor fusion, and machine learning are becoming major engines driving industrial robots' wider application. They increase robots' flexibility and enable them to make smart adaptive solutions, although robots were initially designed only to perform repetitive actions.

As a result of the development of robot control systems, robots' internal structures have also been improved. These improvements typically include the implementation of new mathematical methods for robot control or optimisation of energy consumption [118]. For example, a previous study [119] provided a methodology that allows implementation of a non-typical Denavit–Hartenberg method for a delta robot.

Nonetheless, despite the recent improvements and smart solutions realised in industrial robots, their widespread use in non-typical areas remains limited. The main limitations and guidelines for further research are new intuitive control methods, user-friendly interfaces, specialised software, and real-time control methods.

5. Conclusions

Analysis of robot applications revealed a number of important issues, and showed that the current rare applications of robot implementations are not always limited by technical difficulties.

Some application fields have no tradition in such activities, such as the civil engineering, food, and agriculture industries. Human–robot cooperation in classical industrial robots and in specialised cobot cases still demands an intensive introduction into these industries. However, in this case, the introduction involves non-technical aspects such as human psychology and personal acceptance of the robots in their working place. Another aspect of the subjective attitude to robots is limited by their acceptance by managers and process designers; however, they are also lacking implementation experience and knowledge of cutting-edge achievements in robotic applications.

Many automation cases are still limited by artificial intelligence (AI) issues related to object recognition, object position recognition, and decision generation for object grabbing and manipulating. This issue arises from the process of widening robotic implementation in existing industries, and therefore many technologies should be redesigned. Nevertheless, pressure due to the absence of a skilled labour force has led to new solutions. Many general solutions using machine vision and sensor fusion (camera–lidar scanner, camera–distance sensors, etc.) have been spontaneously implemented in numerous industrial enterprises. These approaches are starting to appear in home appliances, but market penetration of these solutions remains low.

Robot implementations are often subject to systematic difficulties, such as manipulation and orientation of solid objects with non-stable geometrical shapes. These objects are widely used in industry and home appliances, and include textiles, clothes, and cables. At present, this area has few publications and technical solutions, and is in the research stage; presentations of some of the publicly available cases are at the level of scientific publications. Although clamps and templates are currently used for specific industrial cases, general solutions have not yet been achieved. This situation requires rethinking processes and possibly preparing objects for robotic processing, rather than using tremendous computing and multiplying hardware.

The result of this review points to four evident directions in the field of robotics:

- development of intelligent companion equipment for robots (sensors, grippers, and servo-applications);
- AI-based solutions for signal processing and decision making;
- the redesign of general objects and the related features for robotic applications;
- provision of psychological solutions for robot–human collaboration and acceptance of robots in the workplace.

Author Contributions: Conceptualisation, V.B. and A.D.; methodology, U.S.-B.; formal analysis, U.S.-B. and E.Š.; investigation, J.S.-Ž.; resources, V.B.; writing—original draft preparation, U.S.-B. and J.S.-Ž.; writing—review and editing, V.B. and A.D.; visualisation, A.D.; supervision, V.B.; project administration, J.S.-Ž.; funding acquisition, V.B. All authors have read and agreed to the published version of the manuscript.

Funding: This work is part of the AI4DI project, receiving funding from the Electronic Components and Systems for European Leadership Joint Undertaking in collaboration with the European Union's H2020 Framework Programme (H2020/2014-2020) and National Authorities, under grant agreement No 826060.

Institutional Review Board Statement: Not applicable.

Informed Consent Statement: Not applicable.

Data Availability Statement: The data presented in this study are available on request from the corresponding author. The data are not publicly available due to the project consortium agreement.

Conflicts of Interest: The authors declare no conflict of interest.

References

1. ISO—ISO 8373:2012—Robots and Robotic Devices—Vocabulary. Available online: https://www.iso.org/standard/55890.html (accessed on 7 April 2021).
2. IFR Presents World Robotics Report 2020—International Federation of Robotics. Available online: https://ifr.org/ifr-press-releases/news/record-2.7-million-robots-work-in-factories-around-the-globe (accessed on 7 April 2021).
3. ScienceDirect Search Results—Keywords (Industrial Robot). Available online: https://www.sciencedirect.com/search?qs=Industrial%20robot (accessed on 7 April 2021).
4. Dekle, R. Robots and industrial labor: Evidence from Japan. *J. Jpn. Int. Econ.* **2020**, *58*, 101108. [CrossRef]
5. Olivares-Alarcos, A.; Foix, S.; Alenyà, G. On inferring intentions in shared tasks for industrial collaborative robots. *Electronics* **2019**, *8*, 1306. [CrossRef]
6. Smith, R.; Cucco, E.; Fairbairn, C. Robotic Development for the Nuclear Environment: Challenges and Strategy. *Robotics* **2020**, *9*, 94. [CrossRef]
7. Rojas, R.A.; Wehrle, E.; Vidoni, R. A Multicriteria Motion Planning Approach for Combining Smoothness and Speed in Collaborative Assembly Systems. *Appl. Sci.* **2020**, *10*, 5086. [CrossRef]
8. Ivanov, S.; Seyitoğlu, F.; Markova, M. Hotel managers' perceptions towards the use of robots: A mixed-methods approach. *Inf. Technol. Tour.* **2020**, *22*, 505–535. [CrossRef]
9. Colim, A.; Sousa, N.; Carneiro, P.; Costa, N.; Arezes, P.; Cardoso, A. Ergonomic intervention on a packing workstation with robotic aid case study at a furniture manufacturing industry. *Work* **2020**, *66*, 229–237. [CrossRef] [PubMed]
10. Giusti, A.; Guzzi, J.; Ciresan, D.C.; He, F.L.; Rodriguez, J.P.; Fontana, F.; Faessler, M.; Forster, C.; Schmidhuber, J.; Di Caro, G.; et al. A Machine Learning Approach to Visual Perception of Forest Trails for Mobile Robots. *IEEE Robot. Autom. Lett.* **2016**, *1*, 661–667. [CrossRef]
11. Elsisi, M.; Mahmoud, K.; Lehtonen, M.; Darwish, M.M.F. Effective Nonlinear Model Predictive Control Scheme Tuned by Improved NN for Robotic Manipulators. *IEEE Access* **2021**, *9*, 64278–64290. [CrossRef]
12. Elsisi, M.; Mahmoud, K.; Lehtonen, M.; Darwish, M.M.F. An improved neural network algorithm to efficiently track various trajectories of robot manipulator arms. *IEEE Access* **2021**, *9*, 11911–11920. [CrossRef]
13. A Brief History of Collaborative Robots | Material Handling and Logistics. Available online: https://www.mhlnews.com/technology-automation/article/21124077/a-brief-history-of-collaborative-robots (accessed on 8 April 2021).
14. Colgate, J.E.; Peshkin, M.A. Cobots. U.S. Patent 5,952,796, 14 September 1999.
15. Galin, R.; Meshcheryakov, R. Automation and robotics in the context of Industry 4.0: The shift to collaborative robots. *IOP Conf. Ser. Mater. Sci. Eng.* **2019**, *537*, 032073. [CrossRef]
16. Tran, M.Q.; Elsisi, M.; Mahmoud, K.; Liu, M.K.; Lehtonen, M.; Darwish, M.M.F. Experimental Setup for Online Fault Diagnosis of Induction Machines via Promising IoT and Machine Learning: Towards Industry 4.0 Empowerment. *IEEE Access* **2021**, *9*, 115429–115441. [CrossRef]
17. Elsisi, M.; Mahmoud, K.; Lehtonen, M.; Darwish, M.M.F. Reliable Industry 4.0 Based on Machine Learning and IoT for Analyzing, Monitoring, and Securing Smart Meters. *Sensors* **2021**, *21*, 487. [CrossRef]
18. Rao, S.K.; Prasad, R. Impact of 5G Technologies on Industry 4.0. *Wirel. Pers. Commun.* **2018**, *100*, 145–159. [CrossRef]
19. Pérez, L.; Rodríguez-Jiménez, S.; Rodríguez, N.; Usamentiaga, R.; García, D.F.; Wang, L. Symbiotic human–robot collaborative approach for increased productivity and enhanced safety in the aerospace manufacturing industry. *Int. J. Adv. Manuf. Technol.* **2020**, *106*, 851–863. [CrossRef]
20. Song, J.; Chen, Q.; Li, Z. A peg-in-hole robot assembly system based on Gauss mixture model. *Robot. Comput. Integr. Manuf.* **2021**, *67*, 101996. [CrossRef]

21. De Pace, F.; Manuri, F.; Sanna, A.; Fornaro, C. A systematic review of Augmented Reality interfaces for collaborative industrial robots. *Comput. Ind. Eng.* **2020**, *149*, 106806. [CrossRef]
22. Matheson, E.; Minto, R.; Zampieri, E.G.G.; Faccio, M.; Rosati, G. Human-robot collaboration in manufacturing applications: A review. *Robotics* **2019**, *8*, 100. [CrossRef]
23. ISO—ISO/TS 15066:2016—Robots and Robotic Devices—Collaborative Robots. Available online: https://www.iso.org/standard/62996.html (accessed on 2 December 2021).
24. Tannous, M.; Miraglia, M.; Inglese, F.; Giorgini, L.; Ricciardi, F.; Pelliccia, R.; Milazzo, M.; Stefanini, C. Haptic-based touch detection for collaborative robots in welding applications. *Robot. Comput. Integr. Manuf.* **2020**, *64*, 101952. [CrossRef]
25. Tannous, M.; Bologna, F.; Stefanini, C. Load cell torques and force data collection during tele-operated robotic gas tungsten arc welding in presence of collisions. *Data Br.* **2020**, *31*, 105981. [CrossRef]
26. Knudsen, M.; Kaivo-oja, J. Collaborative Robots: Frontiers of Current Literature. *J. Intell. Syst. Theory Appl.* **2020**, *3*, 13–20. [CrossRef]
27. Ghosh, A.; Soto, D.A.P.; Veres, S.M.; Rossiter, A. Human robot interaction for future remote manipulations in industry 4.0. *Proc. IFAC-Pap.* **2020**, *53*, 10223–10228. [CrossRef]
28. Ghosh, A.; Veres, S.M.; Paredes-Soto, D.; Clarke, J.E.; Rossiter, J.A. Intuitive programming with remotely instructed robots inside future gloveboxes. In Proceedings of the ACM/IEEE International Conference on Human-Robot Interaction, Cambridge, UK, 23–26 March 2020; pp. 209–211.
29. Weidemann, A.; Rußwinkel, N. The Role of Frustration in Human–Robot Interaction—What Is Needed for a Successful Collaboration? *Front. Psychol.* **2021**, *12*, 707. [CrossRef]
30. Spezialetti, M.; Placidi, G.; Rossi, S. Emotion Recognition for Human-Robot Interaction: Recent Advances and Future Perspectives. *Front. Robot. AI* **2020**, *7*, 532279. [CrossRef] [PubMed]
31. Ge, S.; Wang, P.; Liu, H.; Lin, P.; Gao, J.; Wang, R.; Iramina, K.; Zhang, Q.; Zheng, W. Neural Activity and Decoding of Action Observation Using Combined EEG and fNIRS Measurement. *Front. Hum. Neurosci.* **2019**, *13*, 357. [CrossRef] [PubMed]
32. Mavridis, N. A review of verbal and non-verbal human–robot interactive communication. *Robot. Auton. Syst.* **2015**, *63*, 22–35. [CrossRef]
33. Dzedzickis, A.; Kaklauskas, A.; Bucinskas, V. Human emotion recognition: Review of sensors and methods. *Sensors* **2020**, *20*, 592. [CrossRef] [PubMed]
34. Shubha, P. International Journal of Engineering Technology Research & Management: A review of multi object recognition based on deep leairning. *Int. J. Eng. Technol. Res. Manag.* **2020**, *2*, 27–33.
35. De Coninck, E.; Verbelen, T.; Van Molle, P.; Simoens, P.; Dhoedt, B. Learning robots to grasp by demonstration. *Robot. Auton. Syst.* **2020**, *127*, 103474. [CrossRef]
36. Fujita, M.; Domae, Y.; Noda, A.; Garcia Ricardez, G.A.; Nagatani, T.; Zeng, A.; Song, S.; Rodriguez, A.; Causo, A.; Chen, I.M.; et al. What are the important technologies for bin picking? Technology analysis of robots in competitions based on a set of performance metrics. *Adv. Robot.* **2020**, *34*, 560–574. [CrossRef]
37. Sughashini, K.R.; Sunanthini, V.; Johnsi, J.; Nagalakshmi, R.; Sudha, R. A pneumatic robot arm for sorting of objects with chromatic sensor module. *Mater. Today Proc.* **2021**, *45*, 6364–6368. [CrossRef]
38. Shaikat, A.S.; Akter, S.; Salma, U. Computer Vision Based Industrial Robotic Arm for Sorting Objects by Color and Height. *J. Eng. Adv.* **2020**, *1*, 116–122. [CrossRef]
39. Chen, P.; Elangovan, V. Object Sorting using Faster R-CNN. *Int. J. Artif. Intell. Appl.* **2020**, *11*, 27–36. [CrossRef]
40. Yu, Y.; Zou, S.; Yin, K. A novel detection fusion network for solid waste sorting. *Int. J. Adv. Robot. Syst.* **2020**, *17*, 172988142094177. [CrossRef]
41. Xiao, W.; Yang, J.; Fang, H.; Zhuang, J.; Ku, Y.; Zhang, X. Development of an automatic sorting robot for construction and demolition waste. *Clean Technol. Environ. Policy* **2020**, *22*, 1829–1841. [CrossRef]
42. Li, M.; Duan, Y.; He, X.; Yang, M. Image positioning and identification method and system for coal and gangue sorting robot. *Int. J. Coal Prep. Util.* **2020**, 1–19. [CrossRef]
43. D'Avella, S.; Tripicchio, P.; Avizzano, C.A. A study on picking objects in cluttered environments: Exploiting depth features for a custom low-cost universal jamming gripper. *Robot. Comput. Integr. Manuf.* **2020**, *63*, 101888. [CrossRef]
44. Ciszak, O.; Juszkiewicz, J.; Suszyński, M. Programming of Industrial Robots Using the Recognition of Geometric Signs in Flexible Welding Process. *Symmetry* **2020**, *12*, 1429. [CrossRef]
45. Minoura, H.; Yonetani, R.; Nishimura, M.; Ushiku, Y. Crowd Density Forecasting by Modeling Patch-Based Dynamics. *IEEE Robot. Autom. Lett.* **2021**, *6*, 287–294. [CrossRef]
46. De Coninck, E.; Verbelen, T.; Van Molle, P.; Simoens, P.; Idlab, B.D. Learning to Grasp Arbitrary Household Objects from a Single Demonstration. *IEEE Int. Conf. Intell. Robot. Syst.* **2019**, 2372–2377. [CrossRef]
47. Kaya, O.; Tağlıoğlu, G.B.; Ertuğrul, Ş. The Series Elastic Gripper Design, Object Detection, and Recognition by Touch. *J. Mech. Robot.* **2022**, *14*, 014501. [CrossRef]
48. Kulkarni, R.G. *Robot Path Planning with Sensor Feedback for Industrial Applications*; Wichita State University: Wichita, KS, USA, 2021.
49. Abdalrahman, M.; Brice, A.; Hanson, L. *New Era of Automation in Scania' s Manufacturing Systems—A Method to Automate a Manual Assembly Process*; Libraries at Lund University: Lund, Sweden, 2021.

50. Thike, A.; Moe San, Z.Z.; Min Oo, D.Z. Design and Development of an Automatic Color Sorting Machine on Belt Conveyor. *Int. J. Sci. Eng. Appl.* **2019**, *8*, 176–179. [CrossRef]
51. Wang, Z.; Xie, S.; Chen, G.; Chi, W.; Ding, Z.; Wang, P. An Online Flexible Sorting Model for Coal and Gangue Based on Multi-Information Fusion. *IEEE Access* **2021**, *9*, 90816–90827. [CrossRef]
52. Sun, Z.; Huang, L.; Jia, R. Coal and gangue separating robot system based on computer vision. *Sensors* **2021**, *21*, 1349. [CrossRef] [PubMed]
53. Fadhil, A.T.; Abbar, K.A.; Qusay, A.M. Computer Vision-Based System for Classification and Sorting Color Objects. *IOP Conf. Ser. Mater. Sci. Eng.* **2020**, *745*, 012030. [CrossRef]
54. Peršak, T.; Viltužnik, B.; Hernavs, J.; Klancnik, S. Vision-Based Sorting Systems for Transparent Plastic Granulate. *Appl. Sci.* **2020**, *10*, 4269. [CrossRef]
55. Sun, L.; Zhao, C.; Yan, Z.; Liu, P.; Duckett, T.; Stolkin, R. A novel weakly-supervised approach for RGB-D-based nuclear waste object detection. *IEEE Sens. J.* **2019**, *19*, 3487–3500. [CrossRef]
56. Albinali, H.; Alzahrani, F.A. Faster R-CNN for detecting regions in human-annotated micrograph images. In Proceedings of the 2021 International Conference of Women in Data Science at Taif University (WiDSTaif), Taif, Saudi Arabia, 30–31 March 2021.
57. Li, S.; Zhao, X.; Li, W. Analysis of Object Detection Performance Based on Faster R-CNN. *J. Phys. Conf. Ser.* **2021**, *1827*, 012085. [CrossRef]
58. Cipta Ramadhan Kete, S.; Darma Tarigan, S.; Effendi, H. Land use classification based on object and pixel using Landsat 8 OLI in Kendari City, Southeast Sulawesi Province, Indonesia. *IOP Conf. Ser. Earth Environ. Sci.* **2019**, *284*, 012019. [CrossRef]
59. Hespeler, S.C.; Nemati, H.; Dehghan-Niri, E. Non-destructive thermal imaging for object detection via advanced deep learning for robotic inspection and harvesting of chili peppers. *Artif. Intell. Agric.* **2021**, *5*, 102–117. [CrossRef]
60. Birglen, L.; Schlicht, T. A statistical review of industrial robotic grippers. *Robot. Comput. Integr. Manuf.* **2018**, *49*, 88–97. [CrossRef]
61. Shim, M.; Kim, J.H. Design and optimization of a robotic gripper for the FEM assembly process of vehicles. *Mech. Mach. Theory* **2018**, *129*, 1–16. [CrossRef]
62. Linghu, C.; Zhang, S.; Wang, C.; Yu, K.; Li, C.; Zeng, Y.; Zhu, H.; Jin, X.; You, Z.; Song, J. Universal SMP gripper with massive and selective capabilities for multiscaled, arbitrarily shaped objects. *Sci. Adv.* **2020**, *6*, eaay5120. [CrossRef] [PubMed]
63. Richter, F.; Orosco, R.K.; Yip, M.C. Open-Sourced Reinforcement Learning Environments for Surgical Robotics. *arXiv* **2019**, arXiv:1903.02090.
64. Kassahun, Y.; Yu, B.; Tibebu, A.T.; Stoyanov, D.; Giannarou, S.; Metzen, J.H.; Vander Poorten, E. Surgical robotics beyond enhanced dexterity instrumentation: A survey of machine learning techniques and their role in intelligent and autonomous surgical actions. *Int. J. Comput. Assist. Radiol. Surg.* **2016**, *11*, 553–568. [CrossRef] [PubMed]
65. Pierson, H.A.; Gashler, M.S. Deep learning in robotics: A review of recent research. *Adv. Robot.* **2017**, *31*, 821–835. [CrossRef]
66. Downey, J.E.; Weiss, J.M.; Muelling, K.; Venkatraman, A.; Valois, J.S.; Hebert, M.; Bagnell, J.A.; Schwartz, A.B.; Collinger, J.L. Blending of brain-machine interface and vision-guided autonomous robotics improves neuroprosthetic arm performance during grasping. *J. Neuroeng. Rehabil.* **2016**, *13*, 28. [CrossRef]
67. Fong, J.; Ocampo, R.; Gross, D.P.; Tavakoli, M. Intelligent Robotics Incorporating Machine Learning Algorithms for Improving Functional Capacity Evaluation and Occupational Rehabilitation. *J. Occup. Rehabil.* **2020**, *30*, 362–370. [CrossRef] [PubMed]
68. Rudovic, O.; Lee, J.; Dai, M.; Schuller, B.; Picard, R.W. Personalized machine learning for robot perception of affect and engagement in autism therapy. *Sci. Robot.* **2018**, *3*, eaao6760. [CrossRef]
69. Grischke, J.; Johannsmeier, L.; Eich, L.; Griga, L.; Haddadin, S. Dentronics: Towards robotics and artificial intelligence in dentistry. *Dent. Mater.* **2020**, *36*, 765–778. [CrossRef]
70. Ma, Q.; Kobayashi, E.; Wang, J.; Hara, K.; Suenaga, H.; Sakuma, I.; Masamune, K. Development and preliminary evaluation of an autonomous surgical system for oral and maxillofacial surgery. *Int. J. Med. Robot. Comput. Assist. Surg.* **2019**, *15*, e1997. [CrossRef]
71. Otani, T.; Raigrodski, A.J.; Mancl, L.; Kanuma, I.; Rosen, J. In vitro evaluation of accuracy and precision of automated robotic tooth preparation system for porcelain laminate veneers. *J. Prosthet. Dent.* **2015**, *114*, 229–235. [CrossRef]
72. Lang, T.; Staufer, S.; Jennes, B.; Gaengler, P. Clinical validation of robot simulation of toothbrushing—Comparative plaque removal efficacy. *BMC Oral Health* **2014**, *14*, 82. [CrossRef]
73. Nelson, C.A.; Hossain, S.G.M.; Al-Okaily, A.; Ong, J. A novel vending machine for supplying root canal tools during surgery. *J. Med. Eng. Technol.* **2012**, *36*, 102–116. [CrossRef] [PubMed]
74. Lepidi, L.; Chen, Z.; Ravida, A.; Lan, T.; Wang, H.L.; Li, J. A Full-Digital Technique to Mount a Maxillary Arch Scan on a Virtual Articulator. *J. Prosthodont.* **2019**, *28*, 335–338. [CrossRef]
75. Zhang, Y.; De Jiang, J.G.; Liang, T.; Hu, W.P. Kinematics modeling and experimentation of the multi-manipulator tooth-arrangement robot for full denture manufacturing. *J. Med. Syst.* **2011**, *35*, 1421–1429. [CrossRef] [PubMed]
76. Spin-Neto, R.; Mudrak, J.; Matzen, L.H.; Christensen, J.; Gotfredsen, E.; Wenzel, A. Cone beam CT image artefacts related to head motion simulated by a robot skull: Visual characteristics and impact on image quality. *Dentomaxillofacial Radiol.* **2013**, *42*, 32310645. [CrossRef]
77. Li, C.; Gu, X.; Xiao, X.; Lim, C.M.; Duan, X.; Ren, H. A Flexible Transoral Robot Towards COVID-19 Swab Sampling. *Front. Robot. AI* **2021**, *8*, 51. [CrossRef]
78. Jose, K.; Pratihar, D.K. Task allocation and collision-free path planning of centralized multi-robots system for industrial plant inspection using heuristic methods. *Rob. Auton. Syst.* **2016**, *80*, 34–42. [CrossRef]

79. Das, P.K.; Jena, P.K. Multi-robot path planning using improved particle swarm optimization algorithm through novel evolutionary operators. *Appl. Soft Comput. J.* **2020**, *92*, 106312. [CrossRef]
80. Fascista, A.; Coluccia, A.; Ricci, G. A Pseudo Maximum likelihood approach to position estimation in dynamic multipath environments. *Signal Processing* **2021**, *181*, 107907. [CrossRef]
81. Karaagac, A.; Haxhibeqiri, J.; Ridolfi, M.; Joseph, W.; Moerman, I.; Hoebeke, J. Evaluation of accurate indoor localization systems in industrial environments. In Proceedings of the 2017 22nd IEEE International Conference on Emerging Technologies and Factory Automation (ETFA), Limassol, Cyprus, 12–15 September 2017; pp. 1–8.
82. Makomo, T.J.; Erin, K.; Boru, B. Real Time Application for Automatic Object and 3D Position Detection and Sorting with Robotic Manipulator. *Sak. Univ. J. Sci.* **2020**, *24*, 703–711. [CrossRef]
83. Hermansson, T.; Carlson, J.S.; Linn, J.; Kressin, J. Quasi-static path optimization for industrial robots with dress packs. *Robot. Comput. Integr. Manuf.* **2021**, *68*, 102055. [CrossRef]
84. Nguyen, V.; Melkote, S. Hybrid statistical modelling of the frequency response function of industrial robots. *Robot. Comput. Integr. Manuf.* **2021**, *70*, 102134. [CrossRef]
85. Jiao, J.; Tian, W.; Zhang, L.; Li, B.; Hu, J.; Li, Y.; Li, D.; Zhang, J. Variable stiffness identification and configuration optimization of industrial robots for machining tasks. *Res. Sq.* **2020**. [CrossRef]
86. Ding, L.; Jiang, W.; Zhou, Y.; Zhou, C.; Liu, S. BIM-based task-level planning for robotic brick assembly through image-based 3D modeling. *Adv. Eng. Inform.* **2020**, *43*, 100993. [CrossRef]
87. Leroux, M.; Raison, M.; Adadja, T.; Achiche, S. Combination of eyetracking and computer vision for robotics control. In Proceedings of the IEEE Conference on Technologies for Practical Robot Applications, TePRA, Woburn, MA, USA, 11–12 May 2015; IEEE Computer Society: Washington, DC, USA, 2015.
88. Xu, Y.; Fang, G.; Lv, N.; Chen, S.; Jia Zou, J. Computer vision technology for seam tracking in robotic GTAW and GMAW. *Robot. Comput. Integr. Manuf.* **2015**, *32*, 25–36. [CrossRef]
89. Rojas, R.A.; Garcia, M.A.R.; Gualtieri, L.; Rauch, E. Combining safety and speed in collaborative assembly systems—An approach to time optimal trajectories for collaborative robots. *Procedia CIRP* **2021**, *97*, 308–312. [CrossRef]
90. Roveda, L.; Magni, M.; Cantoni, M.; Piga, D.; Bucca, G. Human–robot collaboration in sensorless assembly task learning enhanced by uncertainties adaptation via Bayesian Optimization. *Rob. Auton. Syst.* **2021**, *136*, 103711. [CrossRef]
91. Fu, G.; Gu, T.; Gao, H.; Lu, C. A postprocessing and path optimization based on nonlinear error for multijoint industrial robot-based 3D printing. *Int. J. Adv. Robot. Syst.* **2020**, *17*, 172988142095224. [CrossRef]
92. Cvitanic, T.; Nguyen, V.; Melkote, S.N. Pose optimization in robotic machining using static and dynamic stiffness models. *Robot. Comput. Integr. Manuf.* **2020**, *66*, 101992. [CrossRef]
93. Wang, Z.; Zhang, R.; Keogh, P. Real-Time Laser Tracker Compensation of Robotic Drilling and Machining. *J. Manuf. Mater. Process.* **2020**, *4*, 79. [CrossRef]
94. Schultz, U.P. Reversible control of robots. In *Reversible Computation: Extending Horizons of Computing. RC 2020. Lecture Notes in Computer Science)*; Ulidowski, I., Lanese, I., Schultz, U., Ferreira, C., Eds.; Springer: Berlin/Heidelberg, Germany, 2020; Volume 12070, pp. 177–186. [CrossRef]
95. Jiang, J.; Huang, Z.; Bi, Z.; Ma, X.; Yu, G. State-of-the-Art control strategies for robotic PiH assembly. *Robot. Comput. Integr. Manuf.* **2020**, *65*, 101894. [CrossRef]
96. Kumar, S.; Singhal, P.; Krovi, V.N. Computer-vision-based decision support in surgical robotics. *IEEE Des. Test* **2015**, *32*, 89–97. [CrossRef]
97. Bader, F.; Rahimifard, S. Challenges for industrial robot applications in food manufacturing. In Proceedings of the 2nd International Symposium on Computer Science and Intelligent Control, Stockholm, Sweden, 21–23 September 2018.
98. Grobbelaar, W.; Verma, A.; Shukla, V.K. Analyzing human robotic interaction in the food industry. *J. Phys. Conf. Ser.* **2021**, *1714*, 012032. [CrossRef]
99. Sandey, K.K.; Qureshi, M.A.; Meshram, B.D.; Agrawal, A.; Uprit, S. Robotics—An Emerging Technology in Dairy Industry. *Int. J. Eng. Trends Technol.* **2017**, *43*, 58–62.
100. Wang, Z.; Or, K.; Hirai, S. A dual-mode soft gripper for food packaging. *Rob. Auton. Syst.* **2020**, *125*, 103427. [CrossRef]
101. Blöcher, K.; Alt, R. AI and robotics in the European restaurant sector: Assessing potentials for process innovation in a high-contact service industry. *Electron. Mark.* **2020**, *31*, 529–551. [CrossRef]
102. Bader, F.; Rahimifard, S. A methodology for the selection of industrial robots in food handling. *Innov. Food Sci. Emerg. Technol.* **2020**, *64*, 102379. [CrossRef]
103. Boschetti, G.; Carbone, G. *Advances in Italian Mechanism Science*; Springer: Cham, Switzerland, 2017; Volume 18, ISBN 9783030558062.
104. Zhang, B.; Xie, Y.; Zhou, J.; Wang, K.; Zhang, Z. State-of-the-art robotic grippers, grasping and control strategies, as well as their applications in agricultural robots: A review. *Comput. Electron. Agric.* **2020**, *177*, 105694. [CrossRef]
105. Chang, C.-L.; Lin, K.-M. Smart Agricultural Machine with a Computer Vision-Based Weeding and Variable-Rate Irrigation Scheme. *Robotics* **2018**, *7*, 38. [CrossRef]
106. Patrício, D.I.; Rieder, R. Computer vision and artificial intelligence in precision agriculture for grain crops: A systematic review. *Comput. Electron. Agric.* **2018**, *153*, 69–81. [CrossRef]
107. Tankova, T.; da Silva, L.S. Robotics and Additive Manufacturing in the Construction Industry. *Curr. Robot. Rep.* **2020**, *1*, 13–18. [CrossRef]

108. Davila Delgado, J.M.; Oyedele, L.; Ajayi, A.; Akanbi, L.; Akinade, O.; Bilal, M.; Owolabi, H. Robotics and automated systems in construction: Understanding industry-specific challenges for adoption. *J. Build. Eng.* **2019**, *26*, 100868. [CrossRef]
109. Robinson, G. *Global Construction Market to Grow $8 Trillion by 2030: Driven by China, US and India*; Global Construction Perspectives and Oxford Economics: London, UK, 2016; Volume 44, pp. 1–3.
110. Aparicio, C.C.; Balzan, A.; Trabucco, D. Robotics in construction: Framework and future directions. *Int. J. High-Rise Build.* **2020**, *9*, 105–111.
111. Follini, C.; Magnago, V.; Freitag, K.; Terzer, M.; Marcher, C.; Riedl, M.; Giusti, A.; Matt, D.T. Bim-integrated collaborative robotics for application in building construction and maintenance. *Robotics* **2021**, *10*, 2. [CrossRef]
112. Parascho, S.; Han, I.X.; Walker, S.; Beghini, A.; Bruun, E.P.G.; Adriaenssens, S. Robotic vault: A cooperative robotic assembly method for brick vault construction. *Constr. Robot.* **2020**, *4*, 117–126. [CrossRef]
113. Kazemian, A.; Yuan, X.; Davtalab, O.; Khoshnevis, B. Computer vision for real-time extrusion quality monitoring and control in robotic construction. *Autom. Constr.* **2019**, *101*, 92–98. [CrossRef]
114. Gautam, M.; Fagerlund, H.; Greicevci, B.; Christophe, F.; Havula, J. Collaborative Robotics in Construction: A Test Case on Screwing Gypsum Boards on Ceiling. In Proceedings of the 2020 5th International Conference on Green Technology and Sustainable Development, Ho Chi Minh City, Vietnam, 27–28 November 2020; pp. 88–93.
115. Balzan, A.; Aparicio, C.C.; Trabucco, D. Robotics in construction: State-of-art of on-site advanced devices. *Int. J. High-Rise Build.* **2020**, *9*, 95–104.
116. Ghasempourabadi, M.; Taraz, M. Human-robot interaction in construction: A literature review. *Malays. J. Sustain. Environ.* **2021**, *8*, 49–74.
117. Bodea, S.; Mindermann, P.; Gresser, G.T.; Menges, A. Additive Manufacturing of Large Coreless Filament Wound Composite Elements for Building Construction. *3D Print. Addit. Manuf.* **2021**; ahead of print. [CrossRef]
118. Zhang, M.; Yan, J. A data-driven method for optimizing the energy consumption of industrial robots. *J. Clean. Prod.* **2021**, *285*, 124862. [CrossRef]
119. Aksoy, S.; Ozan, E. Robots and Their Applications. *Int. Res. J. Eng. Technol.* **2020**. [CrossRef]

applied sciences

MDPI

Article

Multiple Object Tracking in Robotic Applications: Trends and Challenges

Abdalla Gad [1,†], Tasnim Basmaji [1,†], Maha Yaghi [1], Huda Alheeh [1,2], Mohammad Alkhedher [3] and Mohammed Ghazal [1,*]

1 Electrical, Computer and Biomedical Engineering Department, College of Engineering, Abu Dhabi University, Abu Dhabi 59911, United Arab Emirates
2 Network and Communication Engineering Department, College of Engineering, Al Ain University, Abu Dhabi 112612, United Arab Emirates
3 Mechanical and Industrial Engineering Department, College of Engineering, Abu Dhabi University, Abu Dhabi 59911, United Arab Emirates
* Correspondence: mohammed.ghazal@adu.ac.ae
† These authors contributed equally to this work.

Abstract: The recent advancement in autonomous robotics is directed toward designing a reliable system that can detect and track multiple objects in the surrounding environment for navigation and guidance purposes. This paper aims to survey the recent development in this area and present the latest trends that tackle the challenges of multiple object tracking, such as heavy occlusion, dynamic background, and illumination changes. Our research includes Multiple Object Tracking (MOT) methods incorporating the multiple inputs that can be perceived from sensors such as cameras and Light Detection and Ranging (LIDAR). In addition, a summary of the tracking techniques, such as data association and occlusion handling, is detailed to define the general framework that the literature employs. We also provide an overview of the metrics and the most common benchmark datasets, including Karlsruhe Institute of Technology and Toyota Technological Institute (KITTI), MOTChallenges, and University at Albany DEtection and TRACking (UA-DETRAC), that are used to train and evaluate the performance of MOT. At the end of this paper, we discuss the results gathered from the articles that introduced the methods. Based on our analysis, deep learning has introduced significant value to the MOT techniques in recent research, resulting in high accuracy while maintaining real-time processing.

Keywords: multiple object tracking; MOT; self-driving; autonomous vehicle; autonomous navigation; SLAM; KITTI; MOTChallenges; MOT15; MOT16; MOT17; UA_DETRAC

Citation: Gad, A.; Basmaji, T.; Yaghi, M.; Alheeh, H.; Alkhedher, M.; Ghazal, M. Multiple Object Tracking in Robotic Applications: Trends and Challenges. *Appl. Sci.* **2022**, *12*, 9408. https://doi.org/10.3390/app12199408

Academic Editor: Luis Gracia

Received: 18 August 2022
Accepted: 14 September 2022
Published: 20 September 2022

1. Introduction

Integrating computer vision and deep learning-based systems in the robotics field has led to a massive leap in the advancement of autonomous feature. The utilization of different sensors, such as cameras and LIDARs, and the progress established by the recent research on processing this data, have introduced multiple object tracking techniques in autonomous driving and robotics navigation systems. Multiple object tracking has been one of the most challenging topics researched through computer vision techniques. The reasons behind this are due to: (1) multiple object tracking (MOT), an essential tool that can be used in enhancing security and automating robotics navigation, and (2) occlusion, which is the main obstacle standing in the path of reaching a reliable accuracy and one issue that is difficult to tackle. In this paper, we aim to survey the different approaches of MOT introduced recently in autonomous robotics.

Much research has been done to enhance the performance of tracking in SLAM applications [1,2]. It is simply difficult to navigate through the environment while the positions of the robot itself and other objects in the surrounding are neglected. Tracking is used to estimate the relative location of the robot to other components in the environment. The most challenging part of this process is the existence of highly dynamic objects [3] such as people or vehicles. SLAM-based autonomous navigation in robots has been regarded as essential in development and research primarily because of its potential in many aspects. One example is the autonomous wheelchair systems reviewed in Refs. [4,5]. The authors in Ref. [6] provided a survey of the mobile devices that assist people with disability. Autonomous driving can cause a reduction in the number of accidents that occur due to fatigue and distractions [7]. Although that might be the case, the public opinion about autonomous vehicles is hesitant about whether to consider the technology trustworthy. Providing awareness and understanding of the capabilities of the sensors in autonomous vehicles to the drivers is vital to reaching the proper employment of the technology in our daily lives. These sensors should not be disregarded or become entirely dependable on them [8]. The approach introduced in Ref. [9] aims to reduce the risks firefighters encounter by deploying a team of UAVs with an MOT system to track wildfires and control the situation. The authors in Ref. [10] employ MOT to guide a swarm of drones and control them. Similarly, MOT is utilized with UAV for collision avoidance in Ref. [11]

As has been discussed in Refs. [12–14], the general framework for MOT is shown in Figure 1. The input frame is subjected to an object detection algorithm. Then, the detections from the current frame and the previous frames are used to match the similar trajectories either by motion, appearance, and/or other features. This process would generate tracks presenting the objects through the sequence of frames. Some data association between multiple frames is applied to track an object through multiple frames. A reliable MOT system should be able to handle the new tracks as well as the lost ones. Here, the occlusion issue is where the lost tracks reappear again because they did not move out of the sensor's view but were hidden by other objects.

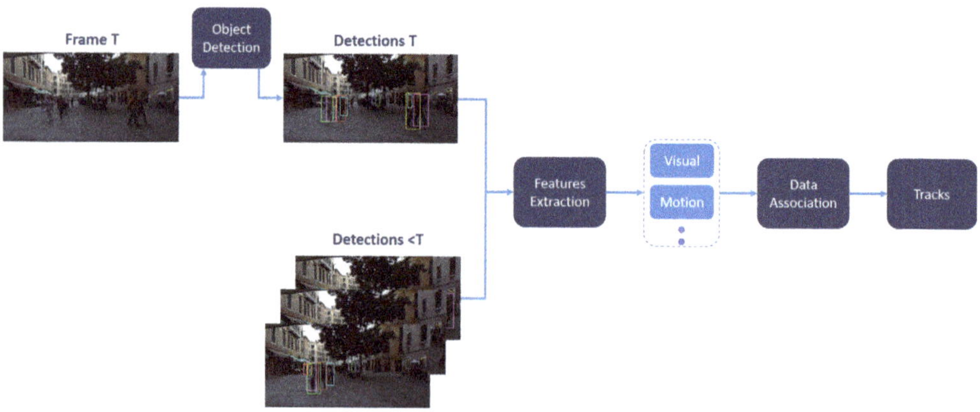

Figure 1. General framework of MOT systems. Visual and motion features of the detected objects at frame T are extracted and compared to those detected from previous frames. A robust data association algorithm would be able to match the features of the same objects. The final output of the system would be tracked with unique IDs identifying the multiple objects detected and tracked over the multiple frames.

1.1. Challenges

To effectively perform object tracking, one must develop a robust and efficient model that the users can effectively use. This section aims to provide a comprehensive overview of the various challenges facing developing and optimizing such models.

The first challenge a model must face is the quality of the input video [15]. If the model cannot process the video properly, it will require additional work to convert it into a clear form so that it can be used to detect objects. The classification system must first identify the objects that fall under a particular class. It then gives them IDs based on their shape and form, which raises the issue that objects of this class come in varying shapes and sizes [16]. After the objects are detected, they must be assigned IDs with a bounding box to identify them to ensure that the model can identify multiple similar objects in the coverage area. The next challenge is to identify the objects that are moving in the ROI of the camera. This phenomenon can cause the classification system to misclassify the objects or even identify them as new ones. Aside from the quality of the video input, other factors that affect the classification of objects are also considered. For instance, the illumination conditions can significantly influence the model's accuracy [12–14,16]. The model may not be able to detect objects that are blunt with the environment or have background conditions. It also needs to be able to identify them at varying speeds. One of the most challenging issues in object tracking is the Occlusion issue, where the object movement gets interrupted by other objects in the scene [12–14,16]. It can be caused by various factors such as natural conditions or the object's movement out of the camera's ROI. Another reason is that other objects might block the visual of the object if the object is in the camera's ROI. Therefore, the system must be trained to identify and track the objects in motion. It also needs to be able to re-identify the IDs of the captured images with the same ones already used by the cameras. Figure 2 shows an example of the occlusion issue. The yellow arrow follows one of the tracks that maintains its ID after experiencing full occlusion. The problem of occlusion is minimized in bird-eye view tracking [17]. However, other challenges arise, such as the low resolution of objects and misclassifications. Another obstacle is related to onboard tracking in self-driving applications. The issue is that the tracking process needs to be quick and accurate for an efficient assistant driving system. The FPS is one of the essential factors determining the tracking quality in this case [18].

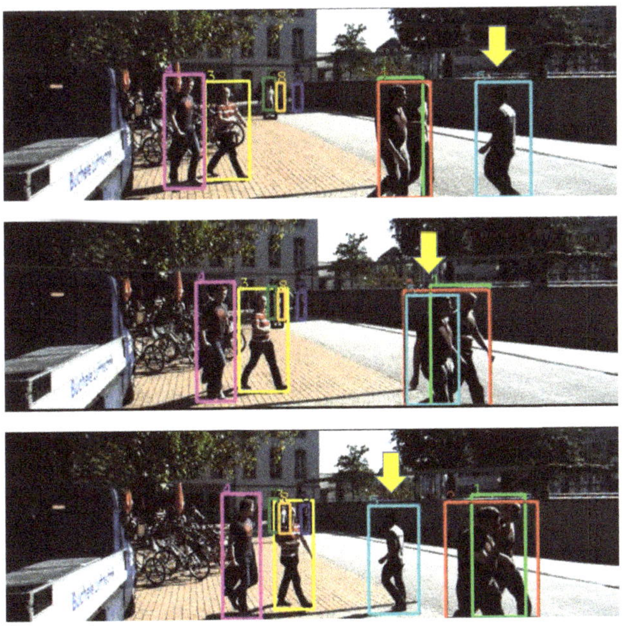

Figure 2. Preserving the ID during full occlusion. The frames are obtained from the MOT15 dataset [19]. The yellow arrow is pointing towards a track (**top image**) that experiences full occlusion (**middle image**). The objective of the MOT system is to preserve the ID of the track (**bottom image**) and matches the previously detected object with the reappeared one.

1.2. Related Work

The authors in Ref. [20] provided a summary of the techniques developed for SLAM-MOT (combination of SLAM and MOT systems) that utilize the dynamic features to construct 3D object tracking and multi-motion segmentation systems. In Ref. [20], 3D tracking of dynamic objects techniques was categorized into trajectory triangulation, particle filter, and factorization-based approaches. The authors in Ref. [20] discussed the data fusion problem in autonomous vehicles. The perception of data from different sensors such as RGB cameras, LIDAR, and depth cameras provides more knowledge and understanding of the surrounding environment and increases the navigation system's robustness. In Ref. [20], a survey is conducted on the techniques used for SLAM in autonomous driving applications and the limitations of the current research. The evaluation in this paper was done on the KITTI dataset.

The authors in Ref. [12] categorized the MOT approaches into three groups. The first is the initialization method, which defines whether the tracking would be detection-based or detection-free. Detection-based tracking or tracking-by-detection is the most common, where a detected object is connected to its trajectories from future frames. This connection can be applied by calculating the similarity based on appearance or motion. Detection-free tracking is where a set of objects is manually localized in the first frame and tracked through the future. This is not optimal in case new objects appear and is rarely applied. The second is based on the processing mode, either online or offline tracking. Online tracking is where objects are detected and tracked in real time. This is more optimal in the case of autonomous driving applications as offline tracking is where a batch of frames are processed at a low FPS. The final one is the type of output where it can be stochastic, in which the tracking varies at different running times, or deterministic, in which the tracking is constant. They would further define the components that are included in the MOT system. Appearance model is used to extract spatial information from the detections and then calculate their similarity. The visual features and representations extracted can be defined either locally or regionally.

Motion models are used to predict the future location of the detected object and hence, reduce the inspection area. A good model would have a good estimation after a certain number of frames as its parameters are tuned towards learning how the object moves. Linear (constant velocity) and non-linear are the two types of motion models. Although there has been a rapid advancement in multiple object detection and tracking for autonomous driving, it is still processing only a few objects. It will be a giant leap forward to have a system capable of tracking all types of objects in real time. This can be achieved by generating a great deal of data that tackles the problem at different perceptions such as camera, LIDAR, ultrasonic, etc. [21]. The issues related to the full deployment of MOT in autonomous vehicles lie in that its reliability heavily depends on many parameters, such as the camera view and the type of background (dynamic or static). This leads to difficulty in being entirely trustful towards MOT in different real scenarios and environments [12]. Tracking pedestrians is a far more difficult task than tracking vehicles whose motion is bounded by the road compared to the motion of people, which is very random and challenging for the system to learn. Another issue is the occlusion, which leads to high fragmentation and ID switches due to losing and re-initializing tracks every time they get lost. There have been very few systems that comprehensively tackle the problem, which leaves a huge space for improvements [22].

In Ref. [16], the tracking algorithms are categorized into two groups. The first is matching-based, which defines how features, such as appearance and motion, are first extracted and used to measure the similarity in the future frames. The second is filtering-based tracking, where Kalman and Particle filters are discussed. The authors in Ref. [13] comprehensively surveyed the deep learning-based methods for MOT. They also provided an overview of the data in MOTChallenges and the type of conditions included. An evaluation of the performance of some methods on this dataset is then listed. In Ref. [14], the deep learning-based methods for MOT were also reviewed. Similarly, the authors also provided an overview of the benchmark datasets, including MOTChallenges and KITTI, and presented the performance of some methods.

In Ref. [23], the vision-based methods used to detect and track vehicles at road intersections were discussed. The authors categorized those methods depending on the sensors used and the approach carried out for detection and tracking. On the other hand, the authors in Ref. [24] presented methods introducing vehicle detection and tracking in urban areas, and an evaluation was then discussed. UAVs' role in civil applications, including surveillance, have been surveyed in Refs. [25,26]. The authors discussed the characteristics and roles of UAVs in traffic flow monitoring. However, there have not been many contributions to vehicle tracking methods using an UAV.

The authors in Refs. [7,8], provided a detailed overview of the types of sensors mounted on autonomous vehicles, such as LIDAR, Ultrasonic, and cameras, for data perception. They also surveyed the current advancement in the autonomous driving field commercially and the type of technology associated with that. In Ref. [21], the authors studied the role of deep learning in autonomous driving including perception and path planning. In addition to deep learning approaches, a general review was introduced in Ref. [27]. In Ref. [28], the methods used to extract and match information from multiple sensors used for perception were reviewed. They also discussed how data association could be an issue in using multiple sensors to achieve reliable multiple object tracking. The authors in Ref. [29], surveyed the methods that utilize LIDAR in data perception and grouped the performance results on the KITTI dataset. Ref. [22] provided a comprehensive overview of the KITTI dataset's role in the autonomous driving application. The dataset can be used for training and testing pedestrians, vehicles, cyclists, and other objects that can be found on the road. Moreover, the dataset was extended to lane and road marks detection by Ref. [30].

Although the techniques mentioned above were very thorough in reviewing techniques, we aim in this paper to provide comprehensive research that surveys the techniques associated with autonomous robotics applications, provides an insight into the different tracking methods, gathers and compares the results from the different methods discussed in the paper, and evaluate the current work and find limitations that require future research. Table 1 lists the recent reviews, the year of publication, and the datasets used for comparing MOT methods.

Table 1. Recent reviews and the data used for evaluation and comparison.

Review	Year	Evaluation Dataset
Ciaparrone et al. [13]	2019	MOT 15, 16, 17
Xu et al. [14]	2019	MOT 15, 16
Luo et al. [12]	2022	PETS2009-S2L1
Ours	**2022**	**KITTI, MOT 15, 16, 17, 20 , and UA_DETRAC**

Section 2 discusses the state-of-art methods and techniques introduced by the literature. Section 3 discusses the benchmark datasets and evaluation metrics popularly used by the research for training and testing. Section 4 presents the evaluation results collected from the literature and discussion. Finally, Sections 5 and 6 provide the current study challenges and the future work that is required.

2. Mot Techniques

In this section, we go through the most recent MOT techniques and the common trends being followed for matching tracks across multiple frames.

Table 2 shows a summary of the components used in MOT techniques. It can be observed that the appearance cue is rarely neglected. Motion cue also shows presence a lot. Most approaches depend on deep learning for extracting visual features. CNNs are vital tools that can extract visual features from the tracks and achieve accurate detections of tracks matching [14]. The approaches introduced in Refs. [31,32] use Long Short Term Memory (LSTM) based networks for motion modeling. LSTM networks are considered in MOT for appearance and motion modeling as they can find patterns by efficiently processing the previous frames in addition to the current ones. On the other hand, The authors in Ref. [33] generated histograms from the detections and used them as the appearance features. As for data association, the Hungarian algorithm is common with MOT techniques, such as Refs. [33–36], for associating the current detections with the previous ones, although the performance of these techniques did not show much potential. Deep learning has rarely been utilized for data association. However, the best performing technique on MOT16 and MOT17 datasets relied on a prediction network to validate that the two bounding boxes are related. For occlusion handling, most approaches rely on feeding the history of tracks into the tracking system to validate the lost ones. The tracks absent for a specific number of frames would be considered lost and deleted from the history. This is to avoid processing a massive number of detections and reducing the FPS.

Table 2. Summary of the components used in MOT techniques.

Tracker	Appearance Cue	Motion Cue	Data Association	Mode	Occlusion Handling
Keuper et al. [37]	-	Optical flow	Correlation Co-Clustering	Online	Point trajectories
Fang et al. [38]	fc8 layer of the inception network	Relative center coordinates	Conditional Probability	Online	Track history
Xiang et al. [32]	VGG-16	LSTM network	Metric Learning	Online	Track history
Chu et al. [39]	CNN	-	Reinforcement Learning	Online	SVM Classifier
Zhu et al. [40]	ECO	-	Dual Matching Attention Networks	Online	Spatial attention network
Zhou et al. [41]	ResNet50	Linear Model	Siamese Networks	Online	Track history
Sun et al. [42]	VGG-like network	-	Affinity estimator	-	Track history
Peng et al. [43]	ResNet-50 + FPN	-	Prediction Network	Online	Track history
Wang et al. [44]	FaceNet	-	Multi-Scale TrackletNet	-	Track history
Mahmoudi et al. [34]	CNN	Relative mean velocity and position	Hungarian Algorithm	Online	Track history
Zhou et al. [35]	ResNet-50	Kalman Filter	Hungarian Algorithm	Online	Track history
Lan et al. [45]	Decoder	Relative position	Unary Potential	Near-online	Track history
Zhou et al. [46]	-	CenterTrack	2D displacement prediction	Online	Track history
Karunasekera et al. [33]	Matching histograms	Relative position	Hungarian Algorithm	Online	Grid Structure + Track history
Chu et al. [47]	Siamese-style network	-	R1TA Power Iteration layer	Online	Track history

Table 2. *Cont.*

Tracker	Appearance Cue	Motion Cue	Data Association	Mode	Occlusion Handling
Zhao et al. [36]	CNN + PCA + Correlation Filter	-	Hungarian Algorithm	Online	APCE + IoU
Chen et al. [48]	Faster-RCNN (VGG16)	-	Category Classifier	Online	Track history
Sadeghian et al. [31]	VGG16 + LSTM network	LSTM network	RNN	Online	Track history
Yoon et al. [49]	Encoder	-	Decoder	Online	Track history
Xu et al. [50]	FairMOT	Kalman Filter	Hungarian Algorithm	Online	Track history
Ye et al. [51]	LDAE	Kalman Filter	Cosine Distance	Online	Track History
Wang et al. [52]	Faster-RCNN (ResNet50)	Kalman Filter	Faster RCNN (ResNet50)	Online	Track history
Yu et al. [53]	DLA-34 + GCD + GTE	-	Hungarian Algorithm	Online	Track history
Wang et al. [54]	PCB	Kalman Filter	RTU++	Online	Track history
Gao et al. [55]	ResNet34	-	Depth wise Cross Correlation	Online	Track history
Nasseri et al. [56]	-	Kalman Filter	Normalized IoU	Online	Track history
Zhao et al. [57]	Transformer + Pixel Decoder	Kalman Filter	Cosine Distance	Online	Track history
Aharon et al. [58]	FastReID's SBS-50	Kalman Filter	Cosine Distance	Online	Track history
Seidenschwarz et al. [59]	ResNet50	Linear Model	Histogram Distance	Online	Track history
dai et al. [60]	CNN	-	Hungarian Algorithm	Online	Track history
Zhang et al. [61]	FairMOT	Kalman Filter	IoU + Hungarian Algorithm	Online	Track history
Hyun et al. [62]	FairMOT	-	Sparse Graph + Hungarian	Online	Track history

Table 2. *Cont.*

Tracker	Appearance Cue	Motion Cue	Data Association	Mode	Occlusion Handling
Chen et al. [63]	DETR Network	-	Hungarian Algorithm	Online	Track history
Cao et al. [64]	-	Kalman Filter + non-linear model + Smoothing	Observation Centric Recovery	Online	Track history
Wan et al. [65]	YOLOx	-	Self and Cross Attention Layers	Online	Track history
Du et al. [66]	-	NST Kalman Filter + ResNet50	AFLink	Online	Track history
Zhang et al. [67]	ResNet50	Kalman Filter	Hungarian Algorithm	Online	Track history

The general framework illustrated in Figure 1 is followed by most of the recent MOT techniques. The most common approach for extracting the visual features of an object is by using CNN. VGG-16 is very popular for this application, as in Refs. [32,42,48]. The issue with deploying CNN is the slow computation time due to the high dimensionality output. Zhao et al. [36] tackled this issue by applying PCA followed by a correlation filter for dimensionality reduction to the output of the CNN. An encoder with fewer parameters is introduced in Ref. [49] for faster computation. Another popular network for extracting appearance features is ResNet-50, as in Refs. [35,41,43], which resulted in a competing accuracy with fast computation. Peng et al. [43] extracted the appearance features from different layers of a ResNet-50 network forming a Feature Pyramid Network, which has the advantage of detecting objects at different scales. The LSTM network is an important concept for architecture design for processing a sequence of movies. It has been used for MOT application in multiple approaches such as Refs. [31,32]. The main approach taken by most current methods is to store the appearance features of the previous frames and retrieve them for comparison with the ones of the current frame. The important factor that affects the reliability of this comparison is the updating of the stored features. The object's appearance varies through the frames but not significantly between two adjacent frames; hence, constant updating can lead to a higher matching accuracy.

The second most common feature used for tracking is that related to the object's motion. This is specifically useful at full occlusion occurrence. In this case, the object's state can change significantly, and the appearance features will not be reliable for matching. A robust motion model can predict the object's location even if it disappears from the scene. The most common approach for motion tracking is the Kalman Filter, as in Refs. [50–52]. The authors in Refs. [34,45] use the relative position between two tracks in two adjacent frames and decide whether or not the two tracks are of the same object. The authors in Refs. [31,32,46], use deep learning approaches for motion tracking. The most common architecture for this approach is the LSTM network. The Kalman filter has a significantly lower computational cost than the deep learning approaches. The issue with utilizing only motion models for tracking is the random motion of objects. For instance, motion models would work better on cars where the motion is limited than on people. Zhou et al. introduced CenterTrack in Ref. [46] for tracking objects as points. The system is end-to-end, taking the current and the previous frames and outputting the matched tracks as illustrated in Figure 3.

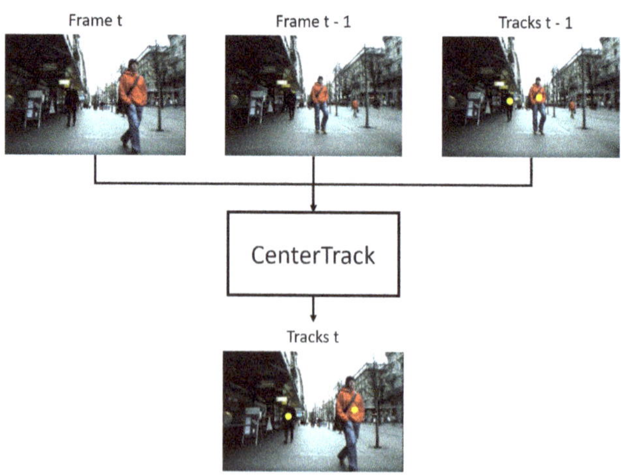

Figure 3. The point tracking approach in Ref. [46]. The current and previous frames are passed into the centerTrack network, which utilize the motion feature to detect and match tracks.

There are other features used for tracking. The authors in Refs. [56,61], added the IoU metric between the adjacent frames' detections to match the two tracks. The tracking of one object in the scene can be affected by other objects. For this reason, some methods have introduced the interactivity feature to the tracking algorithm. In Ref. [31], the interactivity features were extracted from an occupancy map using an LSTM network. The authors in Ref. [54] used a tracking graph and designed a set of conditions to measure the interactivity between two objects. Figure 4 illustrates the overview of tracking graph methods. The approach in Ref. [63], exploited the size and structure as features along with the appearance and motion for tracking. Increasing the number of features can improve the tracking process at the cost of computation time. The main issue would be the processing needed to fuse those features with different dimensionalities. The authors in Refs. [44,68] used IOU in addition to appearance and motion features to increase the reliability of the tracking. The authors in Ref. [44] further improved the model by adding epipolar constraints with the IOU and introduced a tracklenet to group similar tracks into a cluster. Ref. [69] added the deep_sort algorithm to the extracted features to reduce the unreliable tracks, and Ref. [36] added a correlation filter tracker to the CNN. Ref. [47] performed a similar approach of performing feature extraction and matching simultaneously by having affinity estimation and multi-dimensional assignment in one network. The authors in Refs. [70,71] experimented with 3D distance estimation from RGB frames. In Ref. [70], Poisson multi-Bernoulli mixture tracking filter was used to perform the 3D projections. In addition to CNN, Ref. [72] experimented visually with the track's Gaussian Mixture Probability Hypothesis Density. The authors in Ref. [37] introduced a motion segmentation framework using motion cues in addition to the IOU and bounding box clusters for object tracking through multiple frames. An interesting technique is established in Ref. [33], where one network has the current and prior frames as inputs and outputs point tracks. Those tracks are given to a displacement model to measure the similarity. Similarly, another approach to motion modeling is introduced in Ref. [45] to handle overlapping tracks by using the efficient quadratic pseudo-Boolean for optimization.

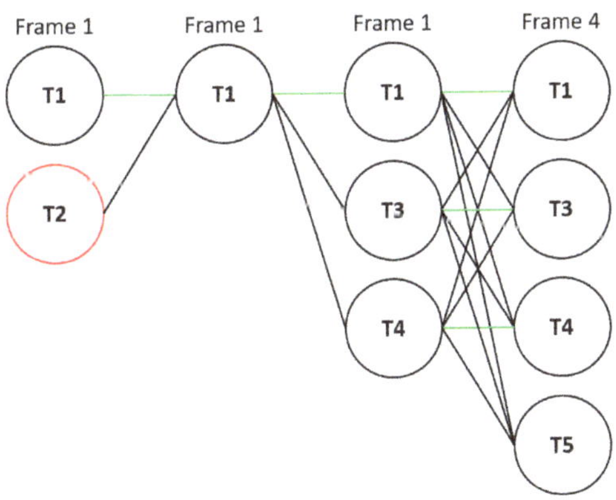

Figure 4. Track tree proposed in Track tree method system overview. The strength of each branch depends on a score evaluated by the matching algorithm. The green lines indicate matched tracks. The red circle indicate a lost track.

The features extracted from every detection in the current frame must be associated with those extracted from the previous frames. The most popular approach taken for data association in recent years would be the Hungarian algorithm, as in Refs. [50,53,60,63,67]. The advantage of this method is the accuracy accompanied by a fast computation time. Zhang et al. proposed ByteTrack in Ref. [50], where the Kalman filter is used for predicting the detection location followed by two levels of association. The first utilizes the appearance features in addition to the Intersection over Union (IoU) for matching tracks using the

Hungarian algorithm. The second level of association deals with the weak detections by utilizing only the IoU with the unmatched tracks remaining from the first level. The authors in Refs. [31,43,52] use deep learning networks for data association. The authors in Ref. [39] introduced a model trained using reinforcement learning. The metric learning concept was used to train the matching model in Ref. [32]. The authors in Ref. [62] take advantage of the object detection network for feature extraction. This would save computational costs from applying an appearance feature network on each detection in the current frame. The approaches in Refs. [43,52] apply the same concept in addition to an end-to-end system that takes the current frame and previous frames as input and outputs the current frame with tracks. The hierarchical single-branch network is an example of an end-to-end system proposed in Ref. [52] as illustrated in Figure 5. The P3AFormer tracker introduced in Ref. [57] uses the simultaneous detection and tracking framework where a decoder and a detector extract pixel-level features from the current and previous frames. The features are passed into a multilayer perceptron (MLP) that outputs the size, center, and class information. The features are then matched using the Hungarian algorithm. The system overview of the P3AFormer is shown in Figure 6.

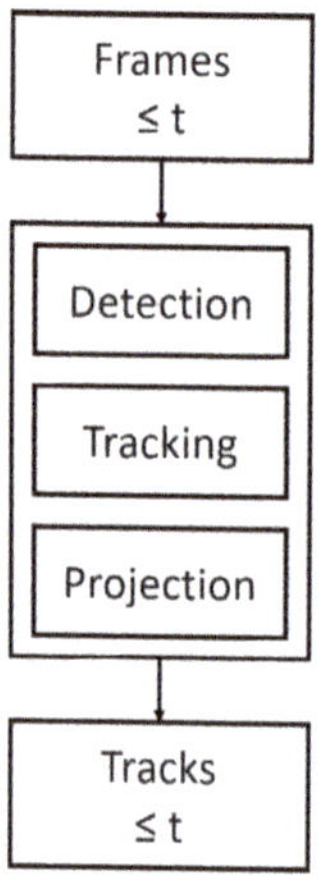

Figure 5. The hierarchical single-branch network proposed in Ref. [52]. The frames from a video source are passed into the network and outputs detections and tracks.

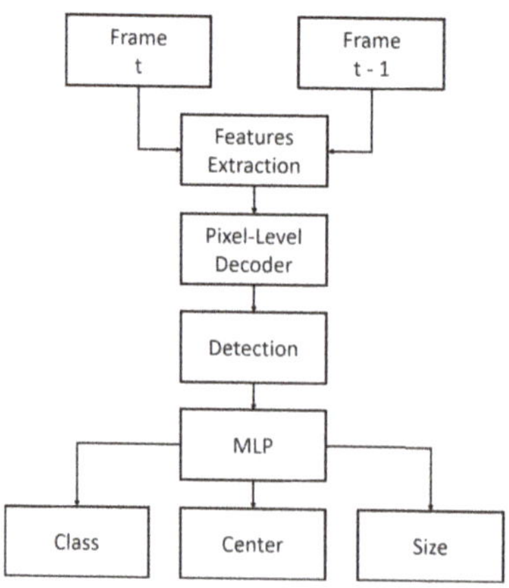

Figure 6. P3AFormer system overview [57]. The current and previous frames are passed for features extraction and detection. The extraction module consists of a backbone network, pixel-level decoder, and a detector, which is illustrated on the right. The features are passed to MLP heads to output the class, center, and size features.

The Recurrent Autoregressive Networks (RAN) approach introduced in Ref. [38] defines an autoregressive model that is used to estimate the mean and variance of appearance and motion features of all associated tracks of the same object stored in the external memory. The internal memory uses the model generated to compare with all upcoming detections. The one with the maximum score and above a certain threshold would then be considered the same object. The external and internal memory would then be updated with the new associated object. There are two types of independence in this approach. The first is that the motion and appearance models include different parameters, so they have different internal and external memories. The second is a new RAN model generated for every new object detected. The lost tracks are terminated after 20 frames. The visual features are extracted using the fully connected layer (fc8) of the inception network, and the motion feature is a 4-dimensional vector representing the width, height, and relative position from the previous detection. The CTracker framework [43] takes two adjacent frames as inputs and matches each detection with the other using Intersection over Union calculations. A constant velocity motion model is used to match the tracks up to a certain number of frames to handle the reappearance of lost tracks. This approach is end-to-end. It takes two frames as input and outputs two frames with detections and matching tracks.

The authors in Ref. [33] introduced an approach in which the detections' dissimilarity measures are computed and then matched. First is the dissimilarity cost computation. Next, the histogram of the H and S channels of the HSV colorspace of the previous detections is compared to the similar histogram of the current detections. A grid structure is used as in Refs. [73,74]. Hence, multiple histograms are used to match the appearance features. Furthermore, Linear Binary Pattern Histogram (LBPH), introduced in Ref. [75] and used for object recognition in Ref. [76], is utilized for computing the structure-based distance. The predicted and the measured position matching using the L2 norm is added as the motion-based distance. Finally, IoU calculates the size difference between the current detections and the previous tracks. The second step is using the Hungarian algorithm [77] to calculate the overall similarity using the four features calculated in the previous step.

The authors in Ref. [68] proposed V-IoU, an extension of IoU [78], for object tracking. The objective here is to reduce the number of ID switches and fragmentation by maintaining the location of lost tracks for a certain number of frames until it appears. A backtracking technique where the reappeared track is projected backward through the frames is implemented to validate that the reappeared track is, in fact, the lost one. In Ref. [46], CenterNet [79] is used to detect objects as points. Centertrack takes two adjacent frames as inputs in addition to the point detections in the previous frame. The tracks are associated using the offset between the current and previous point detections. The authors in Ref. [80] designed a motion segmentation model where the point clusters used for trajectory prediction were placed around the center of the detected bounding box. The approach in Ref. [37] employs optical flow and correlation co-clustering for projecting trajectory points across multiple frames, as illustrated in Figure 7.

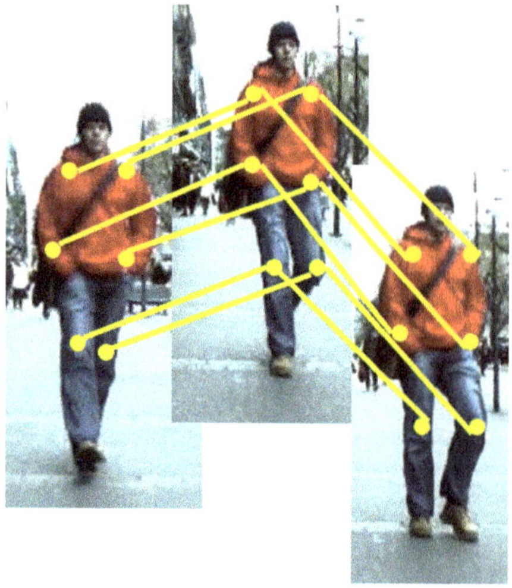

Figure 7. Motion segmentation. The trajectory points are projected across multiple frames.

There has been a recent advancement in multiple object tracking and segmentation (MOTS). This field tackles the issues related to the classic MOT, which are associated with the utilization of bounding box detection and tracking, such as background noises and loss of the shape features. The approach introduced in Ref. [81] used an instance segmentation mask on the extracted embedding. The method in Ref. [82] applies contrastive learning for learning the instance-masks used for segmentation. An offline approach introduced in Ref. [83] exploits appearance features for tracking. This method is currently at the top of the leader board at the MOTS20 challenge [84].

The system can achieve more reliability when utilizing multiple sensors to detect and track targets in addition to understanding their intentions [28]. The deep learning approaches are improving and showing promise in the LIDAR datasets. The issue with this is the bad running time, causing difficulty in real-time deployment [29]. The challenges facing 3D tracking are related to the fusion of the data perceived from LIDAR and RGB cameras. Table 3 lists the recent MOT techniques that utilize LIDAR and camera for tracking.

Table 3. Summary of the sensors fusion approaches used in 3D MOT techniques.

Tracker	RGB Camera	Point Cloud LIDAR	Data Fusion	Target Tracking
Simon et al. [85]	Enet	3D Voxel Quantization	Complex-YOLO	LMB RFS
Zhang et al. [86]	VGG-16	PointNet	Point-wise convolution + Start and end estimator	Linear Programming
Frossard et al. [87]	MV3D	MV3D	MV3D	Linear Programming
Hu et al. [71]	Faster R-CNN	34-layer DLA-up	Monocular 3D estimation	LSTM Network
Weng et al. [88]	ResNet	PointNet	Addition	Graph Neural Network
Sualeh et al. [89]	YOLOv3	IMM-UKF-JPDAF	3D projection	Munkres Association
Shenoi et al. [90]	Mask RCNN	F-PointNet	3-layered fully connected network	JPDA

Simon et al. [85] proposed complexer-YOLO, illustrated in Figure 8, for RGB and LIDAR data detection, tracking, and segmentation. A preprocessing step for the point cloud data input from LIDAR aims to generate a voxalized map of the 3D detections. The RGB frame is passed into ENet [91], which will output a semantic map. Both maps are matched and passed into the Complexer-YOLO network to output tracks. The approach in

Ref. [86] extracts features from the RGB frame and the point cloud data. An end-to-end approach was introduced in Ref. [87] for dealing with features extraction and fusing from RGB and Point Cloud Data, as illustrated in Figure 9. Point-wise convolution in addition to a start and end estimator are utilized for fusing both types of data to be used for tracking.

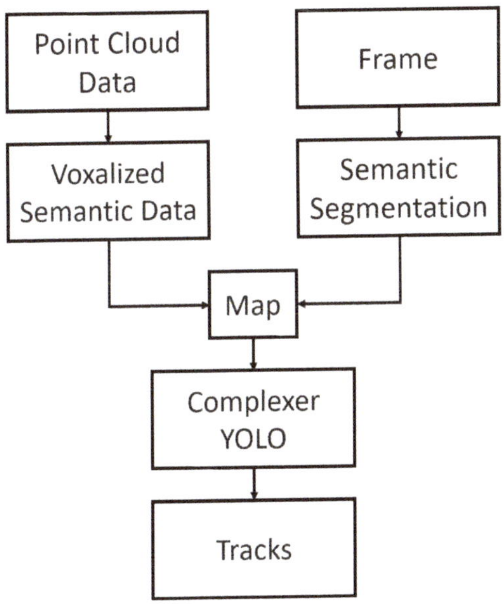

Figure 8. Complexer-YOLO [85]. Data from RGB frame and point cloud data are mapped and passed into Complexer-YOLO for tracking and matching.

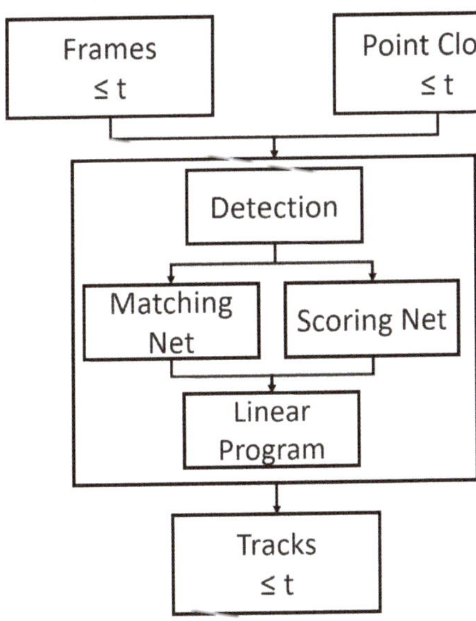

Figure 9. An end-to-end approach for 3D detection and tracking [87]. The RGB and point cloud data are passed into a detection network. Matching and scoring nets are then trained to generate trajectories across multiple frames.

3. Mot Benchmark Datasets and Evaluation Metrics

In this section, we review the most common datasets that are used for training and testing MOT techniques. We also provide an overview of the metrics used to evaluate the performance of these techniques.

3.1. Benchmark Datasets

Most research done in multiple object tracking uses standard datasets for evaluating the state of art techniques. In this way, we have a better view of what criteria, the new methodologies, have shown superiority. For this application, everyday moving objects could be pedestrians, vehicles, cyclists, etc. The most common datasets that provide a variation of those objects in the streets are the MOTChallenge collection and KITTI.

- MOTChallenge: The most common datasets in this collection are the MOT15 [19], MOT16 [92], MOT17 [92], and MOT20 [93]. There is a newly created set, MOT20, but it has not yet become a standard for evaluation in the research community to our current knowledge. The MOT datasets contain some data from existing sets such as PETS and TownCenter and others that are unique. Examples of the data included are presented in Table 4, where the amount of variation included in the MOT15 and MOT16 can be observed. Thus, the dataset is useful for training and testing using static and dynamic backgrounds and for 2D and 3D tracking. An evaluation tool is also given with the set to measure all features of the multiple object tracking algorithm, including accuracy, precision, and FPS. The ground truth data samples are shown in Figure 10.

Table 4. Examples of the types of data included in MOT15 and MOT16.

Dataset	Sequences	Length	Tracks	FPS	Platform	ViewPoint	Density	Weather
MOT 2015	TUD-Crossing	201	13	25	static	horizontal	5.5	cloudy
	PETS09-S2L2	436	42	7	static	high	22.1	cloudy
	ETH-Jelmoli	440	45	14	moving	low	5.8	sunny
	KITTI-16	209	17	10	static	horizontal	8.1	sunny
MOT 2016	MOT16-01	450	23	30	static	horizontal	14.2	cloudy
	MOT16-03	1500	148	30	static	high	69.7	night
	MOT16-06	1194	221	14	moving	low	9.7	sunny
	MOT16-12	900	86	30	moving	horizontal	9.2	indoor

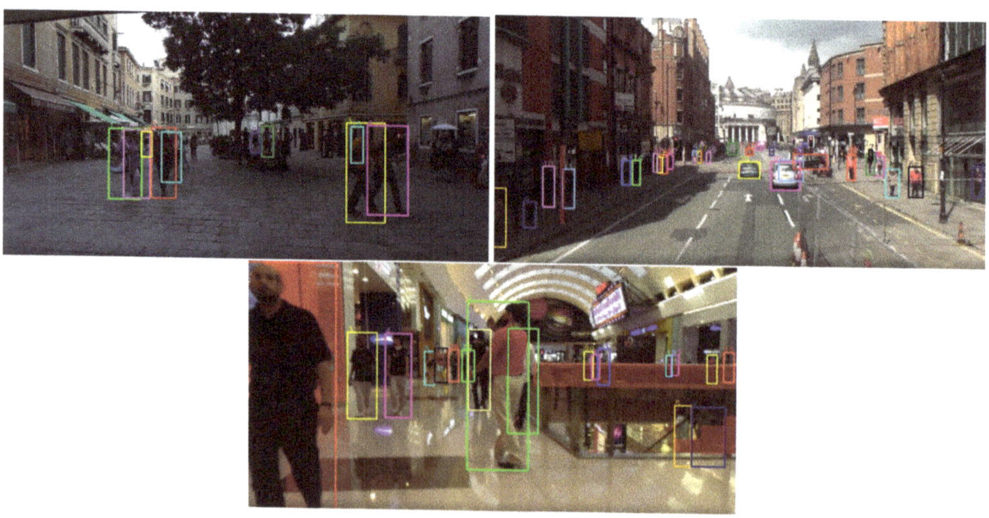

Figure 10. Samples from MOT 15-17 ground truth dataset. Samples of MOT15 (**top image**), MOT16 (**middle image**), and MOT17 (**bottom image**).

- KITTI [94]: This dataset is created specifically for autonomous driving. It was collected by a car driven through the streets with multiple sensors mounted for data collection. The set includes PointCloud data collected using LIDAR sensors and RGB video sequences captured by monocular cameras. It has been included in multiple research related to 2D and 3D multiple object tracking. Samples of the pointcloud and RGB data included in the KITTI dataset are shown in Figure 11.

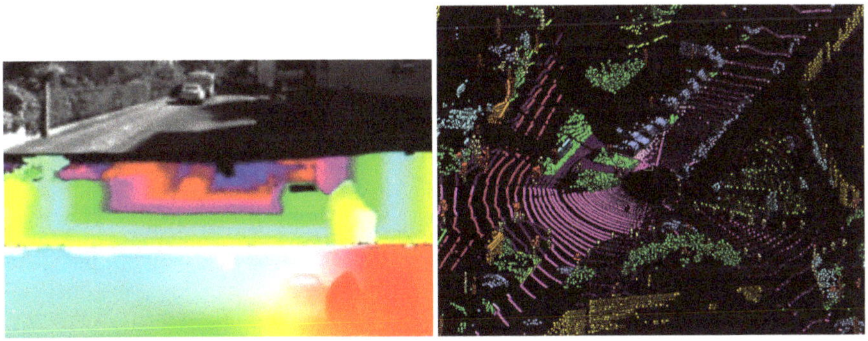

Figure 11. Samples of the KITTI dataset including Pointcloud and RGB. Visual odometry trajectory (**top left**), disparity and optical flow map (**top right**), visualized point cloud data [95] (**middle**), and 3D labels (**bottom**).

- UA-DETRAC [96–98]: The dataset includes videos sequences captured from static cameras looking at the streets at different cities. A huge amount of labeled vehicles can assist in training and testing for static background multiple object tracking in surveillance and autonomous driving. Samples of the UA-DETRAC dataset at different illumination conditions can be shown in Figure 12.

Figure 12. Samples of the UA-DETRAC dataset showing variation of illumination in the environment from a static camera.

3.2. Evaluation Metrics

The most common evaluation metrics are the CLEAR MOT metrics, which were developed by Refs. [22,23]. Mostly tracked objects (MT) and mostly lost objects (ML) in addition to IDF1 are uses to present the leaderboards in MotChallenges. False Positives (*FP*) is the number of falsely detected objects. false Negatives (*FN*) the number of falsely undetected objects. Fragmentation (Fragm) is the number of times a track gets interrupted. ID Switches (IDSW) is the number of times an ID changes. Multiple Object Tracking Accuracy (*MOTA*) is given by (1), whereas Multiple Object tracking Precision (*MOTP*) is given by (2). Finally, frames per second (Hz) and IDF1 given by (3).

$$MOTA = 1 - \frac{(FN + FP + IDSW)}{GT} \quad (1)$$

where GT is the total number of ground truth labels.

$$MOTP = \frac{\sum_{t,i} d_{t,i}}{\sum_t c_t} \tag{2}$$

where c is the number of matches in t frame and d is the correctly overlapping detections and tracks.

$$IdentificationF1(IDF1) = \frac{1}{\frac{1}{IDP} + \frac{1}{IDR}} \tag{3}$$

where identification precision is defined by:

$$IDP = \frac{IDTP}{IDTP + IDFP} \tag{4}$$

and identification recall is defined by:

$$IDR = \frac{IDTP}{IDTP + IDFN} \tag{5}$$

IDTP is the sum of true positives edges weights, *IDFP* is the sum of false positives edges weights, and *IDFN* is the sum of false negatives edges weights

The metrics used for evaluating on the UA_DETRAC dataset utilize the precision recall (PR) for calculating the CLEAR metrics, as introduced in Ref. [96]. In addition to these metrics, the HOTA metric introduced in Ref. [99] is calculated by the formula in (6).

$$\int_{0<\alpha\leq 1} HOTA_\alpha \tag{6}$$

where

$$\sqrt{\frac{\sum_{c \epsilon TP_\alpha}(Ass\_IoU)_\alpha(c))}{|TP_\alpha| + |FN_\alpha| + |FP_\alpha|}} \tag{7}$$

where

$$Ass\_IoU = \frac{|TPA|}{|TPA| + |FNA| + |FPA|} \tag{8}$$

where *TPA*, *FNA*, and *FPA* are the association metrics.

There has been a recent advancement in the multiple object tracking and segmentation (*MOTS*). This field tackles the issues related to the classic *MOT* which are associated with the utilization of bounding box detection and tracking such as background noises and loss of the shape features. The MOTS20 Challenge [84] proposed metrics for evaluating methods that tackle this issue. The multi-object tracking and segmentation accuracy (*MOTSA*) is calculated using the formula in (9). Similarly, multi-object tracking and segmentation precision (*MOTSP*) and soft multi-object tracking and segmentation accuracy (*sMOTSA*) are found by the formulas in (10) and (11), respectively.

$$MOTSA = 1 - \frac{(|FN| + |FP| + |IDSW|)}{|M|} \tag{9}$$

where M is the ground truth masks.

$$MOTSP = \frac{\tilde{TP}}{|TP|} \tag{10}$$

where $\tilde{TP}$ is the soft version true positives (TP).

$$sMOTSA = \frac{\tilde{TP} - |FP| - |IDSW|}{|M|} \tag{11}$$

4. Evaluation and Discussion

In this section, we compare the MOT techniques based on the dataset used for evaluation. Then, analysis and discussion are conducted to provide insight for future work.

The performance of the most recent MOT techniques on MOT15, 16, 17, and 20 datasets are shown in Table 5. The ↑ stands for higher is better, and ↓ stands for lower is better. The protocol indicates the type of detector used for evaluating the results. The dataset provides the public detector and is typical for all methods. The private detector is designed by the method and is not shared. In MOT15, the tracker introduced in Ref. [38] has the highest accuracy and the lowest identity switches (IDs). It also maintained the highest percentage of tracks (MT) and has the lowest percentage of lost ones (ML). On the other hand, it had a significantly higher number of false positives and negatives compared to the method introduced in Ref. [42], which also performed with the highest FPS (Hz). The authors in Ref. [36] evaluated their system using the private detector protocol and have significantly lower fragmentation than all other methods. In Ref. [38], the tracker relied on appearance features extracted from the inception network layer and the position of the detections. The association process was done using conditional probability. The approach in Ref. [48] has the second best accuracy, where the motion features and the appearance features were used, as well as a category classifier for the association. The method with the lowest accuracy [80] only relied on the motion features, and the slowest performing method [39], where reinforcement learning was applied for data association. The approach in Ref. [62] has the highest accuracy on the MOT16 dataset using the private detector protocol. The method in Ref. [53] also used the private detector and has the highest HOTA. Both methods relied on appearance features for tracking and incorporated the Hungarian algorithm for matching. Similarly, a significantly faster-performing method [38] with slightly lower accuracy only used the appearance features and a prediction network for the association. The method in Ref. [35] used the Kalman filter for motion feature prediction and the appearance features extracted from the detection network for tracking. This method has a lower accuracy in comparison to other methods. In MOT17, the approach with the highest accuracy [57] has a significantly higher fragmentation than the one introduced in Ref. [64], which only used the motion feature for tracking. The method in Ref. [67] has slightly lower accuracy but with an acceptable FPS. This method used the appearance and motion features in addition to the Hungarian algorithm for matching. The approaches in Refs. [56,58] have an acceptable accuracy where both used Kalman filter for motion tracking and Ref. [56] neglected the appearance features. The method in Ref. [64] has the lowest number of fragmentation and only uses the motion features in tracking. The method in Ref. [57] has the highest accuracy on MOT20, although it has a significantly higher number of fragmentation compared to the one in Ref. [54]. The approach in Ref. [67] has the highest FPS, followed by the one in Ref. [54]. All of these methods relied on visual and motion features for tracking. The methods in Refs. [58,67] only used the motion features and had an acceptable accuracy. On the other hand, the ones that only relied on the visual features, such as Refs. [60,62,63,65] did not perform well according to the accuracy and other metrics. This evaluation shows that appearance features are essential for high accuracy, and other cues are used to boost performance. Based on the results from Table 5 and the summary presented in Table 2, the utilization of deep learning for data association reduces the processing time, as can be observed from Refs. [43,49]. On the other hand, including motion cues in the system drops the FPS significantly compared to only using the visual features as indicated by the results in Ref. [32]. Although adding complexity to the system drops the FPS, the IDS metric significantly improves when the motion features are included in the system. It can be concluded from these findings that to improve the FPS and the accuracy one should use deep learning in all MOT components. Although that might be the case, the end-to-end approaches introduced in Refs. [32,48] have used deep learning to extract appearance and motion features and data association, and the performance did not compete with other approaches. Deep learning approaches are data driven, which means they are suitable for specific tasks but unsuitable or expected to perform poorly in real scenarios due to the variation from the data used in training [13].

Table 5. The performance using the MOT15, 16, 17, and 20 datasets.

Dataset	Tracker	MOTA ↑	MOTP ↑	IDF1 ↑	MT ↑	ML ↓	FP ↓	FN ↓	IDS ↓	Frag ↓	Hz ↑	HOTA ↑	Protocol
MOT15	Chen et al. [48]	53.0	**75.5**		29.1	20.2	5159	22,984	708	1476			Private
	Chen et al. [48]	38.5	72.6		8.7	37.4	4005	33,204	586	1263			Public
	Sadeghian et al. [31]	37.6	71.7		15.8	26.8	7933	29,397	1026	2024	1		Public
	Keuper et al. [37]	35.6		45.1	23.2	39.3	10,580	28,508	457	**969**			Private
	Fang et al. [38]	**56.5**	73.0	**61.3**	**45.1**	**14.6**	9386	16,921	**428**	1364	5.1		Public
	Xiang et al. [32]	37.1	72.5	28.4	12.6	39.7	8305	29,732	580	1193	1.0		Private
	Chu et al. [39]	38.9	70.6	44.5	16.6	31.5	7321	29,501	720	1440	0.3		Public
	Sun et al. [42]	38.3	71.1	45.6	17.6	41.2	**1290**	**2700**	1648	1515	**6.3**		Public
	Zhou et al. [46]	30.5	71.2	1.1	7.6	41.2	6534	35,284	879	2208	5.9		Private
	Dimitriou et al. [80]	24.2	70.2		7.1	49.2	6400	39,659	529	1034			Private
MOT16	Yoon et al. [49]	22.5	70.9	25.9	6.4	61.9	7346	39,092	1159	1538	**172.8**		Public
	Sadeghian et al. [31]	47.2		**75.8**	14.0	41.6	**2681**	92,856	774	1675	1.0		Public
	Keuper et al. [37]	47.1		52.3	20.4	46.9	6703	89,368	**370**	**598**			Private
	Fang et al. [38]	63	78.8		39.9	22.1	13,663	53,248	482	1251			Public
	Xiang et al. [32]	48.3	76.7		15.4	40.1	2706	91,047	543		0.5		Private
	Zhu et al. [40]	46.1	73.8	54.8	17.4	42.7	7909	89,874	532	1616	0.3		Private
	Zhou et al. [41]	64.8	78.6	**73.5**	**40.6**	22.0	13,470	49,927	794	1050	18.2		Private
	Chu et al. [39]	48.8	75.7	47.2	15.8	38.1	5875	86,567	906	1116	0.1		Public
	Peng et al. [43]	67.6	78.4	57.2	32.9	23.1	8934	48,305	1897		34.4		Private
	Henschel et al. [100]	47.8		47.8	19.1	38.2	8886	85,487	852				Private
	Wang et al. [44]	49.2		56.1	17.3	40.3	8400	83,702		882			Public
	Mahmoudi et al. [34]	65.2	78.4		32.4	21.3	6578	55,896	946	2283	11.2		Public
	Zhou et al. [35]	40.8	74.4		13.7	38.3	15,143	91,792	1051	2210	6.5		Private
	Lan et al. [45]	45.4	74.4		18.1	38.7	13,407	85,547	600	930			Public
	Xu et al. [50]	74.4		73.7	46.1	15.2						60.2	Private
	Ye et al. [51]	62.8	79.4	63.1	34.4	17.2	12,463	54,648	701	983			Private
	wang et al. [52]	50.4		47.5			18,370	69,800	1826				Private
	Yu et al. [53]	75.6	**80.9**	**75.8**	43.1	21.5	9786	**34,214**	448			**61.7**	Private
	Dai et al. [60]	63.8		70.6	30.3	30.6	7412	57,975	629		11.2	54.7	Private
	Hyun et al. [62]	**76.7**		73.1	**49.1**	**10.7**	10,689	30,428	1420				Private

Table 5. *Cont.*

Dataset	Tracker	MOTA ↑	MOTP ↑	IDF1 ↑	MT ↑	ML ↓	FP ↓	FN ↓	IDS ↓	Frag ↓	Hz ↑	HOTA ↑	Protocol
MOT17	Keuper et al. [37]	51.2			20.7	37.4	24,986	248,328	1851	2991			Private
	Zhu et al. [40]	48.2	75.9	55.7	19.3	38.3	26,218	263,608	2194	5378	11.4		Private
	Sun et al. [42]	52.4	53.9	76.9	**49.5**	21.4	25,423	234,592	8431	14,797	6.3		Public
	Peng et al. [43]	66.6	78.2	57.4	32.2	24.2	22,284	160,491	5529		**34.4**		Private
	Henschel et al. [100]	51.3		47.6	21.4	35.2	24,101	247,921	2648				Private
	Wang et al. [44]	51.9		58.0	23.5	35.5	37,311	231,658		2917			Public
	Henschel et al. [101]	52.6		50.8	19.7	35.8	31,572	232,572	3050	3792			Private
	Zhou et al. [46]	61.5		59.6	26.4	31.9	**14,076**	200,672	2583				Private
	Karunasekera et al. [33]	46.9	76.1		16.9	36.3			4478		17.1		Private
	Xu et al. [50]	73.1	73.0	45	16.8							59.7	Private
	Ye et al. [51]	61.7	79.3	62.0	32.5	20.1	32,863	181,035	1809	3168			Private
	Yu et al. [53]	73.8	**81.0**	74.7	41.7	23.2	27,999	118,623	1374			61.0	Private
	Wang et al. [54]	79.5		79.1			29,508	84,618	1302	2046	24.7	63.9	Private
	Gao et al. [55]	70.2		65.5	43.1	15.4	30,367	115,986	3265		10.7		Private
	Nasseri et al. [56]	80.4		77.7			28,887	**79,329**	2325	4689		63.3	Private
	Zhao et al. [57]	**81.2**		78.1	54.5	13.2	17,281	86,861	1893				Private
	Aharon et al. [58]	80.5		**80.2**			22,521	86,037	1212		4.5	**65.0**	Private
	Seidenschwarz et al. [59]	61.7		64.2	26.3	32.2			1639			50.9	Private
	Dai et al. [60]	63.0		69.9	30.4	30.6	23,022	183,659	1842		10.8	54.6	Private
	Zhang et al. [61]	62.1		65.0			24,052	188,264	1768				Public
	Zhang et al. [61]	77.3		75.9			45,030	79,716	3255				Private
	Hyun et al. [62]	76.5		73.6	47.6	**12.7**	29,808	99,510	3369				Private
	Chen et al. [63]	76.5		72.6								59.7	Private
	Cao et al. [64]	78.0		77.5			15,100	108,000	1950	**2040**		63.2	Private
	Wan et al. [65]	67.2		66.5	38.3	20.5	17,875	164,032	2896				Public
	Wan et al. [65]	75.6		76.4	48.8	16.2	26,983	108,186	2394				Private
	Du et al. [66]	79.6		79.5					**1194**		7.1	64.4	Private
	Zhang et al. [67]	80.3		77.3			25,491	83,721	2196		29.6	63.1	Private

Table 5. *Cont.*

Dataset	Tracker	MOTA ↑	MOTP ↑	IDF1 ↑	MT ↑	ML ↓	FP ↓	FN ↓	IDS ↓	Frag ↓	Hz ↑	HOTA ↑	Protocol
	Xu et al. [50]	59.7		67.8	66.5	7.7						54.0	Private
	Ye et al. [51]	54.9	79.1	59.1	36.2	23.5	19,953	211,392	1630	2455			Private
	Wang et al. [52]	51.5		44.5			38,223	208,616	4055				Private
	Yu et al. [53]	67.2	**79.2**	70.5	62.2	8.9	61,134	104,597	4243			56.5	Private
	Wang et al. [54]	76.5		76.8			**19,247**	101,290	971	**1190**	15.7	62.8	Private
	Nasseri et al. [56]	76.8		76.4			27,106	91,740	1446	3053		61.4	Private
	Zhao et al. [57]	**78.1**		76.4	**70.5**	**7.4**	25,413	86,510	1332				Private
	Aharon et al. [58]	77.8		**77.5**			24,638	88,863	1257		2.4	**63.3**	Private
MOT20	Seidenschwarz et al. [59]	52.7		55.0	29.1	26.8			1437			43.4	Private
	Dai et al. [60]	60.1		66.7	46.5	17.8	37,657	165,866	2926		4.0	51.7	Private
	Zhang et al. [61]	55.6		65.0			12,297	216,986	**480**				Public
	Hyun et al. [61]	66.3		67.7			41,538	130,072	2715				Private
	Hyun et al. [62]	72.8		70.5	64.3	12.8	25,165	112,897	2649				Private
	Chen et al. [63]	67.1		59.1								50.4	Private
	Cao et al. [64]	75.5		75.9			18,000	108,000	913	1198		63.2	Private
	Wan et al. [65]	70.4		71.9	65.1	9.6	48,343	101,034	3739				Private
	Du et al. [66]	73.8		77.0			25,491	**83,721**	2196		1.4	62.6	Private
	Zhang et al. [67]	77.8		75.2			26,249	87,594	1223		**17.5**	61.3	Private

Similarly, the performance of the KITTI dataset is shown in Table 6. The dataset is divided into multiple sequences: car, pedestrian, and cyclist. The methods are either evaluated on all of them combined or individually. As it can be observed, the pedestrian data are the most difficult to process with a good performance. Only methods evaluated on all sequences are used for comparison, and accordingly, the values corresponding to the best performance according to the metric are in bold. The rest are listed for observation and analysis. The approach introduced in Ref. [46] showed superiority. Although it did not perform well on the MOT15 dataset, it showed a competing performance on the MOT17 dataset. The authors in Ref. [102] evaluated the proposed technique on each sequence individually. They have superior performance on all of them. The performance on the UA_DETRAC dataset is shown in Table 7. Although the UA_DETRAC dataset is of a static background type and does not include the challenge of dynamic background, the approach introduced in Ref. [47] performed better on the KITTI dataset. This variation in the performance of the MOT techniques on multiple datasets may indicate that the MOT techniques are data driven and difficult to generalize. Similar to 2D Tracking, the deep learning approach is utilized for visual feature extraction in 3D. For processing point cloud data, PointNet is the most popular method. The authors in Refs. [85,89] did not rely on deep learning for techniques for the processing of point cloud data and performed poorly in terms of accuracy on the KITTI dataset shown in Table 6. The approach with the highest accuracy in Table 3 introduced in Ref. [86] creates multiple solutions for data fusion problems. The features extracted from LIDAR and camera sensors are fused by concatenation, addition, or weighted addition and then passed into a custom-designed network to calculate the correlation between the features and output the linked detections. The approaches that depend on deep learning for data fusion, Refs. [85,87,90], have a low MOTA, although [90] has high accuracy on the car set. The localization-based Tracking introduced in Ref. [103] has the best accuracy on the UA_DETRAC dataset. Although the DMM-Net method in Ref. [104] has significantly lower accuracy, it showed superiority in identity switches and fragmentation.

Table 6. The performance using the KITTI dataset. We have marked the highest scores in bold for methods evaluated on all categories.

Tracker	MOTA ↑	MOTP ↑	MT ↑	ML ↓	FP ↓	FN ↓	IDS ↓	Frag ↓	Type
Chu et al. [47]	77.1	78.8	51.4	8.9	760	6998	123		Car
Simon et al. [85]	75.7	78.5	58.0	5.1					All
Zhang et al. [86]	84.8	85.2	73.2	2.77	711	4243	284	753	All
Scheidegger et al. [70]	80.4	81.3	62.8	6.2			121	613	Car
Frossard et al. [87]	76.15	83.42	60.0	8.31			296	868	All
Hu et al. [71]	84.5	85.6	73.4	2.8	**705**	**4242**			All
Weng et al. [88]	82.2	84.1	64.9	6			142	416	Car
Zhao et al. [36]	71.3	81.8	48.3	5.9					All
Weng et al. [102]	86.2	78.43					0	15	Car
Weng et al. [102]	70.9								Pedestrian
Weng et al. [102]	84.9								Cyclist
Luiten et al. [105]	84.8				681	4260	275		Car
Wang et al. [106]	68.2	76.6	60.6	12.3			111	**725**	All
Sualeh et al. [89]	78.10	79.3	70.3	9.1			21	111	Car
Sualeh et al. [89]	46.1	67.6	30.3	38.7			57	500	Pedestrian
Sualeh et al. [89]	70.9	77.6	71.4	14.3			3	24	Cyclist
Shenoi et al. [90]	85.7	85.5					98		Car
Shenoi et al. [90]	46.0	72.6					395		Pedestrian

Table 6. *Cont.*

Tracker	MOTA ↑	MOTP ↑	MT ↑	ML ↓	FP ↓	FN ↓	IDS ↓	Frag ↓	Type
Zhou et al. [46]	**89.4**	85.1	**82.3**	**2.3**			**116**	744	All
Karunasekera et al. [33]	85.0	**85.5**	74.3	2.8			**301**		All
Zhao et al. [57]	91.2		86.5	2.3					Car
Zhao et al. [57]	67.7		49.1	14.5					Pedestrian
Aharon et al. [58]	90.3						250	280	Car
Aharon et al. [58]	65.1						204	609	Pedestrian
Wang et al. [107]	90.4	85.0	84.6	7.38	2322	962			Car
Wang et al. [107]	52.2	64.5	35.4	25.4	1112	2560			Pedestrian
Sun et al. [108]	86.9	85.7	83.1	2.9			271	254	Car

Table 7. The performance using the UA_DETRAC dataset.

Dataset	Tracker	PR-MOTA ↑	PR-MOTP ↑	PR-MT ↑	PR-ML ↓	PR-FP ↓	PR-FN ↓	PR-IDS ↓	PR-FRAG ↓
UA_DETRAC	Bochinski et al. [68]	30.7	37	32	22.6	18,046	179,191	363	1123
	Kutschbach et al. [72]	14.5	36	14	18.1	38,597	174,043	799	1607
	Hou et al. [69]	30.3	36.3	30.2	21.0	20,263	179,317	389	1260
	Sun et al. [42]	20.2	26.3	14.5	18.1	9747	**135,978**		
	Chu et al. [47]	19.8	36.7	17.1	18.2	14,989	164,433	617	
	Gloudemans et al. [103]	**46.4**	**69.5**	**41.1**	16.3				
	Sun et al. [104]	12.2		10.8	14.9	36,355	192,289	**228**	**674**
	Messoussi et al. [109]	31.2	50.9	28.1	18.5	**6036**	170,700	252	
	Wang et al. [110]	22.5	35.2	15.5	**10.1**			1563	3186

Navigation and self-driving applications in robotics depend on the online feature. The system must be able to react in real time. Although most of the techniques discussed can process video sequences online, their FPS is not showing robust performance, according to the Hz metric, to be deployed in an application such as self-driving. The method introduced in Ref. [43] has the highest FPS overall and acceptable accuracy on the MOT16 and MOT17 datasets. However, the research utilizing LIDAR and RGB cameras show potential in robotics navigation and autonomous driving applications.

5. Current Research Challenges

Through this study, we gained insight into the current trends of online MOT methods that can be utilized in robotics applications and the challenges faced. The first challenge would be the online feature. The MOT algorithm should be able to operate in real-time for most robotics applications in order to be able to react to environmental change. The second challenge would be the accurate track trajectory across multiple frames. The lack of this problem can cause multiple identity switches and difficulty keeping a concise description of the surroundings. In addition, the issue is segmenting the detections at pixel level and tracking them. The bounding box provides wrong information about the object in shape and size, along with noises from the background. Finally, the motion feature has proven its value in tracking, but it is not simple to track random moving objects such as people, animals, cyclists, etc.

6. Future Work

The final objective of the research done on deploying MOT algorithms in autonomous robots is to have a reliable system that contributes to reducing accidents and facilitating tasks that might be difficult for humans to carry out. One aspect that we found the current research lacks is the generation of a new benchmark dataset that includes data collected by the standard sensors employed by the current industry. Sensors such as ultrasonic and LiDar are essential in today's autonomous robot manufacturing, and it is necessary to use the same tools to make the research on MOT up-to-date. Moreover, using deep learning algorithms for detection and tracking would face a massive problem due to the risk of meeting a variation that was not included in the training set. Thus, deep learning should be trained on segmenting the road regions and hence, be trained on any area before deployment. This is one of the approaches that can be researched to tackle the problem of dealing with new objects. The current research looks at appearance and motion models as necessary in MOT. They are further going into learning the behavior of objects in the scenes and the interactivity between those objects. For instance, two objects moving towards each other would lead to one of them getting covered, and a track would be lost. As the MOT system's complexity increases, it becomes more challenging to work in real-time. The research on embedded processors that can be utilized in autonomous robots will significantly contribute to increasing the accuracy while maintaining the online feature.

7. Conclusions

This paper aims to review the current trends and challenges related to the MOT for autonomous robotics applications. This area of study has been frequently researched recently due to its high potential and standards, which are difficult to achieve. The paper has discussed and compared the MOT techniques through a common framework and datasets, including MOTChallenges, KITTI, and UA_DETRAC. There is a vast area left to explore and investigate as well as multiple approaches created by the literature that has the potential to build into reliable and robust techniques. A summary of the components utilized in the general MOT framework, including appearance and motion cues, data association, and occlusion handling, has been listed and studied. In addition, the popular methods used for data fusion between multiple sensors, focusing on the camera and LIDAR, have been reviewed. The role that deep learning techniques are utilized in MOT approaches has been investigated thoroughly using quantitative analysis to evaluate its limitations and strong points.

Author Contributions: Conceptualization, M.G., T.B. and A.G.; methodology, M.G., T.B. and A.G.; software, M.G., T.B. and A.G.; validation, M.Y., H.A., and M.A.; formal analysis, M.G., T.B. and A.G.; investigation, M.G., T.B. and A.G.; resources, M.G. and M.A.; data curation, M.G., T.B. and A.G.; writing—original draft preparation, T.B., A.G. and H.A.; writing—review and editing, M.G., T.B., A.G., M.Y., H.A. and M.A.; visualization, M.G., T.B., A.G. and M.A.; supervision, M.G. and M.A.; project administration, M.G. and M.A.; funding acquisition, M.A. All authors have read and agreed to the published version of the manuscript.

Funding: This paper is part of a research project funded by the Office of Research and Sponsored Programs (ORSP) at Abu Dhabi University through a research fund (Grant number 19300564).

Institutional Review Board Statement: Not applicable.

Informed Consent Statement: Not applicable.

Data Availability Statement: Not applicable.

Conflicts of Interest: The authors declare no conflict of interest.

Abbreviations

The following abbreviations are used in this manuscript:

MOT	Multiple Object Tracking
LIDAR	Light Detection and Ranging
KITTI	Karlsruhe Institute of Technology and Toyota Technological Institute
UA-DETRAC	University at Albany DEtection and TRACking
SLAM	Simultaneous localization and mapping
ROI	Region of Interest
IoU	Intersection over Union
HOTA	Higher Order Tracking Accuracy
SLAMMOT	Simultaneous localization and mapping Multiple Object Tracking
RTU	Recurrent Tracking Unit
LSTM	Long Short Term Memory
ECO	Efficient Convolution Operators
PCA	Principal Component Analysis
LDAE	Lightweight and Deep Appearance Embedding
DLA	Deep Layer Aggregation
GCD	Global Context Disentangling
GTE	Guided Transformer Encoder
DETR	DEtection TRansformer
PCB	Part-based Convolutional Baseline

References

1. Runz, M.; Buffier, M.; Agapito, L. MaskFusion: Real-Time Recognition, Tracking and Reconstruction of Multiple Moving Objects. In Proceedings of the 2018 IEEE International Symposium on Mixed and Augmented Reality, ISMAR 2018, Munich, Germany, 16–20 October 2019; pp. 10–20.
2. Zhou, Q.; Zhao, S.; Li, H.; Lu, R.; Wu, C. Adaptive Neural Network Tracking Control for Robotic Manipulators with Dead Zone. *IEEE Trans. Neural Netw. Learn. Syst.* **2019**, *30*, 3611–3620. [CrossRef] [PubMed]
3. Yu, C.; Liu, Z.; Liu, X.; Xie, F.; Yang, Y.; Wei, Q.; Fei, Q. DS-SLAM: A Semantic Visual SLAM towards Dynamic Environments. In Proceedings of the IEEE International Conference on Intelligent Robots and Systems, Madrid, Spain, 1–5 October 2018; pp. 1168–1174.
4. Fehr, L.; Langbein, W.E.; Skaar, S.B. Adequacy of power wheelchair control interfaces for persons with severe disabilities: A clinical survey. *J. Rehabil. Res. Dev.* **2000**, *37*, 353–360.
5. Simpson, R. Smart wheelchairs: A literature review. *J. Rehabil. Res. Dev.* **2005**, *42*, 423–436. [CrossRef]
6. Martins, M.M.; Santos, C.P.; Frizera-Neto, A.; Ceres, R. Assistive mobility devices focusing on Smart Walkers: Classification and review. *Robot. Auton. Syst.* **2012**, *60*, 548–562. [CrossRef]
7. Khan, M.Q.; Lee, S. A comprehensive survey of driving monitoring and assistance systems. *Sensors* **2019**, *19*, 2574. [CrossRef] [PubMed]
8. Van Brummelen, J.; O'Brien, M.; Gruyer, D.; Najjaran, H. Autonomous vehicle perception: The technology of today and tomorrow. *Transp. Res. Part C Emerg. Technol.* **2018**, *89*, 384–406. [CrossRef]
9. Pham, H.X.; La, H.M.; Feil-Seifer, D.; Deans, M.C. A distributed control framework of multiple unmanned aerial vehicles for dynamic wildfire tracking. *IEEE Trans. Syst. Man Cybern. Syst.* **2020**, *50*, 1537–1548. [CrossRef]
10. Cai, Z.; Wang, L.; Zhao, J.; Wu, K.; Wang, Y. Virtual target guidance-based distributed model predictive control for formation control of multiple UAVs. *Chin. J. Aeronaut.* **2020**, *33*, 1037–1056. [CrossRef]
11. Huang, Y.; Liu, W.; Li, B.; Yang, Y.; Xiao, B. Finite-time formation tracking control with collision avoidance for quadrotor UAVs. *J. Frankl. Inst.* **2020**, *357*, 4034–4058. [CrossRef]
12. Luo, W.; Xing, J.; Milan, A.; Zhang, X.; Liu, W.; Kim, T. Multiple object tracking: A literature review. *Artif. Intell.* **2021**, *293*, 103448. [CrossRef]
13. Ciaparrone, G.; Luque Sánchez, F.; Tabik, S.; Troiano, L.; Tagliaferri, R.; Herrera, F. Deep learning in video multi-object tracking: A survey. *Neurocomputing* **2020**, *381*, 61–88. [CrossRef]
14. Xu, Y.; Zhou, X.; Chen, S.; Li, F. Deep learning for multiple object tracking: A survey. *IET Comput. Vis.* **2019**, *13*, 411–419. [CrossRef]
15. Irvine, J.M.; Wood, R.J.; Reed, D.; Lepanto, J. Video image quality analysis for enhancing tracker performance. In Proceedings of the 2013 IEEE Applied Imagery Pattern Recognition Workshop (AIPR), Washington, DC, USA, 23–25 October 2013; pp. 1–9. [CrossRef]
16. Meng, L.; Yang, X. A Survey of Object Tracking Algorithms. *Zidonghua Xuebao/Acta Autom. Sin.* **2019**, *45*, 1244–1260.
17. Jain, V.; Wu, Q.; Grover, S.; Sidana, K.; Chaudhary, D.G.; Myint, S.; Hua, Q. Generating Bird's Eye View from Egocentric RGB Videos. *Wirel. Commun. Mob. Comput.* **2021**, *2021*, 1–11. [CrossRef]

18. Yeong, D.J.; Velasco-Hernandez, G.; Barry, J.; Walsh, J. Sensor and Sensor Fusion Technology in Autonomous Vehicles: A Review. *Sensors* **2021**, *21*, 2140. [CrossRef]
19. Leal-Taixé, L.; Milan, A.; Reid, I.; Roth, S.; Schindler, K. MOTChallenge 2015: Towards a Benchmark for Multi-Target Tracking. *arXiv* **2015**, arXiv:1504.01942.
20. Xu, Z.; Rong, Z.; Wu, Y. A survey: Which features are required for dynamic visual simultaneous localization and mapping? *Vis. Comput. Ind. Biomed. Art* **2021**, *4*, 20. [CrossRef]
21. Grigorescu, S.; Trasnea, B.; Cocias, T.; Macesanu, G. A survey of deep learning techniques for autonomous driving. *J. Field Robot.* **2020**, *37*, 362–386. [CrossRef]
22. Janai, J.; Güney, F.; Behl, A.; Geiger, A. Computer vision for autonomous vehicles. *Found. Trends Comput. Graph. Vis.* **2020**, *12*, 1–308. [CrossRef]
23. Datondji, S.R.E.; Dupuis, Y.; Subirats, P.; Vasseur, P. A Survey of Vision-Based Traffic Monitoring of Road Intersections. *IEEE Trans. Intell. Transp. Syst.* **2016**, *17*, 2681–2698. [CrossRef]
24. Buch, N.; Velastin, S.A.; Orwell, J. A review of computer vision techniques for the analysis of urban traffic. *IEEE Trans. Intell. Transp. Syst.* **2011**, *12*, 920–939. [CrossRef]
25. Otto, A.; Agatz, N.; Campbell, J.; Golden, B.; Pesch, E. Optimization approaches for civil applications of unmanned aerial vehicles (UAVs) or aerial drones: A survey. *Networks* **2018**, *72*, 411–458. [CrossRef]
26. Shakhatreh, H.; Sawalmeh, A.H.; Al-Fuqaha, A.; Dou, Z.; Almaita, E.; Khalil, I.; Othman, N.S.; Khreishah, A.; Guizani, M. Unmanned Aerial Vehicles (UAVs): A Survey on Civil Applications and Key Research Challenges. *IEEE Access* **2019**, *7*, 48572–48634. [CrossRef]
27. Badue, C.; Guidolini, R.; Carneiro, R.V.; Azevedo, P.; Cardoso, V.B.; Forechi, A.; Jesus, L.; Berriel, R.; Paixão, T.M.; Mutz, F.; et al. Self-driving cars: A survey. *Expert Syst. Appl.* **2021**, *165*, 113816. [CrossRef]
28. Wang, Z.; Wu, Y.; Niu, Q. Multi-Sensor Fusion in Automated Driving: A Survey. *IEEE Access* **2020**, *8*, 2847–2868. [CrossRef]
29. Elhousni, M.; Huang, X. A Survey on 3D LiDAR Localization for Autonomous Vehicles. In Proceedings of the IEEE Intelligent Vehicles Symposium, Las Vegas, NV, USA, 9 October–13 November 2020; pp. 1879–1884.
30. Fritsch, J.; Kühnl, T.; Geiger, A. A new performance measure and evaluation benchmark for road detection algorithms. In Proceedings of the 16th International IEEE Conference on Intelligent Transportation Systems (ITSC 2013), The Hague, The Netherlands, 6–9 October 2013; pp. 1693–1700. [CrossRef]
31. Sadeghian, A.; Alahi, A.; Savarese, S. Tracking the Untrackable: Learning to Track Multiple Cues with Long-Term Dependencies. In Proceedings of the 2017 IEEE International Conference on Computer Vision (ICCV), Venice, Italy, 22–29 October 2017; pp. 300–311. [CrossRef]
32. Xiang, J.; Zhang, G.; Hou, J. Online Multi-Object Tracking Based on Feature Representation and Bayesian Filtering Within a Deep Learning Architecture. *IEEE Access* **2019**, *7*, 27923–27935. [CrossRef]
33. Karunasekera, H.; Wang, H.; Zhang, H. Multiple Object Tracking With Attention to Appearance, Structure, Motion and Size. *IEEE Access* **2019**, *7*, 104423–104434. [CrossRef]
34. Mahmoudi, N.; Ahadi, S.M.; Mohammad, R. Multi-target tracking using CNN-based features: CNNMTT. *Multimed. Tools Appl.* **2019**, *78*, 7077–7096. [CrossRef]
35. Zhou, Q.; Zhong, B.; Zhang, Y.; Li, J.; Fu, Y. Deep Alignment Network Based Multi-Person Tracking With Occlusion and Motion Reasoning. *IEEE Trans. Multimed.* **2019**, *21*, 1183–1194. [CrossRef]
36. Zhao, D.; Fu, H.; Xiao, L.; Wu, T.; Dai, B. Multi-Object Tracking with Correlation Filter for Autonomous Vehicle. *Sensors* **2018**, *18*, 2004. [CrossRef]
37. Keuper, M.; Tang, S.; Andres, B.; Brox, T.; Schiele, B. Motion Segmentation & Multiple Object Tracking by Correlation Co-Clustering. *IEEE Trans. Pattern Anal. Mach. Intell.* **2020**, *42*, 140–153. [CrossRef]
38. Fang, K.; Xiang, Y.; Li, X.; Savarese, S. Recurrent Autoregressive Networks for Online Multi-Object Tracking. In Proceedings of the 2018 IEEE Winter Conference on Applications of Computer Vision (WACV), Lake Tahoe, NV, USA, 12–15 March 2018. [CrossRef]
39. Chu, P.; Fan, H.; Tan, C.C.; Ling, H. Online Multi-Object Tracking with Instance-Aware Tracker and Dynamic Model Refreshment. In Proceedings of the 2019 IEEE Winter Conference on Applications of Computer Vision (WACV), Waikoloa, HI, USA, 7–11 January 2019. [CrossRef]
40. Zhu, J.; Yang, H.; Liu, N.; Kim, M.; Zhang, W.; Yang, M.H. Online Multi-Object Tracking with Dual Matching Attention Networks. *arXiv* **2019**, arXiv:1902.00749.
41. Zhou, Z.; Xing, J.; Zhang, M.; Hu, W. Online Multi-Target Tracking with Tensor-Based High-Order Graph Matching. In Proceedings of the 2018 24th International Conference on Pattern Recognition (ICPR), Beijing, China, 20–24 August 2018; pp. 1809–1814. [CrossRef]
42. Sun, S.; Akhtar, N.; Song, H.; Mian, A.; Shah, M. Deep Affinity Network for Multiple Object Tracking. *arXiv* **2018**, arXiv:1810.11780.
43. Peng, J.; Wang, C.; Wan, F.; Wu, Y.; Wang, Y.; Tai, Y.; Wang, C.; Li, J.; Huang, F.; Fu, Y. Chained-Tracker: Chaining Paired Attentive Regression Results for End-to-End Joint Multiple-Object Detection and Tracking. *arXiv* **2020**, arXiv:2007.14557.
44. Wang, G.; Wang, Y.; Zhang, H.; Gu, R.; Hwang, J.N. Exploit the Connectivity: Multi-Object Tracking with TrackletNet. *arXiv* **2018**, arXiv:1811.07258.

45. Lan, L.; Wang, X.; Zhang, S.; Tao, D.; Gao, W.; Huang, T.S. Interacting Tracklets for Multi-Object Tracking. *IEEE Trans. Image Process.* **2018**, *27*, 4585–4597. [CrossRef] [PubMed]
46. Zhou, X.; Koltun, V.; Krähenbühl, P. Tracking Objects as Points. *arXiv* **2020**, arXiv:2004.01177.
47. Chu, P.; Ling, H. FAMNet: Joint Learning of Feature, Affinity and Multi-dimensional Assignment for Online Multiple Object Tracking. *arXiv* **2019**, arXiv:1904.04989.
48. Chen, L.; Ai, H.; Shang, C.; Zhuang, Z.; Bai, B. Online multi-object tracking with convolutional neural networks. In Proceedings of the 2017 IEEE International Conference on Image Processing (ICIP), Beijing, China, 17–20 September 2017; pp. 645–649. [CrossRef]
49. Yoon, K.; Kim, D.Y.; Yoon, Y.C.; Jeon, M. Data Association for Multi-Object Tracking via Deep Neural Networks. *Sensors* **2019**, *19*, 559. [CrossRef]
50. Xu, B.; Liang, D.; Li, L.; Quan, R.; Zhang, M. An Effectively Finite-Tailed Updating for Multiple Object Tracking in Crowd Scenes. *Appl. Sci.* **2022**, *12*, 1061. [CrossRef]
51. Ye, L.; Li, W.; Zheng, L.; Zeng, Y. Lightweight and Deep Appearance Embedding for Multiple Object Tracking. *IET Comput. Vis.* **2022**, *16*, 489–503. [CrossRef]
52. Wang, F.; Luo, L.; Zhu, E.; Wang, S.; Long, J. Multi-object Tracking with a Hierarchical Single-branch Network. *CoRR* **2021**, abs/2101.01984. Available online: http://xxx.lanl.gov/abs/2101.01984 (accessed on 7 September 2022).
53. Yu, E.; Li, Z.; Han, S.; Wang, H. RelationTrack: Relation-aware Multiple Object Tracking with Decoupled Representation. *CoRR* **2021**, abs/2105.04322. Available online: http://xxx.lanl.gov/abs/2105.04322 (accessed on 7 September 2022).
54. Wang, S.; Sheng, H.; Yang, D.; Zhang, Y.; Wu, Y.; Wang, S. Extendable Multiple Nodes Recurrent Tracking Framework with RTU++. *IEEE Trans. Image Process.* **2022**, *31*, 5257–5271. [CrossRef] [PubMed]
55. Gao, Y.; Gu, X.; Gao, Q.; Hou, R.; Hou, Y. TdmTracker: Multi-Object Tracker Guided by Trajectory Distribution Map. *Electronics* **2022**, *11*, 1010. [CrossRef]
56. Nasseri, M.H.; Babaee, M.; Moradi, H.; Hosseini, R. Fast Online and Relational Tracking. *arXiv* **2022**, arXiv:2208.03659.
57. Zhao, Z.; Wu, Z.; Zhuang, Y.; Li, B.; Jia, J. Tracking Objects as Pixel-wise Distributions. *arXiv* **2022**, arXiv:2207.05518.
58. Aharon, N.; Orfaig, R.; Bobrovsky, B.Z. BoT-SORT: Robust Associations Multi-Pedestrian Tracking. *arXiv* **2022**, arXiv:2206.14651.
59. Seidenschwarz, J.; Brasó, G.; Elezi, I.; Leal-Taixé, L. Simple Cues Lead to a Strong Multi-Object Tracker. *arXiv* **2022**, arXiv:2206.04656.
60. Dai, P.; Feng, Y.; Weng, R.; Zhang, C. Joint Spatial-Temporal and Appearance Modeling with Transformer for Multiple Object Tracking. *arXiv* **2022**, arXiv:2205.15495.
61. Zhang, Y.; Wang, C.; Wang, X.; Zeng, W.; Liu, W. Robust Multi-Object Tracking by Marginal Inference. *arXiv* **2022**, arXiv:2208.03727.
62. Hyun, J.; Kang, M.; Wee, D.; Yeung, D.Y. Detection Recovery in Online Multi-Object Tracking with Sparse Graph Tracker. *arXiv* **2022**, arXiv:2205.00968.
63. Chen, M.; Liao, Y.; Liu, S.; Wang, F.; Hwang, J.N. TR-MOT: Multi-Object Tracking by Reference. *arXiv* **2022**, arXiv:2203.16621.
64. Cao, J.; Weng, X.; Khirodkar, R.; Pang, J.; Kitani, K. Observation-Centric SORT: Rethinking SORT for Robust Multi-Object Tracking. *arXiv* **2022**, arXiv:2203.1662.
65. Wan, J.; Zhang, H.; Zhang, J.; Ding, Y.; Yang, Y.; Li, Y.; Li, X. DSRRTracker: Dynamic Search Region Refinement for Attention-based Siamese Multi-Object Tracking. *arXiv* **2022**, arXiv:2203.10729.
66. Du, Y.; Song, Y.; Yang, B.; Zhao, Y. StrongSORT: Make DeepSORT Great Again. *arXiv* **2022**. arXiv:2202.13514.
67. Zhang, Y.; Sun, P.; Jiang, Y.; Yu, D.; Weng, F.; Yuan, Z.; Luo, P.; Liu, W.; Wang, X. ByteTrack: Multi-Object Tracking by Associating Every Detection Box. *arXiv* **2021**, arXiv:2110.06864.
68. Bochinski, E.; Senst, T.; Sikora, T. Extending IOU Based Multi-Object Tracking by Visual Information. In Proceedings of the 2018 15th IEEE International Conference on Advanced Video and Signal Based Surveillance (AVSS), Auckland, New Zealand, 27–30 November 2018; pp. 1–6. [CrossRef]
69. Hou, X.; Wang, Y.; Chau, L.P. Vehicle Tracking Using Deep SORT with Low Confidence Track Filtering. In Proceedings of the 2019 16th IEEE International Conference on Advanced Video and Signal Based Surveillance (AVSS), Taipei, Taiwan, 18–21 September 2019; pp. 1–6. [CrossRef]
70. Scheidegger, S.; Benjaminsson, J.; Rosenberg, E.; Krishnan, A.; Granstrom, K. Mono-Camera 3D Multi-Object Tracking Using Deep Learning Detections and PMBM Filtering. *arXiv* **2018**, arXiv:1802.09975.
71. Hu, H.N.; Cai, Q.Z.; Wang, D.; Lin, J.; Sun, M.; Kraehenbuehl, P.; Darrell, T.; Yu, F. Joint Monocular 3D Vehicle Detection and Tracking. In Proceedings of the 2019 IEEE/CVF International Conference on Computer Vision (ICCV), Seoul, Korea, 27 October–2 November 2019; pp. 5389–5398. [CrossRef]
72. Kutschbach, T.; Bochinski, E.; Eiselein, V.; Sikora, T. Sequential sensor fusion combining probability hypothesis density and kernelized correlation filters for multi-object tracking in video data. In Proceedings of the 2017 14th IEEE International Conference on Advanced Video and Signal Based Surveillance (AVSS), Lecce, Italy, 29 August–1 September 2017; pp. 1–5. [CrossRef]
73. Lowe, D. Distinctive Image Features from Scale-Invariant Keypoints. *Int. J. Comput. Vis.* **2004**, *60*, 91. [CrossRef]
74. Dalal, N.; Triggs, B. Histograms of oriented gradients for human detection. In Proceedings of the 2005 IEEE Computer Society Conference on Computer Vision and Pattern Recognition (CVPR'05), San Diego, CA, USA, 20–25 June 2005; Volume 1, pp. 886–893. [CrossRef]

75. Ojala, T.; Pietikäinen, M.; Harwood, D. A comparative study of texture measures with classification based on featured distributions. *Pattern Recognit.* **1996**, *29*, 51–59. [CrossRef]
76. Ahonen, T.; Hadid, A.; Pietikainen, M. Face Description with Local Binary Patterns: Application to Face Recognition. *IEEE Trans. Pattern Anal. Mach. Intell.* **2006**, *28*, 2037–2041. [CrossRef]
77. Kuhn, H.W. The hungarian method for the assignment problem. *Nav. Res. Logist. Q.* **1955**, *2*, 83–97. [CrossRef]
78. Bochinski, E.; Eiselein, V.; Sikora, T. High-Speed tracking-by-detection without using image information. In Proceedings of the 2017 14th IEEE International Conference on Advanced Video and Signal Based Surveillance (AVSS), Lecce, Italy, 29 August–1 September 2017; pp. 1–6. [CrossRef]
79. Zhou, X.; Wang, D.; Krähenbühl, P. Objects as Points. *arXiv* **2019**, arXiv:1904.07850.
80. Dimitriou, N.; Stavropoulos, G.; Moustakas, K.; Tzovaras, D. Multiple object tracking based on motion segmentation of point trajectories. In Proceedings of the 2016 13th IEEE International Conference on Advanced Video and Signal Based Surveillance (AVSS), Colorado Springs, CO, USA, 23–26 August 2016; pp. 200–206. [CrossRef]
81. Cai, J.; Wang, Y.; Zhang, H.; Hsu, H.; Ma, C.; Hwang, J. IA-MOT: Instance-Aware Multi-Object Tracking with Motion Consistency. *CoRR* **2020**, abs/2006.13458. Available online: http://xxx.lanl.gov/abs/2006.13458 (accessed on 7 September 2022).
82. Yan, B.; Jiang, Y.; Sun, P.; Wang, D.; Yuan, Z.; Luo, P.; Lu, H. Towards Grand Unification of Object Tracking. *arXiv* **2022**, arXiv:2207.07078.
83. Yang, F.; Chang, X.; Dang, C.; Zheng, Z.; Sakti, S.; Nakamura, S.; Wu, Y. ReMOTS: Self-Supervised Refining Multi-Object Tracking and Segmentation. *CoRR* **2020**, abs/2007.03200. Available online: http://xxx.lanl.gov/abs/2007.03200 (accessed on 7 September 2022).
84. Voigtlaender, P.; Krause, M.; Osep, A.; Luiten, J.; Sekar, B.B.G.; Geiger, A.; Leibe, B. MOTS: Multi-Object Tracking and Segmentation. *arXiv* **2019**, arXiv:1902.03604.
85. Simon, M.; Amende, K.; Kraus, A.; Honer, J.; Sämann, T.; Kaulbersch, H.; Milz, S.; Gross, H.M. Complexer-YOLO: Real-Time 3D Object Detection and Tracking on Semantic Point Clouds. *arXiv* **2019**, arXiv:1904.07537.
86. Zhang, W.; Zhou, H.; Sun, S.; Wang, Z.; Shi, J.; Loy, C.C. Robust Multi-Modality Multi-Object Tracking. *arXiv* **2019**, arXiv:1909.03850.
87. Frossard, D.; Urtasun, R. End-to-end Learning of Multi-sensor 3D Tracking by Detection. In Proceedings of the 2018 IEEE International Conference on Robotics and Automation (ICRA), Brisbane, QLD, Australia, 21–25 May 2018; pp. 635–642. [CrossRef]
88. Weng, X.; Wang, Y.; Man, Y.; Kitani, K. GNN3DMOT: Graph Neural Network for 3D Multi-Object Tracking with Multi-Feature Learning. *arXiv* **2020**, arXiv:2006.07327.
89. Sualeh, M.; Kim, G.W. Visual-LiDAR Based 3D Object Detection and Tracking for Embedded Systems. *IEEE Access* **2020**, *8*, 156285–156298. [CrossRef]
90. Shenoi, A.; Patel, M.; Gwak, J.; Goebel, P.; Sadeghian, A.; Rezatofighi, H.; Martín-Martín, R.; Savarese, S. JRMOT: A Real-Time 3D Multi-Object Tracker and a New Large-Scale Dataset. *arXiv* **2020**, arXiv:2002.08397.
91. Paszke, A.; Chaurasia, A.; Kim, S.; Culurciello, E. ENet: A Deep Neural Network Architecture for Real-Time Semantic Segmentation. *CoRR* **2016**, abs/1606.02147. Available online: http://xxx.lanl.gov/abs/1606.02147 (accessed on 7 September 2022).
92. Milan, A.; Leal-Taixé, L.; Reid, I.; Roth, S.; Schindler, K. MOT16: A Benchmark for Multi-Object Tracking. *arXiv* **2016**, arXiv:1603.00831.
93. Dendorfer, P.; Rezatofighi, H.; Milan, A.; Shi, J.; Cremers, D.; Reid, I.D.; Roth, S.; Schindler, K.; Leal-Taixé, L. MOT20: A benchmark for multi object tracking in crowded scenes. *CoRR* **2020**, abs/2003.09003. Available online: http://xxx.lanl.gov/abs/2003.09003 (accessed on 7 September 2022).
94. Geiger, A.; Lenz, P.; Urtasun, R. Are we ready for Autonomous Driving? The KITTI Vision Benchmark Suite. In Proceedings of the IEEE Conference on Computer Vision and Pattern Recognition (CVPR), Providence, RI, USA, 16–21 June 2012; pp. 3354–3361.
95. Behley, J.; Garbade, M.; Milioto, A.; Quenzel, J.; Behnke, S.; Stachniss, C.; Gall, J. SemanticKITTI: A Dataset for Semantic Scene Understanding of LiDAR Sequences. In Proceedings of the IEEE/CVF International Conference on Computer Vision (ICCV), Seoul, Korea, 27 October–2 November 2019.
96. Wen, L.; Du, D.; Cai, Z.; Lei, Z.; Chang, M.; Qi, H.; Lim, J.; Yang, M.; Lyu, S. UA-DETRAC: A New Benchmark and Protocol for Multi-Object Detection and Tracking. *Comput. Vis. Image Underst.* **2020**, *193*, 102907. [CrossRef]
97. Lyu, S.; Chang, M.C.; Du, D.; Li, W.; Wei, Y.; Del Coco, M.; Carcagnì, P.; Schumann, A.; Munjal, B.; Choi, D.H.; et al. UA-DETRAC 2018: Report of AVSS2018 & IWT4S challenge on advanced traffic monitoring. In Proceedings of the 2018 15th IEEE International Conference on Advanced Video and Signal Based Surveillance (AVSS), Auckland, New Zealand, 27–30 November 2018; pp. 1–6.
98. Lyu, S.; Chang, M.C.; Du, D.; Wen, L.; Qi, H.; Li, Y.; Wei, Y.; Ke, L.; Hu, T.; Del Coco, M.; et al. UA-DETRAC 2017: Report of AVSS2017 & IWT4S Challenge on Advanced Traffic Monitoring. In Proceedings of the 2017 14th IEEE International Conference on Advanced Video and Signal Based Surveillance (AVSS), Lecce, Italy, 29 August–1 September 2017; pp. 1–7.
99. Luiten, J.; Osep, A.; Dendorfer, P.; Torr, P.H.S.; Geiger, A.; Leal-Taixé, L.; Leibe, B. HOTA: A Higher Order Metric for Evaluating Multi-Object Tracking. *CoRR* **2020**, abs/2009.07736. Available online: http://xxx.lanl.gov/abs/2009.07736 (accessed on 7 September 2022).
100. Henschel, R.; Leal-Taixé, L.; Cremers, D.; Rosenhahn, B. Fusion of Head and Full-Body Detectors for Multi-Object Tracking. *arXiv* **2017**, arXiv:1705.08314.

101. Henschel, R.; Zou, Y.; Rosenhahn, B. Multiple People Tracking Using Body and Joint Detections. In Proceedings of the 2019 IEEE/CVF Conference on Computer Vision and Pattern Recognition Workshops (CVPRW), Long Beach, CA, USA, 16–17 June 2019; pp. 770–779. [CrossRef]
102. Weng, X.; Kitani, K. A Baseline for 3D Multi-Object Tracking. *CoRR* **2019**, abs/1907.03961. Available online: http://xxx.lanl.gov/abs/1907.03961(accessed on 7 September 2022).
103. Gloudemans, D.; Work, D.B. Localization-Based Tracking. *CoRR* **2021**, abs/2104.05823. Available online: http://xxx.lanl.gov/abs/2104.05823(accessed on 7 September 2022).
104. Sun, S.; Akhtar, N.; Song, X.; Song, H.; Mian, A.; Shah, M. Simultaneous Detection and Tracking with Motion Modelling for Multiple Object Tracking. *CoRR* **2020**, abs/2008.08826. Available online: http://xxx.lanl.gov/abs/2008.08826 (accessed on 7 September 2022).
105. Luiten, J.; Fischer, T.; Leibe, B. Track to Reconstruct and Reconstruct to Track. *IEEE Robot. Autom. Lett.* **2020**, *5*, 1803–1810. [CrossRef]
106. Wang, S.; Sun, Y.; Liu, C.; Liu, M. PointTrackNet: An End-to-End Network For 3-D Object Detection and Tracking From Point Clouds. *arXiv* **2020**, arXiv:2002.11559.
107. Wang, L.; Zhang, X.; Qin, W.; Li, X.; Yang, L.; Li, Z.; Zhu, L.; Wang, H.; Li, J.; Liu, H. CAMO-MOT: Combined Appearance-Motion Optimization for 3D Multi-Object Tracking with Camera-LiDAR Fusion. *arXiv* **2022**, arXiv:2209.02540.
108. Sun, Y.; Zhao, Y.; Wang, S. Multiple Traffic Target Tracking with Spatial-Temporal Affinity Network. *Comput. Intell. Neurosci.* **2022**, *2022*, 1–13. [CrossRef]
109. Messoussi, O.; de Magalhaes, F.G.; Lamarre, F.; Perreault, F.; Sogoba, I.; Bilodeau, G.; Nicolescu, G. Vehicle Detection and Tracking From Surveillance Cameras in Urban Scenes. *CoRR* **2021**, abs/2109.12414. Available online: http://xxx.lanl.gov/abs/2109.12414 (accessed on 7 September 2022).
110. Wang, G.; Gu, R.; Liu, Z.; Hu, W.; Song, M.; Hwang, J. Track without Appearance: Learn Box and Tracklet Embedding with Local and Global Motion Patterns for Vehicle Tracking. *CoRR* **2021**, abs/2108.06029. Available online: http://xxx.lanl.gov/abs/2108.06029 (accessed on 7 September 2022).

applied sciences

MDPI

Article

Drone and Controller Detection and Localization: Trends and Challenges

Jawad Yousaf [1], Huma Zia [1], Marah Alhalabi [1], Maha Yaghi [1], Tasnim Basmaji [1], Eiman Al Shehhi [1], Abdalla Gad [1], Mohammad Alkhedher [2] and Mohammed Ghazal [1,*]

1 Department of Electrical, Computer, and Biomedical Engineering, Abu Dhabi University, Abu Dhabi 59911, United Arab Emirates
2 Mechanical Engineering Department, College of Engineering, Abu Dhabi University, Abu Dhabi 59911, United Arab Emirates
* Correspondence: mohammed.ghazal@adu.ac.ae

Abstract: Unmanned aerial vehicles (UAVs) have emerged as a rapidly growing technology seeing unprecedented adoption in various application sectors due to their viability and low cost. However, UAVs have also been used to perform illegal and malicious actions, which have recently increased. This creates a need for technologies capable of detecting, classifying, and deactivating malicious and unauthorized drones. This paper reviews the trends and challenges of the most recent UAV detection methods, i.e., radio frequency-based (RF), radar, acoustic, and electro-optical, and localization methods. Our research covers different kinds of drones with a major focus on multirotors. The paper also highlights the features and limitations of the UAV detection systems and briefly surveys the UAV remote controller detection methods.

Keywords: Unmanned aerial vehicles (UAVs); detection technologies; radio frequency-based (RF); radar; acoustic; electro optical; hybrid fusion; controller detection

Citation: Yousaf, J.; Zia, H.; Alhalabi, M.; Yaghi, M.; Basmaji, T.; Shehhi, E.A.; Gad, A.; Alkhedher, M.; Ghazal, M. Drone and Controller Detection and Localization: Trends and Challenges. *Appl. Sci.* **2022**, *12*, 12612. https://doi.org/10.3390/app122412612

Academic Editors: Luis Gracia and Carlos Perez-Vidal

Received: 16 September 2022
Accepted: 5 December 2022
Published: 9 December 2022

1. Introduction

In recent years, there has been a significant advancement in unmanned aerial vehicles (UAVs). UAVs are widely used for commercial, civilian, and military applications due to their low cost, spatiotemporal coverage, and remote sensing capability. They have been specifically popular for collecting information in remote and inaccessible areas, such as military surveillance and search and rescue in floods or earthquakes [1–3].

An aircraft without onboard human command and control is called a UAV, also called a drone. Command and control are achieved autonomously by the embedded autopilot or remotely by the operators through a ground station [1]. Moreover, autonomous and remote controls can be integrated as a single UAV control mechanism. Over the years, the technology and features of UAVs have improved tremendously to address the varying requirements of different applications. In addition, ongoing research has been successful in finding ways to improve the performance of the UAV. Various designs and features that support their assigned missions in different fields and sectors have been proposed, such as shape structures, take-off, and landing techniques [2,4].

Surveillance applications use UAV technology to be integrated as a standalone, connected platform for information gathering. The human detection system in [4] was achieved through input from thermal images and videos from a thermal camera connected to a UAV. These images and videos are categorized by reference to a thermal dataset in the system and are processed by sequence operations to achieve the final result. In the military, some geographic areas are difficult to reach for monitoring and detecting unwanted signals or entities. The proposed system in [5] overcomes this demand.

Moreover, smart farming utilizes UAV technology for real-time monitoring and data acquisition of crop parameters, e.g., plant height, presence of weeds, or fungus.

Tsouros et al. [1] discussed different types of UAVs and explored multiple applications of UAVs in precision agriculture, crop health, and growth monitoring. Moreover, this work reviews data acquisition technologies and aerial image processing methods. Further, civil engineering utilizes UAV technologies for seismic risk assessment, transportation management, disaster response, construction management, surveying and mapping, and flood monitoring and evaluation [2].

To summarize, UAV technologies have numerous features that enable their usage in multiple sectors and applications. The benefits of such technologies include: (1) reducing human risk, (2) lower energy consumption, (3) a lower cost, (4) flexibility, and (5) accuracy of data collection. The advances in electronics and sensor technology have widened the scope of UAV applications for their likely invaluable inclusion in police, fire brigades, and disaster management operations. However, in recent years, UAVs have been used to perform malicious actions, such as drug smuggling, intelligence gathering, and suicide attacks [2,3,5]. UAVs also pose a threat to surpassing restricted government or military sites. In addition, with the prevalence of smaller UAVs, concerns over public privacy are rising.

All these threats warrant an urgent need for research into UAV detection methods. It becomes strategic to detect and localize UAVs to prevent such malicious actions. Recently, various detection algorithms have been researched, such as active radar probes, acoustic recognition, infrared spectrum identification, visual recognition, and radio frequency (RF) signal detection [1–7]. This study aims to provide a detailed literature review of these detection methods, identify their strengths, explore various applications where they were used, and compare these methods for the major relevant studies in the open literature. Our study scope includes the detection and localization of multirotor and other UAV types. The study also reviews the techniques for the UAV controller localization of the detected drones. This review aims to survey the quickly evolving field, record what is notable and popular within this sector, and provide recommendations for future investigators. Table 1 summarizes the covered topics in different sections of this study.

Table 1. Summary of reviewed topics for drones and their controller detection.

Detection Technologies		Ref.
UAV Architecture and Security Concerns		[3–8]
UAE Detection Technologies	RF	[7–20]
	Radar	[21–29]
	Acoustic	[30–40]
	Electro-optical	[41–46]
	Hybrid fusion	[40,41,47,48]
Controller Detection and Localization		[12,18,49,50]

The rest of the study is organized as follows: Section 2 details the architecture of UAVs and associated security concerns with drones. A comprehensive review of UAV detection technologies is outlined in Section 3. The studies about drone controller localization are reviewed in Section 4. Lastly, Section 5 concludes the findings of the study.

2. UAV Architecture and Security Concerns

2.1. UAV Architecture

UAVs have multiple subsystems integrated to perform various operations, such as launch, fly, operate, process, transmit, and receive commands from remote or ground stations [3,5]. Four main UAV subsystems should be considered: (1) a power unit, (2) a communication module, (3) the main computing device, and (4) a sensor board. The power unit is designed to provide a longer lifetime for UAV operation without charging it [4]. The high-level architecture of the UAV system is illustrated in Figure 1, including UAV's main computer processes commands based on the collected data from other subsystems or components (GPS, sensors, gyroscopes, accelerometers, antennas, receivers, etc.). These data or commands are transferred through a communication link between the UAV and the ground control station (GCS). This communication is mainly monitored to detect UAVs based on RF and radar-based technologies (details in Sections 3.1 and 3.2).

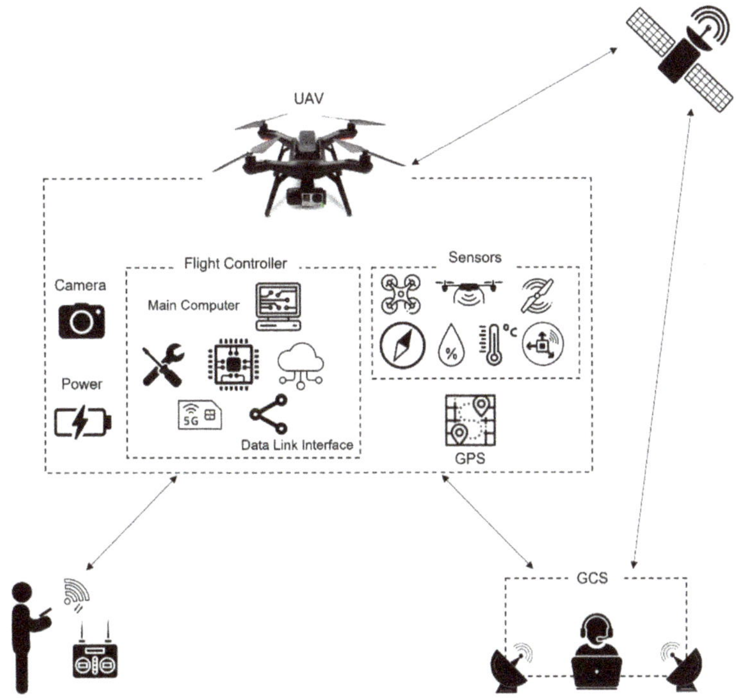

Figure 1. High-level architecture of a UAV.

A brief description of the major components of UAE architecture is as follows:

- UAV's structure/airframe: There are many common features of a UAV's chassis, such as lightweight, small size, endurance, aerodynamic flexibility, etc.
- Main computer: The critical part responsible for autonomous functioning and flight control. The computing subsystem processes sensed information, transmits it back, manages flight operations, and communicates with the control base.
- Sensors/payloads: UAVs can be equipped with a range of possible lightweight sensors per the application's needs, including RGB cameras, thermal sensors, LiDAR sensors, and multispectral and hyperspectral sensors. All of them are connected to the flight controller to gather real-time data and process it for the missions' execution.
- Communication link: UAVs are equipped with a high-quality wireless communication unit, including 5G, WiFi, Bluetooth, and radio-frequency identification (RFID), to facilitate communication with the GCS or the internet.
- Ground control station (GCS): This base station is mainly employed to monitor and control the UAV during its operation. Flight operation is continuously monitored and can be controlled to alter the mission.

2.2. Security Concerns

Regarding UAV security, two main topics are discussed in the literature: the security and safety of UAVs and the potential misuse of UAVs against critical infrastructures and privacy-related issues.

Threats to UAV security are well-researched concerning targeting its hardware, software, and communication module. In [6], threats to various components of the UAV system are discussed. GCS's physical, network, and cloud security have been highlighted as vulnerabilities that can be exploited. Moreover, threats exist against the UAV communication according to the communication medium technology or type (WiFi, cellular network, GPS, and other RF solutions). The possible common attacks are eavesdropping, jamming, replay, denial of service, hijacking, etc. Other threats include mission disruption and itinerary tracking [7,8].

Despite promising application benefits using UAVs, threats also exist by their prevalence in the public domain. UAV security threats and incidents are mainly caused by privacy violations of sensitive sites, airplane flight disruption, damage and explosion in targeted areas, and sensitive data leakage through eavesdropping [7,8].

3. UAV Detection Methods

As mentioned in Section 2, there are many sectors in which UAVs have been explored and adopted, utilizing their practical and advanced features. The continuous development and improvement of UAV's main systems and components, i.e., flight controller, sensors, gyroscopes, cameras, GPS, etc., increased the demand and reliance on UAVs for accomplishing different civilian and military missions. Moreover, they are widely available in the market at a reasonable cost compared to other solutions.

Research has been dedicated to designing, developing, and implementing systems for detecting malicious UAVs. Techniques of these systems are classified into passive and active. RF-based, acoustic, and vision-based techniques are among the passive technologies, whereas radar-based techniques are defined as active technologies. These technologies vary in operational conditions, covering range, consistency, accuracy, and many other parameters. This section focuses on UAV detection technologies and discusses the general framework and related work.

3.1. *RF-Based*

RF is used for UAV remote command and control communication. RF-based detection technologies rely on real-time sensing, capturing, processing, analyzing, and retrieving data from UAV's RF-emitted signals. Acquired RF data are intended to identify, track and classify the detected UAV and localize the controller. RF-based techniques analyze the captured spectrum between the UAV and operators using circular or linear array antennas to detect both the drone and its controller in all-weather environments. As most of the communication between a drone and its controller occurs in the ISM band, around 2.4 GHz, the implementation cost of such a system is much lower compared to a radar-based solution [21–28,51].

In RF-based detection technologies, RF and WiFi-based fingerprinting techniques are major verification systems. RF-based techniques include studying and analyzing the characteristics of the captured transmitted RF signal from UAVs or UAVs' controllers. However, WiFi-based fingerprinting is related to the WiFi links and traffic between the UAV and its remote controller. The reviewed studies include the analysis of RF spectrogram (fingerprinting) [7,9,10,19], angle of arrival (AOA) (MUSIC) [12], and direction of arrival (DOA) [40] methods for the identification and localization of drones using conventional as well machine learning algorithms [9,10,13–15].

In [7], the technique proposes a complete UAV detection and identification system framework designed to work in the 2.4 GHz frequency band. The system starts with capturing the wireless signals in the test area. Then, the captured signal is processed based on a 4-level Haar wavelet transform analysis. The standard deviation of the processed signal is calculated to define the UAV detection condition. After the detection of the UAV, the RF fingerprinting stage is activated, and three main features are extracted: (1) fractal dimension (FD), (2) square integrated bispectra (SIB), and (3) axially integrated bispectra (AIB). These features are adjusted and weighted using principal component analysis (PCA) and neighborhood component analysis (NCA) algorithms. The final RF fingerprints are stored as the training data for a set of machine learning algorithms used to classify the UAV.

Based on indoor and outdoor experimental scenarios, the average identification accuracy of UAVs is summarized with respect to three fingerprinting features. Furthermore, in [8], the WiFi network traffic is monitored, and the UAV detection method is based on WiFi fingerprint analysis. The extracted features are related to the captured traffic's duration, behavior, and distribution. Different scenarios are applied to evaluate the system's performance in UAV detection, where the average precision is about 96%.

The authors of [15,19,20] fed the extracted time-domain characteristics (shape factor, skewness, kurtosis, and variance) of recorded RF signals to the machine learning data processing units to detect and classify UAVs. In [9], indoor experimental testing is conducted for data collection using the RF fingerprints of the transmitted signal from the micro-UAV controller to the UAV for UAV detection and classification. Different micro-UAV controllers (a total of 14) operating at the 2.4 GHz frequency band were used to create the dataset (a total of 100 RF signals) and test the proposed detection and classification technique. Each micro-UAV controller has a different transmitted signal, categorized with its unique transmitter characteristics, excluding the traditional threshold-based detection technique. The Markov model algorithm is later used for UAV detection and energy transient signal approach for feature extraction and UAV classification. The performance and accuracy of the system were found to be 96.3%.

UAVs use the Industrial Scientific and Medical (ISM) frequency bands, i.e., 2.4 GHz and 5.8 GHz bands, to communicate with their remote controllers [11]. Multiple passive RF sensors support these frequency bands and are used for non-invasive surveillance operations, including UAV monitoring, detection, localization, and tracking. In [11], the UAV detection system consists of a sensor node, Keysight RF sensor N6841A, operating in the range of 20 MHz–6 GHz, broadband antenna, and GPS tracker linked with geolocation software, N6854A. The RF signals are detected and collected within a radius of 2 km from the sensor node. A GPS antenna also records the time stamps for these collected signals. The localization of the UAV is performed using a detection algorithm and time difference of arrival (TDOA) measurements. Extended Kalman filter (EKF) framework and fitting motion models (MM) address these errors and improve localization performance.

Furthermore, the research work in [10] illustrates a system model and architecture followed by experimental validation of the proposed direction finding (DF) method of sparse de-noising auto-encoder (SDAE) for UAV surveillance. This method consists of a single channel for a receiver and a directional phased array antenna. The mechanism of the system works as follows. First, the transmitted signal from the drone to its ground controller gets processed using an RF switching mechanism to measure the received signals output power at each directional phased array antenna. Next, the acquired output power values from the N-antennas of the phased directional array are input to the proposed SDAE-based deep neural network (DNN). The first network layer extracts received wattage values. Then the remaining network utilizes sparse representation to categorize UAVs' signal directions. The system diagram of the proposed method in [10] is depicted in Figure 2. To summarize, the wattage power values are passed to the proposed deep network, followed by the DF method, which exploits both the sparsity parameter of the transmitted UAV signal and the gain variation parameters of the directional antenna array.

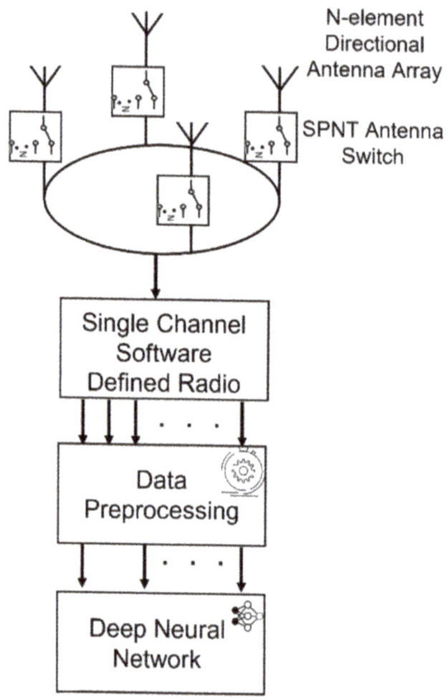

Figure 2. Diagram of the system model in [10].

In [16], the authors discuss UAV detection using RF-transmitted signals between UAVs and their remote controllers. Power spectrum cancellation and multi-hop autocorrelation are developed to achieve RF passive detection of UAVs and controllers to detect emitted signals. The multi-hop autocorrelation method can detect the cross-correlation signal if the Signal-to-Noise (SNR) ratio is small by applying an emitted remote-control signal. A limitation of the multi-hop autocorrelation is the low accuracy in the case of fixed frequency in remote-control signals. The calculated parameters significantly depend on the autocorrelation function, leading to false positives. Hence, the study of [16] used the power spectrum cancellation technique to eliminate the effect of fixed frequency signals. Power spectrum cancellation works by first finding the differences between the control signal power spectrum and fixed frequency signals over time. Once the differences are identified, the fixed frequency signal is eliminated, and the remote control signal is applied to multi-hop auto-correlation to finalize the parameters for UAV detection.

Furthermore, [17] stated that the RF passive detection method has the advantage of low cost, license-free, long-range distance coverage, and early warning capability. They also illustrated an RF passive system architecture, which analyzes the electromagnetic RF spectrum emitted from exchanged signals between the UAV and its controller. The passive RF detection algorithms analyze these signals to sense alternations in the frequency and time domain RF spectrum.

Various studies have reported promising results utilizing different algorithms and techniques for RF-based UAV detection. However, the presence of noise affects the accuracy and detection range. Table 2 summarizes the reviewed papers and tabulates the features and accuracy of the undertaken methodology for RF-based UAV detection.

Table 2. Summary of reviewed RF-based techniques for UAV characterization.

Ref.	Operating Frequency	Functionalities			Performance
		Identification	Classification	Localization/ Tracking	
[7]	2.4 GHz	√	√	-	Average of 97%
[8]	2.4 GHz	√	√	-	Greater than 96%
[9]	2.4 GHz	√	√	-	Average of 96.3%
[11]	20 MHz–6 GHz	√	-	√	-
[13]	1–6 GHz	√	√	-	Average of 99%
[10]	2.401–2.481 GHz	√	√	√	-
[16]	2.4 GHz and 5.8 GHz ISM bands	√	-	-	-
[12]	2.4 GHz	√	-	√	-
[18]	2.4 GHz ISM band	√	-	√	-

3.2. Radar

Radar signal processing is among the classical approach for aircraft and drone detection as it can be used in all weather conditions with 24/7 operation [18,21,52] as compared to acoustic and visual detection methods. In this approach, the received signal is characterized to detect echo, doppler signature, or radar cross-section (RCS) for detecting and tracking the target [21,28,29,53]. The conventional radar signal processing techniques have the limitation of accurate distinction of mini UAVs from birds due to their smaller RCSs. AI-based techniques are proposed [28,29,53,54] to process the extracted features from the radar signals to address this issue to some extent.

In radar-based detection, radio energy is used to detect the target and define its position [21,23,55]. Typically, a radar-based detection system has three main components: RF radar, data acquisition, and signal processing. In RF radar, the electromagnetic energy radiates into space and encounters the UAV's body flying in the monitored area. The UAV's reflected wave is returned and received by the system, measured, and processed in real-time (data acquisition and signal processing). Hence, the UAV is successfully located, and its flight path is tracked by the system [30,36,55,56].

Frequency-modulated continuous wave (FMCW) and continuous wave (CW) radars are preferred to be used in UAV detection and identification, especially for their continuous pulsing, effective cost, and performance [21]. The FMCW radar contains a transmitter and a receiver antenna. The oscillator and the control signal produce the transmitted signal. After the backscattering/reflected signal is received, it gets passed to the I/Q demodulator for filtering. Power is equally distributed into two signals with 90 degrees phase shift to be forwarded to the low pass filter (LPF). The intermediate frequency (IF) signal, resulting from in-phase and quadrature-phase components, is directed to the analog-to-digital converter (ADC) and the digital signal processing (DSP), as depicted in Figure 3. The distance and velocity of the target can be defined by using the time delay and phase information of both the transmitted and received signals [21].

The studies of [24,25,28,54,57,58] employed the principal component analysis (PCA) [24], convolutional neural networks (CNN) [23,28,51,54], long short-term memory (LSTM) [28], and support vector machines (SVM) [57,58] techniques for the processing of extracted features from radar signals such as micro-doppler spectrogram [23,28,54,57,58] and range-doppler signature [24] for the classification of drones. Recently authors in [13] used the hierarchical learning approach for the detection of the presence, type, and flight trajectory of a UAV. Due to the smaller size of most UAVs, wideband, high frequency, expensive radars are required for the accurate detection and tracking of mini UAVs [23,24,52,54], which increases the overall cost of the detection and localization system.

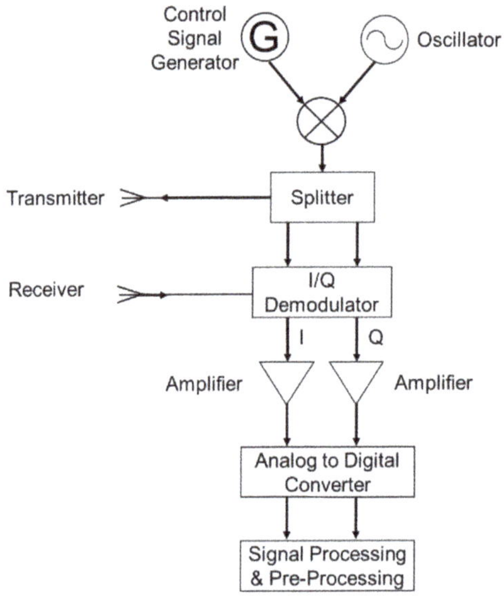

Figure 3. Architecture design of FMCW radar [21].

In [55], the authors proposed research and experiments for evaluating the data acquisition and signal processing algorithm in a CW radar system that supports C and X frequency bands operations. The radar system uses the micro-Doppler principle. The extracted signatures in the frequency and time domains are used in UAV classification for calculating the propeller blades' length and determining the rotation propellers' speed. For performance evaluation, the number of UAV propellers varies during the experiments while fixing the propellers' rotational speed and a maximum distance of 25 m between the radar and UAV. The classification and measurement of UAVs become complex with the increase in propellers.

In another work [56], simulation and analysis of continuous wave radar's echo signals are studied and presented in different conditions at an operating frequency of 35 GHz. Mainly UAV detection is based on the time-frequency characteristics of the Micro Doppler signal produced by the rotor rotation using singular value decomposition. Discrete wavelet transform is also used to remove environmental clutter from the radar echo signal, whereas the support vector machine (SVM) is used as a classifier. The detection accuracy of the developed system achieved 85%.

Another type of radar-based UAV detection mechanism, cylindrical phased array radar, was discussed in [27]. The system performs better for UAV detection when comparing the omnidirectional scanning to planner array radar due to the flexibility of changing the direction of the beam and illumination time to the target after the phased array was used. As for the operational norms, the system's hardware structure and signal processing flow are designed to get a strong clutter suppression specified in the investigation, and the result of the experiment shows potential for UAV detection. Authors in [27] developed a cylindrical phased array radar system and explored signal optimization by specifying signal processing flow with the moving target detection (MTD) based on the maximum signal-to-clutter ratio (SCR) criterion.

Tang et al. [24] explained the type x-band, a small phased array radar based on AD9361, an RF Agile Transceiver. The AD9361 is a highly integrated RF module with a high-performance agile transceiver for 3G and 4G base station applications. The reported radar system consists of a control module controlling the antenna beam pointing through the transmitter/receiver (T/R) module. The signal processor also sends waveforms as transmitted RF signals to AD9361 within the timing sequence. Then the corresponding waveform is generated and established by AD9361. The radar simulation detects a drone with a radar cross section (RCS) of 0.01 m^{-2} within the range of 5 km. For radar detection, enhanced reflected signals are necessary to minimize the effect of noise. An SNR value greater than 14 dB indicates a highly accurate detection.

In the case of a reliable RCS, the chosen wavelength should not reach half of the detected object's dimension. It is critical to use a higher frequency while using Doppler-based detection. As illustrated in [22], radar is used to detect smaller drones; however, it has an ill-prepared standard for UAV detection based on low air-velocity aircraft and weak radar signature. During target detection, the radars would receive reflections from clutter-like objects, landscapes, and precipitation, posing a challenge in detection. A target can only be detected if system noise due to clutter is minimized. A 30 × 30 rectangular phase array used in [22] detects the presence of drones in monostatic radar. It would continuously scan the predefined surveillance region, with the limitation of a 90-degree azimuth sector, to achieve 360 azimuth coverage at a low cost. Doppler estimation discussed in [22] can be described as a spectrum estimation process.

The reference [26] illustrates the new method based on 5G millimeter waves with an end-to-end network. It further explains the detection method done using 5G millimeter-wave radar at rotors of UAVs. The high-resolution range profile (HRRP) can identify a UAV location, while micro-Doppler identifies the UAV. Moreover, the cepstrum method was used to extract any number and speed information of the detected UAV rotor. Multiple UAVs can be identified using the sinusoidal frequency modulation (SFM) parameter optimization method. The proposed method determines the following: the number of detected UAVs, the number of rotors, the rotation speed of all rotors, and the position of the UAVs. The proposed radar detection in [26] presents a UAV identification and detection study by providing a method for UAV tracking using the GPS-independent method, such as GPS signal failure, GPS signal interference, and satellite occlusion areas. HRRP technology and micro-Doppler provide a successful solution to detect and localize any rotating targets regardless of weather conditions. The presented simulated results showed high robustness and performance of the cepstrum method.

Authors in [59] presented a passive radio drone detection system that uses goodness-of-fit (GoF) based spectrum sensing and the MUSIC algorithm to detect the transmitted signal of a drone and its controller and estimate the DOA. Once a signal is detected, the DOA is estimated at the detected frequency. The MDL algorithm detects the number of targets and whether the source is a drone or controller. The detection system detected drones and controllers from different manufacturers with good sensitivity.

A challenge associated with UAV detection is the presence of aircraft and birds in the background [60–62]. Hence, clutter suppression and target detection algorithms are needed to overcome this complex issue, as stated in [26]. Rationally, object detection of possible UAVs comes first, followed by classification to separate UAVs from other detected objects. In addition, the purpose of these classifications and identifications can be used to extract many unique features of these UAVs [23]. As stated previously, the effects of Doppler radar are used to determine the velocity of a distant object more accurately. This is obtained from the radial component of a target velocity in relation to radar. Using stepped frequency waveform (SFW), an HRRP can be obtained. Due to HRRP and Doppler information from a wide-band Doppler radar, detected objects scanned using wide-band are identified and classified. Millimeter wave base stations and 5G network systems can be used as detection network channels for UAV detection using the data from the processing center of 5G base stations. The process includes extracting essential parameters from multipath locations through 5G bases.

Many factors must be considered during the development to enhance the radar systems' performance, such as operating frequency, data acquisition, processing algorithms, classification techniques, and environmental clutter. The summary of reviewed studies of this technique is given in Table 3.

Table 3. Summary of reviewed radar-based techniques for UAV characterization.

Ref.	Detection Technique	Specifications	Functionalities			Performance
			Identification	Classification	Localization/ Tracking	
[21]	FMCW/CW radar	Doppler effect principle	√	-	-	NA
[55]	CW radar	C and X frequency bands, Micro Doppler principle	√	√	-	-
[56]	CW radar	Operating frequency: 35 GHz	√	√	-	Accuracy 85%
[27]	Cylindrical phased array radar	Operating frequency: C band	√	-	√	Performed well under a strong cluttered environment
[24]	Small phased array radar	Based on AD9361	√	-	√	Reliable and stabile
[22]	Rectangular phased array radar	Operating frequency: X band	√	-	-	Mixed up with birds
[26]	5G millimeter wave radar	Starting frequency is 25 GHz, which is in the 5G band	√	-	√	Detected at 300 m with a speed of 157.9 r/s & at 850.2 m with a speed of 88 r/s

3.3. *Acoustic*

Acoustic sensors, such as microphone arrays, capture the generated audio from the rotors and propellers of the drone and then compare the extracted features, including mel- frequency cepstral coefficients (MFCC) and short-time Fourier transform (STFT), with acoustic signature databases for the detection and classification of drones and UAVs using conventional and AI-based architectures. MFCC is a set of reflected human perception features of sounds, which is used in audio classification when paired with machine learning approaches. STFT is considered an intermediate feature compared to MFCC. MFCC compresses signals while representing them with coefficients set. On the other hand, STFT features contain more information and noise than MFCC, giving STFT an advantage. Deep learning models can easily adopt STFT and manage it given more complex data [37].

Authors proposed a machine learning framework in [38], shown in Figure 4, to detect and classify ADr sounds in a noisy environment, among other sounds. The required features are extracted from ADr sound using the feature extraction techniques of MFCC and linear predictive cepstral coefficients (LPCC). Following the feature extraction process, these sounds are then identified using SVMs. The results show that the SVM cubic kernel with MFCC outperforms the LPCC technique by detecting ADr sounds with 96.7% accuracy.

Acoustic-based technologies are effective for detecting UAVs since they are not affected by the UAV's frequency range, weather fluctuations, e.g., fog, environmental disturbance, and noise. Hence, such technologies do not block the acoustic sensors' earshot to detect the UAV's acoustic signals. Acoustic signals produced by the engine and propeller blades of the UAV are collected and processed to classify the UAV and calculate its distance, direction, and location [33].

Authors in [39] proposed a CNN-based system to detect drones using acoustic signals received by a microphone. STFT magnitude is used as the two-dimensional feature in the study since drones' harmonic properties differ from those of other devices that make a similar noise. The dataset comprised 68,931 and 41,958 frames of drone and non-drone sounds collected using DJI Phantom 3 and 4 drones flying outdoors. The proposed approach has a detection rate of 98.97% for the 100-epoch model and a false alarm rate of 1.28. Figure 5 illustrates the system overview of the proposed approach.

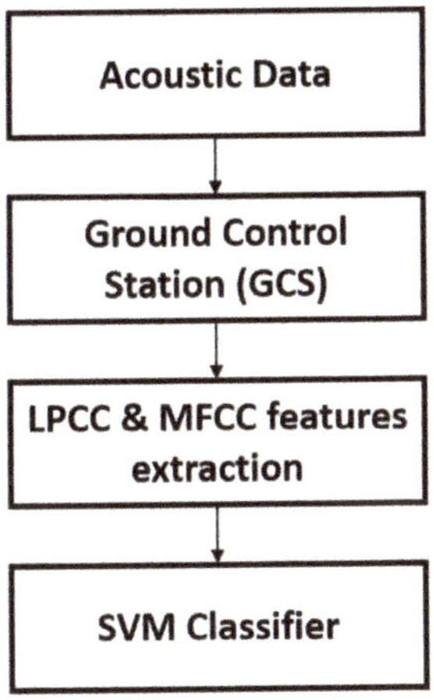

Figure 4. Overall system diagram of the approach presented in [38].

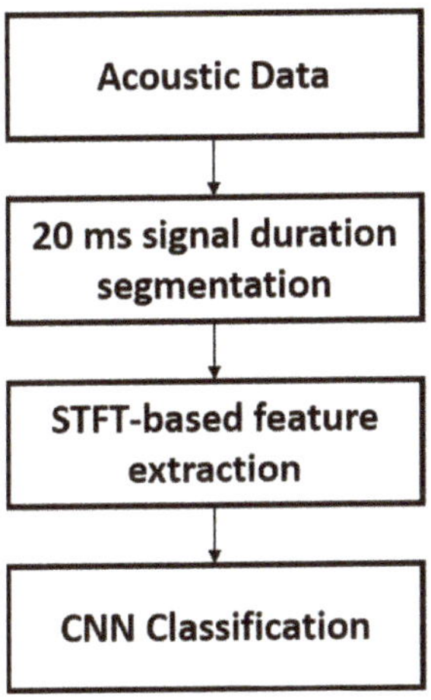

Figure 5. System overview of the method introduced in [39].

An acoustic-based detection system was designed and implemented in [30] to detect and locate the UAVs efficiently. The acoustic sensor array configuration comprises two tetrahedron-shaped microphones. The system uses multiple algorithms for data and features extraction from the collected acoustic signals: cepstral coefficients (CC) for extracting the harmonics' features, SVM to classify and distinguish between the extracted features' vectors related to UAV or background noise, and TDOA based on Bayesian framework. Signal processing is concluded with the temporal and dimensional features' vectors calculations to acquire the accurate UAV classification and localization path. They also study the contribution of the SNR in detecting the UAVs against the detection rate.

In [36], multi-label UAV sound classification is examined using stacked bidirectional long short-term memory (BiLSTM), an advanced, recurrent neural network (RNN) capable of handling sequence or multiple classification tasks and avoiding long-term dependency issues. The proposed BiLSTM model is 94.02% successful in UAVs' sound classification.

Several types of research and studies aim to investigate and evaluate different algorithms used in data acquisition, processing, and classification of the collected acoustic signals. In [31], the system's performance level varies using different audio processing algorithms for characteristic feature vector extraction. The extracted features are inputted into concurrent neural network (CoNN) for classification. The results confirm better accuracy when integrating CoNN with the Wigner-Ville dictionary than MFCC and mean instantaneous frequency (MIF).

The authors in [35] gathered the acoustic data from a local suburban airport for the five samples of commercial multirotor UAVs to establish the performance based on passive acoustic detection. The study characterizes the emitted noise of UAVs of different levels in an anechoic chamber at the airborne time. The microphone array was arranged within two circular tiers, each 1-m in radius, and separated vertically by 1.6 m to collect data from the local airports. The generalized cross-correlation (GCC)-based algorithm is used to find direction by fusing the time difference of both arrivals and steered power response with phase transform (SRP-PHAT). The smallest UAV with a 294 m detection distance was tested and demonstrated. Differential Doppler is used to overcome the decorrelation effect for better accuracy, as stated in [35].

In [32], the authors used classical detection and direction-finding methods using an array of microphones. There had been a physical investigation of the UAVs through experiments on acoustic emission with two signal models presented in harmonic signal and broadband signal for open area and indoor environments, respectively. The spectral signs are used for detecting and recognizing the UAVs in a noisy environment by incorporating the effect of noises in urban transport, speech signals, and environment noises. The result gives the same quality as the MFCC method, where acoustic portraits are unnecessary. The cross-correlation function is efficient in the direction-finding of the UAV. The study of [32] concludes with the following points: (1) high-pass filters are effective in the processing stage of UAV acoustic emission; (2) taking a noisy environment as a background experiment while detecting and recognizing UAVs by spectral signs performs similarly to the MFCC method, excluding acoustic portraits; (3) it is suggested to improve the efficiency of the CCFM algorithm in acoustic signals to filter out low-frequency noise; (4) MFCC and CCFM can be used to create an effective counter-action system against UAVs.

Yang et al. [37] researched the utilization of acoustic nodes in the UAV detection system. The proposed system finds the best configuration of the node for deploying the UAV acoustic detection system using machine learning models. The study was designed to investigate the best combination of acoustic features, STFT and MFCC, machine learning algorithms, SVM and CNN, for node optimization. After integrating the sensing nodes in four different configurations among the test sets, the one that maximizes the detection range without blind spots is selected. A semi-circle by the STFT-SVM model with a 75-m distance between the protected area and node has the best performance for configuration optimization. Demonstrating machine learning in the audio signal domain with different learning algorithms was used for detection module development. The study [37] focused on event sound detection using binary classification with MFCC features in an urban area.

A drone acoustic detection system (DADS) is proposed and demonstrated experimentally to detect, classify, and track airborne objects in [33]. They used a Phantom 4 UAV for testing, which reached 350 m with four degrees as an average precision to track a maneuvering UAV with compact acoustic nodes. This test also implemented the classification algorithm to detect a multirotor UAV based on a specific sound inherent in the flight control mechanism. The Steven Institute of Technology has developed the DADS to detect, track, and classify anonymous UAVs by propeller noise. The proposed system has three or more microphone nodes in a tetrahedron configuration. The communication between the microphone nodes and the central computer is done through WiFi for processing. The orientation calibration for the DADS system is performed by emitting white noise from a speaker and tracking the GPS position for several minutes. Based on the difference between the detected direction and computed ones from the surveyed GPS, the orientation can easily be corrected in the case of detection and tracking. Establishing a tracking process can be predicted using collected data and parameters. Node placement, the direction-finding probability that depends on precision and range for a given target, and ambient conditions with the tracker association threshold are among the collected data.

Two main components affecting the system's overall performance are (1) hardware specifications, including acoustic sensors and data acquisition tools, and (2) software tools and algorithms, including acoustic fingerprints and features extraction, classification, and localization. Table 4 summarizes the recent studies for acoustic-based UAV detection, classification, and localization.

Table 4. Summary of reviewed acoustic-based techniques for UAV characterization.

Ref.	Detection Technique	Functionalities			Performance
		Identification	Classification	Localization/ Tracking	
[30]	Designed for Amateur Drones (200 Hz), SVM (Drone sound identification)	√	-	√	High accuracy
[36]	BiLSTM (UAV sound classification)	√	√	-	UAV sounds 94.02%
[31]	Concurrent Neural Networks	√	-	-	96.3%
[35]	TDoA, SRP-PHAT	√	-	√	SRP-PHAT outperform TDoA
[32]	-	√	-	√	-
[37]	MFCC, STFT, CNN, SVM	√	-	√	Noise affects the detection
[33]	SRP-PHAT	√	-	√	Drone classification algorithm to be improved according to distance
[34]	SRP-PHAT	√	-	√	-

Unlike radar and RF approaches, the acoustic solution does not require a line of sight (LOS). However, this solution has challenges of a short range, the need for an extensive large signature database, and vulnerability to ambient environmental noise and clutters, particularly in urban areas [14,30,31,40], and quiet operation of the drone [9,30,38]. The detection of the drone pilot could be very difficult, too, using acoustic sensors.

3.4. Electro-Optical

The electro-optic sensing system transmits, detects, and examines radiations in the optical spectrum, including visible light, infrared, and ultraviolet radiation. It can handle long-range imaging and has reliable results under different illumination levels. The components associated include optics, laser, detectors, camera, processing unit, etc. Such systems have been used for UAV detection, direction finding, and localization continuously and in all weather conditions.

In [46], the authors proposed using machine learning techniques to automatically detect and track small moving objects in the airfield from their motion patterns, i.e., the ways an object moves. The system utilized remote digital towers with high-resolution cameras covering the 360-degree view of airports to construct a video dataset comprising aircraft in an airfield and drones. Harris detection and convolutional neural network followed by optical flow we applied to the dataset to locate and track very small moving objects in the wide-area scene. The results showed that the system can detect objects with 15×15 pixels in 1080p images with a low miss rate. Motion-based features are extracted from their trajectories, after which a K-nearest neighbor classifier is applied to classify objects into drones or aircraft, with an accuracy of 93%.

The proposed approach in [42] performs the detection by integrating a 3D LADAR sensing system. The study employed voxel-based background subtraction and variable radially bounded nearest neighbor (V-RBNN) techniques to detect small UAVs up to 2 km. During the development phase, this integration is supported with augmented data set to enhance the model's performance. The developed LADAR scanner can be rotated to cover a wide range of areas, e.g., 350 degrees for azimuth direction and 120 degrees for elevation direction. Furthermore, the used clustering algorithm, V-RBNN, has a good impact on the target UAV classification, which may increase the use of this proposed detection system in various applications.

In some electro-optic solutions, the detected data transferred to the analysis phase, including advanced processing, machine vision, or machine learning, are not accurate enough to track small UAVs effectively. Authors in [44] proposed an electro-optic system integrated with an all-sky camera system to get a wider view of the monitored area to improve the detection resolution. Multiple experiments were performed to evaluate the proposed solution and test its integration with other cues, i.e., acoustic. The combination of these three systems, electro-optic, all-sky camera, and acoustic cues, is also evaluated.

The study in [41] improved its outcomes' reliability by using the electro-optical method for small UAV detection and tracking. The actual video stream was used in real-time, and a differential method was employed for analyzing and investigating UAV detection and tracking. The differential method finds the differences in sequenced frames in a video stream. In the case of hovering UAVs or some axially moving and revealing objects near the frames, the contrast was selected only for the displaced part of the object to process the video streams using the DIS algorithm.

The electro-optical detection method needs to consider the following factors: size and movement in 3D, speed of detected airborne objects, the maximal distance of detected objects from the camera position, optical lens descriptions, and linear object image resolution. Since these factors directly relate to image processing methods, detecting distance and outputs of detections get affected negatively if one of the aforementioned factors contains faulty or inaccurate information. Detecting moving objects at a maximal distance from a camera in real-time is the main objective of the method.

A single dynamic vision sensing (DVS) camera, a base station, UAV, and a blinking marker are used in [45] to detect and locate mobile UAVs. During video streaming, the differences among the captured frames are computed and filtered to detect UAVs in the background image using a temporal-filtering algorithm. The triangulation algorithm was also used to help capture UAVs or drones by extracting spatial localization parameters and providing details about the physical size of the detected object.

Similarly, in [43], Seidaliyeva et al. developed an algorithm to detect drones from a video stream. The system overview for the process is illustrated in Figure 6. The input frames are passed into a moving object detector algorithm. The authors relied on background subtraction followed by threshold filtering and morphological operations for detecting moving objects. The background subtraction method describes a model background image that is subtracted from all frames to extract the foreground. This method heavily depends on the background remaining static throughout the operation. A CNN-based classifier is used on the detections to distinguish drones from other objects such as birds.

The limited detection range can pose a challenge while employing the electro-optic technique. The detection performance can be enhanced by incorporating other supporting algorithms. The summary of reviewed studies of this method is depicted in Table 5.

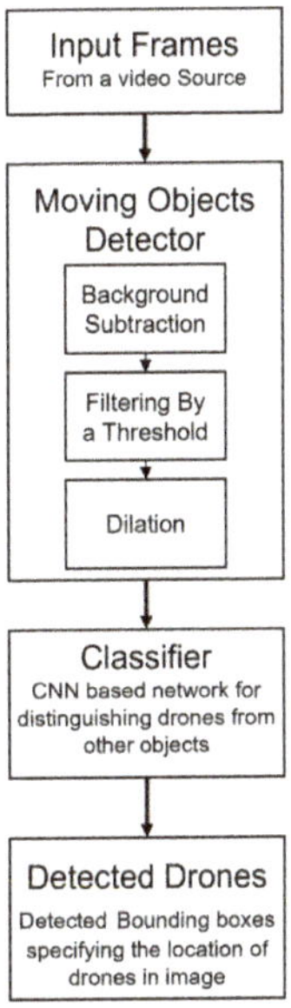

Figure 6. Proposed UAV detection algorithm used in electro-optical method [43].

Table 5. Summary of reviewed electro-optical sensor-based techniques for UAV characterization.

Ref.	Detection Technique	Functionalities			Performance
		Identification	Classification	Localization/ Tracking	
[42]	3D LADAR sensor, 3D background subtraction, V-RBNN	√	-	√	Detection Range 2 km
[44]	Combination of: EO/IR, All-sky, and acoustic cues	√	-	√	Line of sight limitation
[41]	Real stream detection, Differential method	√	-	√	-
[45]	DVS camera, Temporal filtering, Triangulation	√	-	√	Accurate Detection range 30 m
[43]	Background subtraction, CNN's	√	-	-	Moving Background dependency

The performance of vision-based solutions becomes poor with no LOS (angle of camera), bad quality of lenses, in foggy, dark, and dusty environments (weather conditions), and background temperature [9,10,21,63]. The aforementioned limitations could be addressed to some extent by using an IR camera, i.e., detection based on drone component heat, but that increases the system cost significantly and limits the detection range and environment due to the sensibility of the sensors that measure the thermal difference between the drone and the background [14,63].

3.5. Hybrid Fusion Systems

The hybrid fusion of multiple cues, such as radio frequency, radar, acoustic, and visual sensors, improves the performance of detection, classification, and localization of both the drone and its controller. Jovanoska et al. [47] suggested an array of sensors to collect the detected drone's data to be fed to a fusion engine for further analysis using the multiple hypothesis tracker (MHT) techniques, as illustrated in Figure 7. The RF signal is received and processed to compute the detected drone's DOA for drone localization. The captured signals from the integrated acoustic sensors are filtered to remove unwanted noise. After identifying and detecting the drone signatures, the coherent broadband beamforming technique is used to recognize the drone bearing angle and reduce its error by the two-step filter. Finally, the extracted DOA and bearing angle are referred to by the fusion engine for localization purposes. Finally, the GSM passive radar [41] is used for UAV detection and localization, and its output is fed to the fusion engine of the overall system. Combining all these technologies improved the system performance and enhanced the localization accuracy.

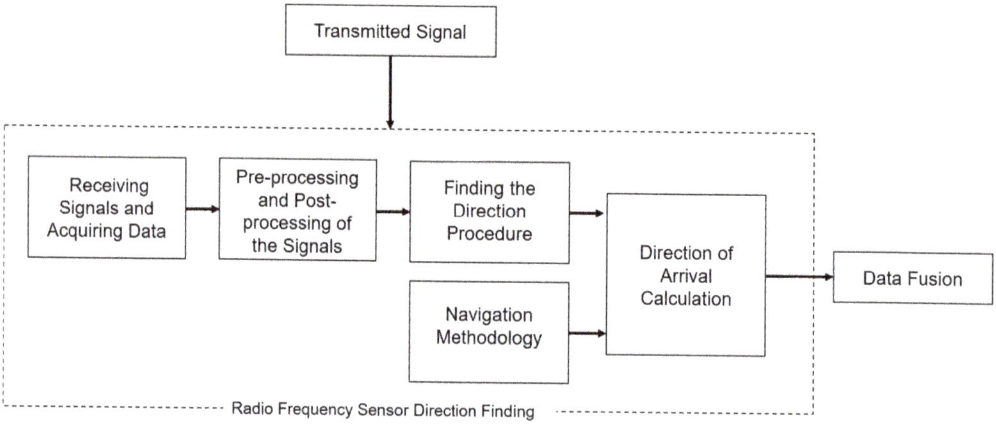

Figure 7. RF sensor direction finding reported in [47].

In [40], the proposed UAV detection and localization system relies on time delay and beamforming of the collected acoustic signal from a set of microphones. Acoustic signal characteristics with such signal processing are used to find the DOA of the UAV's detected recorded signal. Furthermore, Kalman filtering is used to improve the UAV's trajectory. The system is designed to identify and track the RF signal emitted by portable RF devices [48]. The system consists of two parts: (1) RF signal acquisitions achieved by an antenna array followed by a Nyquist ADC converter and (2) signal processing. The RF signal from the first part is passed into FFT to measure the DOA. The DOA is passed into the digital bandpass filter to measure the TDOA, which is used together with the DOA to estimate the location. AOA calculated from the DOA, the location, and past tracking information are used for tracking the drone's position [40].

3.6. Comparison of Detection Technologies

Earlier sections have discussed different techniques for detecting, identifying, and localizing UAVs. Each technique's performance varies according to equipment complexity and cost, coverage range and distance, operation efficiency, accuracy and precision measurements, etc. Table 6 summarizes the techniques cited in this study with their main features and affected factors. Combining the different techniques and integrating different sensors can increase the accuracy and reliability of the UAV detection systems, reduce the possibility of errors, and improve the system's ability to adapt.

Table 6. Summary of all reviewed UAV detection techniques.

Detection Technique	Summary	Limitations	Ref.
Radio Frequency	Real-time analysis for the detected radio communication between UAV and its controller. However, it does not apply to autonomous UAV detection. Low cost and simple architecture and elements: Antennas, Processors, RF sensors. Power and sensitivity of each affect detection system performance and accuracy. Common frequency bands are around 2.4 and 5 GHz Covering a long detection range will perform more efficiently in the less congested RF zones. Referring to RF datasets and integrating with machine learning algorithms are advanced ways to enhance detection, localization, and precise classification.	The RF-based detection technique applies only if the UAV is remotely controlled.	[7–20]
Radar	Transmitting radio signals, then receiving and analyzing the reflection/backscattering/echo radar signals. UAV's detection, tracking (Doppler-based), classification, and localization are based on the analysis of the reflected radio signal. Active sensor (Radar) and data processing modules with high-range detection and accurate localization. Machine learning algorithms and techniques' integration for better performance and results. Less noise and applicable in different weather conditions (fog, dust, rain, etc.). UAVs with small radar cross-sections are difficult to be identified and classified.	UAVs generally have limited Radar Cross Sections similar to birds or pedestrians. The amount of false positives remains high and low-RCS limits the detection range of the radar, especially X-band Radars.	[21–29]
Acoustic	Analyze acoustic signals coming from UAV's engine or propeller blades. Acoustic sensors/microphones arrays combined with data acquisition and signal processing modules Acoustic fingerprint analysis, features extraction, classification, and localization UAV's identification and distinction from other objects Effective in a short distance, however, it's affected by the nearby noise sources and weather. Acoustic dataset and Machine learning techniques integration for higher performance (detection and classification).	The detection of acoustic noise emitted by UAVs is low; thus, the acoustic technique requires a network of sensors deployed around sensitive places.	[30–40]
Electro-optic	Imaging and motion line of sight detection. High-cost equipment Ability to track autonomous UAVs. Controlling false alarms with advanced integration with other methods/algorithms/machine learning. Detection performance can vary with different environmental conditions and weather.	Using different electro optics is required, and the fusion of video streams is required to cope with UAVs' environment and type/size. This increases the cost of the solution.	[41–46]

4. Drone Controller Detection and Localization

Once UAVs are detected, the detection and localization of the drone controller are implemented to monitor their communication and limit illegal use.

UAV detection systems differ according to the technologies and functions performed, such as identification, classification, tracking, localization, interdiction, destruction, and damage. Technology and functions are selected and implemented based on the main requirements of the UAV detection system. In this section, some detection systems that support localization functionality are reviewed.

The process of locating and positioning a UAV is mainly based on collecting the direct measurements of the detected UAV and its emitted signals. These direction measurements and other extracted features are calculated and utilized in the UAV detection system to estimate the geolocation of the UAV. The computed geolocation parameters for the direction-finding methods include angle of arrival (AOA) [40], time of arrival (TOA) [64], direction of arrival (DOA) [18,47], frequency difference of arrival (FDOA) [51], time difference of arrival (TDOA) [11,35], and received signal strength (RSS) [65].

The proposed system in [12] utilizes a low-cost passive RF-based UAV detection and localization method. The system computes AOA for the RF-based signal to determine whether the transmitted signals' peaks correspond to the UAV or its controller. Then, it uses the triangulation technique to estimate the location of RF signal peak sources. The free-space path loss model and triangulation combination is reported in [50] to detect and localize a stationary drone controller. The proposed system contains two direction-finding systems for direction identification and localization for the drone and its controller. Each direction-finding system has an omnidirectional antenna for detecting drone signal occurrence and a mechanically agile directional antenna for directions identification and localization of RF signal peaks for UAV and/or its controller signals. The whole system is depicted in Figure 8.

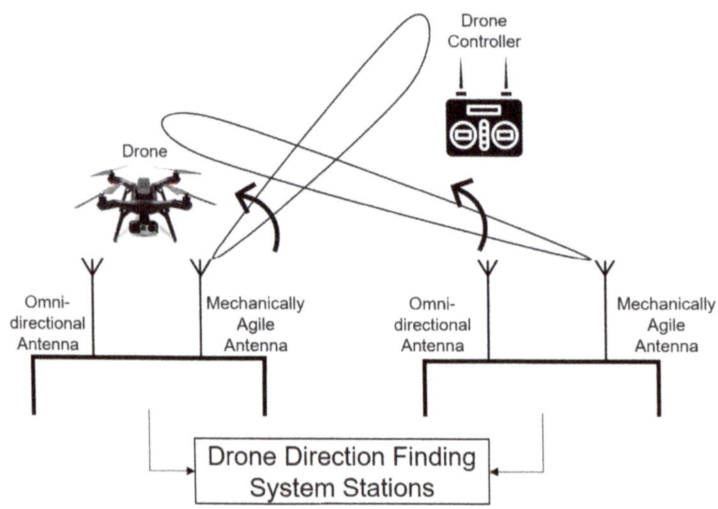

Figure 8. Diagram of direction finding mechanism in the proposed system [12].

The study [12] discussed that distinguishing between RF signals from the UAV and its controller from other RF signals in the surrounding area poses a challenge in drone controller localization. The reported direction-finding station consists of two modules: drone signal analysis to classify drone and remote controller (RC) signals and the direction-finding module. Each direction-finding station extracts acute parameters from detected RF signals and uses a mechanical steering antenna for identification and localization. The precision of the direction-finding function is dependent and affected by the antenna's directivity and gain, drone velocity, scanning velocity, and beam width of the used directional antenna.

The reported detection system in [18] employs frequency hopping spread spectrum (FHSS) to detect and locate UAVs and RCs. The cyclostationarity analysis algorithm is used to identify the FHSS-type drone RC signals and differentiate them from other background signals operating in the same frequency band. After the successful classification of the drone RC signals, STFT and additional re-sampling processing are applied to enhance the detection accuracy of the reconstructed RC signal. Finally, the direction-finding phase is achieved by implementing the subspace algorithms to identify the AOA of the FHSS drone RC signal. The proposed system [49] utilizes a set of a uniform linear array of quasi-Yagi antennas in the experimental setup to enhance the precision of the direction-finding function.

5. Conclusions

This study has reviewed the most recent techniques for UAV/drone and its controller detection, classification, and localization. Cost-effectiveness, precision, accuracy, reliability, and real-time processing are among the factors considered while developing UAV detection systems. After discussing the high-level architecture of UAVs and security concerns, a comprehensive review of radio frequency, radar, acoustic, electro-optic, and hybrid systems for UAV detection is presented. The UAV detection systems employ different algorithms and techniques depending on the applications for detecting, classifying, locating, tracking, and alerting. To address the challenges, meet market needs, and improve reliability, employing a hybrid fusion of multiple cues, such as radio frequency, radar, acoustic, and visual sensors, can enhance detection performance.

Author Contributions: Conceptualization, J.Y. and M.G.; methodology, J.Y., M.A. (Marah Alhalabi), M.Y., T.B., E.A.S. and M.G.; validation, J.Y., H.Z., M.A. (Marah Alhalabi), M.Y., T.B., E.A.S., A.G. and M.G.; formal analysis, J.Y., H.Z., M.A. (Marah Alhalabi), M.Y., T.B., E.A.S., A.G. and M.G.; investigation, J.Y., H.Z., M.A. (Marah Alhalabi), M.Y., T.B., E.A.S., A.G. and M.G.; resources, J.Y., H.Z., E.A.S., M.A. (Mohammad Alkhedher) and M.G.; data curation, J.Y., H.Z., M.A. (Marah Alhalabi), M.Y., T.B., E.A.S., A.G. and M.G.; writing—original draft preparation, J.Y., E.A.S. and M.G.; writing—review and editing, J.Y., H.Z., M.A. (Marah Alhalabi), M.Y., T.B., E.A.S., A.G, M.A. (Mohammad Alkhedher) and M.G.; visualization, J.Y., M.A. (Marah Alhalabi), M.Y., T.B., E.A.S. and A.G.; supervision, J.Y., M.A. (Mohammad Alkhedher) and M.G., project administration, J.Y. and M.G.; funding acquisition, J.Y., M.A. (Mohammad Alkhedher) and M.G. All authors have read and agreed to the published version of the manuscript.

Funding: This paper is part of a research project funded by the Office of Research and Sponsored Programs (ORSP) at Abu Dhabi University through a research fund (Grant number 19300564).

Institutional Review Board Statement: Not applicable.

Informed Consent Statement: Not applicable.

Data Availability Statement: Not applicable.

Conflicts of Interest: The authors declare no conflict of interest.

References

1. Tsouros, D.C.; Bibi, S.; Sarigiannidis, P.G. A review on UAV-based applications for precision agriculture. *Information* **2019**, *10*, 349. [CrossRef]
2. Liu, P.; Chen, A.Y.; Huang, Y.-N.; Han, J.-Y.; Lai, J.-S.; Kang, S.-C.; Wu, T.-H.; Wen, M.-C.; Tsai, M.-H. A review of rotorcraft Unmanned Aerial Vehicle (UAV) developments and applications in civil engineering. *Smart Struct. Syst.* **2014**, *13*, 1065–1094. [CrossRef]
3. Shahmoradi, J.; Talebi, E.; Roghanchi, P.; Hassanalian, M. A Comprehensive Review of Applications of Drone Technology in the Mining Industry. *Drones* **2020**, *4*, 34. [CrossRef]
4. Kannadaguli, P. YOLO v4 Based Human Detection System Using Aerial Thermal Imaging for UAV Based Surveillance Applications. In Proceedings of the International Conference on Decision Aid Sciences and Application (DASA), Sakheer, Bahrain, 8–9 November 2020; pp. 1213–1219.
5. Utsav, A.; Abhishek, A.; Suraj, P.; Badhai, R.K. An IoT Based UAV Network for Military Applications. In Proceedings of the 6th International Conference on Wireless Communications, Signal Processing and Networking (WiSPNET), Chennai, India, 25–27 March 2021; pp. 122–125.
6. Whelan, J.; Almehmadi, A.; Braverman, J.; El-Khatib, K. Threat Analysis of a Long Range Autonomous Unmanned Aerial System. In Proceedings of the 2020 International Conference on Computing and Information Technology (ICCIT-1441), Tabuk, Saudi Arabia, 9–10 September 2020; pp. 1–5.
7. Nie, W.; Han, Z.C.; Zhou, M.; Xie, L.B.; Jiang, Q. UAV Detection and Identification Based on WiFi Signal and RF Finger-print. *IEEE Sens. J.* **2021**, *21*, 13540–13550. [CrossRef]
8. Bisio, I.; Garibotto, C.; Lavagetto, F.; Sciarrone, A.; Zappatore, S. Unauthorized Amateur UAV Detection Based on WiFi Statistical Fingerprint Analysis. *IEEE Commun. Mag.* **2018**, *56*, 106–111. [CrossRef]

9. Ezuma, M.; Erden, F.; Anjinappa, C.K.; Ozdemir, O.; Guvenc, I. Micro-UAV detection and classification from RF finger-prints using machine learning techniques. In Proceedings of the IEEE Aerospace Conference, Big Sky, MT, USA, 2–9 March 2019; pp. 1–13.
10. Abeywickrama, S.; Jayasinghe, L.; Fu, H.; Nissanka, S.; Yuen, C. RF-based Direction Finding of UAVs Using DNN. In Proceedings of the IEEE International Conference on Communication Systems (ICCS), Chengdu, China, 19–21 December 2018; pp. 157–161.
11. Bhattacherjee, U.; Ozturk, E.; Ozdemir, O.; Guvenc, I.; Sichitiu, M.L.; Dai, H. Experimental Study of Outdoor UAV Localization and Tracking using Passive RF Sensing. In Proceedings of the 15th ACM Workshop on Wireless Network Testbeds, Experimental evaluation & CHaracterization, New Orleans, LA, USA, 31 January–4 February 2022; pp. 31–38.
12. Nguyen, P.; Kim, T.; Miao, J.; Hesselius, D.; Kenneally, E.; Massey, D.; Frew, E.; Han, R.; Vu, T. Towards RF-based Localization of a Drone and Its Controller. In Proceedings of the 5th Workshop on Micro Aerial Vehicle Networks, Systems, and Applications, Seoul, Republic of Korea, 21 June 2019.
13. Nemer, I.; Sheltami, T.; Ahmad, I.; Yasar, A.U.-H.; Abdeen, M.A.R. RF-Based UAV Detection and Identification Using Hierarchical Learning Approach. *Sensors* **2021**, *21*, 1947. [CrossRef]
14. Ozturk, E.; Erden, F.; Guvenc, I. RF-based low-SNR classification of UAVs using convolutional neural networks. *ITU J. Futur. Evol. Technol.* **2021**, *2*, 39–52. [CrossRef]
15. Alipour-Fanid, A.; Dabaghchian, M.; Wang, N.; Wang, P.; Zhao, L.; Zeng, K. Machine Learning-Based Delay-Aware UAV Detection and Operation Mode Identification Over Encrypted Wi-Fi Traffic. *IEEE Trans. Inf. Forensics Secur.* **2019**, *15*, 2346–2360. [CrossRef]
16. Lv, H.; Liu, F.; Yuan, N. Drone Presence Detection by the Drone's RF Communication. *J. Phys. Conf. Ser.* **2021**, *1738*, 12044. [CrossRef]
17. Flórez, J.; Ortega, J.; Betancourt, A.; García, A.; Bedoya, M.; Botero, J.S. A review of algorithms, methods, and techniques for detecting UAVs and UAS using audio, radiofrequency, and video applications. *TecnoLógicas* **2020**, *23*, 262–278. [CrossRef]
18. Kaplan, B.; Kahraman, İ.; Ektı, A.R.; Yarkan, S.; Görçın, A.; Özdemır, M.K.; Çirpan, H.A. Detection, Identification, and Direction of Arrival Estimation of Drone FHSS Signals with Uniform Linear Antenna Array. *IEEE Access* **2021**, *9*, 152057–152069. [CrossRef]
19. Ezuma, M.; Erden, F.; Anjinappa, C.K.; Ozdemir, O.; Guvenc, I. Detection and Classification of UAVs Using RF Fingerprints in the Presence of Wi-Fi and Bluetooth Interference. *IEEE Open J. Commun. Soc.* **2019**, *1*, 60–76. [CrossRef]
20. Zhang, H.; Cao, C.; Xu, L.; Gulliver, T.A. A UAV Detection Algorithm Based on an Artificial Neural Network. *IEEE Access* **2018**, *6*, 24720–24728. [CrossRef]
21. Coluccia, A.; Parisi, G.; Fascista, A. Detection and Classification of Multirotor Drones in Radar Sensor Networks: A Review. *Sensors* **2020**, *20*, 4172. [CrossRef]
22. Bouzayene, I.; Mabrouk, K.; Gharsallah, A.; Kholodnyak, D. Scan Radar Using an Uniform Rectangular Array for Drone Detection with Low RCS. In Proceedings of the 2019 IEEE 19th Mediterranean Microwave Symposium (MMS), Hammamet, Tunisia, 31 October–2 November 2019; pp. 1–4. [CrossRef]
23. Martinez, J.; Kopyto, D.; Schutz, M.; Vossiek, M. Convolutional Neural Network Assisted Detection and Localization of UAVs with a Narrowband Multi-site Radar. In Proceedings of the IEEE 19th Mediterranean Microwave Symposium (MMS), Hammamet, Tunisia, 31 October–2 November 2019; pp. 1–4. [CrossRef]
24. Tang, L.; Wang, H.; Feng, Z.; Xu, D.; Wang, Y.; Quan, S.; Xu, W. Small Phased Array Radar Based on AD9361 For UAV Detection. In Proceedings of the IEEE MTT-S International Microwave Biomedical Conference (IMBioC), Nanjing, China, 6–8 May 2019; pp. 1–3.
25. Zhang, P.; Yang, L.; Chen, G.; Li, G. Classification of drones based on micro-Doppler signatures with dual-band radar sensors. In Proceedings of the Progress in Electromagnetics Research Symposium-Fall (PIERS-FALL), Singapore, 19–22 November 2017; pp. 638–643. [CrossRef]
26. Zhao, J.; Fu, X.; Yang, Z.; Xu, F. Radar-Assisted UAV Detection and Identification Based on 5G in the Internet of Things. *Wirel. Commun. Mob. Comput.* **2019**, *2019*, 2850263. [CrossRef]
27. Yang, J.; Lu, X.; Dai, Z.; Yu, W.; Tan, K. A Cylindrical Phased Array Radar System for UAV Detection. In Proceedings of the 6th International Conference on Intelligent Computing and Signal Processing (ICSP), Xi'an, China, 9–11 April 2021; pp. 894–898.
28. Huizing, A.; Heiligers, M.; Dekker, B.; de Wit, J.; Cifola, L.; Harmanny, R. Deep Learning for Classification of Mini-UAVs Using Micro-Doppler Spectrograms in Cognitive Radar. *IEEE Aerosp. Electron. Syst. Mag.* **2019**, *34*, 46–56. [CrossRef]
29. Wang, L.; Tang, J.; Liao, Q. A Study on Radar Target Detection Based on Deep Neural Networks. *IEEE Sens. Lett.* **2019**, *3*, 1–4. [CrossRef]
30. Chang, X.; Yang, C.; Wu, Z.; Wu, J. An Acoustic-Based Surveillance System for Amateur Drones Detection and Localization. *IEEE Trans. Veh. Technol.* **2020**, *69*, 2731–2739.
31. Dumitrescu, C.; Minea, M.; Costea, I.M.; Chiva, I.C.; Semenescu, A. Development of an Acoustic System for UAV Detection †. *Sensors* **2020**, *20*, 4870. [CrossRef]
32. Kartashov, V.; Oleynikov, V.; Koryttsev, I.; Sheiko, S.; Zubkov, O.; Babkin, S.; Selieznov, I. Use of acoustic signature for detection, recognition and direction finding of small unmanned aerial vehicles. In Proceedings of the IEEE 15th International Conference on Advanced Trends in Radioelectronics, Telecommunications and Computer Engineering (TCSET), Lviv-Slavske, Ukraine, 25–29 February 2020; pp. 1–4.

33. Sedunov, A.; Haddad, D.; Salloum, H.; Sutin, A.; Sedunov, N.; Yakubovskiy, A. Stevens Drone Detection Acoustic System and Experiments in Acoustics UAV Tracking. In Proceedings of the IEEE International Symposium on Technologies for Homeland Security (HST), Woburn, MA, USA, 5–6 November 2019; pp. 1–7. [CrossRef]
34. Sedunov, A.; Salloum, H.; Sutin, A.; Sedunov, N. Long-term testing of acoustic system for tracking low-flying aircraft. In Proceedings of the IEEE International Symposium on Technologies for Homeland Security (HST), Woburn, MA, USA, 23–24 October 2018; pp. 1–6.
35. Sedunov, A.; Salloum, H.; Sutin, A.; Sedunov, N.; Tsyuryupa, S. UAV passive acoustic detection. In Proceedings of the IEEE International Symposium on Technologies for Homeland Security (HST), Woburn, MA, USA, 23–24 October 2018; pp. 1–6.
36. Utebayeva, D.; Almagambetov, A.; Alduraibi, M.; Temirgaliyev, Y.; Ilipbayeva, L.; Marxuly, S. Multi-label UAV sound classification using Stacked Bidirectional LSTM. In Proceedings of the 4th IEEE International Conference on Robotic Computing (IRC), Taichung, Taiwan, 9–11 November 2020; pp. 453–458.
37. Yang, B.; Matson, E.T.; Smith, A.H.; Dietz, J.E.; Gallagher, J.C. UAV Detection System with Multiple Acoustic Nodes Using Machine Learning Models. In Proceedings of the 3rd IEEE International Conference on Robotic Computing (IRC), Naples, Italy, 25–27 February 2019; pp. 493–498. [CrossRef]
38. Anwar, M.Z.; Kaleem, Z.; Jamalipour, A. Machine Learning Inspired Sound-Based Amateur Drone Detection for Public Safety Applications. *IEEE Trans. Veh. Technol.* **2019**, *68*, 2526–2534. [CrossRef]
39. Seo, Y.; Jang, B.; Im, S. Drone Detection Using Convolutional Neural Networks with Acoustic STFT Features. In Proceedings of the 15th IEEE International Conference on Advanced Video and Signal Based Surveillance (AVSS), Auckland, New Zealand, 27–30 November 2018; pp. 1–6.
40. Blanchard, T.; Thomas, J.-H.; Raoof, K. Acoustic localization and tracking of a multi-rotor unmanned aerial vehicle using an array with few microphones. *J. Acoust. Soc. Am.* **2020**, *148*, 1456–1467. [CrossRef] [PubMed]
41. Kartashov, V.; Oleynikov, V.; Zubkov, O.; Sheiko, S. Optical Detection of Unmanned Air Vehicles on a Video Stream in a Real-Time. In Proceedings of the International Conference on Information and Telecommunication Technologies and Radio Electronics (UkrMiCo), Odessa, Ukraine, 9–13 September 2019; pp. 1–4.
42. Kim, B.H.; Khan, D.; Bohak, C.; Choi, W.; Lee, H.J.; Kim, M.Y. V-RBNN Based Small Drone Detection in Augmented Datasets for 3D LADAR System. *Sensors* **2018**, *18*, 3825. [CrossRef] [PubMed]
43. Seidaliyeva, U.; Akhmetov, D.; Ilipbayeva, L.; Matson, E.T. Real-Time and Accurate Drone Detection in a Video with a Static Background. *Sensors* **2020**, *20*, 3856. [CrossRef] [PubMed]
44. Siewert, S.B.; Andalibi, M.; Bruder, S.; Rizor, S. Slew-to-Cue Electro-Optical and Infrared Sensor Network for small UAS Detection, Tracking and Identification. In Proceedings of the AIAA Scitech Forum, San Diego, CA, USA, 7–11 January 2019; p. 2264.
45. Stuckey, H.; Al-Radaideh, A.; Escamilla, L.; Sun, L.; Carrillo, L.G.; Tang, W. An Optical Spatial Localization System for Tracking Unmanned Aerial Vehicles Using a Single Dynamic Vision Sensor. In Proceedings of the IEEE/RSJ International Conference on Intelligent Robots and Systems (IROS), Prague, Czech Republic, 27 September–1 October 2021; pp. 3093–3100. [CrossRef]
46. Thai, V.-P.; Zhong, W.; Pham, T.; Alam, S.; Duong, V. Detection, Tracking and Classification of Aircraft and Drones in Digital Towers Using Machine Learning on Motion Patterns. In Proceedings of the Integrated Communications, Navigation and Surveillance Conference (ICNS), Herndon, VA, USA, 9–11 April 2019; pp. 1–8. [CrossRef]
47. Jovanoska, S.; Knoedler, B.; Palanivelu, D.P.; Still, L.; Fiolka, T.; Oispuu, M.; Steffes, C.; Koch, W. Passive Sensor Processing and Data Fusion for Drone Detection. In Proceedings of the NATO STO Meeting Proceedings: MSG-SET-183 Specialists' Meeting on Drone Detectability: Modelling the Relevant Signature, Prague, Czech Republic, 27–29 April 2021; p. 16.
48. Daponte, P.; De Vito, L.; Picariello, F.; Rapuano, S.; Tudosa, I. Compressed Sensing Technologies and Challenges for Aerospace and Defense RF Source Localization. In Proceedings of the 5th IEEE International Workshop on Metrology for AeroSpace (MetroAeroSpace), Rome, Italy, 20–22 June 2018; pp. 634–639. [CrossRef]
49. Yousef, M.; Iqbal, F. Drone forensics: A case study on a DJI Mavic Air. In Proceedings of the IEEE/ACS 16th International Conference on Computer Systems and Applications (AICCSA), Abu Dhabi, United Arab Emirates, 3–7 November 2019; pp. 1–3
50. Huang, X.; Yan, K.; Wu, H.C.; Wu, Y. Unmanned Aerial Vehicle Hub Detection Using Software-Defined Radio. In Proceedings of the IEEE International Symposium on Broadband Multimedia Systems and Broadcasting (BMSB), Jeju, Republic of Korea, 5–7 June 2019; pp. 1–6.
51. Wang, Y.; Wang, W.; Zhang, X.; Wu, L.; Yin, H. The Joint Phantom Track Deception and TDOA/FDOA Localization Using UAV Swarm without Prior Knowledge of Radars' Precise Locations. *Electronics* **2022**, *11*, 1577. [CrossRef]
52. Zhao, Y.; Su, Y. The Extraction of Micro-Doppler Signal with EMD Algorithm for Radar-Based Small UAVs' Detection. *IEEE Trans. Instrum. Meas.* **2019**, *69*, 929–940. [CrossRef]
53. Choi, B.; Oh, D. Classification of Drone Type Using Deep Convolutional Neural Networks Based on Micro- Doppler Simulation. In Proceedings of the International Symposium on Antennas and Propagation (ISAP), Busan, Republic of Korea, 23–26 October 2018; pp. 1–2.
54. Kim, B.K.; Kang, H.-S.; Park, S.-O. Drone Classification Using Convolutional Neural Networks with Merged Doppler Images. *IEEE Geosci. Remote. Sens. Lett.* **2016**, *14*, 38–42. [CrossRef]
55. Mazumder, J.; Raj, A.B. Detection and Classification of UAV Using Propeller Doppler Profiles for Counter UAV Systems. In Proceedings of the5th International Conference on Communication and Electronics Systems (ICCES), Coimbatore, India, 10–12 June 2020; pp. 221–227. [CrossRef]

56. Li, S.; Chai, Y.; Guo, M.; Liu, Y. Research on Detection Method of UAV Based on micro-Doppler Effect. In Proceedings of the 39th Chinese Control Conference (CCC), Shenyang, China, 27–29 July 2020. [CrossRef]
57. Oh, B.-S.; Guo, X.; Wan, F.; Toh, K.-A.; Lin, Z. Micro-Doppler Mini-UAV Classification Using Empirical-Mode Decomposition Features. *IEEE Geosci. Remote. Sens. Lett.* **2017**, *15*, 227–231. [CrossRef]
58. Ren, J.; Jiang, X. Regularized 2-D complex-log spectral analysis and subspace reliability analysis of micro-Doppler signature for UAV detection. *Pattern Recognit.* **2017**, *69*, 225–237. [CrossRef]
59. Basak, S.; Scheers, B. Passive radio system for real-time drone detection and DoA estimation. In Proceedings of the International Conference on Military Communications and Information Systems (ICMCIS), Warsaw, Poland, 22–23 May 2018; pp. 1–6.
60. Solomitckii, D.; Gapeyenko, M.; Semkin, V.; Andreev, S.; Koucheryavy, Y. Technologies for Efficient Amateur Drone Detection in 5G Millimeter-Wave Cellular Infrastructure. *IEEE Commun. Mag.* **2018**, *56*, 43–50. [CrossRef]
61. Han, S.; Huang, Y.; Meng, W.; Li, C.; Xu, N.; Chen, D. Optimal Power Allocation for SCMA Downlink Systems Based on Maximum Capacity. *IEEE Trans. Commun.* **2018**, *67*, 1480–1489. [CrossRef]
62. Han, S.; Zhang, Y.; Meng, W.; Chen, H.-H. Self-Interference-Cancelation-Based SLNR Precoding Design for Full-Duplex Relay-Assisted System. *IEEE Trans. Veh. Technol.* **2018**, *67*, 8249–8262. [CrossRef]
63. Chiper, F.-L.; Martian, A.; Vladeanu, C.; Marghescu, I.; Craciunescu, R.; Fratu, O. Drone Detection and Defense Systems: Survey and a Software-Defined Radio-Based Solution. *Sensors* **2022**, *22*, 1453. [CrossRef] [PubMed]
64. Siva, J.; Poellabauer, C. Robot and Drone Localization in GPS-Denied Areas. In *Mission-Oriented Sensor Networks and Systems: Art and Science*; Springer: Berlin/Heidelberg, Germany, 2019; pp. 597–631. [CrossRef]
65. Sinha, P.; Yapici, Y.; Guvene, I.; Turgut, E.; Gursoy, M.C. RSS-Based Detection of Drones in the Presence of RF Interferers. In Proceedings of the IEEE 17th Annual Consumer Communications & Networking Conference (CCNC), Las Vegas, NV, USA, 10–13 January 2020; pp. 1–6. [CrossRef]

applied sciences

MDPI

Article

Evaluating Levels of Automation in Human–Robot Collaboration at Different Workload Levels

Dana Gutman, Samuel Olatunji * and Yael Edan

Department of Industrial Engineering and Management, Ben-Gurion University of the Negev, P.O. Box 653, Be'er Sheva 8410501, Israel; Danagut@post.bgu.ac.il (D.G.); yael@bgu.ac.il (Y.E.)
* Correspondence: olatunji@post.bgu.ac.il

Abstract: This study explored how levels of automation (LOA) influence human robot collaboration when operating at different levels of workload. Two LOA modes were designed, implemented, and evaluated in an experimental collaborative assembly task setup for four levels of workload composed of a secondary task and task complexity. A user study conducted involving 80 participants was assessed through two constructs especially designed for the evaluation (quality of task execution and usability) and user preferences regarding the LOA modes. Results revealed that the quality of task execution and usability was better at high LOA for low workload. Most of participants also preferred high LOA when the workload increases. However, when complexity existed within the workload, most of the participants preferred the low LOA. The results reveal the benefits of high and low LOA in different workload situations. This study provides insights related to shared control designs and reveals the importance of considering different levels of workload as influenced by secondary tasks and task complexity when designing LOA in human–robot collaborations.

Keywords: human–robot collaboration; assembly task; user studies; user preferences; quality of task execution; usability

Citation: Gutman, D.; Olatunji, S.; Edan, Y. Evaluating Levels of Automation in Human–Robot Collaboration at Different Workload Levels. *Appl. Sci.* **2021**, *11*, 7340. https://doi.org/10.3390/app11167340

Academic Editors: Luis Gracia, Carlos Perez-Vidal and Manuel Armada

Received: 30 May 2021
Accepted: 6 August 2021
Published: 10 August 2021

1. Introduction

Human–robot collaboration (HRC) involves one or more humans working with one or more robots to accomplish a certain task or a specific goal [1]. Significant research has focused on interaction aspects for designing robotic systems for use by or with humans [2–6]. This research, which focuses on factors that affect HRC [1,7] at different levels of automation, specifically evaluates the influence of workload.

The level of automation (LOA) of the system, defined as the degree to which the robot and the human are involved in the collaborative task [8–11], influences the characteristics of the dynamics of the collaboration, the behavior of the robots, actions to be taken, as well as autonomy of the human in the collaboration [12,13]. Workload addresses the actual and perceived amount of work that the human operator experiences as related to the effort invested in the task [14,15]. It can be described in terms of the elements that constitute the cost of accomplishing the goal for the human operator in the HRC [16]. These elements could be task-related (such as mental, temporal, and physical demands [17], operator-related (such as skill, strategy, experience [18]) or machine-related (such as poorly designed controls, feedback, inappropriate, or inadequate automation [15]. Workload consequences could be reflected in the stress, fatigue or frustration experienced by the human operator [16], depletion of attentional, cognitive or response resources [15], as well as in performance changes [19]. Workload can also be influenced by task complexity as characterized in terms of the stimuli involved in the task for inputs, as well as the behavioral requirements the human operator should emit in order to achieve a specific level of performance [20]. It could depend on the objective complexity derived from the task properties and on the subjective complexity which is influenced by the human operator's perception [21]. The task properties include the component complexity—number of distinct actions that the human operator must execute or number of informational cues that should be processed (e.g., the number and type of subtasks to be managed, [22]); coordinative complexity—nature of relationships between task inputs and task products, the strength of these relationships as well as the sequencing of inputs (e.g., timing, frequency, intensity and location requirements [23]), and dynamic complexity—changes in the states of the environment which the human operator should adapt to [20,24].

The influence of LOA on HRC has been intensively investigated [25]. However, there are limited studies that investigated factors influencing workload in relation to the design of LOA modes suitable for different HRC collaboration contexts [26]. Moreover, research has revealed that the alignment between manufacturing strategy and automation decisions are often ad hoc in nature [27]. The current study therefore aims to examine the influence of different levels of workload when operating at different levels of automation (LOA) in a human–robot collaborative system. This is important when introducing robotics in real life situations.

To evaluate the overall performance and interaction in such HRC contexts, many different measures are commonly applied for the assessment [22,28–30]. However, by evaluating each measure separately, a holistic evaluation is lacking. We therefore specially designed two constructs that compile different evaluation measures. These constructs are useful in assessing the preferences, performance, and perception of the users regarding various aspects of the collaboration with the robot as required in a user-centered design [31–33]. The constructs are quality of task (QoT) execution (the user's performance aspects) and usability (performance aspects along with other user perception aspects such as perceived ease of use). Additionally, user preferences were evaluated.

We design, implement and evaluate LOA modes in a user study involving 80 participants working at different workload conditions. Section 2 presents the study hypotheses, system design, LOA modes, task, and experimental evaluations of the design. Section 3 is devoted to the experimental results. Discussion is presented in Section 4 while Conclusions and suggestions for future work are discussed in the last section.

2. Materials and Methods

2.1. Experimental System

The experimental system included a 4 degree of freedom DOBOT Magician robotic arm (https://www.dobot.cc/dobot-magician/product-overview.html, accessed on 30 May 2021) equipped with a suction gripper, user interface (presented on a computer), cubes to be assembled and the human operator (Figure 1). The DOBOT Magician (135 mm high, 158 mm wide with a 320 mm radius and 500 g payload) connects to the computer through a USB connection and was programmed for the two LOA modes using the Python programming language.

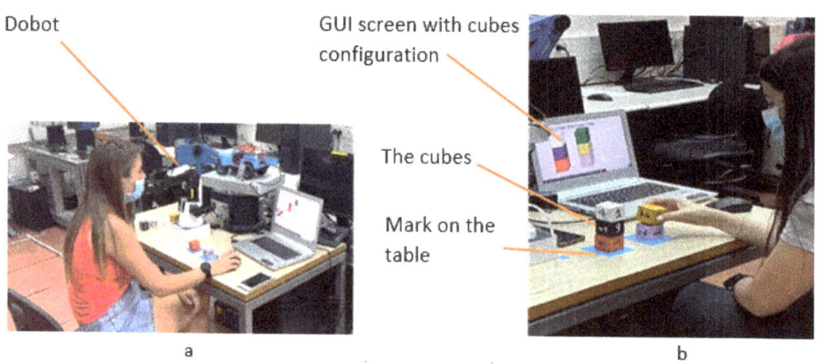

Figure 1. The experimental system.

The HRC assembly task simulates a work scenario where participants are expected to assemble blocks made from cubes brought to them by a robot according to a configuration presented to them through a user interface. The task was performed in two LOA modes, at four workload levels. The workload levels, detailed below, are composed of different combinations of a secondary task and task complexity.

The user communicates with the robot through a user interface implemented on a GUI screen (Figure 2). This was designed to be friendly to promote ease of use as the human interacts with the robot through the GUI [34–36]. The configuration to be assembled is displayed on the GUI screen when starting the task. The robot brings the cubes in a sequence one after another from a predetermined place according to the specific LOA the robot is operating in. The robot releases the cube when it reaches the front of the participant. The participants are expected to assemble the cubes when received from the robot and place these cubes in a marked area on the desk in front of them.

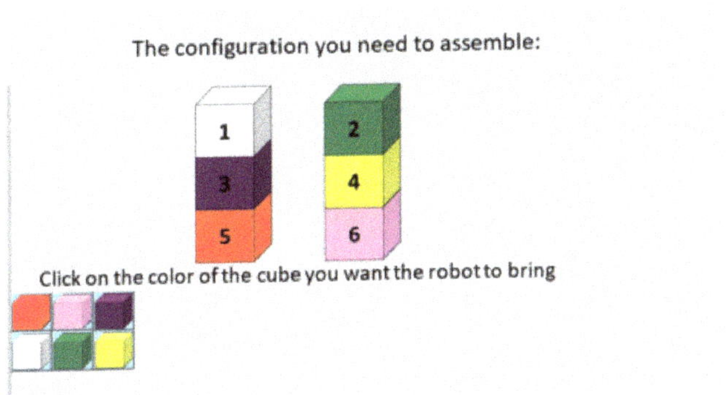

Figure 2. The GUI screen.

2.2. Design of the Experimental Conditions

2.2.1. Levels of Automation (LOA) Modes

The automation design focuses on the *decision* and *action* aspects of the overall process taken either by the robot or the user. This specifies the degree of control the user or robot in the decision of action(s) to be taken and the execution of the actions. It is conditioned in two levels for this study:

(a) Low LOA—the user has autonomy to select the type and order of cubes. The robot supports the user by bringing the type of cube the user selected via the user interface.
(b) High LOA—the robot has autonomy to bring the specific type of cube and in the order preprogrammed in its operation. The user simply demands for a cube through the user interface and the robot brings the type of cube suitable for the specific configuration assembled.

2.2.2. Levels of Workload

The workload design focuses mainly on the physical and cognitive workload induced through the selection of the right cubes to assemble in the minimum possible time. This is the main task. Workload is increased in two ways: through a secondary task and by increasing task complexity.

The secondary task influence was depicted through an off-the shelf well known cognitive game, the "RUSH HOUR" (https://www.thinkfun.com/products/rush-hour/, accessed on 30 May 2021) thinking game (Figure 3). It involves arranging toy cars in a way to get a specific car out of a gridlock. There are tabs at each stage showing how to arrange the cars and finding a way to get the required red car out at different stages.

Figure 3. "RUSH HOUR" game.

In the main task, where cubes are assembled, the default setting is that the cubes for the assembly differ only by color. The users are required to assemble the cubes to match particular configurations characterized by differences in color pattern (Figure 4a).

Figure 4. Sample of cubes configurations in (**a**) without complexity and (**b**) with complexity.

The task complexity influence was depicted by introducing the cubes for the assembly that differ in color and in numbers on a particular side (Figure 4b). The users are required to assemble the cubes in color patterns as done in the low task complexity condition, but in addition, they must ensure that the specific numbers on a particular color of cubes match the required configuration per time. The task complexity is increased by the additional information cue (presence of numbers) and their spatial consideration (position of the number in the configuration). It represents component and coordinative task complexity induced through the number and type of sub-actions to be performed while selecting the right cubes and assembling along with the coordination of the actions in the secondary task.

Four levels of workload were designed using these factors:

(a) Low workload (LWL)—the users perform only the main task, assembling cubes (without reference to the numbers on the cubes) to match the specific configuration required. The workload involves some physical demand of arranging the cubes, mental demand of thinking about the type of cube that would match the required configuration and some temporal demand related to completing the task in the shortest possible time.

(b) Medium workload 1 (MWL1)—the users perform only the main task of assembling the cubes but with reference to the numbers on the cubes. It depicts the LWL level with increased task complexity (or high workload without secondary task).

(c) Medium workload 2 (MWL2)—the users perform the main task of assembling (without references to the numbers on the cubes) simultaneously with the secondary task. It depicts the high workload level without complexity included (or the LWL with a secondary task).

(d) High workload (HWL)—the users perform the main task of assembling the cubes (with reference to the numbers on the cubes) along with a secondary task. This combines both secondary task and increased task complexity.

2.3. Experimental Design

The experimental design includes two independent variables: LOA and levels of workload. A between-within participant experimental design was conducted with the LOA as the within variable while level of workload was the between variable. Four groups were designed depicting the different levels of workload. Each participant was randomly assigned to one of the four groups and experienced both LOA modes (Table 1).

Table 1. Experimental design.

	Workload			
	Low Workload	**Medium Workload 1 Task Complexity**	**Medium Workload 2 Secondary Task**	**High Workload**
Level of Automation (LOA) — **Low LOA**	Condition 1a The **user chooses** via a GUI screen which color of cube the robot will bring him. The user concentrates only on the **main task, without** reference to the numbers written on the cubes.	Condition 2a The **user chooses** via a GUI screen which color of cube the robot will bring him. The user concentrates only on the **main task, which** has increased complexity (through the numbers written on the cubes).	Condition 3a The user chooses via a GUI screen which color of cube the robot will bring him. The user performs a **main + secondary task** simultaneously, **without** reference to the numbers written on the cubes.	Condition 4a The **user chooses** via a GUI screen which color of cube the robot will bring him. The user concentrates on performing a **main + secondary task** simultaneously, **with** an increased task complexity (must refer to the numbers written on the cubes).
Level of Automation (LOA) — **High LOA**	Condition 1b The **robot brings** the cubes to the user in a predefined order. The user concentrates only on the **main task, without** reference to the numbers written on the cubes.	Condition 2b The **robot brings** the cubes to the user in a predefined order. The user concentrates only on the **main task, which** has increased complexity (through the numbers written on the cubes).	Condition 3b The **robot brings** the cubes to the user in a predefined order. The user concentrates on performing a **main + secondary task** simultaneously, **without** reference to the numbers written on the cubes.	Condition 4b The **robot brings** the cubes to the user in a predefined order. The user concentrates on performing a **main + secondary task** simultaneously, **with** increased task complexity (must refer to the numbers written on the cubes).

2.4. Study Hypotheses

The model for the study (Figure 5) and the hypotheses describing the proposed connection between the constructs, user preferences and the study variables (LOA and levels of workload) along for the rationale for the hypotheses are presented as follows:

We suspect that at all workload levels, high LOA will enable the users to perform efficiently and effectively since the high LOA involves the robot carrying out most aspects of the main task which would likely improve performance [37]. Therefore, we propose:

Hypothesis 1. *Quality of task (QoT) execution will be higher with high LOA than with low LOA for all workload levels.*

Several meta-studies conducted regarding levels of automation [38], ref. [39] seem to suggest that the workload experienced by users is influenced by the LOA of the system, particularly in situations of routine performance. This does not discountenance the effect of task complexity but seems to point to the effect level of workload may have in low task complexity. Since a major component of usability is the users' perception of the system use [40] along with effectiveness and efficiency, which high LOA will likely increase, we posit:

Hypothesis 2. *Usability will be higher with high LOA than with low LOA for all workload levels.*

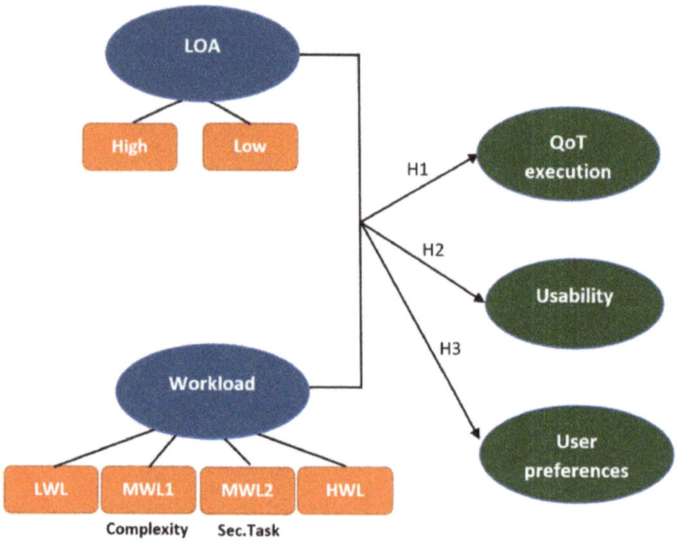

Figure 5. Model for the study and hypotheses.

Research has revealed that as automation increases, workload is expected to decrease, particularly if the automation is properly designed and does not provide new challenges and tasks related to monitoring or other forms of engagement [39]. Moreover, in the design of adjustable robot autonomy in human–robot systems, research shows that as task complexity increases, robot effectiveness is likely to reduce if the robot is operating at higher autonomy [41]. Users seem to intuitively understand that autonomous systems could encounter difficulties in more complex situations with high uncertainty [42] Therefore, in terms of user preferences, we propose:

Hypothesis 3. *Participants will prefer high LOA to low LOA for high workload and low LOA to high LOA when task complexity is increased.*

2.5. Participants

Eighty undergraduate industrial engineering third year students (44 females, 36 males, mean age = 26, SD = 1.4) participated in the study. All students had experience with both computers and robots. Participation was voluntary and every participant received compensation in the form of a bonus point contributing to a credit in an academic course. The participants completed a preliminary questionnaire which included demographics questions for the participants and the negative attitudes towards robots scale (NARS) [43].

The NARS results revealed that 21.06% of the participants had a negative attitude towards situations and interactions with robots while 63.65% were neutral about it. 26.58% had highly negative attitudes towards the social influence of robots, 47.61% had a low attitude and 25.81% were neutral about it. 65.82% had a highly negative attitude towards the concept of robots having emotions, 8.87% were indifferent about it while 25.31% had a low negative attitude towards it.

2.6. Experimental Procedure

Explanation was provided to the participants noting the robot would operate differently in the two trials. To avoid bias, the details of each trial in terms of LOA was not explained to them. They were told that a post-trial and final questionnaire will be provided to express their observations, assessments, and preferences. Then, the participant experienced two experimental trials in which they collaborated with the robot to assemble the configuration that appeared during the GUI in a specific LOA (high/low) in random order. After each trial, they completed a post-trial questionnaire regarding their experience with the robot. At the end of the two trials, each participant completed a final questionnaire where they indicated their preferred level of automation. The experimental design and protocol were approved by the departmental ethical committee.

2.7. Dependent Variables

2.7.1. Objective Measures

Effectiveness: Accuracy of the robot during the task—calculated from the number of times the robot erred in bringing the cubes (e.g., failed to catch a cube, brought an incorrect cube). These are system errors to portray the context of a system whose performance may not be absolutely optimum at all times.

Performance in the secondary task was measured as the number of stages they passed in the secondary task (for the participants that experienced the higher workload).

Efficiency: Total time (in seconds) that it took the participant to complete the task for each trial. In the higher level of automation, the total time was constant since depended on robot motions only.

2.7.2. Subjective Measures

The subjective measures were collected through questionnaires that included questions regarding the participants' experience with the robot. The post-trial questionnaire was prepared as a 5-point Likert scale ranging from "1 = strongly disagree" to "5 = strongly agree" through which participants were expected to express their experience and assessments. The questionnaire included NASA-TLX questions [17] to assess perceived workload in relation to the system efficiency. The raw NASA-TLX scores were added without the weights to provide an estimate of the overall workload (RTLX aggregation technique). The post-trial questionnaire also included questions from the technology acceptance model (TAM) to assess perceived ease of use [44]. The final questionnaire assessed **user preferences** regarding LOA modes and their perceptions as they collaborate with the robot at specific LOA modes.

2.7.3. Constructs

The dependent variables were defined through two constructs: QoT execution and usability. These constructs were derived from the objective and subjective measures explained above (mapping is provided in Figure 6). They were adapted to the context of human–robot collaboration from the ISO 9241-151 guideline [40,45] as follows:

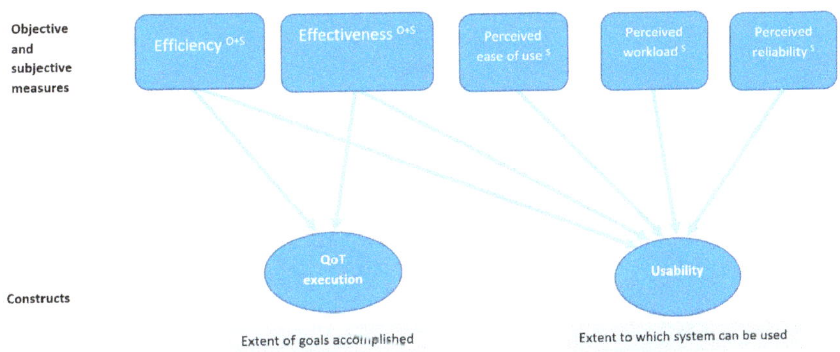

Figure 6. Mapping of the measures into constructs for assessment. (O—objective measures; S—subjective measures).

Quality of task (QoT) execution. The extent to which specific goals in a task are accomplished to a specified degree of accuracy for a specified time period [46]. This construct involves effectiveness and efficiency of the collaboration. Effectiveness of the collaboration was evaluated by the accuracy and completeness of the task which the human and robot cooperate to execute. The efficiency of the collaboration depends on resources such as time and human effort spent to achieve the required goal [47].

Usability. The extent to which the robotic system can be used to achieve specified goals with effectiveness, efficiency and satisfaction in a specified context of use (adapted from [40]). This construct, in this study, is composed of effectiveness, efficiency in addition to satisfaction derived from the perceived ease of use, perceived workload and perceived reliability of the system. All these variables could affect the degree to which the human operator believes that working with the robot will be free of difficulty or great effort. This is an adaption from [44] in the information technology domain to the context of HRC. They constitute the user's perception regarding use of the system and is essential to ensure that the human can successfully team up with the robot to achieve such collaboration [35]. A negative user perception could lead to disuse of the support the robot can provide in the collaboration [48]. In the current study, the usability construct was comprised of the QoT measures, along with other user perceptions on ease of use, workload, and reliability.

2.8. Analysis

A generalized linear mixed model (GLMM) was applied to analyze the data with the LOA, and workload as independent variables. To combine variables for the constructs, multivariate analyses of variance (MANOVA) was used. The analyses considered all the constituent variables within constructs and combined them into a composite variable. Tukey's honestly significant difference (Tukey's HSD) test were used as the post-hoc test for multiple comparison. The tests were designed as two-tailed with a significance level of 0.05. The items in the user preferences questionnaire were analyzed using ANOVA to assess the effect of workload on their preferences for the LOA mode they experienced.

3. Results

Results of the assessments using the constructs (QoT execution and usability), details of the user preference regarding the LOA modes and a comparison within the workload groups are presented below.

3.1. QoT Execution

The interaction of LOA and workload had significant effect (F (3, 152) = 5.198, p = 0.002) on the QoT execution. The QoT execution was higher at the high LOA when the workload was low compared to other LOA-workload combinations, confirming H1. LOA (F (3, 150) = 45.15, $p < 0.001$) and workload (F (3, 152) = 18.725, $p < 0.001$) were also significant as main effects on the QoT execution. The high LOA produced better QoT execution compared to the low LOA. Best results were obtained for low workload as expected. When the workload is high, the high LOA also produced a better QoT execution compared to the low LOA. Details of the constituent variables in the QoT execution (effectiveness and efficiency) are presented below:

3.1.1. Effectiveness

The interaction of LOA and workload did not have a significant effect on accuracy (F (3, 152) = 0.512, p = 0.675) and neither did the LOA (F (1, 152) = 1.024, p = 0.313) and workload (F (3, 152) = 0.376, p = 0.77) as main effects. Workload level however, had a significant effect on the performance in the secondary task (F (1, 32) = 4.23, $p < 0.001$) with MWL2 (M = 2.02, SD = 1.239) resulting in better performance compared to HWL (M = 1.93, SD = 1.047). All of the participants who did the secondary task finished the first stage of the game. The majority (71/80) reached the second stage of the game, 56/80 reached the third stage while only 10/80 reached the fourth stage.

3.1.2. Efficiency

The interaction of LOA and workload had a significant effect on completion time ($F(3, 152) = 4.838$, $p = 0.003$). At high LOA and LWL, participants completed the task at shorter time compared to the other combinations. LOA also had significant effect on the completion time ($F(1, 152) = 136.565$, $p < 0.001$) with the high LOA ($M = 87.3$, $SD = 0$) having lower completion time compared to the low LOA ($M = 107.945$, $SD = 16.547$) as expected, even though the users had the option to stop the robot's operation at any point in the high LOA mode, thereby increasing the completion time. Workload also had significant effect on the completion time ($F(3, 152) = 4.838$, $p = 0.004$) with the LWL ($M = 94.62$, $SD = 9.028$) having less completion time compared to the HWL ($M = 103.158$, $SD = 23.924$). Higher task complexity (MWL1, $M = 96.449$, $SD = 12.766$) resulted in less completion time compared to the workload caused by the secondary task (MWL2, $M = 96.595$, $SD = 11.241$).

3.2. *Usability*

The interaction of LOA and workload on usability was not significant ($F(18.137) = 1.615$, $p = 0.064$). However, the main effects of LOA ($F(18, 135) = 7.768$, $p < 0.001$) and level of workload ($F(18, 137) = 11.905$, $p < 0.001$) was significant. At high LOA, the usability was higher ($M = 4.36$, $SD = 0.83$) compared to the low LOA ($M = 4.31$, $SD = 0.773$), in agreement with H2. At LWL ($M = 4.37$, $SD = 0.633$), usability was higher compared to HWL ($M = 4.25$, $SD = 0.742$). Higher usability was obtained when task complexity increased (MWL1, $M = 4.45$, $SD = 0.959$) as compared to when there was a secondary task (MWL2, $M = 4.29$, $SD = 0.835$).

There was no difference in the workload groups in terms of the perceived ease of use. However, workload level significantly influenced perceived workload as measured through the aggregated raw NASA-TLX scores ($F(3, 152) = 11.767$, $p < 0.001$), with the HWL ($M = 14.6$, $SD = 4.337$) resulting in higher perceived workload compared to the LWL ($M = 12.58$, 3.796) as expected. Between the medium workload groups, MWL2 ($M = 15.33$, $SD = 3.318$) resulted in higher perceived workload compared to MWL1 ($M = 11.18$, $SD = 2.123$).

Workload also had significant effect ($F(3, 152) = 3.646$, $p = 0.014$) on perceived reliability as assessed through the questionnaire. The reliability was perceived as higher by the participants who experienced the LWL ($M = 4.53$, $SD = 0.687$) compared to the HWL ($M = 4.5$, $SD = 0.555$). Between the medium workload levels, MWL1 (4.63, $SD = 0.628$) resulted in higher perceived reliability compared to MWL2 ($M = 4.19$, $SD = 0.634$).

3.3. *User Preferences*

A one-way ANOVA revealed that there was a significant difference between workload groups ($F(3, 76) = 9.276$, $p < 0.001$). When comparing LWL and HWL, high LOA was preferred. However, when comparing between MWL1 and MWL2, low LOA was preferred for the MWL1 (confirming H3). More details regarding user preferences for the LOA modes between the workload groups are depicted in Figure 7.

3.4. *Comparison between Workload Groups for Different LOA Modes*

Multiple comparison made between the different workload groups with details on each LOA mode for groups that were significantly different are presented in Table 2. Results revealed that at low LOA: QoT execution is higher when workload is lower; usability is higher when a secondary task is involved, and user preference tended towards low LOA when complexity increases. However, at high LOA: QoT execution was the same for all workload types except when complexity is involved; usability was higher when a secondary task is involved, and user preference tended towards high LOA when a secondary task is involved.

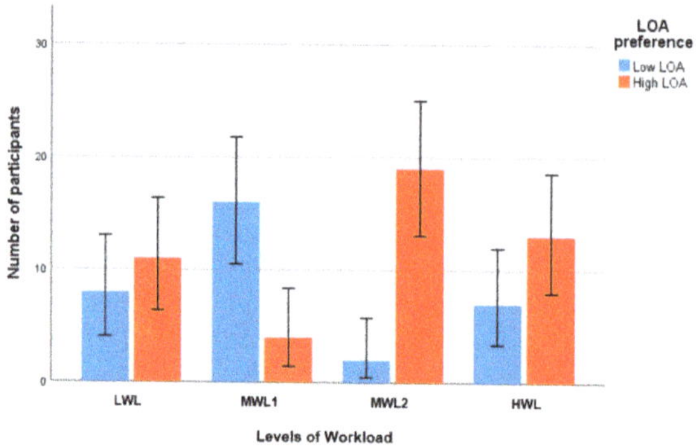

Figure 7. LOA preference for the different workload levels.

Table 2. Comparison of assessment (with p-values) within the workload groups *.

Groups	QoT Execution	Usability	User Preferences
LWL I MWL1	0.858	0.297	0.038 * Low LOA > High LOA
LWL I MWL2	0.88	0.03 * Low LOA: Low < MWL2 High LOA: Low < MWL2	0.089
LWL I High	0.004 * Low LOA: LWL > HWL High LOA: LWL = HWL	0.059	0.956
MWL1 I MWL2	0.1	0 < 0.001 * Low LOA: MWL1 < MWL2 High LOA: MWL1 < MWL2	0 < 0.001 * Low LOA < High LOA
MWL1 I HWL	0.042 * Low LOA: MWL1 > HWL High LOA: MWL1 < HWL	0 < 0.001 * Low LOA: MWL1 < HWL High LOA: MWL1 < HWL	0.008 * Low LOA > High LOA
MWL2 I HWL	0.033 * Low LOA: MWL2 > HWL High LOA: MWL2 = HWL	0.782	0.242

* green depicts comparison with statistical significance; similar trends are marked with identical colors.

4. Discussion

The main influences and interacting influences of LOA in HRC in an assembly task context, considering different levels of workload is summarized in Table 3.

Table 3. Summary of findings.

Metrics	Constituent Measures	Significant Effects	Finding
QoT execution	Efficiency; effectiveness	LOA ($p < 0.001$); workload ($p < 0.001$); LOA*workload ($p = 0.002$)	LOA and workload had significant effect on the QoT execution. The QoT execution was higher at the high LOA.
Usability	QoT execution measures; perceived ease of use, perceived reliability, perceived workload	LOA ($p < 0.001$); Workload ($p < 0.001$)	The usability was higher at high LOA. The workload had more influence on the constituent variables, with the LWL resulting in higher usability.
User preferences	User choices regarding LOA modes	Workload ($p < 0.001$)	Most of the participants preferred the high LOA for both LWL and HWL. In the medium workload levels, the low LOA was preferred for the MWL1 where some task complexity was involved

4.1. Influence of LOA

In HWL situations, where additional resources are needed to complete the task in the least possible time and with minimal effort, high LOA is preferred. This corresponds with the observations made in the meta-analyses conducted in [38,39] where several automation-related data where analyzed. It also agrees with the characteristics of the suggested line of solution in workload demands amidst multiple resources as elaborated upon in [37]. However, in cases where complexity is involved, as seen in the results for the LOA preference of participants in the medium workload category, a low LOA can be considered. Most participants seem to prefer a low LOA when the task complexity is high. This confirms H3, and is also in agreement with previous studies where it was stated that a higher LOA may not always give a positive outcome in situations where uncertainties, and higher probabilities of failure exist [38,39]. In high complex tasks where high component and coordinative complexity increases the probabilities of failure [23,49], humans usually have a higher potential to better manage unknown or unexpected situations [50,51]. This reinforces the significance of evaluating LOA modes alongside different workload situations as emphasized in [52] for various contexts and causes of workload. It also calls for further assessments using these constructs.

4.2. Workload Considerations

Workload had significant influence on most of the measures. The significant effects were seen in effectiveness and efficiency leading to reduced QoT execution in situations where the workload was high. This is consistent with the literature highlighting the contribution of task-related demands (such as mental, temporal, and physical demands, including complexity demands involved in the HRC task) to workload, which could negatively influence resources available to complete task at hand [15].

The medium workload category more clearly reflects some of the differences in additional workload which can be induced by a secondary task or task complexity. Secondary task inclusion (depicted in MWL2) seems to produce a higher perception of workload compared to complexity in the task (depicted through MWL1). This could explain the reason why most users preferred the high LOA (which autonomously executes more aspects of the task) compared to the low LOA for MWL2. The LOA option seems to provide more mental space for the users to execute other tasks, particularly when the automation functioned well, as suggested in [38,39].

This difference in the medium workload category also brings into prominence the relevance of task complexity, specifically the influence reflected through the perceived reliability where MWL1 (reflecting higher complexity) condition was perceived more reliable compared to MWL2 (reflecting secondary task influence). This could be a result of higher uncertainty and failure probabilities which complexity induces as elaborated in [53,54]. It is therefore understandable that users preferred low LOA to the high LOA in this level of workload (where the task complexity exists) where they seem to have an increased sense of control over the operation [55]. This enables them to better manage the higher uncertainties in this condition (through the low LOA) compared to relying on the robot (through the high LOA). The results reveal that both objective and subjective complexity considerations as noted in [21] should be considered along with the suitable LOA modes for such HRC assembly tasks. This consequently affects the QoT execution and usability of the system.

4.3. Limitations

Evaluation was performed with users who had experience with computers and robots. We expect these results to be amplified with users who have experience in real industrial setting. We are also cognizant of potential differences in the subjective assessment of the students in comparison to professionals in an industrial setting since this plays a role in the perception of the users working alongside a robot in a work setting [56]. We therefore consider the results obtained with caution, with the perception that these could be relatively equivalent to assessment with novice operators and different from expert or professional assessments.

The LOA and levels of workload design is simplified for research purpose and not fully representative of the degree of automation, workload levels demanded in more industrial settings. The results obtained, therefore, serve as building blocks and insights for further developments where more detailed automation, workload and complexity conditions are tested in sample industrial settings. Some other social aspects of interacting (such as verbal [57] and non-verbal communication methods [58]) with the robot for the collaborative work were not explicitly investigated in this study. However, further research should also investigate the interplay of the socio-technical aspects of the collaboration while also considering economic and societal issues to understand fuller dimensions of improved HRC in industry [56,59,60].

5. Conclusions

This paper presented the influence of LOA on a human robot collaborative assembly task considering different workload levels. The user study yielded valuable insights into participants' preferences and influence of LOA and workload. The study also introduced two constructs for the evaluation: quality of task (QoT) execution and usability. The evaluation obtained through these constructs highlighted their potential for use in HRI studies. The study has served to provide support tools to further align manufacturing strategies and automation decisions putting into consideration level of workload to further improve productivity.

The QoT execution construct also pointed to the significance of combining efficiency and effectiveness together as a single variable. It revealed the influence of the LOA and workload in the extent to which goal of the task was accomplished under specified degree of accuracy and duration of the task. The usability construct was significant in revealing the combined effect of QoT execution and user perceptions of the ease of use, workload, and system reliability. The interactive effect of LOA and levels of workload on this construct pointed to the added value which user perceptions contribute when combined with the QoT measure.

We recommend a high LOA to support the user when the workload is high. A high LOA could reduce the stress or pressure of additional secondary tasks which the robot could support in. This was observed in the outcome of the user preferences which tended towards higher LOA when the workload was high. It also agrees with the observations of [38] in their meta-analyses considering the influence of LOA on workload. High LOA, when designed effectively, helps to extend the capabilities of the user to attend to other tasks concurrently as noted by [42,61]. However, lower LOA is helpful when high task complexities are involved, for which failure performance may occur as also noted in [39]. An adaptive LOA design that takes these outcomes into consideration is therefore recommended for further investigation.

There may be significant differences in the influence of these variables when observed in other settings, with different forms of robots, tasks and robot feedback modalities [62] and with the perception of different users as emphasized in [63]. Future work should evaluate different forms of increased workload. The workload design can be fine-tuned to portray distinct types of workload demands such as physical, cognitive and temporal demands during the task. Evaluation should also be conducted with other forms of tasks e.g., with a mobile robot delivering items and with other populations. Ongoing research is aimed at performing studies with older adults for daily living tasks and for non-professional users, putting into consideration the influence of demographics on the changes automation brings [64]. LOA has proven to influence performance for older adults [12]. We expect the effect of the levels of workload to amplify with them. The change of preferences and the differences in the reaction and performance of the older adults should be examined with different LOA options for different workload levels.

Author Contributions: Conceptualization, D.G., S.O. and Y.E.; methodology, D.G., S.O. and Y.E.; software, D.G.; validation, D.G., S.O. and Y.E.; formal analysis, D.G., S.O. and Y.E.; investigation, D.G., S.O. and Y.E.; resources, Y.E.; data curation, D.G.; writing—original draft preparation, D.G.; writing—review and editing, D.G., S.O. and Y.E.; visualization, D.G., S.O. and Y.E.; supervision, S.O. and Y.E.; project administration, Y.E.; funding acquisition, Y.E. All authors have read and agreed to the published version of the manuscript.

Funding: This research was supported by the EU funded Innovative Training Network (ITN) in the Marie Skłodowska-Curie People Programme (Horizon2020): SOCRATES (Social Cognitive Robotics in a European Society training research network), grant agreement number 721619. Partial support was provided by Ben-Gurion University of the Negev through the Agricultural, Biological and Cognitive Robotics Initiative, the Marcus Endowment Fund, and the W. Gunther Plaut Chair in Manufacturing Engineering.

Institutional Review Board Statement: This study was approved by the ethical committee of the Department of Industrial Engineering and Management at Ben-Gurion University of the Negev.

Informed Consent Statement: Informed consent was obtained from all subjects involved in the study.

Data Availability Statement: Data supporting the reported results can be found at https://github.com/samuelolatunji/LOA-WorkloadLevels_Analyses.git (accessed on 9 August 2021).

Conflicts of Interest: The authors declare no conflict of interest.

References

1. Bauer, A.; Wollherr, D.; Buss, M.D. Human-Robot collaboration: A survey. *Int. J. Humanoid Robot.* **2008**, *5*, 47–66. [CrossRef]
2. Goodrich, M.A.; Schultz, A.C. *Human-Robot Interaction: A Survey*; Now Publishers Inc: Hanover, MA, USA, 2007.
3. Hentout, A.; Aouache, M.; Maoudj, A.; Akli, I. Human-robot interaction in industrial collaborative robotics: A literature review of the decade 2008–2017. *Adv. Robot.* **2019**, *33*, 764–799. [CrossRef]
4. Bonarini, A. Communication in human-robot interaction. *Curr. Robot. Rep.* **2020**, *1*, 279–285. [CrossRef]
5. Mohebbi, A. Human-Robot interaction in rehabilitation and assistance: A review. *Curr. Robot. Rep.* **2020**, *1*, 131–144. [CrossRef]
6. Prati, E.; Peruzzini, M.; Pellicciari, M.; Raffaeli, R. How to include user experience in the design of human-robot interaction. *Robot. Comput. Integr. Manuf.* **2021**, *68*, 102072. [CrossRef]
7. Johnson, G.I.; Wilson, J.R. Future directions and research issues for ergonomics and advanced manufacturing technology (AMT). *Appl. Ergon.* **1988**, *19*, 3–8. [CrossRef]
8. Kaber, D.; Endsley, M.R. Out-of-the-loop performance problems and the use of intermediate levels of automation for improved control system functioning and safety. *Am. Inst. Chem. Eng.* **1997**, *16*, 126–131. [CrossRef]
9. Kaber, D.; Endsley, M.R. Level of automation effects on performance, situation awareness and workload in a dynamic control task. *Ergonomics* **1999**, *42*, 462–492. [CrossRef]
10. Lindström, V.; Winroth, M.; Stahre, J. Levels of automation in manufacturing. *Int. J. Ergon. Hum. Factors* **2008**, *30*, 1–29.
11. Shi, J.; Jimmerson, G.; Pearson, T.; Menassa, R. Levels of human and robot collaboration for automotive manufacturing. In Proceedings of the Workshop on Performance Metrics for Intelligent Systems, College Park, MD, USA, 20–22 March 2012; pp. 95–100. [CrossRef]
12. Olatunji, S.; Markfed, N.; Gutman, D.; Givati, S.; Sarne-Fleischman, V.; Oron-Gilad, T.; Edan, Y. Improving the interaction of older adults with a socially assistive table setting robot. In *Lecture Notes of Artificial Intelligence, Proceedings of the 11th Interna-tional Conference on Social Robotics, Madrid, Spain, 26–29 November 2019*; Springer: Berlin/Heidelberg, Germany, 2019; Volume 11876, pp. 568–577.
13. Wang, W.; Chen, Y.; Li, R.; Jia, Y. Learning and comfort in human-robot interaction: A review. *Appl. Sci.* **2019**, *9*, 5152. [CrossRef]
14. Xu, J.; Anders, S.; Pruttianan, A.; France, D.; Lau, N.; Adams, A.J.; Weigner, M.B. Human performance measures for the evaluation of process control human-system interfaces in high-fidelity simulations. *Appl. Ergon.* **2018**, *73*, 151–165. [CrossRef] [PubMed]
15. Hart, S.G.; Wickens, C.D. Workload assessment and prediction. In *Manprint*; Springer: Amsterdam, The Netherlands, 1990; pp. 257–296.
16. Hart, S.G. NASA-Task load index (NASA-TLX); 20 years later. *Proc. Hum. Factors Ergon. Soc.* **2006**, *50*, 904–908. [CrossRef]
17. Hart, S.G.; Staveland, L.E. Development of NASA-TLX (Task Load Index): Results of empirical and theoretical research. *Adv. Psychol.* **1988**, *52*, 139–183. [CrossRef]
18. Hilburn, D.; Jorna, P.G. Workload and air traffic control. In *Human Factors in Transportation. Stress, Workload, and Fatigue*; Hancock, P.A., Desmond, P.A., Eds.; Lawrence Erlbaum Associates Publishers: Mahwah, NJ, USA, 2001; pp. 384–394.
19. Yeh, Y.-Y.; Wickens, C.D. Dissociation of performance and subjective measures of workload. *Hum. Factors J. Hum. Factors Ergon. Soc.* **1988**, *30*, 111–120. [CrossRef]
20. Wood, R.E. Task complexity: Definition of the construct. *Organ. Behav. Hum. Decis. Process* **1986**, *37*, 60–82. [CrossRef]

21. Rasmussen, M.; Standal, M.I.; Laumann, K. Task complexity as a performance shaping factor: A review and recommendations in standardized plant analysis risk-human reliability analysis (SPAR-H) adaption. *Saf. Sci.* **2015**, *76*, 228–238. [CrossRef]
22. Olsen, D.R.; Goodrich, M.A. Metrics for Evaluating Human-Robot Interactions. 2003. Available online: https://faculty.cs.byu.edu/~{}mike/mikeg/papers/OlsenGoodrichPERMIS2003.pdf (accessed on 9 August 2021).
23. Campbell, D.J. Task complexity: A review and analysis. *Acad. Manag. Rev.* **1988**, *13*, 40–52. [CrossRef]
24. Braarud, P.Ø. Subjective task complexity and subjective workload: Criterion validity for complex team tasks. *Int. J. Cogn. Ergon.* **2001**, *5*, 261–273. [CrossRef]
25. Neill, T.A.O.; Mcneese, N.J.; Carolina, S.; Barron, A. Human—Autonomy teaming: A Review and analysis of the empirical literature. *J. Hum. Factors Ergon. Soc.* **2017**. [CrossRef]
26. Kolbeinsson, A.; Lagerstedt, E. Lindblom, J. Foundation for a classification of collaboration levels for human-robot cooperation in manufacturing. *Prod. Manuf. Res.* **2019**, *7*, 448–471. [CrossRef]
27. Lindström, V.; Winroth, M. Aligning manufacturing strategy and levels of automation: A case study. *J. Eng. Technol. Manag.* **2010**, *27*, 148–159. [CrossRef]
28. Steinfeld, A.; Fong, T.; Kaber, D.; Lewis, M.; Scholtz, J.; Schultz, A.; Goodrich, M. Common metrics for human-robot interaction. In Proceedings of the 1st ACM SIGCHI/SIGART Conference on Human-Robot Interaction (H 06), Salt Lake City, UT, USA, 2–3 March 2006; pp. 33–40. [CrossRef]
29. Aly, A.; Griffiths, S.; Stramandinoli, F. Metrics and benchmarks in human-robot interaction: Recent advances in cognitive robotics. *Cogn. Syst. Res.* **2017**, *43*, 313–323. [CrossRef]
30. Bensch, S.; Jevtíc, A.; Hellström, T. On interaction quality in human-robot interaction. In Proceedings of the 9th International Conference on Agents and Artificial Intelligence (ICAART 2017), Porto, Portugal, 24–26 February 2017; pp. 182–189. [CrossRef]
31. Efthimiou, E.; Papageorgiou, X.S.; Fotinea, S.E.; Karavasili, A.; Vacalopoulou, A.; Goulas, T. User centered design in practice adapting HRI to real user needs. In Proceedings of the 12th ACM International Conference on Pervasive Technologies Related to Assistive Environments, Rhodes, Greece, 5–7 June 2019; pp. 425–429. [CrossRef]
32. Lindblom, J.; Alenljung, B.; Billing, E. *Evaluating the User Experience of Human-Robot Interaction*; Springer: Cham, Switzerland, 2020; pp. 231–256.
33. Honig, S.S.; Oron-Gilad, T.; Serna-Fleischmann, V.; Olatunji, S.; Edan, Y. A user-needs based approach for designing human-robot interactions. In Proceedings of the Workshop on Robotic Co-Workers 4.0: Human Safety and Comfort in Human-Robot Interactive Social Environments, IEEE/RSJ International Conference on Intelligent Robots and Systems, Madrid, Spain, 1–5 October 2018.
34. Parasuraman, R. Designing automation for human use: Empirical studies and quantitative models. *Ergonomics* **2000**, *43*, 931–951. [CrossRef] [PubMed]
35. Bröhl, C.; Nelles, J.; Brandl, C.; Mertens, A.; Nitsch, V. Human-Robot collaboration acceptance model: Development and comparison for Germany, Japan, China and the USA. *Int. J. Soc. Robot.* **2019**, *11*, 709–726. [CrossRef]
36. Renner, P.; Pfeiffer, T. Model-based acquisition and analysis of multimodal interactions for improving human-robot interaction. In Proceedings of the Symposium on Eye Tracking Research and Applications, Safety Harbor, FL, USA, 26–28 March 2014; pp. 361–362. [CrossRef]
37. Wickens, C.D. Multiple resources and mental workload. *J. Hum. Factors Ergon. Soc.* **2008**, *50*, 449–455. [CrossRef]
38. Wickens, C.D.; Li, H.; Santamaria, A.; Sebok, A.; Sarter, N.B. Stages and levels of automation: An integrated meta-analysis. *Proc. Hum. Factors Ergon. Soc. Ann. Meet.* **2010**, *4*, 389–393. [CrossRef]
39. Onnasch, L.; Wickens, C.D.; Li, H.; Manzey, D. Human performance consequences of stages and levels of automation: An integrated meta-analysis. *J. Hum. Factors Ergon. Soc.* **2014**, *56*. [CrossRef]
40. ISO. *ISO 9241-11:2018(en). ERGONOMICS of Human-System Interaction—Part 11: Usability: Definitions and Concepts.* 2018. Available online: https://www.iso.org/obp/ui/#iso:std:iso:9241:-11:ed-2:v1:en (accessed on 29 February 2020).
41. Ashcraft, C.C.; Goodrich, M.A.; Crandall, J.W. Moderating operator influence in human-swarm systems. In Proceedings of the IEEE International Conference on Systems, Man and Cybernetics (SMC), Bari, Italy, 6–9 October 2019; pp. 4275–4282. [CrossRef]
42. Endsley, M.; Kiris, E.O. The Out-of-the-loop performance problem and level of control in automation. *J. Hum. Factors Ergon. Soc.* **1995**, *37*, 381–394. [CrossRef]
43. Syrdal, D.S.; Dautenhahn, K.; Koay, K.; Walters, M.L. The negative attitudes towards robots scale and reactions to robot behaviour in a live human-robot interaction study. In Proceedings of the 23rd Convention of the Society for the Study of Artificial Intelligence and Simulation of Behaviour, Edinburgh, UK, 6–9 April 2009; pp. 109–115.
44. Davis, F.D. Perceived usefulness, perceived ease of use, and user acceptance of information technology. *MIS Q.* **1989**, *13*, 319–340. [CrossRef]
45. Ekşioğlu, M.; Kiriş, E.; Çakir, T.; Güvendik, M.; Koyutürk, E.D.; Yilmaz, M. Design, user experience and usability. Web, mobile, and product design. In *Lecture Notes on Computer Science, Proceedings of the 2nd International Conference, DUXU 2013 (Part of the HCI International), Las Vegas, NV, USA, 21–26 July 2013*; Springer: Berlin/Heidelberg, Germany, 2013; Volume 8015, pp. 173–182.
46. International Organization for Standardization (ISO). 9000 Store. ISO 9001 Processes, Procedures and Work Instructions. 2020. Available online: https://the9000store.com/iso-9001-2015-requirements/iso-9001-2015-context-of-the-organization/processes-procedures-work-instructions/ (accessed on 5 October 2020).

47. Baraglia, M.; Cakmak, Y.; Nagai, R.; Rao, R.; Asada, M. Initiative in robot assistance during collaborative task execution. In Proceedings of the 11th ACM/IEEE International Conference on Human-Robot Interaction (HRI), Christchurch, New Zealand, 7–10 March 2016; pp. 67–74. [CrossRef]
48. Parasuraman, R.; Riley, V. Humans and automation: Use, misuse, disuse, abuse. *J. Hum. Factors Ergon. Soc.* **1997**, *39*, 230–253. [CrossRef]
49. Hærem, T.; Pentland, B.T.; Miller, K.D. Task complexity: Extending a core concept. *Acad. Manag. Rev.* **2015**, *40*, 446–460. [CrossRef]
50. Monostori, L.; Váncza, J.; Kumara, S.R.T. Agent-based systems for manufacturing. *CIRP Ann.* **2006**, *55*, 697–720. [CrossRef]
51. Wang, X.V.; Kemény, Z.; Váncza, J.; Wang, L. Human-Robot collaborative assembly in cyber-physical production: Classification framework and implementation. *CIRP Ann.* **2017**, *66*, 5–8. [CrossRef]
52. Kaber, D.B. Issues in human-automation interaction modeling: Presumptive aspects of frameworks of types and levels of automation. *J. Cogn. Eng. Decis. Mak.* **2018**, *12*, 7–24. [CrossRef]
53. Murthy, D.N.P. Confiabilidade e garantia de produto: Visão geral e pesquisas futuras. Product reliability and warranty: An overview and future research. *SciELO Braz.* **2007**, *17*, 426–434.
54. Niu, J.; Geng, H.; Zhang, Y.; Du, X. Relationship between automation trust and operator performance for the novice and expert in spacecraft rendezvous and docking (RVD). *Appl. Ergon.* **2017**, *71*, 1–8. [CrossRef] [PubMed]
55. Kaber, D.B.; Endsley, M.R. The effects of level of automation and adaptive automation on human performance, situation awareness and workload in a dynamic control task. *Theor. Issues Ergon. Sci.* **2004**, *5*, 113–153. [CrossRef]
56. Brynjolfsson, E.; McAfee, A. *The Second Machine Age: Work, Progress, and Prosperity in a Time of Brilliant Technologies*; WW. Norton & Company: New York, NY, USA; London, UK, 2016; pp. 15–25.
57. Giuliani, M.; Lenz, C.; Müller, T.; Rickert, M.; Knoll, A. Design principles for safety in human-robot interaction. *Int. J. Soc. Robot.* **2010**, *2*, 253–274. [CrossRef]
58. Lenz, A.; Skachek, S.; Hamann, K.; Steinwender, J.; Pipe, A.G.; Melhuish, C. The BERT2 infrastructure: An integrated system for the study of human-robot interaction. In Proceedings of the 10th IEEE-RAS International Conference on Humanoid Robots, Nashville, TN, USA, 6–8 December 2010; pp. 346–351. [CrossRef]
59. Moniz, A.B.; Krings, B.-J. Robots Working with humans or humans working with robots? Searching for social dimensions in new human-robot interaction in industry. *Societies* **2016**, *6*, 23. [CrossRef]
60. Moniz, A.B. *Human Resource Management and Technological Challenges*; Springer: Berlin/Heidelberg, Germany, 2014; pp. 123–131. [CrossRef]
61. Gutman, D. Levels of Automation in Human-Robot Collaboration. Master's Thesis, Industrial Engineering and Management, Ben-Gurion University of the Negev, Beer Sheva, Israel, 2020.
62. Markfeld, N. Feedback for Human-Robot Collaboration. Master's Thesis, Industrial Engineering and Management, Ben-Gurion University of the Negev, Beer Sheva, Israel, 2020.
63. Broadbent, E. Interactions with robots: The truths we reveal about ourselves. *Annu. Rev. Psychol.* **2017**, *68*, 627–652. [CrossRef]
64. Acemoglu, D.; Restrepo, P. Demographics and Robots. 2018. Available online: http://www.nber.org/papers/w24421 (accessed on 9 August 2021).

applied sciences

MDPI

Article

Dual-Arm Peg-in-Hole Assembly Using DNN with Double Force/Torque Sensor

David Ortega-Aranda [1,*], Julio Fernando Jimenez-Vielma [1], Baidya Nath Saha [2] and Ismael Lopez-Juarez [3]

1 Centro de Ingenieria y Desarrollo Industrial (CIDESI), Apodaca 66628, Mexico; julio.jimenez@cidesi.edu.mx
2 Faculty of Science, Concordia University of Edmonton, Information Technology, Edmonton, AB T5B 4E4, Canada; baidya.saha@concordia.ab.ca
3 Centre for Research and Advanced Studies (CINVESTAV), Ramos Arizpe 25900, Mexico; ismael.lopez@cinvestav.edu.mx
* Correspondence: dortega@cidesi.edu.mx

Abstract: Assembly tasks executed by a robot have been studied broadly. Robot assembly applications in industry are achievable by a well-structured environment, where the parts to be assembled are located in the working space by fixtures. Recent changes in manufacturing requirements, due to unpredictable demanded products, push the factories to seek new smart solutions that can autonomously recover from failure conditions. In this way, new dual arm robot systems have been studied to design and explore applications based on its dexterity. It promises the possibility to get rid of fixtures in assembly tasks, but using less fixtures increases the uncertainty on the location of the components in the working space. It also increases the possibility of collisions during the assembly sequence. Under these considerations, adding perception such as force/torque sensors have been done to produce useful data to perform control actions. Unfortunately, the interaction forces between mating parts produced non-linear behavior. Consequently, machine learning algorithms have been considered an alternative tool to avoid the non-linearity. In this work we introduce an assembly strategy for an industrial dual arm robot based on the combination of a discrete event controller and Deep Neural Networks (DNN) to solve the peg-in-hole assembly. Our results show the difference between the use of DNN with one and with two force/torque sensors during the assembly task and demonstrate a 30% increase in the assembly success ratio when using a double force/torque sensor.

Keywords: robotic assembly; deep neural networks; peg-in-hole; dual-arm

Citation: Ortega-Aranda, D.; Jimenez-Vielma, J.F.; Saha, B.N.; Lopez-Juarez, I. Dual-Arm Peg-in-Hole Assembly Using DNN with Double Force/Torque Sensor. *Appl. Sci.* **2021**, *11*, 6970. https://doi.org/10.3390/app11156970

Academic Editors: Luis Gracia and Carlos Perez-Vidal

Received: 17 June 2021
Accepted: 14 July 2021
Published: 29 July 2021

1. Introduction

Due to rapid changes in production demand and to product changes, new challenges have arisen in the manufacturing of products, a trend that has increased throughout the years and has affected most of the industry. Therefore, more robots that work in an unstructured environment for assembly tasks will be required in factories, where fixtureless operations could be executed, giving high flexibility to the production processes [1].

Considering that current automation systems do not support intelligent solutions for assembly tasks, a great opportunity arises, that is, to develop practical methodologies to include machine learning algorithms in assembly problems with robots. When non-desired contact among objects occurs, the assembly cycle requires assistance from an operator. Moreover, in automated cells, when such conditions appear, the robot is programmed to reject the assembled components and start a new cycle. This represents a delay in the production schedule and additional costs due to scrap generation.

Dual arm robots have already been deployed as research projects. New methods have been proposed to achieve different manufacturing tasks. However, assembly tasks with dual arm robots have not yet been broadly studied. They have been reviewed in [2], among some advantages of using the dual arm configuration and trying to execute more tasks in an unstructured environment. In previous work, we had presented the study of the assembly process of a starter motor using fixtures to position components and applying neural networks for contact state identification [3]. It also has been pointed out that controlling a dual arm robot for an assembly task can be very challenging [4]. Research has also focused on the performance of dual arm systems to assemble an automotive component, trying to imitate human-like activities [5]. Among the advantages, there is higher manipulation capability, flexibility, and stiffness [6]. Moreover, stiffness can be controlled or adjusted when both arms make contact with an object [4]. The advantage of using not only one arm robot but two arms is presented in [7], even avoiding the use of force-torque sensors in [8]. Tele-operation activities using visual servoing and manipulation of a single object with both arms is presented in [9]. To process contact states, direct policy search using linear Gaussian controllers has been proposed in [10]. Regarding the difficulties of programming the dual arm robots, a framework for task oriented programming that decomposes complex activities in simpler ones has been proposed [11]. Another matter of study has been human-like operations and analysis of the peg-in-hole problem, the same way a human executes this task in [12].

Regarding the peg-in-hole problem, as the most studied research case for an assembly task, there is extensive literature that presents different points of view and issues. With regard to how to consider the assembly task, a definition of assembly primitives has been proposed [13]. It has also been described as "the basis of a wide range of component assemblies" where two main strategies can be considered, contact model based and contact model free in [14].

In the case of model free strategies, it is known how reinforcement learning finds a solution without knowing the models of the robot [15]. Another study proposes skill acquisition, where low accuracy of conventional methods is compensated by a learning method without parameter tuning [16]. For the case of peg-in-hole with a dual arm robot, a three step method inspired by humans operations is designed in [17]. Seed works that inspired many recent studies such as force/toque maps [18], self-organizing maps [19], reinforcement learning [20], and event discrete systems [21].

Impedance control and force control methods have been reviewed in [14], but most industrial setups work on positional control methods [22]. Working with industrial robots isn't simple; the lack of easy development tools for programming, integration of perception technologies, and computational power are some challenges which limit the development of technical solutions for complex problems in the industry [23].

This paper presents a novel approach to investigate the peg-in-hole assembly with a dual arm robot. An industrial dual arm robot is programmed to follow a sequence of steps to achieve a peg-in-hole assembly task. A positional error is defined and an assembly strategy to error recovery based on classification of contact states is proposed. Classification is investigated using Deep Neural Networks (DNN) to learn the force/torque (F/T) patterns of defined contact states. The novelty of the investigation lies on the use of a double force/torque sensor (F/T sensor), to increase the number of features the DNN learns.

It also considers the integration of a DNN model as a trigger condition into a discrete event controller. In order to test the proposed strategy, the dual arm robot system is integrated with the Robot Operating System (ROS) and two force/torque sensors mounted on the wrist of each robot arm. To evaluate the performance of the assembly strategy using two scenarios, the first by training the DNN with one F/T sensor and the second with double F/T sensors, experiments were executed and the number of completed assemblies are counted for calculation of Success Ratio for each scenario.

The article is organized as follows: after this introduction, Section 2 presents the related work and original contribution. In Section 3, the methodology that includes the description of the test bed is introduced, while the results are explained in Section 4. Finally, Section 5 provides the conclusions and further work.

2. Related Work

For solving the peg-in-hole with a dual arm robot, Ref. [17] describes the problem of insertion as a search problem of the hole, where force/torque (F/T) sensor data was used to identify the different assembly stages using thresholds. The challenge with thresholds is the continuous tuning due to changes in the assembly conditions. In this research we compare the threshold method with machine learning methods for contact states classification.

In [13] is proposed a pool of assembly primitives and maintaining an unilateral contact constraint, where the need of more advanced force control is needed to prevent the abrupt changes in contact forces. Our approach considers quasi-static motion and discrete events, which excludes complex models of the force feedback.

In [7] there is a definition of a master-slave strategy and a cooperative strategy to evaluate the assembly of the peg-in-hole with one and two arms. The advantage of such work is the use of a compliant robot with low accuracy, which can manage contact forces with less risk of damage. It also defines thresholds as trigger functions in the assembly sequence. In [8] a dual arm robot is also used using a leader-follower strategy and a sensor-less and active compliance admittance method with real time trajectory generation.

Double F/T sensors in robotics have been reported in [24,25] where two robots are used for a bi-manual peg-in-hole assembly task. Discrete event systems (DES) are used to model the assembly process. The F/T data is processed separately for each robot and compared with defined thresholds of forces and torques; qualitative values are defined due to the high variability of force/torque signals in every execution. This eliminates the need of tuning constantly.

Another dual arm system with double sensor is presented in [9], where the main focus is the manipulation of objects in a coordinated way using the F/T signals as the feedback for the impedance controller. There, the authors designed their own software platform. Our approach was using ROS, a developing platform in order to transfer knowledge to different platforms.

The single arm peg-in-hole assembly has been broadly studied in recent years. In [22] a safe learning mechanism is used to avoid damage to the system. This idea was taken for our work to prepare the robot for uncertain and unexpected conditions. In [26] the authors express their concern about their classical programming methods and how position control-based methods are not suitable for high precision assembly. In contrast, in this work we consider that the classical position control-based method can be improved by integration of machine learning.

In [27] a combination of different methods such as Principal Component Analysis (PCA), Hidden Markov Models (HMM) and time series for contact state recognition is used. The authors' approach of analysing the contact states recognition as a classification problem is suitable for our project. Expected Maximization and Gaussian Mixture Models (EM-GMM) for monitoring contact states during the peg-in-hole assembly is presented in [28], where the authors present a clear notation of the contact state classification problem that is adopted in this work.

Different searching techniques are proposed such as spiral search [29], attractive region in environment (ARIE) [30], random search in a plane [7]. Those approaches are not suitable in our project due to the high stiffness of our robot, which can be unsafe during constrained motion.

2.1. Problem Definition

Industrial robots usually lack appropriate external programming tools to design a suitable control system [9]. Moreover, industrial robots programming method is based on defining a sequence of steps in the position space. The robot is moved to a desired position, then this position is saved in order to follow a sequence of movements.

An automated assembly process is programmed to repeat a defined sequence. In case the robot cannot reach the final destination, an error recovery sequence must be programmed. In that sense, when the error position is difficult to estimate, there are few chances that this strategy will work.

When using dual arm robots for a fixtureless assembly process, an uncertain environment is produced, hence, a positioning error among the parts is expected to appear previous to the assembly stage. This positioning error is very difficult to estimate when using F/T sensors due to noisy signals that make it difficult to classify. Moreover, a robot with high accuracy and high stiffness abruptly presents changes during contact states. In such conditions, using F/T data in a qualitative force and moment templates as proposed in [31], making calculations using transformation matrix as proposed in [23], and requires continuous tuning, resulting in less successful assemblies.

By applying an appropriate assembly strategy, the automated process can be strengthened by including machine learning algorithms that drive better decision making on the process and avoid assembly failures.

2.2. Original Work

This research work explores the integration of machine learning algorithms into a discrete event controller as a trigger function in order to accomplish a better performance in the overall assembly cycle.

Three main challenges have been addressed in this work:

- Incorporation of off-line learning capabilities using DNN to an industrial robot system.
- Investigation of the advantages of using a double force/torque sensor in a bi-manual assembly task.
- Undesrtanding how to evaluate the effects of deep neural networks in an automated assembly cycle.

The main contribution of this paper is the methodology for evaluating contact state classification using one and two force/torque sensors as a single feature vector for a single controller. Few works use two sensors to feed two separate controllers, meanwhile most of the literature presents the use of only one sensor. Our approach considers a twelve component vector to increase the number of features to input data into the DNN classifier.

It has been found that the second sensor helps the systems perform better on contact state recognition. This methodology also considers the Success Ratio evaluation of the peg-in-hole assembly cycle. An increase in performance due to the use of machine learning classifiers was demonstrated. An automated robot cell sequence for assembly was modeled as a discrete event system. We found that, using this methodology instead of standard automation logic, the discrete event controller is able to overcome positioning errors in more assembly cycles.

3. Methodology

The methodology used in this research is based on an automated assembly cycle, such as a real manufacturing environment. To simplify the investigation, only the final assembly stage is considered, when the robot has picked up the parts, carried them to approach closely and executed the final assembly.

The problem to be analyzed assumes the existence of a positioning error or misalignment between the peg and the hole. This error during the attempt to assemble causes a high force contact state and a stop command to the robot in order to avoid damage.

Therefore, in a traditional automation system, an error recovery sequence must be programmed to deal with unexpected situations. In this case, when using force/torque sensors, one of the great issues to address is the nonlinear behavior during the constrained motion. This makes it difficult to estimate the positioning error.

The methodology is summarized as follows: an error recovery sequence inspired by human behavior is defined. A positioning error region is defined and selected according to the force/torque measurements caused by the interaction forces. The contact states are defined to simplify the experiment. A classification problem is defined where DNN are proposed.

To train the DNN, a data set is gathered running the error recovery sequence with a finite number of positioning errors. The data sets are used to train the DNN models and other machine learning algorithms for comparison purposes. The DNN trained model is evaluated using different data sets. Finally, the DNN models are integrated into the automatic sequence. The methodology is tested by running the assembly cycle a certain number of times to determine the assembly success ratio.

The error recovery sequence is shown in Figure 1. A similar approach was considered in [7]. A positioning error is defined empirically as in [23,32]. Considering how a human would approach recovery form positioning error, the following actions can be described in six stages:

(a) Move until contact between parts is reached.
(b) Change orientation in the hole center direction.
(c) Move the part until a lateral impact of the peg in the hole edge is reached.
(d) Re-orient the part to align with the hole axis.
(e) Insert the part into the hole.
(f) Insertion with fine adjustment.

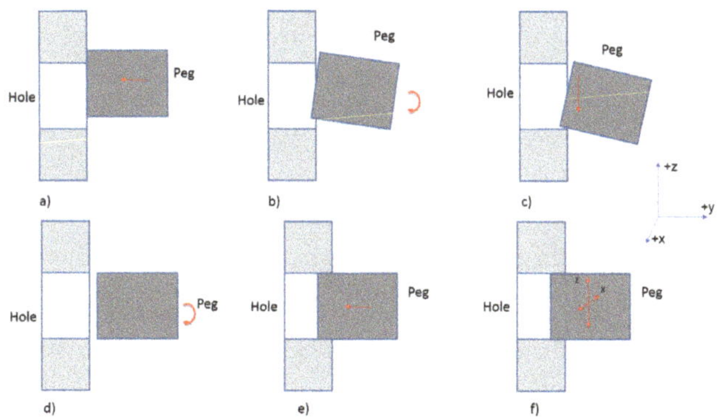

Figure 1. Assembly Sequence. (**a**) Frontal contact. (**b**) New orientation. (**c**) Perform lateral search. (**d**) Orientation for assembly. (**e**) Insertion. (**f**) Adjustment during insertion.

These stages can be represented as an assembly sequence modeled as a discrete event system (DES). This model simplifies the control strategy that will facilitate the constraint motion commands [33]. The discrete events are graphically represented based on Petri-Nets as in [21]. This helps visualize the sequence of actions considered in the control strategy.

3.1. Positioning Error

The positioning error can be described as the sum of positioning inaccuracies during the initial actions of the process [34]. This can be considered a random positioning error that in normal circumstances would cause a fatal error in an automated sequence or rejection of the assembled product. The next equation describes the positioning error:

$$\hat{p}_p = p_p + e \quad (1)$$

where $\hat{p_p}$ is the sum of the random error e and the desired position p_p. e as a position vector according to the deviation from the center of the hole can be described as follows:

$$e = \left[e_x \; e_y \; e_z \; e_\alpha \; e_\beta \; e_\gamma\right]^T \quad (2)$$

The error e has an infinite number of values. To simplify the investigation in this research, a finite number of values are defined. In some works, such as [16] one finds the use of only four error positions around the hole but of different distances along the orthogonal axis in the plane. In [23], eight regions are selected based on experimentation. For this research, eight defined positioning errors in the orthogonal plane to the assembly direction are considered. We will use these defined positions to show how the controller decides in which direction to move after the first impact and classification of the contact state.

The direction of insertion is along the Y axis, so the X-Z plane is where the eight positioning errors are defined. Then the error vector e is simplified as

$$e = [e_x \; e_z]^T \tag{3}$$

The eight defined errors can be expressed in polar notation as $e_r = 4mm$, $\theta = i \times 45°$, where e_r is the magnitude, θ is the angle, and i is a number from 0 to 7 corresponding to one of the defined error positions. Figure 2a shows a graphical representation of the error positions around the hole, and Figure 2b shows the representation of the peg with a positioning error in the $+Z$ axis direction.

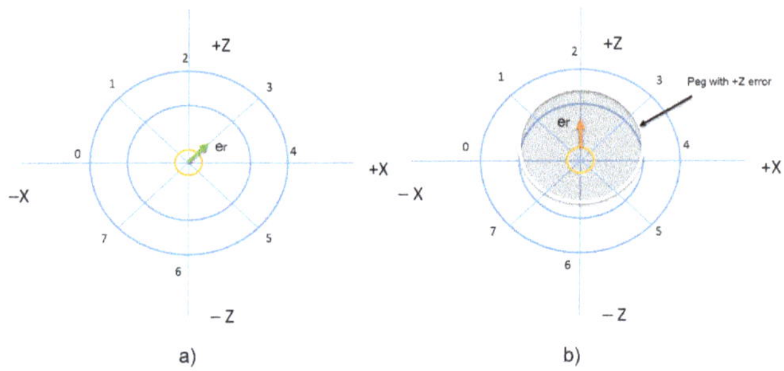

Figure 2. X-Z plane (**a**) Eight locations around the hole. (**b**) Peg with +Z positioning error.

The simplification of the positioning errors leads one to consider this problem a supervised learning classification problem where machine learning algorithms can be used.

3.2. Data Set Acquisition

In order to train the machine learning algorithms, a representative data set needs to be acquired. For this purpose, the error recovery sequence is used to acquire the force/torque signals on every defined error position.

To safely perform constraint movements with the dual arm robot, considering the control strategy for a discrete event system, two primitive movement functions are defined to avoid damage to the system. This can be described as follows:

$$move_{max}(f) = \begin{cases} 1 & f < f_{max} \\ 0 & f \geq f_{max} \end{cases} \tag{4}$$

$$move_{min}(f) = \begin{cases} 1 & f > f_{min} \\ 0 & f \leq f_{min} \end{cases} \tag{5}$$

$$f = \sqrt{f_x^2 + f_y^2 + f_z^2} \tag{6}$$

where $move_{max}(f)$ is a function that starts the robot movement whenever the force f is less than the maximum limit f_{max} and the movement stops whenever the force f is greater than the limit f_{max}. For the case of the $move_{min}(f)$, the function starts or stops the robot movement according to the conditions based on the limit f_{min}. These primitive movements are used for every step of the sequence in Figure 3 according to logic constraints.

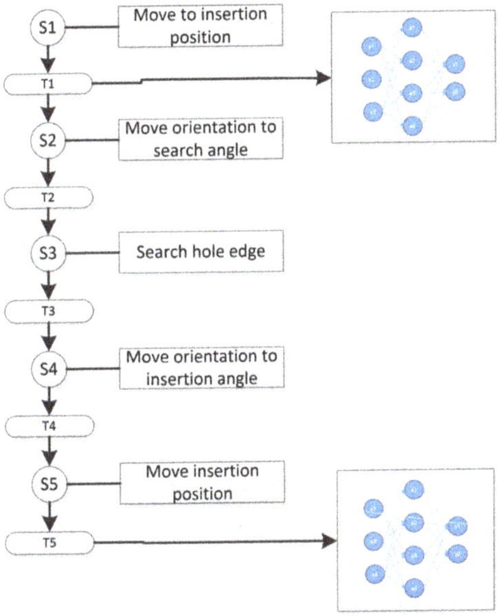

Figure 3. Petri-Net representation of the sequence.

Step 1 moves the robot to the error position as shown in Figure 1a. The error position is selected randomly. On step 2, the orientation is selected according to the error position as shown in Figure 1b. In step 3, the peg is moved along the X-Z plane in the -Z direction until a lateral contact force is reached Figure 1c, step 4 sets the orientation to be aligned with the hole Figure 1d and, finally, step 5 moves the peg towards insertion Figure 1e.

The sequence is repeated for each error position, gathering contact state information to make the data set for training, testing, and to make a validation data set.

3.3. Data Set Analysis

In order to analyze whether the data set can be used for learning the contact states, the t-distributed stochastic neighbor embedding (t-SNE) method is performed. As can be seen in the results, this graphical representation helps find out whether there are separable classes. The t-SNE [35] is a nonlinear dimensionality reduction technique that helps in visualization of data in a low dimensional space.

3.4. Standard Automation Logic Rules

In the case of programming a standard automated program, logic rules are used to trigger transitions among sequence steps. Variable limits or thresholds to classify F/T data are presented in [7,23]. To find whether this method can be used for this research, the data set is analyzed using statistical measures. Thresholds are used as logic conditions to classify the F/T input values in accordance to the next equation:

$$Th_m = \begin{cases} lim_{sup} & \mu_M + \sigma_M \\ lim_{inf} & \mu_M - \sigma_M \end{cases} \tag{7}$$

where μ_M corresponds to the mean of the momentum in any axis and σ_M corresponds to the standard deviation of the momentum in any axis. This defines the value of the superior and inferior limits: lim_{sup} and lim_{inf}.

3.5. Deep Neural Network Definition

Deep Neural Networks can map the force/torque signals to the defined error position class. These classes can also map the corresponding robot actions. Contact states are defined as quasi-static motion, which means that velocity of the motion is not taken into account.

For this investigation, the contact states that are under analysis are those that occur during the first assembly stage: the initial approach between peg and hole. Although the error position during this stage is unknown, a set of error positions can be defined to convert the case into a classification problem.

The sensors deliver x, y, z components for the force and momentum. Instead of using separate sensor signals, a single F/T vector of 12 features is proposed as shown next:

$$FT = [f_{x1}\ f_{y1}\ f_{z1}\ m_{x1}\ m_{y1}\ m_{z1}\ f_{x2}\ f_{y2}\ f_{z2}\ m_{x2}\ m_{y2}\ m_{z2}] \tag{8}$$

where FT represents the force/torque vector, and the sub-indexes $x1$, $x2$ and so on, represent the component from sensor 1 or sensor 2.

Then, a DNN model as expressed in the next equations is defined experimentally by changing parameters and hyper-parameters to reach an acceptable accuracy during the training and testing.

$$y = \sigma(\sum_{i=0}^{n} W_i X_i) \tag{9}$$

$$\sigma = \frac{e^z - e^{-z}}{e^z + e^{-z}} = \tanh \tag{10}$$

$$Loss(\hat{y}, y) = -\sum_{i=1}^{2} y_i \log(\hat{y}_i) \tag{11}$$

$$S(y)_i = \frac{\exp(y_i)}{\sum_{j=1}^{n} \exp(y_i)} \tag{12}$$

where X_i corresponds to the number of features or inputs, y correspond to the number of outputs or classes, h correspond to the number of hidden layers, and n to the number of hidden units.

In order to pre-process the data set for training the DNN, a standardization process corresponding to the following equation is executed.

$$\hat{X} = \frac{X - \mu}{\sigma} \tag{13}$$

where $\hat{X}$ is the normalized sensor features, X is the input feature, μ the mean of each feature in the data set, and σ the standard deviation of each feature in the data set.

A second DNN model is designed to classify basic contact states such as orthogonal contacts in all axes X, Y, Z. That means that the second DNN model will have the same 12 input features and 5 outputs corresponding to the actions: move to $+X$, move to $-X$, move to $+Z$, move to $-Z$, and move to the insertion position $-Y$. This model classifies the contact states during the insertion step. In any case, a small deviation causes a contact state force $f > f_{limit}$ that commands the robot to stop insertion and align according to the second DNN model output.

The trained DNN model helps the system classify the contact states caused by the error position. This classification selects the relative position to which the robot should move next. Other algorithms are also trained and tested with the same data set to compare the performance.

3.6. Dnn Evaluation and Implementation on the Robot Controller

For the purpose of evaluating the DNN performance, different data sets were generated running the same sequence on different dates. The trained model is packed and loaded to run an offline validation test.

The DNN model is then packed as a computational library to be integrated into the robot controller. Thus the robot controller is programmed using the robot operating system (ROS). The libraries were programmed in python and packed as a Service-Node. The force/torque vector is received and processed running the DNN model. The result is sent back to the main control program that processes the data in the sequence logic.

3.7. Error Recovery Sequence

Since the DNN models present misclassifications in a real scenario, the assembly sequence is strengthened by adding logic constraints in every transition between steps. Two repetition cycles are added to the sequence. The first is represented as step S7, which repeats an orientation selection in case that the first DNN model outputs a wrong class, causing different contact states conditions. This can be identified and then triggers transition T6 to start this correction cycle.

For the case of the final insertion step S5, in case the insertion forces exceed the defined limits, the logic conditions will trigger T8 to start the adjustment cycle S8, where the second DNN model outputs a robot action that moves the peg to a center position.

This is one of the strengths of a discrete event system, that any step can be adjusted without modifying the whole control strategy. This new sequence program would lead the system to increase the rate of successful assemblies. Figure 4 shows the petri-net that represents the recovery sequence.

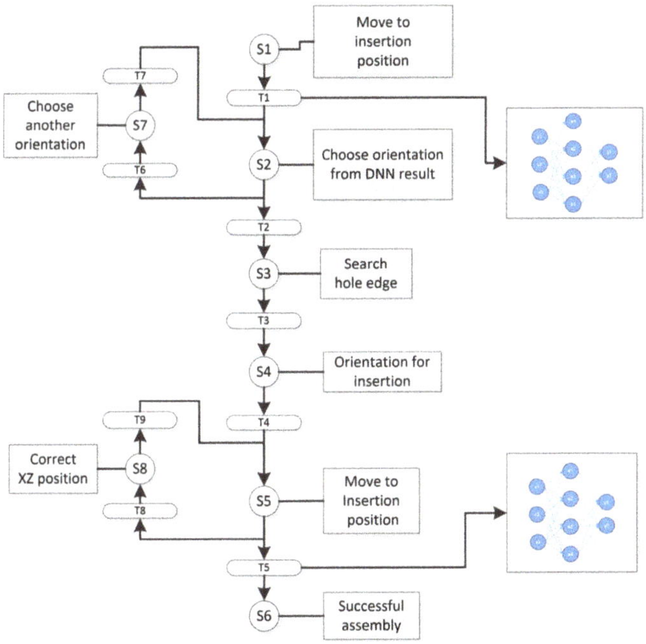

Figure 4. Modified sequence considering recovery steps.

The rule for trying another orientation after the first contact step is defined as follows:

$$o_{n+1} = \begin{cases} o_n + 1 & f > f_{min}, o_{axis} < o_{max}, try = 1 \\ o_n - 1 & f > f_{min}, o_{axis} < o_{max}, try = 2 \end{cases} \tag{14}$$

The rule for aligning the peg during the insertion step uses a DNN trained with basic contact states in order to output a reaction movement according to the predicted F/T pattern. The logic rule is described in the Table 1.

Table 1. Centering Rule for the correction position step.

Pattern	Movement in X	Movement in Z	Movement in Y
1	+X	0	0
2	−X	0	0
3	0	+Z	0
4	0	−Z	0
5	0	0	$-Y$

The first four patterns consider X and Z axis movements for error correction and for the pattern # 5; no error correction is commanded in the X-Z plane but there is a movement in the Y axis to continue insertion.

3.8. Validation Process of the Methodology

A validation process of different data sets is executed and the results are compared with the training set. A comparison of threshold methodology of single sensor data with DNN single sensor data, and then DNN with dual sensor data, is also executed.

A counter variable is set to limit the number of iterations or attempts in the assembly sequence.

In order to evaluate the performance of the machine learning algorithms, the Exact Match Ratio (MR) [36] as presented in the next equation is used to report experimental results.

$$MR(\hat{y}) = \frac{1}{n}\sum_{i=1}^{n} I_n \tag{15}$$

$$I_n = \begin{cases} 1 & if \quad \hat{y} = y \\ 0 & otherwise \end{cases} \tag{16}$$

where n is the sample number and I_n is a logic function of the correct classification of the sample.

Once the DNN models are trained and integrated into the assembly sequence, Success Ratio (SR) is used to evaluate the effects of using deep neural networks with one sensor and a double sensor in the automated cycle. SR is defined as follows:

$$SR = \frac{1}{n}\sum_{i=1}^{n} SA_n \tag{17}$$

$$SA_n = \begin{cases} 1 & if \quad p_{pos} \geq p_{i\_pos} \\ 0 & otherwise \end{cases} \tag{18}$$

where SR is the Success Ratio, SA_n is the successful assembly that is true if the peg position p_{pos} at the end of the cycle is greater or equal than the defined insertion position p_{i_pos} and n is the sample number.

4. Experimental Results

4.1. TestBed

The robotic testbed is composed of an industrial robot assembly cell that includes a dual arm robot SDA20D with two 7 DOF articulated arms and 1 rotating base giving a total of 15 DOF (degrees of freedom) as a whole system Figure 5. The robot controller is a DX100 with Ethernet port.

The maximum payload is 20 kg per arm, and repetitive positioning accuracy ±0.1 mm. Mounted in each arm are the force/torque sensors in a range with ±300 N and Torque

±30 Nm with 3 Cartesian components for each magnitude with a maximum data stream of 100 Hz for all component signals. Noise around 1.2 N and 0.02 Nm. Recommended threshold 5 N and 0.12 Nm maximum.

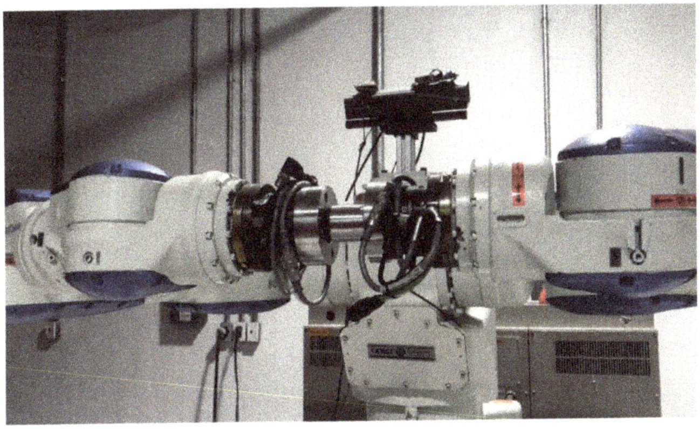

Figure 5. Motoman SDA20D Dual Arm Robot.

Both sensors are connected to an RS-485 to USB converter that sends the data to a Raspberry pi 3 running Robot Operating System (ROS). The parts to be assembled are a circular peg and hole in stainless steel with diameter of 40 mm and 39.9 mm for the peg with 50mm length. Both are fixed to the robot arms.

The main PC controller runs on Linux Ubuntu 18, and ROS Melodic is connected to the robot by the Ethernet 100 MB. A second PC controller runs Linux Ubuntu 20 and ROS Noetic, which runs the Deep Neural Network model as a service node. The robot has installed the MotoROS library that allows transfer of data via TCP/IP to the PC controllers and the Motodriver for ROS. Figure 6 shows a representation of the setup.

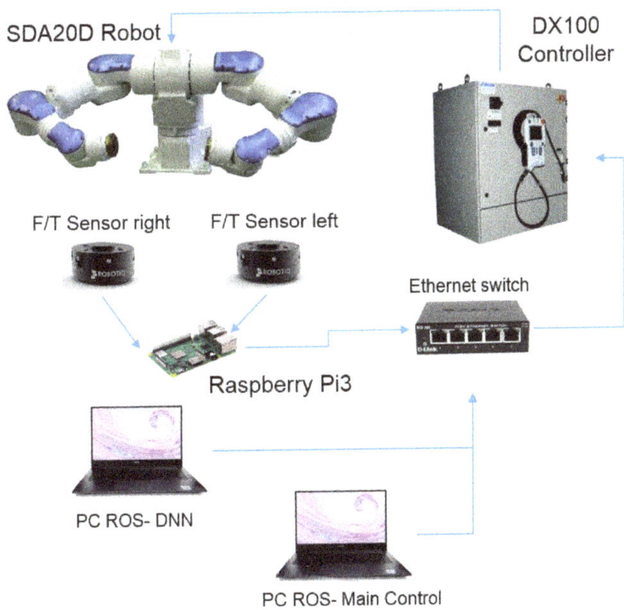

Figure 6. Experimental Setup.

4.2. Sequence Testing

In order to validate the proposed methodology, the previous experimental setup is used. The first goal in the experiment was to generate the data set of the force/torque signals from the two sensors. The assembly sequence is programmed in the computer to follow the 5 stages described in Figure 1.

As described in the graphical representation of the sequence in Figure 3, the robot arm that holds the peg moves towards the hole at a very slow speed; when the force limit is reached due to the first contact state, the robot will stop the movement and the contact forces are stabilized in a range of 1 or 2 s. With stable values of force/torque signals, the program gathers the information and saves a data file with the values of the error classification and the force/torque vectors from both sensors.

4.3. Data Set Creation

The database generated had a total of 2056 contact samples. The data set was analyzed using t-SNE (t- Distributed Stochastic Neighbor Embedded) technique to graphically visualize the separability of the features as shown in Figure 7, where it shows how the data can be separated into the 8 classes. As it can be observed, the classes 0, 6 and 7 show very poor separability characteristics, meaning that most of the misclassifications happen on these corresponding positions, meanwhile the other features seem to be separable.

Additional databases were generated in order to have validation data sets. The number of samples for the validation data sets are reported in Table 4 and start from 35 samples to go to up to 240 samples.

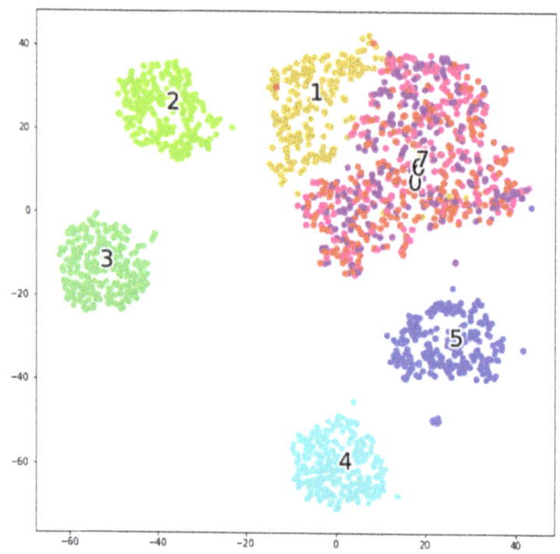

Figure 7. Representation of the data set and the 8 classes (0–7) by applying t-SNE technique.

4.4. Standard Automation Rules Results

For comparison purposes, the threshold is the first classification algorithm, and it consists of statistics metrics that define thresholds based on momentum M_x and M_y as decision constraints for the robot movements during the assembly cycle. Values of the thresholds obtained are presented in Table 2.

Table 2. Resulting thresholds of the data set. Values in Nm.

y	M_{x1low}	M_{y1low}	M_{x1high}	M_{y1high}
0	−1.22821	−1.168189	−0.90158	−0.882675
1	−0.57804	−1.554708	−0.404512	−1.171961
2	0.296703	−1.591266	0.558239	−1.164399
3	0.582325	−0.478162	1.069566	0.018427
4	−1.578333	0.662668	−1.148375	0.967986
5	−1.684699	−0.551533	−1.203931	−0.171751
6	−1.160531	−1.178287	−0.860481	−0.837394
7	−1.132745	−1.172939	−0.820671	−0.825263

A testing algorithm that runs the data set into the threshold comparisons was used to define whether the force/torque information corresponds to a certain defined class. The Equations (10) and (11) defined the threshold limits.

4.5. Deep Neural Network Training and Comparisons

The design of the neural networks was done through iterative testing where the initial configuration was based on empirical estimations. During the iterative design phase, different layer sizes and hyper-parameters were tested. The best result was obtained with the architecture of input layer 12 neurons, 5 hidden layers with unit size of 24, 48, 48, 48, 24 neurons, and, finally an output layer of 8 neurons. All input and hidden layers use an activation function *tanh*. For the output layer, $Softmax$ activation function is selected. The Loss Function is configured as Categorical Cross-entropy, and, as a learning optimizer, the Adam algorithm is selected.

The DNN model is programmed using Python libraries and Tensor-flow Keras [37]. Then, the data set is divided into a 70/30 relationship for training and testing in 180 epochs. The data set contains the force/torque values of the two sensors. It must be considered that every time the DNN model is trained, even with the same data, different results of performance are expected.

The performance of the algorithm for training/testing is shown in Figures 8 and 9. These graphs present the accuracy of the DNN, where the blue line represents the training data and the orange line the testing data. As seen in the one sensor case, the test data fits only up to 0.8 of the value during the 180 epochs, while the training data continues to fit the model. In the case of one sensor, the model implemented in the robot controller will present at least 20% of errors in classification.

On the other hand, the graph that corresponds to the two sensors shows the training and testing data with a high accuracy value. These results show a clear advantage to using two sensors rather than one.

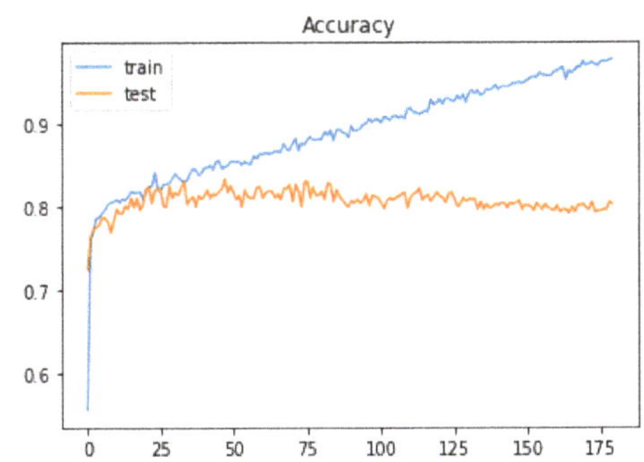

Figure 8. Single Sensor data training results 80% Accuracy.

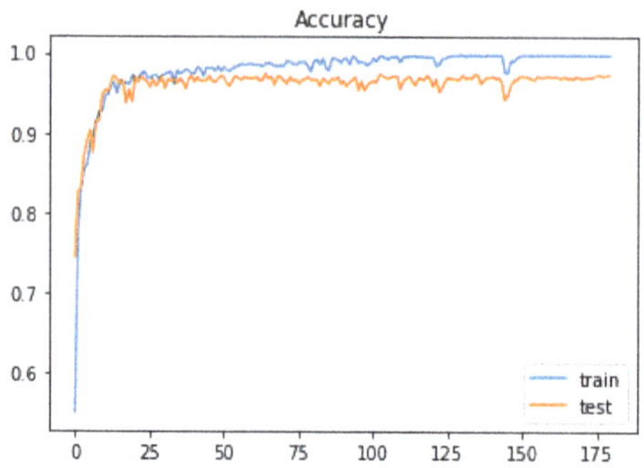

Figure 9. Double Sensor data training results 97.4% Accuracy.

Some other machine learning algorithms based on the sci kit-learn [38] were designed to train/test different classification techniques and compare how well they fit the data. A list of tested machine learning algorithms with this data set is presented as follows:

- Random Forest
- K-Nearest Neighbors
- Decision Tree
- Multi Layer Perceptron
- Deep Neural Networks

In order to demonstrate the difference between using only one F/T sensor versus two F/T sensors, the classification model uses either 6 features of the peg sensor or the 12 features of both peg and hole sensors.

To show the results of the training/testing of the machine learning models, a comparison of their performance is presented in Table 3. It can be seen that the threshold algorithms based on statistical metrics performs worst. The best model corresponds to the Deep Neural Network followed by the Random Forest. The metrics reported in the table consider the accuracy defined as the number of correct predictions divided by the total number of predictions. The second metric is the F1-Score that represents the harmonic mean between precision and recall metrics.

Table 3. Comparison of accuracies and F1-Score on different machine learning algorithms.

Algorithm	Acc One Sen	F1 One Sen	Acc Two Sen	F1 Two Sen
Threshold	22%	NA	31%	NA
Random Forest	80%	80%	94.9%	94.8%
Decision Tree	78%	78%	92%	92%
K Nearest Neighbors	71.9%	72.4%	78.7%	78.4%
Multi Layer Perceptron	79.6%	79.3%	93.5%	93.1%
Deep Neural Networks	91%	91%	98%	98%

4.6. Validating the DNN Models

The model was tested using extra data sets gathered after the main training data set acquisition. Using the same automated program, these new data sets with fewer contact samples were used as validation for the trained DNN model. The data sets start from 35 samples to got to 240 samples as shown in Table 4.

The algorithm to validate the trained model considers the same values of mean and standard deviation of the train data set. The Equation (12) is used to standardize the validation data sets. Then, the validation data set is used on the python Tensorflow Keras program to evaluate the correct classifications versus the incorrect classifications according to the Exact Match Ratio defined in (17).

Table 4 summarizes the accuracy results for the one sensor and two sensors. As shown in the results, there is a clear advantage to using two sensors rather than one sensor.

Table 4. Validation results of DNN Accuracy of one and two sensor.

Samples	Exact Match Ratio One Sensor	Exact Match Ratio Two Sensor
159	63%	93%
130	53%	81%
240	36%	84%
120	30%	85%
35	51%	88%

4.7. Testing on the Dual Arm Robot

In order to test this methodology in a real scenario of operation with the dual arm robot, the automated recovery program in Figure 5 is executed. Two computers were used to run the experimental setup, one computer ran the automatic sequence and sensor acquisition. The whole controller is programmed in C++ running on ROS using Moveit packages and libraries to perform Cartesian commands on the robot.

The second computer that runs the service node receives the F/T data as an array of twelve elements that is computed and, at the end, gives back the class prediction. This prediction $\hat{y}$ is used in the assembly sequence to define the next orientation angle or movement in the corresponding direction to find the next contact states of the sequence. Table 5 shows the corresponding orientation angles according to the predicted class.

Table 5. Classes and angle of movement.

$\hat{y}$	0	1	2	3	4	5	6	7
Angle	0°	45°	90°	135°	180°	225°	270°	315°

The last experiment considers the validation of the full assembly cycle. Logic rules used as transition constraints between the steps. The full cycle F/T data is presented in Figure 10 where the different steps and their corresponding F/T pattern can be seen. From free motion, the sequence evolves to the frontal contact state. The second step changes the orientation. The next step searches for lateral contact of the peg into the hole. At the end of the assembly sequence, the insertion phase shows a little deviation that exceeds a force limit value, which triggers the last DNN model. Reaction motions are selected according to the F/T data pattern and the DNN output. As the cycle continues, the patterns are detected and the position adjustment drives the F/T towards minimum values to lead the final assembly.

For these experiments a very slow motion is considered for safe contacts and to avoid any possible damage on the mechanical systems of the robot. Therefore, the time constraints for these experiments are not taken into account.

Finally, in order to compare the differences between the success ratio of assemblies using one sensor and using two sensors with two DNN's as classifiers, the complete assembly experiment is reported in Table 6.

Table 6. Final results and comparison of one vs two sensors.

Sensor	Examples	Average Steps	Success Ratio
One Sensor	40	17.65	68%
Two Sensors	40	14.975	98%

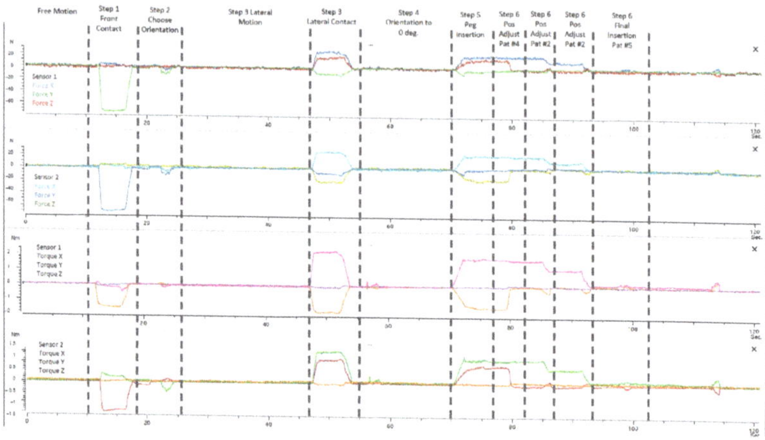

Figure 10. Forces and torques during one assembly cycle.

These results show that the whole experiment executed 40 assembly insertion attempts, where the number of sequence steps needed to finish the insertion is counted, after which the average number of steps for the total of assembly attempts is reported as a comparison metric. The Success Ratio, is the average of the completed insertions in the experiment, is also presented.

The graphical representation of the results are presented in Figure 11a, where one can see a better performance in the Success Ratio and in Figure 11b the Average of steps for the two sensor configurations.

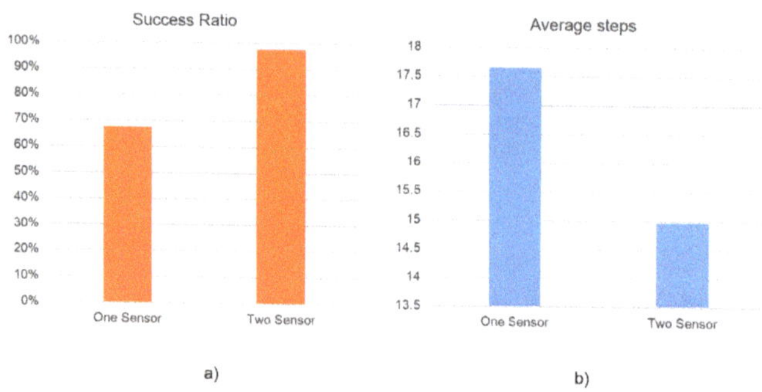

Figure 11. Experiment results. (**a**) Success Ratio and (**b**) Average of Steps.

5. Conclusions

We have presented in this paper a methodology to evaluate the effects of using double force/torque sensors on the peg-in-hole assembly and deep neural networks. An assembly strategy was proposed, which was inspired on how a human would solve the positioning error of a peg-in-hole assembly. A discrete event controller designed to run the assembly sequence was designed as it would be programmed in an industrial automated robot cell. The purpose of this strategy is to achieve the assembly task despite the aleatory positioning error of the peg and hole.

A force/torque signal acquisition program was implemented to generate the data set for training machine learning algorithms. The automated data acquisition proved very efficient if different conditions are to be considered. This allows for testing different parameters before the final experiments. It also ensures that the contact states are gathered with the same assembly conditions as it would be in a normal automated cycle.

Off-line learning of the contact states is achieved by using deep neural networks. It was found that the publicly available machine learning libraries are a great evaluation solution to integrate in a ROS-based system. Despite the good results using DNN, we encourage evaluation other machine learning algorithms in the experimental setup. We noticed how the F/T signals are less separable in a certain geometric region presenting more false positives. Further investigation of this F/T signal behavior is needed.

Additionally, it is shown that the success ratio of the assembly process improved by the use of double force/torque sensors and deep neural networks. Although the comparison between different research works is not simple due to different experimental conditions, such as robot manufacturer, developing software and perception systems among others, the findings of this research presents a contribution on the evaluation of dual force torque sensors for assembly tasks.

The results obtained by using one and two force/torque sensors to classify contact states showed an improvement of 30% of Success Rate. The assembly strategy was based on a discrete event controller represented by a Petri-Net sequence diagram. Because this sequence diagram represents an automation program, we consider it feasible to test the methodology on industrial scenarios. However, this work does not take into account the overall assembly time due to safety considerations such as working with slow speed motions to avoid any damage to the robot. For an industrial application, this needs to be investigated.

There are still many conditions that need to be taken into account to implement this strategy in an industrial application. For example, changing shapes of the mating parts, the effects of adjustments such as speed, force limits and so on. These changes would affect the performance of the classifier due to the changes in the F/T behavior.

Future work considers different challenging scenarios such as testing different machine learning algorithms, the evaluation of different assembly parameters such as higher speed movements, higher force/torque limits, shape of the mating parts, the increase of uncertain conditions such as the number of aleatory error positions and so on. Evaluation of dual arm coordination strategies will also be investigated. Moreover, integration of grippers to manipulate the objects will increase positioning uncertainties that also need to be studied.

Author Contributions: D.O.-A. contributed to the conceptualization, design, implementation, and python programming as well as the DNN design. I.L.-J. conceptualized the use of the DNN for peg-in-hole assembly and wrote-up the first paper draft. B.N.S. and J.F.J.-V. contributed to the experimental design, scikit-learn validation and the revision of the final draft. All authors contributed to writing-up the final paper version and agreed to the published version of the manuscript. All authors have read and agreed to the published version of the manuscript.

Funding: This research was fully funded through CIDESI and CONACYT.

Institutional Review Board Statement: Not applicable.

Informed Consent Statement: Not applicable.

Data Availability Statement: The data presented in this study are openly available in https://data.mendeley.com/datasets/m7g93whs3t/1, doi:10.17632/m7g93whs3t.1. (accessed on 15 June 2021).

Acknowledgments: Thanks are sincerely due to Israel Uribe and Sinai Aranda for his valuable technical support. Hugo Gamez for his kind support on reviewing the writing.

Conflicts of Interest: The authors declare no conflict of interest.

Abbreviations

The following abbreviations are used in this manuscript:

DNN	Deep Neural Networks
DES	Discrete Event Systems
t-SNE	t-distributed Stochastic Neighbour Embedded

F/T	Force and Torque
ROS	Robot Operating System

References

1. Dharmara, K.; Monfared, R.P.; Ogun, P.S.; Jackson, M.R. Robotic assembly of threaded fasteners in a non-structured environment. *Int. J. Adv. Manuf. Technol.* **2018**, *98*, 2093–2107. [CrossRef]
2. Smith, C.; Karayiannidis, Y.; Nalpantidis, L.; Gratal, X.; Qi, P.; Dimarogonas, D.V.; Kragic, D. Dual arm manipulation—A survey. *Robot. Auton. Syst.* **2012**, *60*, 1340–1353. [CrossRef]
3. Ortega-Aranda, D.; Lopez-Juarez, I.; Nath-Saha, B.; Osorio-Comparan, R.; Pena-Cabrera, M.; Lefranc, G. Towards learning contact states during peg-in-hole assembly with a dual-arm robot. In *2017 CHILEAN Conference on Electrical, Electronics Engineering, Information and Communication Technologies (CHILECON)*; Staff, I., Ed.; IEEE: Piscataway, NJ, USA, October 2017; pp. 1–6. [CrossRef]
4. Weng, C.Y.; Tan, W.C.; Chen, I.M. A Survey of Dual-Arm Robotic Issues on Assembly Tasks. In *ROMANSY 22—Robot Design, Dynamics and Control*; Arakelian, V., Wenger, P., Eds.; CISM International Centre for Mechanical Sciences; Springer International Publishing: Cham, Switzerland, 2019; Volume 584, pp. 474–480. [CrossRef]
5. Tsarouchi, P.; Makris, S.; Michalos, G.; Stefos, M.; Fourtakas, K.; Kaltsoukalas, K.; Kontrovrakis, D.; Chryssolouris, G. Robotized Assembly Process Using Dual Arm Robot. *Procedia CIRP* **2014**, *23*, 47–52. [CrossRef]
6. Rojas, J.; Guan, T.; Luo, W.; Harada, K. Strategies, controllers, and coordination: Bi-manual snap assembly automation. In Proceedings of the 2014 IEEE International Conference on Robotics and Biomimetics, IEEE ROBIO 2014, Bali, Indonesia, 5–10 December 2014; pp. 1266–1271. [CrossRef]
7. Zhang, X.; Zheng, Y.; Ota, J.; Huang, Y. Peg-in-hole assembly based on two-phase scheme and F/T sensor for dual-arm robot. *Sensors* **2017**, *17*, 2004. [CrossRef] [PubMed]
8. Polverini, M.P.; Zanchettin, A.M.; Castello, S.; Rocco, P. Sensorless and constraint based peg-in-hole task execution with a dual-arm robot. In Proceedings of the IEEE International Conference on Robotics and Automation, Stockholm, Sweden, 16–21 May 2016; pp. 415–420. [CrossRef]
9. Kruse, D.; Wen, J.T.; Radke, R.J. A Sensor-Based Dual-Arm Tele-Robotic System. *IEEE Trans. Autom. Sci. Eng.* **2015**, *12*, 4–18. [CrossRef]
10. Levine, S.; Abbeel, P. *Learning Neural Network Policies with Guided Policy Search under Unknown Dynamics*; MIT Press: Cambridge, MA, US , 2014; Volume 27, pp. 1071–1079.
11. Makris, S.; Tsarouchi, P.; Surdilovic, D.; Krüger, J. Intuitive dual arm robot programming for assembly operations. *CIRP Ann. Manuf. Technol.* **2014**, *63*, 13–16. [CrossRef]
12. Krüger, J.; Schreck, G.; Surdilovic, D. Dual arm robot for flexible and cooperative assembly. *CIRP Ann.* **2011**, *60*, 5–8. [CrossRef]
13. Almeida, D.; Karayiannidis, Y. Folding assembly by means of dual-arm robotic manipulation. In Proceedings of the 2016 IEEE International Conference on Robotics and Automation (ICRA), Stockholm, Sweden, 16–21 May 2016; Volume 2016, pp. 3987–3993. [CrossRef]
14. Xu, J.; Hou, Z.; Liu, Z.; Qiao, H. Compare Contact Model-Based Control and Contact Model-Free Learning: A Survey of Robotic Peg-in-Hole Assembly Strategies). *arXiv* **2019**, arXiv:1904.05240.
15. Ren, T.; Dong, Y.; Wu, D.; Chen, K. Learning-Based Variable Compliance Control for Robotic Assembly. *J. Mech. Robot.* **2018**, *10*. [CrossRef]
16. Inoue, T.; Magistris, G.D.; Munawar, A.; Yokoya, T.; Tachibana, R. Deep reinforcement learning for high precision assembly tasks. In Proceedings of the IEEE/RSJ International Conference on Intelligent Robots and Systems (IROS 2017), Vancouver, BC, Canada, 24–28 September 2017; pp. 819–825. [CrossRef]
17. Huang, Y.; Zheng, Y.; Wang, N.; Ota, J.; Zhang, X. Peg-in-hole assembly based on master-slave coordination for a compliant dual-arm robot. *Assem. Autom.* **2020**, *40*, 189–198. [CrossRef]
18. Newman, W.S.; Zhao, Y.; Pao, Y.H. Interpretation of force and moment signals for compliant peg-in-hole assembly. *Proc. ICRA IEEE Int. Conf. Robot. Autom.* **2001**, *1*, 571–576. [CrossRef]
19. Cervera, E.; Del Pobil, A.P.; Marta, E.; Serna, M.A. Perception-based learning for motion in contact in task planning. *J. Intell. Robot. Syst.* **1996**, *17*, 283–308. [CrossRef]
20. Gullapalli, V.; Grupen, R.A.; Barto, A.G. Learning reactive admittance control. In *1992 IEEE International Conference on Robotics and Automation*; IEEE Computer Society Press: Los Alamitos, CA, USA, 1992; pp. 1475–1480. [CrossRef]
21. McCarragher, B.J. Task primitives for the discrete event modeling and control of 6-DOF assembly tasks. *IEEE Trans. Robot. Autom.* **1996**, *12*, 280–289. [CrossRef]
22. Beltran-Hernandez, C.C.; Petit, D.; Ramirez-Alpizar, I.G.; Nishi, T.; Kikuchi, S.; Matsubara, T.; Harada, K. Learning Force Control for Contact-Rich Manipulation Tasks With Rigid Position-Controlled Robots. *IEEE Robot. Autom. Lett.* **2020**, *5*, 5709–5716. [CrossRef]
23. Abdullah, M.W.; Roth, H.; Weyrich, M.; Wahrburg, J. An Approach for Peg-in-Hole Assembling using Intuitive Search Algorithm based on Human Behavior and Carried by Sensors Guided Industrial Robot. *IFAC PapersOnLine* **2015**, *48*, 1476–1481. [CrossRef]
24. Chung, S.Y.; Doo Yong, L. Discrete event systems approach to fixtureless peg-in-hole assembly. In Proceedings of the 2001 American Control Conference (Cat. No.01CH37148), Arlington, VA, USA, 25–27 June 2001; Volume 6, pp. 4962–4967. [CrossRef]

25. Hwang, M.J.; Lee, D.Y.; Chung, S.Y. Motion planning of bimanual robot for assembly. In Proceedings of the IEEE International Conference on Systems, Man and Cybernetics, Montreal, QC, Canada, 7–10 October 2007; pp. 240–245. [CrossRef]
26. Zhao, X.; Zhao, H.; Chen, P.; Ding, H. Model accelerated reinforcement learning for high precision robotic assembly. *Int. J. Intell. Robot. Appl.* **2020**, *4*, 202–216. [CrossRef]
27. Al-Yacoub, A.; Zhao, Y.; Lohse, N.; Goh, M.; Kinnell, P.; Ferreira, P.; Hubbard, E.M. Symbolic-Based Recognition of Contact States for Learning Assembly Skills. *Front. Robot. AI* **2019**, *6*, 99. [CrossRef] [PubMed]
28. Jasim, I.F.; Plapper, P.W. Contact-state monitoring of force-guided robotic assembly tasks using expectation maximization-based Gaussian mixtures models. *Int. J. Adv. Manuf. Technol.* **2014**, *73*, 623–633. [CrossRef]
29. Park, H.; Park, J.; Lee, D.H.; Park, J.H.; Baeg, M.H.; Bae, J.H. Compliance-Based Robotic Peg-in-Hole Assembly Strategy without Force Feedback. *IEEE Trans. Ind. Electron.* **2017**, *64*, 6299–6309. [CrossRef]
30. Salem, A.; Karayiannidis, Y. Robotic Assembly of Rounded Parts With and Without Threads. *IEEE Robot. Autom. Lett.* **2020**, *5*, 2467–2474. [CrossRef]
31. Chung, S.Y.; Lee, D.Y. An augmented Petri net for modelling and control of assembly tasks with uncertainties. *Int. J. Comput. Integr. Manuf.* **2005**, *18*, 170–178. [CrossRef]
32. Tang, T.; Lin, H.C.; Tomizuka, M. *A Learning-Based Framework for Robot Peg-Hole-Insertion. Volume 2: Diagnostics and Detection; Drilling; Dynamics and Control of Wind Energy Systems; Energy Harvesting; Estimation and Identification; Flexible and Smart Structure Control; Fuels Cells/Energy Storage; Human Robot Interaction; HVAC Building Energy Management; Industrial Applications; Intelligent Transportation Systems; Manufacturing; Mechatronics; Modelling and Validation; Motion and Vibration Control Applications*; American Society of Mechanical Engineers: New York, NY, US , 2015; doi:10.1115/DSCC2015-9703. [CrossRef]
33. Dietrich, F.; Buchholz, D.; Wobbe, F.; Sowinski, F.; Raatz, A.; Schumacher, W.; Wahl, F.M. On contact models for assembly tasks: Experimental investigation beyond the peg-in-hole problem on the example of force-torque maps. In Proceedings of the IEEE/RSJ 2010 International Conference on Intelligent Robots and Systems, IROS 2010 - Conference Proceedings, Taipei, Taiwan, 18–22 October 2010; pp. 2313–2318. [CrossRef]
34. Park, H.; Park, J.; Lee, D.H.; Park, J.H.; Bae, J.H. Compliant Peg-in-Hole Assembly Using Partial Spiral Force Trajectory with Tilted Peg Posture. *IEEE Robot. Autom. Lett.* **2020**, *5*, 4447–4454. [CrossRef]
35. van der Maaten, L.; Hinton, G. Visualizing Data using t-SNE. *J. Mach. Learn. Res.* **2008**, *9*, 2579–2605.
36. Sorower, M.S. A literature survey on algorithms for multi-label learning. *Or. State Univ. Corvallis* **2010**, *18*, 1–25.
37. Abadi, M.; Agarwal, A.; Barham, P.; Brevdo, E.; Chen, Z.; Citro, C.; Corrado, G.S.; Davis, A.; Dean, J.; Devin, M.; et al. TensorFlow: Large-Scale Machine Learning on Heterogeneous Systems, 2015. Available online: tensorflow.org (accessed on 13 March 2021)
38. Pedregosa, F.; Varoquaux, G.; Gramfort, A.; Michel, V.; Thirion, B.; Grisel, O.; Blondel, M.; Prettenhofer, P.; Weiss, R.; Dubourg, V.; et al. Scikit-learn: Machine Learning in Python. *J. Mach. Learn. Res.* **2011**, *12*, 2825–2830.

applied sciences

MDPI

Article

Compliant Human–Robot Collaboration with Accurate Path-Tracking Ability for a Robot Manipulator

Daniel Reyes-Uquillas *,† and Tesheng Hsiao †

Department of Electrical and Computer Engineering, National Yang Ming Chiao Tung University, Hsinchu 30010, Taiwan; tshsiao@cn.nctu.edu.tw
* Correspondence: danielreyes.04g@g2.nctu.edu.tw
† These authors contributed equally to this work.

Abstract: In this article, we aim to achieve manual guidance of a robot manipulator to perform tasks that require strict path following and would benefit from collaboration with a human to guide the motion. The robot can be used as a tool to increase the accuracy of a human operator while remaining compliant with the human instructions. We propose a dual-loop control structure where the outer admittance control loop allows the robot to be compliant along a path considering the projection of the external force to the tangential-normal-binormal (TNB) frame associated with the path. The inner motion control loop is designed based on a modified sliding mode control (SMC) law. We evaluate the system behavior to forces applied from different directions to the end-effector of a 6-DOF industrial robot in a linear motion test. Next, a second test using a 3D path as a tracking task is conducted, where we specify three interaction types: free motion (FM), force-applied motion (FAM), and combined motion with virtual forces (CVF). Results show that the difference of root mean square error (RMSE) among the cases is less than 0.1 mm, which proves the feasibility of applying this method for various path-tracking applications in compliant human–robot collaboration.

Keywords: path tracking; compliance control; human–robot interaction; sliding mode controller; manual guidance

Citation: Reyes-Uquillas, D.; Hsiao, T. Compliant Human–Robot Collaboration with Accurate Path-Tracking Ability for a Robot Manipulator. *Appl. Sci.* **2021**, *11*, 5914. https://doi.org/10.3390/app11135914

Academic Editor: Luis Gracia

Received: 31 May 2021
Accepted: 23 June 2021
Published: 25 June 2021

1. Introduction

Guidance constraints are a type of active constraints/virtual fixtures, which have been studied in the field of physical human–robot interaction (pHRI) by many researchers [1–7]. From the work of Rosenberg [1], the concept of perceptual overlay to increase task performance has been explored in different types of human–machine manipulation systems, such as telemanipulators and cooperative manipulators [3]. These virtual fixtures are often used to guide the movement of the robot towards desired targets while avoiding obstacles and preventing the robot from entering predefined forbidden regions. Although reference paths for the robot are generated in some applications, the accuracy of path following is not the focus of these techniques. Instead, compliant behavior of the robot that allows deviation of the path is the main concern of most pHRI studies. However, accurate path tracking and compliant behavior are not necessarily antinomies. They can be delicately tailored to meet the requirements for dedicated purposes and achieve better performance. For example, rehabilitation patients can benefit greatly from practicing with equipment that is capable of following a path while still being compliant along the path, because it allows the patient to closely track an optimal path predefined by medical experts, and progress in different levels of assistance, which has been shown to improve rehabilitation progress [8–10]. Another important application is in industrial assembly tasks. Many efforts have been reported in this area towards efficient human–robot collaboration [11–15]. The authors of [13] presented a force-overshoot-free controller for high-accuracy assembly tasks using an industrial robot. Furthermore, in [14] a collaborative task for a homokinetic joint assembly is performed using admittance control. In [15], a method for robotic fixtureless approach of large flexible panels assembly using compliance force control is presented and evaluated with an industrial robot. The aim of these approaches is to relieve the operator from having to manipulate heavy components during the assembly and instead being able to focus only with the dexterous part of the assembly. Moreover, an important factor is that in high-precision assembly tasks static fixtures are necessary to keep in place the parts being joined and task performance depends on these fixtures. These tasks could benefit from a compliant path-tracking system. Furthermore, other direct application of path tracking in human–robot collaboration is for educational and physical training purposes where the robot system can be used as a tool to teach complex tasks that are best learned by hand-in-hand demonstration. Examples include practicing calligraphy, assisted hand–motor skills development, and so on [16,17]. In this sense, the robot system acts as the coach and the user is guided to follow an optimal path that perfectly accomplishes the task while obtaining different levels of assistance to trace the path, from completely passive to completely active. Based on these analyses, we found that in many manual guidance applications, it is beneficial for the robot to follow a path while being compliant along the path. Consequently, the goal of this paper is to investigate the related techniques.

Path-tracking is an important feature of most robotic systems. From manipulators to mobile platforms, the ability to follow a path with high accuracy has been investigated by many researchers [2,18,19]. Moreover, path-tracking in compliant human–robot collaboration has been the topic of recent studies. In [4], virtual fixtures were created and modified through an iterative method based on kinesthetic teaching and Akima splines. The approach was evaluated in a simulated sanding task using a 3-DOF collaborative robot and a user study which showed that virtual guide assistance improves the performance of co-manipulation tasks. Later, Raiola et al. in [20] defined virtual guides as virtual mechanisms using Gaussian mixtures model (GMM) from multiple user demonstrations, which can be refined through incremental training. They proposed to establish a library of multiple virtual guides which can be created, modified and used by the human operator. The performance was experimentally evaluated with a compliant 3-DOF robot and a user study. The algorithms in [4,20] concentrated on virtual guides construction and how to iteratively modify the nominal path, but there was no treatment of the accuracy of path-tracking. Wu et al. in [21] optimally reshaped the desired robot trajectory using an adaptive neural network controller to effectively complete co-manipulation tasks with a robotic exoskeleton. The experimental validation was demonstrated in a 2-DOF motion case but did not describe extensibility of the method to higher DOFs or to different types of manipulators. In other words, the research effort was to conceive an algorithm that allows the human to modify the desired trajectory using shared control; thus, the contributions focus on the path generation aspect of human–robot collaboration. However, the challenge of accuracy in human–robot interaction by regulating robot motion in the predetermined task direction remains.

Previous works on the accuracy aspect for compliant tasks by the use of virtual fixtures include [5–7], where virtual fixtures were used together with admittance control to allow for compliant path tracking tasks. A vision-based virtual fixture algorithm for cooperative manipulation was designed and implemented in [5], where the authors described how to create hard and soft virtual fixtures for two types of reference targets—point positioning and curve following—as well as the extensions to virtual geometries such as tubes or cones to confine motion to a volume for a cooperative system which uses admittance control. The experimental setup was validated using the "Steady Hand Robot" [22] in macroscale and microscale planar tasks. The results showed that the addition of vision-based virtual fixtures assistance increased performance regarding position error and execution time, and that there is a trade-off between performance and user control. However, the experimental validation consisted in using different fixed levels of admittance, but as covered in [23] adaptive admittance control has been proven more effective. Moreover, there is no distinction between the virtual fixture definition and the admittance control parameter tuning. This complicates the definition of the virtual fixtures and directly influences the achievable tracking performance, as shown in their study. In [6], virtual fixtures were used to control the robot for robotic-assisted surgery. The task was defined and projected into a space of orthogonal directions that consisted of preferred directions and non-preferred ones. The authors of [6] explored and compared two types of error compensation—manual and autonomous—to reduce the path-tracking task error. The results showed that manual compensation was not sufficient for correction of all deviations, particularly when a translational virtual fixture was defined and the deviation error for orientation occurred, and vice versa. This was solved with their proposed autonomous compensation method where the user had control of the system along the preferred directions through the forces applied, while the robotic system independently compensated deviations. In [7], a framework for compliant path tracking tasks was proposed. The system used a virtual robot approach to obtain the desired velocities that were treated as the command to the internal position controller. Motion constraints and task space constraints can be defined in this framework. The proposed approach was experimentally evaluated using a 7-DOF KUKA LWR cobot. The results showed that it is easy to define different constraints in the framework by adjusting the properties and parameters in their algorithm. However, all parameters were tuned heuristically and specifically for the evaluated task. The authors of [24] presented a hybrid position-force control oriented to surface treatment tasks, e.g., sanding, deburring, polishing, etc. They used sliding mode concepts for the robot force control to maintain the orientation of the tool always perpendicular to the surface and to obtain the desired tool pressure on the surface. The lower-priority tracking controller was designed to follow the desired trajectory where deviations to satisfy safety constraints are allowed. Finally, redundancy resolution was also included to keep the manipulator near its home configuration for safety purposes. The method showed increased robustness and low computational cost, and experimental validation of the approach using a 7R manipulator proved its applicability and effectiveness.

The common drawbacks of the cited works are that, in many cases, the accuracy aspect of path tracking when compliance is combined for co-manipulation tasks was not explored. Besides, the experimental setup often relies on robots designed to be inherently suitable for manipulation tasks. Furthermore, kinematic constraints such as singularities, joint limits, task space limits, etc., if considered, are part of the virtual fixture definition and therefore task dependent. As a result, the task must be carefully planned to avoid these kinematic constraints, which imposes restrictions on the task definition. Given these recent references in the field and considering their contributions and stated drawbacks, we believe the topic of path tracking is highly relevant to human–robot collaboration applications, and the study of the associated design strategies is of great importance and value.

In this paper, we aim to achieve compliance of a robot manipulator along a predefined path, giving the robot the ability to move forward or backward depending on the direction of an external force applied to the end effector while tracking the path. Figure 1 illustrates this objective, where a desired path is shown and in the case of an external force is applied, the robot is compliant only along the path tangential direction, and not compliant along the normal direction. To accomplish this goal, we propose a dual-loop control structure, including an inner motion control loop and an outer admittance control loop. A modified sliding mode controller (SMC) is designed for the inner motion control, which guarantees desirable state tracking performance and robustness while significantly suppresses chattering. In the outer admittance loop, we define the compliance control problem in the tangential-normal-binormal (TNB) frame associated with the path to find the projection of the external force in the tangential direction and develop an atemporal path generation strategy by parameterizing the path with respect to the arc length to move along the path as desired by the user. Furthermore, an adaptive admittance law that has been experimentally verified in [23] is exploited to allow a timely, intuitive, and simple communication between the human and robot. In addition, the proposed adaptive admittance law considers the most common kinematics constraints that can appear in manual guidance interaction and adapts the admittance parameters appropriately based on the objectives of safety, accuracy, and comfort. In this approach, kinematic constraints are not task dependent and instead are configuration dependent, which means that they can be reused in multiple applications. Furthermore, these constraints are treated simultaneously, considering smooth transitions among them. Combining the TNB frame projection and adaptive admittance law, the outer admittance control loop realizes a safe and intuitive human-robot collaboration along a predefined path while the robust inner motion control loop guarantees that the robot follows the path accurately.

We validate the proposed approach with two experiments including a linear motion test to demonstrate the behavior of the robot to the direction of the external force applied and a general 3D path test. Both tests are performed on a traditional 6-DOF industrial robot equipped with a force/torque (F/T) sensor on its end effector. No dedicated hardware is required. We compare three types of motion—free motion (FM), force-applied motion (FAM), and combined motion with virtual forces (CVF)—to show the performance in both ends of the assistance range spectrum. FM demonstrates the fully assisted case, where no force from the user is needed; FAM is the no-assistance case, where the motion of the robot along the path depends on the force applied by the user; and CVF is the smooth transition between the previous two cases. To achieve the proposed objective of compliant human–robot collaboration with accurate path-tracking, the robot follows a reference path and the tracking error should be small, whether it is subject to external forces or not. Thus, we evaluate the tracking performance of each type of motion by calculating the root mean square error (RMSE). The experiments exhibit not only accurate path tracking with desired compliant behavior but also the applicability of the proposed approach to traditional industrial manipulators with F/T sensors. Therefore, the main contributions of this work are as follows:

- Development of a dual-loop control structure that achieves accurate path tracking for collaborative tasks while maintaining compliance along the path.
- Safe and intuitive human–robot interaction achieved by the outer adaptive admittance control loop.
- Equivalent path tracking performance guaranteed by the inner modified SMC loop under various human–robot collaborative scenarios.
- Experimental validation with a traditional 6-DOF industrial robot proving the generality of the method implementation.

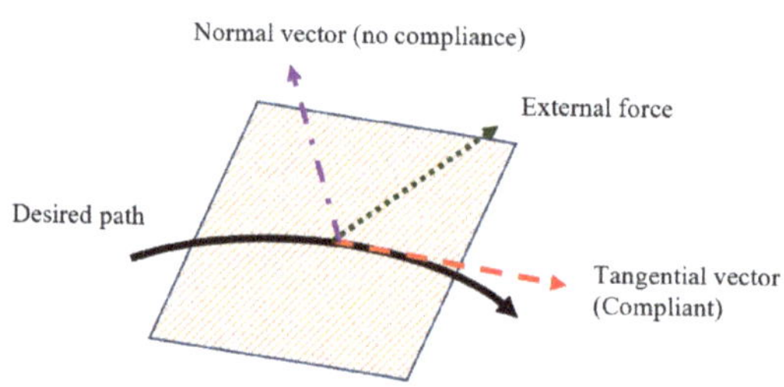

Figure 1. Path tracking for human–robot interaction with compliance along a predefined path.

The paper is structured as follows. Section 2 covers the problem formulation and the methodology, including the inner motion control loop and the outer admittance control loop. In Section 3, we describe the experimental setup and tests performed to validate the proposed approach. Additionally, we present the results and analysis of RMSE, which is used to evaluate the tracking performance. Finally, the conclusions of the work are presented in Section 4.

2. Problem Formulation and Methodology

As we discussed in the previous section, the most effective way to accomplish particular human–robot collaboration in the fields of rehabilitation, industrial assembly, education, and physical training is by guiding the user to closely follow an optimal path with robots, while reserving a certain degree of compliance for safe and smooth human–robot interaction. In these applications, the path is defined in advance by domain experts. The goal of this paper is to design a controller that confines the end effector of a robot on the path but allows it to move forward or backward freely along the path depending on the direction of external forces. If the external force is too weak or null, a virtual force is applied to the end effector from the controller to track the path autonomously and accurately. In other words, the proposed control system endows the robot with directional compliance along a prescribed path, and high stiffness in other directions for accurate path tracking. Moreover, the degree of compliance is online adjustable according to the applied forces and safety operation criteria of the robot such that the interaction between the human and the robot is safe, comfortable, and intuitive.

In this paper, we consider an n-joint robot that carries out tasks in the 3D task space, where $n \geq 3$. An F/T sensor is mounted on the end effector of the robot to measure external forces. To accomplish the goal of this paper, we propose a dual-loop control structure as shown in Figure 2. The inner loop is dedicated to precision motion control in the joint space. Namely, given the desired trajectory in the joint space $q_d \in \mathbb{R}^n$, the motion controller in the inner loop calculates the corresponding joint torque such that the actual joint angle follows q_d accurately. q_d is determined by the outer admittance control loop based on the desired task space path and the external force $F_{ext} \in \mathbb{R}^3$. The filtered external force $f_{ext} \in \mathbb{R}^3$ is projected to the tangent direction of the path and then converted to the corresponding task space velocity $\dot{r}_d \in \mathbb{R}^3$ through an admittance function whose parameters are online adjustable. Then, $\dot{r}_d$ is transformed to the joint space to obtain q_d. In the following subsections, we present the details of the techniques used in the proposed dual-loop control structure.

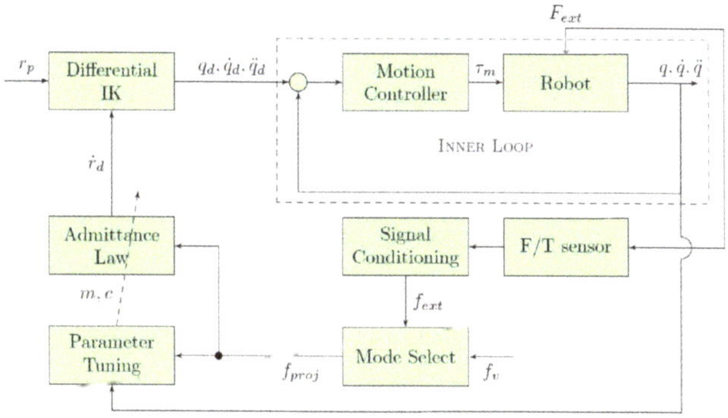

Figure 2. Block diagram of adaptive admittance control with path tracking.

2.1. Inner Motion Control Loop

The dynamic equation of an n-joint robot is given as follows:

$$M(q)\ddot{q} + C(q,\dot{q})\dot{q} + G(q) + F(\dot{q}) = \tau + J^T(q)F_{ext} \quad (1)$$

where $q, \dot{q}, \ddot{q}, \tau \in \mathbb{R}^n$, denote the joint angle, angular velocity, angular acceleration, and torque, respectively. $M(q), C(q,\dot{q}) \in \mathbb{R}^{n\times n}$ are the inertial matrix, Coriolis, and centrifugal matrix, respectively. $G(q), F(\dot{q}) \in \mathbb{R}^n$ are the gravitational torque and frictional torque, respectively. $J(q) \in \mathbb{R}^{3\times n}$ is the Jacobian matrix of the end effector, and $F_{ext} \in \mathbb{R}^3$ is the external force applied to the end effector.

Given the desired joint space trajectory $q_d \in \mathbb{R}^n$, we would like to design a robust motion control law that calculates the required joint torque τ for keeping the tracking error $\tilde{q} = q - q_d$ as small as possible, even in the presence of model uncertainties and external disturbance. It is well known that SMC has desirable robustness with respect to matched uncertainties, but high-frequency chattering causes degradation of performance and should be suppressed carefully. Many variations of SMC have been investigated in the past for alleviating chattering with minor cost in performance deterioration [25–29]. Among all these techniques, we adopt the modified SMC developed and experimentally verified by our research group in [30].

Let $M_0(q), C_0(q,\dot{q}), G_0(q)$, and $F_o(\dot{q})$ be the nominal models of $M(q), C(q,\dot{q}), G(q)$, and $F(\dot{q})$, respectively. Then, (1) can be rewritten as

$$\ddot{q} = -M_0^{-1}(q)[C_0(q,\dot{q})\dot{q} + G_0(q) + F_0(\dot{q})] + (I + \Delta_M)M_0^{-1}(q)\left(\tau + J^T(q)F_{ext}\right) + d \quad (2)$$

where d is the lumped additive model uncertainties resulting from all model parameters in $M(q), C(q,\dot{q}), G(q)$ and $F(\dot{q})$. Δ_M is the multiplicative model uncertainty of $M^{-1}(q)$. We assume that $\|\Delta_M\| \leq \kappa < 1$ and $\|d\| \leq \bar{d}$ for all q and $\dot{q}$.

Define the sliding variable $\sigma \in \mathbb{R}^n$ as

$$\sigma = [\sigma_1, \cdots, \sigma_n]^T = \dot{\tilde{q}} + C_1\tilde{q} + \eta \quad (3)$$

where $C_1 \in \mathbb{R}^{n\times n}$ is positive definite and $\eta \in \mathbb{R}^n$ is an auxiliary vector which will be determined shortly. The proposed control law is

$$\tau = [C_0(q,\dot{q})\dot{q} + G_0(q) + F_0(\dot{q})] + M_0(q)(\ddot{q}_d - C_1\dot{\tilde{q}} + \Lambda_0 u_1) \quad (4)$$

$$\dot{u}_1 = -\Lambda_1 u_1 + \eta \quad (5)$$

$$\dot{\eta} = -\Lambda_0 u_1 - K_s\sigma - \rho \quad (6)$$

Note that the control law in (4) cancels the nominal nonlinear terms and introduces an auxiliary control input u_1 to enhance robustness. $\Lambda_1, \Lambda_0, K_s \in \mathbb{R}^{n\times n}$ are constant gain matrices and

$$\rho = \left[sgn(\sigma_1)\bar{d}_{s1}, \cdots, sgn(\sigma_n)\bar{d}_{sn}\right]^T \quad (7)$$

where $sgn(\cdot)$ denotes the sign function, and $\bar{d}_{si} = \bar{d} + \kappa\|\ddot{q}_d - C_1\dot{\tilde{q}} + \Lambda_0 u_1\|, i = 1, \cdots, n$, is the bound of the uncertain terms. Each component of ρ switches its sign across the sliding surface $\sigma_i = 0$ and is used to suppress the model uncertainty. However, the high-frequency switching behavior of ρ causes significant chattering if it is directly added to the joint torque, just as the traditional SMC does.

Let K_s be positive definite, if we substitute (4)–(6) into (2), it can be shown that [30]

$$\sigma^T\dot{\sigma} \leq -\sigma^T K_s\sigma \quad (8)$$

This implies that $\sigma \to 0$ as $t \to \infty$. In other words, the system will reach and stay on the sliding surface $\sigma \equiv 0$. Moreover, (5) and (6) can be combined into one equation:

$$\ddot{u}_1 + \Lambda_1\dot{u}_1 + \Lambda_0 u_1 = -(K_s\sigma + \rho) \quad (9)$$

where $\Lambda_1, \Lambda_0 \in \mathbb{R}^{n\times n}$ are chosen such that the left-hand side of (9) is a stable system. For example, Λ_1, Λ_0 can be chosen as diagonal matrices with positive diagonal elements.

From (4) and (9) we see that, unlike traditional SMC, the switching term ρ does not appear in the joint torque directly. Instead, it is filtered by (9) before being incorporated into the joint torque. Therefore, the chattering phenomenon is significantly alleviated.

Once the sliding surface is reached, i.e., $\sigma \equiv 0$, the switching term is equivalent to the lumped effect of model uncertainty and external disturbance, and the closed-loop system can be expressed as

$$\ddot{u}_1 + \Lambda_1 \dot{u}_1 + \Lambda_0 u_1 = -\rho \quad (10)$$

$$\dot{\tilde{q}} + C_1 \tilde{q} = -\eta = -(\dot{u}_1 + \Lambda_1 u_1) \quad (11)$$

For ease of illustration, let us temporarily assume that Λ_1, Λ_0, and C_1 are diagonal matrices, i.e., $\Lambda_i = diag(\lambda_{i1}, \cdots, \lambda_{in}),\ i = 0, 1$, and $C_1 = diag(c_{11}, \cdots, c_{1n})$. Then, the closed-loop system (10) and (11) can be expressed as Figure 3 below. It is clear from Figure 3 that the closed-loop system is stable. In addition, the switching term, which is equivalent to uncertainty and disturbance in the sliding mode, is filtered by a second-order system $\frac{s+\lambda_{1j}}{s^2+\lambda_{1j}s+\lambda_{0j}}$ and affects the tracking error through a first-order system $\frac{1}{s+c_{1j}}$. Although the tracking error is not guaranteed to converge to zero, we can choose appropriate gain matrices Λ_1, Λ_0, and C_1 such that the tracking error is as small as desired.

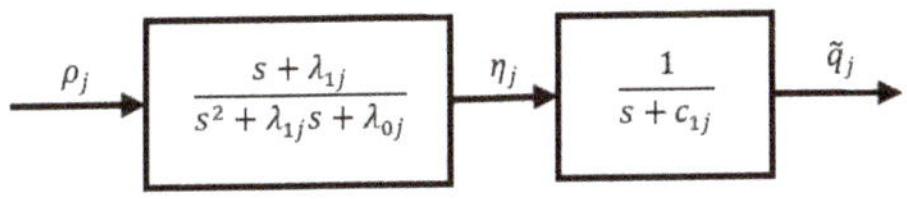

Figure 3. The jth component of the closed-loop system when the sliding surface $\sigma_j = 0$ is reached, $j = 1, \cdots, n$.

Besides the chattering attenuation property, another appealing feature of the proposed modified SMC is that if the initial tracking error is known, we can set $\eta(0) = -\dot{\tilde{q}}(0) - C_1\tilde{q}(0)$. As a result, the system is on the sliding surface as it starts, which eliminates the reaching phase of the traditional SMC and reduces the transient response time.

Experimental verification conducted in [30] showed that the proposed modified SMC has the best tracking performance in comparison with traditional first-order SMC (FOSMC), second-order SMC (SOSMC), and integral SMC (ISMC). Its chattering reduction ability is much better than FOSMC and ISMC, and similar to SOSMC. However, the proposed modified SMC does not need to find bounds to the time derivative of uncertainties as SOSMC does. Therefore, the proposed control law induces less vibration and achieves the best tracking performance with a simple structure.

2.2. Outer Admittance Control Loop

Suppose that the reference path with respect to an inertial frame in the 3D task space is

$$r_p(s) = \begin{bmatrix} x_p(s) & y_p(s) & z_p(s) \end{bmatrix}^T \quad (12)$$

which is parameterized by a single parameter $s \in [0, 1]$. We assume that r_p has finite length and choose the normalized arc length as the parameter s. Thus, $s = 0$ and $s = 1$ correspond to the start point and the end point of the reference path, respectively. Furthermore, we assume that r_p is twice differentiable with respect to s. At any point on $r_p(s)$, we can find the unit tangent vector $t(s)$, unit normal vector $n(s)$ and unit binormal vector $b(s)$ at that point as follows:

$$t(s) = \frac{r_p'(s)}{\|r_p'(s)\|}, \quad n(s) = \frac{t'(s)}{\|t'(s)\|}, \quad b(s) = t(s) \times n(s) \quad (13)$$

Note that $r_p'(s)$ and $t'(s)$ denote the first derivatives of $r_p(s)$ and $t(s)$ with respect to s, and $\times$ is the cross product of two vectors. These vectors comprise the TNB frame, which moves with the end effector along r_p. Whenever an external force F_{ext} is applied to the end effector, they are measured by the F/T sensor equipped on the end effector, and the measurements are preprocessed to alleviate noise, obtaining f_{ext} (see the output of the signal conditioning block in Figure 2). If the magnitude of f_{ext} exceeds a threshold f_{th}, the f_{ext} is projected to the unit tangent vector of the path. On the other hand, if f_{ext} is too weak or null, i.e., $\|f_{ext}\| < f_{th}$, then a virtual force with magnitude $f_v > 0$ is added along the tangent direction of the path. In other words,

$$f_{proj} = \begin{cases} (f_{ext}^T t(s))t(s), & \|f_{ext}\| \geq f_{th} \\ f_v t(s), & \|f_{ext}\| < f_{th} \end{cases} \tag{14}$$

Note that (14) is represented by the mode select block in Figure 2. Then, the desired velocity of the end effector in the task space, denoted by $\dot{r}_d$, is determined as $\dot{r}_d = g(t) * f_{proj}$, where $*$ denotes convolution, and $g(t)$ is the impulse response of the desired admittance function $G(s)$, which has the following first-order form:

$$G(s) = \frac{\frac{1}{c}}{\frac{m}{c}s + 1} \tag{15}$$

Parameters m and c denote the virtual mass and damping of the end effector, respectively, and they are online adjustable for avoiding common kinematic constraints such as singularities, joint limits, and task space limits. These constraints are directly related to the safety of the interaction because it may cause damage to the environment or the nearby human operators if any of these constraints is violated. Furthermore, providing the robot with awareness of these constraints and a way to preemptively deal with them relieves the operator from the mental load required to avoid them, which increases performance of the task [1,5–7,14,20]. If none of these constraints are going to be violated, the admittance parameters are adjusted to achieve safe, comfortable, and intuitive human–robot interaction. The online parameter tuning algorithm will be introduced shortly. Once the desired task space velocity $\dot{r}_d$ is obtained, the desired joint space velocity is calculated by

$$\dot{q}_d = J^{\dagger}(q)[\dot{r}_d + K(r_d - r)] \tag{16}$$

where r is the actual position of the end effector and $K \in \mathbb{R}^{3\times 3}$ is a positive definite gain matrix which guarantees that $r \to r_d$ as $t \to \infty$. $J^{\dagger}(q)$ is the pseudo-inverse of $J(q)$. We assume that the reference path r_p does not pass through any singularity of the robot; therefore, $J(q)$ is full rank and $J^{\dagger}(q)$ exists for all points on r_p. Existence of singularities is an inherent limitation for industrial robots, regardless of the method used. However, one desirable feature of the proposed method is that it forces the robot to stop before reaching singularities and allows the robot to retract along the path without getting stuck when guided by the human operator. This is accomplished by the outer admittance control loop. Therefore, violation of the no-singularities-on-the-reference-path assumption does not fail the normal operation of the robot. It just restricts its motion on the portion of the reference path without singularities.

Note that only the tangential component of f_{ext} is used to generate $\dot{r}_d$ which is in the tangent direction of r_p. In other words, the robot moves along r_p when the end effector is pushed by an external tangential force. However, if a large force, i.e., $\|f_{ext}\| \geq f_{th}$, is applied in the perpendicular direction of r_p, then f_{proj} is zero. Consequently, $\dot{r}_d$ reduces to zero exponentially with time constant $\frac{m}{c}$, and the robot stops almost immediately if the time constant is very small. In this way, the robot rejects perpendicular forces while complies with tangential forces based on an adaptive admittance law.

The first-order admittance function $G(s)$ is widely used in many compliance control problems [31–34], but the appropriate admittance parameters are task dependent and are often set by heuristics. To deal with this problem, we have proposed an online parameter tuning approach in [23] that considers the safety, comfort, and intuition of manual guidance. The underlying idea is to reduce compliance without significantly changing the dynamic behavior of the admittance function whenever the robot is close to constrained areas such as singularities, joint limits, and task space limits. If no constraints are going to be breached, the compliance varies with the external force, allowing the user to guide the motion of the robot more comfortably and intuitively. In this paper, the desired task space path r_p is defined in advance and the end effector does not deviate from this path; thus, we can assume that the robot will not pass through or go near any constrained areas. Therefore, the admittance parameters are adapted based on the applied forces.

To tune the admittance parameters, we first set up the varying range for the virtual damping c. The maximum allowable motion speed is considered when selecting the limits of the range for damping c. This ensures that any motion is below the safety relative speed limit defined in the Annex A of ISO/TS 15066:2016 [35] for transient and quasi-static contact. Let $c \in [c_{min}, c_{max}]$, where c_{min} and c_{max} are the pre-established minimum and maximum limits of c, respectively. Then, the projected force f_{proj} is normalized with respect to an estimated maximum value $\bar{f}$, which could be determined either experimentally or heuristically. Define

$$\hat{f} = min\left\{\frac{||f_{proj}||}{\bar{f}}, 1\right\} \tag{17}$$

Therefore, $\hat{f} \in (0, 1]$. The proposed parameter adaptive law is

$$c = (c_{min} - c_{max})\hat{f} + c_{max} \tag{18}$$

Notice that the largest force $\hat{f} = 1$ is mapped to the minimum damping c_{min}, which corresponds to the largest DC gain of $G(s)$, while the smallest force $\hat{f} \approx 0$ is mapped to the maximum damping c_{max} corresponding to the smallest DC gain of $G(s)$. Consequently, the user can guide the robot to move faster and reach the destination quickly just by applying a larger force. On the other hand, the user slows down the robot to accomplish fine work with high accuracy when a smaller force is applied. Once the virtual damping c is determined, the virtual mass m is adjusted accordingly to keep the ratio $\frac{m}{c}$ unaltered. The ratio $\frac{m}{c}$ is the time constant of $G(s)$, which is related to the response time of $G(s)$. Fixed time constant gives consistent dynamic behavior of the admittance function and allows the user to quickly get accustomed to the way for operating the robot.

Experimental verification of the proposed adaptive admittance law and comparison with other parameter tuning methods were conducted in [23] by implementing a 3D shape tracing task. The admittance laws in comparison included constant parameters and adaptive virtual damping with fixed virtual mass. Objective evaluation of the test results was done by analyzing the path length errors and task execution time. Meanwhile, a subjective evaluation was also conducted by collecting questionnaires that asked each test user to score the performance of each parameter tuning method in terms of vibration of the system, required efforts, and ease of completing the task. The results showed that the proposed adaptive admittance law has the best performance among all compared methods for both objective and subjective evaluation.

The outer admittance control loop is summarized as Algorithm 1 below.

Algorithm 1 Compliant path tracking using force projection

Inputs: f_{ext}, f_v, s
Outputs: $\dot{r}_d$

From path $r_p(s)$, calculate $t(s)$, $n(s)$, $b(s)$ using (13)
Initialize f_{th}, f_v, c and ratio m/c
for $t = 0$ to ∞ **do**
  if ($s \leq 1$) **then**
    Select mode to choose force input and project it to the tangent direction.
    if ($\|f_{ext}\| < f_{th}$) **then**
      Update $f_{proj} = f_v\, t(s)$ ▷ f_v is the task-dependent constant virtual force
    else
      Update $f_{proj} = (f_{ext}^T t(s))t(s)$
    end if
  else
    Robot at goal, stop motion. $f_{proj} = 0$
  end if
  Obtain $\dot{r}_d$ from passing f_{proj} through the admittance law (15).
  Normalize f_{proj} using (17) to obtain $\hat{f}$
  Update c using (18) and calculate m to maintain the fixed ratio m/c
end for

3. Experiments and Results

3.1. Experimental Setup

The tests are performed using a HIWIN RA605 6-DOF robot arm with a Robotiq 2F-85 gripper attached to the end effector. An ATI Gamma F/T sensor with analog output sampled at 1 kHz was used to measure the external force applied to the end effector. See Figure 4.

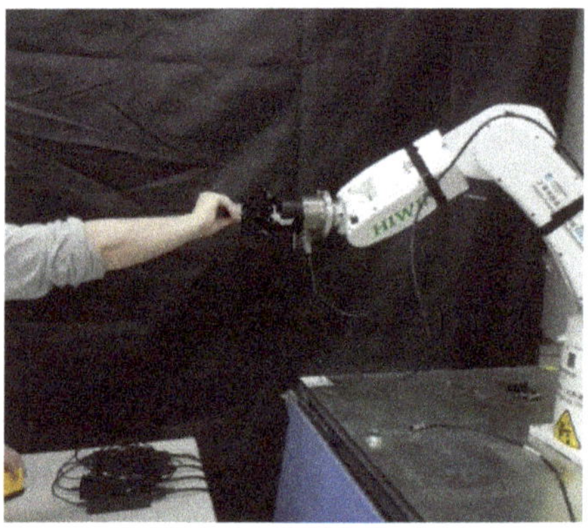

Figure 4. Experimental setup used for the tests showing the robot equipped with the force/torque (F/T) sensor and a gripper on the end-effector.

In this section, the two experiments conducted to validate the proposed approach are detailed. They include a linear motion test and a 3D path test. For these tests, the robot moves with constant orientation. Furthermore, the robot tool actual position is obtained from the robot joint encoders. The sliding mode control parameters used are summarized in Table 1. Regarding the admittance law parameters, the limits for damping c were chosen as $c_{min} = 300$ N·s·m^{-1} and $c_{max} = 1200$ N·m·s^{-1}. From these values, the maximum speed is 266 mm·s^{-1}, which allows us to comply with the relative speed limits defined in Annex A of ISO/TS 15066:2016 [35] for applications where contact between body areas and the robot system is possible. In our case, these body areas include the lower arm and hand/finger, where the speeds are restricted to 1300 and 650 mm·s^{-1} for transient and quasi-static contact, respectively, considering an estimated robot mass of 20 kg. In addition, the user has an emergency button at hand throughout the experiments conducted. The time constant of $G(s)$ is the ratio $\frac{m}{c} = 0.0167$ s. Finally, $f_{th} = 1.2$ N and $f_v = 40$ N.

Table 1. Sliding mode controller (SMC) parameters.

Parameter	Value
Λ_0	$diag(1936, 3884, 8100, 1764, 6400, 4900)$
Λ_1	$diag(44\sqrt{2}, 62\sqrt{2}, 90\sqrt{2}, 42\sqrt{2}, 80\sqrt{2}, 70\sqrt{2})$
C_1	$diag(125, 65, 32, 190, 187, 275)$
K_s	$diag(90, 100, 100, 100, 100, 100)$

Illustrated in Figure 5 are all the coordinate frames and their corresponding relations with each other in the environment used for the experiments. In the linear motion test, we define a horizontal line in front of the robot to demonstrate the behavior as external forces are applied to the end effector from different directions. The line is defined from points $A = [0.34,\ 0.45,\ 0.15]^T$ to $B = [0.0,\ 0.6,\ 0.15]^T$ in robot base frame, see Figure 5 and Figure 6a, and parametrized with respect to s as in (19), where $\overrightarrow{AB} = B - A$ is the 3×1 direction vector from A to B and $s \in [0, 1]$.

$$r(s) = A + s\overrightarrow{AB} \tag{19}$$

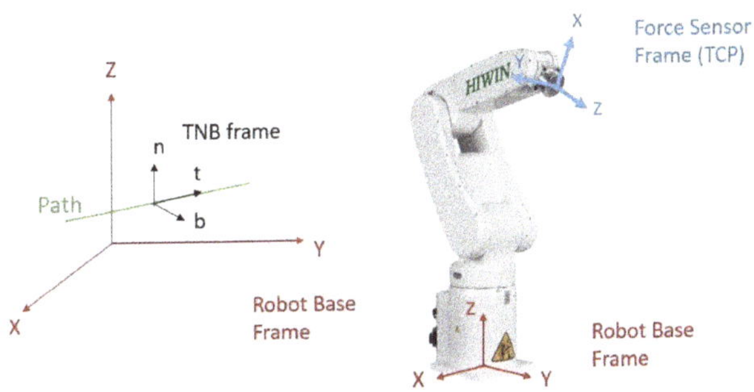

Figure 5. Coordinate frames description. Force sensor frame in blue, robot base frame in red, tangential-normal-binormal (TNB) frame in black, and predefined path in green.

For the second test, the 3D path we chose to demonstrate the path-tracking capability of our system is a helix, defined in (20)–(22) and parametrized with respect to $s \in [0, 1]$, where $s = 0$ marks the initial point of the path and $s = 1$ marks the end point of the path. $R = 0.1$ m is the helix radius centered at $o = [0.24,\ 0.45,\ 0.15]^T$, $n = 2$ is the number of rounds for the helix, $b_o = 0.05$ m is the increase in z per round.

$$x(s) = o_x + R\cos(2\pi ns) \tag{20}$$

$$y(s) = o_y + R\sin(2\pi ns) \tag{21}$$

$$z(s) = o_z + nb_o s \tag{22}$$

3.2. Linear Motion Test

We performed a linear motion test to show the behavior of the robot to the external forces applied from different directions. The detailed data are presented in Figure 6, including the desired and actual 3D paths in the robot base frame, desired and actual Cartesian positions, path parameter s, desired velocities in the TNB frame, and external forces in the robot frame and the TNB frame. Each subfigure of Figure 6 is annotated with labels using numerals from 1 to 5 to separate the data into regions where forces are applied in different directions: 1 and 5 for free-motion, 2 for force applied along the y-axis of the robot frame, 3 for force applied along the x-axis of the robot frame, and 4 for force applied along the z-axis of the robot frame. The differences between the desired and actual Cartesian positions in Figure 6a,b are the tracking errors represented in the task space which demonstrate the performance of the inner loop SMC controller. Figure 6d shows the evolution of the s parameter, which defines the position of the robot in the path. In region 1, from $t = 0$ s to $t = 16$ s, there is no external force applied to the robot, and the robot moves to the target point B at $s = 1$ with a constant velocity based on f_v as in (14) and Algorithm 1, (see also Figure 6c–f). Next, in region 2, from $t = 16$ s to $t = 24$ s a negative external force on the y-axis of the robot frame is applied, see F_y in Figure 6c, which is reflected in the TNB projected force as F_t and F_b in Figure 6f. This causes the robot to move backward along the line reaching $s = 0$ in Figure 6d.

Likewise, from $t = 24$ s to $t = 35$ s a positive external force on the y-axis of the robot frame was applied to move the robot forward along the path. During this interval there was a pause from $t = 29$ s to $t = 31$ s where no force was applied, and the slope of the s parameter changed to the same value as in the previous free-motion interval, see Figure 6c–f. For region 3, from $t = 35$ s to $t = 59$ s, the same tests were repeated but this time applying the external force on the x-axis of the robot frame showing a similar behavior in the s parameter, see F_x in Figure 6c and F_t, F_b in Figure 6f. In region 4, from $t = 59$ s to $t = 78$ s, an external force normal to the path along the z-axis of the robot frame was applied three times with alternating signs, see F_z in Figure 6c and F_n in Figure 6f. As expected, the applied force was not considered for robot motion because the projection to the tangent direction of the path is null. This can also be seen in Figure 6d where three stair patterns are created, meaning the s parameter remained constant during the three periods where the perpendicular force was applied, see Figure 6c,f. For region 5, from $t = 78$ s to $t = 84$ s, no force was applied to the robot; thus, the robot moved towards the goal with constant velocity obtained from the virtual force f_v until $s = 1$.

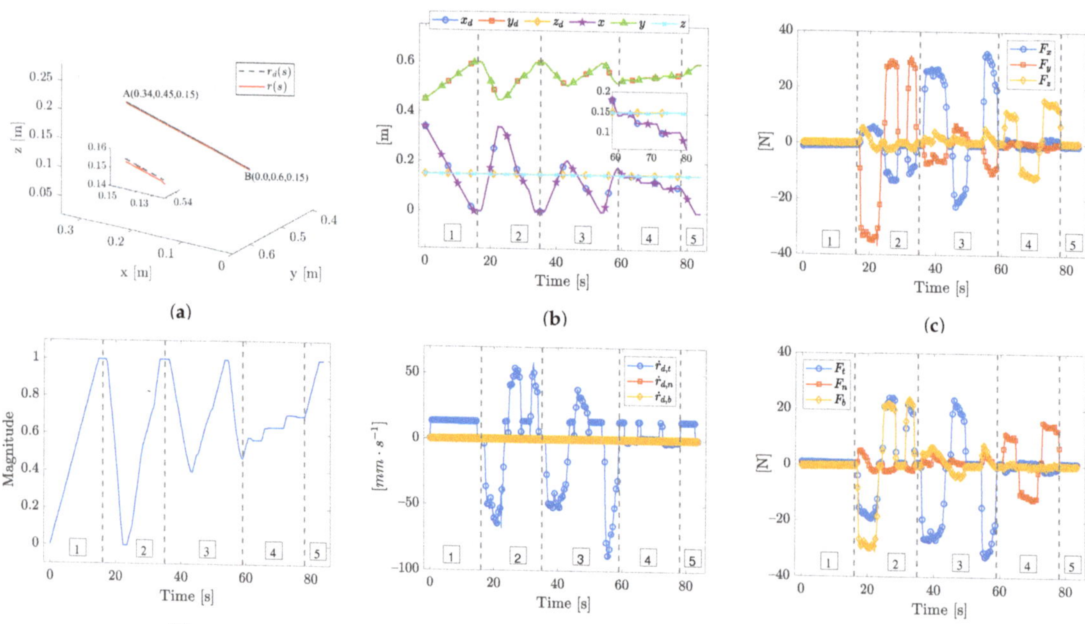

Figure 6. Linear motion path tracking test. Numerals 1–5 are used to separate regions where forces are applied in different directions: 1 and 5—free-motion, 2—force applied along y-axis of robot frame, 3—force applied along x-axis of robot frame, 4—force applied along z-axis of robot frame. (**a**) Desired and actual 3D path in robot base frame. (**b**) Desired and actual Cartesian positions. (**c**) Forces in robot frame. (**d**) Parameter s. (**e**) Desired Velocities in TNB frame. (**f**) Forces in TNB frame.

This linear test proves the capability of our proposed method to customize the compliance behavior of the robot for a path tracking task. Furthermore, for clear presentation, the tracking performance of the modified SMC controller in both the joint space and task space is summarized in Table 2.

Table 2. Linear motion task tracking root mean square error (RMSE).

Joint	RMSE (mrad)	Cart. Pos.	RMSE (mm)
1	0.4102	x	0.3458
2	0.4793	y	0.4253
3	2.9438	z	1.4255
4	0.7535		
5	4.5410		
6	2.0088		

3.3. 3D Path-Tracking Tests

Three tests were executed using the helix path for comparison to show the behavior of the proposed algorithm in both ends of the assistance range spectrum, namely, fully assisted motion or free motion (FM), and not assisted motion, i.e., force-applied motion (FAM). The combined motion with virtual forces (CVF) is the smooth transition between FM and FAM. Note that for the FAM test we defined $f_v = 0$ N to test the case of no assistance individually. This causes that at any instant when there is no external force applied, the robot stops and maintains the last position, meaning that only the user can make the robot move in this case. Moreover, the proposed virtual force scheme provides the means for a smooth transition between FM and FAM. The detailed data of the tests are shown in Figure 7a–i, including the external forces in the TNB frame, the s parameter variation throughout the task, and the desired velocity $\dot{r}_d$ in the TNB frame obtained from the constant virtual force f_v.

3.3.1. Free Motion (FM)

For this test as shown in Figure 7a–c, the robot follows the defined helix path, without external forces applied to the end effector as presented in Figure 7a. This demonstrates the case of fully-assisted interaction because the robot moves from the starting point $s = 0$ to the final point $s = 1$, see Figure 7b. This case serves as the reference to evaluate the error in the results of Section 3.4. The desired velocities $\dot{r}_d$ in Figure 7c, which are used in (16) to obtain the desired joint velocities, are calculated from passing the constant virtual force f_v projected to the tangent direction of motion of the path through the admittance law as described in Section 2.

3.3.2. Force-Applied Motion (FAM)

Figure 7d–f illustrates this test, which consists of applying an external force, see Figure 7d, to the robot end-effector to move it along the predefined helix path. This force is projected to the tangential direction of the path, which gives the magnitude of the desired velocity in task space to start or stop robot motion, see Figure 7f. No virtual force is generated for this case because the magnitude of $f_v = 0$ N. In this test, the robot moves from the initial point $s = 0$ to the final point $s = 1$ in a nonlinear way opposed to the FM case because the desired velocity depends on the external force applied to the end effector, as can be seen in Figure 7e,f.

3.3.3. Combined Test Using Virtual Forces (CVF)

In this test, we merge the two previous cases to demonstrate the capabilities of the proposed algorithm for multiple applications and the smooth transition between collaborative and non-collaborative interaction. The robot will track the helix path with constant velocity obtained from f_v if no external forces are applied, moving from the initial point $s = 0$ to the final point $s = 1$. This can be seen from $t = 0$ s to $t = 30$ s in Figure 7g–i. At any point of this free motion towards the goal, the user can apply an external force that will cause the robot to retract or advance while tracking the path, as seen in Figure 7g,h from $t = 34$ s to $t = 46$ s. If the user stops applying the external force, the robot will continue to move to the goal in FM case as shown in the same figures from $t = 46$ s to $t = 49$ s. The desired velocity in each step of the test in the TNB frame is calculated accordingly, see Figure 7i.

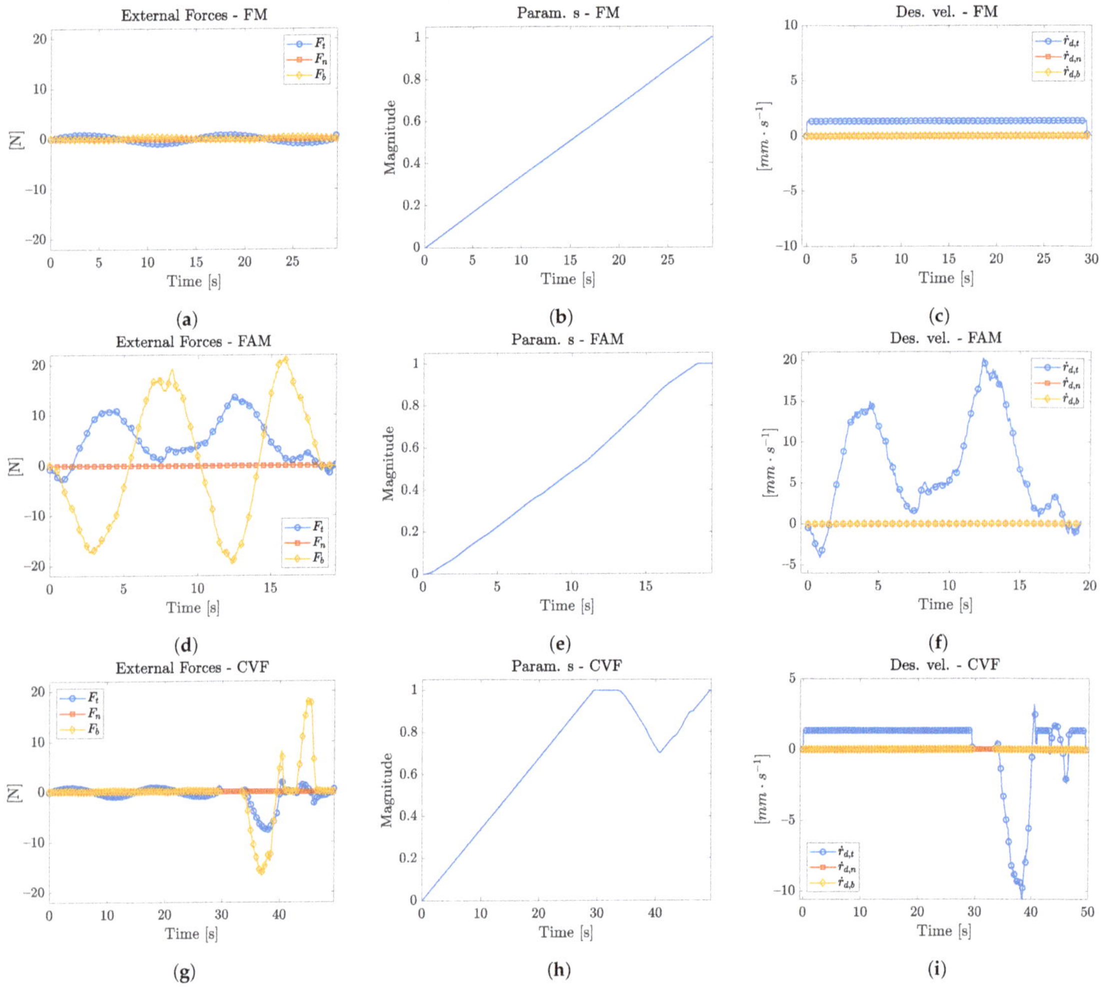

Figure 7. Helix path-tracking test detailed data for free motion (FM), force-applied motion (FAM), and combined motion with virtual forces (CVF) in the TNB frame. (**a**) External Forces-FM. (**b**) Parameter s-FM. (**c**) Desired Velocities-FM. (**d**) External Forces-FAM. (**e**) Parameter s-FAM. (**f**) Desired Velocities-FAM. (**g**) External Forces–CVF. (**h**) Parameter s–CVF. (**i**) Desired Velocities-CVF.

3.4. RMSE Evaluation and Comparison

Following our objective of accurate path tracking in human–robot collaboration tasks, we evaluate the error between the reference path and the actual recorded path. It is important for the robot to continue tracking the path accurately, despite forces being applied or not during the human–robot interaction. In this context, the addition of directional compliant motion should have as small influence as possible on path tracking accuracy.

Thus, we must ensure the error difference is small among the proposed cases that represent the interaction assistance range.

Figure 8 shows the reprojection of the RMSE residuals to the 3D helix path calculated between the reference helix path $r_p(s)$ in black dashed line and the recorded test data $r(s)$ in red solid line, also green dotted lines connect the points from $r(s)$ to $r_p(s)$ to display the error every 10 samples. Figure 9 presents for each of the three evaluated cases, the RMSE values in green dashed line and the tracking error between the actual and reference samples in solid blue line. For every joint position data collected in the test, we find the shortest distance $d_{p,i\perp r_p}$ between the ith point and the reference curve $r_p(s)$, which was sampled from (20)–(22) to have the same number of points as the test data, see (23). Then, we find the RMSE as in (24), where $p_{j,i}$ is the ith test point of the jth component, $j = x, y, z$, and N is the total number of points in the test.

$$d_{p,i\perp r_p} = \min_s \left(\sqrt{(r_{p,x}(s) - p_{x,i})^2 + (r_{p,y}(s) - p_{y,i})^2 + (r_{p,z}(s) - p_{z,i})^2} \right), s \in [0,1] \quad (23)$$

$$RMSE = \sqrt{\frac{\sum_{i=1}^{i=N} \left(d_{p,i\perp r_p} \right)^2}{N}} \quad (24)$$

In Table 3, we can see that the RMSE values obtained between the FM and FAM motion have a very small difference of 0.021 mm. This shows path tracking has a good performance even when external forces are applied to guide the motion. In the other case for the combined test with virtual forces, the error difference with the free motion test is slightly higher, 0.038 mm; however, it is still less than 0.1 mm, which is the pose accuracy value reported for commercial robots used in collaborative applications [36–38].

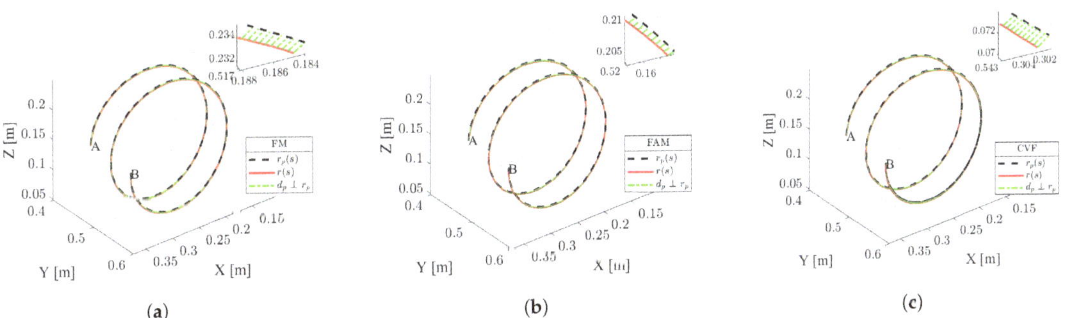

Figure 8. Helix path tracking. $r_p(s)$-black dashed line, $r(s)$-red solid line, distance from each point of $r(s)$ to $r_p(s)$-green dotted line. (**a**) FM. (**b**) FAM. (**c**) CVF.

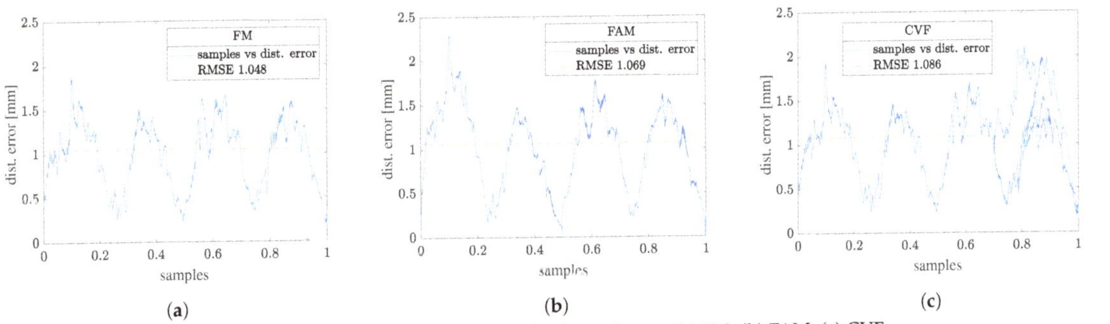

Figure 9. RMSE and distance error plot for each test. (**a**) FM. (**b**) FAM. (**c**) CVF.

Table 3. RMSE error differences among cases: FM vs. FAM, and FM vs. CVF.

Cases Compared	Error Difference (mm)
FM vs FAM	0.021
FM vs CVF	0.038

4. Conclusions

We successfully achieved compliant path tracking motion using an industrial 6-DOF robot manipulator by employing the proposed dual-loop control structure including an inner motion control loop and an outer admittance control loop. The safety of the application is guaranteed by adhering to the constraints presented in our previous work [23], which include singularity avoidance, joint limits, and workspace limits. Additionally, the modified SMC designed and implemented as the inner motion controller showed a satisfactory tracking performance. A linear motion test successfully demonstrated the system ability to modify the robot compliance with respect to the external force direction applied to the end effector. The results from this linear test show the capability of our proposed method to customize the compliance behavior of the robot for a path tracking task. Furthermore, a 3D helix path was used to test the path-tracking performance and display the generality to many applications such as rehabilitation, assisted drawing, assisted hand–motor skills development, and so on. For this test, three cases were compared depending on the level of the interaction; FM, FAM, and CVF motion. The error difference between cases is less than 0.1 mm, which is suitable for path tracking applications where the user can dynamically interact with the robot. Our future research directions include testing the framework in more specific applications such as rehabilitation or assisted hand–motor skills development as well as in different types of robot manipulators, given that our framework implementation can be easily ported to other manipulators with the addition of an F/T sensor. Moreover, a user study will be carried out to further explore the perceived performance in each specific application.

Author Contributions: Conceptualization, D.R.-U. and T.H.; Data curation, D.R.-U.; Formal analysis, D.R.-U.; Funding acquisition, T.H.; Investigation, D.R.-U.; Methodology, D.R.-U. and T.H.; Project administration, T.H.; Resources, T.H.; Software, D.R.-U.; Supervision, T.H.; Validation, D.R.-U. and T.H.; Visualization, D.R.-U.; Writing–original draft, D.R.-U. and T.H.; Writing–review & editing, D.R.-U. and T.H. All authors have read and agreed to the published version of the manuscript.

Funding: This research was funded by the Ministry of Science and Technology, Taiwan, grant number MOST 109-2221-E-009-092-.

Conflicts of Interest: The authors declare no conflict of interest. The funders had no role in the design of the study; in the collection, analyses, or interpretation of data; in the writing of the manuscript; or in the decision to publish the results.

References

1. Rosenberg, L.B. Virtual fixtures: perceptual tools for telerobotic manipulation. In Proceedings of the IEEE Virtual Reality Annual International Symposium, Seattle, WA, USA, 18–22 September 1993; pp. 76–82. [CrossRef]
2. Bowyer, S.A.; Davies, B.L.; Baena, R.F.Y. Active constraints/virtual fixtures: A survey. *IEEE Trans. Robot.* **2014**, *30*, 138–157. [CrossRef]
3. Abbott, J.J.; Marayong, P.; Okamura, A.M. Haptic virtual fixtures for robot-assisted manipulation. In *Robotics Research*; Thrun, S., Brooks, R., Durrant-Whyte, H., Eds.; Springer: Berlin/Heidelberg, Germany, 2007; Volume 28, pp. 49–64. [CrossRef]
4. Restrepo, S.S.; Raiola, G.; Chevalier, P.; Lamy, X.; Sidobre, D. Iterative virtual guides programming for human-robot comanipulation. In Proceedings of the 2017 IEEE International Conference on Advanced Intelligent Mechatronics (AIM), Munich, Germany, 3–7 July 2017; pp. 219–226. [CrossRef]
5. Bettini, A.; Marayong, P.; Lang, S.; Okamura, A.M.; Hager, G.D. Vision-assisted control for manipulation using virtual fixtures. *IEEE Trans. Robot.* **2004**, *20*, 953–966. [CrossRef]
6. Castillo-Cruces, R.A.; Wahrburg, J. Virtual fixtures with autonomous error compensation for human-robot cooperative tasks. *Robotica* **2010**, *28*, 267–277. [CrossRef]

7. Žlajpah, L.; Petrič, T. Virtual Guides for Redundant Robots Using Admittance Control for Path Tracking Tasks. In *Advances in Service and Industrial Robotics*; Aspragathos, N.A., Koustoumpardis, P.N., Moulianitis, V.C., Eds.; Springer International Publishing: Cham, Switzerland, 2019; pp. 13–23.
8. Barri, M.H.; Widyotriatmo, A.; Suprijanto. Path Reference Generation for Upper-Limb Rehabilitation with Kinematic Model. In Proceedings of the 2018 IEEE International Conference on Robotics, Biomimetics, and Intelligent Computational Systems (Robionetics), Bandung, Indonesia, 8–10 August 2018; pp. 38–43. [CrossRef]
9. Wang, W.W.; Tsai, B.C.; Hsu, L.C.; Fu, L.C.; Lai, J.S. Guidance-control-based exoskeleton rehabilitation robot for upper limbs: Application to circle drawing for physiotherapy and training. *J. Med. Biol. Eng.* **2014**, *34*, 284–292. [CrossRef]
10. Zhao, Y.; Liang, C.; Gu, Z.; Zheng, Y.; Wu, Q. A new design scheme for intelligent upper limb rehabilitation training robot. *Int. J. Environ. Res. Public Health* **2020**, *17*, 2948. [CrossRef] [PubMed]
11. Villani, V.; Pini, F.; Leali, F.; Secchi, C. Survey on human–robot collaboration in industrial settings: Safety, intuitive interfaces and applications. *Mechatronics* **2018**, *55*, 248–266. [CrossRef]
12. Lefebvre, T.; Xiao, J.; Bruyninckx, H.; de Gersem, G. Active compliant motion: A survey. *Adv. Robot.* **2005**, *19*, 479–499. [CrossRef]
13. Roveda, L.; Pedrocchi, N.; Beschi, M.; Molinati Tosatti, L. High-accuracy robotized industrial assembly task control schema with force overshoots avoidance. *Control Eng. Pract.* **2018**, *71*, 142–153. [CrossRef]
14. Cherubini, A.; Passama, R.; Crosnier, A.; Lasnier, A.; Fraisse, P. Collaborative manufacturing with physical human-robot interaction. *Robot. Comput. Integr. Manuf.* **2016**, *40*, 1–13. [CrossRef]
15. Peng, Y.C.; Chen, S.; Jivani, D.; Wason, J.; Lawler, W.; Saunders, G.; Radke, R.J.; Trinkle, J.; Nath, S.; Wen, J.T. Sensor-Guided Assembly of Segmented Structures with Industrial Robots. *Appl. Sci.* **2021**, *11*, 2669. [CrossRef]
16. Tsumugiwa, T.; Yokogawa, R.; Hara, K. Variable impedance control based on estimation of human arm stiffness for human-robot cooperative calligraphic task. In Proceedings of the 2002 IEEE International Conference on Robotics and Automation, Washington, DC, USA, 11–15 May 2002; Volume 1, pp. 644–650. [CrossRef]
17. Bingham, G.P.; Snapp-Childs, W. Training children aged 5–10 years in manual compliance control to improve drawing and handwriting. *Hum. Mov. Sci.* **2019**, *65*, 42–50. [CrossRef]
18. Ravankar, A.; Ravankar, A.A.; Kobayashi, Y.; Hoshino, Y.; Peng, C.C. Path smoothing techniques in robot navigation: State-of-the-art, current and future challenges. *Sensors* **2018**, *18*, 3170. [CrossRef]
19. Hsiao, T.; Huang, P.H. Iterative Learning Control for Trajectory Tracking of Robot Manipulators. *Int. J. Autom. Smart Technol.* **2017**, *7*. [CrossRef]
20. Raiola, G.; Restrepo, S.S.; Chevalier, P.; Rodriguez-Ayerbe, P.; Lamy, X.; Tliba, S.; Stulp, F. Co-manipulation with a library of virtual guiding fixtures. *Auton. Robot.* **2018**, *42*, 1037–1051. [CrossRef]
21. Wu, X.; Li, Z.; Kan, Z.; Gao, H. Reference Trajectory Reshaping Optimization and Control of Robotic Exoskeletons for Human-Robot Co-Manipulation. *IEEE Trans. Cybern.* **2020**, *50*, 3740–3751. [CrossRef] [PubMed]
22. Taylor, R.; Jensen, P.; Whitcomb, L.; Barnes, A.; Kumar, R.; Stoianovici, D.; Gupta, P.; Wang, Z.X.; DeJuan, E.; Kavoussi, L. Steady-hand robotic system for microsurgical augmentation. *Int. J. Rob. Res.* **1999**, *18*, 1201–1210. [CrossRef]
23. Reyes-Uquillas, D.; Hsiao, T. Safe and intuitive manual guidance of a robot manipulator using adaptive admittance control towards robot agility. *Robot. Comput. Integr. Manuf.* **2021**, *70*, 102127. [CrossRef]
24. Gracia, L.; Solanes, J.E.; Muñoz-Benavent, P.; Valls Miro, J.; Perez-Vidal, C.; Tornero, J. Adaptive Sliding Mode Control for Robotic Surface Treatment Using Force Feedback. *Mechatronics* **2018**, *52*, 102–118. [CrossRef]
25. Hung, J.Y.; Gao, W.; Hung, J.C. Variable Structure Control: A Survey. *IEEE Trans. Ind. Electron.* **1993**, *40*, 2–22. [CrossRef]
26. Tseng, M.L.; Chen, M.S. Chattering reduction of sliding mode control by low-pass filtering the control signal. *Asian J. Control* **2010**, *12*, 392–398. [CrossRef]
27. Bartolini, G.; Pisano, A.; Punta, E.; Usai, E. A survey of applications of second-order sliding mode control to mechanical systems. *Int. J. Control* **2003**, *76*, 875–892. [CrossRef]
28. Capisani, L.M.; Ferrara, A.; Magnani, L. Design and experimental validation of a second-order sliding-mode motion controller for robot manipulators. *Int. J. Control* **2009**, *82*, 365–377. [CrossRef]
29. Pan, Y.; Yang, C.; Pan, L.; Yu, H. Integral Sliding Mode Control: Performance, Modification, and Improvement. *IEEE Trans. Ind. Inform.* **2018**, *14*, 3087–3096. [CrossRef]
30. Wang, P.H. Comparisons of a Novel Low Chattering Sliding Mode Control and Other Sliding Mode Controls with Application to Trajectory Tracking of a 6-Axis Robot Manipulator. Master's Thesis, National Chiao Tung University, Hsinchu City, Taiwan, 2019.
31. Lecours, A.; Mayer-St-Onge, B.; Gosselin, C. Variable admittance control of a four-degree-of-freedom intelligent assist device. In Proceedings of the 2012 IEEE International Conference on Robotics and Automation, Saint Paul, MN, USA, 14–18 May 2012; pp. 3903–3908. [CrossRef]
32. Ficuciello, F.; Villani, L.; Siciliano, B. Variable Impedance Control of Redundant Manipulators for Intuitive Human-Robot Physical Interaction. *IEEE Trans. Robot.* **2015**, *31*, 850–863. [CrossRef]
33. Reyes-Uquillas, D.A.; Hsiao, T. Online motion adjustment using compliance control for a multi-axis robot manipulator. In Proceedings of the 2017 International Automatic Control Conference (CACS), Pingtung, Taiwan, 12–15 November 2017; pp. 1–6. [CrossRef]
34. Dimeas, F.; Moulianitis, V.C.; Aspragathos, N. Manipulator performance constraints in human-robot cooperation. *Robot. Comput. Integr. Manuf.* **2018**, *50*, 222–233. [CrossRef]

35. International Organization for Standardization. *ISO/TS 15066:2016*; International Organization for Standardization: Geneva, Switzerland, 2016.
36. Pollák, M.; Kočiško, M.; Paulišin, D.; Baron, P. Measurement of unidirectional pose accuracy and repeatability of the collaborative robot UR5. *Adv. Mech. Eng.* **2020**, *12*, 1–21. [CrossRef]
37. Płaczek, M.; Piszczek, Ł. Testing of an industrial robot's accuracy and repeatability in off and online environment. *Eksploat. Niezawodn. Maint. Reliab.* **2018**, *20*, 455–464. [CrossRef]
38. Morozov, M.; Pierce, S.G.; MacLeod, C.N.; Mineo, C.; Summan, R. Off-line scan path planning for robotic NDT. *Meas. J. Int. Meas. Confed.* **2018**, *122*, 284–290. [CrossRef]

Short Biography of Authors

Daniel Reyes-Uquillas obtained his bachelor's in mechatronic engineering from the Armed Forces University (ESPE), Ecuador, in 2013, and his master's degree in mechatronic engineering from the National Taipei University of Technology (Taipei Tech), Taiwan, in 2015. He is currently a Ph.D. candidate in the field of robotics at the National Yang Ming Chiao Tung University, Taiwan. He is working as a part-time researcher of service robot navigation with the Mechanic and Mechatronic Systems Lab at the Industrial Technology Research Institute (ITRI), Taiwan. His research interests include compliance control for robotic systems and physical human-robot interaction, as well as autonomous mobile robots and SLAM.

Tesheng Hsiao (M'05) received his B.S. and M.S. degrees in control engineering from National Chiao Tung University, Taiwan, in 1995 and 1997 respectively, and the Ph.D. degree in mechanical engineering from the University of California, Berkeley, in 2005. He is currently an associate professor with the Department of Electrical and Computer Engineering at National Yang Ming Chiao Tung University, Taiwan. His research interests include advanced vehicle control systems, robot motion and compliance control, and CNC precision motion control.

applied sciences

MDPI

Article

Creating a Novel Mathematical Model of the Kv10.1 Ion Channel and Controlling Channel Activity with Nanoelectromechanical Systems

Jasmina Lozanović Šajić [1,2,*], Sonja Langthaler [1] and Christian Baumgartner [1]

[1] Institute of Health Care Engineering with European Testing Center for Medical Devices, Graz University of Technology, A-8010 Graz, Austria; s.langthaler@tugraz.at (S.L.); christian.baumgartner@tugraz.at (C.B.)
[2] Innovation Center of the Faculty of Mechanical Engineering, University of Belgrade, 11000 Belgrade, Serbia
* Correspondence: j.lozanovicsajic@tugraz.at

Featured Application: Nanoelectromechanical systems and nanorobots can be used to treat cancers associated with the Kv10.1 voltage-gated ion channel activity. The Kv10.1 model was developed by applying the control engineering theory. Nanoelectromechanical systems play the role of a PID regulator.

Abstract: The use of nanoelectromechanical systems or nanorobots offers a new concept for sensing and controlling subcellular structures, such as ion channels. We present here a novel method for mathematical modeling of ion channels based on control system theory and system identification. We investigated the use of nanoelectromechanical devices to control the activity of ion channels, particularly the activity of the voltage-gated ion channel Kv10.1, an important channel in cancer development and progression. A mathematical model of the dynamic behavior of the selected ion channel Kv10.1 in the Laplace (s) domain was developed, which is given in the representation of a transfer function. In addition, we addressed the possibilities of controlling ion channel activity by nanoelectromechanical devices and nanorobots and finally presented a control algorithm for the Kv10.1 as a control object. A use case demonstrates the potential of a Kv10.1 controlled nanorobot for cancer treatment at a single-cell level.

Keywords: nanoelectromechanical system (NEMS); nanorobots; ion channel Kv10.1; mathematical modeling; system identification; control algorithm

Citation: Šajić, J.L.; Langthaler, S.; Baumgartner, C. Creating a Novel Mathematical Model of the Kv10.1 Ion Channel and Controlling Channel Activity with Nanoelectromechanical Systems. *Appl. Sci.* **2022**, *12*, 3836. https://doi.org/10.3390/app12083836

Academic Editors: Luis Gracia and Carlos Perez-Vidal

Received: 14 March 2022
Accepted: 8 April 2022
Published: 11 April 2022

1. Introduction

Nanoelectromechanical systems, nanomachines, and nanorobots represent a challenging and future-orientated research area in biomedical engineering. A nanoelectromechanical system (NEMS) is a device that combines electrical and mechanical system behavior at the nanoscale level. The NEMS has electromechanical parts that have been developed on the nanoscale level, such as sensors, actuators, controllers, and drives [1]. Nanorobots are nanoelectromechanical or nanomechatronic systems, which are innovative devices expected to have revolutionary applications in health care, cancer therapy, monitoring, and drug delivery [2]. These nanoelectromechanical systems and nanorobots represent miniaturizations of microelectromechanical systems and microrobots that travel in the human body and can be used in applications to monitor, interact, and control processes at a cellular level. In particular, these robotic applications involve sensing, control, actuation and propulsion, communications, interfacing, programming and coordination at macro, micro, and nano-levels [2].

In the field of biomedical engineering and medicine, microelectromechanical/nanoelectromechanical systems (MEMS/NEMS) are also called bioMEMS/bioNEMS [3], with almost unlimited practical applications, e.g., in surgery, drug delivery, or gene therapy [4–8]. Figure 1 shows a schematic nanoscale comparison of different items in nature, adapted based on material published in [3,9].

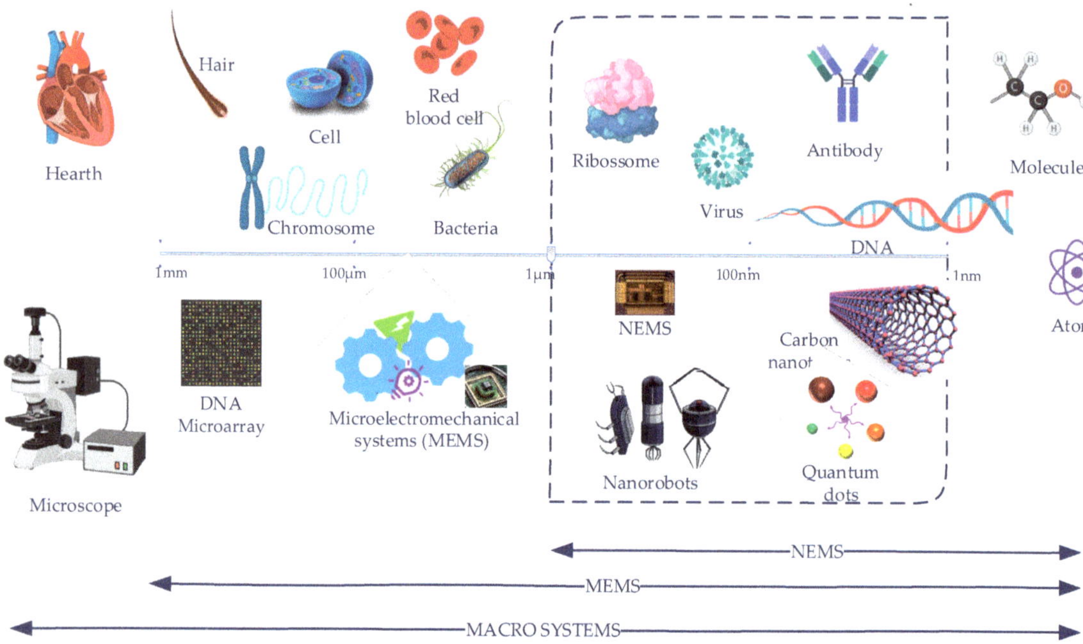

Figure 1. Scales in nature, ranging from micro-systems to nano-systems.

In this work, we chose a voltage-gated potassium ion channel as an example of a biological system at a nanoscale. Ion channels are membrane proteins the size of about 4 nm that enable the passive transport of ions through the cell membrane [10,11]. Based on the scale depicted in Figure 1, we assumed that the dynamic behavior of the ion channel could be measured and controlled with MEMS/NEMS and/or nanorobots. Depending on the gating mechanism, three main types of ion channels are classified as: voltage-gated, receptor/ligand-gated, and second messenger gated channels [12]. Cells usually express a variety of ion channel types, e.g., there are more than 300 different ion channels in an inner ear cell alone [13].

We here selected a single channel from a voltage-gated ion channel family representing the biological system. Voltage-gated ion channels open (activate) and close depending on the cell membrane potential. Subtypes of voltage-gated ion channels (Nav, Cav, Kv, CLC, and Hv) are specifically selective for sodium (Na^+), calcium (Ca^{2+}), potassium (K^+), chloride (Cl^-), and proton (H^+) channels. Voltage-gated potassium channels are referred to as the giant subtype and denoted as Kv. These Kv channels encode more than 100 human genes [14]. Potassium channels represent the most complex class of voltage-gated channels [15] and are found in basically all types of mammalian cells, such as cells in the nervous, muscular, and immune systems, among others [15–18].

Each ion channel exhibits a specific dynamic behavior with different physiological and pharmacological properties [15]. In particular, aberrant expression and functionality of various Kv channels in cancer cells have been associated with tumor development and progression. The voltage-gated Kv10.1 potassium channel, encoded by the gene KCNH1 (subfamily H member 1, known as EAG1 or Ether-à-go-go 1), is implicated in various cellular processes, including, for example, cell proliferation [15].

An analysis of Kv channel expression in human cancer cells [18] revealed an up-regulation of Kv10.1 in a variety of tumors and can be found in blood, bone, brain, breast, stomach, colon, cervix cancer, and prostate cancers cells, thus providing a novel biomarker candidate and potential oncological target for cancer [5,19,20].

For single-channel modeling, we finally selected the Kv10.1 ion channel, as it represents a significant ion channel in cancer development and progression. The Kv10.1 model is based on control system theory, which is similarly based on the modeling concept of the Kv1.1 voltage-gated channel recently presented in [21].

2. Methods

For mathematical modeling of voltage-gated ion channels, the method of system identification known from control engineering was used. We considered the voltage-gated ion channel Kv10.1 as a system, object, or process in accordance with the control system theory. Kv10.1 was treated as a separate object (or system/process) within a space that can interact or connect with other elements. If a system is interpreted as a combination of elements that act together and perform specific objectives [22], then a single ion channel can be considered as a system or, in more precise terms, as an object in a control loop. As such, the channel can achieve the desired dynamics under nominal conditions and display acceptable behaviors under random conditions that deviate from the maximum prescribed boundary conditions.

System identification is a scientific methodology that is used to develop mathematical models of a dynamical system. This method is based on observed or measured data of the system [23]. A block diagram of a general system ($G(s)$) represents a transfer function with a complex variable s in the Laplace domain, specified with an input $c(t)$, output $r(t)$, measured disturbances $dc(t)$, and unmeasured disturbances $d(t)$ (Figure 2a). An illustration of a voltage-gated ion channel as a system is shown in Figure 2b. The block diagram of the system shows its unilateral property, where system inputs and outputs, in general, are vectors. If there are multiple inputs and multiple outputs, then the system is called a MIMO system. In the case of a single input and a single output, however, the control system is called a SISO system. For example, if a voltage-gated ion channel has a voltage stimulus as input and a measured current as output, we assume that it can be described as a SISO system.

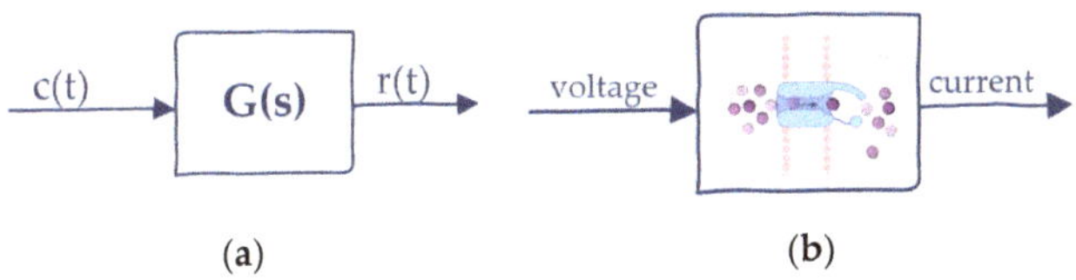

Figure 2. Block diagram of a system (**a**); illustrated voltage-gated ion channel as a system (**b**).

Furthermore, we assume that the voltage-gated ion channel Kv10.1 is a linear time-invariant system based on the experimental results given in the following sections. A linear time-invariant system is defined by

$$\begin{aligned} &a_n r^{(n)}(t) + a_{n-1} r^{(n-1)}(t) + \ldots + a_1 \dot{r}(t) + a_0 r(t) = \\ &b_0 c(t) + b_1 \dot{c}(t) + \ldots + b_{n-1} c^{(n-1)}(t) + b_n c^{(n)}(t), \quad n \geq m. \end{aligned} \tag{1}$$

where $r(t)$ is the system's output and $c(t)$ is the system's input. The transfer function, $G(s)$, is defined as the ratio of the left Laplace-transformed output and the left Laplace-transformed input under zero initial conditions, where s is a complex variable in the Laplace (S) domain.

$$G(s) = \frac{\mathcal{L}\{r(t)\}}{\mathcal{L}\{c(t)\}} = \frac{R(s)}{C(s)} = \frac{b_0+b_1 s+\ldots+b_{n-1}s^{n-1}+b_n s^n}{a_n s^n+a_{n-1}s^{n-1}+\ldots+a_1\dot{y}(t)+a_0 y(t)} = \\ = \frac{\sum_0^m b_m s^m}{\sum_0^n a_n s^n} \quad (2)$$

The application of the transfer function concept makes it possible to represent the system's dynamics in the form of algebraic equations. If the highest power of s in the denominator of a transfer function is equal to n, the system is called an nth-order system [22]. The transfer function of a system is thus a mathematical model of this system. If the system's transfer function is unknown, we can estimate it based on known inputs and measured outputs for the system. This resulting transfer function provides a complete description of the system's dynamic behavior [22,24,25].

A transfer function can be determined experimentally based on measured input and output signals if a mathematical model cannot be developed using physical correlations and equations. Therefore, the process of developing a model from measured input and output data is known as system identification [23,26]. The main elements in the system identification process cycle are experimental design, the experiment itself, data preprocessing, fitting the model to the data, testing the model structure, validating and auditing the model [26].

The process of system identification has been greatly simplified due to the availability of modern software tools, such as MATLAB or LabVIEW. In this work, we used the System Identification toolbox provided by MATLAB [27,28]. This toolbox includes tools to identify a modeling approach and to define the model structure properties. Measured data can be imported as time-domain or frequency-domain data, and functions for preprocessing, representing, filtering, and estimating the model can be applied. Furthermore, a model can be estimated as a transfer function, a state-space, process, polynomial, or nonlinear model by the System Identification toolbox.

3. Results

We considered experimental patch-clamp data [15] from CHO (Chinese hamster ovary) host cells, stably expressing rat Kv10.1 channels, which exhibit a highly comparable electrophysiological behavior to human Kv10.1 channels [29]. The data provided comprises voltage-clamp measurements at three different temperature levels of 15, 25, and 35 °C. The dynamic behavior of the voltage-gated ion channels displays three main characteristics: activation, deactivation, and inactivation (Figure 3). In the case of the depolarization of the cell membrane, the Kv channel transits from the resting (closed) to the active (open) state. During prolonged depolarization, Kv channels switch to an inactivated state [30]. Figure 3 shows the whole-cell current response to applied voltage-step protocols for the determination of the activation, deactivation, and inactivation characteristics of the channel. In developing the dynamic behavior of the system, the three different stimulus signals, as shown in Figure 3, were used to provoke these activities. In addition, these activation, deactivation, and inactivation characteristics could also be stimulated alternatively by ramp, action potential (AP), and recovery protocols [15].

For model development, the dynamic behavior of Kv10.1 was considered at a temperature of 25 °C, which corresponds to standard experimental conditions in in vitro electrophysiological studies. The activation or opening is stimulated with appropriate depolarizing input voltage step functions. Nevertheless, the Kv10.1 response corresponds to the transient response of the system. Based on the experimentally measured response to given stimuli, we concluded that Kv10.1 behaves as a first-order system. We assumed all initial conditions as equal to zero and at a nominal voltage of 70 mV. It was possible to obtain an average model based on averaged data for all cells, as given in [21] for the Kv1.1 voltage-gated ion channel. We also considered a randomly chosen active cell with high-quality recording data from the database available via Channelpedia, a web-based freely accessible information management network and electrophysiology data repository [15].

The chosen cell ID is 9514 at 25 °C, and the host cell is CHO_FT; the species is rat. Input and output experimental data from patch-clamp measurements are given in Figure 3.

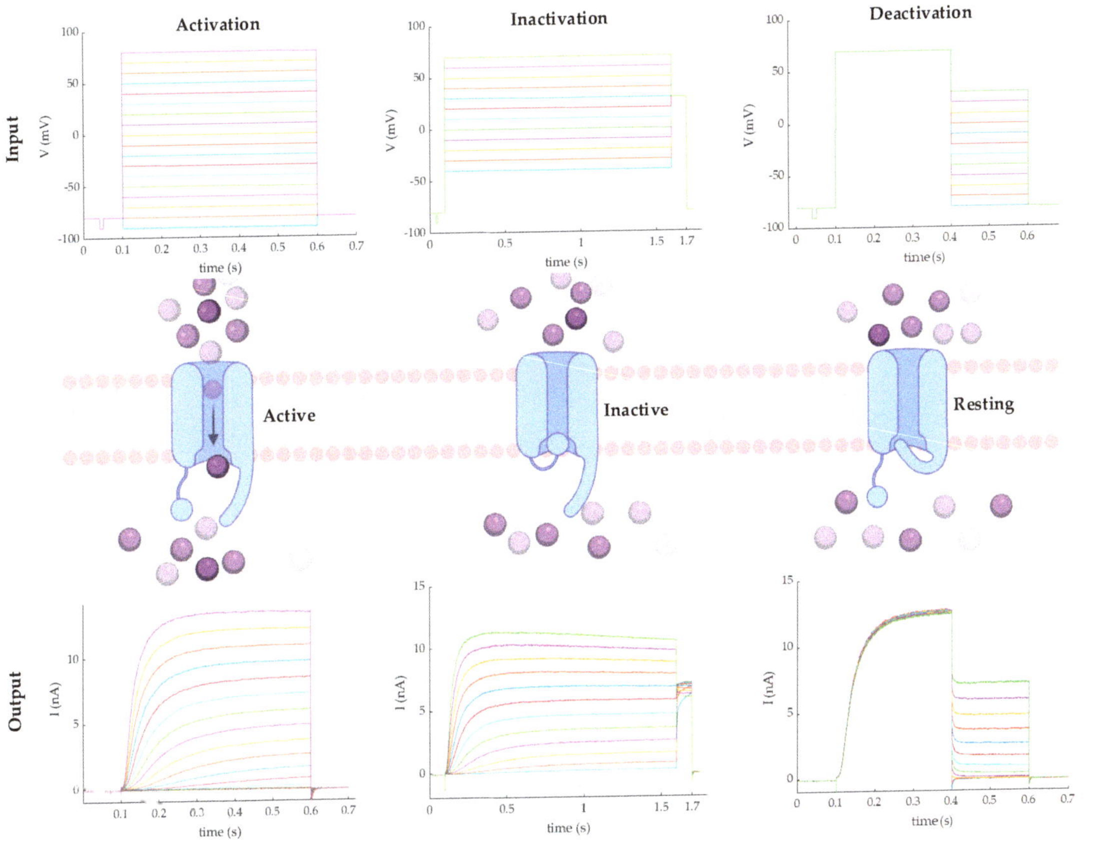

Figure 3. Dynamic behavior of the voltage-gated ion channel Kv10.1. Created with BioRender.

Figure 4 shows the activation voltage step protocol to determine the opening or activation of the ion channel and the measured macroscopic ion current as input and output for system identification. We assumed a nominal operating point at an input voltage of 70 mV, based on the experimental design and input functions for inactivation and deactivation. Input-output data from the patch-clamp experiments for the nominal operating point are shown in Figure 5a. Time-domain experimental data were imported into the System Identification toolbox in MATLAB. The data were not preprocessed and, based on the measured output signal and input step signals, we adopted a first-order linear time-invariant system. A comparison between measured and simulated model output is shown in Figure 5b.

The transfer function was estimated with one pole and null zeros in the continuous-time domain, the i/o delay was fixed to zero, and we set the initial condition to zero. The initialization method implies that the algorithm is used to initialize the numerator and denominator. Algorithms applicable only for estimating continuous-time transfer functions using time-domain data are the instrument variable (IV), state variable filters (SVF), generalized Poisson moment functions (GPMF), subspace state-space estimation (N4SID), and a combination of all of the preceding (ALL) approaches [27,28]. We chose to use the initialization method ALL. This method of systems modeling is known as black-box modeling.

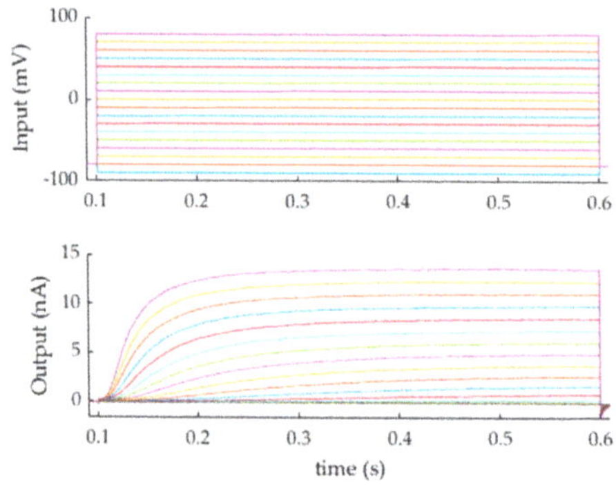

Figure 4. Input: activation voltage step protocol (**above**); output: measured macroscopic ion-current (**below**) for system identification.

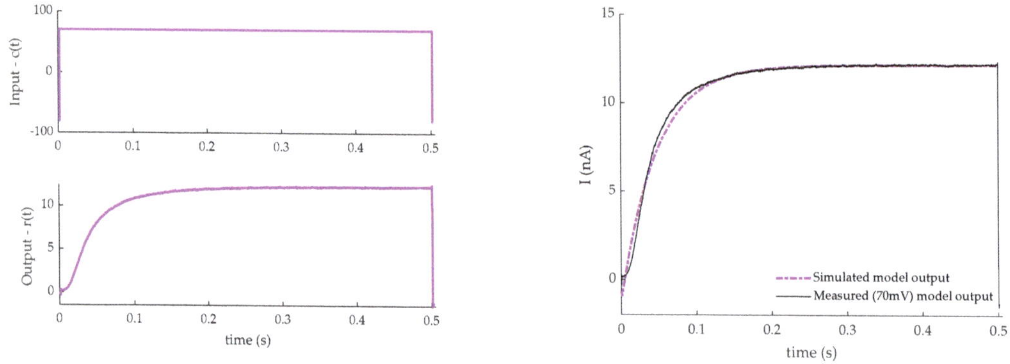

(**a**) (**b**)

Figure 5. Input-output data for nominal operating point at 70 mV (**a**); output: macroscopic ion current (**b**) used for system identification.

The estimated transfer function model for the assumed nominal step input of 70 mV and measured output macroscopic current in nA is given by

$$G(s) = \frac{3.687}{s + 21.22} \quad (3)$$

where units of the transfer function *G*(*s*) are ratio units of output (*s*)—the output macroscopic current (nA), and units of input (*s*)—the input voltage step (mV).

The model, Equation (3), is written in a general form of a first-order system in the Laplace domain by

$$G(s) = \frac{K}{Ts + 1} \quad (4)$$

where *K* is the gain with *K* = 0.1738 and *T* the time constant, *T* = 0.0471. The system's gain is the ratio between the input signal and the steady-state value of the output.

4. Discussion

4.1. System Analysis

The mathematical model of the system is given by Equations (3) and (4) in the Laplace domain. The unit step response of the system is shown in Figure 6. Besides the gain and time constant, other essential characteristics of the system are the rise time, transient time, settling time, and steady-state. These system characteristics are calculated from the transfer function or are found by plotting the step response in MATLAB.

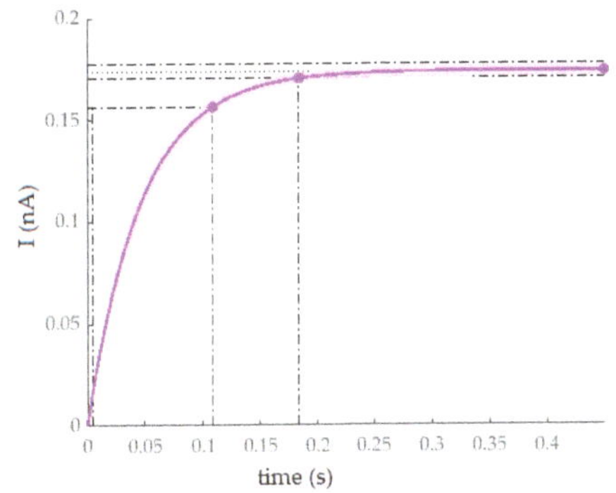

Figure 6. Unit step response of the system $G(s)$ with confidence region.

In Figure 6, the rise time is 0.104 s, the transient time is 0.184 s, and the final value corresponds to a gain of 0.174 nA; thus, the system does not have an overshoot. Experimental output data and the mathematical model indicate that stability is enforced. The same conclusion was drawn by conducting a pole location analysis. The model has one pole with a negative real part, and the system was determined to be stable [22,24,25].

The model in the Laplace domain corresponds to the differential equation under the assumption that all initial conditions are zero [22]. The inverse Laplace transformation gives a solution of a differential equation in the time domain, and the system described with the solution to Equation (4) in the time domain is

$$g(t) = 0.1738\left(1 - e^{\frac{t}{0.0471}}\right)h(t) \quad (5)$$

where $g(t)$ corresponds to $G(s)$, representing the unit step response of the system $G(s)$ (see Figure 6) when all initial conditions are equal to zero, and $h(t)$ is the input function, which corresponds to the unit step or Heaviside function $h(t)$.

We simulated outputs of the model using Equation (4) with inactivation and deactivation stimuli. The compared results of the measured outputs and simulated outputs are shown in Figure 7. The model satisfies the dynamic behavior of the Kv10.1 voltage-gated ion channel based on the adopted assumptions.

Since different cellular mechanisms can affect the opening behavior of ion channels, we believe that time-dependent transfer functions can be used to provide a more accurate description of the dynamic characteristics of ion channels. This hypothesis is supported by the fact that all mathematical systems are non-stationary and that the mathematical origin of non-stationarity can be proved [31].

4.2. Control Algorithm

The model (Equation (3)) is stable, observable, and controllable because the rank of the controllability and observability matrix is equal to n, where $n = 1$, and the system is a first-order system [22,24,25]. We considered a classical control algorithm for the Kv10.1 model. The controller is designed using the Sisotool in MATLAB. We chose an automated tuning method for a PID compensator in a feedback loop, the tuning method is a robust time response, and the controller type is PID. The controller is given by

$$C(s) = 316.85\left(\frac{1 + 0.0026s}{s}\right) \quad (6)$$

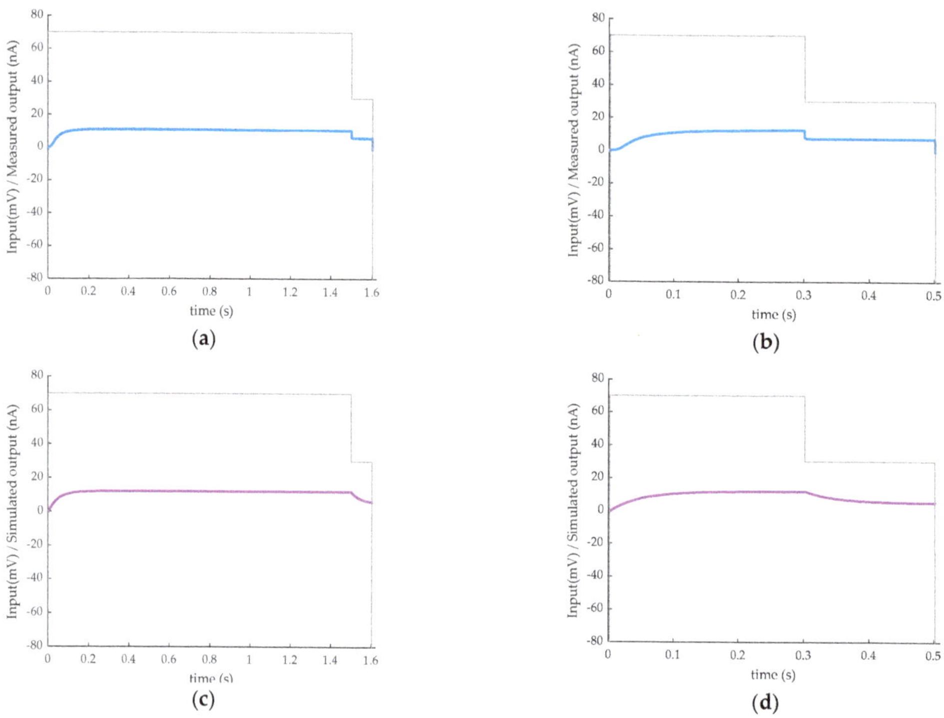

Figure 7. Measured and simulated output macroscopic current for the inactivation and deactivation input stimuli (grey lines). Input stimulus inactivation: measured output (**a**), simulated output (**b**); input stimulus deactivation: measured output (**c**), simulated output (**d**).

The unit step response of the controlled Kv10.1 voltage-gated ion channel in a closed-loop, also called a feedback loop, is shown in Figure 8. The Kv10.1 in the feedback loop displays a rise time of 0.043 s, a peak amplitude of 1.06 nA, an overshoot of 6.08%, a transient time of 0.149 s, and a final value of 1 nA. The controlled model has a faster rise time and transient response, and the final value corresponds to the desired value.

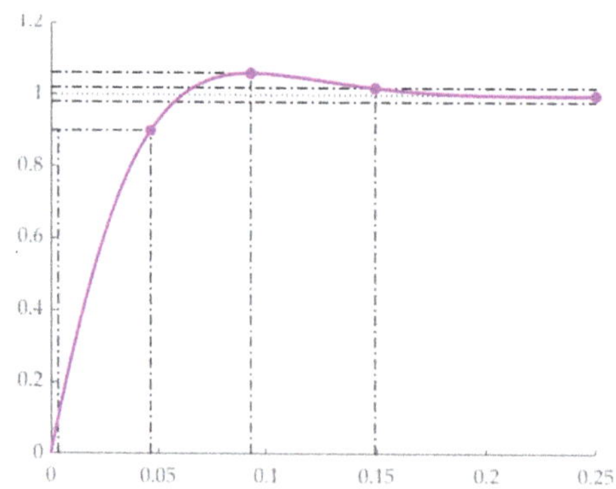

Figure 8. Unit step response of the Kv10.1 model in a control loop.

The feedback control system with the Kv10.1 model, $G(s)$, and controller $C(s)$ in a loop was implemented in Simulink (Figure 9a). A sine waveform with unit amplitude and frequency of 1 Hz was chosen as a simulation input. The simulated output is shown in

Figure 9b. Nevertheless, using this approach of ion channel modeling, the ion channel could be simulated as a single system or a system in a control loop with any randomly chosen input, enabling the obtained output to be analyzed.

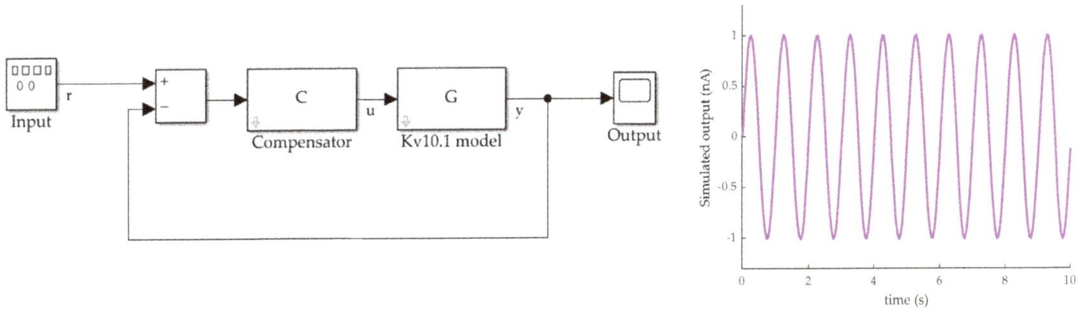

(**a**) (**b**)

Figure 9. Simulink Kv10.1 model in a control loop (**a**); simulated output for sine input (**b**).

Microelectromechanical (MEMS) devices and nanoelectromechanical (NEMS) devices can be designed as MEMS PID controllers and NEMS PID controllers. In a clinical context, nanorobots are already used for the mechanical opening of a cell membrane [32], microscopic surgical procedures, drug delivery, gene therapy, or selective destruction of cancer cells [4]. Therefore, our vision is that free-floating and guided nanorobots could have promising applications in cancer detection and therapy. Research has shown that the up-regulation of the selected channel Kv10.1 can promote cell proliferation and thus tumor development in different organs and human body systems [18], whereas the downregulation of Kv10.1 could potentially have the opposite effect, providing a potential oncological target. However, since the same ion channels are also expressed in normal, healthy cells, albeit to minor expression levels, one of the biggest challenges is to specifically block or manipulate them only in cancer cells. Nanorobots could be programmed to target certain cell clusters, tissues or organs (e.g., prostate, stomach, or breast) and specifically control the activity of selected ion channels in cancer cells without affecting the same channels expressed in surrounding healthy cells or other tissues and thus impairing their function. A controlled down-regulation of the ion channel current in specific cells only would also allow for a systemic application, such as the use of traveling nanorobots for the treatment of cancer in various regions of the body or targeted tissues and cell clusters, as shown in Figure 10, for example.

4.3. Potential Use Case for Treating Breast Cancer Using a Kv10.1 Controlled Nanorobot

Our hypothesis for the use of swimming nanorobots is to control ion-channel activity to specifically target cancer cells in breast tumors. We know that Kv10.1 channels in the nervous system contribute to the control of neuronal excitability. However, since activation requires strong depolarization, a single action potential would not be sufficient for activation [33]. In contrast, cancer cells are usually depolarized so that the channels are predominantly open, providing a potential target for pinpoint destruction by nanorobots. For example, overexpression of Kv10.1 channels is significantly up-regulated in invasive breast carcinoma, whereas healthy breast tissue shows hardly any expression or activity of the Kv10.1 channel [30,33–36]. Floating nanorobots applied via the blood circulation can be guided to the targeted sites and measure and control the Kv10.1 channel activity of the cells in the area of interest. Since nanorobots act as sensors and actuators, if the sensor interface detects an altered behavior of Kv10.1 channel activity, as is the case in breast cancer cells, the channels could be controlled accordingly by the Kv10.1 feedback control system of the robot (see Figures 9 and 10). Alternatively, after detecting a cancer cell based on aberrant or abnormal channel activity, such MEMS/NEMS-based nanorobot devices may also act as a nano-scalpel that mechanically destroys the targeted cell.

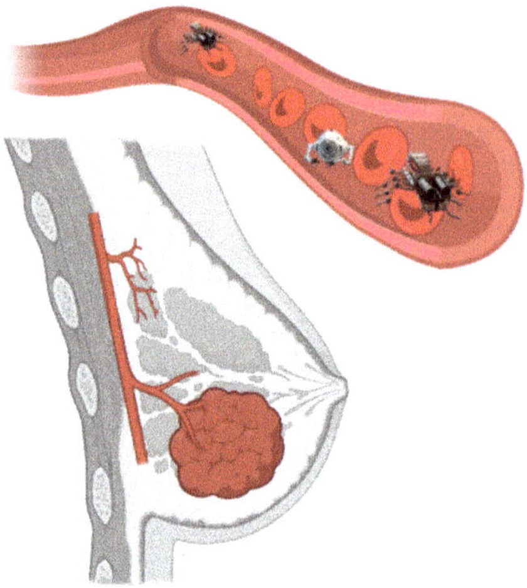

Figure 10. Illustration of possible application of Kv10.1-controlled nanorobots in breast cancer detection and treatment. The cancerous area will be reached by the nanorobots through the blood vessel system. Created with BioRender.

Further examples of possible applications are discussed in [34,37–39]. However, although the practical implementation of this use case has not been performed and validated, this work demonstrates a potential new, forward-looking direction for cancer detection and therapy.

5. Conclusions

In this work, voltage-gated ion channels were modeled as control objects or systems by applying the control system theory and a system identification method. The Kv10.1 voltage-gated ion channel exhibits dynamic behavior as a first-order dynamic system. We presented a classical control algorithm and described how this algorithm could theoretically be used to control the response of the Kv10.1 voltage-gated ion channel. Other control algorithms, such as robust control or fuzzy control, could also be used. Research has shown that controlling the Kv10.1 ion channel under up-regulated conditions plays a vital role in cancer prevention and treatment. Nanoelectromechanical systems, such as nanorobots, could be used in the treatment of cancer as they are able to travel through the blood vessel system to the target cells and bind to the ion channels. The robot's control unit can then be activated based on the implemented control algorithm, and the desired ion channel regulation can be performed to treat channel activity by the down or upregulation of the targeted channels.

It is, therefore, expected that such models, which could also describe the activity of multiple ion channels and their interactions, e.g., described by a MIMO system, will enable the use of nanoelectromechanical systems and nanorobots, as demonstrated for the Kv10.1 voltage-gated ion channel, and for other ion channel families, with the goal of establishing new tools for cancer detection and treatment at the single-cell level.

Author Contributions: Conceptualization, J.L.Š.; methodology, J.L.Š.; validation, J.L.Š.; formal analysis, J.L.Š.; investigation, J.L.Š. and S.L.; resources, C.B.; data curation, J.L.Š. and S.L.; writing—original draft preparation, J.L.Š.; writing—review and editing, J.L.Š., S.L. and C.B.; visualization, J.L.Š.; supervision, J.L.Š. and C.B. All authors have read and agreed to the published version of the manuscript.

Funding: Open Access Funding by the Graz University of Technology.

Institutional Review Board Statement: Not applicable.

Informed Consent Statement: Not applicable.

Data Availability Statement: https://channelpedia.epfl.ch/ionchannels/ (accessed on 15 February 2022).

Acknowledgments: We acknowledge Open Access Funding by Graz University of Technology.

Conflicts of Interest: The authors declare no conflict of interest.

References

1. Tan, A.; Jeyaraj, R.; De Lacey, S.F. Nanotechnology in Neurosurgical Oncology. In *Nanotechnology in Cancer*; Elsevier: Amsterdam, The Netherlands, 2017; pp. 139–170. ISBN 978-0-323-39080-4.
2. Requicha, A.A.G. Nanorobots, NEMS, and Nanoassembly. *Proc. IEEE* **2003**, *9*, 1922–1933. [CrossRef]
3. Simovic-Pavlovic, M.; Bokic, B.; Vasiljevic, D.; Kolaric, B. Bioinspired NEMS—Prospective of Collaboration with Nature. *Appl. Sci.* **2022**, *12*, 905. [CrossRef]
4. Giri, G.; Maddahi, Y.; Zareinia, K. A Brief Review on Challenges in Design and Development of Nanorobots for Medical Applications. *Appl. Sci.* **2021**, *11*, 10385. [CrossRef]
5. Asher, V.; Sowter, H.; Shaw, R.; Bali, A.; Khan, R. Eag and HERG Potassium Channels as Novel Therapeutic Targets in Cancer. *World J. Surg. Oncol.* **2010**, *8*, 113. [CrossRef]
6. García-Becerra, R.; Díaz, L.; Camacho, J.; Barrera, D.; Ordaz-Rosado, D.; Morales, A.; Ortiz, C.S.; Avila, E.; Bargallo, E.; Arrecillas, M.; et al. Calcitriol Inhibits Ether-à Go-Go Potassium Channel Expression and Cell Proliferation in Human Breast Cancer Cells. *Exp. Cell Res.* **2010**, *316*, 433–442. [CrossRef]
7. Wang, G.J.; Chen, C.L.; Hsu, S.H.; Chiang, Y.L. Bio-MEMS Fabricated Artificial Capillaries for Tissue Engineering. *Microsyst. Technol.* **2005**, *12*, 120–127. [CrossRef]
8. Tsuda, S.; Zauner, K.-P.; Gunji, Y.-P. Robot Control with Biological Cells. *Biosystems* **2007**, *87*, 215–223. [CrossRef]
9. Mendonça Munhoz, A.; Santanelli di Pompeo, F.; De Mezerville, R. Nanotechnology, Nanosurfaces and Silicone Gel Breast Implants: Current Aspects. *Case Rep. Plast. Surg. Hand Surg.* **2017**, *4*, 99–113. [CrossRef]
10. Ion Channel | Learn Science at Scitable. Available online: https://www.nature.com/scitable/topicpage/ion-channel-14047658/ (accessed on 7 April 2022).
11. Marchesi, A.; Gao, X.; Adaixo, R.; Rheinberger, J.; Stahlberg, H.; Nimigean, C.; Scheuring, S. An Iris Diaphragm Mechanism to Gate a Cyclic Nucleotide-Gated Ion Channel. *Nat. Commun.* **2018**, *9*, 3978. [CrossRef]
12. Shad, K.F.; Salman, S.; Afridi, S.; Tariq, M.; Asghar, S. Introductory Chapter: Ion Channels. In *Ion Channels in Health and Sickness*; Shad, K.F., Ed.; InTech: London, UK, 2018; ISBN 978-1-78984-227-2.
13. Gabashvili, I.S.; Sokolowski, B.H.A.; Morton, C.C.; Giersch, A.B.S. Ion Channel Gene Expression in the Inner Ear. *J. Assoc. Res. Otolaryngol.* **2007**, *8*, 305–328. [CrossRef]
14. Sands, Z.; Grottesi, A.; Sansom, M.S.P. Voltage-Gated Ion Channels. *Curr. Biol.* **2005**, *15*, R44–R47. [CrossRef] [PubMed]
15. EPFL Blue Brain Project, Channelpedia. Available online: https://channelpedia.epfl.ch/ (accessed on 7 April 2022).
16. Coetzee, W.A.; Amarillo, Y.; Chiu, J.; Chow, A.; Lau, D.; McCormack, T.; Morena, H.; Nadal, M.S.; Ozaita, A.; Pountney, D.; et al. Molecular Diversity of K+ Channels. *Ann. N. Y. Acad. Sci.* **1999**, *868*, 233–255. [CrossRef] [PubMed]
17. Jan, L.Y.; Jan, Y.N. Voltage-Gated and Inwardly Rectifying Potassium Channels. *J. Physiol.* **1997**, *505*, 267–282. [CrossRef] [PubMed]
18. Serrano-Novillo, C.; Capera, J.; Colomer-Molera, M.; Condom, E.; Ferreres, J.; Felipe, A. Implication of Voltage-Gated Potassium Channels in Neoplastic Cell Proliferation. *Cancers* **2019**, *11*, 287. [CrossRef] [PubMed]
19. Lastraioli, E.; Iorio, J.; Arcangeli, A. Ion Channel Expression as Promising Cancer Biomarker. *Biochim. Biophys. Acta BBA-Biomembr.* **2015**, *1848*, 2685–2702. [CrossRef]
20. Asher, V.; Khan, R.; Warren, A.; Shaw, R.; Schalkwyk, G.V.; Bali, A.; Sowter, H.M. The Eag Potassium Channel as a New Prognostic Marker in Ovarian Cancer. *Diagn. Pathol.* **2010**, *5*, 78. [CrossRef]
21. Langthaler, S.; Lozanović Šajić, J.; Rienmüller, T.; Weinberg, S.H.; Baumgartner, C. Ion Channel Modeling beyond State of the Art: A Comparison with a System Theory-Based Model of the Shaker-Related Voltage-Gated Potassium Channel Kv1.1. *Cells* **2022**, *11*, 239. [CrossRef]
22. Ogata, K. *Modern Control Engineering*, 5th ed.; Prentice-Hall Electrical Engineering Series. Instrumentation and Controls Series; Prentice-Hall: Boston, MA, USA, 2010; ISBN 978-0-13-615673-4.
23. Schoukens, J.; Pintelon, R.; Rolain, Y. *Mastering System Identification in 100 Exercises*; Wiley: Hoboken, NJ, USA, 2012; ISBN 978-0-470-93698-6.
24. Dorf, R.C.; Bishop, R.H. *Modern Control Systems*, 13th ed.; Pearson: Hoboken, NJ, USA, 2016; ISBN 978-0-13-440762-3.
25. Nise, N.S. *Control Systems Engineering*, 7th ed.; Wiley: Hoboken, NJ, USA, 2015; ISBN 978-1-118-17051-9.
26. Ljung, L. *System Identification: Theory for the User*, 2nd ed.; Prentice Hall Information and System Sciences Series; Prentice Hall PTR: Upper Saddle River, NJ, USA, 1999; ISBN 978-0-13-656695-3.
27. Ljung, L. *System Identification Toolbox^{TM} User's Guide*; The MathWorks, Inc.: Natick, MA, USA, 2014.

28. Option Set for Tfest-MATLAB-MathWorks Switzerland. Available online: https://ch.mathworks.com/help/ident/ref/tfestoptions.html (accessed on 7 April 2022).
29. Delgado-Ramírez, M.; De Jesús-Pérez, J.J.; Aréchiga-Figueroa, I.A.; Arreola, J.; Adney, S.K.; Villalba-Galea, C.A.; Logothetis, D.E.; Rodríguez-Menchaca, A.A. Regulation of Kv2.1 Channel Inactivation by Phosphatidylinositol 4,5-Bisphosphate. *Sci. Rep.* **2018**, *8*, 1769. [CrossRef]
30. Martínez, R.; Stühmer, W.; Martin, S.; Schell, J.; Reichmann, A.; Rohde, V.; Pardo, L. Analysis of the Expression of Kv10.1 Potassium Channel in Patients with Brain Metastases and Glioblastoma Multiforme: Impact on Survival. *BMC Cancer* **2015**, *15*, 839. [CrossRef]
31. Lazic, D. *Nelinearni Sistemi (Nonlinear Systems)*; Faculty of Mechanical Engineering, University of Belgrade: Belgrade, Serbia, 2005; ISBN 978-86-7083-530-6.
32. Wang, W.; Wu, Z.; He, Q. Swimming Nanorobots for Opening a Cell Membrane Mechanically. *View* **2020**, *1*, 20200005. [CrossRef]
33. Ouadid-Ahidouch, H.; Ahidouch, A.; Pardo, L.A. Kv10.1 K+ Channel: From Physiology to Cancer. *Pflüg. Arch.-Eur. J. Physiol.* **2016**, *468*, 751–762. [CrossRef] [PubMed]
34. Yang, R.; Xi, N.; Lai, K.W.C.; Patterson, K.C.; Chen, H.; Song, B.; Qu, C.; Zhong, B.; Wang, D.H. Cellular Biophysical Dynamics and Ion Channel Activities Detected by AFM-Based Nanorobotic Manipulator in Insulinoma β-Cells. *Nanomed. Nanotechnol. Biol. Med.* **2013**, *9*, 636–645. [CrossRef] [PubMed]
35. Pardo, L.A. Oncogenic Potential of EAG K+ Channels. *EMBO J.* **1999**, *18*, 5540–5547. [CrossRef] [PubMed]
36. Cázares-Ordoñez, V.; Pardo, L.A. Kv10.1 Potassium Channel: From the Brain to the Tumors. *Biochem. Cell Biol.* **2017**, *95*, 531–536. [CrossRef]
37. Li, S.; Jiang, Q.; Ding, B.; Nie, G. Anticancer Activities of Tumor-Killing Nanorobots. *Trends Biotechnol.* **2019**, *37*, 573–577. [CrossRef]
38. Hariharan, R.; Manohar, J. Nanorobotics as Medicament: (Perfect Solution for Cancer). In Proceedings of the INTERACT-2010, Chennai, India, 3–5 December 2010; IEEE: Chennai, India, 2010; pp. 4–7.
39. Manjunath, A.; Kishore, V. The Promising Future in Medicine: Nanorobots. *Biomed. Sci. Eng.* **2014**, *2*, 42–47. [CrossRef]

applied sciences

MDPI

Article

The Spherical Inverted Pendulum: Exact Solutions of Gait and Foot Placement Estimation Based on Symbolic Computation

Giuseppe Menga

Department of Control and Computer Engineering, Politecnico di Torino, Corso Duca degli Abruzzi 24, 10129 Torino, Italy; menga@polito.it; Tel.: +39-011-090-7272

Abstract: The gait and the Foot Placement Estimation (FPE) has recently been extended to 3-D spaces by adopting a specific form of a spherical inverted pendulum (SIP). The approach is very attractive, as it does not involve dynamics, but it is based solely on energies and momenta, however the authors (DeHart et al.) introduced several questionable approximations, in order to reach a manageable solution. The scope of the present paper is to revisit this spherical inverted pendulum applied to biped walking, offering an exact solution to the gait and the FPE by using symbolic computation. This is facilitated by exploiting the Kane's approach to dynamical modelling, and his software environment for symbolic manipulation, called Autolev. It generates explicit formulas describing the energies and angular momenta before/after the impact, along with the mechanics of the impact. As the resulting equations, function of (measurable) angular positions and velocities, are very compact, embedded in a numerical nonlinear solver, are suitable to be implemented in real time and used in practice to control biped robots or lower limb exoskeletons. The two main contributions of the paper are: the recovery of the balance by stepping, in the presence of a push in an arbitrary direction and omnidirectional walking. In this last respect, this specific form of SIP emphasizes the expenditure of energy in the walk. For the first time, at our knowledge, the walk of the SIP, based on energy, has been compared to the simulation of a 12 degrees of freedom biped robot tracking preview signals using the Zero Moment Point (ZMP) of the Linear Inverted Pendulum (LIPM). This quantitatively shows the inefficiency, in terms of energy, of the ZMP-based walk, and the gain due to the recovery of the collision of the flying foot. Similarity in the sagittal plane and differences in the frontal plane of the center of mass trajectories of the two approaches are shown, to open the road to an integration of fully actuated and underactuated controls, for an efficient full-dimensional robot gait to be developed in a future paper

Keywords: humanoid and bipedal locomotion; legged robots; passive walking; foot placement estimation

Citation: Menga, G. The Spherical Inverted Pendulum. Exact Solutions of Gait and Foot Placement Estimation Based on Symbolic Computation. *Appl. Sci.* **2021**, *11*, 1588. https://doi.org/10.3390/app11041588

Received: 29 December 2020
Accepted: 4 February 2021
Published: 10 February 2021

1. Introduction

1.1. Background

In the last decade, especially driven by the robotic school of the University of Waterloo, the inverted pendulum model, as an evolution of the passive pantograph walker, has been proposed for foot placement estimation (FPE) and the related biped gait design. First, the problem was solved in two dimensions [1,2], with a first extension and test in 3-D in [3]. More recently, the approach has been transferred to 3-D adopting a specific form of a spherical inverted pendulum (SIP) and called SFPE [4,5]. However, in order to make the problem manageable several questionable simplifications were introduced, i.e. the projections of the central inertia on the two rotation axes of the SIP are considered constant, the rotation velocity does not change before and after the contact of the swing foot with the ground, the impact is approached approximatively, so it is the computation of the angles after the impact.

The solution of the problem depends on three phases: pre-impact, impact and post-impact of the flying foot. The total energy and the angular moment on the pivot point projected on the vertical axis are constant in the periods before and after the impact, the velocities and loss of energy after the impact can be calculated, finally, the equilibrium is reached imposing zero to both angular velocities at the erect standing balance point.

Instead of introducing approximations in these three steps, an exact solution is possible by using explicitly the mathematical equations of all involved variables, and processing them with a numerical solver, to find a solution.

1.2. Symbolic Computation

The programming of the exact expressions of the previously mentioned three steps is simplified by adopting a method to describe dynamical systems introduced at the end of the last century by Prof. Kane of Stanford, known as the Kane's method [6]. He also developed a symbolic manipulation software environment, called Autolev (now MotionGenesis) [7], to support his method and to generate fragments of very efficient code of all needed mathematical expressions to be embedded into a nonlinear numerical solver. The approach allows to represent unitarily, either holonomic and non-holonomic systems, and to handle explicitly kinetic energy, momenta, impact, impulsive forces, and generalized momenta.

1.3. State of the Art of Balance, Stepping and Walking

In humanoid robotics the generality of the techniques covers basic walking (flat-footed) on flat surfaces in the absence of disturbances. They mostly track for the whole stride a preview signal based on the ZMP of the LIPM [8–10].

At difference, human-like gait, with its mix of fully actuated and underactuated phases (where walking during one of the phases is a "controlled falling") is more complex [11].

Push recovery, walking on rough terrain, and agile footstep control are active research topics [12].

For push recovery in 2-D, along with [2], using the LIPM, see also [13], and, by adding a flywheel [14]. In 3-D the concept of "N step-capturabiliy" has been introduced in [15]. However, the LIPM is used, i.e., the hight of the COG is assumed constant, and no inertia is accounted for.

Ref. [11] discusses theoretically the problems of underactuation and collision in the walk, applied in practice in this paper. The very recent reference [12] contains a comprehensive review of the literature in this field and approaches the walk, as is done here, through foot placement.

1.4. Aims and Organization of the Paper

The novel approach of processing with a numerical solver the basic energy equations of the SFPE does not involve dynamics and offers several advantages in generating the walk. It does not follow an a priory trajectory, the gait style can be changed at each step, so also cadence and step length, allowing aperiodic walk. It is robust to disturbances. Flat ground has been considered here for simplicity; however, looking one step ahead (watch your step!), also non-flat ground can be accounted for. Maneuvering has been demonstrated in the last section.

This paper has not the intention to offer a solution for the control of a complete robot, but to pave the road, with a motion generation, for a future integration in the gait of three aspects: a realist stride with finite double stance periods, energy efficiency, and a mixture of fully actuated and underactuated phases. The idea to add a small energy at each step to maintain the walk is similar to passive-dynamic walkers [16], and in the line of the frameworks of hybrid zero dynamics [17–19].

The paper is organized as follows: Section 2 discusses the existing results of the SFPE; Section 3 introduces the Kane's method and the symbolic environment Autolev; Section 4 presents the spherical inverted pendulum (SIP) model used in this paper; Section 5 describes the equations needed to estimate the foot placement to reach the balance point; Section 6 applies the approach to find the balance point by stepping in the presence of disturbances in any direction; Section 7 applies the SIP to generate a gait with arbitrary trajectory and pace, and compares the gait generated from the SIP with the classical gait of an equivalent biped robot based on the ZMP of the LIPM and previews signals [20]; Section 8 concludes the paper, and outlines future works to embed the SIP in the generation of the gait for a complete biped.

2. The Spherical Foot Placement Estimation

Traditionally, the two degrees of freedom of the SIP (sometime three degrees of freedom are defined, involving, also, the length of the pendulum) were obtained with two rotations on the horizontal axes of the inertial frame [8]. As the final objectives were the two projections of the COG trajectory on the ground, this model originated the celebrated ZMP expressions of the LIPM [9].

In the present case, in order to exploit energies and momenta, rotations along the vertical and one of the horiziontal axes have been chosen. Indicated with γ and ω the rotation velocities of the SIP, on the vertical and the horizontal axis on the frontal plane of the biped, the approach of [4,5] is based on the projection of the angular momentum on the pivot point of the pendulum on these two directions, and expressing the kinetic energy as function of these two projections. Cleverly, noting that the total energy and the momentum projected on the vertical axis remain constant during the periods before and after the collision of the swing foot with the ground, approaching the collision and the switching of the pivot foot, the problem is solved by writing the equations of the pre-impact, impact and post-impact phases.

Discussion

To render the problem manageable the two projections of the angular momentum are expressed as a function exclusively of γ or ω and the two projections of the central inertia are considered constants. i.e., the central inertial matrix is assumed diagonal, and I_{11} and I_{33} have identical values (see Appendix A). The impact is solved approximatively. Moreove, no importance is given to the angle of rotation on the vertical axis.

3. The Kane's Method and Autolev

In this work, the so-called Kane's method [6] was adopted to model the spherical inverted pendulum. This method is particularly interesting in this case because it is equally applicable to either holonomic and non-holonomic systems and, for non-holonomic systems, without the need to introduce Lagrangian multipliers. Briefly, the main contribution of the Kane's method is that, through the concepts of motion variables (later called generalized speeds), the vectors of partial velocities and partial angular velocities, generalized active forces and generalized inertia forces, the dynamical equations are automatically determined, enabling forces and torques with no influence on the dynamics to be eliminated early in the analysis. Early elimination of these noncontributing forces and torques greatly simplifies the mathematics and enables problems with greater complexity to be handled.

3.1. Generalized Coordinates and Speeds

A multi-body system, which possesses n degrees of freedom, is represented by a state with a n-dimensional vector $\mathbf{q}$ of configuration variables (*generalized coordinates*) and an identical dimension vector $\mathbf{u}$ of *generalized speeds* called also *motion variables*, that could be any nonsingular combination of the time derivatives of the generalized coordinates that describe the configuration of a system. These are the kinematical differential equations:

$$u_r = \sum_{i=1,\cdots,n} Y_{ri}\dot{q}_i, r = 1,\cdots,n \tag{1}$$

Y_{ri} may be in general nonlinear in the configuration variables so that the equations of motion can take on a particularly compact (and thus computationally efficient) form with the effective use of generalized speeds.

3.2. Partial Velocities and Angular Velocities

Partial velocities of each point (partial angular velocity of each body) are the n three-dimensional vectors expressing the velocities of that point (angular velocity of that body) as a linear combination of the generalized speeds. Let be $\mathbf{v}^B$ the translational velocity of a point B and $\boldsymbol{\omega}^P$ the rotational velocity of a body P with respect to the inertial reference frame, then

$$\begin{aligned} \mathbf{v}^B &= \textstyle\sum_{r=1...n} \mathbf{v}_r^B u_r \\ \boldsymbol{\omega}^P &= \textstyle\sum_{r=1...n} \boldsymbol{\omega}_r^P u_r \end{aligned} \tag{2}$$

where $\mathbf{v}_r^B$ and $\boldsymbol{\omega}_r^P$ are the rth partial velocity and partial angular velocity of B and P, respectively.

3.3. Generalized Active and Inertia Forces

The n generalized forces acting on a system are constructed by the scalar product (projection) of all contributing forces and torques on the partial velocities and partial angular velocities of the points and bodies they are applied to.

Let us consider a system composed by N bodies P_i, where the torque $\mathbf{T}^{P_i}$, and force $\mathbf{R}^{B_i}$ applied to a point B_i of P_i are the equivalent resultant (*"replacement"* [6]) of all active forces and torques applied to P_i. Then

$$F_r^{P_i} = \boldsymbol{\omega}_r^{P_i} \cdot \mathbf{T}_i^{P_i} + \mathbf{v}_r^{B_i} \cdot \mathbf{R}_r^{B_i} \tag{3}$$

is the rth generalized active force acting on P_i and

$$F_r = \sum_{i=1,\cdots,N} F_r^{P_i} \tag{4}$$

the rth generalized active force acting on the whole system. Identically for the inertia forces, indicated as F_r^*.

The dynamical equations for an n degree of freedom system are formed out from generalized active and inertial forces F_r^*

$$F_r + F_r^* = 0, r = 1, \cdots, n. \tag{5}$$

These are known as Kane's dynamical equations.

They result in a n-dimensional system of second order differential equations ($2n$ order state variable representation) on generalized coordinates and speeds

$$\bar{\mathbf{M}}(\mathbf{q})\dot{\mathbf{u}} + \bar{\mathbf{C}}(\mathbf{q}, \mathbf{u})\mathbf{u} + \bar{\mathbf{G}}(\mathbf{q}) - \bar{\mathbf{\Gamma}}(\mathbf{q}, \mathbf{u}, \boldsymbol{\tau}) = \mathbf{0}, \tag{6}$$

where the parameter definitions are similar but not identical of the classical Lagrangian form and more efficient computationally [21].

3.4. Non-Holonomic Constraints

When m constraints on the motion variables are added to the model, only $n - m$ generalized speeds are independents. The system is, then, called a non-holonomic system. The *non-holonomic* constraints are expressed as a set of m linear relationships between dependent and independent generalized speeds of the type

$$u_r = \sum_{i=1,\cdots,p} A_{ri} u_i, r = p + 1, \cdots, n, \tag{7}$$

with $p = n - m$. In this case, selected the independent speeds, the Kane's method immediately offers the minimal $2p$ order state variable representation from

$$\widetilde{F}_r + \widetilde{F}_r^* = 0, r = 1, \cdots, p, \tag{8}$$

where Kane calls $\widetilde{F}_r$ and $\widetilde{F}_r^*$ non-holonomic generalized active and inertial forces, while the remaining m original redundant equations resolve themselves in the expressions of the m reaction forces/torques returned by the constraints. Because the Kane's method is fundamentally based on the projection of forces on a tangent space on which the system dynamics are constrained to evolve, spanned by the partial velocities, reaction forces/torques result from the projection on its null-space.

Moreover, it is always possible to handle an holonomic (configuration) constraint as if it is non-holonomic, that is, to treat it as a motion constraint. This is particularly advantageous to represents the spherical inverted pendulum with a pantograph during a step, where in the first phase non-holonomic constrains allow pivoting on the supporting leg, and in the second phase, releasing the non-holonomic constraints the impact of the swing leg with the ground can be represented.

3.5. Unilateral Constraints and Collision

As a consequence of switching between different non-holonomic models during gait, unilateral constraints and collisions cannot be ignored.

Clearly, adopting non-holonomic dynamics assuming points of the feet fixed to the ground is valid for bilateral constraints (ignoring eventual detachment from the ground and slipping). In the approaches known as hybrid complementarity dynamical systems based on forward dynamics [22] the necessary conditions for satisfying unilateral constraints are directly embedded into the model. Vice versa, a minimalistic view is adopted here, noting that in a physiological gait, normally, bilateral constraints on the feet are not assumed to be violated. Hence, we design a priori walking strategies and we test through the simulator that this effectively occurs, by monitoring, a posteriori, reaction forces for the conditions:

$$F_{z_{foot_i}} > 0, i = 1,2 \tag{9}$$

and

$$|F_{j_{foot_i}}| < \mu F_{z_{foot_i}}, j = x, y, i = 1,2. \tag{10}$$

Obviously, the control we propose cannot adapt itself to pathological conditions, such as a slipping surface.

For the second point, mechanics of the collision of the swing foot to the ground has to be considered, when switching to the next step causes the transfer of final conditions of the generalized speeds of one phase to the initial conditions of the successive. With reasonable assumptions of non-slipping and anelastic restitution the reaction impulsive force $\mathbf{F}^B$ at the impact point B and the initial conditions of the generalized speeds for the new phase $\mathbf{u}(t^+)$ can be computed. Also for this aspect, Autolev offers all needed mechanical expressions.

The following analysis is based on two concepts: *generalized impulse* and *generalized momentum* [6,23]. Indicate, as usual, with $\mathbf{v}_r^B$ the r-th component of the partial velocity vectors of the point B (the swing foot), the *generalized impulse* at the point B at the contact with the ground at instant t^- is defined as the scalar product of the integral of the reaction impulsive force $\mathbf{F}^B\delta(t-\tau)$ in the time interval $t^- \div t^+$ with the corresponding partial velocities

$$I_r \approx \mathbf{v}_r^B(t^-)^T \cdot \mathbf{F}^B, r = 1, \cdots, n, \tag{11}$$

the *generalized momentum* is defined as the partial derivative of the kinetic energy K with respect to the r-th generalized speed

$$p_r(t) = \partial K / \partial u_r, r = 1, \cdots, n, \tag{12}$$

then, Kane proves that

$$I_r \approx p_r(t^+) - p_r(t^-). \tag{13}$$

Indicate the matrices

$$\mathbf{V}^B = (\mathbf{v}_1^B(t^-) \cdots \mathbf{v}_n^B(t^-)) \tag{14}$$

$$\mathbf{P} = \{\partial p_i(t^-)/\partial u_j\}, i, j = 1, \cdots, n \tag{15}$$

of vectors of partial velocities, and of partial derivatives of $p_r(t)$ with respect to the generalized speeds, and the vectors

$$\mathbf{I} = [I_1 \cdots I_n]^T = \mathbf{V}^{B^T} \cdot \mathbf{F}^B \tag{16}$$

$$\mathbf{u}(t) = [u_1(t) \cdots u_n(t)]^T \tag{17}$$

$$\mathbf{v}^B(t) = \mathbf{V}^B \cdot \mathbf{u}(t) \tag{18}$$

$$[p_1(t), \cdots, p_n(t)]^T = \mathbf{P} \cdot \mathbf{u}(t) \tag{19}$$

of *generalized impulses*, of *generalized speeds*, of the velocity of point B and of *generalized momenta*, respectively.

Then, taking into account from (16) to (19), considering that $\mathbf{v}^B(t^-)$ is known and $\mathbf{v}^B(t^+)$ is zero, assuming non-slipping condition and inelastic collision, the following system of equations is solved to derive the unknown $\mathbf{F}^B$ and $\mathbf{u}(t^+)$:

$$\begin{bmatrix} -\mathbf{Pu}(t^-) \\ 0 \end{bmatrix} = \begin{bmatrix} \mathbf{V}^B(t)^T & -\mathbf{P} \\ 0 & \mathbf{V}^B(t) \end{bmatrix} \cdot \begin{bmatrix} \mathbf{F}^B \\ \mathbf{u}(t^+) \end{bmatrix} \tag{20}$$

An essentially similar equation was discussed in [11]. At the solution, along with the velocity $\mathbf{u}(t^+)$ after the impact, it must be verified that the impulsive force $\mathbf{F}^B$ satisfies the conditions of unilateral constraint (9) and (10).

4. The Spherical Inverted Pendulum Model

In describing the spherical inverted pendulum the same notation used in [4,5] was adopted. In addition, explicitly, θ_z indicates the angle of rotation with respect to the vertical axis, and two degrees of freedom of the swing leg relative to the pendulum were introduced, with the angles α_z and α, as shown in Figure 1, with the kinematics of the joints in Figure 2a. The configuration Figure 2b will be used in special situations, only to perform side shuffle.

The angle position and velocity of the pendulum on the z and y axes of the inertial frame, and the two rotations of the swing leg with respect to the local axis z, and y of the supporting leg are $\theta_z, \theta, \gamma, \omega, \alpha_z, \alpha$, respectively. The configuration variables of the model are $\theta_z, \theta, x, y, z$, as the swing leg is considered without mass and inertia, where x, y, z are the coordinates of the pivot foot. The motion variables are $\gamma, \omega, u_1, u_2, u_3$, where imposing a non-holonomic constraint to the pivot foot, the velocities , $u_1 = \dot{x}, u_2 = \dot{y}, u_3 = \dot{z}$, are zero during the swing phase, but are released at the impact of the swing foot with the ground.

In the next sections the SFPE and the gait based on this model are described.

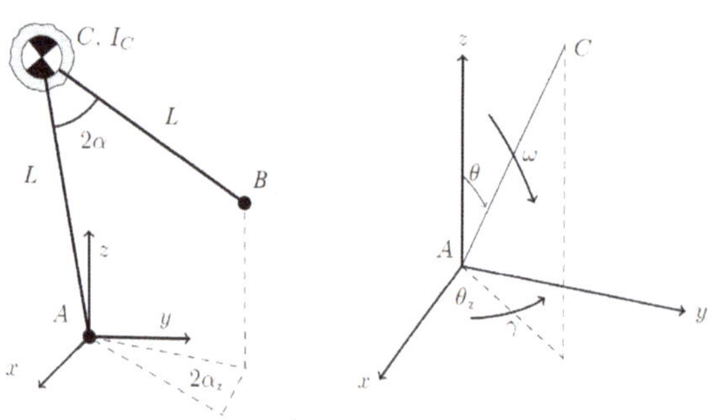

Figure 1. The spherical inverted pendulum.

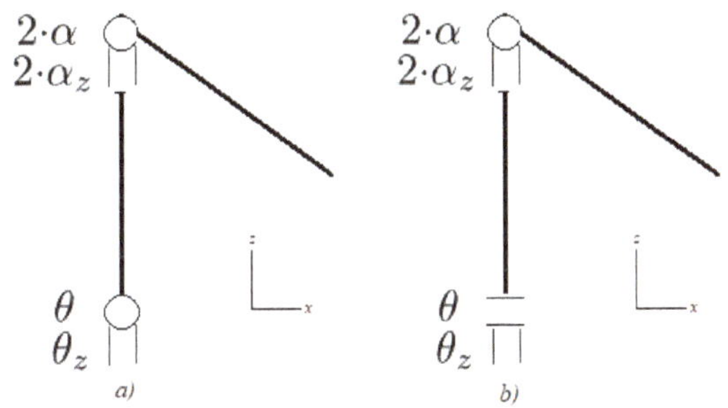

Figure 2. The kinematics of the joints—(**a**) forward/bacward motion of the flying leg, (**b**) side shuffle.

5. The Estimation of the Balance Point

Before the impact the motion variables have value $\gamma^-, \omega^-, 0, 0, 0$ and after γ^+, $\omega^+, u_1^+, u_2^+, u_3^+$. The total energy and the projection on the vertical axis of the angular momentum, k^γ (the notation of [4] is maintained, even if it is shown in Appendix A that in general both speeds are present in this projection, and here an explicit dependency on γ is no more needed), are constant before and after the impact, however, they have a reduction during the impact.

To simplify the computations and to avoid spurious solutions, the next procedure is started after the pendulum reaches the vertical position and $\theta > 0$. Let us say that at time t_0 the state variables assume the values $\theta_{z0}, \gamma_0, \theta_0, \omega_0$, the total energy T_0, and the moment on the vertical axis $k^\gamma{}_0$ (these last two values are the same, also, at the unknown instant of the impact t^-). This gives the first equation, linking all state variables at the pre-impact.

$$T_0 = T(\theta_z{}^-, \gamma^-, \theta^-, \omega^-) \tag{21}$$

At the impact the swing foot, indicated with the point B, touches the ground. The vertical coordinate of B offers the second equation, linking the pre-impact angle θ^- to α and α_z

$$B_z(\theta^-, \alpha, \alpha_z) = 0 \tag{22}$$

The constant momentum k^γ offers the third equation, linking γ^- to the other pre-impact motion variable

$$k^\gamma{}_0 = k^\gamma(\gamma^-, \theta^-, \omega^-) \tag{23}$$

Switching the pivot foot after the impact, the relationship between the angles $\theta_z{}^+, \theta^+$ of the new pivot leg from $\theta_z{}^-, \theta^-, \alpha, \alpha_z$ is obtained equating the three projections, with respect to the inertial axes, of swing and support legs (i.e., the swing leg becomes the new support leg)

$$SU(\theta_z{}^+, \theta^+) = SW(\theta_z{}^-, \theta^-, \alpha, \alpha_z) \tag{24}$$

The solution of the impact equation (performed symbolically) (20) gives the motion variables after the impact, hence the total energy and the angular momentum k^γ. The total energy after the impact can be evaluated before the switching of the pivot foot so it does not require the value of θ^+ and θ_z^+ after switching. The angular momentum is computed on the new pivot point, so it requires the new value of θ^+ and θ_z^+ after the switching. This gives the third and fourth equations.

$$\begin{aligned} TE^+ &= TE(\theta_z{}^-, \theta^-, \gamma^+, \omega^+, u_1+, u_2+, u_3+) \\ k^{\gamma+} &= k^\gamma(\theta_z{}^+, \gamma^+, \theta^+, \omega^+) \end{aligned} \tag{25}$$

Moreover, by imposing velocity zero of the swing foot, after the impact, angles before the impact can be related to motion variables after, with a further relationship

$$[\dot{B_x}, \dot{B_y}, \dot{B_z}]^T = 0 = F(\theta_z{}^-, \theta^-, \alpha, \alpha_z, \gamma^+, \omega^+, u_1{}^+, u_2{}^+, u_3{}^+) \quad (26)$$

To estimate the foot placement to reach the balance in an erect posture after the impact, with $\omega = 0, \gamma = 0$, and $\theta = 0$, noting that $k^{\gamma+}$ is zero by the last condition (28), it is imposed that the total energy after the impact is equal to the maximal potential energy

$$TE^+ = m \cdot g \cdot L \quad (27)$$

Finally, to impose that γ be zero at the balance point (but not necessarily after the impact, as it will be seen in the next Figure 4), from the impact the last equation is set

$$k^{\gamma+}(\theta_z{}^+, \gamma^+, \theta^+, \omega^+) = 0 \quad (28)$$

From the previous relationships, the unknown variables $\theta_z{}^-, \theta_z{}^+, \gamma^-, \theta^-, \theta^+, \omega^-, \alpha, \alpha_z$ are determined, using non-linear least squares, with some numerical solver such as the Levenberg-Marquardt algorithm [24,25].

6. Recovering Balance

The first application of the SFPE is to recover balance in the presence of an impulsive disturbance. It is assumed that the biped is in quiet standing balance, conventionally oriented in the direction of the x axis. An impulsive horizontal force in an arbitrary direction generates a velocity of the COG. The biped, in reorienting himself, will react in two different ways, with a minimal rotation around the z axis, according to the relative direction of the pulse: if the direction is closer to his sagittal plane he will mostly move the free leg forward (or backward), if it closer to the frontal plane the motion will be mostly laterally with a side shuffle. To emulate these two distinct situations if the direction is less than 45° to the sagittal plane the model adopted is the standard one of Figure 2a, otherwise Figure 2b. When the angle θ falls below a safety value and the direction is detected from the falling velocity, let say $[v_x, v_y, v_z]^T$, the model (a) or (b) is selected according to $atan2(v_y, v_x)$, and initial velocities are assigned to ω and γ from the inverse of the first two rows of the matrix of partial velocities (29)

$$\mathbf{V}^{COG} = [\mathbf{v}_{\omega}^{COG} \mathbf{v}_{\gamma}^{COG}]. \quad (29)$$

Then, the SFPE algorithm is run. This defines, among other variables, the swing foot angles α, α_z for the impact that allows recovering the balance in one step. The numerical algorithm does not converge in two cases: when the total energy after the impact is lower than the maximum potential energy, or when one step is not enough to recover the balance. In the first case the condition (27) cannot be satisfied. In the second case the bound on the maximum allowable α is not satisfied.

Two examples are presented using the two models (Figure 3). The situation of a push perfectly aligned with either the sagittal or the frontal plane is not considered, as it can be simply solved with the classical 2-D approach. In both cases the initial velocity is of 0.5 m/s and it is detected, at instant 0.5 s, when the falling angle reaches 0.1 rad. The first example, using model (a), presents the response to a push at an angle of 20° from the x axis, in the second, with model (b), the angle is of 110°.

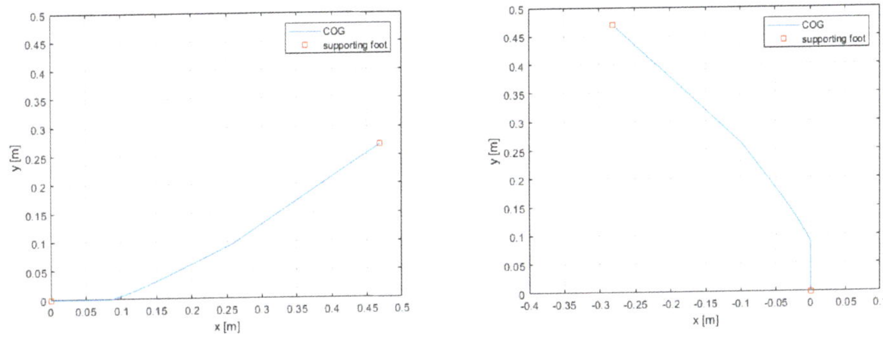

(**a**) model (a) with $\alpha = -0.316, \alpha_z = 0.087$ rad (**b**) model (b) with $\alpha = -0.32, \alpha_z = 0.88$ rad

Figure 3. Recovering an impulsive disturbance-COG behaviours.

Apart from a difference in the sign of the angles, and the velocities the two behaviours are very similar, so the case of model (b) is detailed, only.

The original central inertia matrix of the example, the same of the biped of [20], given in the Appendix A, had the elements I12 and I23 equal zero, the correct responses of angles (speeds and positions) in Figure 4a,b are represented with solid lines. To see the differences, fictitious values different from zero have been assigned to I12 and I23, and the example rerun. The responses are indicated in the legends with a X, and plotted with dashed lines.

From the SFPE, the flying foot final position can be forecasted and the SIP COG trajectory used as a reference, knowing the kinematics, to control the joints of a real biped robot or exoskeleton to recover balance.

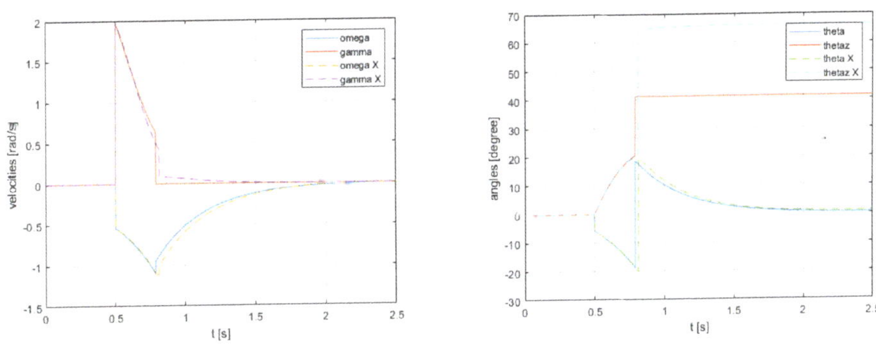

(**a**) model (b) angular velocities (**b**) model (b) angles

Figure 4. Angle velocity and position behaviours-with the effect of off-diagonal elements different from zero in the inertial matrix.

7. The Gait

At difference of other works, only some of the expressions of Section 5 are exploited for generating the gait, the SFPE is only used to impose a halt at the end of the walk.

α is chosen to achieve the desired step length, α_z to achieve the COG sway from $-\theta_{zMax}$ to θ_{zMax} and to control the offset with respect to the baseline of walk. The gait is initiated giving an initial condition to $\theta_z, \gamma, \theta, \omega$, or, simply, from a standing up balance by leaving the pendulum to fall forward.

Each step is concluded when the swing foot touches the ground (the vertical coordinate of point B becomes zero, Equation (22)). From the impact Equation (20) the new motion variables $\gamma^+, \omega^+, u_1^+, u_2^+, u_3^+$ are determined, and from Equation (24) the starting values of $\theta_z{}^+, \theta^+$ for a new step are computed. The gait is maintained by increasing at each step, after impact, the resulting ω^+ to compensate for the reduction of kinetic energy due to the impact, controlling, also, the gait cadence, and perturbing γ^+ to correct the angle of direction of the walk.

In the present model, the two legs have no mass and inertia. Therefore, the motions of the angles α and α_z are instantaneous and energy free. The only energy contribution to maintain the walk is given by proper impulsive forces and torques just after the impact to modify the velocities ω^+ and γ^+ resulting from the impact. This emulates, in a real walk, the contribution given by the biped in the brief double support phase and in the period of single support when the foot is flat and able to transfer torques.

Five control variables are identified to control the five objectives of the walk: *cadence*, *step length*, *distance between feet*, *y offset* with respect to the baseline of walk, and *direction of walk* (even if interacting each other, each of the five variables predominantly controls one of the five objectives). After each impact, at the start of step k they are

$$
\begin{gathered}
\delta_\omega(k) \Rightarrow \omega(k) = \omega^+ + \delta_\omega(k)(cadence) \\
\delta_\alpha(k) \Rightarrow \alpha(k) = \alpha_0 + \delta_\alpha(k)(step\,length) \\
\delta_{sway}(k) \cdot u + \delta_y(k) \Rightarrow \alpha_z(k) = (\alpha_{z0} + \delta_{sway}(k)) \cdot u + \delta_y(k) \\
(spacing\ between\ feet\ and\ y\ offset) \\
\delta_\gamma(k) \Rightarrow \gamma(k) = \gamma^+ + \delta_\gamma(k)(direction)
\end{gathered} \tag{30}
$$

where u assume the values $+1, -1$ according to the right or left foot support.

It must be noted that no periodic reference is tracked. The whole gait style (cadence, length of the step, offset with respect to the baseline of walk-through a side shuffle, spacing between the two feet and direction) can be changed at each step. The energy consumption of the gait is measured by the difference, after the foot collision, of the kinetic energies before and after the application of the control (in particular $\delta_\omega(k)$ and $\delta_\gamma(k)$).

The next Figures 5 and 6 show a sample of a typical rectilinear walk, terminating with a halt. In particular, the Figure 7 show the energy needed at each step to maintain the gait, provided by $\delta_\omega(k)$ and how the gait collapses after two steps if no maintenance is performed.

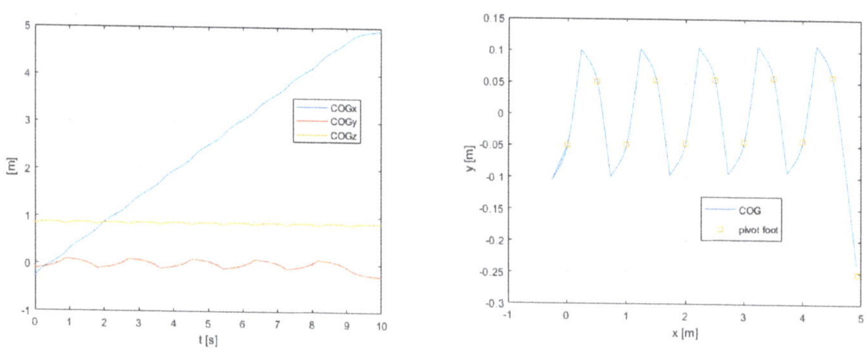

(**a**) COG along the x and y axes

(**b**) Details of the projection of the COG on the ground and pivot foot posizion

Figure 5. The COG behaviour.

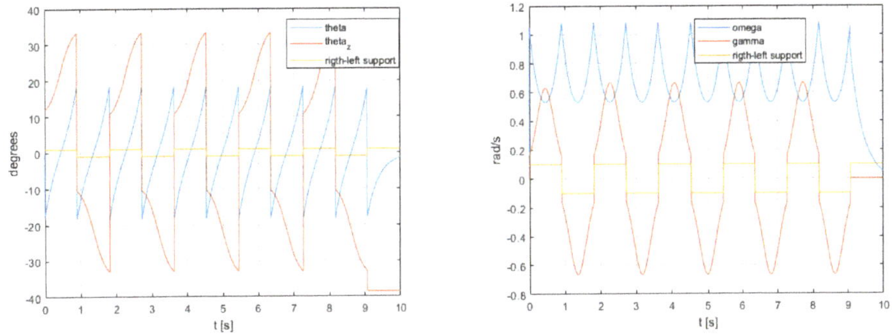

(**a**) The angles θ and θ_z.

(**b**) The velocities ω and γ.

Figure 6. Angle position and velocity behaviors.

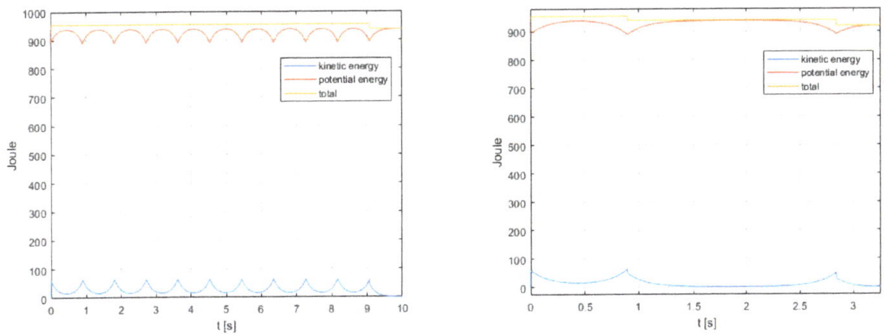

(**a**) The energies when the gait is maintained.

(**b**) The decay of the energies in absence of maintainment.

Figure 7. The energies.

7.1. Comparison between SIP and ZMP Based Gaits

An interesting question is how the SIP gait compares with the classical one based on the ZMP of a linear inverted pendulum. In previous papers, a 10/12 degrees of freedom biped model was simulated in rectilinear and curved trajectories [20]. The technique adopted was classical, by controlling the model to follow a preview trajectory based on the ZMP.

The total mass, the central inertia and the COG height of the biped were used to model the SIP. The parameters that control the gait of the SIP were adjusted to synchronize the two walks. To compare the center of pressure (COP) on the shoes in the two cases, the SIP is mounted on the ankle of the same foot of the biped (see Appendix A), no torque is transferred from the joints of the SIP, but the force on the ankle, returned from the non-holonomic constraint, are balanced by an identical force in the ZMP on the sole. The comparison is shown in the next figures.

In Figures 8a,b and 9 the two trajectories are superimposed. In particular, the right and left supports of the SIP and the periods of double support of the preview based simulation are also indicated.

The COG behaviours along the x a z axes are very similar, but not along the y axis. In fact, the details of the two COGs projected on the ground, and the relationships between the COP and the foot support placement, compared in the Figure 10a,b, show marked differences.

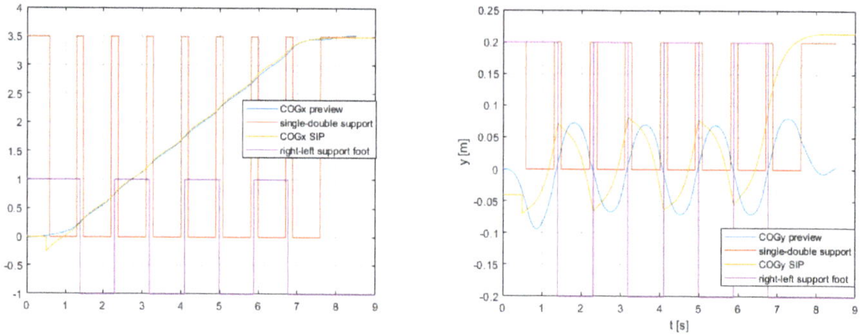

(**a**) COGx, comparisons (**b**) Cogy, comparisons

Figure 8. Comparison of COGx and COGy.

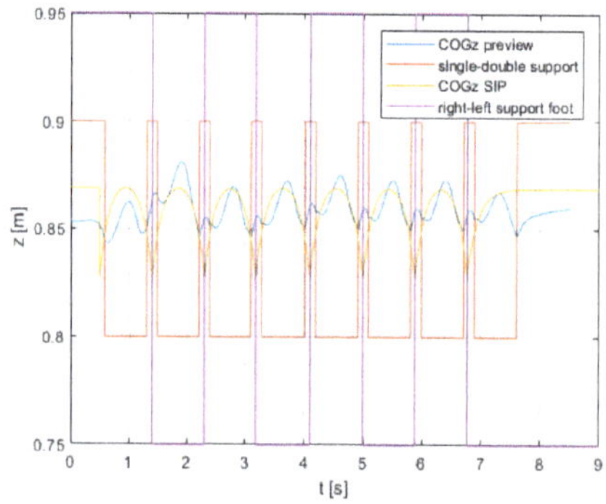

Figure 9. COGz, comparisons.

To achieve a similar sway of the *COG* on the *y* direction in the two cases, the SIP keeps the feet closer than in the preview case. Let consider that the SIP, passed the erect position is in free fall and the COP jumps suddenly on the new foot. Vice versa, in the preview-based gait the biped is always controlled to maintain the COP close to the supporting foot, and to transfer it, almost continuously from single to double support.

Finally, the energies were compared in Figure 11. As expected, the preview based on the ZMP demands a greater expenditure of energy.

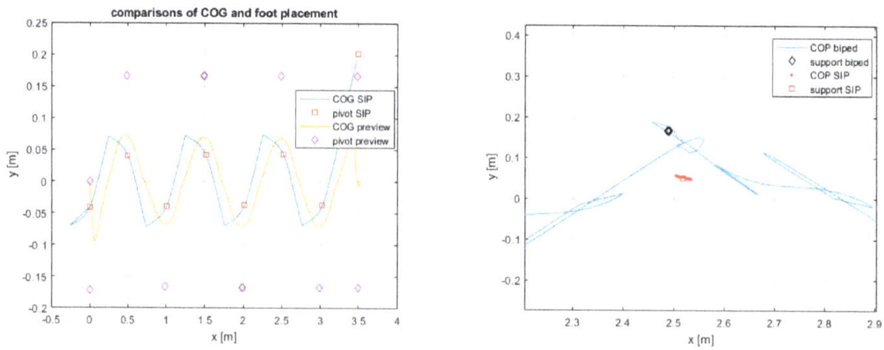

(**a**) The COG, and the foot support (**b**) Relationship between COP and foot support

Figure 10. Comparison of COG and COP.

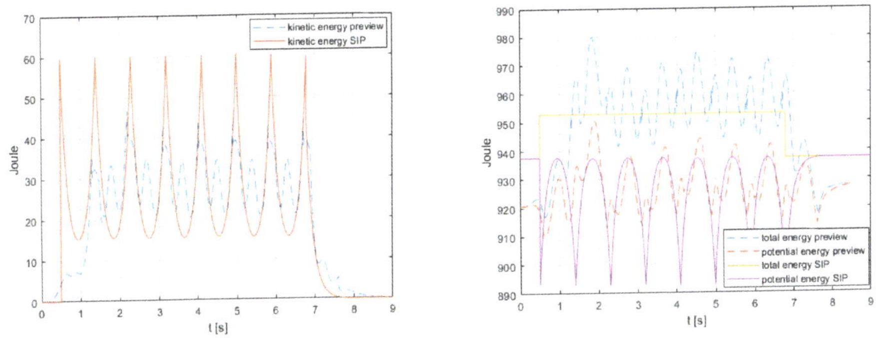

(**a**) Kinetic energies (**b**) Total and Potential energies

Figure 11. Comparison of energies.

7.2. Turning While Walking

This last subsection shows how to generate turning while walking using the SIP model.

The walking trajectory is described in part III of [20]. It is obtained concatenating arcs of circle, whose ray can be changed at the beginning of each step. The control is based on a running local reference frame with respect to which the SIP is constrained to maintain a rectilinear walk. The local running frame, at each instant, rotates along the z axis with respect to the world space to have its x axis tangent to the trajectory, and moves its origin to follow an involute of the path curve. Let $s(t)$ and $\dot{s}(t)$ be the curvilinear coordinate and its velocity on the path, their values also represent the motion on the abscissa of the local frame, and $\Theta_z(t)$ the orientation on the z axis of the local frame at each instant t.

At time $T(k)$, immediately before the step k starts, let indicate with $A_{xL}(k), A_{y_L}(k)$, $B_{xL}(k), B_{y_L}(k)$ the x and y coordinates of the last supporting foot, and of the swing foot at the contact (it is going to be the next support), in the local frame, respectively, and with $\theta_z(k) = {\theta_z}^+$ the current orientation angle after the contact.

δ_ω is fixed to obtain the desired average *cadence* of the gait, δ_α the average *step length*, and δ_{sway} the average *distance between feet*, averaging $1/(T(j) - T(j-1)), B_{xL}(j) - A_{xL}(j)$, $(B_{y_L}(j) - A_{y_L}(j)) \cdot u$, respectively, over the period $1 \leq j \leq k$. $(B_{y_L}(k) + A_{y_L}(k))/2$ measures the *offset* of the trajectory with respect to the centerline in the local frame and $(\theta_z(k-1) + \theta_z(k))/2$ the mean *direction* of walk at time $T(k)$.

From those measures a mild proportional, integral feedback is set to $\delta_y(k)$ and $\delta_\gamma(k)$ to maintain *offset* to zero, and *direction* to follow $\Theta_z(k)$. The result is the following Figure 12.

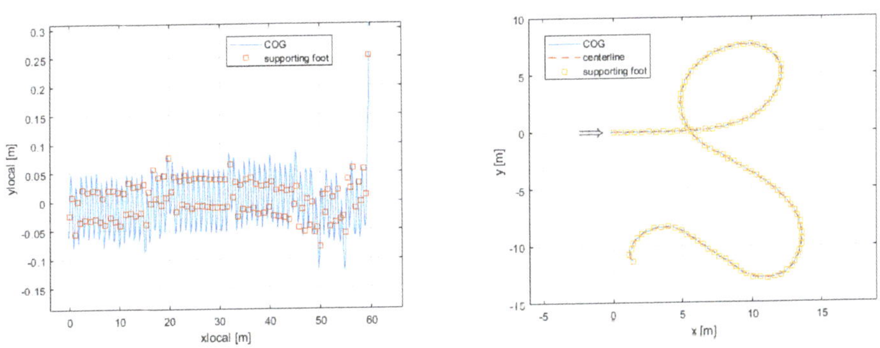

(**a**) Trajectory in the local frame (**b**) Trajectory in the world frame

Figure 12. Trajectory-turn while walking.

8. Conclusions

Starting from the ideas of [4,5] of using a SIP model that emphasizes energies, the gait and the SFPE were reviewed with a solution based on the exact computation of the kinetic energy, the momentum projected on the z axis and the foot collision. Let note, that with this approach of SFPE the complicate handling of the projection of the angular momentum on the horizontal axis, contained in the original development, is not any more needed and arbitrary central inertia matrices are allowed. Consider that the solution of the impact (Equation (20)) offers the reaction forces at the collision, also. They are not used in this paper, but they will be in future extensions.

Very compact code for the equations have been obtained adopting the Kane's method and using his software environment. The solution of these equations, through a non-linear least square algorithm such as Levenberg-Marquardt, allows to introduce box bounds of the variables, to exploit the Jacobian, and to test for the feasibility of the solution. Moreover, the dependency of these equations on measurable angle positions and velocities and knowledge one step ahead of the ground level consents a real-time implementation to control robots or exoskeletons.

The introduction of the angle α_z, taking into account the angle θ_z, has extended the approach to non-periodic and omnidirectional gait and to recover the balance from a push in an arbitrary direction.

This paper, originally motivated by the need to help to recover balance, when disturbances are present, in a lower limb exoskeleton [26], is just the starting point. Future exentions, in order of complexity, are: introducing in the pantograph model the width of the pelvis in the distance between the two legs, adding a mass to the legs (sometime balance is recovered by inertia, just extending one leg), running and jumping.

The extension of the real-time control of the gait in the Cartesian space to all five control variables (30) with Model Predictive Control [27] will be also considered. So far we did not succeed, and only two variables have been controlled with classical PI techniques.

The approach was tested on a dynamic simulation of the SIP. Moreover, as the final objective is to offer a flexible Motion Generator for biped robot walking, the results have, also, been compared with the simulation of a fully actuated 12 DOF of a biped robot. In the comparison, the similarity of the COG behaviour in the sagittal plane suggests that actuated and underactuated controls can be intermixed [28], transforming the instantaneous increment of rotational velocities after the flying foot collision into a proper torque of the ankle of the supporting foot during double support and in the phase of single support, when the foot is flat. However, to obtain a nontrivial finite period of double support, compliance has to be added to the actual anelastic collision. As compliance is essential for that [11,29].

The behaviour of the COG in the frontal plane, that is consequence of a pure ballistic trajectory, will require a further investigation, as the resulting sway of the COG in the frontal plane seems excessive.

Funding: This research received no external funding.

Conflicts of Interest: The author declares no conflict of interest.

Appendix A. A Kinetic Energy, Projection of the Angular Momentum on the Vertical Axis and Biped Parameters

The expressions of the kinetic energy and the angular moment on the vertical axis of the model in Figure 2a are the following

$$\begin{aligned} KE &= 0.5 \cdot (I22 + m \cdot L^2) \cdot \omega^2 \\ &+ (I23 \cdot cos(\theta) - I12 \cdot sin(\theta)) \cdot \gamma \cdot \omega \\ &+ 0.5 \cdot (I33 \cdot cos(\theta)^2 + I11 \cdot sin(\theta)^2) \cdot \gamma^2 \\ &- I31 \cdot sin(\theta) \cdot cos(\theta) \cdot \gamma^2 \\ &+ 0.5 \cdot m \cdot L^2 \cdot sin(\theta)^2 \cdot \gamma^2 \end{aligned} \tag{A1}$$

$$k^{\gamma} = -2 \cdot I31 \cdot cos(\theta) \cdot sin(\theta) \cdot \gamma$$
$$(I23 \cdot cos(\theta) - I12 \cdot sin(\theta)) \cdot \omega$$
$$I33 \cdot cos(\theta)^2 \cdot \gamma \quad \text{(A2)}$$
$$I11 \cdot sin(\theta)^2 \cdot \gamma$$
$$m \cdot L^2 \cdot sin(\theta)^2 \cdot \gamma$$

The parameters of the model used in this paper, taken from [20], and representing a patient wearing an exoskeleton are shown in Figure A1.

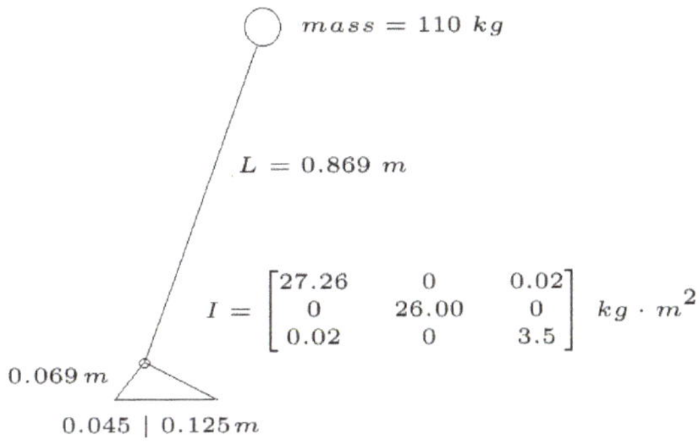

Figure A1. the biped parameters.

References

1. Wight, D.L. A Foot Placement Strategy for Robust Bipedal Gait Control. Ph.D. Thesis, University of Waterloo, Waterloo, ON, Canada, 2008.
2. Wight, D.L.; Kubica, E.G.; Wang, D.W.L. Introduction of the Foot Placement Estimator: A Dynamic Measure of Balance for Bipedal Robotics. *J. Comput. Nonlinear Dyn.* **2008**, *3*, 011009-1. [CrossRef]
3. Choudhury, S.; Kulić, D. Gait Generation via the Foot Placement Estimator for 3D Bipedal Robots. In Proceedings of the 2013 IEEE International Conference on Robotics and Automation, Karlsruhe, Germany, 6–10 May 2013. [CrossRef]
4. DeHart, B.J. Dynamic Balance and Gait Metrics for Robotic Bipeds. Ph.D. Thesis, University of Waterloo, Waterloo, ON, Canada, 2019.
5. DeHart, B.J.; Gorbet, R.; Kulić, D. Spherical Foot Placement Estimator for Humanoid Balance Control and Recovery. In Proceedings of the 2018 IEEE International Conference on Robotics and Automation (ICRA), Brisbane, Australia, 21–25 May 2018.
6. Kane, T.R.; Levinson, D.A. *Dynamics: Theory and Applications*; McGraw-Hill: New York, NY, USA, 1985.
7. Mitiguy, P. MotionGenesis: Advanced Solutions for Forces, Motion, and Code-Generation. Available online: http://www.motiongenesis.com/ (accessed on 17 March 2016).
8. Kajita, S.; Kanehiro, F.; Kaneko, K.; Yokoi, K.; Hirukawa, H. The 3D Linear Inverted Pendulum Mode: A simple modeling for a biped walking pattern generation. In Proceedings of the 2001 IEEE/RSJ International Conference on Intelligent Robots and Systems, Maui, HI, USA, 29 October–3 November 2001.
9. Kajita, S.; Kanehiro, F.; Kaneko, K.; Fujiwara, K.; Harada, K.; Yokoi, K.; Hirukawa, H. Biped Walking Pattern Generation by using Preview Control of Zero-Moment Point. In Proceedings of the 2003 IEEE International Conference on Robotics & Automation Taipei, Taiwan, 14–19 September 2003.
10. Tedrake, R.; Kuindersma, S.; Deits, R.; Miura, K. A closed-form solution for real-time ZMP gait generation and feedback stabilization. In Proceedings of the 2015 IEEE-RAS International Conference on Humanoid Robots, Seoul, Korea, 3–5 November 2015.
11. Grizzle, J.W.; Chevallereau, C.; Ames, A.D.; Sinnet, R.W. 3D Bipedal Robotic Walking: Models, Feedback Control, and Open Problems. *IFAC Proc.* **2010**, *43*, 505–532.
12. Missura, M.; Bennewitz, M.; Behnke, S. Capture Steps: Robust Walking for Humanoid Robots. *Int. J. Humanoid Robot.* **2020**, *16*, 1950032. [CrossRef]
13. Aftab, Z.; Robert, T.; Wieber, P.B. Comparison of Capture Point estimation with human foot placement: Applicability and Limitations. In *5èmes Journées Nationales de la Robotique Humanoïde*; Université de Poiters: Poitiers, France, 2010. [CrossRef]
14. Pratt, J.E.; Carff, J.; Drakunov, S.; Goswami, A. Capture Point: A Step toward Humanoid Push Recovery. In Proceedings of the 6th IEEE-RAS International Conference on Humanoid Robots, Genova, Italy, 4–6 December 2006. [CrossRef]
15. Koolen, T.; Boer, T.D.; Rebula, J.; Goswami, A.; Pratt, J. Capturability-based analysis and control of legged locomotion, part 1: Theory and application to three simple gait models. *Int. Robot. Res.* **2012**, *31*, 1094–1113,. [CrossRef]

16. Collins, S.; Ruina, A.; Tedrake, R.; Wisse, M. Efficient bipedal robots based on passive-dynamic walkers. *Science* **2005**, *307*, 1082–1085. [CrossRef]
17. Ames, A.D.; Cousineau, E.A.; Powell, M.J. Dynamically stable bipedal robotic walking with NAO via human-inspired hybrid zero dynamics. In Proceedings of the 15th ACM International Conference on Hybrid Systems: Computation and Control (HSCC '12), Beijing, China, 13–15 April 2012; pp. 135–144. [CrossRef] [PubMed]
18. Hereid, A.; Cousineau, E.A.; Hubicki, C.M.; Ames, A.D. 3D dynamic walking with underactuated humanoid robots: A direct collocation framework for optimizing hybrid zero dynamics. In Proceedings of the 2016 IEEE International Conference on Robotics and Automation (ICRA), Stockholm, Sweden, 16–21 May 2016. [CrossRef]
19. De Oliveira, A.; Vicinansa, G.; da Silva, P.; Ang'elico, B. Frontal Plane Bipedal Zero Dynamics Control. *arXiv* **2019** arXiv:1904.12939.
20. Menga, G.; Ghirardi, M. Modelling, Simulation and Control of the Walking of Biped Robotic Devices—Part I : Modelling and Simulation Using Autolev; Part II: Rectilinear Walking; Part III: Turning while Walking. *Inventions* **2016**, *1*, 6.
21. Gillespie, R.B. Kane's equations for haptic display of multibody systems. *Haptics-e* **2003**, *3*, 144–158. Available online: http://www.haptics-e.org (accessed on 17 March 2016). [CrossRef]
22. Hurmuzlu, Y.; Genot, F.; Brogliato, B. Modeling, stability and control of biped robots—A general framework. *Automatica* **2004**, *40*, 1647–1664.
23. Bajodah, A.H.; Hodges, D.; Chen, Y. Nonminimal generalized Kane's impulse-momentum relations. *J. Guid. Control Dyn.* **2004**, *27*, 1088–1092. [CrossRef]
24. Levenberg, K.A. A Method for the Solution of Certain Non-linear Problems in Least Squares. *Q. Appl. Math.* **1944**, *2*, 164–168. [CrossRef]
25. Marquardt, D. An Algorithm for the Least-Squares Estimation of Nonlinear Parameters. *Siam J. Appl. Math.* **1963**, *11*, 431–441. [CrossRef]
26. Menga, G.; Ghirardi, M. Lower Limb Exoskeleton for Rehabilitation with improved Postural Equilibrium. *Robotics* **2018**, *7*, 28. [CrossRef]
27. Saraf, N.; Bemporad, A. An efficient non-condensed approach for linear and nonlinear model predictive control with bounded variables. *arXiv* **2019**, arXiv:1908.07247v1. [CrossRef]
28. Liu, Y.; Zang, X.; Heng, S.; Lin, Z.; Zhao, J. Human-Like Walking with Heel Off and Toe Support for Biped Robot. *J. Appl. Sci.* **2017**, *7*, 499.
29. Miossec, S.; Aoustin, Y. Walking gait composed of single and double supports for a planar biped without feet. In Proceedings of the 5th International Conference on Climbing and Walking Robots, Paris, France, 25–27 September 2002.

applied sciences

MDPI

Article

Robotic-Assisted Hand Therapy with Gloreha Sinfonia for the Improvement of Hand Function after Pediatric Stroke: A Case Report

Federica Bressi [1,*], Fabio Santacaterina [1], Laura Cricenti [1], Benedetta Campagnola [1], Francesca Nasto [1], Carla Assenza [2], Daniela Morelli [2], Francesca Cordella [3], Martina Lapresa [3], Loredana Zollo [3], Silvia Sterzi [1] and Marco Bravi [1]

1 Physical Medicine and Rehabilitation Unit, Campus Bio-Medico University of Rome, 00128 Rome, Italy; f.santacaterina@policlinicocampus.it (F.S.); l.cricenti@unicampus.it (L.C.); b.campagnola@unicampus.it (B.C.); f.nasto@unicampus.it (F.N.); s.sterzi@policlinicocampus.it (S.S.); m.bravi@policlinicocampus.it (M.B.)
2 I.R.C.C.S. Santa Lucia Foundation, 00179 Rome, Italy; c.assenza@hsantalucia.it (C.A.); d.morelli@hsantalucia.it (D.M.)
3 Unit of Advanced Robotics and Human-Centred Technologies, Campus Bio-Medico University of Rome, 00128 Rome, Italy; f.cordella@unicampus.it (F.C.); m.lapresa@unicampus.it (M.L.); l.zollo@unicampus.it (L.Z.)
* Correspondence: f.bressi@policlinicocampus.it; Tel.: +39-06-2254-1624

Abstract: Background: Stroke in childhood presents a serious rehabilitation challenge since it leads to physical, cognitive and psychosocial disability. The objective of our study was to describe the effectiveness of robot-mediated therapy (RMT) with Gloreha Sinfonia in addition to a conventional treatment in the recovery of the sensory-motor capabilities of the paretic hand and the quality of life in a ten-year-old child after a stroke. Methods: The girl was enrolled to undergo 10 sessions of RMT with Gloreha Sinfonia. She was evaluated with functional scales and with upper limb kinematic analysis at pre-treatment (T0) and at the end of treatment (T1). Outcome measures were Fugl-Meyer Assessment-Upper Extremity (FMA-UE), Visual Analogic Scale (VAS) and Activities and Participation of Daily Life (ADL). In addition, a Force Assessment System based on Virtual Reality games was used to assess the force control and modulation capability at T0 and T1. Results: At the end of treatment, the patient improved in functional scales and in quality of life for greater involvement in some activity of daily living. Force control and modulation capability significantly increased after the treatment. Conclusions: This clinical case highlights possible positive effects of a combined (conventional plus robotic) rehabilitation treatment for the upper limb in pediatric stroke outcomes from both a sensorimotor and functional point of view, also improving the motivational and affective aspects of the patient and of family members. Further studies are needed to validate these results and to identify the most appropriate modalities and doses

Keywords: pediatric stroke; robotics; upper limb; hand rehabilitation; device; Gloreha Sinfonia; case report

Citation: Bressi, F.; Santacaterina, F.; Cricenti, L.; Campagnola, B.; Nasto, F.; Assenza, C.; Morelli, D.; Cordella, F.; Lapresa, M.; Zollo, L.; et al. Robotic-Assisted Hand Therapy with Gloreha Sinfonia for the Improvement of Hand Function after Pediatric Stroke: A Case Report. *Appl. Sci.* **2022**, *12*, 4206. https://doi.org/10.3390/app12094206

Academic Editor: Enrico Vezzetti

Received: 11 March 2022
Accepted: 18 April 2022
Published: 21 April 2022

1. Introduction

Even though stroke has long been accepted as an adult health problem causing substantial morbidity and mortality, it is also an important cause of acquired brain injury in young patients, occurring most commonly in the neonate and during childhood. Pediatric stroke incidence rates, including both neonatal and childhood and both ischemic and hemorrhagic stroke, range from 3 to 25 per 100,000 children in developed countries. Newborns have the highest risk ratio: 1 in 4000 live births [1,2]. The incidence of arterial ischemic stroke is highest in children aged under 1 year [3]. Children also have a more diversified and larger number of risk factors for stroke that differ significantly from adults [1,4,5]. The risk factors for AIS in the pediatric population are arteriopathy, cardiac disease, cardiac surgery/interventions, sickle cell disease, infections, thrombophilia and perinatal factors [1,6]. The clinical presentation of childhood stroke varies depending on the age of the child, and the most common clinic presentations are functional neurological disorders (20.0%), transient neurological deficits (17.8%), migraine (15.6%) and seizure (11.1%) [6,7]. The recovery of the arm function is one of the main goals of rehabilitation attempts after childhood stroke; the upper limb function is essential in the performance of everyday activities and has a significant impact on independent functioning and the overall quality of life of the affected children. The aim of rehabilitation of the upper limbs is to prevent the disuse of the impaired side of the body. Many studies in the literature have shown that therapy involving sensorimotor exercises to simulate meaningful tasks used in daily life increases the functional recovery of the affected upper limb [8–10]. Realistic contexts of functional activities, such as reaching or pointing towards an everyday object, help patients to acquire control strategies to compensate muscle weakness and inaccuracies [11]. In order to rehabilitate the upper limbs, there are different treatments available. Constraint Induced Movement Therapy (CIMT) is a rehabilitative methodology widely used currently, even on infants under one year of age (named baby-CIMT); it is considered feasible and without adverse effects [12]. Kwakkel et al. demonstrated that high-intensity and task-specificity are two of the main features of any successful stroke rehabilitation program [13,14]. In the last few years, therapy aided by a robotic exoskeleton is noted for its capability of supporting repetitive and high-intensity training tasks [15]; a rehabilitation robot can assist the physiotherapist in administering programmable and customizable rehabilitation procedures according to the type of treatment required. When combined with interactive programs such as virtual reality (VR), robot-aided therapy can assign functional meaning to the therapy, creating a motivating environment [16,17]. Active and very inspiring practice, with intensive and oriented repetitions, is essential to induce changes in neuroplasticity within the sensory-motor system and improves performance in motor tasks [17]. The use of robotics and video gaming within a pediatric population can improve motivation and attention while focusing on the practice of specifically difficult motor tasks [18,19].

In the pediatric field, it is possible to combine rehabilitative treatments in any stroke phase, exploiting the potential of innovative technology for sensor-motor and cognitive recovery and for the continuum of care also at home [8]. In pediatric strokes, studies currently available are mainly aimed at the use of virtual reality and tele-rehabilitation. The use of robotic devices in the pediatric population is still under study. In fact, to date, no studies have investigated the efficacy of Gloreha in the motor recovery of hand functions in children after stroke. Therefore, this case report aims to describe the feasibility and the effectiveness of an RMT with Gloreha Sinfonia plus a conventional treatment in the recovery of the sensory-motor capabilities of the paretic hand and the quality of life in a ten-year-old child after stroke.

The paper structure is as follows: Section 2 describes the materials, i.e., Gloreha Sinfonia, the developed rehabilitation exercises, the experimental setup and protocol, the developed assessment system based on VR and force measurement, and the evaluated outcome measures; Section 3 reports the results of the experimental validation, which are discussed in Section 4. Conclusions and future work are provided in Section 5.

2. Materials and Methods

2.1. Trial Design

In this case report, a ten-year-old girl with right hemiparesis following ischemic stroke caused by probable primary central nervous system angioitis was enrolled to undergo 10 sessions of RMT with Gloreha Sinfonia.

The aim of this study was to provide an account of the efficacy of RMT using Gloreha Sinfonia in terms of improvements in the range of motion, spasticity, force modulation capability and functionality of the hemiplegic hand in pediatric stroke patients. This paper was organized according to CARE guidelines [20]. Figure 1 presents the timeline.

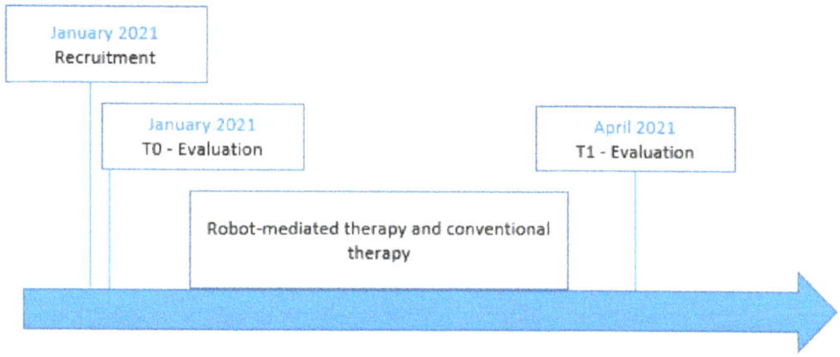

Figure 1. Timeline.

2.2. Participant

A nine-year and eleven-month-old girl on 24 March 2020, who apparently was in a state of full wellbeing, fell to the ground and lost consciousness after feeling a severe headache while she was playing with her brother; then, she had generalized hypertonus, trismus and emission of noises for about 10 min with apparent resolution. When the ambulance arrived, the little girl had regained consciousness and was able to carry out simple orders with her 4 limbs. When she arrived in the Umberto I ER, her speech appeared fluid and there were no evident cranial nerve deficits. Later, the girl complained of headaches again, and she had a reduced ability to move on the right side, deviation of the buccal rim and dysarthria. She underwent brain MRI in urgency, which highlighted "ischemic acute lesion in the left Rolandic area". Her parents reported a history of mild and sporadic headache in the previous days. Transferred for competence to the ER of the Pediatric Hospital Bambino Gesù of Rome, the girl appeared alert, oriented and responsive to verbal stimulus. The right side was compromised, with deviation of the buccal rim and a deficit of tone and strength that was greater in the upper limb than in the lower limb; moreover, she had difficulty in trunk control in sitting positions and had no problems with dysphagia and desaturation. She started low molecular weight heparin, as per the hematological prescription and according to Pediatric Stroke Guidelines [8]. As soon as possible, the child was transferred to the Department of Neurology of the same hospital to complete the investigations planned for this case. During neurological hospitalization, the girl underwent another brain MRI, which confirmed recent ischemic lesions. In particular, they were in the territory of the left Middle Cerebral Artery (MCA) with greater extension compared to the first MRI examination, associated with a reduction in the left Internal Carotid Artery (ICA) caliber and left M1 and Al flow profile irregularities (left MCA and Anterior Cerebral Artery—ACA—arteritis). The baby was subjected to ultrasound of the neck vessels, evaluation and cardiological examinations, immuno-infectious examinations, thrombophilia screening, trace oligoclonal bands and platelet aggregation tests. In consideration of the exclusion of secondary causes and of vasculitis alterations in other parts of the body, the diagnostic suspicion of primary CNS angiitis was conducted, for which, to complete the diagnosis, and further in-depth examinations were carried out according to the national protocol for primary vasculitis [8]. By 27 March 2020 to 4 April 2020, the child had also started treatment with intravenous cortisone. In a short period of time, the child showed an improvement in clinical conditions, in particular in dysarthria and in the ability to make minimal movements with the right lower limbs. Upon discharge from neurology, she was prescribed aspirin, heparin suspension and cortisone scaling. On 4 April 2020, the child was transferred to the Pediatric Neurorehabilitation Department of the same hospital located in Palidoro (near Rome) to begin an intensive neuro-motor rehabilitative treatment. She was admitted to Bambino Gesù Rehab Department from 4 April 2020 to 12 June 2020.

Therefore, after 15 days from the stroke onset, she began intensive rehabilitation treatment consisting of 5 sessions/week of neuro-motor therapy, 2 sessions/week of speech therapy and 1 session/week of occupational therapy. In addition to conventional motor treatments, she performed robotic motor training for the upper limb (Mit Manus). The proposals were recognition activities of different textures, heights and shapes to stimulate the sensitivity of the right upper limb, activate the intrinsic muscles of the hand, and to improve the pronation–supination of the forearm and coordination of the two upper limbs in simple gestures of daily life. At the last MRI examinations on 10 April and 19 May 2020, respectively, there was a reduction in the known ischemic lesions in the left hemisphere and a reduction in the size of ICA, the M1/M2 sections of the MCA and the A1 section of the ACA. Upon discharge, the condition of the child improved overall, reached a moderate static and dynamic control of the trunk and started ambulatory training with the help of an ankle-foot orthosis, but right hemiparesis and dysarthria persisted.

Followed a period of non-in-patient rehabilitation at the Fondazione Santa Lucia in Rome, there she performed a neuro-motor rehabilitative treatment of 180 days, 4 times a week, including speech therapy and occupational therapy as well.

After 180 days of treatment and with respect for the recovery of the right upper limb, the girl was able to perform movements against gravity flexo-extension and abduction–adduction of the shoulder with extended elbow and spinal compensation. In addition, she was able to perform movements against gravity elbow flexo-extension, while hinting at muscle recruitment at the level of the wrist and hand extensor muscles and first finger opponent. However, the girl was able to find strategies to solve practical problems of daily life and involving the right upper limb in base activities.

In January 2021, the child began, in association with the conventional treatment and in progress at the Fondazione Santa Lucia, a treatment with the Gloreha robotic glove for the right upper limb at the Campus Bio-Medico University. The objective of the robotic treatment was to recover the movements of extension of the fingers and wrist and improve hand-wrist bending and manual dexterity movements. The use of Gloreha aimed also at recovering the sensory skills of the hand, improving its muscle tone and functionality. Moreover, the use of non-immersive virtual reality through interactive video games had the objective of stimulating cognitive abilities, such as attention, visual-spatial analysis, working memory and executive functions as well.

2.3. Gloreha Sinfonia

The Gloreha Sinfonia is a robotic device for the neuromotor rehabilitation of the upper limb, which can facilitate the patient in all phases of recovery. It can support the movement of the finger joints in passive, active-assisted and active modes. It consists of a complete set of gloves, braces and accessories for finger mobilization; a dynamic support to compensate the weight of the arm; a stimulating software equipped with 3D animation; a voice guide and audio video effects; a touchscreen PC and an ergonomic table for performing functional exercises and to allow the use with a wheelchair.

The device allows the execution of the following exercises:

1. Passive mobilization exercises (the movements are carried out entirely by the device);
2. Active-assisted exercises with graphic interface (the patient trains in flexion–extension of the fingers thanks to motivating games; the motors support and integrate the patient's voluntary movements only to the extent necessary);
3. Active-assisted functional exercises with real objects (the patient trains fine grip);
4. Interactive games (the patient can improve dexterity);
5. All possible combinations of flexion–extension of the fingers;
6. Therapies based on the action–observation approach (the patient performs a task that he first observed in a preview video, supported by the robot);
7. Exercises with partial or total compensation of the weight of the upper limb, free to float in space interacting with real objects (the system is able to self-adjust based on the residual abilities of the patient).

Moreover, the robot allows the performance of therapies on patients who are bedridden or in an upright position or seated in a chair/wheelchair. It also allows a constant measurement of motor performance and condition of the patient's hand (active/passive ROM, movement speed, coordination and improvement in the execution of the various tasks). The robotic glove records all data associated with each patient and allows the operator to monitor the performance of each subject treated. Graphs show the trend of the obtained results exercise by exercise and session by session so that the patient can have immediate feedback on the progress achieved.

Since this is a new technology, there are a limited number of studies in the literature on the use of Gloreha in pediatric and adult stroke patients, which show that the main clinical benefits related to its use are as follows: the maintenance and improvement of the joint range of the fingers of the hands; the prevention of adhesions, contractures and damage from immobilization; the reduction in pain, edema and hypertonia; proprioceptive stimulation, improvement of joint metabolism and lymphatic and blood circulation; the maintenance of functional afferences and the perception of the body; the increase in coordination, dexterity and functional independence; the increase in gripping and gripper strength; and the improvement of visual-spatial and attentional skills [21].

2.4. Force Assessment System Based on Virtual Reality Games

A Force Assessment System based on VR games was developed to assess the child's improvements in grip force control and modulation. The system was composed of force sensors positioned at the fingertips of the Gloreha Sensor Glove and of custom-designed VR games in the form of tracking tasks purposely developed to assess different aspects of force control and modulation.

Previous works in the literature proposed tracking tasks for the assessment and training of grip force control in mildly to severely affected hemiparetic stroke patients. In the study of Kurillo et al. [22], the authors presented and tested a grip force training system that enabled the improvement of grip force control in 8 out of 10 post-stroke patients. In the study of Lindberg et al. of 2012 [23], the authors proved that the power grip force tracking tasks developed to assess grip force modulation capability were feasible to quantify the accuracy of grip force control.

The force sensors to be embedded in the Gloreha Sensor Glove were chosen according to the following technical specifications: (i) maximum dimension of 2×10^{-2} m diameter; (ii) thickness lower than 5×10^{-4} m; (iii) force range comparable to the one detected during hand rehabilitation training (0–25 N) [24]. The piezoresistive sensor FSR® Model 402 Short Tail (Interlink Electronics) was selected to meet the specifications. It has a diameter of 1.83×10^{-2} m, a thickness of 4×10^{-4} m, a force sensitivity range of 0.2–20 N and continuous force resolution. Piezoresistive sensors were selected because they are low cost, suitable for wearable applications, require a very simple conditioning electronics and have a good shock resistance.

Three sensors were embedded, respectively, in the fingertips of the first, second and third finger of the Gloreha Sensor Glove, i.e., the fingers mostly involved in daily life grasps, as shown in Figure 2. The interface between the sensor and the glove was modified by fixing a thin 3D printed PLA plate with double-sided tape (1×10^{-3} m thickness and 2×10^{2} m diameter) to uniformly distribute the force on the sensitive area of the sensor. A 3D-printed PLA support was positioned to interlock the plastic components of the glove that allow connecting Gloreha Sinfonia flexible transmissions to the glove itself. The sensors are secured on the fingertips by means of an elastic band fastened at the top of the PLA support.

The sensors with the developed PLA plate were calibrated using the Instron® testing machine to relate the force and output voltage.

The VR game was developed with Unity, using Microsoft Visual Studio as the pre-set editor and C# as the programming language. Attention was paid to provide an intuitive, clear and engaging visual feedback to the user. In fact, providing the patient with biofeedback can improve the outcome of the treatment and promote neuroplasticity [22].

The aim of the VR game in the form of tracking tasks was to assess the patient capability to finely control the sub-maximal forces exerted when grasping real objects, the ability to balance and release the grip and the general accuracy of force control.

The average force exerted by the three fingers was given in input to the VR game to move the avatar according to the force exerted on the object grasped by the participant (i.e., a wood parallelepiped 1.35×10^{-1} m $\times$ 6.5×10^{-2} m $\times$ 4×10^{-2} m). In the proposed VR game, shown in Figure 3, the patient was asked to move the avatar of the game vertically, according to the exerted force, in order to track the three proposed waveforms moving on the screen. The three waveforms were a "Ramp", a "Square Wave" and a "Sinusoidal Wave" [25]. Each waveform was developed to assess different aspects of grip force control: the Ramp aimed at assessing the capability to gradually increase and decrease the grip force; the Square Wave aimed at assessing the capability to exert discrete force levels and stabilize the force; and the Sinusoidal Wave aimed at assessing the overall force modulation capability. "Ramp" and "Square Wave" were composed of 10 discrete force levels to be reached (and held for the "Square Wave") and are uniformly distributed between the maximum and minimum forces recorded at the beginning of the trial, whereas the "Sinusoidal Wave" had a peak-to-peak amplitude that corresponded to the range between the minimum and maximum forces.

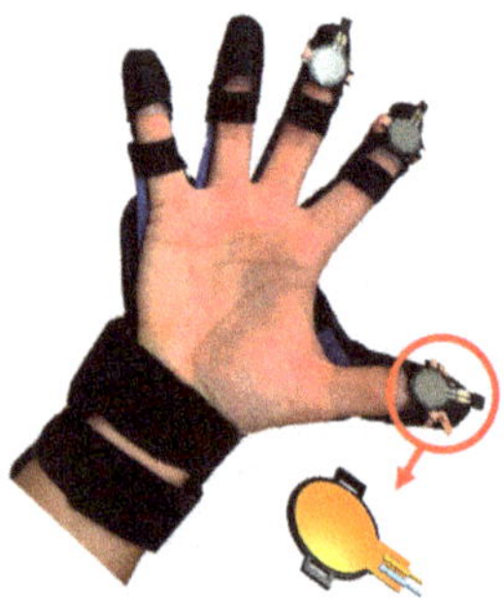

Figure 2. Force sensors positioned at the fingertips of Gloreha Sensor Glove.

To assess the child's improvements in grip force control and modulation, she was instructed to move the avatar (i.e., a turtle) and to follow the proposed waveform pattern in order to collect the maximum number of "bubbles" of the VR game with the avatar. In the first assessment session (T0), she performed 5 one-minute repetitions of the "Ramp" and 3 one-minute repetitions of the "Square Wave" and "Sinusoidal Wave" exercises, respectively. In the second assessment session (T1), she performed 3 one-minute repetitions of the "Ramp" and 2 one-minute repetitions of the "Square Wave" and "Sinusoidal Wave" exercises. Before each session, the maximum and minimum forces applicable by the child were recorded with a custom-made graphical user interface to set the force range in which the avatar could be moved. Then, the maximum force to be reached was set at 90% of the maximum recorded value.

Figure 3. VR game for force assessment.

2.5. Intervention

The Gloreha treatment lasted from January to April 2021, and the commitment was once a week for a total of 10 weeks; each session lasted 60 min. During each session, the child performed passive, active-assisted and active movements with gradually increasing complexity exercises supported and stimulated by sensory feedback. In addition, the child was also offered exercises in bi-dexterity to improve both the quality of the recovery of the paretic limb and its coordination and functionality in daily activities.

2.6. Outcome Measures

The child was evaluated with functional scales, with upper limb kinematic analysis and with an ad hoc developed force assessment tool (FSR model 402 Short Tail, Interlink Electronics) at pre-treatment (T0) and at the end of the treatment (T1).

She was evaluated with Fugl-Meyer Assessment-Upper Extremity (FMA-UE) for upper-extremity motor impairment and the Visual Analogic Scale (VAS) for pain intensity and activities and participation of daily life with ADL scale [26].

Improvements in force control and modulation capability were assessed by means of the Force Assessment System. Two main performance indicators were computed for each repetition of the three VR exercises: the Root Mean Square Error (RMSE) (N) between the target force pattern and the exerted force pattern and the Peak Performance (%), computed as the percentage of reached force peaks, for "Ramp" and "Sinusoid", and the percentage of reached and held discrete force levels, for the "Square Wave". A statistical analysis was performed to evaluate improvements of the child from one session to the other. The non-parametric one-way ANOVA test (i.e., Kruskal–Wallis test) was conducted between RMSE and Peak Performance indicators in the two sessions. The significance level was set at 5%.

3. Results

Results are reported in Table 1. At the end of treatment (T1), the patient improved in functional scales: FMA-UE had a percentage variation ($\Delta\%$) of 44% from T0 (34/66) to T1 (49/66). Moreover, at the end of treatment, her quality of life was better for greater involvement in some little but significative activities of daily living such as food, dress and undress (ADL scale: 4/6 at T0 vs. 6/6 at T1) and had greater interest in the surrounding world with less fear of feeling different from others.

Table 1. Results of outcome measurement.

	T0	T1
ADL	4/6	6/6
FMA-UE	34/66	49/66
VAS	0/10	0/10

ADL: Activity of Daily Living; FMA-UE: Fugl-Meyer Assessment for Upper Limb; VAS: Visual Analogic Scale.

It is important to highlight that the child had never complained of pain (VAS scale 0/10) during both evaluations.

The Kruskal–Wallis test performed on RMSE and Peak Performance indicators between sessions at T0 and T1, allowed the assessment of improvements in force control and modulation. RMSE was significantly reduced from one session to the other (p value = 0.0018), meaning that training with the Gloreha improved the child's capacity to follow target force patterns. Peak Performance significantly increased from the first to the second evaluation session (p value = 0.0120). Boxplots for the two performance indicators are shown in Figure 4, where the central mark is the median and the box edges are the 25th and 75th percentiles.

No adverse events occurred during the entire treatment.

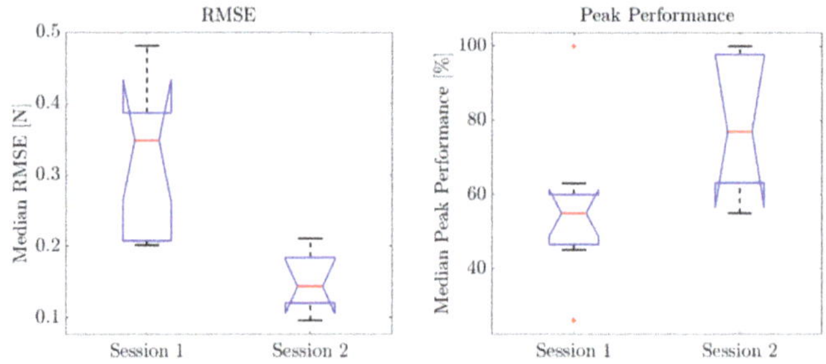

Figure 4. Boxplots of RMSE and Peak Performance for session 1 (T0) and session 2 (T1).

4. Discussion

The aim of this study is to describe the effects on the range of motion, muscle tone and functionality of the paretic upper limb, particularly the hand, when Gloreha Sinfonia combined with conventional therapy was used for rehabilitation treatments in a pediatric stroke patient. The results revealed an improvement in FMA-UE, a significant reduction in RMSE and a significant improvement in Peak Performance.

To the best of our knowledge, this is the first study using Gloreha Sinfonia in pediatric strokes. Previous rehabilitation studies on pediatric stroke patients were mostly dedicated to the robotic treatment of the proximal portion of the upper limb, while there are still very few studies on the distal extremity, especially the hand. Gloreha Sinfonia allows the performance of tasks that combine the activity of the entire upper limb. In fact, unlike other robots designed exclusively for shoulder and elbow movements, Gloreha permits the improvement of distal control through the implementation of exercises focused on the use of hands combined with the involvement of the entire upper limb, reproducing activities of daily living. As other studies have previously mentioned, when a patient uses the distal part of the paretic upper limb, at the same time, the proximal segment is also trained, albeit the upper arm is supported or restrained in the distal group [27]. Furthermore, Gloreha Sinfonia is well suited for use in combination with other traditional rehabilitation activities because it can integrate rehabilitation treatments with highly stimulating and interactive exercises for patients.

Our results show an important enhancement in FMA-UE, with values that proceed from 34/66 to 49/66 between T0 and T1; they indicate an overall improvement of the upper limb due to a clinical advance both in the shoulder district, implemented thanks to reaching exercises, and to the hand control, achieved thanks to manipulation exercises. These results were significantly higher than the minimal clinically important difference (MCID) seen in adults by Page et al. [28], but it is not possible to make a comparison with a pediatric population, because their MCID is not present in the literature. In addition, FMA-UE results are in line with ADL records, which report improvements in activities of daily life due to a better use of the upper limb.

From Figure 4, it is evident that the child significantly improved her capacity of reaching force peaks and holding force levels. Furthermore, the results show that force control and modulation capability significantly increased after treatment.

We know from the girl's parents that she has reported a psycho-emotional improvement and presented new interest in socializing with other children without fear of being judged and perceived less pronounced motor deficits and less disability at the end of therapy. This statement can be considered as another positive effect of the combined (robot + conventional rehab therapy) and continuative therapy. It is important to underline that the girl never complained of pain, demonstrating a good ad safe adaptability of the robot with respect to the child.

Another aspect that makes this robot an interesting tool in pediatric applications is the possibility of performing specific repetitive task-oriented exercises, placed in the form of playing a videogame; as already reported, robot-mediated therapy in children with acquired or congenital brain injury appears to be beneficial: enhancing motivation and improving perception, it incorporates the advantages of the enjoyable game-like experience [15,29]. Furthermore, the use of a virtual and motivational environment in which the patient has immerged has a positive effect to facilitate motor and sensory and cognitive relearning by stimulating the patient to challenge himself to perform better and to obtain a higher score in the game. Moreover, as recorded also by Mirkowski et al. in their systematic review [18], robotic therapy seems to significantly improve the upper extremity function and spasticity in children after strokes. This theory is also supported by the study of Wann et al., which considers motor learning theory and reports how the repetition of motor patterns is seen as a key factor in improving movement [30]. Moreover, Rizzo et al. reports how a rehabilitation treatment that uses feedback, either visual or auditory, contributes to gains made in motor learning [31]. It follows that Virtual Reality is a powerful medium for providing stimulation in the form of visual and auditory events to increase the motivation and desire to continue practicing [32].

To date, there are few studies in the literature on this specific population. Most pediatric upper extremity rehabilitation studies are aimed at patients with cerebral palsy, spinal cord injury or quadriplegia. As reported also in the case report of Čolović et al. [33], there is still no clear indication on how robotic therapy can be combined with conventional therapy and for how many times; however, combining robotic and traditional rehabilitation can improve the functional motor performance of the arm involved in the chronic recovery phase after a pediatric stroke.

Our results, being relative to a single case, cannot find a generalization. Further studies on larger samples, with control and randomization groups and with adequate follow-ups, are needed in order to reach meaningful conclusions. Furthermore, we have not assessed patient's emotional changes and the level of social integration before and after the robotic treatment with specific and objective measures; the administration of scales that could quantify these changes could be useful to determine social aspects in further studies.

5. Conclusions

Given the results and given the literature evidence, Gloreha Sinfonia seems to be suitable for the treatment of post-stroke hand disabilities in the pediatric age, but further studies on larger populations, with stratification of the sample for clinical characteristics, as well as clinical scales that are more sensitive to any change are needed to support this hypothesis. It would be interesting in the future to consider how this type of technology can also support cognitive difficulties. In addition, this type of rehabilitative approach facilitates the need for personalized and easily monitored rehabilitation protocols. Clinical trials with follow-ups and on large populations could confirm the results obtained.

Author Contributions: Conceptualization, F.B., C.A. and D.M.; methodology, F.S. and F.C.; software, L.Z., F.C. and M.L.; formal analysis, F.S.; investigation, F.N.; data curation, L.Z. and F.C.; writing—original draft preparation, F.N., B.C. and L.C.; writing—review and editing, F.B. and L.C.; supervision, S.S. and M.B.; project administration, F.B. All authors have read and agreed to the published version of the manuscript.

Funding: This research received no external funding.

Institutional Review Board Statement: The study was conducted in accordance with the Declaration of Helsinki and approved by the Ethics Committee of Fondazione Policlinico Campus Bio-Medico of Rome (protocol code PT-O 20.22; date of approval: 29 March 2022).

Informed Consent Statement: Informed consent was obtained from all subjects involved in the study.

Data Availability Statement: Not applicable

Conflicts of Interest: The authors declare no conflict of interest.

References

1. Ferriero, D.M.; Fullerton, H.J.; Bernard, T.J.; Billinghurst, L.; Daniels, S.R.; DeBaun, M.R.; Deveber, G.; Ichord, R.N.; Jordan, L.C.; Massicotte, P.; et al. Management of Stroke in Neonates and Children: A Scientific Statement from the American Heart Association/American Stroke Association. *Stroke* **2019**, *50*, e51–e96. [CrossRef] [PubMed]
2. Mallick, A.A.; Ganesan, V.; Kirkham, F.J.; Fallon, P.; Hedderly, T.; McShane, T.; Parker, A.P.; Wassmer, E.; Wraige, E.; Amin, S.; et al. Diagnostic Delays in Paediatric Stroke. *J. Neurol. Neurosurg. Psychiatry* **2015**, *86*, 917–921. [CrossRef] [PubMed]
3. Mallick, A.A.; Ganesan, V.; Kirkham, F.J.; Fallon, P.; Hedderly, T.; McShane, T.; Parker, A.P.; Wassmer, E.; Wraige, E.; Amin, S.; et al. Childhood Arterial Ischaemic Stroke Incidence, Presenting Features, and Risk Factors: A Prospective Population-Based Study. *Lancet Neurol.* **2014**, *13*, 35–43. [CrossRef]
4. Lanthier, S.; Carmant, L.; David, M.; Larbrisseau, A.; de Veber, G. Stroke in Children: The Coexistence of Multiple Risk Factors Predicts Poor Outcome. *Neurology* **2000**, *54*, 371–378. [CrossRef]
5. Riela, A.R.; Roach, E.S. Etiology of Stroke in Children. *J. Child Neurol.* **1993**, *8*, 201–220. [CrossRef] [PubMed]
6. Royal College of Paediatrics and Child Health Stroke in Childhood—Clinical Guideline for Diagnosis, Management and Rehabilitation | RCPCH. Available online: https://www.rcpch.ac.uk/resources/stroke-in-childhood-clinical-guideline (accessed on 10 March 2022).
7. DeLaroche, A.M.; Sivaswamy, L.; Farooqi, A.; Kannikeswaran, N. Pediatric Stroke and Its Mimics: Limitations of a Pediatric Stroke Clinical Pathway. *Pediatr. Neurol.* **2018**, *80*, 35–41. [CrossRef]
8. Ictus Cerebrale: Linee Guida Italiane Di Prevenzione e Trattamento Raccomandazioni e Sintesi. Available online: https://www.iso-stroke.it/wp-content/uploads/2017/02/LIBRO-SPREAD-VIII-ED-13-09-16.pdf (accessed on 10 March 2022).
9. Bütefisch, C.; Hummelsheim, H.; Denzler, P.; Mauritz, K.H. Repetitive Training of Isolated Movements Improves the Outcome of Motor Rehabilitation of the Centrally Paretic Hand. *J. Neurol. Sci.* **1995**, *130*, 59–68. [CrossRef]
10. Hendricks, H.T.; van Limbeek, J.; Geurts, A.C.; Zwarts, M.J. Motor Recovery after Stroke: A Systematic Review of the Literature. *Arch. Phys. Med. Rehabil.* **2002**, *83*, 1629–1637. [CrossRef]
11. Thielman, G.; Kaminski, T.; Gentile, A.M. Rehabilitation of Reaching after Stroke: Comparing 2 Training Protocols Utilizing Trunk Restraint. *Neurorehabil. Neural Repair* **2008**, *22*, 697–705. [CrossRef]
12. Eliasson, A.C.; Nordstrand, L.; Ek, L.; Lennartsson, F.; Sjöstrand, L.; Tedroff, K.; Krumlinde-Sundholm, L. The Effectiveness of Baby-CIMT in Infants Younger than 12 Months with Clinical Signs of Unilateral-Cerebral Palsy; an Explorative Study with Randomized Design. *Res. Dev. Disabil.* **2018**, *72*, 191–201. [CrossRef]
13. Kwakkel, G.; Wagenaar, R.C.; Koelman, T.W.; Lankhorst, G.J.; Koetsier, J.C. Effects of Intensity of Rehabilitation after Stroke. A Research Synthesis. *Stroke* **1997**, *28*, 1550–1556. [CrossRef] [PubMed]
14. Feys, H.M.; de Weerdt, W.J.; Selz, B.E.; Cox Steck, G.A.; Spichiger, R.; Vereeck, L.E.; Putman, K.D.; van Hoydonck, G.A. Effect of a Therapeutic Intervention for the Hemiplegic Upper Limb in the Acute Phase after Stroke: A Single-Blind, Randomized, Controlled Multicenter Trial. *Stroke* **1998**, *29*, 785–792. [CrossRef] [PubMed]
15. Frascarelli, F.; Masia, L.; Di Rosa, G.; Cappa, P.; Petrarca, M.; Castelli, E.; Krebs, H.I. The Impact of Robotic Rehabilitation in Children with Acquired or Congenital Movement Disorders. *Eur. J. Phys. Rehabil. Med.* **2009**, *45*, 135–141. Available online: https://www.minervamedica.it/it/riviste/europa-medicophysica/articolo.php?cod=R33Y2009N01A0135 (accessed on 10 March 2022). [PubMed]
16. Doerbeck, V.; Nallinger, A.; Burger, H.; Rosner, V.; Machuta, K.; Gruender, S.; Nastulla, T.; Schroeder, S.; Berweck, S. Integration of Armeo Spring Pediatrics in Inpatient Rehabilitation of Children and Adolescents with Hemiparesis. *Neuropediatrics* **2012**, *43*, PS14_11. [CrossRef]
17. Friel, K.M.; Kuo, H.C.; Fuller, J.; Ferre, C.L.; Brandão, M.; Carmel, J.B.; Bleyenheuft, Y.; Gowatsky, J.L.; Stanford, A.D.; Rowny, S.B.; et al. Skilled Bimanual Training Drives Motor Cortex Plasticity in Children with Unilateral Cerebral Palsy. *Neurorehabil. Neural Repair* **2016**, *30*, 834–844. [CrossRef]
18. Mirkowski, M.; McIntyre, A.; Faltynek, P.; Sequeira, N.; Cassidy, C.; Teasell, R. Nonpharmacological Rehabilitation Interventions for Motor and Cognitive Outcomes following Pediatric Stroke: A Systematic Review. *Eur. J. Pediatr.* **2019**, *178*, 433–454. [CrossRef]
19. Gordon, A.M.; Okita, S.Y. Augmenting Pediatric Constraint-Induced Movement Therapy and Bimanual Training with Video Gaming Technology. *Technol. Disabil.* **2010**, *22*, 179–191. [CrossRef]
20. Gagnier, J.J.; Kienle, G.; Altman, D.G.; Moher, D.; Sox, H.; Riley, D.; Group, C. Standards & Guidelines CARE: Linee Guida per Il Reporting Di Casi Clinici. *Evidence* **2016**, *8*, e1000147.
21. Miccinilli, S.; Bravi, M.; Foti, C.; Morrone, M.; Maselli, M.; Santacaterina, F.; Scotto Di Luzio, F.; Zollo, L.; Sterzi, S.; Bressi, F. Robotic Hand Treatment of Patients Affected by Chronic Stroke: A Monocentric Longitudinal Pilot Study. *J. Biol. Regul. Homeost. Agents* **2020**, *34*, 79–86.
22. Kurillo, G.; Gregoric, M.; Goljar, N.; Bajd, T. Grip Force Tracking System for Assessment and Rehabilitation of Hand Function. *Technol. Health Care* **2005**, *13*, 137–149. Available online: https://pubmed.ncbi.nlm.nih.gov/15990417/ (accessed on 6 April 2022). [CrossRef]
23. Lindberg, P.G.; Roche, N.; Robertson, J.; Roby-Brami, A.; Bussel, B.; Maier, M.A. Affected and Unaffected Quantitative Aspects of Grip Force Control in Hemiparetic Patients after Stroke. *Brain Res.* **2012**, *1452*, 96–107. [CrossRef] [PubMed]
24. de Pasquale, G.; Mastrototaro, L.; Pia, L.; Burin, D. Wearable System with Embedded Force Sensors for Neurologic Rehabilitation Trainings. In Proceedings of the Symposium on Design, Test, Integration and Packaging of MEMS/MOEMS, DTIP 2018, Rome, Italy, 22–25 May 2018; pp. 1–4. [CrossRef]

25. Lapresa, M.; Tamantini, C.; Scotto Di Luzio, F.; Cordella, F.; Bravi, M.; Miccinilli, S.; Zollo, L. A Smart Solution for Proprioceptive Rehabilitation through M-IMU Sensors. In Proceedings of the 2020 IEEE International Workshop on Metrology for Industry 4.0 and IoT, MetroInd 4.0 and IoT 2020, Rome, Italy, 3–5 June 2020; pp. 591–595. [CrossRef]
26. Katz, S.; Ford, A.B.; Moskowitz, R.W.; Jackson, B.A.; Jaffe, M.W. Studies of illness in the aged. The index of adl: A standardized measure of biological and psychosocial function. *JAMA* **1963**, *185*, 914–919. [CrossRef] [PubMed]
27. Hsieh, Y.W.; Lin, K.C.; Wu, C.Y.; Shih, T.Y.; Li, M.W.; Chen, C.L. Comparison of Proximal versus Distal Upper-Limb Robotic Rehabilitation on Motor Performance after Stroke: A Cluster Controlled Trial. *Sci. Rep.* **2018**, *8*, 2091. [CrossRef] [PubMed]
28. Page, S.J.; Fulk, G.D.; Boyne, P. Clinically Important Differences for the Upper-Extremity Fugl-Meyer Scale in People with Minimal to Moderate Impairment Due to Chronic Stroke. *Phys. Ther.* **2012**, *92*, 791–798. [CrossRef] [PubMed]
29. Reid, D.T. Benefits of a Virtual Play Rehabilitation Environment for Children with Cerebral Palsy on Perceptions of Self-Efficacy: A Pilot Study. *Pediatr. Rehabil.* **2002**, *5*, 141–148. [CrossRef]
30. Wann, J.P.; Rushton, S.K.; Smyth, M.; Jones, D. Virtual Environments for the Rehabilitation of Disorders of Attention and Movement. *Stud. Health Technol. Inform.* **1997**, *44*, 157–164. [CrossRef]
31. Rizzo, A.A.; Buckwalter, J.G.; Neumann, U.; Kesselman, C.; Thiebaux, M. Basic Issues in the Application of Virtual Reality for the Assessment and Rehabilitation of Cognitive Impairments and Functional Disabilities. *CyberPsychol. Behav.* **2009**, *1*, 59–78. [CrossRef]
32. Alt Murphy, M.; Häger, C.K. Kinematic Analysis of the Upper Extremity after Stroke—How Far Have We Reached and What Have We Grasped? *Phys. Ther. Rev.* **2015**, *20*, 137–155. [CrossRef]
33. Čolović, H.; Dimitrijević, L.; Đurić, V.; Janković, S. Upper Limb Robotic Neurorehabilitation after Pediatric Stroke. *Srp. Arh. Za Celok. Lek.* **2020**, *148*, 368–376. [CrossRef]

applied sciences

MDPI

Article

A Smart Modular IoT Sensing Device for Enhancing Sensory Feedbacks in Surgical Robotics

Mafalda Rosa [1,2], Rongrong Liu [1,2], Giorgio Pitruzzello [1,2,3] and Giuseppe Tortora [3,*]

1 BioRobotics Institute, Scuola Superiore Sant'Anna, 56127 Pisa, Italy
2 Department of Excellence in Robotics & AI, Scuola Superiore Sant'Anna, 56127 Pisa, Italy
3 Smart Medical Theatre Laboratory, ABzero, 56124 Pisa, Italy
* Correspondence: smartmedicaltheatre@abzero.it

Abstract: This paper proposes a device of sensing that could be integrated into the instruments of any surgical robot. Despite advances in robot-assisted laparoscopic surgery, the tools currently supplied to surgical robots have limited functions, due to the absence of sensorization. With this motivation, we present a preliminary work based on the design, development, and early stages of experimentation with smart and multifunctional devices of sensing for surgical tools. The proposed device of sensing has a proximity sensor, colorimetric sensor, and BLE connection for different surgical instruments to connect to each other. The proximity feedback allows the surgeon to know the distance of the instrument from a particular tissue, to operate in conditions of greater safety. With the colorimetric feedback, on the other hand, we intend to proceed to the identification of specific tissue areas with characteristics that are not typical of the physiological tissue. The results show that the device is promising and can be further developed for multiple clinical needs in robotic procedures. This system can effectively increase the functionality of surgical instruments by overcoming the sensing limitations introduced by using robots in laparoscopic surgery.

Keywords: robot-assisted laparoscopic surgery; surgical robot; da Vinci Research Kit; IoT

Citation: Rosa, M.; Liu, R.; Pitruzzello, G.; Tortora, G. A Smart Modular IoT Sensing Device for Enhancing Sensory Feedbacks in Surgical Robotics. *Appl. Sci.* **2022**, *12*, 8083. https://doi.org/10.3390/app12168083

Academic Editor: Alessandro Gasparetto

Received: 11 July 2022
Accepted: 10 August 2022
Published: 12 August 2022

1. Introduction

The past few decades have seen an exponential growth in medical technology, particularly with regard to the application of robotics to surgery. Robotic surgery, the latest evolution of minimally invasive surgery, overcoming the limits of traditional surgery, has allowed the broadening of therapeutic horizons and represents the gold standard for various clinical applications.

Robotics is the center of modern health engineering. The first robot used in the clinical setting to obtain neurosurgical biopsies was the Puma 560 robot in 1985. Since then, more and more advanced surgical robots have been developed [1,2]. In general, the use of robotic surgery increased significantly from 2012 to 2018, with an increase from 1.8% to 15.1% for all general surgery procedures. Over the same period, the use of both laparoscopic and open surgery declined. For example, the proportional use of open surgery was 42.4% in 2012, compared to 32.4% in 2018 [3]. It has also been witnessed that the use of robotic surgery has increased rapidly and spread widely in numerous procedures during the years following the adoption of this practice in hospitals. Therefore, for most surgeons, it was already considered a safe and effective approach when clinically feasible.

Current robotic platforms are designed to incorporate advanced features that allow for increased accuracy by making the execution of operator tasks easier and safer. Additionally, surgical robots have retained the ability to perform surgical operations through smaller incisions. These characteristics aim to improve the results compared to those obtainable through traditional surgical methods. The adoption and diffusion of robotic surgery shows a positive trend in some geographical areas, especially in countries with advanced economies. This is shown by the widespread use of the da Vinci Surgical System (Intuitive Surgical Inc., Sunnyvale, CA, USA), in a multitude of application areas [4]. Currently, the da Vinci Surgical System represents the most widespread surgical system, with over 5000 models implemented worldwide, performing over 7 million surgical procedures in different anatomical areas. The da Vinci Research Kit (dVRK) research platform fits into this context, developed through a collaboration among academic institutions to address the challenges in starting research on surgical robotics. This has led to a significant boost in the development of surgical robotics research over the past decade, and has generated new opportunities for collaboration and linking of a surgical robot to other technologies.

Among the advantages introduced by robot-assisted surgery are the reduction in tissue trauma thanks to small incisions, less bleeding, and less need for transfusions, a reduction in hospital stays and post-operative pain, a reduction in recovery times, and a quicker recovery rate in carrying out daily activities and greater ease in the execution of complex surgical tasks, which entails greater safety for the patient. On the other hand, the disadvantages of robotic surgery are mainly linked to the cost of the robotic system, the instrumentation, the system maintenance, and to the fact that, to operate the robot, very high-level skills are required on the part of the surgeon and room staff, to be acquired through specific training [5].

During an open surgery, surgeons can use their hand to locate and diagnose abnormal tissue by direct palpation; instead, in laparoscopic surgery, direct palpation is not feasible, due to the limits of the incision [6,7]. For this reason, one aspect that many studies are focusing on is the lack of haptic feedback during the surgical procedure, in addition to visual feedback [8,9]. Nevertheless, surgeons using robotic technologies could benefit from other types of feedback, such as feedbacks of color, speed, and proximity, to further broaden the fields of the application of surgical robots [10,11].

In this work, an attempt is made to restore the functions that are lost in robotic laparoscopic surgery, by using the sensorization of the surgical instrumentation. The design intention is to develop an intelligent and multifunctional sensing device to improve the performance of surgical robots, interconnect instruments, and enable, in the future, the development of AI algorithms.

2. Materials and Methods

2.1. Device Overview

In this section, the overall design of the proposed device of sensing is introduced. For the implementation, the following needs have been considered and analyzed:

- The design of a compact mechanical concept;
- A circuit that integrates the functional components to detect color and proximity;
- A machine learning model that allows the classification of tissues based on their color;
- A BLE (Bluetooth Low Energy) communication to allow the interconnection of the various arms and the introduction of the IoT to surgical robotics.

The main purpose of the first prototype proposed in this manuscript is to carry out the sensorization of the surgical instrument used during procedures with surgical robots, to ensure a safer interaction with the organs. For the first prototyping phase, an Arduino Nano 33 BLE Sense board is used, due to its compact size and the presence of integrated sensors. Reference is made to the colorimetric and proximity sensor integrated inside the APDS9960 unit of the Arduino board considered. In particular, the proximity sensor provides feedback on the distance between the tip of the surgical instrument and the organs, to ensure safe interaction. In addition, the colorimetric feedback allows, with the implementation of a simple neural network, to identify specific tissue areas with characteristics that are not typical of physiological tissue, such as cancerous tissue structures. This kind of recognition leads the way, in the future, to the classification of healthy tissue and diseased tissue. A 3D printed mechanical support is used to mount the electronic board to the robotic instruments, consisting of two main parts held together using a magnetic anchor. The magnets are integrated into the mechanical structure of the device, and allow for a quick alignment of the two parts, and a stable anchoring to the instrument. The dimensions of the holder are compatible with the size of conventional da Vinci EndoWrist tools; as this device of sensing is designed for robot-assisted laparoscopic surgery, it can be integrated into instruments of the Patient Side Manipulator (PSM) of the da Vinci Research Kit (dVRK) and can be teleoperated by the Master Tool Manipulator (MTM) [12] (Figure 1).

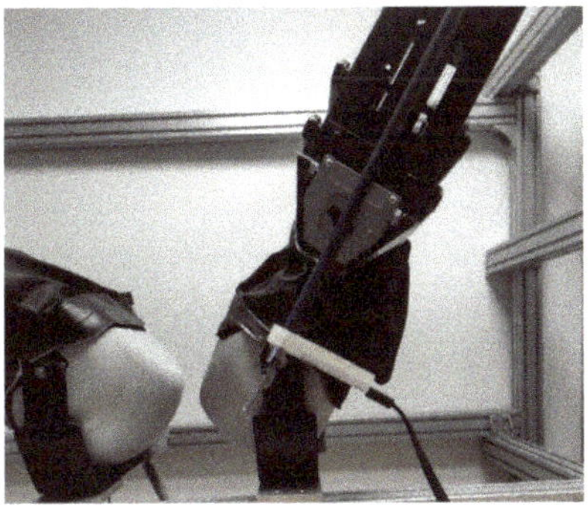

Figure 1. The developed sensing device mounted on the PSM of the dVRK system.

2.2. Design and Implementation of the Sensing Device

2.2.1. Mechanical Design

A mechanical support is created to allow to the Arduino Nano 33 BLE Sense board to be kept in position on the dVRK instrumentation during the execution of the validation tasks. It is conceived in two different versions to adapt to the different positioning along the instruments of the dVRK. The first version of the prototype, shown in Figure 2, allows the lying positioning on the robot instrument. This arrangement is ideal during BLE communication between the various robot arms, as there is no need to keep the sensor integrated on the board in a specific position. Furthermore, the first position is characterized by a reduced size and a better flexibility during the movements.

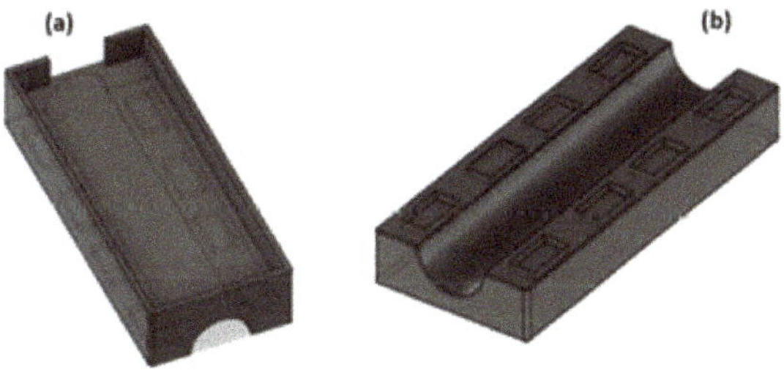

Figure 2. The CAD design of the first version of the mechanical support. (**a**) The housing for the electronic board. (**b**) The connection part for the positioning of the device along the EndoWrist tools.

The second version, shown in Figure 3, is designed for cases in which it is necessary to hold the embedded sensors that look towards the work surface. This is useful in the experimental validation phases during the detection of the color and proximity of tissues placed on the work surface.

This support, for both versions, is made up of two main parts. The first part represents the housing for the electronic board, while the second part connects to the first for the positioning of the device along the EndoWrist tools, as shown in Figures 2 and 3. The two structural portions of the mechanical support are held together using four K&J MAGNETICS neodymium magnets (NdFeB), for each side. The magnetic blocks have dimensions of (0.64 × 0.32 × 0.08) mm and grade N42. The magnetic anchoring enables the quick alignment of the two mechanical parts and the perfect integration of the sensing device with the dVRK instruments. At the same time, this magnetic mechanism makes it possible to separate the two structural portions of the support after performing the tasks using the tip of the surgical instrument itself [13,14]. This operation not only saves time, but can be performed without the assistance of an assistant surgeon.

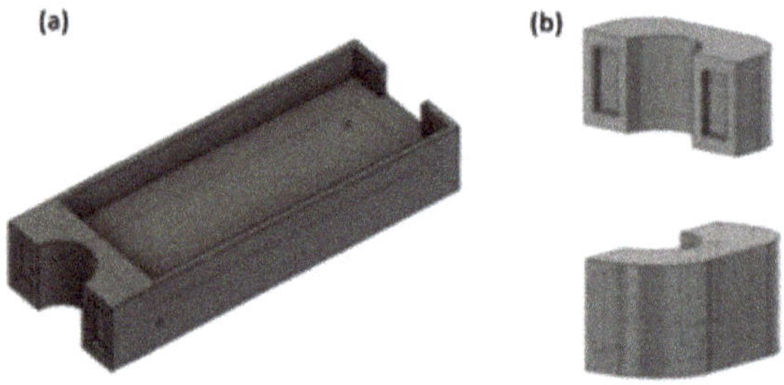

Figure 3. The CAD design of the second version of the mechanical support. (**a**) The housing for the electronic board. (**b**) The connection part for the positioning of the device along the EndoWrist tools.

2.2.2. Hardware Components

In this research, we employ the Arduino Nano 33 BLE Sense board composed of embedded sensors to detect color, proximity, motion, temperature, humidity, audio, and more. The presence of the embedded sensors allows the board to manage numerous IoT and AI applications without requiring the presence of external sensors. This board is built upon the nRF52840 microcontroller and runs on Arm® Mbed™ OS. The processor has other important features such as Bluetooth® pairing via NFC and ultra-low mode energy consumption. For the sensorization of the device, reference made is to the APDS-9960 unit built into the Arduino Nano33 BLE Sense board, which features advanced gesture sensing, proximity sensing, digital ambient light sensing (ALS), and color sensing (RGBC). This modular unit has dimensions 3.94 × 2.36 × 1.35 mm, and incorporates an IR LED and a factory-calibrated LED driver.

The proximity detection function provides the measurement of distance via the photodiode sensing of reflected IR energy from built-in LEDs. Detect/release events are interrupt-driven and occur whenever the proximity result crosses the upper and/or lower threshold settings. The IR LED intensity is factory trimmed to eliminate the need for end-equipment calibration due to component variations. The proximity results are further improved by automatic ambient light subtraction. The proximity results are affected by three basic factors: IR LED emission, IR reception, and environmental factors, including distance to the target and the surface reflectivity. The photodiode signal is combined, amplified, and offset adjusted to optimize performance. The colour and ALS detection feature provides red, green, blue, and clear light intensity data. Each of the R, G, B, C channels has a UV and IR blocking filter and a dedicated data converter producing 16-bit data simultaneously. This architecture allows applications to accurately measure ambient light and sense colour, which enables devices to calculate colour temperature and control display backlight.

2.2.3. IoT and Bluetooth Low Energy Connection

In recent years, we have seen a significant advance in digital technologies, which contributes to the current concept of the Internet of Things (IoT). At the basis of the IoT, there are "intelligent" objects that are interconnected to the exchange information owned, collected, and/or processed. The smart object must first be identifiable; that is, with a unique identifier in the digital world, and then it must be connected to transmit and receive information. These are smart connected devices that process and share all kinds of data with each other, and that can be controlled via the Internet. Into this context fit energy-efficient short-range wireless communication technologies such as Bluetooth Low Energy (BLE) [15].

This section shows how is possible to exchange information between two Arduino Nano 33 BLE Sense boards. With this communication, the various arms of the da Vinci surgical robot can be interconnected.

When a Bluetooth® connection is established, the central device will scan the surrounding devices and "listen" for the devices that transmit information, and at the same time, the device will advertise or transmit its data or information to any nearby device. As soon as the central device collects information from the peripheral device, an attempt is made to connect the peripheral device. Once the connection is established, the central device will interact with the available information to the peripheral device. This information exchange takes place using so-called services. By grouping the various device capabilities into services, central devices allow peripheral devices to quickly find, select, and interact with the desired services. Any service has a unique identifier called UUID. This code can be 16 or 32 bits long, for services with Bluetooth® specifications. One of the two Arduino Nano BLEs is configured as a central device, while the other as a peripheral device. The information shared between the two boards comes from the proximity sensor of the integrated APDS-9960 unit of the Arduino Nano 33 BLE Sense board.

For this purpose, the ArduinoBLElibrary library was used, and a service called proximityService was created with a feature called proximity_type, as shown in Figure 4. The central device tries to establish a connection with the peripheral device, and tries to discover the service and the feature that we have specified when implementing the code. If the connection is made successfully, the Nano 33 BLE Sense board's built-in proximity sensor is activated. When a proximity value is detected by the sensor, the central device gives us feedback, through the serial monitor, on the type of distance detected (FAR, MIDDLE, or CLOSE). The value is written to the proximity_type feature of the proximity service in the peripheral device. In addition, the on-board LED in the peripheral device lights up according to the detected value. If a distance is detected that exceeds the threshold set as FAR, the green LED will turn on; in the case of MIDDLE distance, the blue LED will be on; finally, if the object is close to the sensor, the red LED lights up.

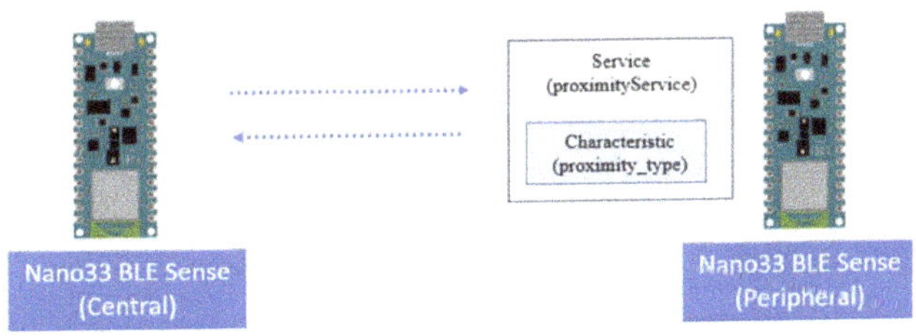

Figure 4. BLE communication between two Arduino Nano 33 BLE Sense units.

2.2.4. Complete Prototype

All the mechanical parts that make up the sensing device were produced using a 3D printer. After that, the two mechanical parts and the electronic board described above were assembled with the use of the magnetic anchor and positioned on the EndoWrist tool of the dVRK system. Figure 5 clearly shows how the first version of the prototype guarantees a minimum footprint and better compactness. This support configuration is to be taken into consideration in cases where there is no need to detect objects placed on the workspace using the integrated sensors. For example, in the case in question, it is possible to refer to this first version to demonstrate the potential of the BLE connection between two boards placed on two arms of the dVRK. Furthermore, during the validation phase of the device on dVRK instrumentation, it was found that the friction between the surface of the instrument and the inside of the device prevents any movement of the support; therefore, the stability of the electronic component is guaranteed.

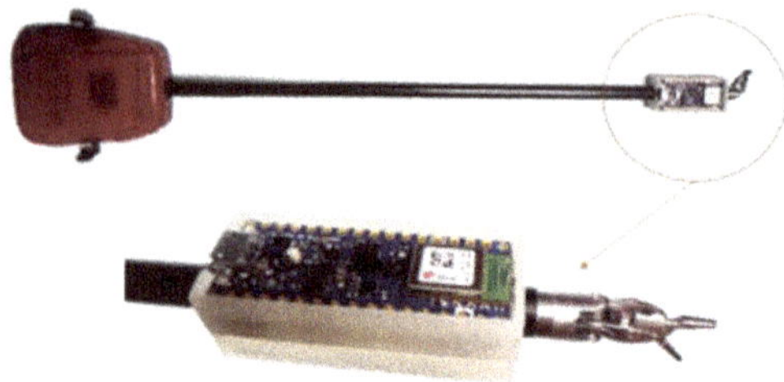

Figure 5. Complete prototype of the first version of the detection device connected to the EndoWrist tool of dVRK system.

The second assembled version of the prototype, unlike the first, allows the board to be kept parallel to the work surface. The main limitation of this configuration is the larger footprint, as shown in Figure 6, which could constitute an obstacle for the other tools of the robotic system during the execution of the tasks. Furthermore, it may seem less stable than the first configuration, but it has been tested that using magnets, the sensing device remains perfectly in position during the entire procedure carried out with the dVRK system.

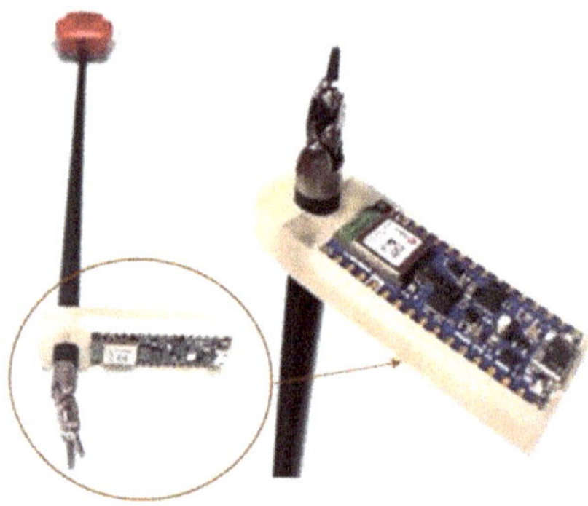

Figure 6. Complete prototype of the second version of the detection device connected to the EndoWrist tool of dVRK system.

2.3. *Experimental Evaluations*

This section presents the experiments conducted to verify the possibility of introducing a proximity and color sensor on the EndoWrist instrument.

2.3.1. Proximity Detection and Experiments with Organs

Proximity calibration was performed using the embedded APDS9960 unit of the Arduino Nano 33 BLE Sense. The path of the IR receive signal begins with the IR detection using four photodiodes, and ends with an 8-bit proximity result (256 values) in the PDATA register. So, we focused on proximity readings, which were based on sensing a tissue on photodiodes, and were then converted to millimeters within the sensor for our use. The board has been programmed to print out simple proximity detections and control the RGB LED accordingly, and to change the colors of the RGB LED according to the proximity of a tissue to the sensor. To take advantage of the functions of the APDS9960 detection unit, the <Arduino_APDS9960.h> library was used. In particular, the LED lights up green if the object is far from the sensor (proximity value >150), blue in the case of intermediate distance (60 < proximity value < 150), and red if the object is very close to the sensor (proximity value < 60). The threshold values shown in parentheses have been implemented in the code. The first test performed concerns the evaluation of the proximity between the tissues and the sensor. This allows the surgeon to work in a safer environment with the sensorization of the instrument, which gives a sound or visual feedback based on the distance of the instrument to the tissue. This aspect is very important during abdominal surgery procedures, where the various organs are very close to each other. Before getting to the heart of the testing phase, a connection was created between the Arduino Nano 33 BLE Sense board and the Matlab software through a serial communication protocol to manage the data collected by the proximity sensor directly in the Matlab environment. Subsequently, we moved on to the real-time display of the proximity values read by the sensor, which allowed for an instant evaluation of the trend of the data collected in relation to the distance in centimeters. Proximity data acquisition was performed with the MELFA RV3-SB industrial robot. For the execution of the tests, a support for the Arduino board to be fixed to the end effector of the robot was designed and built using a 3D printer. The support was fixed to the robot using two M5 screws, as shown in Figure 7.

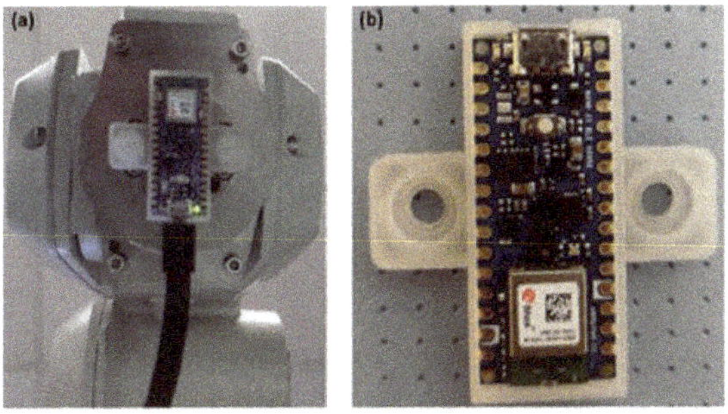

Figure 7. (**a**) Support for the Arduino board fixed to the end effector of the MELFA RV3-SB industrial robot by means of two M5 screws. (**b**) View of the support with an integrated board made with a 3D printer.

The robot was programmed to perform movements along the z axis, allowing the sensor to move away from and approach the tissue sample. Once we acquired the initial position (minimum distance from the tissue) desired, we decided to make the robotic arm move by 13 cm by making vertical movements of 0.5 mm intervals. All the positions necessary for the desired movement were implemented manually using the robot programming language with a structure that is very simple and intuitive; for example, the command MOV was used to manage the movement of the robot from one position to the next. Slices of tissues with a thickness of about six millimeters were obtained from the organ. Obviously, since these are soft and/or spongy organs, it is not possible to consider a precise and absolute measure of thickness, as the tissue surfaces have protuberances and depressions. This condition was considered relevant for an optimal and faithful representation of the operational reality; therefore, no corrections were made. These slices of tissue were placed on a rigid plate on the workstation below the sensor. As shown in Figure 8, a safety space was left between the sensor and the surface of the tissues to ensure the cleanliness and integrity of the detector. During the testing phase, the same environmental lighting conditions were always maintained. Furthermore, the tests on the samples were carried out on the same day that the tissues were taken from the slaughterhouse; in this way, the freshness of the sample was guaranteed to preserve its color and consistency, to better respect the conditions in vivo.

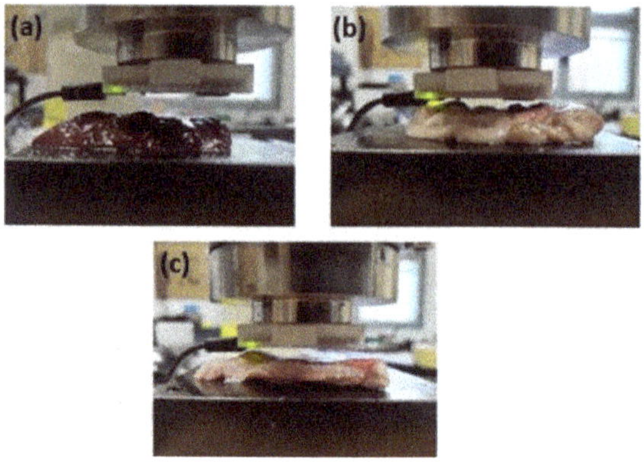

Figure 8. Tissue samples from the pig, placed on the plate below the sensor for the testing phase with the robot MELFA RV3-SB. (**a**) Liver. (**b**) Gut. (**c**) Stomach.

2.3.2. Classification of Tissues by Color

This subsection presents the implementation of a TensorFlow model to classify tissues, based on color detection. The main objective is to be able to distinguish normal tissue from one with characteristics that are not typical of physiological tissue, evaluating the color differences that characterize the two types of tissue. For example, it is possible to define the area of the tissue affected by cancerous manifestations that lead to an alteration of the color, compared to that of normal tissue.

To demonstrate this concept, we limited ourselves to verifying that it is possible, based on color, to distinguish two different tissues, i.e., liver and stomach. In particular, the procedure for the identification and classification of tissues was based on the use of the TensorFlow Lite Micro library and the Arduino Nano 33 BLE Sense colorimetric sensor. To do this, a simple neural network was implemented on the board, combining machine learning with integrated systems to develop an intelligent device. For the first phase of collecting the color data from the two portions of tissues, the Arduino Nano 33 BLE Sense board was first programmed. The colorimetric sensor was integrated inside the APDS9960 unit. This allowed the detection of the intensity of the red, green, and blue colors of each tissue subjected to the sensor. For correct data acquisition, it is advisable to move the sensor around the surface of the tissue to capture the variations of color, as shown in Figure 9. RGB color values were captured as comma separated data grouped in CSV format. This procedure was repeated for both organs, which we were decided to classify by capturing the color data.

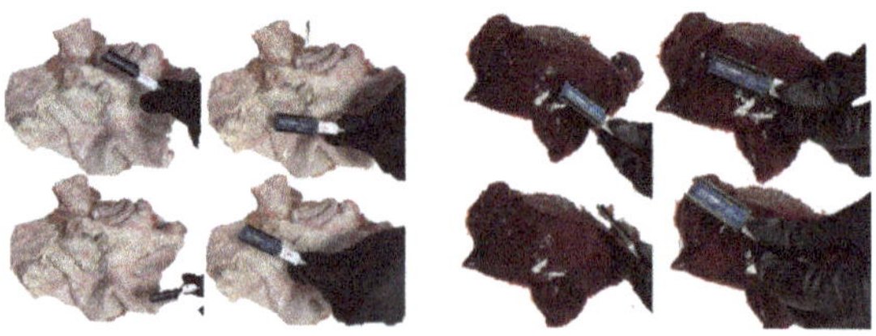

Figure 9. Frames concerning the acquisition of color data from the stomach and liver to detect the intensity of the red, green, and blue colors of each tissue subjected to the sensor.

Following the data acquisition phase, the model training phase was implemented. The model training phase represents the process by which a model learns to produce the correct output for a given set of inputs. This phase involved feeding training data through the model, making small changes until the most accurate predictions were possible. A model was a network of simulated neurons represented by arrays of numbers arranged in various layers. As data are fed into the network, it is transformed by successive mathematical operations involving weights and distortions at each level. The output of the model was the result of executing the input through these operations. Training stopped when the model's performance stopped improving. For the training phase of the machine learning model, reference was made to the Google Colaboratory, an interactive environment that provides a notepad that allows the writing and execution of code in Python, using the data collected in the previous phase. Before TensorFlow Lite ran the model that was trained, it needed to be converted to the TensorFlow Lite format, and then a model.h file would be generated to be downloaded and included in the final Arduino code, to classify fabrics based on color. Finally, after loading this code on the Arduino board, it was possible, by approaching the RGB sensor close to the object to be classified with which the model was trained, to view the percentage relating to the two classes implemented in the model.

2.3.3. Validation by dVRK

The proposed prototype has been tested on the dVRK system. The experimental protocol followed for the tasks performed with the aid of the dVRK system, and the various phases, tools, and objects used are presented below. As shown in Figure 10, an MTM from the dVRK console was used to teleoperate the instrument mounted on the PSM. The activities were carried out by three users with the support of the visual feedback of the dVRK endoscope, designed with two separate optical channels capable of recreating the most important aspect of stereopsis: binocular disparity. Inside the viewer, therefore, users followed the operations thanks to the stereoscopic vision that provided three-dimensional images of the work area [16].

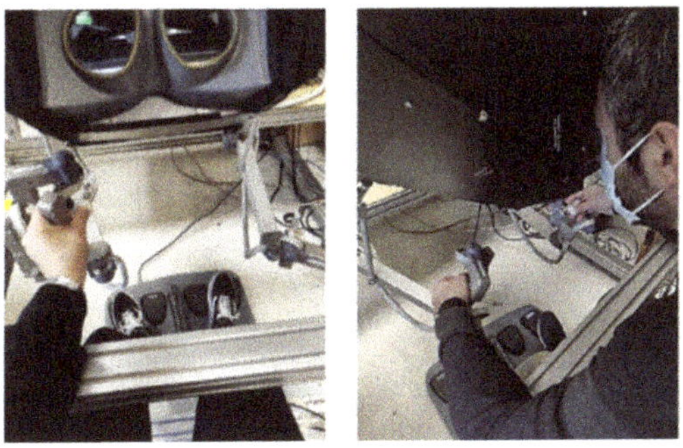

Figure 10. Functionality demonstration. MTM from the dVRK console was used to teleoperate the instrument mounted on the PSM.

The Arduino board with the integrated proximity sensor was positioned, using the support made by a 3D printer, on the EndoWrist instrument two centimeters from the tip of the tool, as shown in Figure 11. The positioning of the board at this distance from the tip of the instrument was in line with the results obtained during the sensor calibration phase; from the characterization curve, it is possible to see that the operating area of the liver sensor occurred after two centimeters, a distance at which the sensor returned the first non-null value.

Users were asked to perform a simple transfer task; a suitably sized ball was picked up from the worktable with the ProGaspr tool supplied with the dVRK, and it was brought within eight centimeters of the liver simulation object. The sequence of images below shows the steps just mentioned in performing the task (Figure 12). Finally, the user placed the ball on the object by approaching it slowly (Figure 13). During the execution of the operations, the user was able to estimate the distance of the instrument from the object based on what could be perceived by the viewer. During the entire execution phase of the task, the proximity values read by the sensor were recorded and compared with the values visually estimated by the users.

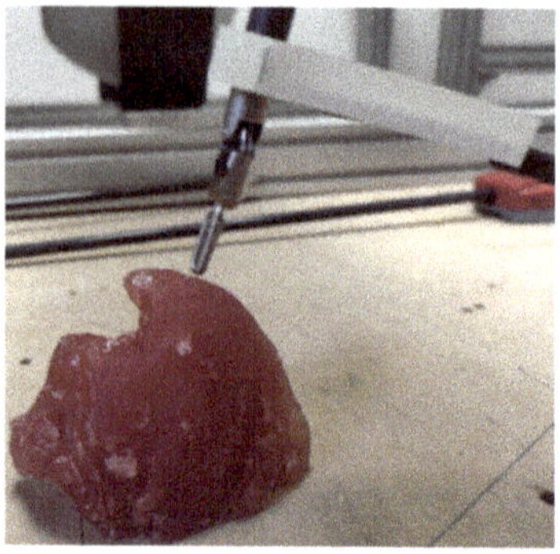

Figure 11. Positioning of the sensing device prototype on the PSM tool of the dVRK system.

Figure 12. Three frames concerning the phases of the performed transfer task, taking the ball from the workspace and positioning it about 8 cm from the tissue.

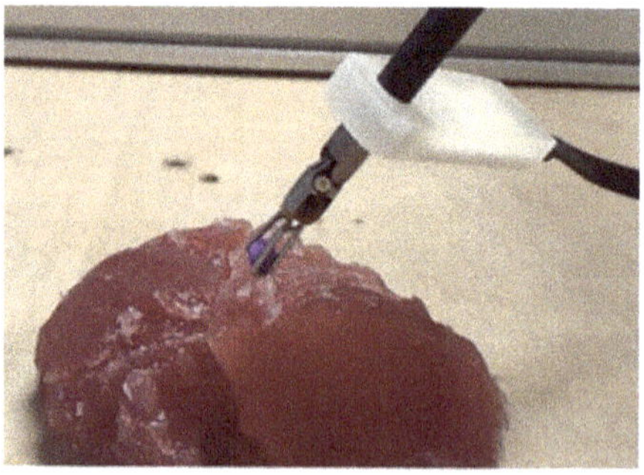

Figure 13. Positioning of the ball on the object placed on the work surface.

3. Results

The results related to the proximity measurements on the various tissue samples, obtained by experimental steps described in the previous section, are shown in this section. Due to the lack of synchronization between the start signal of the robotic arm and the Arduino board containing the proximity sensor, it was necessary to proceed with a manual evaluation of the first useful point at which the sensor returns a non-zero proximity value. To do this, the tissue sample was positioned on the plate as shown in section III, and the robotic arm was moved manually to detect the first significant data by acting on the coordinates of the individual joints. As shown in Figure 14, the sensor returned a series of null proximity values when it was too close to the sample. Starting from the fixed initial position P1 (position of the robotic arm closest to the sample), the sensor, positioned at the end-effector of the robot, was slowly moved away from the tissue sample by taking steps equal to 0.5 mm. In this way, it was possible to evaluate and display on the real-time graph relating to the sensor the first non-zero proximity value detected, and the corresponding distance of the sensor from the sample. The real distance was known to the user who programmed the robotic arm. By repeating this experiment for all three types of tissues, it was possible to precisely define the first useful proximity value for each sample. Regarding the liver, the first useful proximity value was detected at two centimeters from the sample, while for the stomach and intestine, the first non-zero proximity value was shown at 3.5 cm.

The characteristic curves for each tissue are shown in Figure 14. In this way, it was possible to relate the real distance due to the movement of the robotic arm with the proximity values derived from the sensor. As previously mentioned, the proximity measurement by the sensor took place thanks to the detection of the IR energy reflected by the tissue on the photodiode. Consequently, since this measurement was influenced by the reflectivity of the sample surface, it was reasonable to think that tissues of different colors could give a different characterization curve. This concept is clearly shown in Figure 14d, where the stomach and intestines with a very similar color returned the first non-zero value at the same distance. The liver characterization curve, on the other hand, had a different trend than that of the stomach and gut. It should be mentioned that for all three types of tissues, it was possible to note that as the robotic arm moved away from the sample, the readings of the proximity sensor were accompanied by a part of noise.

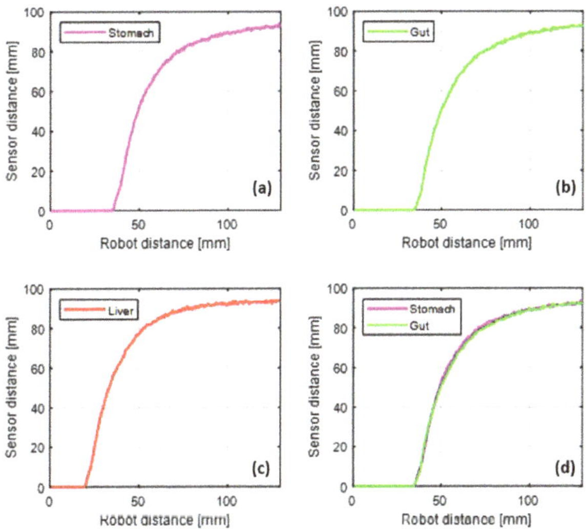

Figure 14. Characteristic curves relating to the proximity of the three tissues analyzed. (**a**) Stomach. (**b**) Gut. (**c**) Liver. (**d**) Superposition of the curves of the intestine and stomach.

To ensure the repeatability of the experiment, four different tests were carried out for each of the three tissues examined. For each organ, as shown in Figures 15–17, the sensor characterization curves were practically the same for the four tests. This allows us to conclude that it is possible to extend this argument to any tissue sample, and therefore, to be able to consider the curves obtained as being representative for the tissue to which they refer in relation to the sensor used.

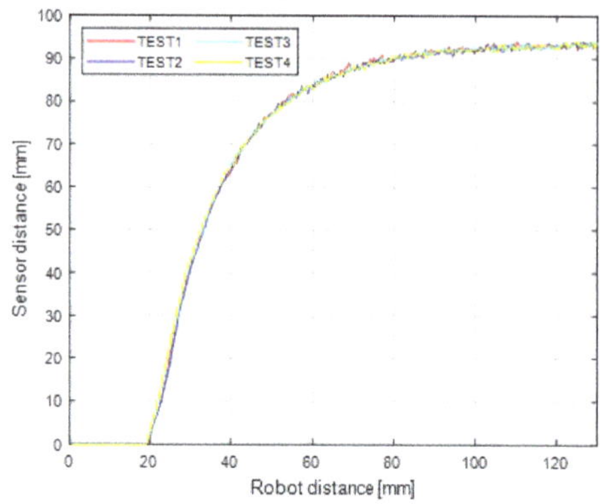

Figure 15. The proximity values recorded for the liver during the movement of the robot in four tests.

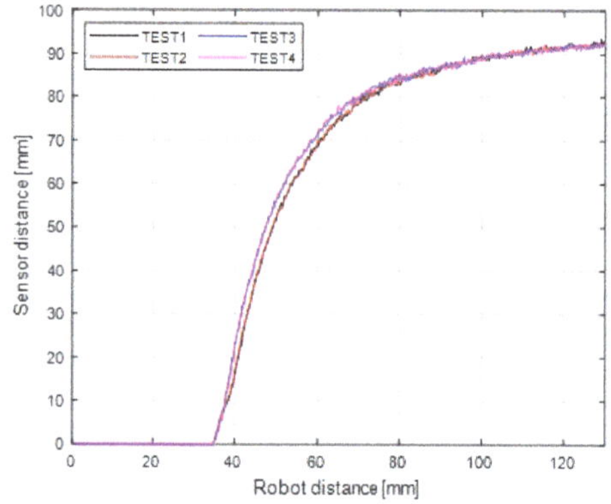

Figure 16. The proximity values recorded for the stomach during the movement of the robot in four tests.

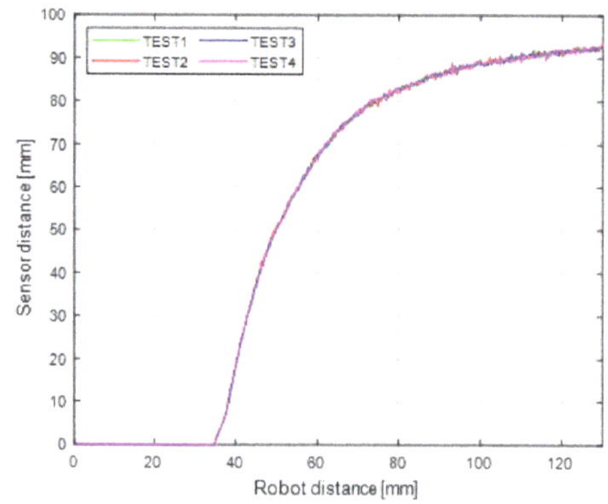

Figure 17. The proximity values recorded for the gut during the movement of the robot in four tests.

During the following analysis, the stomach curve was taken as a reference, but similar results can also be easily extended to the other two tissues. By analyzing the curve, an initial flat area that describes the locations where the sample was too close to the sensor, and therefore, was unable to return proximity values, was pointed out. Continuously, a linear area was highlighted by the green rectangle in Figure 18.

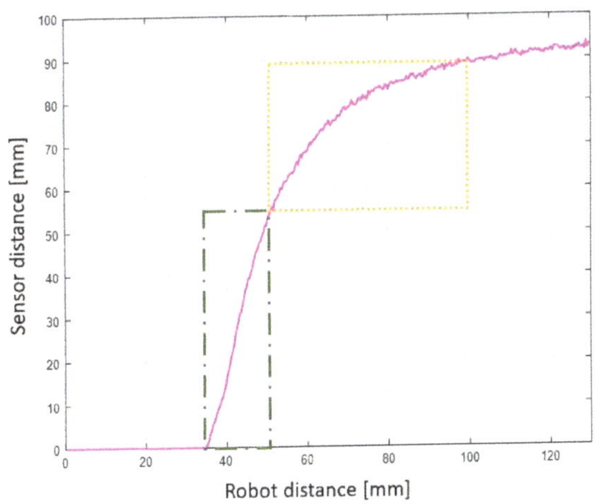

Figure 18. Characteristic curve relative to the stomach, with highlighted areas.

This linear part just mentioned was well approximated by a first-degree polynomial, as demonstrated with the help of MATLAB's Curve Fitting Toolbox (Figure 19). The linear polynomial model that best approximates this data distribution was from the equation:

$$f(x) = p1 \times x + p2 \quad (1)$$

where x was normalized by mean 7.667 and std 4.378. The coefficients $p1$ = 16.88, and $p2$ = 27.9 were estimated by the model with 95% confidence bounds.

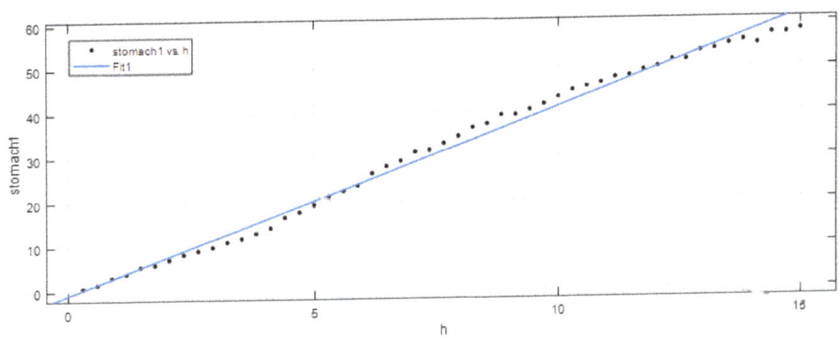

Figure 19. Fitting of the values of the portion of the graph highlighted by the green rectangle with the polynomial model.

In the same way, it was possible to highlight and analyze the next portion of the curve about the stomach, shown by the yellow rectangle in Figure 18. In the second case, it was possible to make an approximation described by a third-degree polynomial in the form:

$$f(x) = p1 \times x3 + p2 \times x2 + p3 \times x + p4 \quad (2)$$

where x was normalized by mean 25.17 and std 14.48. The coefficients $p1$ = 1.116, $p2$ = −3.359, $p3$ = 6.919 and $p4$ = 82.15 were estimated by the model with 95% confidence bounds (Figure 20).

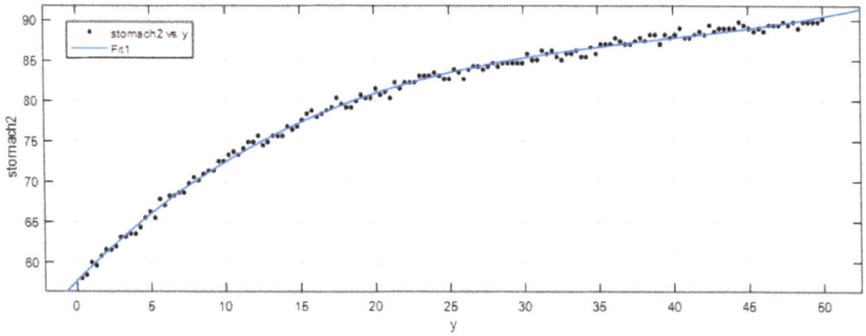

Figure 20. Fitting of the values of the portion of the graph highlighted by the yellow rectangle with the polynomial model.

For both portions of the graph highlighted, the goodness of the fitting was evaluated with appropriate parameters, such as R-squared, to understand how strong the predictive power of a linear regression model is. These measurements evaluate how much difference there is between the observed values in the sample and the values that the model has estimated. The case examined shows small discrepancies between the expected and observed values, and this indicates that the model fits well with the data. In fact, the value of R-squared, respectively, for the first and second case examined, was equal to $R_1 = 0.994$ and $R_2 = 0.997$.

4. Discussion

This work aims to design and develop a sensing device that can be integrated with the EndoWrist instrument of surgical robots, in order to provide the surgeon with colorimetric feedback and information on the distance between the tip of the instrument and the organs. This is the first version of our device of sensing, and hence, the aspects are still preliminary. Despite this, the results obtained allow us to conclude that it is possible to use color to distinguish two different types of tissue with the final goal of making tumor diagnoses of tissue portions hidden from the human eyes. At the same time, through proximity feedback, the surgeon can work in conditions of greater control and safety. Furthermore, thanks to the integration of BLE communication, a connection between the various robotic arms is possible for the exchange of information during surgical procedures. These capabilities will facilitate the opening to the world of IoT applied to the world of surgical robotics, and will enable the development of AI algorithms for automatic or semi-automatic procedures in the future.

Author Contributions: Conceptualization, G.T.; methodology, M.R., R.L., G.P. and G.T.; software, M.R.; validation, M.R.; data analysis, M.R.; investigation, M.R., R.L., G.P. and G.T.; resources, M.R., R.L., G.P. and G.T.; data curation, M.R., R.L., G.P. and G.T.; writing—original draft preparation, M.R.; writing—review and editing, R.L., G.P. and G.T.; visualization, M.R.; supervision, G.T.; project administration, G.T.; funding acquisition, G.T. All authors have read and agreed to the published version of the manuscript.

Funding: This research received no external funding.

Institutional Review Board Statement: Not applicable.

Data Availability Statement: Not applicable.

Conflicts of Interest: The authors declare no conflict of interest.

References

1. Lane, T. A short history of robotic surgery. *Ann. R. Coll. Surg. Engl.* **2018**, *100*, 5–7. [CrossRef] [PubMed]
2. Morrell, A.L.G.; Morrell-Junior, A.C.; Morrell, A.G.; Mendes, J.; Freitas, M.; Tustumi, F.; Morrell, A.; et al. The history of robotic surgery and its evolution: When illusion becomes reality. *Rev. do Col. Bras. de Cir.* **2021**, *48*, 1–9. [CrossRef] [PubMed]
3. Sheetz, K.H.; Claflin, J.; Dimick, J.B. Trends in the adoption of robotic surgery for common surgical procedures. *JAMA Netw. Open* **2020**, *3*, e1918911. [CrossRef] [PubMed]
4. Tan, A.; Ashrafian, H.; Scott, A.J.; Mason, S.E.; Harling, L.; Athanasiou, T.; Darzi, A. Robotic surgery: Disruptive innovation or unfulfilled promise? A systematic review and meta-analysis of the first 30 years. *Surg. Endosc.* **2016**, *30*, 4330–4352. [CrossRef]
5. Yu, H.Y.; Hevelone, N.D.; Lipsitz, S.R.; Kowalczyk, K.J.; Hu, J.C. Use, costs and comparative effectiveness of robotic assisted, laparoscopic and open urological surgery. *J. Urol.* **2012**, *187*, 1392–1399. [CrossRef] [PubMed]
6. Bergeles, C.; Yang, G.Z. From passive tool holders to microsurgeons: Safer, smaller, smarter surgical robots. *IEEE Trans. Biomed. Eng.* **2013**, *61*, 1565–1576. [CrossRef] [PubMed]
7. Koh, F.H.; Tan, K.K.; Lieske, B.; Tsang, M.L.; Tsang, C.B.; Koh, D.C. Endowrist versus wrist: A case-controlled study comparing robotic versus hand-assisted laparoscopic surgery for rectal cancer. *Surg. Laparosc. Endosc. Percutaneous Tech.* **2014**, *24*, 452–456. [CrossRef] [PubMed]
8. Liu, H.; Selvaggio, M.; Ferrentino, P.; Moccia, R.; Pirozzi, S.; Bracale, U.; Ficuciello, F. The MUSHA hand II: A multifunctional hand for robot-assisted laparoscopic surgery. *IEEE/ASME Trans. Mechatron.* **2020**, *26*, 393–404.
9. Puangmali, P.; Liu, H.; Seneviratne, L.D.; Dasgupta, P.; Althoefer, K. Miniature 3-axis distal force sensor for minimally invasive surgical palpation. *IEEE/ASME Trans. Mechatron.* **2011**, *17*, 646–656. [CrossRef]
10. Jeong, S.H.; Seo, K.W.; Min, J.S. Intraoperative tumor localization of early gastric cancers. *J. Gastric Cancer* **2021**, *21*, 4. [CrossRef]
11. Bințințan, V.; Calborean, A.; Mocan, M.; Macavei, S.; Cordoș, A.; Ciuce, C.; Bințințan, A.; Chira, R.; Nagy, G.; Surlin, V.; et al. New inductive proximity sensor platform for precise localization of small colorectal tumors. *Mater. Sci. Eng. C* **2020**, *106*, 110146. [CrossRef] [PubMed]
12. D'Ettorre, C.; Mariani, A.; Stilli, A.; Valdastri, P.; Deguet, A.; Kazanzides, P.; Taylor, R.H.; Fischer, G.S.; DiMaio, S.P.; Menciassi, A.; et al. Accelerating surgical robotics research: Reviewing 10 years of research with the dvrk. *arXiv* **2021**, arXiv:2104.09869.
13. Tognarelli, S.; Salerno, M.; Tortora, G.; Quaglia, C.; Dario, P.; Schurr, M.O.; Menciassi, A. A miniaturized robotic platform for natural orifice transluminal endoscopic surgery: In vivo validation. *Surg. Endosc.* **2015**, *29*, 3477–3484. [CrossRef] [PubMed]
14. Tortora, G.; Dario, P.; Menciassi, A. Array of robots augmenting the kinematics of endocavitary surgery. *IEEE/ASME Trans. Mechatron.* **2014**, *19*, 1821–1829. [CrossRef]
15. Nikodem, M.; Slabicki, M.; Bawiec, M. Efficient communication scheme for Bluetooth low energy in large scale applications. *Sensors* **2020**, *20*, 6371. [CrossRef] [PubMed]
16. Palep, J.H. Robotic assisted minimally invasive surgery. *J. Minimal Access Surg.* **2009**, *5*, 1. [CrossRef] [PubMed]

applied sciences

MDPI

Review

Research Progress on Synergistic Technologies of Agricultural Multi-Robots

Wenju Mao [1,2,3,4], Zhijie Liu [1,2,3,4], Heng Liu [1,2,3,4], Fuzeng Yang [1,2,3,4,*] and Meirong Wang [1,2,3,4]

1 College of Mechanical and Electronic Engineering, Northwest A&F University, Yangling 712100, China; vincentblack@nwafu.edu.cn (W.M.); liuzhijie@nwafu.edu.cn (Z.L.); hengliu@nwafu.edu.cn (H.L.); 2020050991@nwafu.edu.cn (M.W.)
2 Scientific Observation and Experimental Station of Agricultural Equipment for the Northern China Ministry of Agriculture and Rural Affairs, Yangling 712100, China
3 Ministry of Agriculture and Rural Affairs Apple Full Mechanization Research Base, Yangling 712100, China
4 State Key Laboratory of Soil Erosion and Dryland Agriculture on Loess Plateau, Yangling 712100, China
* Correspondence: yangfzkm@nwafu.edu.cn; Tel.: +86-180-4942-0386

Abstract: Multi-robots have shown good application prospects in agricultural production. Studying the synergistic technologies of agricultural multi-robots can not only improve the efficiency of the overall robot system and meet the needs of precision farming but also solve the problems of decreasing effective labor supply and increasing labor costs in agriculture. Therefore, starting from the point of view of an agricultural multiple robot system architectures, this paper reviews the representative research results of five synergistic technologies of agricultural multi-robots in recent years, namely, environment perception, task allocation, path planning, formation control, and communication, and summarizes the technological progress and development characteristics of these five technologies. Finally, because of these development characteristics, it is shown that the trends and research focus for agricultural multi-robots are to optimize the existing technologies and apply them to a variety of agricultural multi-robots, such as building a hybrid architecture of multi-robot systems, SLAM (simultaneous localization and mapping), cooperation learning of robots, hybrid path planning and formation reconstruction. While synergistic technologies of agricultural multi-robots are extremely challenging in production, in combination with previous research results for real agricultural multi-robots and social development demand, we conclude that it is realistic to expect automated multi-robot systems in the future.

Keywords: agriculture; cooperative robots; key technology; control

Citation: Mao, W.; Liu, Z.; Liu, H.; Yang, F.; Wang, M. Research Progress on Synergistic Technologies of Agricultural Multi-Robots. *Appl. Sci.* **2021**, *11*, 1448. https://doi.org/10.3390/app11041448

Received: 31 December 2020
Accepted: 1 February 2021
Published: 5 February 2021

1. Introduction

With the constant progression of urbanization and industrialization, the mobility of rural young and middle-aged laborers has intensified [1]. The sustainable transfer of non-agricultural labor has led to a decline in agricultural labor, and the problem of aging of the agricultural labor force has become more serious [2]. For example, the proportion of agricultural production and management personnel aged 55 and over is as high as 33.6% according to the main data of the Third Agricultural Census Bulletin of China in 2017. Furthermore, as aging continues, the physical health of the elderly labor force continues to decline, which results in a significant reduction in the supply of effective rural labor and an adverse effect on agricultural output [3,4]. Most agricultural production tasks are labor-intensive and seasonally oriented projects that exacerbate the constraint of seasonal labor shortages [3–5] and increase the cost of agricultural labor. For example, according to a survey conducted by Zhen et al. [6], during the rural busy season, the labor cost of agricultural planting increased from 80 CNY per person per day in 2015 to 90 CNY in 2016 to 100 CNY per person per day in 2017, and the labor cost of technical agricultural labor is even higher [7]. Some statistics show that the agricultural unit labor cost in developed countries such as Germany, Japan, and the United States decreased by 30.17%, 44.22%, and 23.44%, respectively, while agricultural labor productivity increased by 64.78%, 81.22%, and 34.83%, respectively, from 2005 to 2014. On the contrary, China's agricultural unit labor cost increased by 45.17%, and agricultural labor productivity, which is much smaller than that of developed countries, increased by 50.28% [8]. Therefore, the increasing cost of labor will lead to an increase in the cost of agriculture, which will result in a relative decrease in agricultural productivity and international competitiveness [9]. It is urgent to enhance the innovation of agricultural science and technology and replace extensive and expensive repetitive manual operations with intelligent agricultural machines or robots [10].

Due to the growing maturity of computer technology, sensor technology, and control theory, different types of agricultural robots have been developed based on characteristics of agronomy, such as fruit- or vegetable-picking robots, spraying robots, and harvesting robots. The agricultural robot can replace traditional human efforts to engage in all kinds of labor-intensive and complicated agricultural production activities and reduce the decline of output caused by improper human operation, negligence, inaccurate operation, and other reasons, as well as major physical injuries and even casualties of operators [11]. However, the operation efficiency of a single agricultural robot is too low and cannot meet the operation demand in busy seasons without coordination and cooperation by artificial auxiliary resources or other robots [12]. As early as 2009 and 2012, Johnson et al. [13], Moorehead et al. [14] and others in the United States replaced a single robot with a group of agricultural mobile robots to complete mud moss harvesting and orchard spraying successively with appropriate cooperative operation mechanism, which can reduce production costs and improve operational efficiency [15]. Therefore, to adapt to the increasing scale of production, meet the needs of social development, and narrow the gap with other international world powers with advanced scientific capabilities, it is necessary to research the relevant technology of agricultural multiple robot systems. This work focuses on the research progress of the cooperative operation, one of the key technologies of agricultural multiple robot systems.

There are many types of agricultural multi-robots, and this article mainly focuses on unmanned aerial vehicles (UAVs), unmanned ground vehicles (UGVs), and unmanned underwater vehicles (UUVs). This paper summarizes the research progress on synergistic technologies of cooperative operation of the abovementioned multiple robots and characterizes the expected research of related technologies.

2. Problems with Multi-Robot Applied to Agricultural Environments

Crops are fixed in cultivation season and time, which determines the labor demand pattern of agricultural production in a year, and agricultural operators need to make flexible responses and treatments according to the growth pattern of crops, such as plowing, planting, management, and harvesting [16]. At the same time, to adapt to the development of agricultural intensification, scale, and industrialization, and to reduce the economic losses caused by untimely processing, a collaboration of multiple farm machinery operated by people has widely appeared in agricultural production (as shown in Figure 1).

In Figure 1a multiple rotary tillers are being used to plow rice fields on sloping land to safeguard food production and mitigate the impact of the phenomena of lack of labor resources, which is brought about by the New Crown epidemic [17]. In Figure 1b, multiple corn planters are employed to sow seeds in a large field, which saves labor and ensures the quality of seeding, and directly improves the yield and quality of corn planting [18]. In Figure 1c, multiple drones are used to spray pesticides in cotton fields, which could be targeted according to the types of pests and diseases, and also prevented their rapid spread in the early stages of infestation [19]. In Figure 1d, it has become a trend to manually operate multiple combines simultaneously during the wheat harvesting season to avoid the effects of rainfall on wheat quality and yield [20].

Figure 1. Man-operated farm machinery worked in a large field. (**a**) Multiple rotary tillers plowing rice fields in spring. (**b**) Multiple seeders sowing corn in spring. (**c**). Multiple drones spraying cotton fields in the summer. (**d**) Multiple harvesters harvesting wheat fields in summer.

By manually operating the machines, the quality of operation of the implements relies heavily on human experience, while the use of robots instead of manual operation can free up manpower and ensure the quality of operation. However, the topography, soil, light, and climate conditions of the crop growing environment are different from those of indoor and urban transportation environments, and these conditions pose a challenge to the application of multiple robots in agriculture.

Take the example of multiple harvesters harvesting grain in farmland. First, farmland is an unstructured environment, which means that the road conditions are undulating, there are various types of obstacles, and there are missing or blurred lane lines on the ground, and agricultural machines both share the same resource and interact with each other to become dynamic obstacles to each other. Secondly, agricultural operation tasks have strong requirements for operation time, such as harvesting grains in a fixed short time frame. Furthermore, the amount of grain output varies from plot to plot, and the number of agricultural machines needed and the number of operational tasks assigned to them is dynamically changing (such as harvesters with large loading capacity should match the plots with large grain output). Finally, even if the same type of farm machines work together, the characteristics of the machines are not the same (for example, the harvester with the same loading capacity, the fuel consumption is different, the harvester with high fuel consumption should be assigned the operation task more than its work cost, and its operation path should be as short as possible but the harvesting volume should be as large as possible). To ensure that multiple machines can cooperate, a multi-robot system is required to be able to organize multiple robots flexibly, quickly, and efficiently according to the changes in the environment and tasks, and to fully utilize the capabilities of each robot to finally complete the given task with high quality [21]. At the technical level, in addition to being accurately informed of the positioning information of the swarm and the environmental information of the operation, and solving collision and obstacle avoidance, it is also necessary to assign operational tasks to multiple machines, plan operational paths (such as in areas where multiple robots work together), coordinate the formation control of multiple robots, and maintain the information interaction between multiple robots.

3. Research Progress on Synergistic Technologies of Agricultural Multi-Robots

To solve the above problems, it is necessary to study technologies of collaborative operation, such as environment perception, task allocation, path planning, formation control, and communication-based on multi-robot architecture. Since each technology does not solve the same agricultural problems, this section first classifies the types of development of these technologies, then describes and reviews each of these cooperative technologies in terms of research methods or problem solving, and finally summarizes their research development status and characteristics.

3.1. Architecture of Agricultural Multiple Robot Systems

A reasonable architecture can guarantee information flow and control flow in the agricultural multi-robot system and make effective cooperation among multiple robots possible [22]. At present, the architecture of agricultural multi-robots can be divided into centralized architecture and distributed architecture. It is found that the earliest recorded structure of agricultural multiple robot systems comes from hay harvesting and transportation robots in farmland [23,24], for which these multi-robot systems were operated under the principle of centralized architecture. As shown in Figure 2a, in leader-follower mode, a relatively powerful robot is selected as the "leader" of the swarm robots, performing specific motion planning for the remaining robots after analyzing and processing the sensory information, but these remaining robots are just executors, without the ability to choose their actions or coordinate with each other. Alternatively, as shown in Figure 2b, in the central controller mode, each robot can perform tasks independently and is commanded by a central controller [25]. The advantage of this centralized architecture is that the theoretical background is clear, and the implementation is intuitive, but the flexibility, fault tolerance, and adaptability are poor [26].

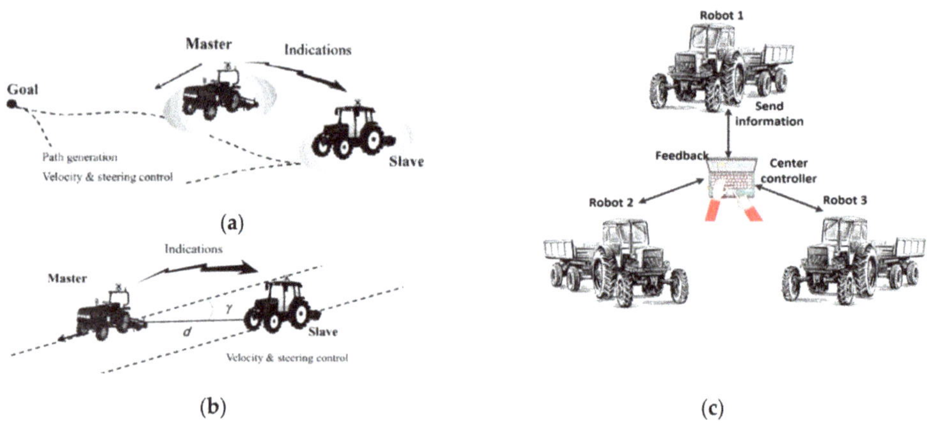

Figure 2. The centralized architecture of an agricultural mobile multi-robot system. (**a**) One of the guided following modes, where the slave robot follows the travel path taken by the active robot [27]. (**b**) Another of the guided following modes, where the slave robot is ordered by the master robot to go the other way [27]. (**c**) The centralized control mode is usually a back-end computer that monitors, plans, and controls the robot's tasks and operating paths.

Compared with the operating environment of UGVs, the operating area of UAVs has the advantage of no obstacles, so these systems generally adopt the distributed structure [28]. As shown in Figure 3, the three UAVs in the multi-robot system carry out agricultural situation monitoring in the individual workspaces independently; each robot of this system had a high degree of autonomous operation ability and can complete a given task according to its aims; robots can communicate with each other, exchange information, and coordinate their behaviors equally and independently to complete a given task [28]. This structure has strong scalability and certain advantages in real-time operation, fault tolerance, and reliability, etc. [29] and is suitable for handling tasks related to spatial states [30]. However, "it costs a lot in terms of the coordination mechanism, such as task allocation and motion planning" [29].

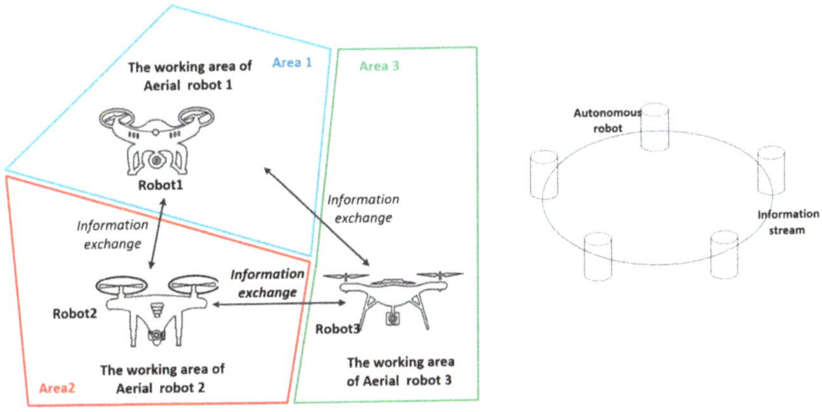

Figure 3. The distributed architecture of an agricultural mobile multi-robot system (**a**) Agricultural multi-robot agricultural condition monitoring in different areas of the vineyard [28]. And the robots exchange their information for autonomous work. (**b**) Distributed structure diagram of a multi-robot. Each robot can exchange information with other robots through communication and make decisions autonomously [26].

The centralized architecture that can be divided into a leader and follower robots is suitable for highly coordinated tasks and is advantageous in a fully known environment. The distributed architecture, in which there is no affiliation among robots, is suitable for weakly coordinated tasks and is advantageous in large-scale, complex, and varying environments.

3.2. *Environment Perception*

Environmental perception is the premise of the cooperative operation of multiple robots in agriculture. That is, the mobile carrier can use the sensors carried by itself (these sensors include internal sensors and external sensors [31], where the internal sensors include odometers, magnetic compasses, inertial navigation, and global positioning systems to determine the speed, position, and direction of the robot in the environment; external sensors include ultrasound, infrared, laser and vision, used to sense surrounding information [26]) to obtain the information of the surrounding environment, extract the effective feature information within the environment for processing and analysis, and finally establish the environment model [32]. This technology mainly involves collaborative positioning, data fusion, and environmental construction.

3.2.1. Co-Location Technology

The concept of co-location was originally proposed in 1994 by Kurazume and Nagata in Japan [33] The concept refers to a robot "sharing" its positioning results with other mobile robots and to other robots using this shared information to integrate their calculation results to improve the accuracy of positioning themselves and, in turn, sharing their positioning results with other mobile robots, repeatedly achieving the precise positioning of mobile robots [34]. According to the collaborative positioning method, this approach can be divided into active positioning, passive positioning, and interactive positioning [35].

However, no research on interactive localization has been reported in the literature of agricultural multi-robots.

- Active positioning

In the absence of information exchange, the robot relies on its sensors to obtain relative distances and angles by observing the neighboring robots for self-positioning [35,36]. As shown in Figure 4, the leading robot guides the follower robot to steer along the leading robot reference route [37].

Figure 4. Active positioning model.

For example, the black-and-white checkerboard feature board was fixed to the leader robot as the following feature, and the 3D information of each corner point on the black-and-white checkerboard feature board was obtained by the binocular vision camera fixed to the following robot, and the information was analyzed to finally obtain the longitudinal spacing, lateral offset and heading declination of the following robot relative to the leader robot. Using this navigation information to realize the automatic following of the following vehicle, the following system of master-slave orchard operation vehicle is established [38].

- Passive positioning

In an environment where information exchange exists, the robot indirectly obtains the relative distance and angle through the "observed" data provided by the friendly neighboring robot to perform its positioning [35]. As shown in Figure 5, the follower robot dynamically creates a reference heading for itself from the position point of the leading robot [39].

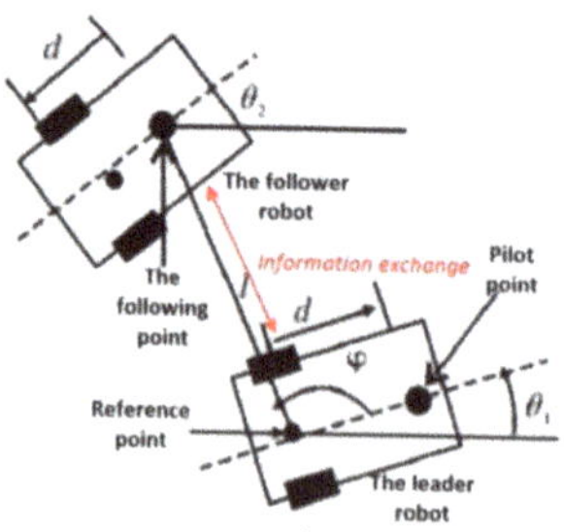

Figure 5. Passive positioning model [39].

Passive positioning mode is usually used in combination with kinematic control models and is one of the most used in master-slave robots. For example, GPS positioning was used for multiple robots, and the travel trajectory of the pilot robot was the main one. Under the premise of communicable, the pilot robot estimated the motion trajectory of other robots using the kinematic model, and the difference between the sensor positioning and the estimated positioning value was calculated by the other robots or the pilot robot, and the lateral and longitudinal displacement of each robot was adjusted according to the difference [40].

However, the above positioning methods are highly dependent on the positioning accuracy of the leader robot, which requires good stability and robustness. When the leader robot fails, it is easy to cause the localization accuracy of the whole team to drop, or even the localization fails. In contrast, the interactive localization approach has an environment of information exchange, where robots achieve joint localization through these steps of mutual observation, data exchange, and information fusion [41]. In this case, the robots are in the same position in the team and there is no master-slave distinction, namely, they do not depend on the positioning accuracy of a fixed robot, and this type of phenomenon is reduced. The above co-location methods are summarized as shown in Table 1.

Table 1. Comparison of co-location methods.

Co-Location Methods	Active Positioning	Passive Positioning	Interactive Positioning [41]
Data measurement object	Robot itself	Neighbors	The robot itself; Neighbors
Observation reference object	Neighbors	Robot itself	The robot itself; Neighbors
Measuring object	Single	Single	Mutual observation
Measurement data	The distance and angle of the robot itself relative to the reference object		
Data exchange	None	Exists	Exists
Data Fusion	None	None	Exists
Advantages	Little computation	Information interaction	Algorithm of high precision and robustness
Disadvantages	Affected by the environment Lower algorithm accuracy and robustness No information interaction	Affected by the environment Lower algorithm accuracy and robustness	A huge amount of information Large complexity of the algorithm
Applications	Drove in orchards [38]	Drove on farmland [42] Drove on sloping land (11°) [43] Plowing [44] Lawn [45] Drove in fields [46] Plowed [47] Harvested farmland [48]	None

It can be seen from Table 1 that the cooperative localization technology for agricultural multi-robots mainly adopts active localization and passive localization, which are computationally small and easy to implement. The real concept of "co-location" should be interactive positioning; through communication among robots, information sharing can be realized, and then the robot's positioning error can be corrected to achieve accurate positioning. However, this method incurs a large amount of calculation and a large complexity of the algorithm, making it difficult to implement. No examples have been found in the literature for agricultural multi-robots.

More practical applications indicate that the cooperative positioning of multiple robots in agriculture is replaced by a central controller or task manager coordination mechanism [49,50]. That is, the working area and travel path of each robot are planned by a task manager or central controller based on the established environment map. The work areas and work paths of these multiple robots usually do not intersect and each robot only needs to localize and navigate based on its sensors. For example [27], the vineyard area was divided into three UAV-monitored vineyard areas using the task manager according to set rules, and path planning was performed for the sub-areas of individual UAV operations, with each UAV flying at a different altitude to avoid collisions between the UAVs.

3.2.2. Data Fusion Technology

A single sensor has certain limitations. For positioning accuracy and reliability, it is necessary to utilize the advantageous features of each sensor, that is, data fusion of multiple sensors. Data fusion technology, also known as multi-sensor information fusion technology (MSIF), essentially involves the comprehensive processing of target information originating from different sensors at different times or multi-target information simultaneously, to obtain more accurate positioning, identification information of the measured environment or object, and comprehensive and timely assessment of the current situation [51,52], thus facilitating the subsequent planning and decision making of the robot. The strength of the data fusion capability directly affects whether the robot can effectively achieve mutual coordination and collaborative work.

The internal sensors in the agricultural robot mainly include a global positioning system (GPS), inertial measurement unit (IMU), steering angle potentiometer, and encoder, which are used to provide the robot with the position, heading, and steering angle information; the external sensors mainly include various LIDAR and cameras, which are used to avoid obstacles and collect environmental information, as shown in Figure 10. Among them, GPS can provide a unified coordinate system and accurate position information for field robots and is used most frequently [53]. With the promotion of satellite positioning technology, agricultural robots equipped with GPS positioning and navigation systems will become increasingly popular. Take the automatic navigation system System150 researched by TOPCON company of Tokyo in Japan as an example [54], the system adopts GPS-based advanced inertial guidance and terrain compensation technology, which can realize navigation in complex terrain environments with ± 2.5 cm accuracy for straight line and turn. However, data fusion techniques are still very important for robot localization when GPS cannot obtain accurate position information in greenhouses or forests [55], or when the robot is too small to install high-precision sensors.

For example, to obtain information on the ambient temperature, humidity, light, and CO_2 concentration in a greenhouse [56], a human remotely operated UAV was first operated to obtain a map of the greenhouse environment, and then the ground robot fused IMU, GPS and odometer information to output the actual location information of the robot through the extended Kalman filter (EKF) algorithm. In practical applications, because of the poor GPS signal in the greenhouse, the EKF was used to fuse the odometer and IMU. EKF is used to linearize the nonlinear system at the reference point using Taylor expansions, and then Kalman filtering theory is used to achieve the prediction and correction of the system. But the EKF still cannot solve the global localization problem [57].

Another example is an agricultural spraying multi-robot [58] that used particle filtering to fuse information from multiple low-cost sensors of the odometer, IMU, wheel encoder sensors, and GPS, which incorporates the open-source library RTKLIB and correction signals. The particle filter could also determine robot attitude based on a series of particles in noisy environments and when the GPS was offline. Particle filtering [59] is a basic method based on Bayesian filtering theory (the robot can determine the poses with a certain degree of confidence based on all available information [60]) and differs from Kalman in using particle sets to describe the probability distribution. However, a large number of particles need to be maintained for higher localization accuracy, which will consume a large number of computational resources, especially as the walking distance gets farther, which will put greater pressure on the computing platform with limited memory resources.

According to the sensor information fusion processing hierarchy, the technique is divided into three levels, namely the data layer, the feature layer, and the decision layer [61]. Their specific scopes, characteristics, and fusion algorithms are shown in Table 2.

From Table 2, it can be seen that information fusion in agricultural multi-robots is focused on multi-sensor information fusion of single robots, and there is almost no research on sensor information fusion between homogeneous or heterogeneous agricultural multi-robots. To obtain more information, the information collected by different types of sensors on a single robot is mostly different (e.g., GPS collects position information, IMU collects robot heading, odometer, etc.), but some information has a mutual transformation relationship (e.g., the integration of velocity from odometer can give distance), which can be regarded as the same kind of sensors at this time. The fusion algorithms are mainly based on classical EKF and particle filtering.

Table 2. Comparison of multi-sensor information fusion processing levels [62,63].

Fusion Level	Data Level	Feature Level	Decision Level
Processing level	Bottom layer	Middle layer	High level
Sensor types	Homogeneous	Heterogeneous	Homogeneous/Heterogeneous
Degree of dependence on the sensor	Maximum	Medium	Minimum
Advantages	Less information loss; High precision;	Takes into account the advantages of data and decision; Fusion is more convenient; Wide range of applications;	Less calculation; Good fault tolerance; Good anti-interference; Low algorithm difficulty; A small amount of communication data; Good real-time performance;
Disadvantages	A large amount of calculation; Poor fault tolerance; Poor anti-interference; The algorithm is complex; A large amount of communication data; Poor real-time performance;	Global decision; Strict requirements for sensor pretreatment;	A large amount of information loss; Low precision; Need to balance constraints (speed, reliability, cost, etc.);
Fusion algorithm	Wavelet transform; Artificial neural networks; Weighted average; Production rules, etc.	Kalman filter; Extended Kalman filter, etc;	Bayesian theory (Particle filter); The statistical decision, etc.;
Applications	None	Drove in a greenhouse [56]	Seeded fields [58]

3.2.3. Mapping

Once the multi-robot has determined its position, it also needs to determine information about the multiple robots' surrounding environment, for instance, the presence of obstacles. Mapping is the task of accurately describing the spatial position of the robot working environment and various objects (such as obstacles and road signs) in the environment, that is, to establish a spatial model (two-dimensional or three-dimensional) or map [26]. The purpose of creating this map is to provide path planning for the robot, so the map must be easy for the robot to understand and computationally manipulate and accommodate revision when new environmental information is detected [64].

At present, many methods have been developed for constructing environmental models for multiple robots, which are mainly summarized in the three types of grid-based model, geometric mode, and topological mode [22]. Probably due to the low environmental information of the topological model, no literature was found on the use of multiple robots in agriculture. On the contrary, the grid model and the geometric model provide abundant information about the agricultural environment, the purpose of multi-robot operations is clear, and more literature is applied to agricultural operations. And the detailed environment model facilitates the task allocation to multiple robots, real-time observation of multiple robots' motion, effective coordination mechanisms, and detection of robot motion faults.

- Grid model

The uniformly distributed grid represents the working environment of the robot, and each grid corresponds to the position of whether an obstacle exists or not [22]. For example, if a grid has an obstacle, it is marked as 1; otherwise, it is 0.

As shown in Figure 6, Bouzouita et al. [28] obtained georeferenced pictures through UAVs and then created a full map of a vineyard. The processing entails first dividing the UAV workspace into multiple regular grids and defined borders (line segments of workspace in the grid-based environment) using Bressenham's linear algorithm (BLA). Then, the procedure to distinguish the subareas in the grid-based environment is a recursive flood-fill algorithm that picks an empty cell (not marked as occupied) and floods in four directions while there are empty cells, and each flooded cell is marked as occupied. This mapping procedure is repeated until all nodes of the grid are occupied by georeferenced pictures. However, the drawback of this mapping is a tradeoff between acquiring an equally sized set of images in cells and inefficient aerial sampling with UAVs.

- Geometric model

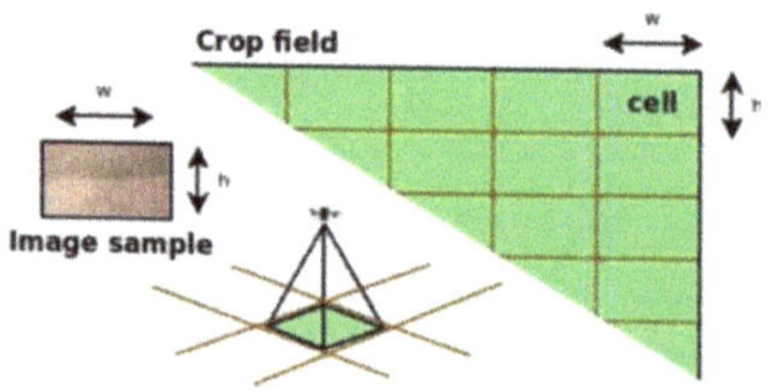

Figure 6. Grid model with rasterized areas [28].

Geometric features such as line segments or curves are extracted from the environmental perception information and used by robots to construct an environmental map [22]. As shown in Figure 7, Ball et al. [65] defined the boundary of a 55-hectare sorghum stubble field by manually selecting appropriate latitude and longitude coordinates but omitted the headland where the UGVs turned around at the end of each row. Besides, the positions of multiple robots and obstacle information (human, pole, vehicle, etc.) were periodically sent to the central controller through sensors of a real robot and 12 simulated robots. Finally, the central controller constructed a 2D map of the environment surrounding the robot.

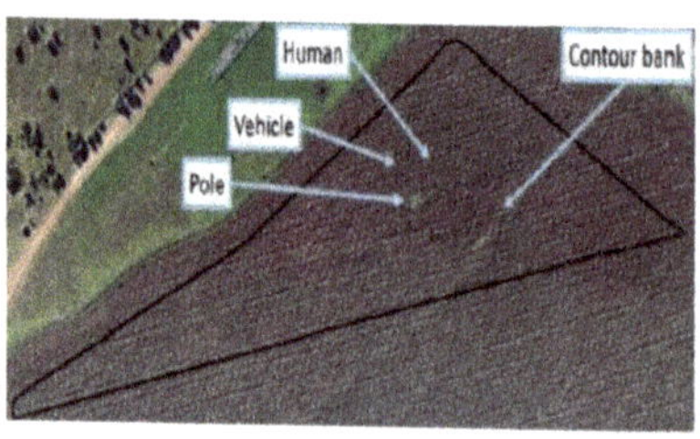

Figure 7. Geometric models were constructed with points and lines [65].

A more scientific approach than manually labeling is the Voronoi diagram [66]. For example, a fruit tree on the robot's driving path was taken as a point in the orchard map, and the two-dimensional coordinates of this point were found in the image acquired by the UAV. At the same time, the data of planting interval and density of fruit trees were learned from this image, and the two-dimensional coordinates of other fruit trees were obtained, and then the K-means clustering algorithm was used to find the center of clusters of multiple fruit trees, and the orchard was divided into multiple sub-regions for robot operation according to the principle that the fruit tree point is most adjacent to this center.

But the experimental results after mapping are to be further validated on a real agricultural spraying multi-robot.

- Topological model

The topological relationship among objects in the environment is represented by the environment model [21]. The topological model consists of multiple unrelated points and lines with a simple model. Some researchers proposed to use this model in agricultural multi-robot path planning, but it is only an idea and has not been applied [67].

The specific application scope and characteristics of the multi-robot environmental model are shown in Table 3.

Table 3. Comparison of mapping methods [68–70].

Mapping Model	Grid Model	Geometric Feature Model	Topological Model
Algorithm	Bressenham's linear algorithm, Bayesian; D-S theory, etc.	Voronoi Diagram; Extended Kalman filter; Particle filter, etc;	Baum-Welch, etc.
Scope of application	Suitable for processing systems such as laser radar and sonar priority ranging;	Local area modeling; Specific environment;	Large-scale, simple environment;
Advantages	Simple and intuitive, easy to create and maintain;	More compactness;	Model is simple and easy to change, and motion planning is fast;
	Good spatial consistency;	Modeling in local areas can achieve higher accuracy;	Better robustness against position errors;
	Better robustness;	Conducive to the estimation of the position of the robot and the identification of the target;	A high degree of abstraction, easy to store;
Disadvantages	A large amount of calculation;	Difficult to address and extract irregular geometric features;	Less information;
	Takes up substantial storage space; Long storage time;	Image data are processed cumulatively;	The topology network establishment process is complicated;
Application	Harvested muddy moss in farmland [13]	Plowed [47]	None
	Monitored vineyards [28] Environmental monitored [49,50]	Drove in a greenhouse [56] Harvested farmland [48]	
	Collected farmland information [71]	Precision irrigation in the vineyards [72] Sprayed in the Orchard [66]	

Compared with 2D maps [14,73–75], 3D maps can provide more information. By fusing the original measurement or small local maps generated from multiple robots to construct global maps at the same time, or matching the 3D maps constructed by heterogeneous robots, more and more abundant data are obtained, which is a new research direction of mapping.

For example [71], the point clouds of the two images acquired by the UAVs and UGVs respectively were first represented by a grid model, and each cell stored excess vegetation index information and ground height information. After that, the relative displacement and rotation parameters between the two grid maps were extracted using the data provided by GPS and Attitude and Heading Reference Systems (AHRS), and the maps were matched by these parameters. Then the correspondence of point clouds between the matched maps was calculated using the Large Displacement Dense Optical Flow (LDOF) system and a voting scheme was used to select a coherent correspondence dataset with high data similarity to infer the initial conversion relationship between the maps, and a non-rigid alignment algorithm was used to eliminate orientation errors. Finally, the input point clouds were brought in for a point-to-point alignment, and non-vegetation points were removed to obtain the 3D maps. The experimental results showed that the constructed maps matched well, but the time taken to construct each such image was quite long.

As the farming environment is often dynamic, learning specific farming information while shortening the time to build maps will help improve the efficiency of multi-robot operations, which should be a direction for further research. For example, simultaneous localization and composition (SLAM) allows the robot to get information about the image while localizing it, and this technique is currently used more often for indoor multi-robots, especially for catastrophic search and rescue where high responsiveness and mobility are required [76]. In the literature [56] it was mentioned that this technique was used on a ground robot for heterogeneous multi-robot in a greenhouse and that the robot was motion controlled using an Augmented Monte Carlo Localization (AMCL) methods. The experimental results showed that the UGV using SLAM takes less time to construct the map than the remotely operated UAV. And based on the map, the ground robot can complete collision-free autonomous movement between the initial and target points. It is obvious that SLAM technology has great potential for multi-robot mapping [77], and the application of multi-SLAM on agricultural multi-robots will be one of the future research directions.

3.3. Multi-Robot Task Allocation

Multi-robot task allocation (MRTA) provides evaluation indicators of a multi-robot system, a task set, and system performance and finds a suitable robot for each subtask to perform, bringing the benefit of a robot system to perform mostly task collection. Thus, the quality of the MRTA results directly affects the efficiency of the entire system and whether each robot of the system can maximize its capabilities [78].

Solve the MRTA problem [79] involves many aspects, such as the capability attributes of the system members, the structural attributes of the tasks, the robot coordination mechanism, and the strategy of task allocation [80]. This approach divides the assignment of agricultural multi-robots into centralized components based on the decision of the central controller in this part.

- Centralized Allocation

The centralized allocation means that the leading robot or control center of the system decomposes the global task and then sends the decomposed subtask to each robot according to the corresponding allocation method.

In 2012, reference [14] reported the work area division and task assignment of two UGVs by a person through remote monitoring based on a citrus orchard map. However, as the number of robots increased, manual task assignment stops when a robot hits a tree or fails to make a reliable turn. Some scholars [66] solved the manual assignment of tasks to multiple robots from the perspective of dividing maps, namely, as many sub-regions as there are robots needed. The edges of these sub-regions are generated from discrete fruit tree points and K-mean clustering points on the map boundary using an integer programming approach. However, this view is too ideal and does not consider the case where the number of robots is less or more than the total number of tasks.

A similar study on the number of robots over the number of tasks is multiple robots and a small number of refueling stations [81]. In this reference, an approximate arbitrary time algorithm based on the branch delimitation method was used to obtain the task sequence of multi-robot collaborative refueling. The limit value of the path length distance was first calculated and used as the upper and lower bounds of the algorithm nodes based on the rules for robots walking infield and the total time cost function of refueling and waiting in the queue. Then the optimal solution was obtained by deleting the sub-nodes whose lower limits were greater than or equal to the optimal upper limits during iteration. Simulation experiments showed that the optimal approximate solution on resource utilization can be found by this method, but it is difficult to apply the method to other aspects of agriculture, such as spraying.

A reinforcement learning-based method (Dyna-Q +) was used to find the optimal search path from the current point to the endpoint, that is, multiple robots randomly selected actions (front, back, left, right) and each action was recorded, and the optimal path was obtained by rewarding and punishing the actions selected by the robots according to the presence or absence of obstacles [62]. Then a weighted graph was used to represent the Gird model, including parameters such as the current position of the robot, the set of grids, and whether there is a path between the grids. The minimum cost time for the UAVs to fly at different speeds under the optimal path was calculated using Dijkstra's algorithm, and finally, the search space was obtained based on the optimal path and time and allocated to multiple robots in proportion to the size of the space. A reinforcement learning [82] is a Markovian decision process where the basic idea is all about modeling or fitting a strategy using a function for more complex decision problems. However, the method requires a large number of samples and a long time when used.

Also, multiple robots working in an agricultural environment are often subject to resource-sharing conflicts. For predictable conflicts, relying only on a central task allocation approach to avoid conflicts, the adaptability of multiple robots is very limited, while adding a decision support system (DDS) to provide options for multi-robot collaboration, i.e., identifying problems and building or modifying decision models to avoid resource conflicts based on explicit goals. For example, as shown in Figure 8, in reference [58], the authors adopted a central entity (OptiVisor) to build a multi-robot seeding map based on inputs such as the location of static obstacles in a large field, the seeding method, the seed density, the location of the central controller, and the number of robots. Based on this map, the location and density of each seed are precisely located and the sowing task is assigned to multiple robots. When the robot finished the job, the path from the robot location to the Central Logistic Unit (CLU) was recorded and the robot was allowed to return to the CLU. Especially when one or more robots failed, the task of the failed robot was reassigned to the other robots of the failed robot, and the sowing path was updated for these replacement robots. OptiVisor could also stop a robot's motion when a multi-robot collision is imminent, and define restricted motion areas for the faulty robot. However, this task assignment method was implemented in a simulation environment and needs to be further tested in real applications.

- Distributed allocation

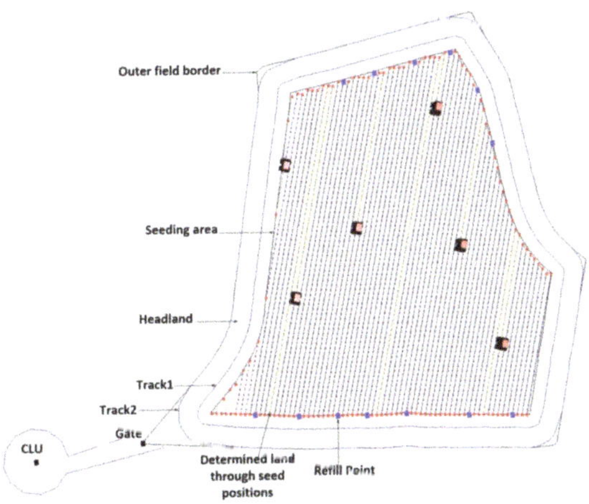

Figure 8. A centralized entity was used to plan the seeding task, seeding patterns, and seed densities for different robots in the simulation environment. At the same time, the centralized entities were used to monitor the whole seeding process to avoid collisions between robots [58]. The small blue dots are the locations of the seeds planned to be sown, the yellow dots are the seeds that have been successfully planted, and the larger red dots are the border seed sowing locations.

In general, the distributed task allocation method has the feature that each robot does not have global knowledge of the task but calculates and plans individually according to the local information obtained by the sensor. The performance of the whole system not only is closely related to the individual but also depends on the combined effects of individuals [83].

It was reported in the literature [84] that when the number of tasks is more than the number of farm machines, the ant colony algorithm can be used to find a suitable task sequence for multiple robots. First, the distance between plots was calculated based on the specified coordinate information of each plot, and the relevant parameters of the ant colony algorithm were set. Then randomly generated the starting points of multiple parcels, according to the state transfer probability formula for path selection, and put the generated path parcels into the forbidden table. Finally, the path distance of the plots was calculated, and the pheromones on the path were continuously updated according to the set rules, and iterations were repeated until the optimal task planning for multiple robots was found. The method is mainly used in simulated farmland environments and has not yet seen the practical application.

In reference [85], researchers adopted Semantic MozartSpaces to describe a task allocation data model based on a resource description framework (RDF) and SPARQL (a query language and data acquisition protocol developed by the RDF) in task storage where the RDF was used to construct nested blank nodes and SPARQL was used for querying, updating and interactions of the entry. As shown in Figure 9a, a task was mapped to a nested blank node to generate a semantic tuple (entry) in the task-allocation model. The entry was stored in the task storage with an internal ID that concluded the URL (uniform resource locator). Then, the entry could be selected with a URL according to the relationship between the robot's function and the task requirement. The results of the simulation experiment suggested that the execution time increases correspondingly with an increasing number of tasks, followed by a gradual decrease in production efficiency. It would be necessary to add new robots temporarily to ensure productivity, but the production cost would also increase, so the tasks need to be set in advance.

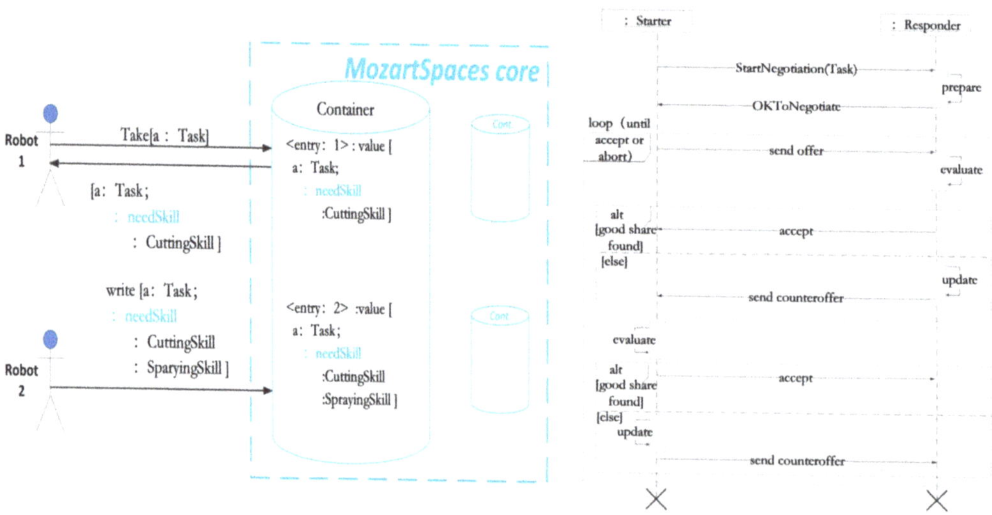

Figure 9. Distributed task allocations: (**a**) An entry was stored in task storage with ID numbers. And robots could select ID to work according to the relationship between the robot's function and the task requirement [85]; (**b**) A task was split into several subtasks. And each robot proposed the largest possible task allocation initially, then gradually decreased its offers based on negotiation mechanism until a deal was done [28].

As shown in Figure 9b, researchers developed market and auction-based approaches for task subdivision and allocation based on the Rubinstein negotiation protocol [28]. The advantage of such a protocol was that the auctioneer robot had to split the task into subtasks during the negotiation. In each round of negotiation, each robot initially started proposing the largest possible task allocation for itself and decreased its offers based on a negotiation of cost functions (discount factors) at each round until the other party indicated acceptance. With distributed task assignment under this protocol, each robot comes with a DDS, and each robot can dynamically adjust to the actual situation to get a suitable task. However, the task allocation result was generally an approximate optimal solution because of the discount factors, such as the area of the mission area, the distance of the robot from the goal, the area beyond the target, and the overlapping work area.

Combined with the above task allocation technology, the research progress of agricultural multi-robots in task allocation technology in the past 10 years is summarized in Table 4.

Table 4. Task allocation classification comparison of agricultural multi-robots [86,87].

Classification	Centralized Distribution	Distributed Distribution
Advantages	Algorithm design is simpler; It has the potential to generate globally optimal solution;	Communication is more dispersed, which can avoid the situation that affects the efficiency of the algorithm due to communication congestion; Good real-time performance;
Disadvantages	The communication is concentrated and susceptible to congestion, affecting the algorithm efficiency; Poor real-time performance;	The solution result can easily fall into a local optimum; Consultation increases the communication burden of the system;
Scope of the application	The task is known and determined in a static environment; The environment model is fully known; The scale of this system is relatively small;	The task is unknown and uncertainly determined in the dynamic environment; The environmental model is partially known or unknown; The scale of this system is medium or large
Allocation method	Mixed integer programming (branch-and-bound); Heuristic search algorithm (reinforcement learning);	A method based on behavioral motivation; A market-based approach; Group intelligence algorithm (greedy algorithm, ant colony algorithm [84]);
Application	Sprayed and weeding in an orchard (manned) [14]; Seeded in a field [58]; Sprayed in the Orchard [66] Refill scheduling [78]	Weeded farmland [85]; Agricultural monitored in vineyards [28]; Harvested fields (simulation) [88];

It can be seen from Table 5 that the current task allocation methods for agricultural multi-robots are mainly centralized, and most of them are implemented in the simulation experiment.

The centralized task allocation mainly adopts the integer programming method, which describes the nature of the task allocation problem by establishing the objective function and constraints. Integral programming (IP) and mixed-integer programming (MIP) problems are an important branch in the field of operations research, which includes branch and bound method [89], cutting plane algorithm [90], graph theory method, etc [91]. The idea is to determine the transfer method and transfer relationship from one search point to another, and the result is a unique optimal solution, which is suitable for small scale; when the scale is extended, the computation is considerable and the computation time will be greatly increased [92]. Thus, the computation of the algorithm grows exponentially with increases in the number of tasks and robots. In general, it is often difficult to meet real-time requirements in task allocation issues.

Distributed task allocation mainly uses a method based on behavioral motivation and a market-based approach. The former applies to multi-robot systems with strong autonomy, but the method has low system allocation efficiency; the latter has a wide range of applications based on the resource optimization configuration idea of economics, but the difficulty of this research method is how to design the negotiation mechanism and reasonably determine the cost-income models of the task [44].

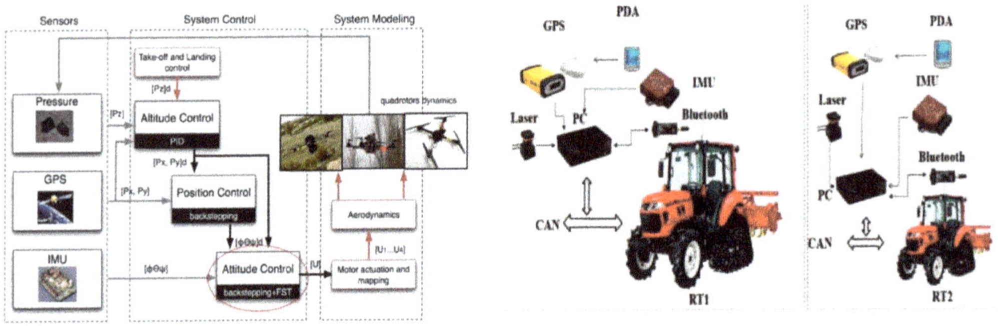

(a) (b)

Figure 10. Heterogeneous sensors carried by agricultural multi-robots. (**a**) Typical sensors for aerial robots: inertial measurement unit (IMU), Global Positioning System (GPS), and pressure gauge [28]. (**b**) Typical sensor for ground robots RT1 and RT2: GPS, Laser, and IMU [45].

3.4. Path Planning

Path planning is the fundamental guarantee for multiple robots to accomplish tasks together. This technology refers to using the known static environment information, or the dynamic environment information obtained by the sensor, to autonomously plan a collision-free optimal path for each robot from a known starting point to a target point, which requires not only a single robot to avoid obstacle but also a plan to satisfy collision avoidance among multiple robots [93]. The path planning methods for single robots to avoid obstacles mainly include traditional methods, intelligent methods, and other methods. The traditional methods include the construction space method, V-Graphic (visibility graph), Voronoi diagram, grid method, A* algorithm, and artificial potential field [94]. Intelligent methods include the ant colony algorithm, particle swarm algorithm (PSO), reinforcement learning algorithm, immune algorithm, genetic algorithm (GA), neural network, and fuzzy logic algorithm [95]. Other methods include dynamic programming (DP) and optimal control algorithms. Collision avoidance strategies among multiple robots include the priority method, rate adjustment method, traffic management rule, and consultation method.

The path planning method can be divided into centralized path planning and distributed path planning according to the ability classification of path planning [96].

- Centralized path planning

The centralized path planning method uses a centralized control unit to plan the optimal path for swarm robots. This method can improve the ability of "close coordination and optimal coordination" among mobile robots [97]. However, it encounters other problems such as "dimensions", "computational complexity", and "non-deterministic polynomial problem (NP) difficulties" with the increase in the number of robots, task difficulty, and space complexity. In particular, the "NP difficulties" problem [98,99], in theory, has not yet received a simple or fast solution.

Connecting known path points into a line, which is simulated to a topological model, and letting multiple robots walk along their respective paths is one of the simplest and easiest path planning methods to implement. But, this method requires obtaining accurate known point information in advance, which is a large preliminary job and is not suitable for situations with a large number of robots or a large operating area. A method similar to the point-to-point method is the visible map method, which aims to reduce collisions. That is, the information of the edge projection points of each obstacle is obtained in advance, and the robot free walking path, path points that can be combined or disconnected are represented by edges and nodes, respectively. Then the starting node is connected to the target node, or the starting node is connected to the raised point of each obstacle edge until it reaches the target node, forming a multi-robot walking path. Finally, depending on the size of the robot, the path width is increased or decreased appropriately. Similarly, the workload of measuring points is larger and does not apply when there are more obstacles.

Planning multi-robot paths on mapping is another method of global path planning. For example, in the grid method, the map was divided into multiple grid cells, and the paths were extended in eight directions with each grid cell as the center, and the path segments were formed by connecting the center of the grid vertically and diagonally with the centers of other grid cells. To get the globally optimal path in the grid, the A* algorithm was used to search for the path segment with the lowest travel cost, in which the cost of the free space cell was set to 0, the cost of the cell with obstacles was set to the maximum, and the travel cost was the sum of all grid cells on the travel path segment, and the globally optimal path was set when the sum was the smallest.

Since information about the farmland changes dynamically, multiple robots operating with precision need to re-plan to create multiple paths each time based on different information. This multi-robot path planning problem with time windows has also been solved as a multi-objective optimization problem. As shown in Figure 11, where a multi-robot system including two aerial robots and three ground robots was jointly developed in Spain and other countries [49]. This system adopted a centralized control unit to provide global path planning for the multiple robots on a grid map with weed information, which divides the sequence of operations for the multiple robots in advance. The No dominated Sorting Genetic Algorithm II (NSGA-II) algorithm [100,101] was then used to coordinate the relationship between the distance of the robot travel path, the number of turns, the number of robots, the amount of weed killer used, and the capacity function to obtain an approximate optimal solution between the time and money spent by the multiple robots and the cost of weed treatment. It is a type of genetic algorithm, which mainly focuses on the simulation of crossover, variation, and hereditary phenomena occurring during natural selection and genetic inheritance, incorporating the natural law of superiority and inferiority, and deriving the candidate solutions for each generation based on the results, and finally deriving the optimal solution from the derived candidate solutions. However, this method is more computationally intensive and the experimental results are not suitable for fields that are unstructured or d fields without a fixed column or row lengths.

- Distributed path planning

The distributed path planning method requires little calculation and is robust but exhibits low efficiency and can provide only a suboptimal solution [102,103]. In a fully known environment, it is necessary to consider each robot obstacle avoidance method and collision avoidance strategy among robots [104], that is, selecting a robot for path planning first, then broadcasting its path to other robots, and finally planning paths of other robots by themselves. However, this method is difficult to achieve [97] for large numbers of robots. In an unknown environment, the preferred method is to plan a path for each robot to avoid static obstacles based on neglecting the movement of other mobile robots in the environment and then using the multi-robot collision avoidance strategy to solve the conflict problem among mobile robots.

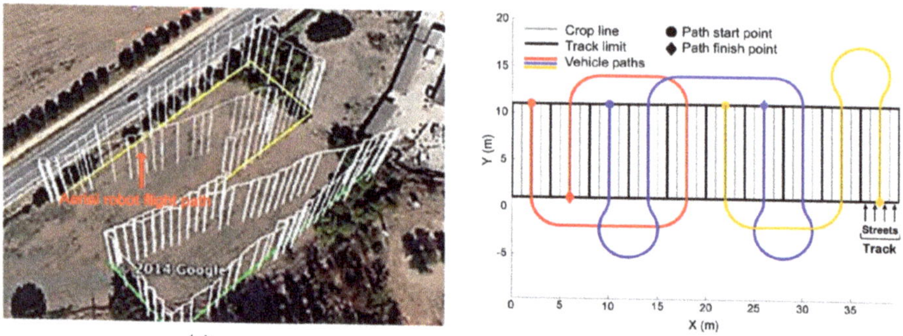

(a) (b)

Figure 11. The central controller plans the driving path for the unmanned aerial vehicles (UAVs) and unmanned ground vehicles (UGVs) [49]: (**a**) The path planning of two UAVs; (**b**) The path planning of three UGVs.

Table 5. Research progress of path planning method for agricultural multi-robots.

Planning Method	Environmental Circumstances	Head-Turning Mode	Coordination Strategy	Obstacle	Application
Point-to-point planner	Fully known	Right-angle turn;	None	None	Drove in farmland [42] Plowed [105,106] Seeded in farmland (simulation) [107]
Visibility graph (V-Graphic)	Fully known	Zigzag; U-shape; Right-angle turn	Priority; Rate adjustment;	Vehicle	Harvested muddy moss in the farmland [13]; Bee pollination (simulation) [108];
Grid method	Partially known Fully known	Zigzag; U-shape;	Others Rate adjustment	Pole Human Vehicle	Seeded in a field [58] Harvested muddy moss in the farmland [13];
A * algorithm [1]	Partially known	None	Others	None	Monitored in a vineyard [28]
Ballistic method	Unknown	None	Others	None	Weeded in a rice field [109]
Breadth-first search algorithm (BFS)	Partially known	None	Others	None	Monitored in a vineyard [28]
Genetic algorithm (GA)	Partially known	U-shape The light bulb shape	Rate adjustment	None	Weeded in farmland [101]
Others	Fully known	Right-angle turn;	Rate adjustment	None	Drove in a greenhouse [56]

[1] The A * algorithm is a direct search method for solving the shortest path in a road network.

At present, there is little research on the distributed path planning of agricultural multi-robots. As shown in Figure 12, Bouzouita et al. [28] developed UAVs to monitor agricultural information in the vineyard. The path planning of the UAVs is based on the grid map and A*. On the map, the UAV path planning function is constructed according to constraint conditions, such as the number of UAV turns, the number of covered grid visits, and the time to complete the single partition. And using a heuristic algorithm like A*, the next best node to be expanded is obtained by partying the generation value of each node. Then, the breadth-first search (BFS) algorithm is used to find the local maximum of the function by the distance between the cells. The result is the path from any starting point in the environment to the target unit. The practice shows that the method can find the approximate optimal solution, reduce the possibility of repeated access to the same cell, and facilitate the avoidance of known obstacles. However, it needs to consider the local environmental conditions to find wide applicability.

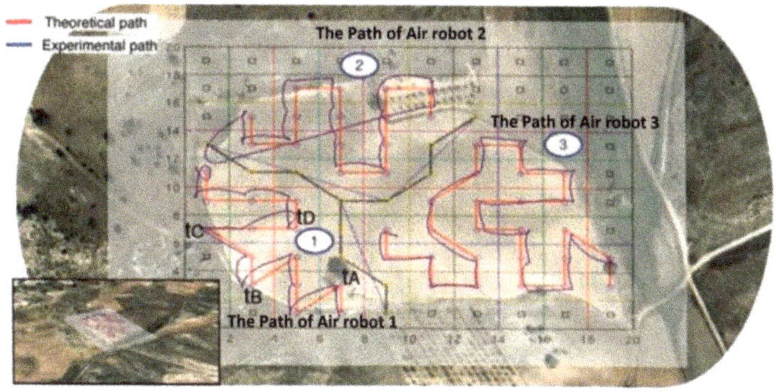

Figure 12. Distributed path planning [28], 1, 2, and 3 are the number of the UAVs, the red line is the planned path of the robot in the delimited area, and the blue line is the actual flight path of the robot.

Combining the above path planning techniques, we summarize the research progress of agricultural multi-robots in path planning methods over the past 10 years, as shown in Table 5.

It can be seen from Table 5 that agricultural mobile robots mainly use ground robots in farmland, and the number of robots is no more than four. The path planning method of multi-robots is mainly conducted in fully known conditions, and the grid method and V-Graphic under centralized planning are the methods most commonly used.

The collision avoidance strategy between agriculture robots usually does not incorporate changes in the path [110], and the obstacle avoidance strategy of a single robot comprises mainly speed adjustment and the priority principle. The path planning of multi-robot generally does not consider the presence of obstacles in agricultural production.

When the UGVs turn around at the headland, they need to consider the relationship between the minimum turning radius and the headspace. The general head-turning method has a bulb shape (as shown in Figure 11b), zigzag or U-shape, etc. The U-shape predominates in practice. As shown in Figure 13, the zigzag (forward-reverse-forward) is used in smaller spaces; in contrast, the U-shape turn (turn-straight- turn) is used in larger spaces. However, the difficulty of robot control is increased with the zigzag turning shape, and the task allocation and path planning of the multi-robot are prepared for U-shape turning in advance.

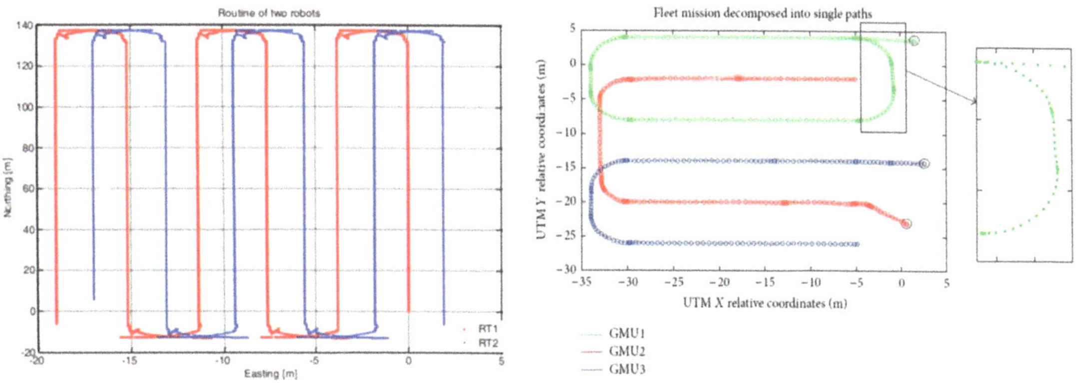

(a) (b)

Figure 13. Agricultural ground multiple robot head-turning mode: (**a**) Multi-robot turned on the ground with a zigzag way, that is, first forward, then backward, then forward again [47]. (**b**) Multi-robot turned on the ground with a U-shape (GMU is the abbreviation for Ground Mobile Unit) [25].

3.5. Formation Control

Multi-robot formation control technology means that multiple robots maintain a certain formation to the target while still adapting to environmental constraints such as obstacle blocking or spatial physical limitations [111]. This technology can improve the robustness of the multi-robot system, and the robots can complete the task with higher efficiency and shorter time [112]. At present, the formation control technology of agricultural multi-robots is divided into forming formation and formation control.

- Formation forming

As shown in Figure 14, the formation of agricultural multi-robots generally has five types: column, I-shape, linear, V-shape, and W-shape, and the circular nodes in the formation structure represent robots. Each robot is represented by RID, such as R1 and R2, and the black arrow indicates the direction of robot movement.

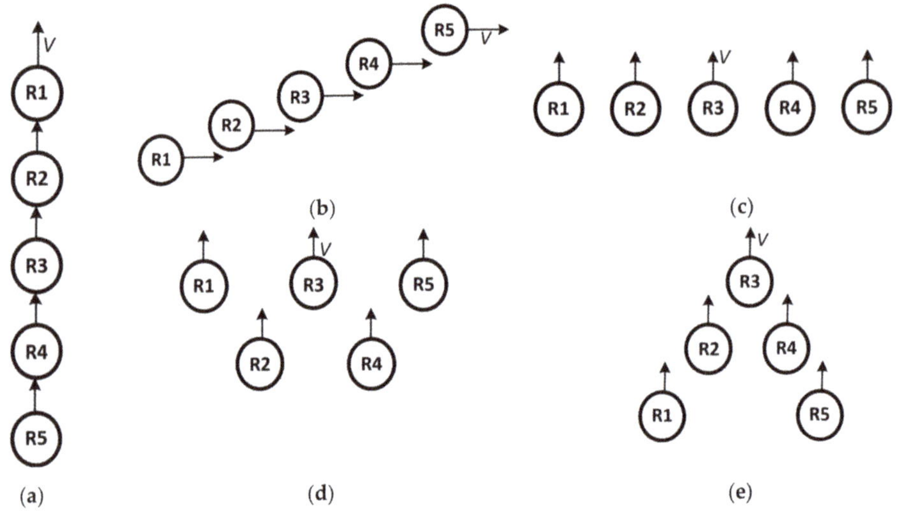

Figure 14. Multi-robot team arrangement, the arrow points to the direction of robot movement, and the circle represents the robot: (**a**) The robots of R1 to R5 formed a longitudinal linear queue. (**b**) The robots of R1 to R5 formed an I-shape queue. (**c**) The robots of R1 to R5 formed a transverse linear queue. (**d**) The robots of R1 to R5 formed a W-shape queue (**e**) The robots of R1 to R5 formed a V-shape queue.

It is necessary to achieve the desired formation by determining the formation position reference point after determining the root formation. There are usually three reference points: center, neighbors, and leading robot; as shown in Figure 15, the position of each root node is represented by PID, such as P1 and P2, and the arrows indicate the relationship between robot dependence and information transfer.

- Formation control

From the perspective of a multi-robot system control framework, formation control is divided into two types: centralized control and distributed control. The former uses a centralized control unit to make decisions, optimize robot coordination, accommodate individual robot failures, and supervise the entire group of robots. The latter does not have a unified control unit, and a single robot makes decisions based on its local information [112,113].

At present, the method of centralized formation control of agricultural multi-robots includes the virtual structure, graph-theoretic approach [114], and model predictive control [115,116]. The method of distributing the formation control of agricultural multi-robots includes leader-follower [34,41,75,117] and the artificial potential field [113].

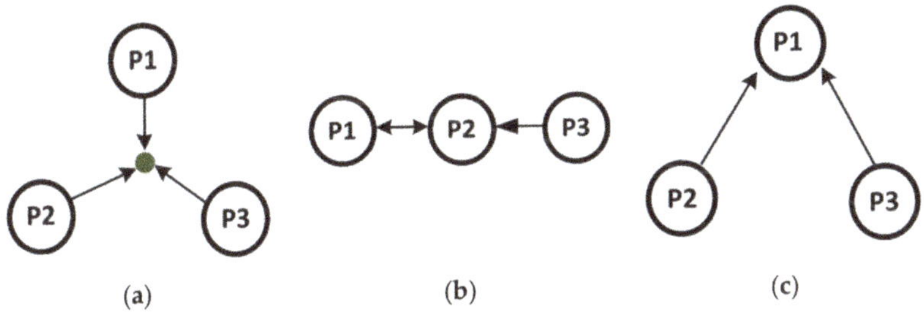

Figure 15. Selection of formation reference points: (**a**) Take the green dot, in the center of P1 P2 P3, as a reference. (**b**) P1 and P3 take neighbor P2 as a reference, P2 takes the nearest neighbor P1 as a reference. (**c**) Take the leading robot P1 of P2 and P3 as a reference.

Guillet et al. in France [44] adopted the bidirectional control strategy based on the virtual structure method. As shown in Figure 16a, each robot of the whole queue is a fixed point on the virtual structure. In this structure, the queue also increases two virtual leaders' interaction with the extreme robots and carries the desired global velocity for the whole fleet. The advantage of this method is simpler communication protocols and lower communication costs; however, the reaction of the robots is slower because of different acceleration performances.

Figure 16. Formations of agricultural multi-robots: (**a**) the head and tail robots in the formation are used to guide UGVs in I-shape operation. And the robot in the middle of the formation plows with farming tools in the field [44]. (**b**) Leader-follower method to control the formation in V-shape operation [105]

Berman et al. in the USA adopted the graph theory approach in bee pollination [108]. When a beehive was opened, the swarm robot flew radially from a moving beacon at a constant speed. And once it encountered the edge of the graph, it flew eastward at a fixed speed. As the robot flies over the plant, it acquires at least one flower within its range through sensors and hovers over the flower with unit time probability to pollinate it and record the location of the pollinated flower, returning to the hive after pollination and starting the next flight until complete coverage of the whole field is achieved. However, this method takes a long time and the model used in the simulation is too idealized. Whether they can be used for practical production needs to be further explored.

Smith et al. in Korea adopted model predictive control (MPC) and nonlinear feedback control respectively in fish tracking (simulation) [117]. MPC is a finite-domain rolling optimal control strategy with three parts: model, prediction, and decision, sacrificing optimality to some extent [118]. The fish population location was first divided into discrete points, and the discrete points were clustered to get the vertices of fish population density, and the transition model was constructed by transforming the movement of the fish population into the movement between the vertices. The transition model and nonlinear feedback were used to obtain the transition matrix, and the underwater robot was guided to the vertices with high fish population density according to the transition matrix. The simulation results showed that the model-based control of the underwater robot could approach the nearest point, while the feedback control made the underwater robot approach the target point. However, in practice, the underwater robot movement speed is smaller than the fish population movement speed, and the method needs further improvement when applied in practice. The leader-follower method [48,105,117,119,120] is also another classic model and widely used in the formation control of agricultural multi-robots. Japan's Zhang et al. [106] used the leader-follower method to control UGV formation. As shown in Figure 16b, the relative positional relationship between the leader and the follower is determined according to the lateral and longitudinal safety distances (l−l) between the robots first, and then the distances are dynamically adjusted with feedback linearization technology to assemble different formations. Based on the leader-follower model, Bai et al. in China also combined slide mode control with the harvester swarm [48]. The kinematic model of the farmland leader-follower harvester swarm was established first, and based on this model; the asymptotically stable path-following control law and the formation-keeping control law were designed by combining feedback linearization and sliding-mode control theory. The advantage of this leader-follower model is that the behavior of the fleets can be controlled through the determined trajectory of the leading robot. The method decouples the cooperative navigation control problem into lateral distance keeping control and longitudinal distance keeping control. The formation control is mainly accomplished by establishing the location and gesture of the following robot relative to the leading robot, such as $(l-\varphi)$, $(l-l)$ first, then obtaining the formation information through feedback linearization, and finally adjusting the formation according to the threshold value. The leader-follower test results show that the real paths of robots can achieve centimeter-level average error with the planned path based on the safe distance of the vehicle. But this method is only applicable for environments involving a single-tasking of agricultural production and a fixed site. The adaptability to the headland turns is not strong. The question of how to maintain robot formation in encountering static or dynamic obstacles is not considered. If the leading robot malfunctions, the formation of the fleets cannot be maintained. Once the leading robot fails, the multi-robot system is susceptible to deadlock, and the formation cannot be maintained. The "leader" replacement method was proposed [121] to overcome this shortcoming, but the method has not been applied to agricultural multi-robots.

Ju and Son in Korea adopted Ramadge-Wonham theory in supervisory control to solve the above deadlock problem [122]. Supervisory control is a feedback control theory for discrete-event systems, where the control goal is achieved by observing the occurrence of events or states and using allowable or prohibited controllable events. Finally, a time-driven system is combined with a low-level controller and an event-driven system with a high-level controller with the criterion of satisfying the behavior specification and maximizing the allowable events. Time-driven is used when there is no fault, and once the queue encounters a fault, the control outcome is selected based on event-driven. Simulation results demonstrate that the method can be used to control complex dynamic systems, but it has not been tested in practical applications.

The characteristics and formation process of the formation control methods are shown in Table 6.

From Table 6, it can be found that more complex or hybrid control methods are mostly used in simple or simulation environments, and the application in actual agricultural production is still dominated by the leader-follower method, and the research is also mainly focused on multiple machines traveling in a straight line in a fixed column. Further research should be conducted on how to continue driving, maintain the formation, or adjust the formation after multiple robots encounter obstacles.

Table 6. Comparison of formation control methods [34,41,75,86,104,106,109–111,113,114,117,123,124].

Formation Control Method	Advantages	Disadvantages	Steps	Application
Bidirectional Weighted Constraints Approach	Simplified description and assignment of tasks; High formation control accuracy; No complicated communication protocols, low communication costs;	Difficult to handle fault tolerance; Poor reliability; Low mobility limits range of motion;	Defining the dynamic model of the virtual structure; Transforming the overall movement of the structure into the motion of the robot; Obtain the tracking control law of each robot;	Plowed [44];
Graph theory approach	Suitable for large-scale robot formation; Easy addition and deletion of robot nodes; Easy to change between different formations;	The implementation is more complicated, mainly limited to simulation research;	Define the formation right map G = (set of points V, edge E, weight W); Specify the ideal distance of each edge in the formation;	Bee pollination (simulation) [108]; Seeded in a field (12 simulation robots) [107]
Artificial potential field	Little calculation; Easy to implement real-time control; Easy to handle collision avoidance problems in obstacle spaces;	The design of the potential field function is difficult; Problem with local extreme points;	Design artificial potential field (environment and constraints between robots in the formation); Establishing a potential field function;	Precision irrigation in the vineyards [72]
Model predictive control (MPC)	Has a strong theoretical foundation; Adds multiple constraints in the control process, and optimizes the control sequence by online scrolling optimization combined with feedback correction of real-time information;	A large amount of calculation, mostly used for simulation;	Real-time planning formation retains the reference center and target control amount of each robot; Building a linear programming model with multiple constraints;	Tracked fish (simulation) [117] Harvested farmland [48];
Leader-follower	Simplified system control;	The ability to adapt to a dynamic environment is not strong. If the leader fails, the entire system crashes;	Identify leading robot and formations; A follower follows the leader Control the spacing between the leading robot and the following robot–angle ($l - \varphi$) or lateral spacing-longitudinal spacing ($l - l$);	Collaborative air-ground surveillance [122] Drove in a field [46]; Lawn [119]; Plowed [106];

3.6. Communication

Communication is the basis of information interaction and collaboration among multiple robots. In agricultural production, many factors affect the fine operation of agricultural robots, and to maintain coordination and cooperation among multiple robots and to gain a more comprehensive understanding of the environment in which multiple robots perform tasks, robots need to interact with each other through information to better perform a given task [29]. Balch and Arkin concluded that even a small amount of communication can improve the performance of multi-robot systems tremendously through experiments [125].

At present, the communication technology of agricultural multi-robots mainly involves three parts: multi-robot communication mode, communication network, and communication protocol.

- Communication mode

The multi-robot communication mode is divided into three categories from a macro perspective: explicit communication, implicit communication, and explicit and implicit communication, as shown in Figure 17. Explicit communication is an interactive mode through communication as a medium, requiring a clear communication protocol between interacting parties. This method is often used for concordant communication among robots, but it incurs fairly large costs. Implicit communication is the acquisition of the required information through the external environment and internal sensors without an explicit exchange of data, so some advanced coordination strategies cannot be used, which affects the capacity to perform certain complex tasks.

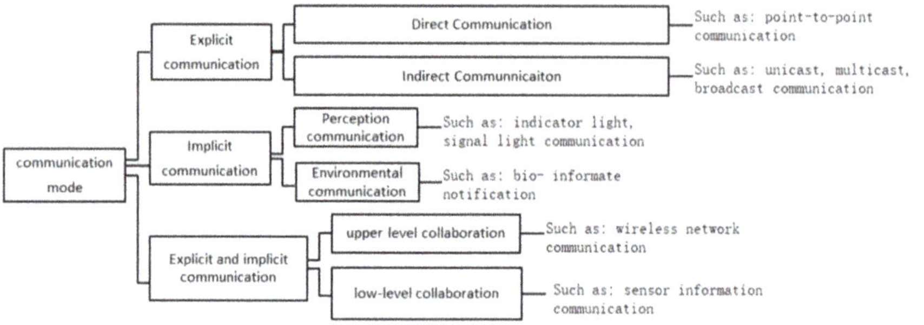

Figure 17. Communication modes of multiple robots.

Since explicit communication and implicit communication have their advantages [120], explicit communication is used for the integrated control of robots in the upper layer, and implicit communication is used for integrated control of robots in the bottom level. Explicit communication means that the robot communicates directly or indirectly with other robots via wireless networks. For example, robot 1 sends a message to all robots in the communication range in broadcast communication, that is, without specifying a particular robot, robot 2, which does not need the message, receives the message. In implicit communication, the intermediary for inter-robot communication is often the surrounding environment. For example, the UAV can be informed about the farmland in advance and build a model of the farmland environment, and the ground robots operate on the ground based on this farmland model [122]. The combination of both communication modes can be used to develop their advantages, improve the flexibility to confront the various dynamic and unknown environments, and complete many complex tasks in agricultural production.

For the implicit communication of multiple robots, you can refer to Section 3.2 environment perceptions, here we focus on robot explicit communication techniques.

In literature [25], two aerial drones were equipped with GPS, visible and near-infrared spectral cameras, which took pictures of the farmland at a set series of ordered waypoints and uploaded them to the backend, which sends the processed information of weeds in the farmland to the ground robot. The ground robots were equipped with RTK-GPS, RGB camera, and LIDAR. RTK-GPS provided accurate heading for the ground robot, RGB camera detects weeds and crop rows, and LIDAR detects obstacles on the vehicle trajectory. While the ground robots were safely walking along their respective set paths, weeding operations start if the weeds detected by the cameras were the same as the weed information in the farmland. In this multi-robot system, the aerial drones and ground robots did not communicate directly but completed the cooperative operation through the interaction of environmental information.

- Wireless communication net of multiple robots

As shown in Figure 18, the agricultural multi-robots need to adjust their pose in real-time. Therefore, the data exchange of communication among multiple robots is mainly based on wireless communication technology in agricultural production. This technology mainly involves a wireless local area network (WLAN) and a wireless personal area network (WPAN), such as WI-FI, Bluetooth, ZigBee, and IRDA (infrared data association). Among them, WI-FI technology has been developed most rapidly in agricultural multi-robots.

- The wireless communication protocol of multiple robots

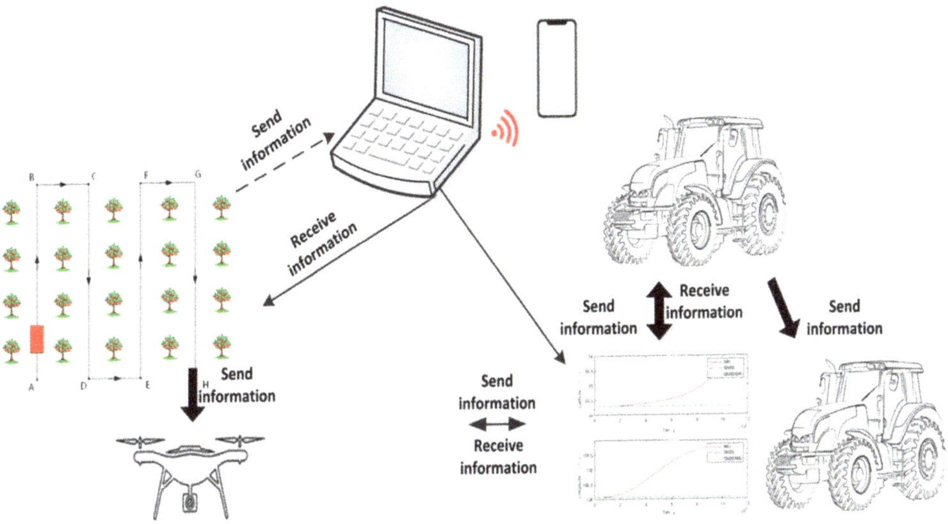

Figure 18. The computer as the center controller was used to send initial paths for UGVs and UAVs control units through communication. And perception data were exchanged between multi-robots and computers.UGV.

The wireless communication protocols are primarily used based on wireless communication standards and the unlicensed band. Taking the WLAN as an example, the IEEE 802.11 series standards and the 2.4 GHz or 5 GHz bands are used in this communication. The IEEE 802.15 series of transmission technology protocols are selected in WPAN.

Combined with the above communication technology, the research progress of agricultural multi-robots in communication in the past 10 years is summarized, as shown in Table 7.

Table 7. Comparison of wireless communication technologies of agricultural multi-robots in the past 10 years.

Communication Technology	WI-FI	Bluetooth	ZigBee
Transmission distance [m]	10~300	10~100	10~75
Theoretical transmission speed [bps]	54~300 M	100 k	10 k
Wireless communication network	Wireless local area network (WLAN)	Wireless personal area network (WPAN)	WPAN
Working frequency [Hz]	2.4~5.4 G	2.4 G	2.4 G (global); 868 M (Europe); 915 M (United States);

Table 7. *Cont.*

Communication Technology	WI-FI	Bluetooth	ZigBee
Communication protocol	IEEE 802.11 a; [1] IEEE 802.11 b [2]; IEEE 802.11 g [3]; IEEE 802.11 p [4];	IEEE 802.15	IEEE 802.15.4
Advantage	Fast transfer speed; Long effective distance; Reliable connection Wide coverage;	High transmission rate; Strong anti-interference ability; Flexible networking; Low power consumption; Small volume;	Low power consumption; Low cost; Large transmission range; Good scalability Strong anti-interference ability; Good security;
Disadvantages	High power consumption; Expensive; Protocol coding complexity;	Slow; Short distance; Weak networking ability Poor security and confidentiality;	Low transmission rate; Short distance; The mobile phone cannot communicate directly;
Application	Plowed in wheat fields or orchards [126,127]; Plowed [44,48];	Plowed [47,105,106];	Monitored vineyards [28]; Monitored and weeded farmland [49]; Plowed [46];

[1] IEEE802.11a standard, operating in the 5 GHz band, has a data transmission rate of 54 Mb/s. [2] IEEE802.11b standard, operating in the 2.4 GHz band, has a data transmission rate of 11 Mb/s, and is not compatible with IEEE 802.11 a. [3] IEEE802.11g is a standard that increases the transmission speed of 802.11b from 11 Mb/s to 54 Mb/s. [4] IEEE 802.11p standard is a communication protocol expanded by IEEE 802.11 standard, which is mainly used in wireless communication of automotive electronics.

In addition to the above wireless communication technologies, Albani et al. adopted a mobile ad hoc (peer-to-peer) network [128,129], which regarded the UAV as a communication node in the network and used three communication strategies (simple, flooding, geo-aware) to solve the communication problem of UAVs flying in the field. The simplest communication strategy is a single broadcast mode, that is, the source node sends information to the nearest node. Flooding constitutes a multi broadcast mode, that is, the source node sends information to multiple agents. Geo-aware employs a source node with the highest utilization rate, and this node sends the messages. All three communication strategies ignore communication errors and focus on the impact of the communication range and protocol on work efficiency. The simulation results show that the effective information of weed monitoring can be transmitted with a minimum number of UAVs under the geo-aware approach. However, the communication strategy discards new information obtained by UAVs of the distributed architecture, and messages cannot be effectively transmitted with a wide range of communication (such as over wide areas of farmland). Agricultural multi-robots working in the farmland often encounter signal occlusion, atypical weather, etc.

Large agricultural multi-robots working in agricultural fields rarely encounter problems such as signal occlusion and atypical weather. However, in other agricultural products, such as greenhouse and orchard, when the size of the multi-robot is smaller than the height of the crop, its communication signal strength is extremely attenuated by factors such as crop planting, growth characteristics, planting scale, and weather (natural wind and rain). Previous references [126,127,130] showed that the test results of the WI-FI communication system of agricultural multi-robots suffered from WI-FI signal intensity attenuation largely because of the reflection and scattering effects of crops, and the effective communication distance was less than 50 m (far less than the theoretical communication distance of 300 m) in mature wheat fields, cornfields, and peach gardens. Therefore, it is a future research direction to select suitable multi-robot communication technology according to the characteristics of crops and to carry out research on multi-robot communication patterns based on crop shape characteristics.

4. Discussion

In general, in the past 10 years, the synergistic technologies of agricultural multi-robots have achieved some research results, and multi-robot collaborative operations have been realized in specific agricultural scenarios. But with the increasing demand for agricultural operations, the following challenges in the application of multi-robots in agriculture still exist to be solved:

- Flexible agricultural multi-robot system architecture

Multi-robot architecture is the basis for collaborative operations of multi-robot systems. In the last decade, agricultural multi-robot systems have mainly focused on centralized or distributed architectures to accomplish collaborative operations under pre-defined conditions. Both architectures have their advantages and disadvantages, but as the number of robots increases (such as multiple aerial robots cooperating with multiple ground robots) and new agricultural operational needs increase (such as sampling in marine environments [131], cargo handling in hilly mountainous areas, pest control in orchards, etc.), it is clear that the scalability and flexibility of multi-robot systems relying on only one architecture are limited. The advantages of centralized and distributed architectures are combined to form a hybrid architecture, or the application architecture is dynamically selected according to the task attributes, which can overcome the low performance caused by the self-centeredness in the distributed architecture and reduce the lack of control flexibility in the centralized architecture.

- Fast and precise environmental perception

In environmental awareness, positioning and sensor fusion answer the question of "where am I" and building a map answers the question of "what's around me", and the answers to these two questions are the prerequisites for robots to start their operations. The positioning and sensor fusion technologies of agricultural multi-robot are mostly used in large fields with unobstructed outdoor signals, where the communication between robots is normal and the robots can get accurate positioning, heading, speed, obstacles, and other information based on their sensors. However, considering the severe compaction of soil by large agricultural machines, the compression of application costs, and the promotion of this concept of refined agriculture, light, and small agricultural robots will be the trend of future development, which will make multi-robot positioning unable to continue to rely on the high-precision positioning of a particular robot or a particular sensor (e.g., GPS). Especially in case of robot failure or communication failure, how to ensure the accurate positioning of the remaining individuals and make the multi-robot system with good robustness is a problem that needs to be solved urgently.

Mapping not only can accurately learn the information of detailed agricultural information, static obstacles, and the location of other robots but also can assign tasks and plan paths for multiple robots. The more accurate the agricultural information, the more accurate the operation objects will be, but this contradicts agricultural tasks that urgently need a fast response, which means that the time spent on the subsequent processing of information data reduces the real-time and flexibility of multi-robot operations. How to obtain dynamic agricultural information quickly and accurately and match it with the precise location of the operation object is another urgent problem in environment sensing.

- Reasonable task assignment in real-time

The task assignment is related to the multi-robot coordination and collaboration mechanism, and the simple zoning assignment of robots cannot adapt to the dynamically changing operational tasks. Also, the number of robots, operating time, and cost of robot operations need to be dynamically adjusted to the operating task. Even for the same type of robots, items such as fuel or electricity, fertilizers, herbicides, and pharmaceuticals can change differently depending on the target of the operation. It is impossible to obtain fast and reasonable response results based on the changes of these uncertainties even depending on a priori knowledge. How to enable multi-robots to timely self-adjust to dynamic task changes and obtain reasonable operational tasks or operational task sequences through real-time interaction with dynamic environments is an urgent problem for multi-robot task assignments in agriculture.

- Dynamic and reliable path planning

The path of agricultural multi-robot operations is designed to accomplish dynamic operational tasks, and the robot's travel rules are usually fixed. The global path of multi-robot offline planning only considers fixed travel rules, such as the point-to-point method and image method, which can avoid static obstacles smoothly, but cannot be extended to be applied to similar agricultural scenarios. In particular, if the dynamics of the agricultural environment change rapidly (e.g., weeds are growing in the field after the rainy season) and the agricultural information is not fully known (e.g., the constructed mapping usually does not contain dynamic obstacles), fixed path planning cannot meet the needs of complex tasks (e.g., weeds are not on the planned path). Therefore, how to perform reliable path planning for multiple robots based on operational tasks with distinct temporal characteristics is a problem that needs to be solved for multi-robot path planning in agriculture.

- Flexible and robust formation control

Multi-robot formation control currently focuses mostly on robot swarms walking steadily along a straight line in a fixed formation. However, when a multi-robot system encounters unexpected events, such as robot failure, communication failure, or stopping travel due to dynamic obstacles, how to mitigate the impact on other robots, respond quickly, and adjust the robot formation shape to continue the task is a concern for agricultural multi-robot formations. Although some studies have shown that multiple robots can be selectively controlled based on time or event drivers, or by replacing the "leader" in the queue, none have been applied in real production.

- Communication system based on plant characteristics

Communication is the basis of multi-robot collaboration in agriculture, whether it is multi-robot positioning, collaborative control, or remote supervision, communication is indispensable. The agricultural environment lacks communication infrastructure construction, and most of them directly used industrial communication systems do not consider the relationship between outdoor plant growth and communication signals, and their communication range and signals will be attenuated to different degrees in the agricultural environment. Therefore, the construction of a communication system adapted to the agricultural multi-robot operating environment is a problem that needs to be solved for multi-robot communication.

5. Conclusions

Given the current challenges in agricultural multi-robot research, this paper points out future research directions in six areas to enhance the application of agricultural multi-robots in practice. Firstly, to build a flexible and changeable agricultural multi-robot system architecture based on hybrid architecture so that the multi-robot system has good environmental adaptability and robustness. Secondly, to develop sensor information fusion technology among agricultural multi-robots based on mutual positioning methods to improve the positioning accuracy of multi-robots in agricultural environments without GPS. Meanwhile, SLAM technology for agricultural multi-robots is studied to rapidly build environment models to adapt to the dynamically changing agricultural information. Third, to introduce deep learning mechanisms in agricultural multi-robot task assignment enables multi-robots to self-identify, evaluate, compare, remember and adjust during their interaction with the environment, and adjust the way they interact with other individuals according to specific tasks so that the group as a whole is equipped with the ability to complete multiple types of tasks. Fourth, dynamic planning of multi-robot paths based on a combination of centralized and distributed path planning methods enables multi-robot systems to adapt to real-time changing operational tasks and avoid obstacles and other robots promptly. Fifth, to modify the reference points of multi-robot formations flexibly according to changing events, adjust the distance and direction between formation members, reduce the impact on other mobile robots, and complete operational tasks. Sixth, to study the relationship between plant growth characteristics and communication system, establish a communication signal attenuation model, and design an agricultural multi-robot communication protocol based on this model to build a communication system.

In summary, the multiple robot system represents the future of robot development. The synergistic technologies for the research of agricultural multi-robots have a great value and bright prospects but are also extremely challenging. Therefore, it requires participation by researchers to combine the former research results, recognize the developing trends, and use practicality as the ultimate goal to drive forward the coordination technology of agricultural multi-robots.

Author Contributions: Writing—review and editing, W.M.; supervision, H.L., Z.L., F.Y. and M.W.; project administration, F.Y. All authors have read and agreed to the published version of the manuscript.

Funding: This research received funding from the Major Science and Technology Project of Shaanxi Province of China (Program No. 2020zdzx03-04-01).

Institutional Review Board Statement: The study in the paper did not involve humans or animals. The project supporting the research of this paper is to study agricultural machinery and equipment for apple orchards, and the research does not involve humans or animals.

Informed Consent Statement: The study in the paper did not involve humans.

Data Availability Statement: This paper is a scientific review paper that provides a detailed analysis and summary of agricultural multi-robots, independent of the data.

Acknowledgments: Authors thank the funding received from the Major Science and Technology Project of Shaanxi Province of China (Program No. 2020zdzx03-04-01). We also thank the critical comments and suggestions from the anonymous reviewers for improving the manuscript.

Conflicts of Interest: The authors declare no conflict of interest.

References

1. Xia, L.Y. Negative Effects of Labor Force Drain on Rural Economy. *J. Nanjing Agric. Univ. (Soc. Sci. Ed.)* **2009**, *9*, 14–19. [CrossRef]
2. Wu, L.L.; Li, G.C.; Zhou, X.S. Change of Factor Endowments and China Agricultural Growth Path Selection. *China Popul. Resour. Environ.* **2015**, *25*, 144–152. [CrossRef]
3. Chen, X.W.; Chen, Y.Y.; Zhang, J.J. A Quantitative Study on the Impact of the Aging of Rural Population on Agricultural Output in China. *Chin. J. Popul. Sci.* **2011**, 26–39. Available online: http://www.cnki.com.cn/Article/CJFDTotal-ZKRK201102005.htm (accessed on 11 March 2019).
4. Zhang, S.W.; Tian, X.; Wang, S.G. Impact of Aging Agricultural Labor on Mechanization and Technical Efficiency in Wheat Production: A prospective analysis based on landforms. *J. China Agric. Univ.* **2018**, *23*, 174–182.
5. Hu, X.P.; Zhu, Y.; Ge, D.Q. Analysis on the Aging of the Agricultural Labor Force in China. 23 December 2011. Available online: http://www.chinanews.com/sh/2011/12-23/3554966_3.shtml (accessed on 11 March 2019).
6. Chen, Z.; Li, C.Y.; Zou, X.J. Supportive Technologies and Research Progresses of Agricultural Multi-robot System. *J. Hua Zhong Agric. Univ.* **2007**, *26*, 914–919. [CrossRef]
7. Zhang, Y.Z.; Chen, L.D.; Tang, T.Z. Analysis on the Causes and Effects of the Rising Labor Cost of Agricultural Planting. *Mod. Agric. Sci. Technol.* **2017**, 259–261. [CrossRef]
8. An International Comparative Study of Labor Productivity and Unit Labor Costs in Three Industries in China. Available online: https://www.sohu.com/a/274944214_481741 (accessed on 24 January 2021).
9. Zhang, Y.H. The "High Cost" Era of Agriculture Needs to Improve Competitiveness. *Finance Community* **2017**, *16*, 74–77. [CrossRef]
10. Zhou, X.S. Labor Migration, and Agricultural Mechanization. *J. South China Agric. Univ. (Soc. Sci. Ed.)* **2017**, *3*, 49–57. [CrossRef]
11. Shamshiri, R.R.; Weltzien, C.; Hameed, I.A.; Yule, I.J.; Grift, T.E.; Balasundram, S.K.; Pitonakova, L., Ahmad, D.; Chowdhary, G. Research and Development in Agricultural Robotics: A perspective of Digital Farming. *Int. J. Agric. Biol. Eng.* **2018**, *11*, 1–14. [CrossRef]
12. Sun, F.K.; Xing, Z.B. Summary and Improvement Measures for Soil Compaction Caused by Machinery. *Agric. Sci. Technol.* **2014**, *15*, 205–206. [CrossRef]

13. Johnson, D.A.; Naffin, D.J.; Puhalla, J.S.; Sanchez, J.; Wellington, C.K. Development and Implementation of A Team of Robotic Tractors for Autonomous Peat Moss Harvesting. *J. Field Robot.* **2010**, *26*, 549–571. [CrossRef]
14. Moorehead, S.; Wellington, C.; Gilmore, B.; Vallespi, C. Automating Orchards: A System of Autonomous Tractors for Orchard Maintenance. In Proceedings of the 2012 IEEE/RSJ International Conference on Intelligent Robots & Systems (IROS), Vilamoura, Algarve, Portugal, 7–12 October 2012.
15. Noreils, F.R. Toward a Robot Architecture Integrating Cooperation Between Mobile Robots. *Appl. Indoor Environ.* **1993**, *12*, 79–98. [CrossRef]
16. Kun, S. Production Time and Labor Time of Agriculture: Academic Debate and Value Implication. *J. China Agric. Univ. (Soc. Sci. Ed.)* **2019**, *36*, 11–23. [CrossRef]
17. This Spring Plowing, Modern Agricultural Machinery "Take the Big Beam" My City Spring Plowing Preparation Plows to Invest 149 Thousand Sets of Agricultural Machinery Tools. Available online: https://app.cqrb.cn/economic/2020-02-27/130410_pc.html (accessed on 30 November 2020).
18. The Agricultural Machinery Plays the Leading Role, Helps the Farmer Spring Plows the Busy. Available online: http://baijiahao.baidu.com/s?id=1665918730118414764 (accessed on 30 November 2020).
19. Xinjiang Plant Protection Drone Doubles Farmers' Income. Available online: https://kuaibao.qq.com/s/20190903A01ZHK00?refer=spider (accessed on 30 November 2020).
20. Summer Grain Harvest Nearly 80%, the Harvest Has Become a Foregone Conclusion. Available online: http://www.xinhuanet.com/fortune/2019-06/12/c_1124613101.htm (accessed on 30 November 2020).
21. Yao, J.W.; Huang, C.S. Review on Multi-robot Coordination and Cooperation Control Technology. *J. Huangshi Inst. Technol.* **2007**, *23*, 1–6. [CrossRef]
22. Zhang, Z.Y. Research on Multi-Robot Collaboration and Environmental Modeling. Ph.D. Thesis, Harbin Engineering University, Harbin, China, 2009.
23. Hao, Y.; Laxton, B.; Agrawal, S.K.; Benson, E.R. Differential flatness-based formation following of a simulated autonomous small grain harvesting system. *Trans. ASAE* **2004**, *47*, 933–941. [CrossRef]
24. Hao, Y.; Laxton, B.; Benson, E.R.; Agrawal, S.K. Robotic Simulation of the Docking and Path Following of an Autonomous Small Grain Harvesting System. In Proceedings of the 2003 ASAE Annual International Meeting, Las Vegas, NV, USA, 27–30 July 2003. [CrossRef]
25. Emmi, L.; Gonzalez-de-Soto, M.; Pajares, G.; Gonzalez-de-Santos, P. New Trends in Robotics for Agriculture: Integration and Assessment of a Real Fleet of Robots. *Sci. World J.* **2014**, 1–21. [CrossRef] [PubMed]
26. Tan, M.; Wang, S.; Cao, Z.Q. *Multi-Robot System*; Tsinghua University Press: Beijing, China, 2005; p. 307. ISBN 978-7-3021-0095-9.
27. Noguchi, N.; Will, J.; Reid, J.; Zhang, Q. Development of A Master-Slave RobotSystem for Farm Operations. *Comput. Electron. Agric.* **2004**, *44*, 1–19. [CrossRef]
28. Bouzouita, A.; Larbi, H.; Cherif, M.; Selmi, M.S.; Belhadj, K.; Derouiche, A.; Ben Slama, M.R.; Chebil, M. Aerial Remote Sensing in Agriculture: A Practical Approach to Area Coverage And Path Planning For Fleets of Mini Aerial Robots. *J. Field Robot.* **2011**, *28*, 667–689. [CrossRef]
29. Cai, Z.X.; Chen, B.F.; Liu, L.Y.; Yu, L.L. *Cooperative Principle and Technology of Multi-Mobile Robot*; National Defense Industry Press: Beijing, China, 2011; p. 286. ISBN 978-7-1180-6854-2.
30. Şahin, E. Swarm Robotics: From Sources of Inspiration to Domains of Application. In Proceedings of the 2004 International Conference on Swarm Robotics, Berlin, Germany, 10–20 July 2004. [CrossRef]
31. Fan, R.F.; Xie, G.M.; He, C.G. *Robot Perception and Application*; Harbin Engineering University Press: Harbin, China, 2013; p. 197. ISBN 978-7-5661-0688-9.
32. Wang, D.S.; Wang, J. Research Review of Environmental Cognition Techniques of Mobile Robots in Unknown Environment. *Mach. Tool Hydraul.* **2013**, *41*, 187–191. [CrossRef]
33. Jiang, H.Z. A Review of Collaborative Positioning Technology. *Telecom. Power Technol.* **2017**, *34*, 48–50. [CrossRef]
34. Rekleitis, I.M.; Dudek, G.; Milios, E.E. Multi-robot cooperative localization: A study of trade-offs between efficiency and accuracy. In Proceedings of the 2002 IEEE/RSJ International Conference on Intelligent Robots and Systems (IROS), EPFL, Lausanne, Switzerland, 30 September–4 October 2002; pp. 2690–2695. [CrossRef]
35. Cai, Y.F. Research on Multiple Robots Cooperative Localization and the Architecture. Ph.D. Thesis, Nanjing University of Science & Technology, Nanjing, China, 2011.
36. Stroupe, A.W.; Balch, T. Collaborative probabilistic constraint-based landmark localization. In Proceedings of the 2002 IEEE/RSJ International Conference on Intelligent Robots & Systems (IROS), EPFL, Lausanne, Switzerland, 30 September–4 October 2002; pp. 447–453. [CrossRef]
37. Wang, Z.Q. Design and realization on the autonomous following control system of agricultural vehicles. Master's Thesis, Nanjing Agricultural University, Nanjing, China, 2014.
38. Bi, W.P.; Zhang, H.; Qu, Z.L.; Ding, Y.Q.; Yu, H.F.; Wang, B. Design of an autonomous following system for master-slave vehicles operating in orchard based on binocular stereo vision. *J. Hunan Agric. Univ. (Nat. Sci.)* **2016**, *42*, 344–348. [CrossRef]
39. Zhao, M.; Lin, M.S.; Huang, Y.Q. Leader-following Formation Control of Multi-robots Based on Dynamic Value of φ. *J. Southwest Univ. Sci. Technol.* **2013**, *28*, 57–61. [CrossRef]

40. Guillet, A.; Lenain, R.; Thuilot, B.; Martinet, P. Adaptable robot formation control: Adaptive and predictive formation control of autonomous vehicles. *IEEE Robot. Autom. Mag.* **2014**, *21*, 28–39. [CrossRef]
41. Wu, X.; Jing, Y.; Sun, F.; Chen, H.; Huang, S. An approach to multi-robot cooperative SLAM. In Proceedings of the 2012 31st Chinese Control Conference (CCC), Hefei, China, 25–27 July 2012; pp. 4904–4913.
42. Zhu, Z.; Takeda, J.-i.; Torisu, R.; Chen, J.; Song, Z.; Mao, E. Control system for tractor-platooning. Proceedings of 2007 International Conference on Mechatronics and Automation (ICMA), Harbin, China, 5–8 August 2007; pp. 3173–3178. [CrossRef]
43. Zhu, Z.; Takeda, J.-I.; Xie, B.; Song, Z.; Torisu, R.; Mao, E. Tractor platooning system on sloping terrain at low speed. *Trans. ASABE* **2009**, *52*, 1385–1393. [CrossRef]
44. Guillet, A.; Lenain, R.; Thuilot, B.; Rousseau, V. Formation Control of Agricultural Mobile Robots: A Bidirectional Weighted Constraints Approach. *J. Field Robot.* **2017**, *34*, 1260–1274. [CrossRef]
45. Cartade, P.; Braconnier, J.-B.; Lenain, R.; Thuilot, B. Adaptive and predictive control of a mobile robots fleet: Application to off-road formation regulation. In Proceedings of the 2013 IEEE International Conference on Robotics and Automation (ICRA), Karlsruhe, Germany, 6–10 May 2013; pp. 1836–1842. [CrossRef]
46. Xi, Z.; Marcus, G.; Patrick, O.N.; Ludwig, G. Development of an intelligent master-slave system between agricultural vehicles. In Proceedings of the 2010 Intelligent Vehicles Symposium(IV), La Jolla, CA, USA, 21–24 June 2010; pp. 250–255. [CrossRef]
47. Chi, Z.; Noboru, N. Development of Leader-follower System for Field Work. In Proceedings of the 2015 IEEE/SICE International Symposium on System Integration (SII), Meijo University, Nagoya, Japan, 11–13 December 2015; pp. 364–368. [CrossRef]
48. Bai, X.; Wang, Z.; Hu, J.; Gao, L.; Xiong, F. Harvester Group Corporative Navigation Method Based on Leader-Follower Structure. *Trans. Chin. Soc. Agric. Mach.* **2017**, *48*, 14–21. [CrossRef]
49. Gonzalez-de-Santos, P.; Ribeiro, A.; Fernandez-Quintanilla, C.; Lopez-Granados, F.; Brandstoetter, M.; Tomic, S.; Pedrazzi, S.; Peruzzi, A.; Pajares, G.; Kaplanis, G.; et al. Fleets of robots for environmentally-safe pest control in agriculture. *Precis. Agric.* **2017**, *18*, 574–614. [CrossRef]
50. Gonzalez-de-Soto, M.; Emmi, L.; Perez-Ruiz, M.; Aguera, J.; Gonzalez-de-Santos, P. Autonomous systems for precise spraying - Evaluation of a robotized patch sprayer. *Biosyst. Eng.* **2016**, *146*, 165–182. [CrossRef]
51. Zhou, H.M.; Qian, Z. *Intelligent Sensing Technology and System (11th Five-Year Plan)*; Beijing University of Aeronautics and Astronautics Press: Beijing, China, 2008; p. 301. ISBN 978-7-8112-4366-6.
52. Qing, Z.Q. Data Fusion Technique and Its Application. *Ordnance Industry Autom.* **2003**, *22*, 25–28. [CrossRef]
53. Perez-Ruiz, M.; Upadhyaya, S.K. GNSS in precision agricultural operation. In *New Approach of Indoor and Outdoor Localization Systems*; Intech: Brisbane, Australia, 2012; pp. 1–25. [CrossRef]
54. High Precision TOPCON Autopilot System for Agricultural Seeding, Land Preparation and Spraying. Available online: http://www.app17.com/supply/offerdetail/9628838.html (accessed on 25 January 2021).
55. Juan-Carlos, T.; Rodrigo, M.; Edmundo, G.; Antoni, G. Cooperative monocular-based SLAM for multi-UAV systems in GPS-denied environments. *Sensors* **2018**, *18*, 1351. [CrossRef]
56. Roldán, J.J.; Garcia-Aunon, P.; Garzón, M.; León, J.D.; Cerro, J.D.; Barrientos, A. Heterogeneous Multi-Robot System for Mapping Environmental Variables of Greenhouses. *Sensors* **2016**, *7*, 1018. [CrossRef] [PubMed]
57. Fang, Z.; Tong, G.; Xu, X. Study of Autonomous Robot Self-localization Methods Based on Bayesian Filter Theory. *Control Decis. Mak.* **2006**, *21*, 841.
58. Blender, T.; Buchner, T.; Fernandez, B.; Pichlmaier, B.; Schlegel, C. Managing a Mobile Agricultural Robot Swarm for a seeding task. In Proceedings of the 2016 42nd Annual Conference of the IEEE Industrial Electronics Society (IES), Florence, Italy, 24–27 October 2016.
59. Dellaert, F.; Fox, D.; Burgard, W.; Thrun, S. Monte Carlo localization for mobile robots. In Proceedings of the 1999 IEEE International Conference on Robotics and Automation, Detroit, MI, USA, 10–15 May 1999. [CrossRef]
60. Thrun, S. Probabilistic Algorithms in Robotics. *AI Mag.* **2000**, *21*, 9–109. [CrossRef]
61. Thrun, S.; Burgard, W.; Fox, D. *Probabilistic Robotics (Intelligent Robotics and Autonomous Agents)*; MIT Press: Cambridge, MA, USA, 2005; p. 668. ISBN 978-0-2622-0162-9.
62. Guan, Y.; Yang, X.J.; Jiang, T. Research Advances on the Multisensor Information Fusion of the Agricultural Robot. *J. Anhui Agric. Sci.* **2010**, *25*, 14127–14128.
63. Yang, L.; Yu, H. *Multi-Source Information Fusion Theory and Application*; Beijing University of Posts and Telecommunications Press: Beijing, China, 2011; p. 234. ISBN 978-7-5635-2740-3.
64. Wang, W.H.; Chen, W.D.; Xi, Y.G. Uncertain Information Based Map-Building Of Mobile Robots In Absolutely Unknown Environment. *Robot* **2001**, *23*, 563–568. [CrossRef]
65. Ball, D.; Ross, P.; English, A.; Patten, T.; Upcroft, B.; Fitch, R.; Sukkarieh, S.; Wyeth, G.; Corke, P. Robotics for Sustainable Broad-Acre Agriculture. In *Field and Service Robotics*; Springer: Cham, Switzerland, 2015; pp. 439–453. [CrossRef]
66. Kim, J.; Son, H.I. A Voronoi Diagram-Based Workspace Partition for Weak Cooperation of Multi-Robot System in Orchard. *IEEE Access* **2020**, *8*, 20676–20686. [CrossRef]
67. Zhang, H. Research on path planning of multiple agricultural robots based on ant colony algorithm. *Digital Technol. Appl.* **2017**, *6*, 147–149. [CrossRef]
68. Moravec, H. High Resolution Maps from Wide Angle Sonar. In Proceedings of the 1985 IEEE International Conference on Robotics and Automation, St. Louis, MO, USA, 25–28 March 1985. [CrossRef]

69. Kortenkamp, D.; Weymouth, T.E. Topological Mapping for Mobile Robots Using a Combination of Sonar and Vision Sensing. In Proceedings of the 1994 12th National Conference on Artificial Intelligence (AAAI), Seattle, WA, USA, 31 July–4 August 1994; Volume 2, pp. 979–984.
70. Mouaddib, E.M.; Marhic, B. Geometrical matching for mobile robot localization. *IEEE Trans. Robot. Autom.* **2000**, *16*, 542–552. [CrossRef]
71. Potena, C.; Khanna, R.; Nieto, J.; Siegwart, R.; Nardi, D.; Pretto, A. AgriColMap: Aerial-Ground Collaborative 3D Mapping for Precision Farming. *IEEE Robot. Autom. Lett.* **2019**, *4*, 1085–1092. [CrossRef]
72. Faryadi, S.; Mohammadpour Velni, J. A reinforcement learning-based approach for modeling and coverage of an unknown field using a team of autonomous ground vehicles. *Int. J. Intell. Syst.* **2020**, 1–16. [CrossRef]
73. Dong, W.; Zhu, K.; Liang, S.H.; Wen, W.S.; Guo, Y.; Tan, Y. Path planning algorithm of field robot based on topological map and robot control. *China Sci. Paper* **2016**, *11*, 2525–2530. [CrossRef]
74. Shalal, N.; Low, T.; Mccarthy, C.; Hancock, N. A preliminary evaluation of vision and laser sensing for tree trunk detection and orchard mapping. In Proceedings of the 2013 Australasian Conference on Robotics and Automation (ACRA), UNSW, Sydney, Australia, 2–4 December 2013; pp. 1–10.
75. Dos Santos, F.N.; Sobreira, H.M.P.; Campos, D.F.B.; Morais, R.; Moreira, A.P.G.M.; Contente, O.M.S. Towards a Reliable Monitoring Robot for Mountain Vineyards. In Proceedings of the 2015 IEEE International Conference on Autonomous Robot Systems & Competitions (IROS), Gateway to the Era of Robots, Hamburg, Germany, 28 September–2 October 2015; pp. 37–43. [CrossRef]
76. Wang, H.L.; Zhang, C.J.; Song, Y.; Pang, B. Master-followed Multiple Robots Cooperation SLAM Adapted to Search and Rescue Environment. *Int. J. Control Autom. Syst.* **2018**, *16*, 2593–2608. [CrossRef]
77. Schmuck, P.; Chli, M. CCM-SLAM: Robust and efficient centralized collaborative monocular simultaneous localization and mapping for robotic teams. *J. Field Robot.* **2019**, *36*, 763–781. [CrossRef]
78. Zhang, Y.; Liu, S.H. Survey of multi-robot task allocation. *CAAI Trans. Intell. Syst.* **2008**, *3*, 115–120. [CrossRef]
79. Gerkey, B.P.; Matarić, M.J. A Formal Analysis and Taxonomy of Task Allocation in. *Int. J. Robot. Res.* **2004**, *23*, 939–954. [CrossRef]
80. Zan, J. Research On Key Technologies of Multiple Autonomous Mobile Robot Cooperation. Ph.D. Thesis, Chang'an University, Xi'an, China, 2014.
81. D'Urso, G.; Smith, S.L.; Mettu, R.; Oksanen, T.; Fitch, R. Multi-vehicle refill scheduling with queueing. *Comput. Electron. Agric.* **2018**, *144*, 44–57. [CrossRef]
82. Wan, L.P.; Lan, X.G.; Zhang, H.B.; Zheng, N.N. A Review of Deep Reinforcement Learning Theory and Application. *Pattern Recognit. Artif. Intell.* **2019**, *32*, 67–81.
83. Zhou, J.; Mu, D.J. Study of multi-robot system on task allocation. *J. Northwest Univ. (Nat. Sci. Ed.)* **2014**, *44*, 403–410. [CrossRef]
84. Ruyue, C.; Shichao, L.; Yuhan, J.; Hongzhen, X.; Man, Z.; Minzan, L. Multi-machine Cooperation Task Planning Based on Ant Colony Algorithm. *J. Agric. Mach.* **2019**, *50*, 34–39.
85. Drenjanac, D.; Klausner, L.; Kühn, E.; Tomic, S.D.K. Semantic Shared Spaces for Task Allocation in a Robotic Fleet for Precision Agriculture. In Proceedings of the 2013 7th Research Conference on Metadata and Semantic Research Conference (MTSR), Thessaloniki, Greece, 19–22 November 2013; pp. 440–446. [CrossRef]
86. Wamg, Y.F. Research on Task Allocation of a Multi-Robot System for Intelligent Manufacturing. Ph.D. Thesis, Nanjing University, Nanjing, China, 2016.
87. Tang, S.Y.; Zhu, Y.F.; Li, Q.; Lei, Y.L. Survey of task allocation in multi Agent systems. *Syst. Eng. Electron.* **2010**, *32*, 2155–2161. [CrossRef]
88. Song, Z.; Vaughan, R.T. Sustainable robot foraging: Adaptive fine-grained multi-robot task allocation for a maximum sustainable yield of biological resources. In Proceedings of the 2013 IEEE/RSJ International Conference on Intelligent Robots & Systems (IROS), Tokyo, Japan, 3–7 November 2013; pp. 3309–3316. [CrossRef]
89. Sherali, H.D.; Driscoll, P.J. Evolution and state-of-the-art in integer programming. *J. Comput. Appl. Math.* **2000**, *124*, 319–340. [CrossRef]
90. Gomory, R.E. Outline of an algorithm for integer solutions to linear programs. *Bull. Am. Math. Soc.* **1958**, *64*, 275–278. [CrossRef]
91. Tan, S.W.; Zhang, D.L. Graphic solution for a type of integer programming. *J. Pet. Univ. (Nat. Sci. Ed.)* **2003**, *27*, 124–127.
92. Zhao, Y. Survey on intelligent optimization algorithms for solving integer programming problems. *Comput. Appl. Res.* **2010**, *27*, 408–412.
93. Fu, M.J.; You, X.M. A review of multi-robot systems and their path planning methods. *Softw. Guide* **2017**, *16*, 177–179. [CrossRef]
94. Du, L.C.; Qian, H.; Xiao, A.P. Path Planning Technology and Its Application in Greenhouse Robot. *Hubei Agric. Sci.* **2010**, *49*, 1205–1208. [CrossRef]
95. Liao, S.H.; Zhang, L.D. Present Situation and Developing Trend of Robot Path Planning Technology. *Mech. Eng.* **2007**, *7*, 13–16. [CrossRef]
96. Lei, Y.M. Research on Dynamic Path Planning Method for Multi-Robot Systems. Ph.D. Thesis, Harbin Engineering University, Harbin, China, 2011.
97. Ma, Y. Research on Path Planning Problems for Multiple Moblie Robots. Ph.D. Thesis, Huazhong University of Science & Technology, Wuhan, China, 2012.

98. Kolendo, P.; Smierzchalski, R.; Jaworski, B. Fitness function scaling in the evolutionary method of path planning. In Proceedings of the 2011 16th International Conference on Methods & Models in Automation & Robotics (MMAR), Miedzyzdroje, Poland, 22–25 August 2011; pp. 1989–1994. [CrossRef]
99. Chiddarwar, S.S.; Babu, N.R. Conflict free coordinated path planning for multiple robots using a dynamic path modification sequence. *Robot. Auton. Syst.* **2011**, *59*, 508–518. [CrossRef]
100. Conesamuñoz, J.; Emmi, L.A.; Gonzálezdesoto, M.; Ribeiro Seijas, Á. Effective supervision of a tractor fleet. In Proceedings of the 2014 2nd International Conference on Robotics & Associated High-Technologies & Equipment for Agriculture and Forestry (RHEA), Madrid, Spain, 30–31 January 2014.
101. Conesa-Munoz, J.; Ribeiro, A.; Andujar, D.; Fernandez-Quintanilla, C.; Dorado, J. Multi-path planning based on an NSGA-II for a fleet of robots to work on agricultural tasks. In Proceedings of the 2012 IEEE Congress on Evolutionary Computation (CEC), Brisbane, Australia, 10–15 June 2012; pp. 1–8. [CrossRef]
102. Mudgal, C.; Vassileva, J. An influence diagram model for multi-agent negotiation. In Proceedings of the 2000 4th International Conference on Multiagent Systems (ICMAS), Washington, DC, USA, 10–12 July 2000; pp. 451–452. [CrossRef]
103. Hasegawa, J.; Kurihara, K.; Nishiuchi, N. Collision-free path planning method for a mobile robot. In Proceedings of the 2002 IEEE International Conference on Systems, Man and Cybernetics (SMC), Yasmine Hammamet, Tunisia, 6–9 October 2002; Volume 3, pp. 323–327. [CrossRef]
104. Kolushev, F.A.; Bogdanov, A.A. Multi-agent Optimal Path Planning for Mobile Robots in Environment with Obstacles. In Proceedings of the 1999 Perspectives of System Informatics—3rd International Andrei Ershov Memorial Conference, Akademgorodik, Novosibirsk, Russia, 6–9 July 1999; pp. 503–510. [CrossRef]
105. Zhang, C.; Noguchi, N. Development of a multi-robot tractor system for agriculture field work. *Comput. Electron. Agric.* **2017**, *142*, 79–90. [CrossRef]
106. Zhang, C.; Noguchi, N. Cooperation of Two Robot Tractors to Improve Work Efficiency. *Adv. Robot. Autom.* **2016**, *5*, 11. [CrossRef]
107. Ball, D.; Ross, P.; English, A.; Patten, T.; Upcroft, B.; Fitch, R.; Sukkarieh, S.; Wyeth, G.; Corke, P. Robotics for Sustainable Broad-Acre Agriculture. In Proceedings of the 2013 9th International Conference on the Field and Service Robotics (FSR), Brisbane, Australia, 9–11 December 2013. [CrossRef]
108. Berman, S.; Nagpal, R.; Halász, Á.M. Optimization of Stochastic Strategies for Spatially Inhomogeneous Robot Swarms: A Case Study in Commercial Pollination. In Proceedings of the 2011 IEEE/RSJ International Conference on Intelligent Robots & Systems(IROS), San Francisco, CA, USA, 25–30 September 2011; pp. 3923–3930. [CrossRef]
109. Tamura, K.; Naruse, K. Unsmooth Field Sweeping by Balistic Random Walk of Multiple Robots in Unsmooth Terrain. In Proceedings of the 2014 Joint 7th International Conference on Soft Computing and Intelligent Systems (SCIS) and 15th International Symposium on Advanced Intelligent Systems (ISIS), Kita-Kyushu, Japan, 3–6 December 2014; pp. 585–589. [CrossRef]
110. Noguchi, N.; Will, J.; Ishii, K.; Reid, J. Development of Master-slave Robot System—Obstacle avoidance algorithm. In Proceedings of the 2002 Automation Technology for Off-road Equipment Conference, Chicago, IL, USA, 26–27 July 2002; pp. 432–441. [CrossRef]
111. Dong, S.L.; Chen, W.D.; Yugeng, X. An distributed control system for multi-mobile robot formation. *Robot* **2002**, *22*, 433–438. [CrossRef]
112. Qing, Y.Q. Research on Motion Control of Multiple Mobile Robots System. Ph.D. Thesis, Huazhong University of Science and Technology, Wuhan, China, 2007.
113. Liu, Y.P.; Yang, Y.M. Research review on formation control of multiple robots. *Control Eng.* **2010**, *53*, 182–186. [CrossRef]
114. Zhang, Q.H. *Graph Theory with Applications*; Tsinghua University Press: Beijing, China, 2013; p. 184. ISBN 978-7-3023-2997-8.
115. Gong, J.W.; Jiang, Y.; Xu, W. *Model Predictive Control for Driverless Vehicles*; Beijing Institute of Technology Press: Beijing, China, 2014; p. 196. ISBN 978-7-5640-9084-5.
116. Wang, D.Z.; Wu, S.J. *Model Tracking Control System Design*; East China University of Science and Technology Press: Shanghai, China, 2016; p. 127. ISBN 978-7-5628-4703-8.
117. Smith, K.D.; Hsiung, S.C.; White, C.; Lowe, C.G.; Clark, C.M. Stochastic modeling and control for tracking the periodic movement of marine animals via AUVs. In Proceedings of the IEEE/RSJ International Conference on Intelligent Robots & Systems (IROS), Daejeon, Korea, 9–14 October 2016; pp. 3101–3107. [CrossRef]
118. Zhang, C.; Zhao, Y. Development and application of model-based predictive control theory. *Electron. Technol. Softw. Eng.* **2014**, *4*, 256–257.
119. Guillet, A.; Lenain, R.; Thuilot, B. Off-road path tracking of a fleet of WMR with adaptive and predictive control. In Proceedings of the 2013 IEEE/RSJ International Conference on Intelligent Robots and Systems (IROS), Tokyo, Japan, 3–7 November 2013; pp. 2855–2861. [CrossRef]
120. Pitla, S.K.; Luck, J.D.; Shearer, S.A. Multi-Robot System Control Architecture (MRSCA) for Agricultural Production. *Br. J. Learn. Disabil.* **2010**, *11*, 68–69. [CrossRef]
121. Chaimowicz, L.; Sugar, T.; Kumar, V.; Campos, M.F.M. An architecture for tightly coupled multi-robot cooperation. In Proceedings of the 2001 IEEE International Conference on Robotics & Automation (ICRA), Seoul, Korea, 21–26 May 2001; Volume 3, pp. 2992–2997. [CrossRef]
122. Ju, C.; Son, H.I. Modeling and control of heterogeneous agricultural field robots based on ramadge-wonham theory. *IEEE Robot. Autom. Lett.* **2020**, *5*, 48–55. [CrossRef]

123. Bochtis, D.D.; Sorensen, C.G. The vehicle routing problem in field logistics part I. *Biosyst. Eng.* **2009**, *104*, 447–457. [CrossRef]
124. Bochtis, D.D.; Sorensen, C.G. The vehicle routing problem in field logistics: Part II. *Biosyst. Eng.* **2010**, *105*, 180–188. [CrossRef]
125. Arkin, R.C.; Balch, T.; Nitz, E. Communication of Behavioral State in Multi-Agent Retrieval Tasks. In Proceedings of the 1993 IEEE International Conference on Robotics & Automation (ICRA), Atlanta, GA, USA, 2–6 May 1993; pp. 588–594. [CrossRef]
126. Zhou, Y. Research of Wireless Communication Technology for Agricultural Multi-Robot Systems. Master's Thesis, Northwest A&F University, Yangling, China, 2014.
127. Zhu, H. Research and Design of Wireless Communication Platform For Agricultural Multi-Robot. Master's Thesis, Northwest A&F University, Yangling, China, 2015.
128. Albani, D.; Ijsselmuiden, J.; Haken, R.; Trianni, V. Monitoring, and Mapping With Robot Swarms for Agricultural Applications. In Proceedings of the 2017 14th IEEE International Conference on Advanced Video & Signal Based Surveillance (AVSS), Lecce, Italy, 29 August–1 September 2017; pp. 1–6. [CrossRef]
129. Albani, D.; Nardi, D.; Trianni, V. Field Coverage and Weed Mapping by UAV Swarms. In Proceedings of the 2017 IEEE/RSJ International Conference on Intelligent Robots & Systems (IROS), Vancouver, BC, Canada, 24–28 September 2017; pp. 4319–4325. [CrossRef]
130. Zhou, Y.; Yang, F.Z.; Zhu, H.; Wang, B.B.; Qing, H.W.; Wang, Z. Test of Propagation Performance of Wheat Field Wireless Channel based on Wi-Fi. *Agric. Mech. Res.* **2015**, *2*, 155–160. [CrossRef]
131. Deusdado, P.; Pinto, E.; Guedes, M.; Marques, F.; Rodrigues, P.; Lourenco, A.; Mendonca, R.; Silva, A.; Santana, P.; Corisco, J.; et al. An Aerial-Ground Robotic Team For Systematic Soil and Biota Sampling in EstuarineMudflats. In Proceedings of the 2nd Iberian Robotics Conference, Lisbon, Portugal, 15–26 November 2015.

applied sciences

MDPI

Article

Bio-Inspired Model-Based Design and Control of Bipedal Robot

David Polakovič, Martin Juhás *, Bohuslava Juhásová and Zuzana Červeňanská

Institute of Applied Informatics, Automation and Mechatronics, Faculty of Material Science and Technology in Trnava, Slovak University of Technology in Bratislava, 91724 Trnava, Slovakia
* Correspondence: martin_juhas@stuba.sk

Abstract: The article aims to describe stages of development of a mobile, anthropomorphic robot. The individual phases of the bio-inspired model-based design of a two-legged robot are described, from the analysis of human walking patterns, through the creation of a simulation model and a control system to the creation and testing the physical model of such a robot. The bio-inspired design is based on an analysis of the movements of the individual parts of a lower human body during walking locomotion. Based on this analysis, the chosen motion model is described as a combination of passive and dynamic walking with a controlled, linear inverted pendulum model. When creating the simulation model, an open kinematic chain consisting of a base (the frame) and two effectors (the legs) is used. The simulation of the virtual model is realized using the software tool Matlab with its toolboxes Simulink and Simscape, and the results confirm the correctness of the design. The feasibility of the design is confirmed by creation of a physical robot skeleton using 3D printing and by a commissioning of the control system based on Atmel ATmega2560 and Raspberry Pi Zero W microcontrollers.

Keywords: bipedal robot; bio-inspired design; simulation; model-based design; LIPM; passive walker

Citation: Polakovič, D.; Juhás, M.; Juhásová, B.; Červeňanská, Z. Bio-Inspired Model-Based Design and Control of Bipedal Robot. *Appl. Sci.* **2022**, *12*, 10058. https://doi.org/10.3390/app121910058

Academic Editors: Luis Gracia and Carlos Perez-Vidal

Received: 8 August 2022
Accepted: 30 September 2022
Published: 6 October 2022

1. Introduction

The trend of using robots in all spheres of life, especially in industry, is constantly growing. We are witnessing an increasingly massive scale of production processes robotization in order to relieve people from performing periodic operations or directly creating processes that are physically impossible for humans. In addition to industrial robotics, we can currently observe an increase in interest in experimental robotics. Humanoid robots such as Atlas [1], Digit [2] or many others [3,4] represent the current pinnacle of human effort and intelligence in this technical field.

More and more progress in the field of human walking emulation also brings the seeds of practical use in terrain inaccessible to other mobile robots. Apart from uneven natural surfaces, these are mainly urban and industrial environments designed for the movement of people, not wheeled, or flying robots.

The area of interest in our contribution is to provide an analytical view of human walking from a mechanical point of view and, based on it, to design robot models at different levels of abstraction. The different levels cover models from the most primitive model—a mathematical graph of the robot, which provides information about the degrees of freedom of individual joints and the links between them, and this model is a source of data for the creation of another—a kinematic model. The next one—three-dimensional model containing the physical and material properties of the future real model, such as the weight and density of the construction material, can be considered as another level of abstraction. This model serves as a source of information about specific dimensions for the kinematic model. Based on these dimensions, a movement model of the robot is created with utilization of the trajectories of individual joints in space during walking. A simulation model is a combination of motion and three-dimensional models. The diagram of the simulation model represents the physical relationships acting on individual parts of the robot, as well as the control interface. The final product after the simulation is finished is the physical model. The key design model is a motion model of the robot, and the model is designed based on the linear inverse pendulum method. An important part of the design is the control of the robot and the construction of its physical model, based on which the main idea—identification and transfer of the key features of human walking into the engineering design of technical equipment—can be verified.

2. Analysis of the Problem Area

2.1. Dynamic Side of the Problem—Human Walking

Human biomechanics deals with the analysis of gait as part of the movements of human beings. In order to analyze human gait, the human walking was divided into the unique atomic periodic movements called step phases (Figure 1). These phases are further divided into two primary step phases, which are made up of eight secondary step phases, referred to as BAC 1–8 (Basic Action Concept). A set of two primary or eight secondary phases constitutes a gait cycle. The cycle can be measured from the start of any phase back to the start of the same phase of the same leg. A cycle is defined by two basic parameters: the cycle time and the spatial measures of the step. Based on the parameters of the cycle, the symmetry, variability and quality of the step are further determined. It is obvious that the step quality parameters are variable for each individual [5,6].

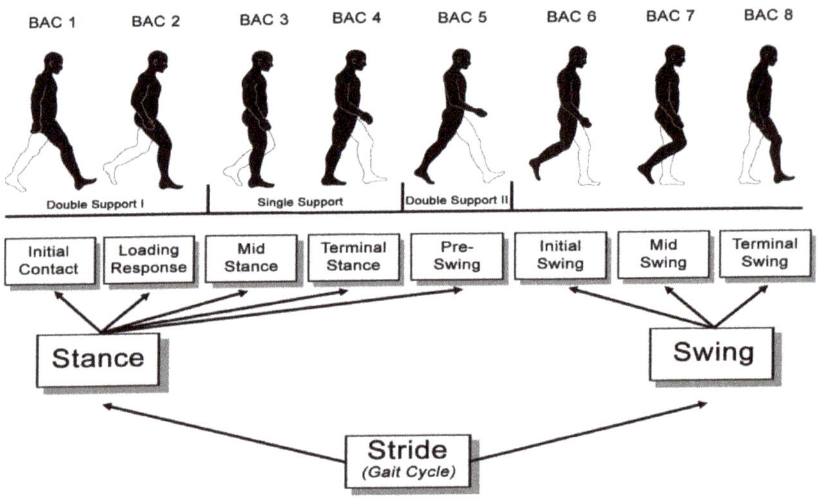

Figure 1. Visual representation of the gait cycle [7].

Two primary phases of the step are identified as the stance and the swing. During the stance phase, the tracked leg is in contact with the ground at all times. In the swing phase, the tracked leg is in the air all the time. In secondary phases, it is possible to observe three basic events [5].

The weight transfer makes up 12% of the total gait cycle. This event can be observed in phases BAC 1 (3%) to BAC 2 (9%). During this event, the initial contact of the heel with the ground takes place and damping also occurs. Damping factors are in ascending order of total contribution: bending at the knee, rolling over the heel and elasticity of the skin, muscles and tendons in the leg. During weight transfer, there is an initial increase in the moment of force M acting on the person's center of gravity [5,6].

The support of one leg makes up 38% of the total gait cycle. This event can be observed in phases BAC 3 (19%) to BAC 4 (19%). During this action, there is a movement of the upper part of the body and an overall increase in stability. At the end of the BAC 3 phase, the system (human body) acquires the greatest stability during the walking cycle. At the end of BAC 4, the person's center of gravity is so deviated that it is not possible to finish the step stably. In general, single-limb stance as an event is best defined by ankle rotation [5].

The swing makes up 50% of the total gait cycle. This event can be observed at BAC 5 (12%), BAC 6 (13%), BAC 7 (12%) and BAC 8 (13%) phases. Three sub-events occur during this event.

- Lifting the heel off the ground while keeping the toe in contact with the ground. This action is included in the BAC 5 phase. It forms the transition between the one-limbed stance and the swing itself [5].
- The movement (swaying) of the leg in front of the person's gravity center is described in the events of BAC 6 and BAC 7. The distance that the leg travels from the beginning of the BAC 6 phase to the end of BAC 7 is defined as the step length [5].
- Preparing for impact. During the BAC 8 phase, the center of gravity acquires the greatest moment of force (while reducing the acceleration) and the system (human being) prepares to absorb the impact [5,6].

Based on detailed standardized video footage of real human walking cycle considering previous movement analysis, data were experimentally collected from the figure's walking cycle, which describes the movement of the hip, knee and ankle in space and time. A virtual measurement unit for distance *j* was involved in this video footage. During one phase, three measurements were taken, in which the position of the joint was monitored on the x and y axes (α-plane—side view of the figure) and in the z and y axes (β-plane—front view of the figure). The beginning of the measurement is in phase BAC 3, because the measurement also dealt with starting from the rest position, when the joints are located on one vertical axis.

From the graphical representation of the data in Figures 2 and 3, the critical region and the damping region were identified. Relatively complicated events take place in these areas and cause significant deviations from the otherwise simply describable trajectory, especially in the ankle. The activity in this joint has a direct effect on the stability of the step. Furthermore, a deviation between the cycle lengths in different parts of the gait was observed from the data. Based on these deviations, the cycle stability function for the given figurant was visualized. The observed trajectory forms a set of desired values of joint positions in time. According to the deviations between the observed trajectory and the generated trajectory of the robot, it is possible to clearly indicate the complexity of the created model.

In order to control the gait cycle of the robot, the zero-movement point (ZMP) concept is implemented in the linear inverse pendulum movement model (LIPM).

The place of operation of the pressure force is directly in the place of contact with the surface on which the robot walks. Therefore, it is a point on the robot foot that occurs during the physical interaction of the foot with the ground. The principle of the ZMP concept is that a new walking cycle is planned based on a reference trajectory. In the case of uncontrolled passive-walking robots, the ZMP forms the point of contact of the foot arch with the ground. The position of the ZMP during the stance phase is not stationary, but changes from heel to toe. This phenomenon causes complications in the simulation of flat-footed robots because the stability changes by leaps and bounds after the application of the pressure force moves beyond the tip of the flat edge. From the point of view of control, movement based on the ZMP concept forms one of several optimal trajectories. Walking appears visually fluid and energetically efficient [8].

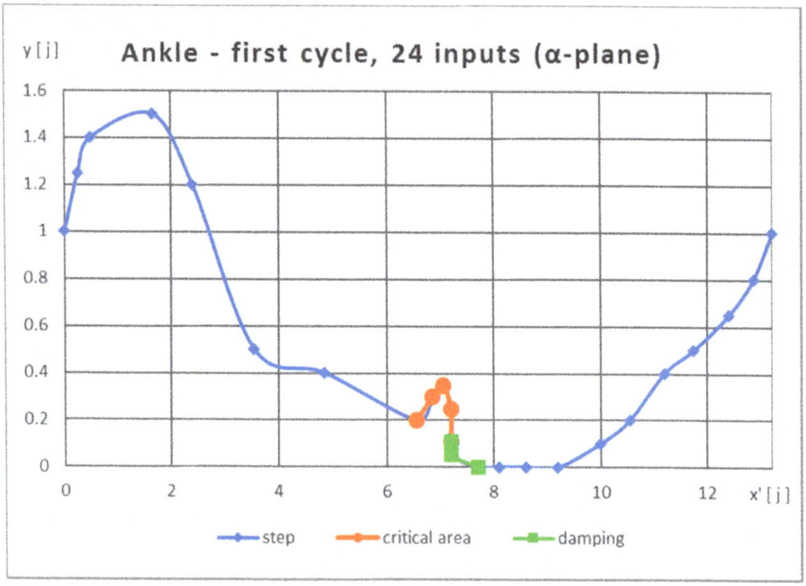

Figure 2. Ankle trajectory (the first step).

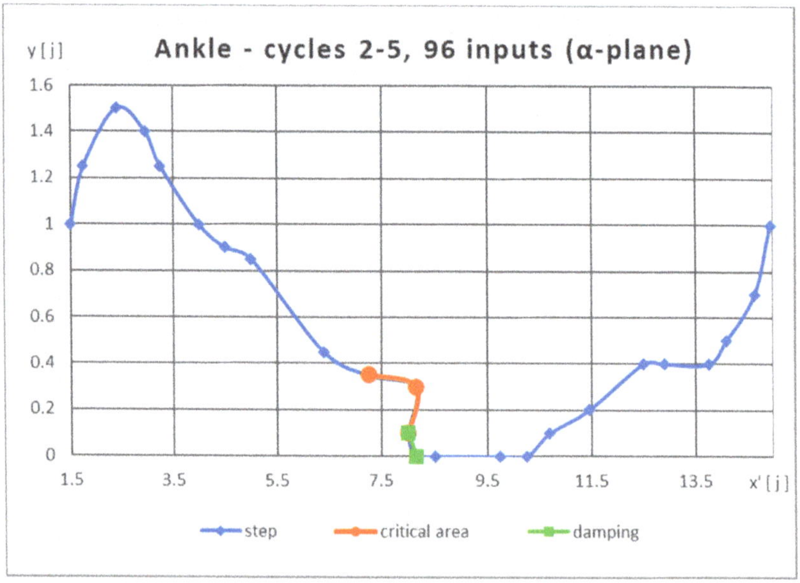

Figure 3. Ankle trajectory (the second step).

By default, the robot step is modeled according to the inverted pendulum model (LIPM). Unlike an ordinary pendulum, an inverted pendulum does not make an oscillating motion, but tends to fall to a certain side. The moving base of the pendulum tries to compensate for the fall and keep it in a vertical position. The vertical position is the unstable equilibrium position of the pendulum [9–11]. The body of the robot represents the weight of the pendulum, which is supported by the leg, which represents the arm of the pendulum, and rotates around the ankle, which represents the base of the pendulum. The step cycle is thus formed by the controlled fall of the robot body in a certain range, after exceeding which the second arm of the pendulum—the other leg—hits. A system modeled in this way is considered passive, since the dynamics of the system depends on the force of gravity and the momentum of the system. To complete one walking cycle, the momentum of the system is required to produce forward motion. A system modeled by LIPM is inherently unstable, but for each input there are initial conditions from which the system will converge to a stable output. Likewise, there is a stabilization input for all initial conditions [9,10].

The LIPM model provides some variability in the robot design process, as it is independent of the dimensions and kinetic model of the structure, or the number of robot legs. However, it assumes that the weight of the robot's body is much greater than the weight of the legs. In an ideal scenario, the weight of the legs is negligible, which is highly unlikely with a physical model. The LIPM model does not further limit the robot construction in the mutual position of the center of gravity and ZMP—the length of the pendulum arm. The implementation of the model is also possible with a higher weight of the legs, which, however, significantly reduces the stability of the gait cycle. LIPM limits the physical model only in the number of degrees of freedom in the leg, which must be at least six [12].

2.2. Static Side of the Problem—Construction Solution

The primary goal of the robot construction was to maintain an anthropomorphic shape and to copy the original as best as possible in terms of physical properties. For the proper functioning of the robot, it was necessary that:

- the ratio of the dimensions of the frame and individual arms, including the joints of the robot, was as similar as possible to the ratio of the dimensions of human limbs,
- the center of gravity of the robot was placed as close as possible to the position of the human body.

In order to fill the requirement that the construction design describes the shape of the human body as reliably as possible, one of the proportional models of the human body was chosen (Table 1). The selected model divides the distances between individual points on the human body into units of length. By observing the given proportions, it was possible to design a construction on a different scale while preserving the properties of the original. Since the shape of the body itself is not important in the LIPM model, but only the position of the center of gravity, the construction of the entire body of the robot was not absolutely necessary [13].

Table 1. Proportions of the human body according to the R.B. Hale model [13].

Point 1	Point 2	Number of Units
right hip joint	left hip joint	2
hip joint	knee	3
knee	ground (bottom of heel)	3

Subsequently, based on published information [6,14,15], an analysis of individual joints in the lower part of the human body was carried out (Table 2). A rotational degree of freedom was granted to a joint if its angle range exceeded 10° [16].

Table 2. Analysis of joints in the lower part of the human body.

Coupling	DOF	Type of Kinematic Pair
hip joint	3	spherical
knee	1	rotary
ankle	2	two cylindrical

The design of the robot also depends on the level of system control. According to this criterion, walking robots are divided into controlled and passive walking robots. Controlled walking robots are those in which the range of each robot movement is given by a controlled quantity. Passive walking robots are kinematic chains that move stably with a step without control and without an energy source. The passive step concept can be applied to both quadrupedal and bipedal robots. The movement is achieved by an initial impulse relative to the system and an inclined pad along which the robot moves downwards. A key feature of passive walking robots is the curved foot. As a rule, the bending radius is equal to the distance of the center of gravity from the ground when the robot is in a rest state. Some controlled robots are also based on this concept, where the bending of the foot—completely or in the position of the heel and toe—improves the energy efficiency of the step and eliminates a step change in stability. A combination of controlled and passive joints in walking robots is also frequent. A typical example is the reduction of the knee joint in the robot's leg in exchange for a shock absorber. Another example is a robot leg composed of controlled joints in the hip and knee, or the other knee on the same leg, and a passive foot containing a spring in the toe. The stiffness of this spring ensures a flat foot shape in the secondary phases of the gait cycle, BAC 1 to 4 and later in BAC 7 to 8. In BAC phases 5 and 6, the foot bends under its own weight to such an extent that it directs part of the robot's weight into the toe, similar to human walking. Of course, both mentioned methods can be replaced by a controlled process. Usually, a combination of controlled and passive elements of the robot is used to simplify complicated movements that are more difficult to control [17,18].

3. Modeling and Simulation of Biped Robot

The modeling process was based directly on the analysis results and involved designing the robot at different levels of abstraction and generating walking trajectories. The individual levels created a simulation model for which a control system was proposed in the next stages.

3.1. Design of a Simulation Model of a Bipedal Robot

The starting point for the design of the robot was the kinematic model of the robot, which can be obtained in different ways depending on the principles used [19,20]. In this case, the modeling process was dependent on the analyzed walking model LIPM and ZMP, which was adapted to the number of degrees of freedom of the robot. Since it is a mobile robot, no part of the robot is firmly attached to the ground, so we cannot talk about a mechanism, but about a kinematic chain. The kinematic chain was divided into three main parts.

- **The frame** is a mobile base on which two manipulators are attached. The center of gravity of the robot is also located in this part and forms the material part (the head) of the inverse pendulum. The frame represents the human pelvis.
- **The left leg** is the manipulator of the robot, one end of which is attached to the left side of the frame, and the other end is terminated by an effector. The manipulator consists of four arms and six joints. The manipulator represents the arm of the inverted pendulum and the human leg, where the effector fulfills the role of the foot.
- **The right leg** is a manipulator functionally identical to the left leg, but it is fixed on the right side of the frame.

Based on the movement analysis, it follows that there are two degrees of freedom in the ankle and up to three in the hip joint. The implementation of several degrees of freedom in one place—such as in a spherical kinetic pair—was realized by dividing the joint into three rotary kinematic pairs with the smallest possible arms. These rotation pairs rotate around different axes of the coordinate system. In this way, the three degrees of freedom of the spherical pair were preserved with a simpler implementation in practice. One manipulator has six degrees of freedom, just like the analyzed template. Five degrees of freedom are controlled and one degree of freedom at both ends of the chain is uncontrolled. The robot has two arms on a frame with six degrees of freedom. The additional six degrees of freedom were granted because it is a mobile robot which can freely move and rotate in space in all directions. The number of freedom degrees of the kinetic chain

$$n_{robot} = 6 + \sum_{i=1}^{j} n_i = 6 + 12 = 18 \geq 6, \quad (1)$$

where n_i represents the number of degrees of freedom of the i-th kinematic pair and j is the total number of kinematic pairs, is greater than the minimum required number for the LIPM model.

The graph of the robot in Figure 4 shows the relationships between individual kinematic pairs. In the vertices of the graph, the number of degrees of freedom in each pair is indicated. The mobility of the robot is represented by the absence of a link between the ground and the kinematic chain. Furthermore, it can be observed that the graph of the robot is acyclic; therefore, we speak of an open kinematic chain.

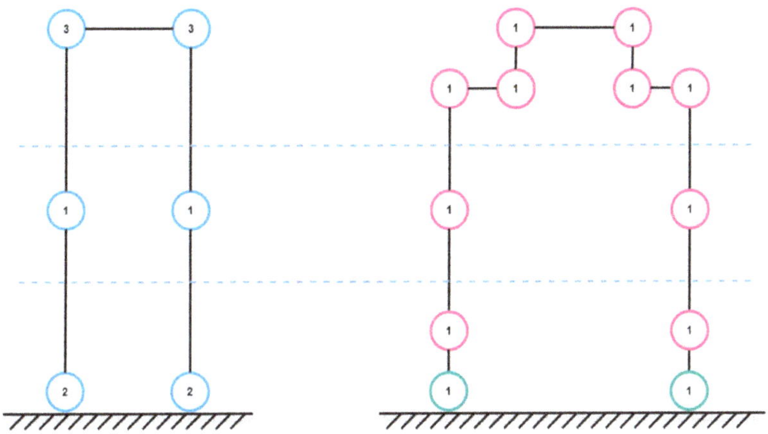

Figure 4. Robot graph based on robot analysis (left), robot graph adapted for robot construction (right).

The kinematic model of the robot was derived from the robot graph (Figure 5). Kinematic pairs were defined in space and therefore in this model we are already talking about rotational joints with one degree of freedom. Relationships between joints are now represented by arms. This is a model with a combined type of walking (passive and controlled).

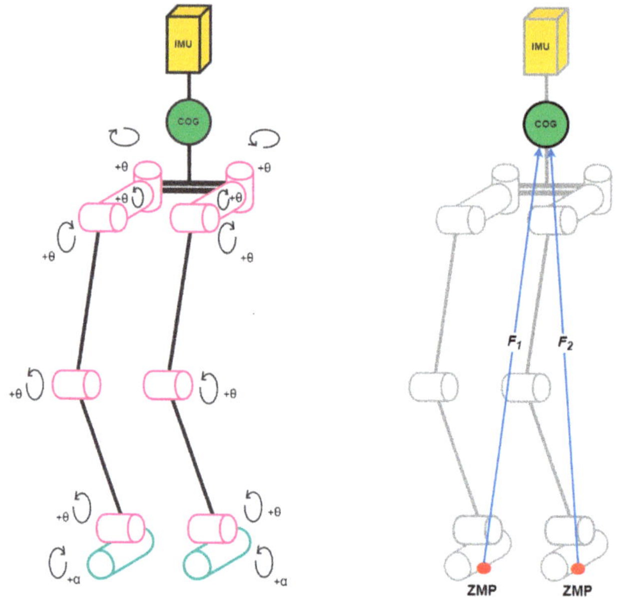

Figure 5. Kinematic model of the robot (left), visualization of LIPM and ZMP principles on the kinematic model (right).

Denavit–Hartenberg (DH) parameters were determined in the kinematic model (Figure 6). Conventional DH parameterization uses a uniform method for marking the axes of coordinate systems in individual joints. It uses two rules:

$$\text{DH1}: x_i \perp z_{i-1};\ \text{DH2}: |x_i \cap z_{i-1} \vee 1. \quad (2)$$

Therefore, in DH, the representation of the homogeneous transformation contains two parameters less than the usual homogeneous transformation, which facilitates the analysis of the n-arm manipulator [21].

In the process of identifying the DH parameters, the axes of rotation were first marked depending on the types of joints. Convention states that the axis of rotation is the z-axis. Based on the positive direction of rotation, the direction of the axis was determined. For some opposing joints, the direction of the z-axis was chosen differently to preserve the natural direction of rotation. Subsequently, two different origins of zero coordinate systems were chosen, specifically for the right and left leg. In both cases, the origin was chosen at the joint that connects the leg to the frame. Unlike the rest of the centers of the coordinate systems, these were not numbered. The center of the coordinate system in the joint connecting the left leg and the frame was marked O_0, and the center in the joint connecting the right leg and the frame was marked O_0'. A different zero-system naming convention for walking robots led to this decision. After finding the remaining origins of the coordinate systems, the x and y axes were determined by convention according to relations (3) and (4).

$$x = z_{i-1} \times z_0 \quad (3)$$

$$y = z \times y \quad (4)$$

In effector coordinate systems, the z-axis is generally oriented in the same way as the axis of the last arm. Thus, the two non-controlled joints at the ends of the kinematic chain will rotate along the y-axis, not along the z-axis like the controlled joints. In both cases, the origins of the effectors coordinate systems were chosen at the point created by the intersection of the axis perpendicular to the ground, passing through the center of gravity, and the straight line created when the given effector touches the ground when the robot is in the starting position. Under ideal conditions, this point is identical to the ZMP. The analysis of the DH parameters of the robot's legs is shown in Figure 6. The specific values of the DH parameters of the kinematic model of the robot are listed in Table 3.

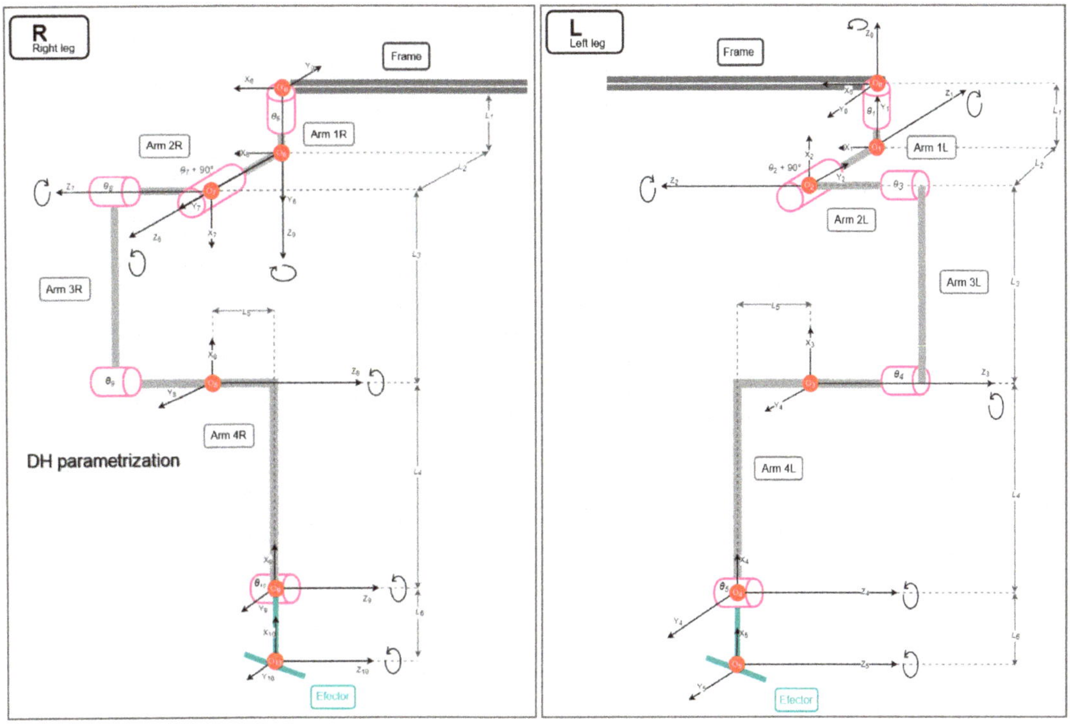

Figure 6. DH parameterization of the right (R) and the left (L) legs.

Table 3. DH parameters of the robot kinematic model.

Left a_i [mm] [1]	**α_i [deg]** [2]	**d_i [mm]** [3]	**θ_i [deg]** [4]
0	90	L_1	θ_1
0	−90	L_2	$\theta_2 + 90$
L_3	180	0	θ_3
L_4	0	L_5	θ_4
L_6	0	0	θ_5
Right a_i [mm] [1]	**α_i[deg]** [2]	**d_i [mm]** [3]	**θ_i[deg]** [4]
0	90	L_1	θ_6
0	90	L_2	θ_7—90
L_3	180	0	θ_8
L_4	0	L_5	θ_9
L_6	0	0	θ_{10}

[1] perpendicular distance between axes z_{i-1} and z_i, measured along the axis x_i.
[2] the angle between z_{i-1} and z_i, measured from z_{i-1} in a plane perpendicular to x_i.
[3] perpendicular distance between axes x_{i-1} and x_i, measured along the axis z_{i-1}.
[4] the angle between x_{i-1} and x_i, measured from x_{i-1} in a plane perpendicular to z_{i-1}.

In the DH parameterization, the centers of the coordinate systems and their relative positions were found. Despite the fact that the effectors seem to fulfill the role of joints in the kinematic chain, this phenomenon is not included in the DH parameters, since they are uncontrolled "joints" between the ground and the robot. The parameterization of the effector was developed by creating a three-dimensional model of the entire kinetic chain based on the analyzed kinematic model.

When modeling the effector, knowledge from the materials for the construction of the foot of a passive walking robot was used. In the starting position, when in contact with the ground, the effectors create support in the form of two parallel lines. Since the position of the effector itself is controlled, the length of the foot was designed to be as small as possible in order to maintain relative stability in the controlled direction of rotation. The foot width was determined to be as large as possible for the longest arch of its lower part. The resulting shape of the foot design is shown in Figure 7.

Figure 7. Proposal of a robot effector.

Due to the chosen ratio of the weights of the legs together with the effectors, with respect to the weight of the robot frame, the requirement for the use of the LIPM walking model for the designed three-dimensional model is fulfilled. The basic parameters of the robot are listed in Table 4.

Table 4. Parameters of the three-dimensional model.

Property	Value	Unit
the height of the robot's center of gravity	445	mm
height of the center of gravity of the frame	645	mm
foot height	30	mm
the length of the foot arch	144	mm
the weight of the robot	2.755	kg

The resulting dimensions from Table 4 were used in motion modeling using the inverse pendulum model. In the modeling process, the center of gravity of the robot represented the head of the pendulum in the form of a mass point p defined in space by default as $p = (x, y, z)$ according to (5)–(7). This position is described in this model as a system of state variables $q = (\theta r1, \theta p1, z)$ in which r represents the length of pendulum arm, or robot leg in this specific case [22].

$$x = rS_p; S_p \equiv sin\theta_p \tag{5}$$

$$y = -rS_r; S_r \equiv sin\theta_r \tag{6}$$

$$z = rD;\ D \equiv \sqrt{1 - Sr^2 - Sp^2} \tag{7}$$

τ_r, τ_p and f, as vectors acting on the center of gravity, arm and head of the pendulum in Formula (8), represent the inputs to the system of equations describing the movement of the inverse pendulum in space, in the Cartesian right-handed system [22] as

$$m\begin{pmatrix} \ddot{x} \\ \ddot{y} \\ \ddot{z} \end{pmatrix} = \left(J^T\right)^{-1}\begin{pmatrix} \tau_r \\ \tau_p \\ f \end{pmatrix} + \begin{pmatrix} 0 \\ 0 \\ -mg \end{pmatrix}, \tag{8}$$

where m is the mass of the center of gravity. Since the arm of the pendulum is considered as a link, the variable m belongs to the mass point p. The Jacobian J is an antisymmetric matrix dependent on the current configuration $q(t)$ and for motion analysis, according to [21,22], has the form

$$J = \frac{\partial p}{\partial q} = \begin{pmatrix} 0 & rC_p & S_p \\ -rC_r & 0 & -S_r \\ \frac{-rC_rS_r}{D} & \frac{-rC_pS_p}{D} & D \end{pmatrix}; C_r \equiv cos\theta_r, C_p \equiv cos\theta_p \tag{9}$$

A similar Jacobian was later used in the recalculation of joint velocities of the robot in order to determine the inverse kinematics. In this case, it was used to create the dynamic equations of the pendulum, and therefore it is multiplied by the vector $(\ddot{x},\ddot{y},\ddot{z})$ from the left to create a mathematical model of the inverse pendulum in space [21,22] according to the relation

$$m \begin{pmatrix} 0 & rC_p & S_p \\ -rC_r & 0 & -S_r \\ \frac{-rC_rS_r}{D} & \frac{-rC_pS_p}{D} & D \end{pmatrix} \begin{pmatrix} \ddot{x} \\ \ddot{y} \\ \ddot{z} \end{pmatrix} = \begin{pmatrix} \tau_r \\ \tau_p \\ f \end{pmatrix} - mg \begin{pmatrix} -rC_rS_r \\ \frac{-rC_pS_p}{D} \\ D \end{pmatrix} \tag{10}$$

In order to use this model to generate step trajectories, restrictions on the movement of the pendulum along the z-axis were defined. These constraints ensure that the height of the pendulum head does not change under the influence of the inputs $q(t)$. However, this height can be changed manually. This restriction is represented in a form of help vector $(k_x, k_y, -1)$ of controlled inputs of height of the robot and the intersection z_c of the z-axis. The step width was determined as the length of the arch of the foot; i.e., the maximum possible. The parameters for the design of the pendulum are listed in Table 5 [22].

Table 5. Three-dimensional model parameters for the LIPM model.

Parameter	Model	Value	Unit
spread of the legs in the starting position	$L_r + L_5$	0.091	m
height of the frame in the starting position	h	0.644	m
frame height while walking	h_W	0.570	m
spread of the legs during gait	L_a	0.144	m
step length in walking direction	L_s	0.020	m

Step generation took place in three stages. In the first stage, the movement of the center of gravity and the ZMP in space was identified (Figure 8). According to the limitations of the model, the height of the center of gravity was anchored. This phenomenon simulated the fact that the legs are not extended when walking as in a fully passive walking robot. A step with bent knees reduces the movement of the center of gravity along the z-axis. At this stage, the pendulum consisted of a head and one arm, which alternately played the role of the robot's left and right legs. The weight of the pendulum head is preset to 1 and the weight of the arm is assumed to be zero.

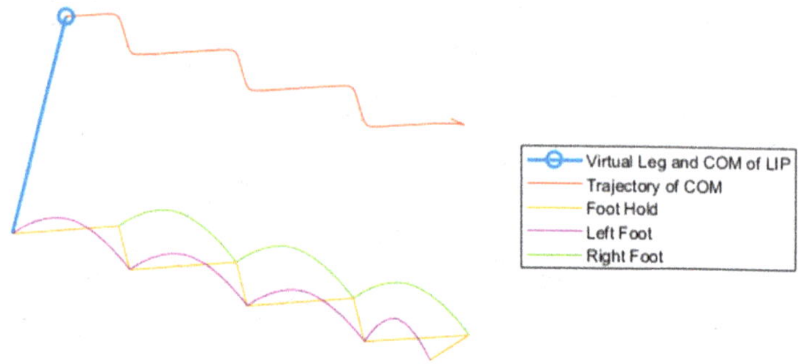

Figure 8. Generated trajectories of the center of mass and the position of the ZMP.

In the second stage, the foot trajectory was generated based on the mutual position of the ZMP and the movement of the center of gravity. Trajectories were generated for eight cycle runs. The analysis showed that during the first two runs of the cycle, stride length and movement were less similar than during the rest of the runs. This phenomenon can also be observed in the first two steps of the model, when the feet in the initial position are located at a spacing of 91 mm from the robot axis, given by the construction. The model reaches the predetermined step width of 144 mm and length of 20 mm only during the third run of the cycle. The trajectories were stored in the data-structure *footinfos* 1×8 (Figure 9), where the column number determined the number of cycle run. One data element contained a 1×200 vector (*timevec*) with time values sampled for a one-second period. Furthermore, this element also contained the joint positions in space stored in two 6×200 matrices for the left leg (*footleft*) and the right leg (*footright*). The joints positions are represented by the amount of each joint rotation with respect to the defined direction using DH parameterization.

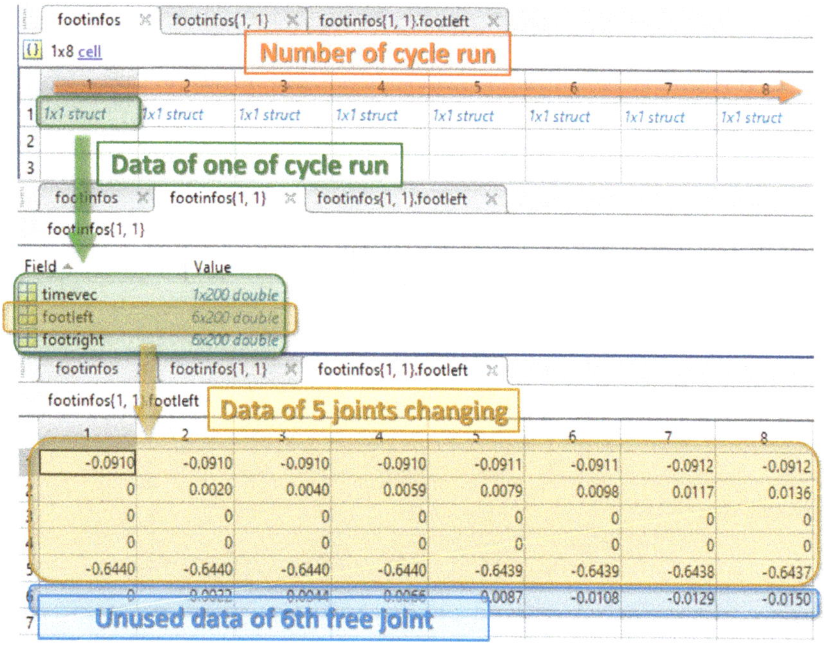

Figure 9. Data-structure covering the changes of DH parametrization in kinematic, 3D and simulation models.

In the third stage, the trajectories were recalculated to the controlled variables of the DH parameters using the inverse kinematics task, and the result is the motion model of the robot made from a 3D material model, which was created with Autodesk Inventor software.

3.2. *Simulation of a Bipedal Robot Walking*

The simulation model of the biped robot was created by exporting the three-dimensional model to the Matlab-Simscape simulation environment. The three-dimensional model was divided into 27 parts, which created links corresponding to the three-dimensional model in the simulation environment. This process was automated using the Simscape Multibody Link toolbox. Six degrees of freedom were added to the model for kinematic chain simulation. Contact forces on the lower part of the effectors were also created. These forces ensure that the bottom of the foot behaves like an ideal rigid body when in contact with another object.

The axes of rotation in the joints were adjusted to the z-axis in order to follow the convention of the DH parameterization and the positive direction of rotation was chosen according to the kinematic model. The resulting block diagram of the simulation model is shown in Figure 10.

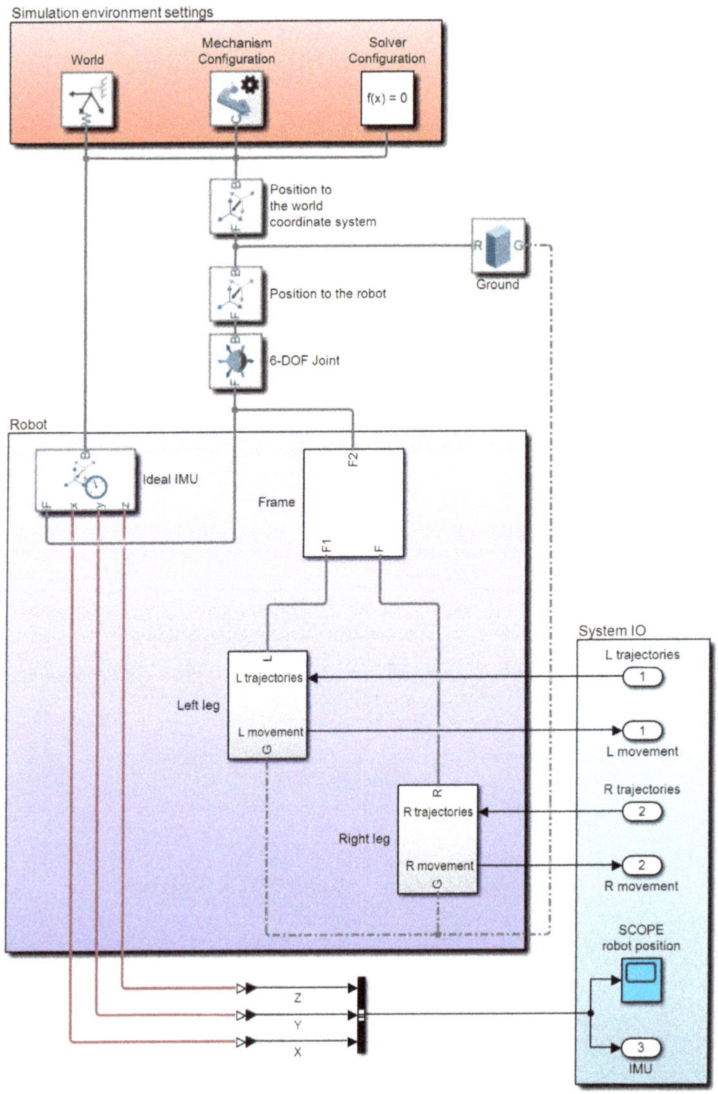

Figure 10. Simulation model of bipedal robot walking.

The output of the inverse kinematic task is the required quantity in the unmodified form W′. The quantity W′ must be filtered before entering the controller, because the generated trajectory is in the form of a function that the simulation solver cannot process correctly. Sharp local extremes of the trajectories were evaluated as noise and removed with a low-pass filter. The filter was implemented using the Filter Design Toolbox as a part of Matlab R2022a software, which allows the user to design the required filter in a short time even without expert experience in the field of signal processing.

From the diagram of the simulation model shown in Figure 11, it follows that the control system is based on the principle of regulating the speeds and positions of the servomotors. To increase the level of robot autonomy, an ideal orientation sensor was included in the system. The sensor detects the acceleration and position of the frame in three axes. The value on the z-axis was monitored at the output of the system. If the deviation from the planned position is exceeded by 30 mm, the robot will be turned off in an emergency. The reason is to prevent damage to the robot due to redundant movements in an unwanted position on the ground.

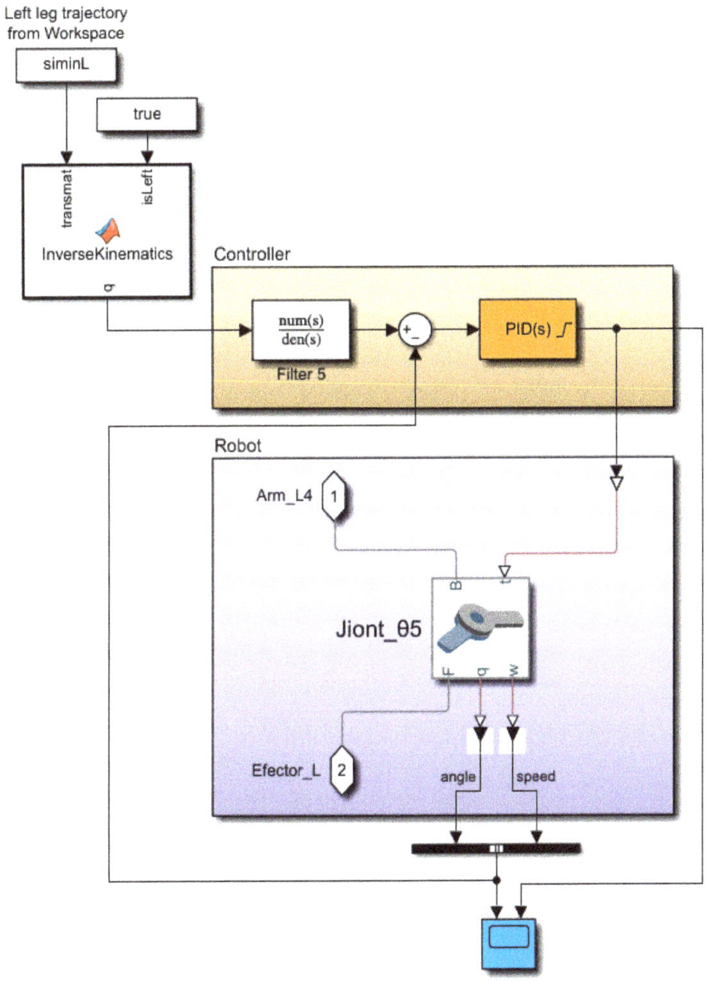

Figure 11. Closed control loop for joint number 5.

The result of the simulation at this stage is the fact that the designed model of the robot can perform all eight planned steps from the starting position. The length and width of the first three steps contained negligible deviations from the movement model, which may be caused by using passive joints. The remaining seven steps were identical to the motion model. Figure 12 shows the simulation model in motion in different design phases from different perspectives.

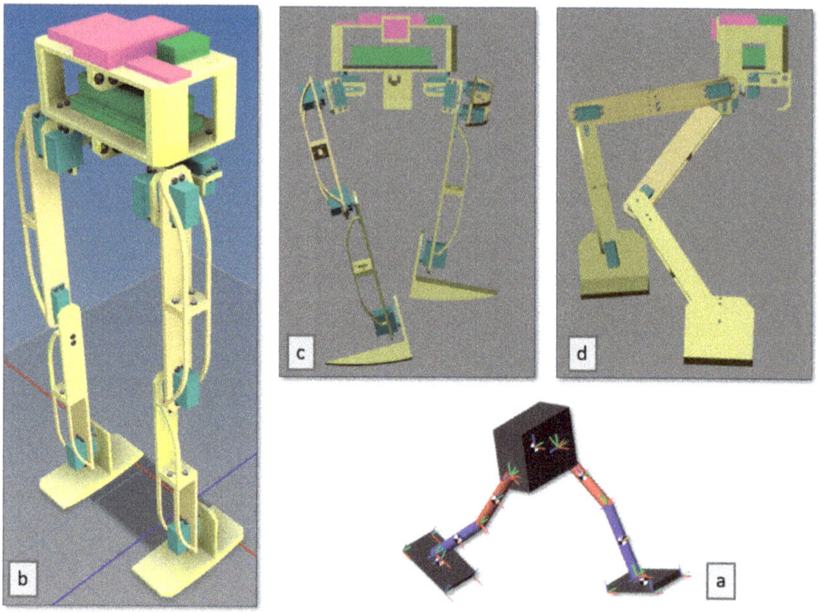

Figure 12. Simulation of walking of the bipedal robot (**a**) basic kinematic model; (**b**) static 3D model; (**c**) walking 3D model—front view; (**d**) walking 3D model—side view.

4. Realization of a Physical Model of a Bipedal Robot

The supporting structure of the robot was made using 3D printing technology. This technology has proven itself as an effective method of manufacturing parts, mainly due to its sufficient accuracy and speed for prototyping purposes. The advantage of the material used in printing the parts is the low weight compared to other parts of the robot and the possibility to modify the parts even after printing. For most parts, the wall thickness was set to four layers and the material filling between the walls was 70%. The Micro-Electro-Mechanical Systems (MEMS) storage components and the component for stabilizing the robot in a seated position were printed with only three layers of wall thickness and 50% infill, as these components do not represent the load-bearing elements of the structure. On the contrary, the effectors were printed with a wall thickness of six layers and a filling of 80%.

4.1. Construction of a Bipedal Robot

The construction of the robot consists of three main parts.

The robot frame is made up of eleven parts and is divided into two floors. The first floor is intended for energy sources and is closed from above and below. The second floor is for the robot's control interface and is open at the top. For this reason, simple handling of the interface is possible without the need to disassemble the frame. Each floor consists of two parts that are connected by a detachable joint. After connecting the two floors, two components were attached to the front and back of the upper floor to hold the two MEMS systems. In the space on the first floor, there is a part that allows the robot to sit without support.

The robot's legs are made up of twelve parts, five of which are unique. The two parts are used to secure the servos together to create three degrees of freedom for the 3R and 4L arms. The connection of these three motors simultaneously creates arms 1L, 1R, 2L and 2R. The rest of the arms are structurally very similar. Each of them is composed of two parts connected by a detachable joint.

Therefore, a new frame height was derived from the simulation model and compared with the measured value. The resulting deviation is largely attributed to the resulting deviations between the lengths of L6, L3, L4 and L1.

All the differences identified in this phase of testing can be involved in the next phases of the spiral model of the engineering design of mechatronic systems in order to achieve an ideal overlap between the design, simulation and final implementation of the solution.

The second part of the testing was focused on monitoring of stability of the robot at different step lengths. The measurement in which the robot passed all eight steps from the generated trajectory was considered successful. Five measurements were taken at each step length. The robot walked on a horizontal concrete pad in a closed room. The robot walked independently without physical contact with another object. A successful step is considered the one after which the walking cycle is completed without falling. The beginning of the measurement of all cycles was in phase BAC 3. The results of the second part of the test experiments are shown in Table 7.

Table 7. Functionality testing results of a bipedal robot.

Test No.	Step Length [mm]	Number of Successful Measurements	Number of Failed Measurements	The Largest Number of Steps Taken in a Failed Measurement (X/8)
1	50	5	0	-
2	60	5	0	-
3	70	5	0	-
4	80	5	0	-
5	90	5	0	-
6	95	5	0	-
7	100	5	0	-
8	105	5	0	-
9	110	5	0	-
10	115	5	0	-
11	120	5	0	-
12	125	5	0	-
13	130	4	1	3
14	135	2	3	3
15	140	1	4	2
16	145	1	4	2
17	150	0	5	2
18	155	0	5	0
-	-	$\sum = 90$	MAX = 3	

From the test results it is clearly evident that the first three steps are critical, after which the gait is stabilized. The test experiments show that the robot is able to walk stably with steps up to 12.5 cm long. The number of all measurements during the second phase of testing was 90. The measurements took place without interference and external influences. It is necessary to emphasize that this testing stage served for verification of control system design proposal. The correctness of the intended idea in defined form was confirmed, but at the same time, it is necessary to declare that the designed solution is not ready for independent continuous infinite walking yet, except for some stable steps. The constructed and tested prototype of the two-legged robot is shown in Figure 14.

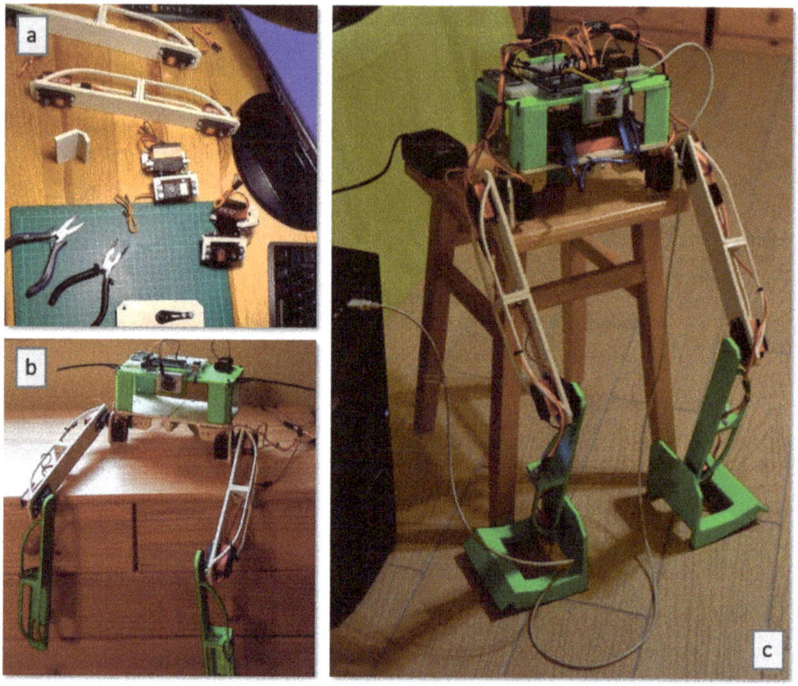

Figure 14. A tested prototype of bipedal robot MaRoŠ (**a**) construction parts; (**b**) partial construction; (**c**) final version.

6. Conclusions

The topic of the development of humanoid robots is very extensive. Several possible approaches to the way of controlling bipedal robots in order to ensure movement as similar as possible to human walking have been published. This article is focused on one of the possible methods using the analysis of real human walking. The principle of model-based design is used, while the bio-inspired design is based on the analysis of the movements of individual parts of the human body while walking. The control system is designed based on a combination of passive and dynamic walking with a controlled, linear inverted pendulum model. Validation of the designed model using the Matlab simulation tool confirmed the feasibility of the design. The result of this article is the finding that the proposed concept can be developed further, after fine-tuning the details, as one of the real options for the design of walking humanoid robots. This statement is based on the course and results of the final experiments with the physically constructed prototype of the bipedal robot MaRoŠ. The development of the prototyping solution is constantly ongoing, inspired by the spiral model of software product development, and the goal is to completely reconcile the simulation and physical model of the robot, together with the optimization of energy use in the mechatronic subsystems of the robot.

Author Contributions: Conceptualization, D.P. and M.J.; methodology, D.P., M.J. and B.J.; software, D.P., M.J. and B.J.; validation, D.P., M.J. and Z.Č.; formal analysis, D.P. and Z.Č.; resources, M.J., B.J., D.P. and Z.Č.; data curation, D.P.; writing—original draft preparation, D.P. and M.J.; writing—review and editing, M.J., D.P. and B.J.; visualization, D.P. and M.J.; supervision, M.J.; funding acquisition, M.J., B.J. and Z.Č. All authors have read and agreed to the published version of the manuscript.

Funding: This research was funded by VEGA agency, grant number 1/0176/22 "Proactive control of hybrid production systems using simulation-based digital twin".

Conflicts of Interest: The authors declare no conflict of interest. The funders had no role in the design of the study; in the collection, analyses, or interpretation of data; in the writing of the manuscript, or in the decision to publish the results.

References

1. Boston Dynamics. ATLAS^{TM}. Available online: https://www.bostondynamics.com/atlas (accessed on 1 April 2022).
2. IEEE. Robots—Your Guide to the World of Robotics. Available online: https://robots.ieee.org/robots/digit (accessed on 1 April 2022).
3. Yang, X.; She, H.; Lu, H.; Fukuda, T.; Shen, Y. State of the Art: Bipedal Robots for Lower Limb Rehabilitation. *Appl. Sci.* **2017**, *7*, 1182. [CrossRef]
4. Gulletta, G.; Erlhagen, W.; Bicho, E. Human-Like Arm Motion Generation: A Review. *Robotics* **2020**, *9*, 102. [CrossRef]
5. Perry, J.; Burnfield, J.M. *Gait Analysis: Normal and Pathological Function*; Slack Incorporated: West Deptford, NJ, USA, 2010; p. 551, ISBN 978-1556427664.
6. Čapek, L.; Hájek, P.; Henyš, P. *Biomechanika Člověka*; Grada Publishing: Prague, Czech Republic, 2018; ISBN 978-80-271-0367-6.
7. Stöckel, T.; Jacksteit, R.; Behrens, M.; Skripitz, R.; Bader, R.; Mau-Moeller, A. The Mental Representation of the Human Gait in Young and Older Adults. *Front. Psychol.* **2015**, *6*, 943. [CrossRef]
8. Mitobe, K.; Capi, G.; Takayama, H.; Yamano, M.; Nasu, Y. A ZMP Analysis of the Passive Walking Machines. In Proceedings of the SICE 2003 Annual Conference (IEEE Cat. No.03TH8734), Fukui, Japan, 4–6 August 2003; Volume 1, pp. 112–116.
9. Omer, A.; Hashimoto, K.; Lim, H.-O.; Takanishi, A. Study of Bipedal Robot Walking Motion in Low Gravity: Investigation and Analysis. *Int. J. Adv. Robot. Syst.* **2014**, *11*, 139. [CrossRef]
10. Lanari, L.; Hutchinson, S.; Marchionni, L. Boundedness Issues in Planning of Locomotion Trajectories for Biped Robots. In Proceedings of the 2014 IEEE-RAS International Conference on Humanoid Robots, Atlanta, GA, USA, 18–20 November 2014; pp. 951–958.
11. Bakaráč, P. Konštrukcia a Riadenie Inverzného Kyvadla. Master's Thesis, Slovak University of Technology in Bratislava, Bratislava, Slovakia, 2017.
12. Erbatur, K.; Kurt, O. Natural ZMP Trajectories for Biped Robot Reference Generation. *IEEE Trans. Ind. Electron.* **2009**, *56*, 835–845. [CrossRef]
13. Prokopenko, S. Human Figure Proportions—Cranial Units—Robert Beverly Hale. Available online: https://www.proko.com/human-figure-proportions-cranium-unit-hale (accessed on 1 April 2022).
14. Vachálek, J.; Takács, G. *Robotika*; STU Publishing: Bratislava, Slovakia, 2014; p. 166. ISBN 978-80-227-4163-7.
15. Somisetti, K.; Tripathi, K.; Verma, J.K. Design, Implementation, and Controlling of a Humanoid Robot. In Proceedings of the 2020 International Conference on Computational Performance Evaluation (ComPE), Shillong, India, 2–4 July 2020; pp. 831–836.
16. Hamilton, N.; Weimar, W.; Luttgens, K. *Kinesiology: Scientific Basis of Human Motion*, 12th ed.; McGraw Hill: New York, NY, USA, 2012; p. 640, ISBN 978-0078022548.
17. McGeer, T. Passive Dynamic Walking. *Int. J. Robot. Res.* **1990**, *9*, 62–82. [CrossRef]
18. Yamane, K.; Trutoiu, L. Effect of Foot Shape on Locomotion of Active Biped Robots. In Proceedings of the 2009 9th IEEE-RAS International Conference on Humanoid Robots, Paris, France, 7–10 December 2009; pp. 230–236.
19. Virgala, I.; Miková, Ľ.; Kelemenová, T.; Varga, M.; Rákay, R.; Vagaš, M.; Semjon, J.; Jánoš, R.; Sukop, M.; Marcinko, P.; et al. A Non-Anthropomorphic Bipedal Walking Robot with a Vertically Stabilized Base. *Appl. Sci.* **2022**, *12*, 4108. [CrossRef]
20. Che, J.; Pan, Y.; Yan, W.; Yu, J. Leg Configuration Analysis and Prototype Design of Biped Robot Based on Spring Mass Model. *Actuators* **2022**, *11*, 75. [CrossRef]
21. Babinec, A. *Riadenie Robotických Manipulátorov. Lectures*; Faculty of Electrical Engineering and Information Technology STU: Bratislava, Slovakia, 2019.
22. Kajita, S.; Kanehiro, F.; Kaneko, K.; Yokoi, K.; Hirukawa, H. The 3D Linear Inverted Pendulum Mode: A Simple Modeling for a Biped Walking Pattern Generation. In Proceedings of the 2001 IEEE/RSJ International Conference on Intelligent Robots and Systems. Expanding the Societal Role of Robotics in the the Next Millennium (Cat. No.01CH37180), Maui, HI, USA, 29 October–3 November 2001; Volume 1, pp. 239–246.
23. Robotshop. Arduino Mega 2560 Datasheet. Available online: https://www.robotshop.com/media/files/pdf/arduinomega2560datasheet.pdf (accessed on 1 April 2022).
24. Raspberry Pi Foundation. Raspberry Pi Hardware. Available online: https://www.raspberrypi.org/documentation/hardware/raspberrypi (accessed on 1 April 2022).
25. Adafruit. Adafruit BNO055 Absolute Orientation Sensor. Available online: https://learn.adafruit.com/adafruit-bno055-absolute-orientation-sensor (accessed on 1 April 2022).

applied sciences

MDPI

Article

A Symmetry Evaluation Method, Using Elevation Angle, for Lower Limb Movement Patterns during Sitting-to-Standing †

Kensuke Nakamura and Norihiko Saga *

Department of Engineering, Kwansei Gakuin University, 1 Gakuenuegahara, Sanda 669-1330, Hyogo, Japan
* Correspondence: saga@kwansei.ac.jp

† This paper is an extended version of the proceedings of the IMEKO XXIII World Congress held in Yokohama, Japan, 30 August–3 September 2021. *Measurement: Sensors* **2021**, *18*, 100195. https://doi.org/10.1016/j.measen.2021.100195.

Abstract: In this paper, we propose a method that uses the femoral and tibial elevation angles to quantitatively evaluate the symmetry of lower limb movement during the transition from a sitting position to a standing position. In kinematic analysis of the transition from sitting to standing, the angles of the three joints of the lower limb are often measured. However, due to the large number of variables, it is difficult to evaluate the symmetry of the lower limb movement by comparing data from the six joints of the left and right lower limbs. In this study, therefore, we measured the femoral and tibial elevation angles of healthy participants and rehabilitation patients and visually and numerically evaluated the symmetry and asymmetry of the movement of the left and right lower limbs. We were able to identify the kinematically major lower limbs in the transition from sitting to standing and quantify the symmetry of the movement patterns of the left and right lower limbs. Furthermore, we examined the possibility that the method could be effectively used in the rehabilitation field to evaluate the motor co-ordination that constitutes the lower limb movement pattern in the transition from the sitting to standing position, such as the gait plane rule.

Keywords: symmetry; sit-to-stand; elevation angle; coordination

Citation: Nakamura, K.; Saga, N. A Symmetry Evaluation Method, Using Elevation Angle, for Lower Limb Movement Patterns during Sitting-to-Standing. *Appl. Sci.* **2022**, *12*, 9454. https://doi.org/10.3390/app12199454

Academic Editors: Luis Gracia and Carlos Perez-Vidal

Received: 17 August 2022
Accepted: 17 September 2022
Published: 21 September 2022

1. Introduction

In our daily lives, we repeatedly travel to and from our destinations to accomplish our goals, often choosing walking as our means of transportation. However, disabled or elderly people who cannot walk safely may choose a wheelchair as their means of transportation [1]. When we walk or sit in a wheelchair, we need to perform a standing movement (sitting to standing (STS)). However, disabled and elderly people may not be able to perform the STS movement and may not be able to continue to live independently. Therefore, sometimes, the goal of rehabilitation in hospitals and nursing homes is to ensure that the patient achieves the STS ability [2]. Post-hip-fracture and post-stroke patients targeted for rehabilitation have been reported to have asymmetric STS movements due to muscle weakness and impaired sensory integration in the lower limbs [3,4]. Pao-Tsai et al. reported that post-stroke patients who had experienced falls had greater asymmetry in the weight distribution to the lower limbs during the STS movement [5]; the asymmetry of the STS movements is shaped by the associated muscle forces and joint moments. Previous studies examining weight bearing in the lower limbs of post-stroke patients have reported that patients often shift their center of gravity to the nonparalyzed lower limb [6]. However, mechanical asymmetry of the lower limb during the STS movement may reduce safety when the person is performing an even more advanced and dynamic gait [7]. Ryoichiro et al. reported an STS rehabilitation system that improves the functional asymmetry of the lower limb caused by disease [8]. These previous studies indicate that it is important for rehabilitation to improve the muscle strength, movement, and asymmetric STS movement of the asymmetric lower limb.

In previous studies on STS symmetry, lower limb joint torque and floor reaction force were selected as mechanical parameters and joint angle as a kinematic parameter parameters that can be observed in measurement experiments [9–11]. These parameters are observed as a result of the human central nervous system controlling the muscles (the locomotion organs). In Japan today, cerebrovascular disorders are the leading cause of the need for long-term care [12] and normalization of muscle activity patterns and center-of-gravity trajectories during movement through rehabilitation-incorporating robots is becoming the treatment of choice [13]. In other words, rehabilitation of the central nervous system, which controls muscle activity temporally and spatially, rather than rehabilitation to increase motor muscle mass and the force exerted by the muscles, is attracting attention Ningjia et al. analyzed muscle synergy during the STS movement in stroke patients [14] However, this report mentions asymmetry based on the analysis of muscle synergy but does not propose a quantitative method for evaluating symmetry and asymmetry. Wendy et al proposed a symmetry index in their report on STS symmetry evaluation [9]. This method measures the angles of the right and left (R–L) lower limb joints during the STS movement and calculates the ratio of the angles as a symmetry index. Michalina et al. selected ankle, knee, and hip joint angles, joint torque, and floor reaction force as symmetry indices [11] These previous studies converted the observed information into a single value as an index of symmetry and cannot represent the motion patterns during the STS movement.

The planar law of gait (PLG) is a well-known evaluation method that allows us to observe the coordination patterns of the neurophysiological body movement based on changes in joint angles observed over time during movement [15,16]. The PLG involves the three-dimensional plotting of the elevation angles (EAs) of three segments (thigh, shin, and foot) during walking. Gianluca et al. used the PLG of the L–R lower limbs to present the symmetry of the walking motion as visual information [17]. They evaluated the symmetry; if the PLG could be applied to the STS movement and the STS movement symmetry could be quantitatively evaluated, it would be possible to geometrically evaluate the symmetry of the control by the central nervous system during the STS movement. However, there are different factors involved in the STS movement and gait. In the gait, the EAs of three segments (thigh, shank, and foot) are observed, but in the STS movement the EA of the foot is always constant because the foot is restrained above the floor and only changes in the EAs of two segments (thigh and shank) are observed. Therefore, the observed changes in the EAs of the two body segments in the STS movement cannot be plotted on 3D coordinates as in the gait. Furthermore, since the STS movement is not a periodic motion similar to a gait, the PLG cannot be directly adapted to the STS movement.

2. Methods

2.1. Participants

This study included 4 healthy participants (22.3 $\pm$ 0.5 years) and 4 rehabilitation patients (82.8 $\pm$ 3.1 years). The selection criterion for healthy participants was the absence of trauma or a disability that would affect the STS motion at the time of participation. In the case of patients, those who were judged by a rehabilitation specialist to be able to complete the STS movement unaided were included. The study was conducted after the participants were informed about the study and after they signed a consent form. Table 1 provides physical information about the healthy participants and the rehabilitation patients.

Table 1. Summary of the participants' physical information.

Group	Code	Gender	Age	Height (cm)	Weight (kg)	BMI	Muscle Mass (kg)	Body Fat (%)
Health (N [1] = 4)	H-1	Male	22	169.0	54.0	18.9	42.3	17.3
	H-2	Male	22	176.0	81.8	26.4	59.7	23.1
	H-3	Male	22	172.0	68.4	23.1	49.8	23.3
	H-4	Male	22	177.0	67.6	21.6	49.2	23.2
	Mean ± SD [2]		22.3±0.5	173.5±3.7	67.9±11.4	22.5±3.1	50.2±7.2	21.7±3.0
Patient (N = 4)	P-1	Male	86	139.0	48.8	25.2	27.7	40.3
	P-2	Female	78	165.5	40.3	14.7	35.6	6.5
	P-3	Female	81	156.0	56.0	23.0	33.9	36.0
	P-4	Female	86	134.5	33.0	18.2	23.1	27.2
	Mean ± SD		82.8 ± 3.1	148.8 ± 11.2	44.5 ± 7.8	20.3 ± 3.7	30.1 ± 4.5	27.5 ± 11.6

[1] Number of participants. [2] Standard deviation

2.2. Experiment

In this study, motion capture experiments were conducted with healthy participants at Kwansei Gakuin University from 20 February 2019 to 24 February 2019 and with rehabilitation patients at Toyonaka Heisei Hospital on 7 March 2020. The 40 motion capture markers were placed on landmarks throughout the body [18]. In this study, four markers (the lateral ankle, the lateral 1/3 of the lower tibia, the lateral epicondyle of the knee, and the lateral 1/3 of the lower thigh) were placed on one lower extremity, and participants sat in a chair with a height of 43.5 cm and without a backrest. Participants were instructed to not allow their upper limbs to come into contact with their body or the external environment during the STS movement. The motion capture experiment was conducted as follows (Figure 1).

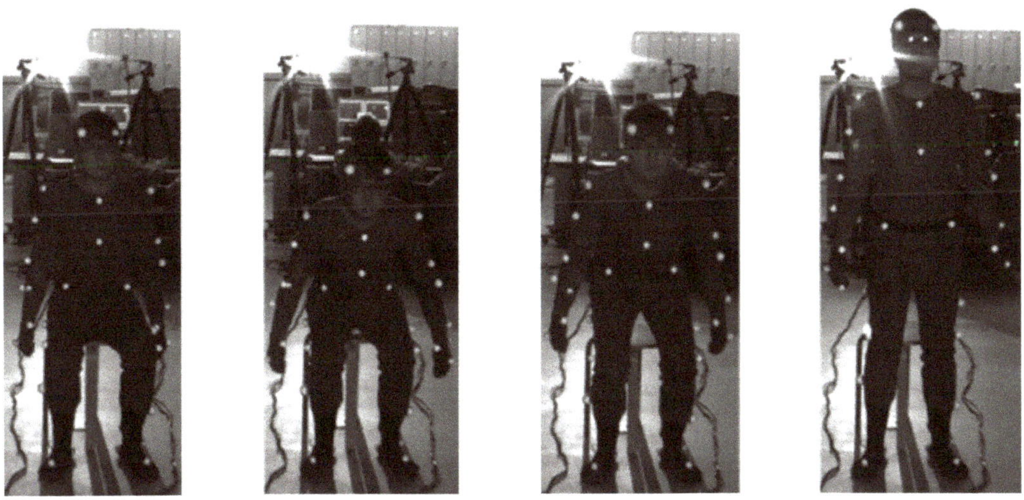

(a) (b) (c) (d)

Figure 1. STS measurement experiment. The upper extremity did not touch the body or the outside environment. The plantar was fixed in an arbitrary position that was not interchanged from the beginning to the end of the STS motion. (**a**) The start of the STS motion from the stationary starting posture at the signal, (**b**, **c**) the STS motion in progress, and (**d**) the end of the STS motion when the ending posture is stationary.

1. Each participant sat in a chair and maintained their posture to initiate the STS movement.
2. When we judged that the participant's sitting posture was stationary, we signaled the start of STS.
3. On hearing the cue, the participant initiated the STS movement.
4. When the participant stood up and we judged that the standing posture was stationary, we signaled the end of the STS movement.
5. The video images from the cue for the start of the movement to the cue for the end of the movement were included in the analysis.

A calibration space with an area of 150 cm × 150 cm and a height of 195 cm centered on the participant's feet was set up (Figure 2). The captured images were recorded using Capture-Ex (Library Co., Ltd., Tokyo, Japan) at a sampling frequency of 50 Hz. Move-tr/3D (Library Co., Ltd., Tokyo, Japan) was used to convert the marker information in the video into coordinate data, and KineAnalyzer (KISSEI COMTEC Co., Ltd., Nagano, Japan) was used to measure the EAs of the thigh and shank. A 2 Hz low-pass filter was applied to all marker data. The EA formed by the shank and the vertical axis was defined as θ_S, and the EA formed by the thigh and the vertical axis was defined as θ_T (Figure 3).

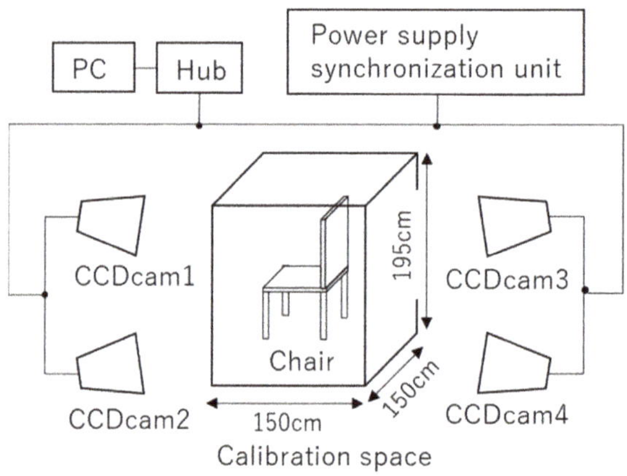

Figure 2. Experimental setup. A calibration space (150 cm × 150 cm × 195 cm) photographed with four CCD cameras.

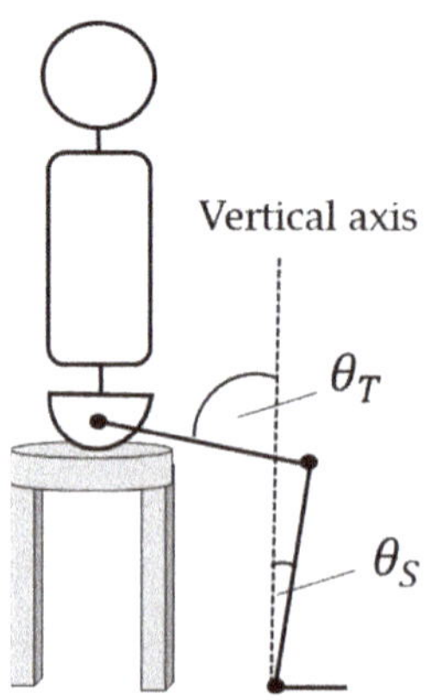

Figure 3. Definition of EAs for the thigh and the lower leg. The EA was measured using KineAnalyzer. The angle between the vertical axis and the thigh axis was defined as θ_T, and the angle between the vertical axis and the lower leg was defined as θ_S.

2.3. Analysis

Data for each of the healthy and rehabilitation patient groups were analyzed, and statistical tests were performed on the results. Details of the analysis are provided below.

2.3.1. Definition of the EA error

During the STS motion, the shank EA was defined as θ_S and the thigh EA was defined as θ_T. In the first experiment of this study, the L–R errors of θ_S and θ_T were calculated. The indices of the L–R error were the shank EA error SE_{ave}, the thigh EA error TE_{ave}, and the overall error E_{sum}. Formulas (1)–(3) provide the calculation of each index.

$$E_{sum} = SE_{ave} + TE_{ave}, \tag{1}$$

$$SE_{ave} = \frac{\sum|Right\ \theta_{Si} - Left\ \theta_{Si}|}{n}, \tag{2}$$

$$TE_{ave} = \frac{\sum|Right\ \theta_{Ti} - Left\ \theta_{Ti}|}{n}, \tag{3}$$

where n is the total number of samples for each participant and i is the i-th sample. The total number of samples was calculated by dividing the STS measurement time for each participant by the sampling frequency of 50 Hz.

2.3.2. Analysis of Measured EA changes

Two-dimensional Cartesian coordinates C_1 were created with the shank EA θ_S as the y-axis and the thigh EA θ_T as the x-axis, and n points F_i (θ_{Si}, θ_{Ti}) were recorded on these coordinates. Point F_i was recorded on coordinate C_1 for the R–L lower limbs. The position vector norm pointing from the origin to point F_i was then calculated. Another two-dimensional Cartesian coordinate C_2 was created, with the position vector norm $\|a_{Ri}\|$ of the right lower limb as the x-axis and the position vector norm $\|a_{Li}\|$ of the left lower limb as the y-axis. Thus, a line (the R–L vector line) consisting of n points f_i (a_{Ri}, a_{Li}) on the two-dimensional coordinate C_2 was illustrated. Finally, a symmetrical reference line y = x was created on the two-dimensional coordinate C_2 and the average error ME and the average error sum of squares MSE at each point of the target reference line and the R–L vector line were calculated using Equations (4) and (5):

$$ME = \frac{\sum(\|a_{Li}\| - \|a_{Ri}\|)}{n}, \tag{4}$$

$$MSE = \frac{\sum\sqrt{ME^2}}{n}. \tag{5}$$

3. Results

3.1. Symmetrical Comparison Using Measured EAs

Figures 4 and 5 display the changes in shank EA θ_S and thigh EA θ_T during the STS motion. Figure 4 shows the results for each participant in the healthy group, and Figure 5 shows the results for each participant in the rehabilitation patient group.

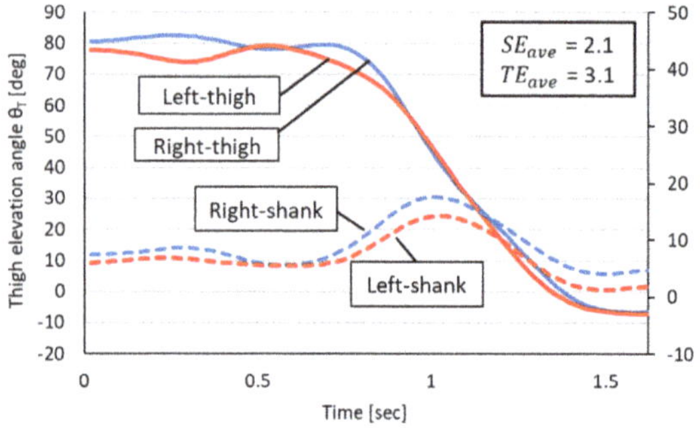

(**a**) H-1

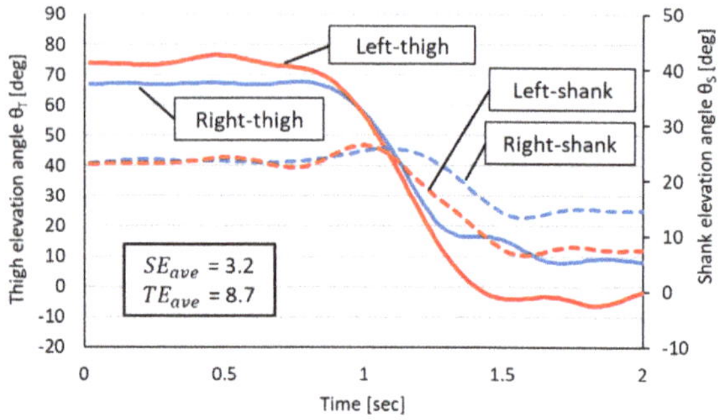

(**b**) H-2

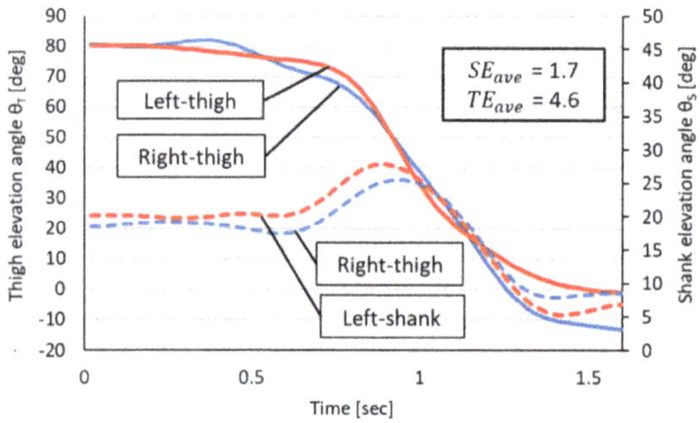

(**c**) H-3

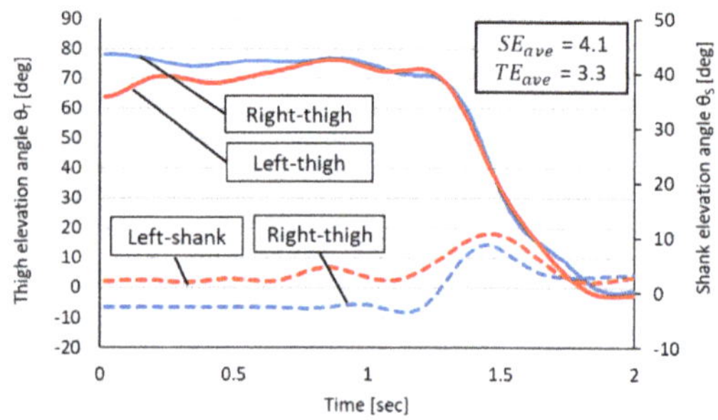

(**d**) H-4

Figure 4. Measured angular changes in thigh and shank EAs in the healthy group. The horizontal axis is the STS time (in seconds) of each participant. H-1 through H-4 in the figure refer to each healthy participant in Table 1.

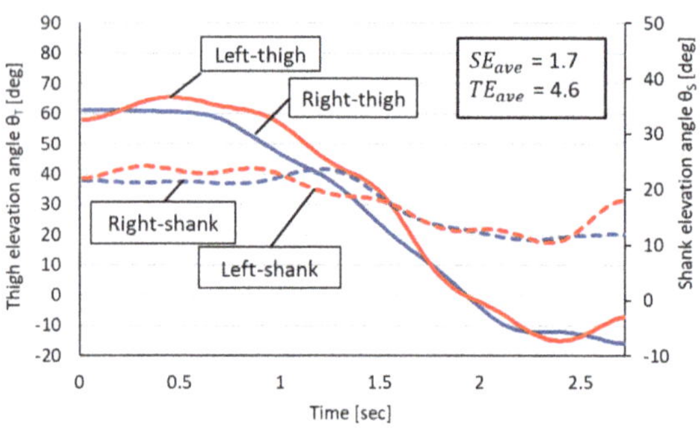

(**a**) P-1

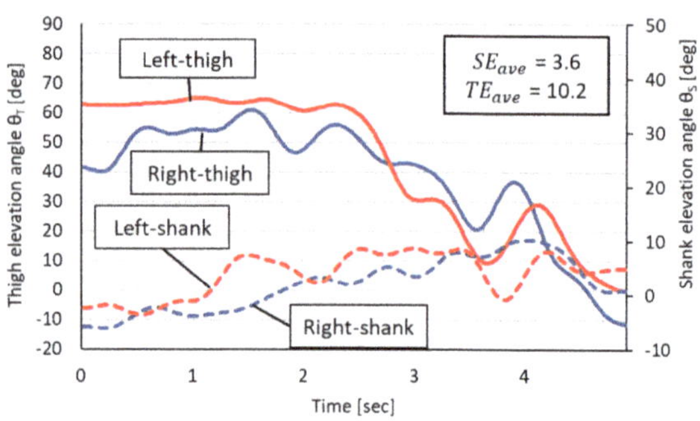

(**b**) P-2

Figure 5. *Cont.*

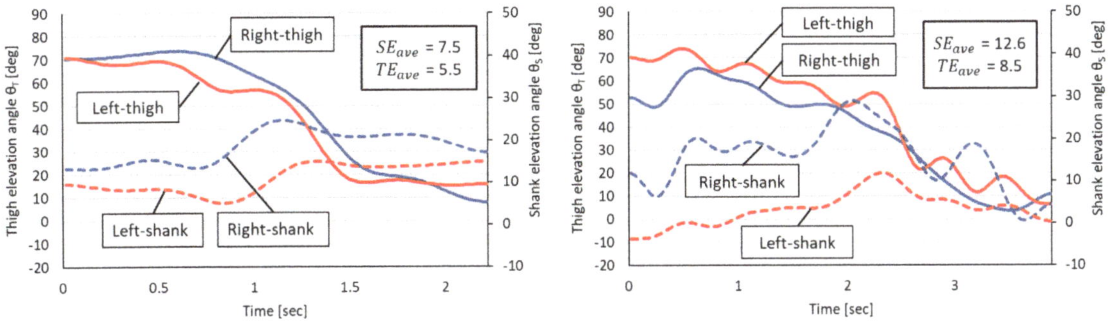

(**c**) P-3 (**d**) P-4

Figure 5. Measured angular changes in thigh and shank EAs in the rehabilitation patient group. The horizontal axis is the STS time (in seconds) of each participant. P-1 through P-4 in the figure refer to each rehabilitated patient in Table 1.

On the basis of the measured EA changes in the healthy group, the following four phases were commonly observed in the healthy participants:

- Phase 1: The STS start angle;
- Phase 2: A decrease in the thigh EA and an increase in the shank EA;
- Phase 3: A decrease in the thigh EA and a decrease in the shank EA;
- Phase 4: The STS end angle.

A common characteristic of the healthy group was a small L–R error in the decrease in the thigh EA corresponding to Phases 2 and 3 but a trend toward larger L–R errors in thigh and shank EAs for Phases 1 and 4.

The EA changes in the rehabilitation patient group were characterized by two features. The first was that irregular EA changes occurred with shifting STS phases, making it impossible to identify the Phases 1–4 observed in the healthy group. The second was that the R–L thigh EAs were not as close as in the healthy group during Phases 2 and 3, even though the EA changes were similar in appearance to those in the healthy group (e.g., Patients 1 and 3).

Table 2 summarizes the R–L EA errors in the healthy group and the rehabilitation patient group. In this experiment, the mean and the standard deviation of each EA error SE_{ave}, TE_{ave}, and the error sum E_{sum} of the rehabilitation patient group were larger than those of the healthy group. A two-tailed Student's t-test (R—4.0.2) at the 5% level of significance for the above three errors showed no significant difference between the two groups.

Table 2. EA error in both groups.

Group	Code	Sample (n [2])	SE_{ave} (deg/n)	TE_{ave} (deg/n)	E_{sum} (deg/n)
Health (N [1]=4)	H-1	81	2.1	3.1	5.2
	H-2	101	3.2	8.7	11.9
	H-3	81	1.7	4.6	6.3
	H-4	101	4.1	3.3	7.4
Mean ± SD		91.0 ± 11.5	2.8 ± 1.1	4.9 ± 2.6	7.7 ± 3.0
Patient (N = 4)	P-1	136	1.7	4.6	6.4
	P-2	246	3.6	10.2	13.9
	P-3	111	7.5	5.5	13.0
	P-4	195	12.6	8.5	21.2
Mean ± SD [3]		172.0 ± 60.6	6.4 ± 4.8	7.2 ± 2.6	13.6 ± 6.1

[1] Number of participants. [2] Number of samples. [3] Standard deviation. A significance test for the above three errors showed no significant difference between the two groups.

3.2. Comparison of R–L Symmetry by the R–L Vector Line

Figures 6 and 7 present the R–L vector lines for the healthy participants and the rehabilitation patients, respectively. Here, the numbers 1–4 assigned to each participant refer to the same participant's results for the measured EAs (Figures 4 and 5). Table 3 shows the *ME* and the *MSE* obtained from the R–L vector line and the symmetric reference line; the *ME* is the error in the position vector norm of the R–L lower limbs, which approaches 0 if the motion patterns of the R–L lower limbs are symmetric ($\|a_{Ri}\| = \|a_{Li}\|$).

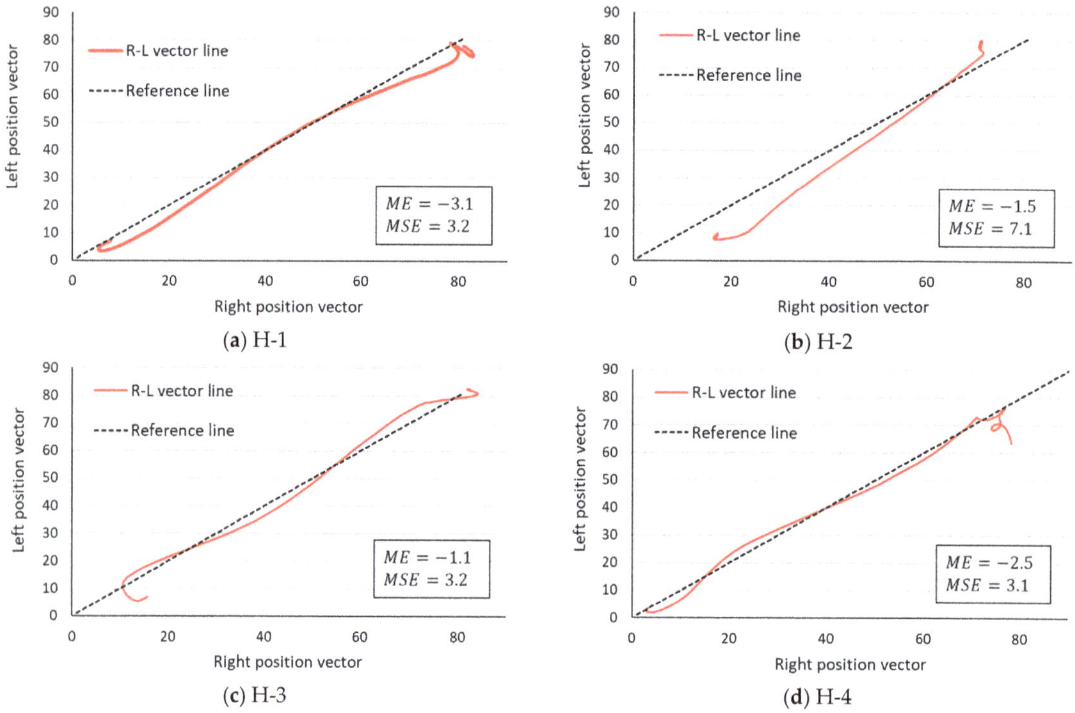

Figure 6. R–L vector line in the healthy group. The lower limb movement pattern is symmetrical enough to be drawn near the central symmetry reference line. H-1 through H-4 in the figure refer to each healthy participant in Table 1.

Table 3. EA errors in both groups.

		Sample	ME	MSE
Health (N = 4)	H-1	81	−3.1	3.2
	H-2	101	−1.5	7.1
	H-3	81	−1.1	3.2
	H-4	101	−2.5	3.1
Mean ± SD		91.0 ± 11.5	−2.0 ± 0.9	4.2 ± 2.0
Patient (N = 4)	P-1	136	3.1	3.6
	P-2	246	3.8	9.6
	P-3	111	−6.9	7.2
	P-4	195	−4.1	6.6
Mean ± SD		172.0 ± 60.6	−1.0 ± 5.3	6.8 ± 2.5

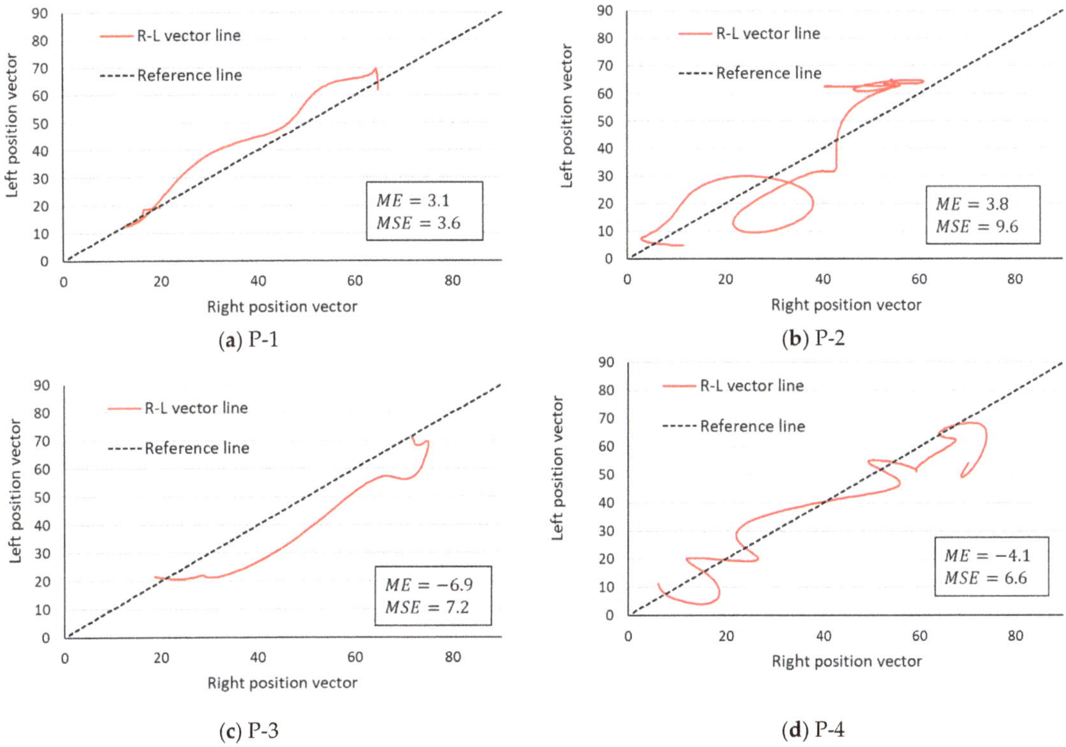

Figure 7. R–L vector line in the rehabilitation patient group. The lower limb movement pattern is symmetrical enough to be drawn near the central symmetry reference line. P-1 through P-4 in the figure refer to each rehabilitated patient in Table 1.

If $ME > 0$, then the point f_i (a_{Ri}, a_{Li}) is distributed more on the y-axis side of the symmetry reference line, and if $ME < 0$, then the point f_i is distributed more on the x-axis side of the symmetry reference line. In the healthy participants, four *ME* values were negative, while in the rehabilitation patient group, two were positive and the remaining two were negative. Next, the *MSE* is a parameter that quantifies the asymmetry of movement patterns. The *MSE* of the healthy participants was 4.2 ± 2.0, while the *MSE* of the rehabilitation patient group was 6.8 ± 2.5. The *MSE* of the rehabilitation patient group was approximately 1.6 times that of the healthy group, but a two-tailed Mann–Whitney U test (R—4.0.2) at the 5% significance level showed no significant difference between the two groups.

4. Discussion

In post-total-hip-arthroplasty patients, the weight bearing on the lower limb during the STS motion is asymmetric because it is biased toward the healthy lower limb. In post-stroke patients, the center of foot pressure during the STS movement is asymmetric because it is biased toward the nonparalyzed side. Thus, previous studies have shown that muscle torque and joint angle changes exerted in the gait and during the STS movement are asymmetric due to disability or muscle weakness in one lower limb [7,19,20]. It was assumed that the rehabilitation patient group in this study would also have asymmetric lower limb joint angle changes during the STS motion due to disability and muscle weakness.

In patients with post-stroke syndrome, the muscle synergy of the R–L lower limbs is reportedly related to the asymmetry of the lower limb joint angle changes during the STS motion [21]. It is already known that the muscle synergy in such patients is altered compared with normal muscle synergy [22,23], and it was inferred that the patients in this study also had altered motor synergy of the unilateral lower limb due to disability and muscle weakness, resulting in asymmetric R–L lower limb movement during the STS motion. Therefore, we hypothesized that it is important to bring the synergy of the impaired lower limb of the patient closer to the normal state for the patient to recover the STS ability [21] and that improving the R–L lower limb coordination during the STS movement means that the muscle or motor synergy of the impaired lower limb used to perform the STS movement is closer to the synergy expressed by the healthy lower limb. We hypothesized that this would be the case. For this reason, we thought that a method for evaluating the coordination of R–L lower limbs during the STS movement was necessary and sought a method for evaluating this coordination from the perspective of both EAs and the R–L vector line in this study.

The R–L differences in shank and thigh EAs were also observed in the healthy participants. The reason for the observed differences in EAs in the healthy participants, despite the absence of a functional impairment, is presumably the habitual use of the dominant leg. Although there is no difference in muscle strength between the dominant and nondominant leg [24,25], the dominant leg has priority in postural control [26]. In addition, placing one leg posteriorly at the start of the STS movement can reduce the external tension moment of the hip joint [27], which may have caused the difference in EAs in the healthy group. In the rehabilitation patients, both EA values and temporal–spatial differences were observed. These temporal–spatial differences were visually complex and difficult to analyze. Therefore, we performed an analysis using R–L vector lines.

The PLG is a geometric method used to identify the cooperative structure of lower limb movement patterns in the gait [15,16]. However, the PLG represents the coordinated motion of one lower limb and it cannot represent the coordinated movement of the R–L lower limbs. Therefore, in this paper, we proposed a method that presents information on the changes in the EA of the R–L thighs and tibia as a single line (the R–L vector line). This method suggests that the closer the drawn R–L vector line is to the symmetry reference line, the more parallel it is to the target reference line, and the higher is the degree of synchronization of the EA and the change per time point between the R–L segments. In other words, the symmetry of the movement pattern can be evaluated as high. The STS movement has a muscle coordination structure that is reportedly similar to that of walking [28,29], and the movements of the R–L lower limbs during the STS motion are considered to be the result of coordination. Therefore, we hypothesized that the symmetry of the coordinated movements of the R–L lower limbs during the STS motion could be evaluated by the R–L vector line. The *ME* of the R–L vector line analysis results provides information that can be used to identify which motion (left or right) is dominant based on the sign of the value (positive or negative) and its magnitude. In a previous study analyzing the STS motion of patients with femoral neck fractures, it was reported that the angular displacement of the knee joint and the hip joint is greater in the nonaffected lower limb, where the peak joint moment values are greater [30]. Therefore, by observing the *ME*, the lower limb that is the primary source of force used to perform the STS motion can be estimated, helping to interpret the visually complex information on elevation displacement. Previous studies have reported asymmetry in the STS motion in patients after a lower limb fracture or a stroke [4,5,7,19,20]. According to the R–L vector line and the sign of the *ME* value, two trends were observed in the rehabilitated patient group. The first is patients with increased use of one lower limb during STS movements (Figure 7, P-1 and P-3), where the R–L vector line was also drawn on the side indicated by the sign of the *ME* value. The second was a patient with alternating and irregular increases in R–L lower limb use (Figure 7, P-2 and P-4). In this case, it was difficult to determine whether the predominantly used lower limb was the right or left one only by observing the R–L vector line, but the direction of the predominantly used lower limb could be determined from the sign of the *ME* value. On the other hand, compared with the rehabilitation patients, the R–L vector line of the healthy participants passed near the symmetric reference line. In other words, the R–L lower limbs of the healthy participants showed the possibility of coordinated temporal and spatial movement. Chun et al. reported that a robotic rehabilitation intervention for stroke patients improves the muscle synergy asymmetry between the affected and nonaffected lower limbs [31]. In the study by Chun et al., the symmetry between R–L lower extremities was assessed by the correlation coefficient of muscle synergy, but it only provides information on the similarity of muscle synergy changes between R–L lower extremities, not explicit symmetry. Another weakness of R–L symmetry evaluation using only muscle synergy is that the number of synergies must be the same in the R–L lower limb as a condition for comparison, and the results of muscle synergy analysis vary depending on the number and types of muscles being investigated [32].

The R–L vector line proposed in this study will clearly demonstrate the symmetry of STS R–L lower limb movements when the R–L vector line, which plots STS movements, approaches the symmetry reference line through rehabilitation of the patient. In addition, since the EAs of the thigh and lower leg are used, the problem of differences in the measurement target affecting the analysis results can be avoided. This makes the determination of the effectiveness of the rehabilitation of R–L symmetry reliable and easy to perform.

The MSE, which excludes the positive and negative signs of the ME and calculates the magnitude of the error between the symmetric reference line and the R–L vector line as a numerical value, is the mean error value obtained by dividing this value by the measurement time of each participant. Previous studies have reported that hemiplegia and pain reorganize the cooperative structure of muscles differently from the healthy side [14,33,34]. The R–L asymmetry of the lower limb movement pattern during the STS motion in the rehabilitation patient group could be quantified by MSE values. In the present study, the number of participants in both groups was small (four in each) and there was no significant difference in the statistical test for the MSE. Therefore, we can only mention the possibility that the MSE is an indicator that can determine asymmetry. If the sample size is too small, the power of the test is estimated to be small [35]. In a previous study on the comparison of peak muscle synergy values, 21 participants, 12 with mild stroke sequelae and 7 with severe stroke sequelae, were compared [21]. A study examining the accuracy of the perception of the asymmetry of lower limb weight bearing during standing movements compared 19 stroke survivors and 15 healthy participants [36]. In contrast to these previous studies, our study had a short duration, which did not allow us to have a large number of participants, particularly rehabilitation patients. By increasing the duration of the study and obtaining the cooperation of several medical institutions, we could increase the number of rehabilitation patients. If we could increase the number of participants, it would be possible to study more clearly the magnitude of the MSE error between healthy participants and rehabilitation patients and to calculate a cutoff value. If we could calculate a cutoff value for the MSE, we would have an R–L vector line that would make it easier to visually determine the effectiveness of rehabilitation and treatment, aiding the staff working in clinical settings.

A limitation of this study is the large age difference between healthy participants and rehabilitation patients. The typical rehabilitation patient admitted to a medical facility is an elderly person. To determine the effectiveness of rehabilitation, it is desirable to compare the STS performance of patients and healthy participants of a similar age. However, since it cannot be said that elderly patients who are considered healthy do not experience musculoskeletal or cardiovascular diseases, close attention must be paid to the definition of STS performance as a normal model. In addition, because this study evaluated R–L lower extremity coordination during the STS motion based on EAs, the theory was limited to a kinematic perspective. If we could compare the muscle synergies of the same participants during the STS motion, errors between the neurological assessment and the kinematic assessment obtained from the results of this study could be identified.

5. Conclusions

In this study, we proposed a method for evaluating the improvement in the R–L synergy of the impaired lower limb during the STS movement, i.e., the measurement of the EA and the measurement of the R–L vector line, which is a secondary type of information obtained from the EA. The information obtained from the EAs of multiple body segments of the R–L lower limbs is complex, with many variables, making it difficult to evaluate coordination. In contrast, the R–L vector line can be represented by a single line in two-dimensional coordinates, and the ME and the MSE facilitate comparison with a symmetrical reference line by a numerical representation. This means that the R–L vector line utilizes the kinematic synergy of the R–L lower limbs and can visually represent the difference in the kinematic coordination of the R–L lower limbs. The conventional assessment of R–L differences using muscle synergy consisting of multiple muscles can be depicted with fewer variables.

Because of the small number of participants in this study and the fact that the MSE did not yield useful conclusions from the statistical tests, we were unable to quantitatively determine the presence or absence of R–L differences using the MSE cutoff values. In the future, we hope to calculate the cutoff value of the MSE by increasing the number of participants and establish this method as a symmetry evaluation method. Furthermore, we would like to apply this method to the PLG, which can be expressed in two degrees of freedom, to verify whether this method can be applied not only to STS but also to the symmetry evaluation of the gait cycle of the R–L lower limbs in the gait.

Author Contributions: K.N., conceptualization and writing of the original draft; N.S., discussion of the original draft, review, and editing. All authors have read and agreed to the published version of the manuscript.

Funding: This research was partially funded by Kwansei Gakuin University.

Institutional Review Board Statement: Not applicable.

Informed Consent Statement: Not applicable.

Data Availability Statement: Not applicable.

Conflicts of Interest: The authors declare no conflict of interest.

References

1. Susan, S. The long-term impact of wheelchair delivery on the lives of people with disabilities in three countries of the world. *Afr. J. Disabil.* **2017**, *6*, a344. [CrossRef]
2. Sui, F.H.; Avril, T.; Tricia, M.; Julie, M.; Andrew, K. Automated Movement Feedback for Recovering Independence in the Sit-to-Stand Movement in an Older Population: A Pilot Randomised. *OBM Geriatr.* **2019**, *3*, 1904089. [CrossRef]
3. Janet, A.K.; Tiffany, N.H.; Josh, T.; Jeff, R.H. Weight-Bearing Asymmetry in Individuals PostHip Fracture During the Sit to Stand Task. *Clin. Biomech.* **2015**, *30*, 14–21.
4. Bivalkar, A.; Varadharajulu, G. Establishment and Analysis of Ideal Lifting Technique Based on Kinesiological Principles for Outpatient Stroke Patients: An Experimental Study. *Int. J. Curr. Res. Rev.* **2020**, *12*, 14–18.
5. Cheng, P.-T.; Liaw, M.-Y.; Wong, M.-K.; Tang, F.-T.; Lee, M.-Y.; Lin, P.-S. The sit-to-stand movement in stroke patients and its correlation with falling. *Arch. Phys. Med. Rehabil.* **1998**, *79*, 1043–1046. [CrossRef]
6. Pai, Y.-C.; Rogers, M.W.; Hedman, L.D.; Hanke, T.A. Alterations in Weight-Transfer Capabilities in Adults With Hemiparesis. *Phys. Ther.* **1994**, *74*, 647–657. [CrossRef]
7. Ryan, L.M.; Lynn, S.-M. Altered loading during walking and sit-to-stand is affected by quadriceps weakness after total knee arthroplasty. *J. Orthop. Res.* **2005**, *23*, 1083–1090. [CrossRef]
8. Ryoichiro, S.; Hiroaki, K.; Yoshiyuki, S. Development of sit-to-stand and stand-to-sit training system for hemiplegie patients. In Proceedings of the 38th Annual International Conference of the IEEE Engineering in Medicine and Biology Society (EMBC), Orlando, FL, USA, 16–20 August 2016.
9. Wendy, G.; Jack, C.; Richard, S. Rising to stand from a chair: Symmetry, and frontal and transverse plane kinematics and kinetics. *Gait Posture* **2008**, *27*, 8–15.
10. Helga, H.; Maria, T.; Elisabeth, O. Coordinated Ground Forces Exerted by Buttocks and Feet are Adequately Programmed for Weight Transfer During Sit-to-Stand. *J. Neurophysiol.* **1999**, *82*, 3021–3029.

11. Michalina, B.; Ida, W.; Andrzej, W. A new method of determination of phases and symmetry in stand-to-sit-to-stand movement. *Int. J. Occup. Med. Environ. Health* **2014**, *27*, 660–671.
12. Cabinet Office. White Paper on Aging Society. 2018. Available online: https://www8.cao.go.jp/kourei/whitepaper/w-2021/zenbun/03pdf_index.html (accessed on 5 September 2021).
13. Yosra, C.; Maryam, H.; Michael, B.; Nicolas, T. Muscle coordination during robotic assisted walking using Lokomat. *Comput. Methods Biomech. Biomed. Eng.* **2020**, *22*, S216–S218.
14. Ningjia, Y.; Qi, A.; Hiroki, K.; Hiroshi, Y.; Yusuke, T.; Kouji, T.; Kinomoto, M.; Yamasaki, H.; Itkonen, M.; Shibata-Alnajjar, M.; et al. Temporal Features of Muscle Synergies in Sit-to-Stand Motion Reflect the Motor Impairment of Post-Stroke Patients. *IEEE Trans. Neural Syst. Rehabil. Eng.* **2019**, *27*, 2118–2127.
15. Lacquaniti, F.; Grasso, R.; Zago, M. Motor Patterns in Walking. *Physiology* **1999**, *14*, 168–174. [CrossRef] [PubMed]
16. Francesco, L.; Yury, I.; Mattia, Z. Kinematic control of walking. *Arch. Ital. Biol.* **2002**, *140*, 263–272.
17. Gianluca, U.S.; Philippe, S.A.; Joyce, F. The Effects of a Virtual Environment and Robot-Generated Haptic Forces on the Coordination of the Lower Limb During Gait in Chronic Stroke Using Planar and 3D Phase Diagrams. In Proceedings of the International Conference on Virtual Rehabilitation (ICVR), Tel Aviv, Israel, 21–24 July 2019.
18. Plug-in-Gait Maker Placement. Available online: http://www.idmil.org/mocap/Plug-in-Gait+Marker+Placement.pdf (accessed on 14 August 2022).
19. Talos, V.L.; Grishin, A.A.; Solopova, I.A.; Oslanyan, T.L.; Blenky, V.E.; Ivanenko, Y.P. Asymmetric leg loading during sit-to-stand, walking and quiet standing in patients after unilateral total hip replacement surgery. *Clin. Biomech.* **2008**, *23*, 424–433. [CrossRef]
20. Cyril, D.; Sylvia, N.; Julie, L. Lateral Trunk Displacement and Stability During Sit-to-Stand Transfer in Relation to Foot Placement in Patients With Hemiparesis. *Neurorehabilit. Neural Repair* **2008**, *22*, 715–722.
21. Kogami, H.; An, Q.; Yang, N.; Wang, R.; Yoshida, K.; Hamada, H.; Yamakawa, H.; Tamura, Y.; Shimoda, S.; Yamasaki, H.; et al. Analysis of muscle synergy and kinematics in sit-to-stand motion of hemiplegic patients in subacute period. *Adv. Robot.* **2021**, *35*, 867–877. [CrossRef]
22. Jinsook, R.; William, Z.R.; Eric, J.P.; Seng, B.Y.; Randall, F.B. Alterations in upper limb muscle synergy structure in chronic stroke survivors. *J. Neurophysiol.* **2013**, *109*, 768–781.
23. van Criekinge, T.; Vermeulen, J.; Wagemans, K.; Schröder, J.; Embrechts, E.; Truijen, S.; Hallemans, A.; Saeys, W. Lower limb muscle synergies during walking after stroke: A systematic review. *Disabil. Rehabil.* **2020**, *42*, 2836–2845. [CrossRef]
24. Damholt, V.; Termansen, N.B. Asymmetry of Plantar Flexion Strength in the Foot. *Acta Orthop. Scand.* **1977**, *49*, 215–219. [CrossRef]
25. Yoshihiro, K.; Shin, M.; Shinichi, T. Comparision of Foot Gripping Strength and Quadriceps Femoris Muscle Strength of the Dominat Foot and Non-Dominant Foot. *Phys. Ther.* **2007**, *22*, 365–368.
26. Tomote, Y.; Fuyuko, I.; Yoshihiro, I.; Tositaka, T.; Masahiko, Y.; Mitsuya, S. The dominant foot affects the postural control mechanism: Examination by body tracking test. *Acta Oto-Laryngol.* **2014**, *134*, 1146–1150.
27. Jason, C.G.; Catherine, A.S. The effects of symmetric and asymmetric foot placements on sit-to-stand joint moments. *Gait Posture* **2012**, *35*, 78–82.
28. Ningjia, Y.; Qi, A.; Hiroshi, Y.; Yusuke, T.; Atsushi, Y.; Hajime, A. Muscle synergy structure using different strategies in human standing-up motion. *Adv. Robot.* **2017**, *31*, 40–54.
29. Qi, A.; Hiroshi, Y.; Atsushi, Y.; Hajime, A. Different Temporal Structure of Muscle Synergy Between Sit-to-Walk and Sit-to-Stand Motions in Human Standing Leg. In *Converging Clinical and Engineering Research on Neurorehabilitation II*; Springer: Cham, Switzerland, 2016; pp. 933–937.
30. Takahiro, U.; Kouji, T.; Takuya, Z.; Eiji, S.; Kazuhiko, H. Kinetic and Electromyographic Analysesof Sit-to-StandMotion in Patients with Proximal Femoral Fracture: Comparison of Patients with Pain and without Pain. *Phys. Ther.* **2015**, *42*, 228–236.
31. Chun, K.T.; Hideki, K.; Hiroki, W.; Aiki, M.; Masashi, Y.; Yoshiyuki, S.; Kenji, S. Lateral Symmetry of Synergies in Lower Limb Muscles of Acute Post-stroke Patients After Robotic Intervention. *Front. Neurosci.* **2018**, *12*, 276. [CrossRef]
32. Katherine, M.S.; Matthew, C.T.; Eric, J.P. The number and choice of muscles impact the results of muscle synergy analyses. *Front. Comput. Neurosci.* **2013**, *7*, 105. [CrossRef]
33. Silvia, M.; Deborah, F.; Dario, F. Reorganization of muscle synergies during multidirectional reaching in the horizontal plane with experimental muscle pain. *Neurophysiology* **2014**, *111*, 1615–1630.
34. Michael, R.P.; Peter, M.T.; Victoria, G. Women with fibromyalgia walk with an altered muscle synergy. *Gait Posture* **2005**, *22*, 210–218.
35. Takamitsu, H. Sample size determination given data of preliminary experiment for Student's t-test, ANOVA and Tukey's multiple comparison. *Jpn. J. Psychol.* **2004**, *75*, 213–219. (In Japanese)
36. Anabèle, B.; Séléna, L.; Denis, G.; Sylvie, N. Perception of Weight-Bearing Distribution During Sit-to-Stand Tasks in Hemiparetic and Healthy Individuals. *Stroke* **2010**, *41*, 1704–1708.

applied sciences

MDPI

Article

Face Recognition and Tracking Framework for Human–Robot Interaction

Aly Khalifa *, Ahmed A. Abdelrahman, Dominykas Strazdas, Jan Hintz, Thorsten Hempel and Ayoub Al-Hamadi *

Neuro-Information Technology, Otto-von-Guericke-University Magdeburg, 39106 Magdeburg, Germany; ahmed.abdelrahman@ovgu.de (A.A.A.); dominykas.strazdas@ovgu.de (D.S.); jan.hintz@ovgu.de (J.H.); thorsten.hempel@ovgu.de (T.H.)
* Correspondence: aly.khalifa@ovgu.de (A.K.); ayoub.al-hamadi@ovgu.de (A.A.-H.)

Abstract: Recently, face recognition became a key element in social cognition which is used in various applications including human–robot interaction (HRI), pedestrian identification, and surveillance systems. Deep convolutional neural networks (CNNs) have achieved notable progress in recognizing faces. However, achieving accurate and real-time face recognition is still a challenging problem, especially in unconstrained environments due to occlusion, lighting conditions, and the diversity in head poses. In this paper, we present a robust face recognition and tracking framework in unconstrained settings. We developed our framework based on lightweight CNNs for all face recognition stages, including face detection, alignment and feature extraction, to achieve higher accuracies in these challenging circumstances while maintaining the real-time capabilities required for HRI systems. To maintain the accuracy, a single-shot multi-level face localization in the wild (RetinaFace) is utilized for face detection, and additive angular margin loss (ArcFace) is employed for recognition. For further enhancement, we introduce a face tracking algorithm that combines the information from tracked faces with the recognized identity to use in the further frames. This tracking algorithm improves the overall processing time and accuracy. The proposed system performance is tested in real-time experiments applied in an HRI study. Our proposed framework achieves real-time capabilities with an average of 99%, 95%, and 97% precision, recall, and F-score respectively. In addition, we implemented our system as a modular ROS package that makes it straightforward for integration in different real-world HRI systems.

Keywords: face recognition; face tracking; face detection; face alignment; person identification; human–robot interaction; intelligent robots; interactive systems

Citation: Khalifa, A.; Abdelrahman, A.A.; Strazdas, D.; Hintz, J.; Hempel, T.; Al-Hamadi, A. Face Recognition and Tracking Framework for Human–Robot Interaction. *Appl. Sci.* **2022**, *12*, 5568. https://doi.org/10.3390/app12115568

Academic Editors: Luis Gracia and Carlos Perez-Vidal

Received: 9 May 2022
Accepted: 27 May 2022
Published: 30 May 2022

1. Introduction

Robots have an increasing involvement in real-world contexts, such as homes, schools, hospitals, labs and workplaces. As a result, the field of human–robot Interaction (HRI) presents new challenges in security, automation, and recognition [1]. Robots need social intelligence to interact effectively with and assist humans. Furthermore, a reasonable difference between humans and robots is that humans can recognize and remember individuals by perceiving their facial features smoothly, while robots pose significant challenges in perception [2]. This is an essential part of social cognition and represents a key element for improving human–robot interaction. Moreover, the recent advances in face detection and face recognition (FR) through deep neural networks make it possible to make robots rapidly approach human-level performance and handle several challenging conditions, including large pose variations and occlusions, difficult lighting conditions, and poor-quality images with large motion blur [3,4]. However, there are still unresolved challenges for real-world applications to operate in unconstrained circumstances, including computing power limitations and the lack of training data for user-wise face identification.

As the field of HRI advances, the levels of interaction between humans and robotics become more complex. In order to better understand the critical aspects that influence the human–robot interaction behavior, we conducted a Wizard-of-Oz study [5] to analyze common communication intuitions of new human interaction partners. Figure 1 shows the different interactions between the subjects and the industrial robot.

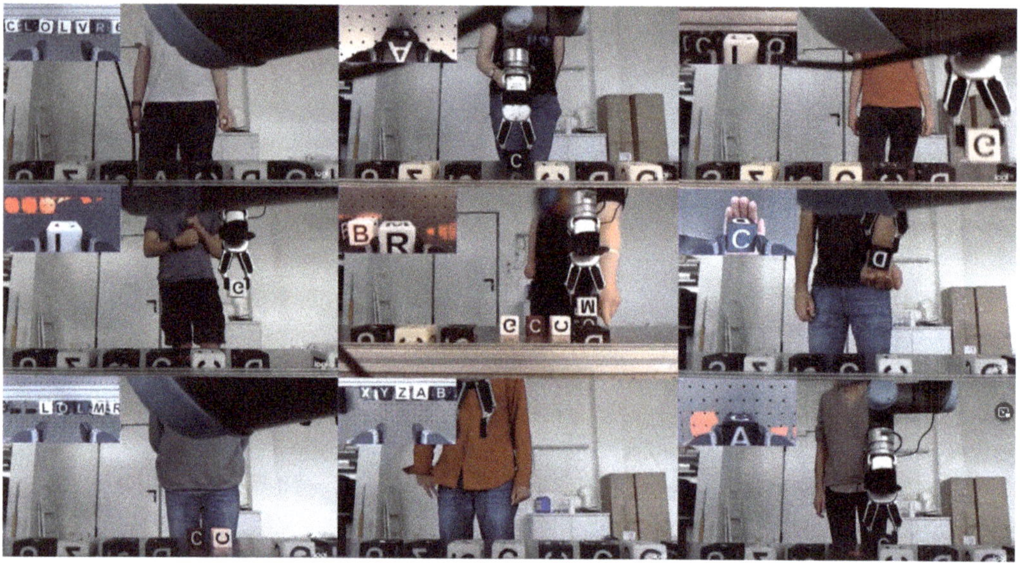

Figure 1. Previous field study Wizard-of-Oz [5]. A video summary can be found here: https://youtu.be/JL409R7YQa0 (accessed on 29 May 2022).

Based on the key results and conclusions of the study, we implemented a multi-modal robotic system called "RoSA" (Robot System Assistant) [6]. This way "RoSA" tackles the challenge of intuitive and user-centered human–robot interaction by integrating different interaction streams such as speech, gesture, object, body, and face recognition.

During the interactions in RoSA, the subjects were not cooperating with the face recognition module as they needed to look down to do the required tasks efficiently as illustrated in Section 3, i.e., the face pitch angle is far away from the camera, preventing the upside camera from capturing the best face pose that fit with the face recognition module. Moreover, face recognition is viewpoint dependent for rotations about all axes (pitch, yaw, and roll) and had the worst accuracy for rotations in pitch [7] as shown in Figure 2. To make the interaction smooth and to increase recognition accuracy for this scenario, we propose a face recognition system that is improved with a tracking capability to handle the subjects' continuous changes in appearance and illumination, in addition, providing the robot with the capability to learn new faces features and recognize them in real-time to participate in social cognition.

In this paper, a typical face recognition framework enhanced with a tracking capability is built by integrating a light-weight RetinaFace-mobilenet [3] with Additive Angular Margin Loss (ArcFace) [4]. Furthermore, to improve the processing speed and accuracy, we propose a tracking algorithm that combines the tracked faces with the actual user identity to improve the recognition performance and accuracy. Finally, we packaged the proposed recognition framework as a real-time Robot Operating System (ROS) node for an easy plugin into other real-world HRI systems.

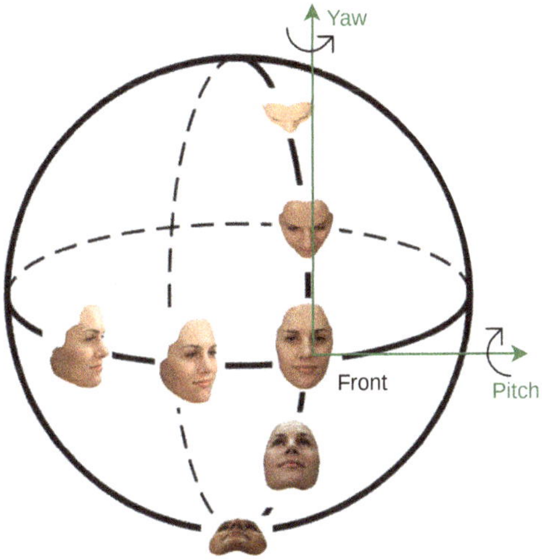

Figure 2. Influence of yaw and pitch angles variations on the head pose, showing that pitch angle have the great impact on the face features. Moving away from the front pose resultant on less distinctive features.

The remainder of the paper is organized as follows: Section 2 reviews recent related work on face detection, face alignment, face recognition, and face tracking algorithms. The RoSA system is illustrated in Section 3. Our proposed methodology and framework are presented in Section 4. Experiments and results are presented in Section 5. Finally, Section 7 concludes this paper.

2. Related Works

Most current deep face recognition systems can be decomposed into three main stages: *face detection*, where faces are localized in an input image, *face alignment*, where the detected faces are warped into a 2D or 3D canonical face model; and *face recognition*, where the aligned faces are classified into different identities. Each part has been actively studied in the field, and near-human performances have been achieved over many benchmark datasets [3,4,8]. In the following, we give a brief overview of recent works on each stage.

2.1. Face Detection Algorithms

Face detection algorithms aim to locate the main face area in input images or video frames. Furthermore, they help robots discriminate between humans and other objects in the scene.

Before the deep learning era, the cascade-based methods and deformable part models (DPM) dominated the face detection field with limitations in unconstrained face images due to considerable variations in resolutions, illumination, expression, skin color, pose, and occlusions [9].

In recent years, deep learning methods have shown their power in computer vision and pattern recognition. As a result, many deep convolutional neural networks (CNN or DCNN)-based face detection methods have been proposed to overcome the limitations mentioned above [3,10–14]. The CNN-based face detection approaches generally have two stages: a feature extraction stage by utilizing a CNN-backbone network to generate the feature map, and a stage for predicting the bounding box locations [15]. They can be divided into two categories: (1) multi-stage; and (2) single-stage detection algorithms.

Two-stage algorithms: Most two-stage algorithms are typically based on Faster R-CNN [12] and generate several candidate boxes and then refine the candidates with a subsequent stage. The first stage utilizes a sliding window to propose the candidate bounding boxes at a given scale, and the second stage rejects the false positives and refines the remaining boxes [16–18]. The advantage of this type of model is that they reach the highest accuracy rates, on the other hand they are typically slower.

Single-stage algorithms: Most single-stage algorithms are typically based on the single shot multi-box detector (SSD) [11]. These algorithms treat object detection as a simple regression problem by performing the candidate classification and bounding box regression from the feature maps directly in only one stage, without the dependence on an extra proposal stage [3,13]. The advantage of this type of model is that they are much faster than two-stage algorithms, but they have lower accuracy rates.

Among the many variants using the single-stage structure, state-of-the-art face detection performance was achieved by RetinaFace [3]. RetinaFace is the latest one-stage face detection model, which is based on the structure of RetinaNet [19] and uses deformable convolution and dense regression loss. We utilized the lightweight version of RetinaFace based on the mobilenet backbone to enhance the detection speed to achieve real-time performance.

2.2. Facial Landmarks and Face Alignment Algorithms

Face alignment plays a vital role in many computer vision applications. It is necessary to improve the robustness of face recognition against in-plane rotations and pose variations [20]. Meanwhile, facial landmarks are essential for most existing face alignment algorithms because they are involved in the similarity transformation for finding the closest shape of the face. So, facial landmark localization is a prerequisite for face alignment.

Face alignment aims to identify the geometric structure of the detected face and calibrate it to the canonical pose, i.e., determining the location and shape of the face elements, such as the mouth, nose, eyes, and eyebrows.

From an overall perspective, face alignment methods can be divided into model-based and regression-based methods [21]. However, the regression methods show superior accuracy, speed, and robustness when compared to model-based methods [22]. Furthermore, model-based methods show difficulties to express the very complex individual landmark appearance.

Trigeorgis et al. [23] further optimize regression-based methods by introducing a single convolutional recurrent neural network architecture that combines all stages' training through facilitating a memory unit that shares information across all levels. The importance of the initialization strategies for face alignment is demonstrated in [24]. Despite that, Valle et al. [25] handled the sensitivity problem of initialization strategies by introducing the Deeply-initialized Coarse-to-Fine Ensemble (DCFE) approach. DCFE refines a CNN-based initialization stage with Ensemble of Regression Trees (ERT) to estimate probability maps of landmarks' locations. Cascade of experts is used by Feng et al. in [26] to improve the face alignment accuracy versus the different face shape poses. Feng et al. proposed Random Cascaded Regression Copse (R-CR-C) method that utilizes three parallel cascaded regressions. Furthermore, Zhu et al. [27] used a probabilistic approach to adopt coarse-to-fine shape searching.

There have been significant improvements in face alignment using deep learning methods. As in [28], Kumar and Chellapa introduced a single dendritic CNN, termed the Pose Conditioned Dendritic Convolution Neural Network (PCD-CNN). Furthermore, they combine a classification network with a second and modular classification network to predict landmark points accurately. In addition, Wu et al. [28] proposed a boundary-aware face alignment algorithm that interpolates the geometric structure of a human face as boundary lines to improve landmark localization.

In a later work, a more efficient compact model has been recently proposed by Guo et al. named practical facial landmark detector (PFLD) [29]. They used a branch of the network to estimate the geometric information for each face sample to make the model more robust. PFLD achieved a size of 2.1 Mb and over 140 fps per face on a mobile phone with high accuracy against complex faces, including unconstrained poses, expressions, lighting, and occlusions, which makes it more suitable for HRI applications.

2.3. Face Recognition Algorithms

A face recognition system is a system that can identify or verify a person in an input image or a video frame. With the current advances in machine learning, the deep face recognition systems based on the CNN models have been the most common due to their remarkable results, and several deep face recognition models have been proposed [4,30–34]. These models work by localizing the face in the input image, extracting the face embeddings, and comparing them to other face embeddings pre-extracted and stored in a database. Every embedding creates a unique face signature and the identity of a specific human face.

Taigman et al. proposed a multi-stage approach called DeepFace [30] based on AlexNet architecture [35]. The faces are first aligned to a generic 3D shape model, and then facial representation is derived from a nine-layer deep neural network. In addition, the authors used a Siamese network trained by standard cross-entropy loss for face verification. Inspired by the work of DeepFace, Sun et al. introduced a high-performance deep convolutional neural network called DeepID2+ [36] for face recognition. DeepID2+ achieved a better performance by adding supervision to early convolutional layers and increasing the dimension of hidden representations. Schroff et al. proposed FaceNet [31] based on the GoogleNet architecture [37]. FaceNet directly optimizes the face embedding by a deep convolutional network trained using a triplet loss function at the final layer. He et al. proposed a Wasserstein convolutional neural network (WCNN) approach [38] that optimizes face recognition by learning invariant features between near-infrared and visual face images.

Recently, different loss functions for face recognition have been proposed [4,32,33,39,40] to enhance discriminative feature learning and representation. Sphereface presents the importance of the angular margin and its advantage in feature separation, but the training is unstable and hard to converge. CosFace defines the decision margin in the cosine space by directly adding the cosine margin penalty to the target logit, which results in better performance than SphereFace with easier implementation and stable training. The ArcFace or Additive Angular Margin Loss [4] is one of the most potent loss functions designed for deep face recognition [41–43]. It enhances discriminative learning by introducing an additive angular margin. In contrast with SphereFace and CosFace which have a nonlinear angular margin, ArcFace has a constant linear angular margin.

The evaluation of single face recognition requires high computational power. Furthermore, multiple faces in a single scene need to be recognized and identified in practice. This makes recognizing multiple faces another challenge, as it requires more computing power to process multiple faces per scene. The accuracy and processing time are the main criteria for any face recognition system. Nevertheless, especially for the HRI, accuracy and real-time recognition are a challenge in scenes with subjects that do not co-operate with the recognition system.

2.4. Face Tracking Algorithms

Visual object tracking has always been a research hotspot in computer vision, and face tracking is a special case. Face tracking is primarily a process of determining the position of the human face in a digital video or frame based on the detected face. This is challenging as the face is not the same during the time (video frames), but it may vary in pose and view. Moreover, other factors affected the face tracking in the actual scene and made it more complex, such as illumination, occlusion, and posture changes. On the other hand, face tracking has many advantages, such as counting the number of human faces in a digital video or camera feed and following a particular face as it moves in a video stream to predict the person's path or direction. Moreover, it can reduce the processing time needed for face detection and recognition.

Many visual object tracking algorithms have been presented; however, Kalman filter [44] and template matching [45] are the most popular methods. In [46], Bewley et al. proposed simple online and real-time tracking (SORT) for multiple object tracking. SORT is a simple approach that associates objects efficiently for online and real-time applications by utilizing the Kalman filter and the Hungarian method. It achieves a favorable performance at high frame rates of 260 Hz. In [47], Wojke et al. integrates SORT with the appearance information by employing a trained CNN to discriminate pedestrians on a large-scale person re-identification dataset, and called it Deep-SORT. This technique has improved the performance and reduced the number of identity switches through longer periods of occlusions.

Recently, deep learning-based face tracking algorithms have been dominant, where the face tracking problem is solved as a binary classification problem for predicting a face or a non-face. Lian et al. [48] proposed a multiple objects tracking algorithm that utilizes a multi-task CNN network (MTCNN) for face detection and fuses multiple features (appearance, motion, and shape features) for tracking. Despite the promising results achieved by deep learning-based face tracking algorithms, SORT has a higher frame rate with favorable accuracy due to its simplicity and ease of implementation.

3. Human–Robot Interaction System

We developed RoSA, a multi-modal system for contactless human–machine interaction based on speech, facial, and gesture recognition [6]. In order to make the interaction smooth and to increase recognition accuracy in RoSA, we propose a face recognition framework that is improved with a tracking capability to handle subjects' continuous changes in appearance and illumination.

The RoSA setup is illustrated in Figure 3, and has two workstations, workstations 1 (WS1) and workstation 2 (WS2), with different designs and purposes [6]. In addition to seven modules (face, speech, gesture, attention, robot, cube, and scene) were designed and implemented. The modules utilize the ROS, ROS network, and ROS messages for communications with the workstations and each other.

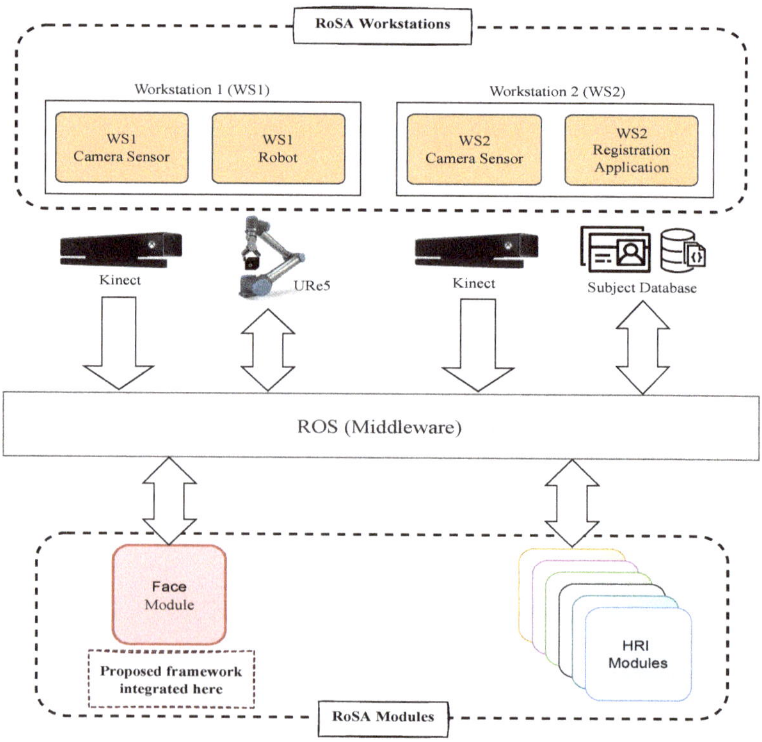

Figure 3. The system setup of the Robot System Assistant (RoSA) framework, showing the communication between the RoSA modules and workstations, is performed via the ROS, and the proposed framework is integrated as the face module.

WS1 is dedicated to all the human–robot interactions and collaborative tasks with the robot. It consists of an industrial robot *UR5e* provided with a gripper *RG6* for easy handling of the required tasks and securely fixed on a metal table. A top camera sensor is used for a live stream of all the human–robot interactions; a time of flight (ToF) Kinect V2 camera is selected for this task. A set of black and white cubes with letters are available for the tasks and under the robot's gripper control. For visual feedback, a projector was utilized to illuminate the cubes and the metal table. The primary purpose of WS2 is for subject registration, and it consists of a smart touch screen with built-in speakers.

In the experiments of the RoSA Study, the subjects enter the required information through a graphical user interface. At the same time, the face embedding is extracted by asking the subject to look at WS2 camera in frontal and profile postures. The collected information and embeddings are stored on the subject database. After the completion of the registration, RoSA asks the subject to go to WS1 to do the practical experiment and the collaborative tasks with the robot. Finally, RoSA asks the subject to answer the questionnaires at WS2. These questionnaires include evaluation questions about RoSA during the interaction. Furthermore, RoSA assists the subject to collect extra data for a module assessment and a benchmark.

An active session is required to enable the interaction between the current subject and the robot. This active session can only be achieved if the face module can effectively recognize and track the identity of the subject during the experiment. Regardless, due to the nature of the collaborative tasks and the unrestricted environment, face recognition is a challenging process and is required to handle the different lighting conditions, pose angles, partially occluded, and sometimes, completely hidden faces. This would sometimes lead to the loss of tracking and active session. The proposed face recognition system enables RoSA to recognize and track subjects robustly.

Using face tracking for user recognition and identification also improves on common problematic situations when implementing body tracking in multi-user scenarios: body tracking mix-up and false body detection in inanimate objects. While the coat hangers and office chairs do sometimes get detected as a person and assigned a body posture for further processing, it is very unlikely that the false body would also have a valid face that could also be detected. By fusing the detected faces to the detected bodies—to which we refer as "fused bodies"—we make sure that each body has a valid face for detection and thus a unique ID, determined by that face.

This approach also reduces the unintentional mix-up of tracked bodies, which occurs when two persons are standing close to each other or pass one another while restricting the view of the body tracker. After the loss of one of the tracked bodies due to occlusion or ambiguities, the body tracker estimation can jump over to the other subject and continue under the wrong ID. By constantly checking for integrity, between the user's skeleton and face with the help of the fused body, the mix-up can be detected right away and the error corrected. This way, it is sufficient to track only the face ID for interaction purposes and sort the detected bodies accordingly. After a mix-up, the information corresponding to the tracked body would be updated in the user's fused body entity, so the system would now be aware of which tracked body and its inputs correspond to the face ID.

4. Methodology and Proposed Framework

The proposed framework is a face recognition system improved with a tracking algorithm. Firstly, the current frame is fed to the face detection module to localize faces in each video frame. Then, a face tracker is created for each detected face across the video frame. Meanwhile, the detected faces are aligned to the canonical face using the detected landmarks and sent to the face recognition module. Finally, the face recognition module gets each detected face identity and associates this identity with the face tracker, and then publishes these identities to the other RoSA modules. The framework is illustrated in Figure 4 and consists of three main modules: *face detection and alignment*, *multi-face tracking*, and *deep face recognition* modules. In the following sections, the details of each module will be discussed.

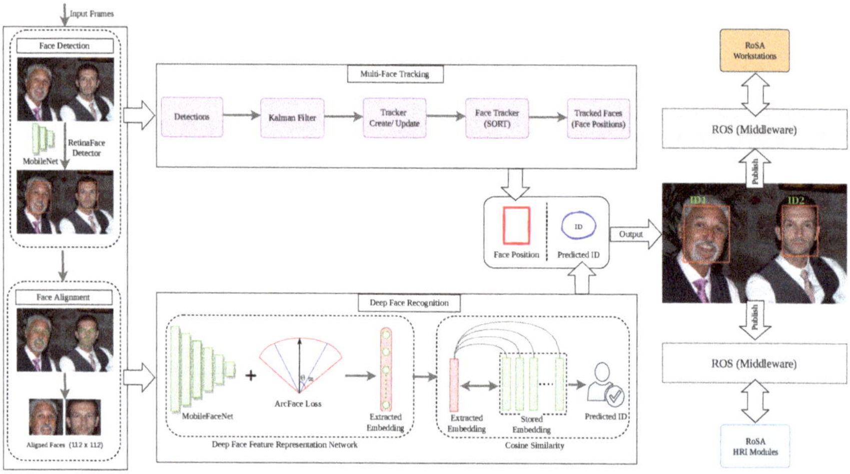

Figure 4. An overview of the proposed face recognition and tracking framework. The predicting face locations and identities are published to the ROS network for broadcasting to RoSA workstations and modules.

4.1. Face Detection and Alignment

For face detection tasks, we use a deep CNN-based face detector by employing a single-shot, multi-level face localization method, called RetinaFace [3]. RetinaFace unifies three different face localization tasks together under one single shot framework: face box prediction, 2D facial landmark localization, and 3D vertices regression. Additionally, all points for these three tasks are regressed on the image plane. RetinaFace proposes a single-shot, multi-level face localization model, which consists of three components: the feature pyramid network, the context head module, and the cascade multi-task loss. First, the feature pyramid network generates five feature maps of different scales. Then, the feature map of a particular scale is fed to the context head module to compute the multi-task loss, i.e., the first context head module predicts the bounding box from the regular anchor. Afterward, the second context head module predicts a more accurate bounding box using the regressed anchor generated by the first context head module. Finally, the anchors are matched to ground-truth boxes if the Intersection over Union (IoU) is greater than 0.7 and 0.5 for the first and second context head respectively, and are matched to the background if IoU is less than 0.3 and 0.4 for the first and second context head, respectively. Furthermore, the unmatched anchors are ignored during training. For any training anchor i, RetinaFace minimizes the following multi-task loss [3]:

$$\mathcal{L} = \mathcal{L}_{cls}(p_i, p_i^*) + \lambda_1 p_i^* \mathcal{L}_{box}(t_i, t_i^*) + \lambda_2 p_i^* \mathcal{L}_{pts}(l_i, l_i^*) + \lambda_3 p_i^* \mathcal{L}_{mesh}(v_i, v_i^*), \tag{1}$$

where t_i, l_i, v_i are box, five landmarks and 1k vertices predictions, t_i^*, l_i^*, v_i^* is the corresponding ground-truth, p_i is the predicted probability of anchor i being a face, and p_i^* is 1 for the positive anchor and 0 for the negative anchor. The classification loss $\mathcal{L}_{cls}$ is the softmax loss for binary classes (face/not face). The loss-balancing parameters λ_1 and λ_2 are set to 0.25 and 0.1, respectively.

For the face landmarks and alignment task, we use a deep CNN-based network by utilizing a practical facial landmark detector (PFLD) by Gue et al. [29]. PFLD employs a branch of the network to estimate the geometric information for each face in order to regularize the landmark localization. Moreover, it adds a multi-scale fully connected (MS-FC) layer to enlarge the receptive field, catch the global structure, and precisely localize landmarks on faces. For predicting landmark coordinates, it utilizes the MobileNet network as a backbone to enhance the processing speed and model size. As a result, it achieved a size of 2.1 Mb and over 140 fps per face on a mobile phone with high accuracy against complex faces, including unconstrained poses, expressions, lighting, and occlusions.

In the face detection and alignment module, all the faces in the images or the video frames are detected with RetinaFace. RetinaFace outputs bounding boxes and five landmarks (2 eyes, nose, and mouth) with a confidence score. For real-time constraints, we select MobileNet-0.25 [49] as a lightweight backbone network, which achieves the real-time speed of 40 fps at GPU for 4K images (4096 × 2160) with outstanding performance.

Next, the filtered faces, i.e., the detection boxes with high confidence scores are sent to face alignment for calibrating to the canonical view and for cropping it to a size of 112 × 112 to be suitable for the subsequent task of face feature extraction. For the face landmarks and alignment task, we used the compact model of the PFLD.

4.2. Face Recognition

For the face recognition task, we utilize the additive angular margin loss (ArcFace) model by Deng et al. [4] to extract the feature embeddings of the faces. ArcFace introduces an additive angular margin penalty m between the deep feature x_i and the target weight W_{y_i} to simultaneously enhance the intra-class compactness and inter-class discrepancy. It provides a more clear geometric interpretation due to its exact correspondence to geodesic distance on a hypersphere. ArcFace is inherited from the most common loss function, Softmax, and is defined as follows [4]:

$$L_{arc} = -\frac{1}{N}\sum_{i=1}^{N}\log\frac{e^{s(\cos(\theta_{y_i}+m))}}{e^{s(\cos\theta_{y_i}+m))}+\sum_{j=1,j\neq y_i}^{n} e^{s\cos\theta_j}}. \quad (2)$$

In Equation (2), n denotes the number of classes in the training database, while N denotes the batch size. ArcFace model starts with extracting the face features x_i by utilizing a DCNN backbone. The backbone network is the bottleneck in terms of processing speed and model size; as in the testing, only this branch is involved so we selected the lightweight MobileFaceNet network [50] as a backbone. Then, based on the feature x_i and weight W normalization, we obtain the logit $cos\,\theta_j$ for each class as $W_j^T x_i$, and get the angle between the feature x_i and the ground truth weight W_{y_i} as $arccos\,\theta_{y_i}$. After that, the angular margin penalty m is added to the target angle θ_{y_i}. Finally, we calculate $cos(\theta_{y_i}+m)$ and multiply all logits by the feature scale s. The logits then go through the softmax function and contribute to the cross-entropy loss. The results of the ablation study by Deng et al. [4] showed that the performance comparison on the LFW, CALFW, and CPLFW datasets for the Arcface loss function outperformed others with 99.82%, 95.45%, and 92.08% accuracies respectively. It was performed against 11 other loss functions, including Softmax, Center Loss, SphereFace, and CosFace. This is the main reason why we selected Arcface as a loss function for the face recognition module.

In the face recognition module, after the filtered faces are aligned, a deep face feature representation network transforms the aligned faces into a feature space. MobileFaceNet [50] was selected as a backbone for this task to handle the real-time constraints. Loss function optimization is challenging for large-scale face classification, as it is needed to strengthen the intra-class compactness and inter-class discrepancy for highly similar individual faces. For that, we used ArcFace as it outperforms the state-of-the-art functions. In addition, it enhances the discriminative power for learning deep features and maximizes the separability between face classes.

Finally, the face recognition module outputs a 512-dimensional feature embedding, and then the predicted identity is calculated by comparing the generated embedding against the stored embeddings by calculating the cosine similarity [51]. The ArcFace model is trained on the MS1M database [14]. Given a face image, the image is aligned, scaled, and cropped before being passed to one of the models. This preprocessing is performed as described in [13] for ArcFace.

4.3. Improved Face Recognition Using Face Tracking

For the face tracking task, we build the face tracking algorithm based on a simple online and real-time tracking algorithm (SORT) [46]. SORT uses a Kalman filter for estimating the location of the face in the current frame given the location in the previous frame. It starts with detecting the target face in the initial frame i. After that, predict the future location $i+1$ of the target face from the initial frame using the Kalman filter. Noting that the Kalman filter just approximates the face's new location, which needs to be optimized. Finally, the Hungarian algorithm is used for face location optimization and association.

The main problem we are targeting is the speed/accuracy trade-offs. Continuous face detection and face recognition processing are time-consuming. Moreover, the quality of the face features depends on the face pose, where the frontal face pose generates the best facial features and degrades in a departure from the frontal pose. Therefore, instead of detecting all faces around all input video frames, we assign only each newly detected face a tracker and start the tracking instead of detection. Furthermore, for each new tracker only, the face embedding will be inferred and compared against the stored embeddings by calculating the cosine similarity to generate the user identity (ID), then add the ID to the tracker metadata for fast recognition, i.e., retrieve the ID from the tracker in the successive frames without the need for recognition. These will improve the processing time, recognition rate, and reduce the recognition errors caused by variations from frontal face poses.

In the proposed tracking Algorithm 1, for each input frame, we are detecting faces using face detection and alignment in Section 4.1. Initially, a new tracker for each detector box will be created by applying SORT [46]. SORT analyzes previous and current frames and predicts face locations on the fly by utilizing the Kalman filter and Hungarian algorithm. Then, the user ID will be obtained using face recognition in Section 4.2 and assigned with the face tracker for use in fast recognition in the further frames. Finally, the tracker will be associated with the detected faces and maintained throughout tracking, and the user ID is assigned for each face tracker. We update the tracker in each frame to validate if a face is there inside the box to improve the tracking quality. If not, we are deleting the tracker to prevent unbounded growth in the number of trackers. Moreover, the actual user identity is attached to the face tracker instead of a unique face tracker ID to improve the face recognition speed.

Algorithm 1: The Proposed Face Tracking Algorithm.

```
Inputs : Video, Detections, KalmanFilter, HeadJoints, SubjectIDs
Output: Recognized Tracked Faces
Initialize KalmanFilterTracker;
foreach frame f_i ∈ Video do
    Trackers ← Predict();
    Trackers ← Assign(Detections, Trackers);
    TrackersID ← Attach(Trackers, SubjectIDs);
    TrackersID ← Assign(TrackersID, HeadJoints);
    Update KalmanFilterTracker;
    foreach tracker t_i ∈ TrackersID do
        ROS ← Publish(t_i);
    end
end
```

To improve the proposed face tracking algorithm and minimize the tracking error, we obtain the head joint from the tracked skeleton provided by the WS1 Kinect V2 camera and try to assign it with the face center. If the assignment is successful, we update the face tracker with the fine location. Otherwise, the tracker will be deleted. Further, this reduces the number of identity switches through longer periods of occlusions.

5. Experiments and Analysis

The most effective parts of the face recognition and tracking framework are the face detection and face recognition models. In order to well evaluate the effectiveness of the introduced tracking approach, we trained and evaluate the two models separately.

5.1. Face Detection

For the face detection, the RetinaFace is trained on the WIDER FACE dataset [52]. It contains 32,203 images and 393,703 face bounding boxes with a high degree of variability in scale, pose, expression, occlusion, and illumination. The evaluation is performed on the WIDER FACE validation set, with Average Precision (AP) of 0.83 for the hard subset.

5.2. Face Recognition

For the face recognition network, the ArcFace is trained on the MS1MV2 dataset [4,53] for 30 epochs with a batch size of 512, feature scale s of 64, and angular margin m of 0.5. MS1MV2 is a semi-automatic refined version of the MS-Celeb-1M dataset [53] which contains about 100k identities with 10 million images. The evaluation is performed on large-pose CPLFW and large-age CALFW datasets and achieved performance of 95.45% and 92.07% respectively.

5.3. Results

The metrics used to measure the overall system performance are precision, recall, F-score, and recognition rate. We classify the predictions into True Positives (TP), False Positives (FP), False Negatives (FN), and True Negatives (TN). A *True Positive* can be obtained in recognition when the model correctly predicted the subject class (i.e., subject ID), which means that it matches the ground truth. Otherwise, the prediction is considered a *False Positive*.

A *True Negative* can be obtained in recognition when the model is not supposed to predict a subject that is not in the database. Otherwise, the prediction is considered a *False Negative*.

Precision is the matching probability of the predicted subject identity relative to the ground truth identity, which shows the results of a correctly recognized subject. It can be calculated as follows:

$$Precision = \frac{TP}{TP + FP}. \tag{3}$$

Recall measures the probability of the subjects that were correctly recognized among ground truth subjects, which is the total number of true positives relative to the sum of true positives and false negatives, as follows:

$$Recall = \frac{TP}{TP + FN}. \tag{4}$$

F-score is evaluated as the harmonic mean of precision and recall to see which model best performs. It can be calculated as follows:

$$F\text{-}score = \frac{Precision * Recall}{Precision + Recall} * 2. \tag{5}$$

The recognition performance can be obtained by the face recognition rate FR_R, and it is the ratio between the total number of correctly recognized faces and the total detected/tracked faces. It can be calculated as follows:

$$FR_R = \frac{TP}{Total faces} * 100. \tag{6}$$

In order to evaluate the proposed framework, we tested it for two different evaluations dataset, and online evaluations.

5.3.1. Dataset Evaluation

We use the ChokePoint dataset [54] to evaluate the proposed framework. This dataset is a video dataset that was collected and designed for experiments on person identification/verification under real-world surveillance conditions. It contains videos of 25 subjects (six female and 19 male). In total, the dataset consists of 48 video sequences and 64,204 face images with variations in terms of illumination conditions, pose, sharpness, as well as misalignment due to automatic face localization/detection.

The experimental results show the performance of tracking for 25 subjects of the ChokePoint dataset. To show recognition refinements, we have tested the proposed face recognition framework with tracker-assisted and without. The average results are shown in Table 1. Furthermore, the Receiver Operating Characteristic (ROC) curve is obtained in Figure 5, which shows that the tracking approach improves the recognition rate for high false positive rates and reduces the false classification rate.

Table 1. The average results of precision, recall, and F-score on ChokePoint dataset.

Tracking	Precision	Recall	F-Score
No	0.83	0.79	0.81
Yes	0.96	0.93	0.94

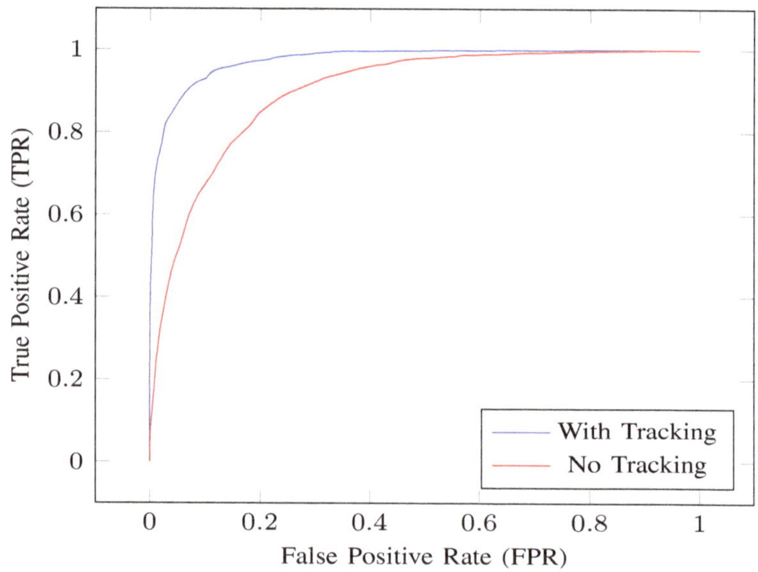

Figure 5. ROC Curve of ChokePoint Dataset for the Proposed Framework.

5.3.2. Online Evaluation

We employ the proposed framework on a real HRI study [6], to further evaluate the framework in real-time HRI and show its robustness. During the experiments in the study, the data for evaluation were collected from 11 subjects (two female and nine male) aged between 20 and 34 years.

The experimental results show the performance of tracking and recognition rate for 11 subjects during the interactions with RoSA [6]. To show recognition refinements, we have tested the proposed face recognition framework with tracker-assisted and without. The proposed framework achieved a face recognition rate of 94% and 76% with tracking and without tracking, respectively. Figure 6 shows the impact of tracking on the *Precision* of the proposed framework, and the impact of tracking on *Recall* of the proposed framework is shown in Figure 7. Furthermore, Figure 8 shows the *F-score* results of the proposed framework with tracker-assisted and without tracking.

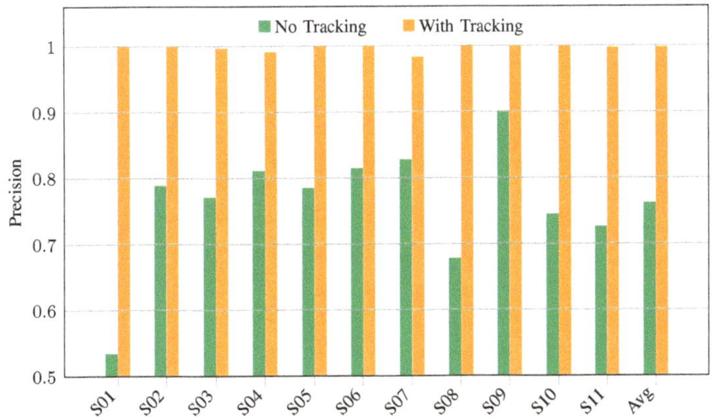

Figure 6. Impact of Tracking on Precision of Face Recognition.

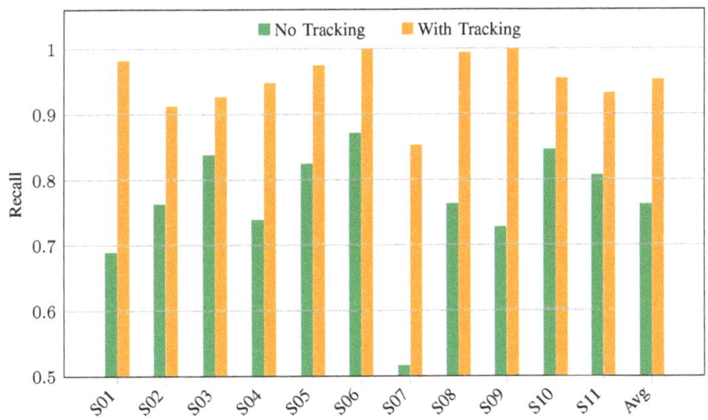

Figure 7. Impact of Tracking on Recall of Face Recognition.

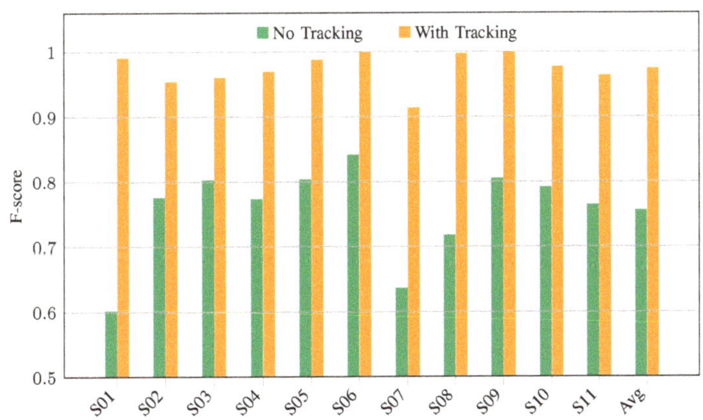

Figure 8. F-score results of the proposed framework with tracker-assisted and without tracking.

Compared to the standard face recognition framework, the proposed framework performance is faster in terms of processing time with frame rates of 25–40 fps. Some results of the proposed framework during the real HRI in our RoSA system [6] are shown in Figure 9.

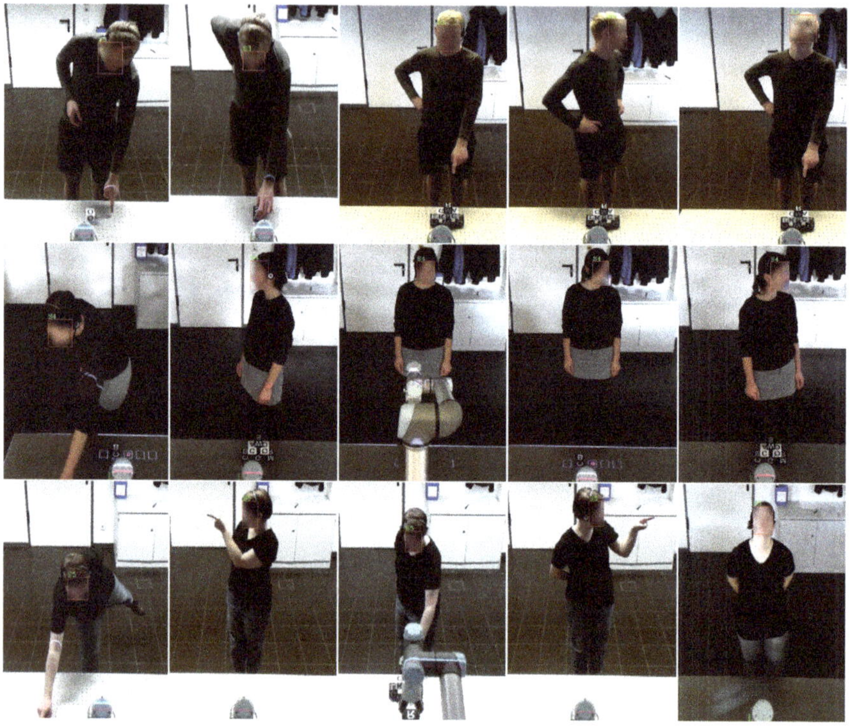

Figure 9. Experimental results of the proposed framework that shows the robustness of the framework against various head posture and illumination conditions.

To confirm the obtained results, we run the experiments again on the recorded videos from the Wizard-of-Oz study [5] with the same results. It contains videos of 36 subjects doing the same tasks on the RoSA study, which were collected on different days with different lighting conditions. For every subject (video), we selected three exemplar face images with different poses and added the extracted embedding to the database to match with video faces. Table 2 shows the precision and recall results for 37 subjects separated by the top ten results.

Table 2. Result of precision and recall for the proposed framework.

No	ID	Precision		Recall	
		Tracking	No Tracking	Tracking	No Tracking
1	4	0.97	0.76	0.81	0.70
2	7	1	0.84	0.92	0.59
3	11	1	0.68	1	0.73
4	16	1	1	1	0.66
5	18	0.96	0.88	1	0.81
6	24	0.98	0.65	0.95	0.77

Table 2. *Cont.*

No	ID	Precision		Recall	
		Tracking	No Tracking	Tracking	No Tracking
7	25	0.89	0.53	0.92	0.63
8	29	0.98	0.83	0.98	0.79
9	32	0.99	0.68	0.98	0.60
10	36	0.95	0.76	0.97	0.75

5.4. Computational Efficiency Assessment

In general, lightweight face networks provide promising results for face recognition. They are able to perform comparably to state-of-the-art very deep face models in most face recognition scenarios. In particular, ResNet100-ArcFace by Deng et al. [4] is one of the best performing state-of-the-art models in the different evaluated scenarios, however, it demands high computational resources. For example, the biggest difference in accuracy between ResNet100-ArcFace and MobileFaceNet (our used network), is 8% in the very large-scale DeepGlint-Image dataset (one of the most challenging databases), while in the remaining databases it is less than 3%. However, regarding the computational complexity, ResNet100-ArcFace requires 19X more storage space and involves 26X more FLOPs and 32X more parameters than MobileFaceNet.

Applying face tracking provides us the advantage of no need to apply face detection and recognition for all input frames. However, to increase the accuracy of our framework and minimize the tracking error, we apply the whole recognition process in each fifth frame.

To calculate the computational efficiency assessment of the proposed framework, we tested it on the collected videos (total of 47 videos) during the RoSA study [6] and the Wizard-of-Oz study [5] and obtained the average processing time for each face recognition module. The hardware setting used was a NVIDIA GeForce GTX 1080 Ti Desktop GPU (12 Gb GDDR5, 3584 CUDA cores). Table 3 shows the average execution time of individual methods used in the proposed framework. To summarize, the average execution time per frame for the whole process takes about 6.7 ms, and the average number of frames per second is $\sim$35 frames.

Table 3. Average execution time of individual methods used in the proposed framework.

Method	Average Time (ms)
Detection	3.2
Alignment	1.4
Tracking	0.8
Recognition (Embedding Inference)	1.3
Identification (Similarity)	0.08
Visualization & Delays	7.5

6. Limitations, and Future Work

The study conducted has a complex setting that contains two workstations (WS1 and WS2) synchronized together using the ROS operating system. In addition, extracting face features during the experiments is a challenging task due to the illumination conditions, extreme deviation in head pose angles, and occlusion. However, the aforementioned performance evaluation showed the effectiveness of the proposed framework in recognizing the subject's identity in a multi-person environment.

Few subjects caused a wrong identification during the experiments due to the lack of the registration process and good face feature embedding, which lead to the re-registration of the mentioned subjects.

The advantage of our framework is that it depends on lightweight CNNs for all face recognition stages, including face detection, alignment and feature extraction, to meet the real-time requirements in HRI systems. Furthermore, the developed framework can simultaneously recognize the faces of the cooperating subjects in various poses, face expressions, illumination, and other outdoor-related factors. Although two of the subjects were wearing face masks for the whole experiment, our model succeeded to recognize their identity with reasonable confidence.

Future work would involve a new study with a large number of subjects with different human–robot interaction scenarios to effectively assess the performance of the framework and overcome the limited number of subjects in the RoSA study. In addition future work would involve designing an end-to-end trainable convolutional network framework for all the face recognition stages.

7. Conclusions

We propose a face recognition system for human–robot interaction (HRI) boosted by face tracking based on deep convolutional neural networks (CNNs). To ensure that our framework can work in real-time HRI systems, we developed our framework based on lightweight CNNs for all face recognition stages, including face detection, alignment, tracking, and feature extraction. Furthermore, we implemented our approach as a modular ROS package that makes it straightforward for integration in different HRI systems. Our results suggest that the use of face tracking alongside face recognition increases the recognition rate.

We utilize the state-of-the-art loss function *ArcFace* for the face recognition task and the *RetinaFace* method for face detection combined with a simple online and real-time face tracker. Furthermore, we propose a face tracker to tackle the challenges faced by the existing face recognition methods including various illumination conditions, continuous changes in head posture, and occlusion.

The face tracker is designed to fuse the tracking information with the recognized identity and associate it with the faces once they are detected for the first time. For the updated trackers, the last recognized identity will be kept alongside the tracker. Despite what preceded, a new identity prediction is required for the new trackers. This method improved the overall processing time and face recognition accuracy and precision for the unconstrained face.

The proposed framework is tested in real-time experiments applied in our real HRI system "RoSA" with 11 participants interacting with the robot to accomplish different tasks. Furthermore, to confirm the obtained results, we tested it on the recorded videos from the Wizard-of-Oz study, which contains videos of 36 subjects doing the same tasks on "RoSA" with the same results. The results showed that the framework can improve the robustness of face recognition effectively and boost the overall accuracy by an average of 25% in real-time. It achieves an average of 99%, 95%, and 97% precision, recall, and F-score respectively.

Author Contributions: Conceptualization, A.K. and A.A.-H.; methodology, A.K., A.A.A., D.S., J.H. and T.H.; software, A.K., D.S., J.H. and T.H.; validation, A.K., D.S., J.H. and A.A.A.; investigation, A.K., D.S., J.H. and A.A.A.; resources, A.A.-H.; writing—original draft preparation, A.K., A.A.A., D.S., J.H. and T.H.; writing—review and editing, A.K., A.A.A., D.S., J.H., T.H. and A.A.-H.; visualization, A.K., A.A.A., D.S., J.H. and T.H.; supervision, A.A.-H.; project administration, A.K. and A.A.-H.; funding acquisition, A.A.-H. All authors have read and agreed to the published version of the manuscript.

Funding: This work is funded by the Federal Ministry of Education and Research of Germany (BMBF) RoboAssist No. 03ZZ0448L and Robo-Lab No. 03ZZ04X02B within the Zwanzig20 Alliance 3Dsensation.

Institutional Review Board Statement: The study was conducted according to the guidelines of the Declaration of Helsinki. Ethical approval was done by Ethik Kommision der Otto-von-Guericke Universtiät (IRB00006099, Office for Human Research) 157/20 on 23 October 2020.

Informed Consent Statement: Informed consent was obtained from all subjects involved in the study.

Data Availability Statement: Not applicable.

Conflicts of Interest: The authors declare no conflict of interest. The funders had no role in the design of the study; in the collection, analyses, or interpretation of data; in the writing of the manuscript, or in the decision to publish the results.

References

1. Zacharaki, A.; Kostavelis, I.; Gasteratos, A.; Dokas, I. Safety bounds in human robot interaction: A survey. *Saf. Sci.* **2020**, *127*, 104667. [CrossRef]
2. Mukherjee, D.; Gupta, K.; Chang, L.H.; Najjaran, H. A survey of robot learning strategies for human–robot collaboration in industrial settings. *Robot. Comput. Integr. Manuf.* **2022**, *73*, 102231. [CrossRef]
3. Deng, J.; Guo, J.; Ververas, E.; Kotsia, I.; Zafeiriou, S. RetinaFace: Single-Shot Multi-Level Face Localisation in the Wild. In Proceedings of the IEEE/CVF Conference on Computer Vision and Pattern Recognition (CVPR), Seattle, WA, USA, 13–19 June 2020.
4. Deng, J.; Guo, J.; Xue, N.; Zafeiriou, S. Arcface: Additive angular margin loss for deep face recognition. In Proceedings of the IEEE/CVF Conference on Computer Vision and Pattern Recognition, Long Beach, CA, USA, 15–20 June 2019; pp. 4690–4699.
5. Strazdas, D.; Hintz, J.; Felßberg, A.M.; Al-Hamadi, A. Robots and Wizards: An Investigation Into Natural Human–Robot Interaction. *IEEE Access* **2020**, *8*, 207635–207642. [CrossRef]
6. Strazdas, D.; Hintz, J.; Khalifa, A.; Abdelrahman, A.A.; Hempel, T.; Al-Hamadi, A. Robot System Assistant (RoSA): Towards Intuitive Multi-Modal and Multi-Device human-robot Interaction. *Sensors* **2022**, *22*, 923. [CrossRef] [PubMed]
7. Favelle, S.; Palmisano, S. View specific generalisation effects in face recognition: Front and yaw comparison views are better than pitch. *PLoS ONE* **2018**, *13*, e0209927. [CrossRef] [PubMed]
8. Albiero, V.; Chen, X.; Yin, X.; Pang, G.; Hassner, T. img2pose: Face alignment and detection via 6dof, face pose estimation. In Proceedings of the IEEE/CVF Conference on Computer Vision and Pattern Recognition, Nashville, TN, USA, 20–25 June 2021; pp. 7617–7627.
9. Minaee, S.; Luo, P.; Lin, Z.; Bowyer, K. Going deeper into face detection: A survey. *arXiv* **2021**, arXiv:2103.14983.
10. Girshick, R. Fast r-cnn. In Proceedings of the IEEE International Conference on Computer Vision, Santiago, Chile, 7–13 December 2015; pp. 1440–1448.
11. Liu, W.; Anguelov, D.; Erhan, D.; Szegedy, C.; Reed, S.; Fu, C.Y.; Berg, A.C. Ssd: Single shot multibox detector. In Proceedings of the European Conference on Computer Vision, Amsterdam, The Netherlands, 11–14 October 2016; Springer: Berlin/Heidelberg, Germany, 2016; pp. 21–37.
12. Ren, S.; He, K.; Girshick, R.; Sun, J. Faster R-CNN: Towards Real-Time Object Detection with Region Proposal Networks. *IEEE Trans. Pattern Anal. Mach. Intell.* **2017**, *39*, 1137–1149. [CrossRef] [PubMed]
13. Najibi, M.; Samangouei, P.; Chellappa, R.; Davis, L.S. Ssh: Single stage headless face detector. In Proceedings of the IEEE International Conference on Computer Vision, Venice, Italy, 22–29 October 2017; pp. 4875–4884.
14. Fiedler, M.A.; Werner, P.; Khalifa, A.; Al-Hamadi, A. SFPD: Simultaneous Face and Person Detection in Real-Time for human–robot Interaction. *Sensors* **2021**, *21*, 5918. [CrossRef]
15. Jiao, L.; Zhang, F.; Liu, F.; Yang, S.; Li, L.; Feng, Z.; Qu, R. A survey of deep learning-based object detection. *IEEE Access* **2019**, *7*, 128837–128868. [CrossRef]
16. Zhang, K.; Zhang, Z.; Li, Z.; Qiao, Y. Joint face detection and alignment using multitask cascaded convolutional networks. *IEEE Signal Process. Lett.* **2016**, *23*, 1499–1503. [CrossRef]
17. Zhang, C.; Xu, X.; Tu, D. Face detection using improved faster rcnn. *arXiv* **2018**, arXiv:1802.02142.
18. Najibi, M.; Singh, B.; Davis, L.S. Fa-rpn: Floating region proposals for face detection. In Proceedings of the IEEE/CVF Conference on Computer Vision and Pattern Recognition, Long Beach, CA, USA, 16–17 June 2019; pp. 7723–7732.
19. Zhang, H.; Chang, H.; Ma, B.; Shan, S.; Chen, X. Cascade retinanet: Maintaining consistency for single-stage object detection. *arXiv* **2019**, arXiv:1907.06881.
20. Huang, G.B.; Mattar, M.; Berg, T.; Learned-Miller, E. Labeled faces in the wild: A database forstudying face recognition in unconstrained environments. In Proceedings of the Workshop on Faces in 'Real-Life' Images: Detection, Alignment, and Recognition, Marseille, France, 17 October 2008.
21. Wu, Y.; Ji, Q. Facial landmark detection: A literature survey. *Int. J. Comput. Vis.* **2019**, *127*, 115–142. [CrossRef]
22. Gogić, I.; Ahlberg, J.; Pandžić, I.S. Regression-based methods for face alignment: A survey. *Signal Process.* **2021**, *178*, 107755. [CrossRef]
23. Trigeorgis, G.; Snape, P.; Nicolaou, M.A.; Antonakos, E.; Zafeiriou, S. Mnemonic descent method: A recurrent process applied for end-to-end face alignment. In Proceedings of the IEEE Conference on Computer Vision and Pattern Recognition, Las Vegas, NV, USA, 27–30 June 2016; pp. 4177–4187.
24. Zhu, H.; Sheng, B.; Shao, Z.; Hao, Y.; Hou, X.; Ma, L. Better initialization for regression-based face alignment. *Comput. Graph.* **2018**, *70*, 261–269. [CrossRef]
25. Valle, R.; Buenaposada, J.M.; Valdes, A.; Baumela, L. A deeply-initialized coarse-to-fine ensemble of regression trees for face alignment. In Proceedings of the European Conference on Computer Vision (ECCV), Munich, Germany, 8–14 September 2018; pp. 585–601.

26. Feng, Z.H.; Huber, P.; Kittler, J.; Christmas, W.; Wu, X.J. Random cascaded-regression copse for robust facial landmark detection. *IEEE Signal Process. Lett.* **2014**, *22*, 76–80. [CrossRef]
27. Zhu, S.; Li, C.; Loy, C.C.; Tang, X. Face alignment by coarse-to-fine shape searching. In Proceedings of the IEEE Conference on Computer Vision and Pattern Recognition, Boston, MA, USA, 7–12 June 2015; pp. 4998–5006.
28. Kumar, A.; Chellappa, R. Disentangling 3d pose in a dendritic cnn for unconstrained 2d face alignment. In Proceedings of the IEEE Conference on Computer Vision and Pattern Recognition, Salt Lake City, UT, USA, 18–23 June 2018; pp. 430–439.
29. Guo, X.; Li, S.; Yu, J.; Zhang, J.; Ma, J.; Ma, L.; Liu, W.; Ling, H. PFLD: A practical facial landmark detector. *arXiv* **2019**, arXiv:1902.10859.
30. Taigman, Y.; Yang, M.; Ranzato, M.; Wolf, L. Deepface: Closing the gap to human-level performance in face verification. In Proceedings of the IEEE Conference on Computer Vision and Pattern Recognition, Columbus, OH, USA, 23–28 June 2014; pp. 1701–1708.
31. Schroff, F.; Kalenichenko, D.; Philbin, J. Facenet: A unified embedding for face recognition and clustering. In Proceedings of the IEEE Conference on Computer Vision and Pattern Recognition, Boston, MA, USA, 7–12 June 2015; pp. 815–823.
32. Wang, H.; Wang, Y.; Zhou, Z.; Ji, X.; Gong, D.; Zhou, J.; Li, Z.; Liu, W. Cosface: Large margin cosine loss for deep face recognition. In Proceedings of the IEEE Conference on Computer Vision and Pattern Recognition, Salt Lake City, UT, USA, 18–23 June 2018; pp. 5265–5274.
33. Zhong, Y.; Deng, W.; Hu, J.; Zhao, D.; Li, X.; Wen, D. SFace: Sigmoid-constrained Hypersphere Loss for Robust Face Recognition. *IEEE Trans. Image Process.* **2021**, *30*, 2587–2598. [CrossRef]
34. Li, L.; Mu, X.; Li, S.; Peng, H. A Review of Face Recognition Technology. *IEEE Access* **2020**, *8*, 139110–139120. [CrossRef]
35. Krizhevsky, A.; Sutskever, I.; Hinton, G.E. Imagenet classification with deep convolutional neural networks. *Adv. Neural Inf. Process. Syst.* **2012**, *25*, 1097–1105. [CrossRef]
36. Sun, Y.; Wang, X.; Tang, X. Deeply learned face representations are sparse, selective, and robust. In Proceedings of the IEEE Conference on Computer Vision and Pattern Recognition, Boston, MA, USA, 7–12 June 2015; pp. 2892–2900.
37. Szegedy, C.; Liu, W.; Jia, Y.; Sermanet, P.; Reed, S.; Anguelov, D.; Erhan, D.; Vanhoucke, V.; Rabinovich, A. Going deeper with convolutions. In Proceedings of the IEEE Conference on Computer Vision and Pattern Recognition, Boston, MA, USA, 7–12 June 2015; pp. 1–9.
38. He, R.; Wu, X.; Sun, Z.; Tan, T. Wasserstein cnn: Learning invariant features for nir-vis face recognition. *IEEE Trans. Pattern Anal. Mach. Intell.* **2018**, *41*, 1761–1773. [CrossRef] [PubMed]
39. Liu, W.; Wen, Y.; Yu, Z.; Li, M.; Raj, B.; Song, L. Sphereface: Deep hypersphere embedding for face recognition. In Proceedings of the IEEE Conference on Computer Vision and Pattern Recognition, Honolulu, HI, USA, 21–26 June 2017; pp. 212–220.
40. Deng, J.; Zhou, Y.; Zafeiriou, S. Marginal loss for deep face recognition. In Proceedings of the IEEE Conference on Computer Vision and Pattern Recognition Workshops, Honolulu, HI, USA, 21–26 June 2017; pp. 60–68.
41. Khalifa, A.; Al-Hamadi, A. A Survey on Loss Functions for Deep Face Recognition Network. In Proceedings of the 2021 IEEE 2nd International Conference on human–machine Systems (ICHMS), Magdeburg, Germany, 8–10 September 2021; IEEE: Piscataway, NJ, USA, 2021; pp. 1–7.
42. Fuad, M.T.H.; Fime, A.A.; Sikder, D.; Iftee, M.A.R.; Rabbi, J.; Al-Rakhami, M.S.; Gumaei, A.; Sen, O.; Fuad, M.; Islam, M.N. Recent Advances in Deep Learning Techniques for Face Recognition. *IEEE Access* **2021**, *9*, 99112–99142. [CrossRef]
43. Hsu, G.S.J.; Wu, H.Y.; Yap, M.H. A comprehensive study on loss functions for cross-factor face recognition. In Proceedings of the IEEE/CVF Conference on Computer Vision and Pattern Recognition Workshops, Seattle, WA, USA, 14–19 June 2020; pp. 826–827.
44. Hu, W.C.; Chen, C.H.; Chen, T.Y.; Huang, D.Y.; Wu, Z.C. Moving object detection and tracking from video captured by moving camera. *J. Vis. Commun. Image Represent.* **2015**, *30*, 164–180. [CrossRef]
45. Liu, F.; Gong, C.; Huang, X.; Zhou, T.; Yang, J.; Tao, D. Robust visual tracking revisited: From correlation filter to template matching. *IEEE Trans. Image Process.* **2018**, *27*, 2777–2790. [CrossRef] [PubMed]
46. Bewley, A.; Ge, Z.; Ott, L.; Ramos, F.; Upcroft, B. Simple online and realtime tracking. In Proceedings of the 2016 IEEE International Conference on Image Processing (ICIP), Phoenix, AZ, USA, 25–28 September 2016; IEEE: Piscataway, NJ, USA, 2016; pp. 3464–3468.
47. Wojke, N.; Bewley, A.; Paulus, D. Simple online and realtime tracking with a deep association metric. In Proceedings of the 2017 IEEE International Conference on Image Processing (ICIP), Beijing, China, 17–20 September 2017; IEEE: Piscataway, NJ, USA, 2017; pp. 3645–3649.
48. Lian, Z.; Shao, S.; Huang, C. A real time face tracking system based on multiple information fusion. *Multimed. Tools Appl.* **2020**, *79*, 16751–16769. [CrossRef]
49. Howard, A.G.; Zhu, M.; Chen, B.; Kalenichenko, D.; Wang, W.; Weyand, T.; Andreetto, M.; Adam, H. Mobilenets: Efficient convolutional neural networks for mobile vision applications. *arXiv* **2017**, arXiv:1704.04861.
50. Chen, S.; Liu, Y.; Gao, X.; Han, Z. Mobilefacenets: Efficient cnns for accurate real-time face verification on mobile devices. In Proceedings of the Chinese Conference on Biometric Recognition, Urumchi, China, 11–12 August 2018; Springer: Berlin/Heidelberg, Germany, 2018; pp. 428–438.
51. Nguyen, H.V.; Bai, L. Cosine similarity metric learning for face verification. In Proceedings of the Asian Conference on Computer Vision, Queenstown, New Zealand, 8–12 November 2010; Springer: Berlin/Heidelberg, Germany, 2010; pp. 709–720.

52. Yang, S.; Luo, P.; Loy, C.C.; Tang, X. Wider face: A face detection benchmark. In Proceedings of the IEEE Conference on Computer Vision and Pattern Recognition, Las Vegas, NV, USA, 27–30 June 2016; pp. 5525–5533.
53. Guo, Y.; Zhang, L.; Hu, Y.; He, X.; Gao, J. MS-Celeb-1M: A Dataset and Benchmark for Large-Scale Face Recognition. In Proceedings of the European Conference on Computer Vision, Amsterdam, The Netherlands, 11–14 October 2016.
54. Wong, Y.; Chen, S.; Mau, S.; Sanderson, C.; Lovell, B.C. Patch-based Probabilistic Image Quality Assessment for Face Selection and Improved Video-based Face Recognition. In Proceedings of the IEEE Biometrics Workshop, Computer Vision and Pattern Recognition (CVPR) Workshops, Colorado Springs, CO, USA, 20–25 June 2011; IEEE: Piscataway, NJ, USA, 2011; pp. 81–88.

applied sciences

MDPI

Article

Dynamic Nonprehensile Manipulation of a Moving Object Using a Batting Primitive

Hyun-Min Joe [1,*], Joonwoo Lee [2] and Jun-Ho Oh [3]

1 Humanoid Robotics LAB, Department of Robot & Smart System Engineering, Kyungpook National University, Daegu 41566, Korea
2 Robotics & Intelligent Systems LAB, Department of Electrical Engineering, School of Electronic and Electrical Engineering, Department of Robot and Smart System Engineering, Kyungpook National University, Daegu 41566, Korea; jwl@knu.ac.kr
3 HUBO LAB, Department of Mechanical Engineering, Korea Advanced Institute of Science and Technology, Daejeon 34141, Korea; jhoh@kaist.ac.kr
* Correspondence: hmjoe@knu.ac.kr; Tel.: +82-10-3869-5170

Abstract: To achieve human-level object manipulation capability, a robot must be able to handle objects not only with prehensile manipulation, such as pick-and-place, but also with nonprehensile manipulation. To study nonprehensile manipulation, we studied robotic batting, a primitive form of nonprehensile manipulation. Batting is a challenging research area because it requires sophisticated and fast manipulation of moving objects and requires considerable improvement. In this paper, we designed a batting system for dynamic manipulation of a moving ball and proposed several algorithms to improve the task performance of batting. To improve the recognition accuracy of the ball, we proposed a circle-fitting method that complements color segmentation. This method enabled robust ball recognition against illumination. To accurately estimate the trajectory of the recognized ball, weighted least-squares regression considering the accuracy according to the distance of a stereo vision sensor was used for trajectory estimation, which enabled more accurate and faster trajectory estimation of the ball. Further, we analyzed the factors influencing the success rate of ball direction control and applied a constant posture control method to improve the success rate. Through the proposed methods, the ball direction control performance is improved.

Keywords: nonprehensile manipulation; robotic batting; high-speed object manipulation; ball recognition; trajectory estimation; motion control; weighted least square

Citation: Joe, H.-M.; Lee, J.; Oh, J.-H. Dynamic Nonprehensile Manipulation of a Moving Object Using a Batting Primitive. *Appl. Sci.* **2021**, *11*, 3920. https://doi.org/10.3390/app11093920

Academic Editor: Luis Gracia

Received: 18 January 2021
Accepted: 20 April 2021
Published: 26 April 2021

1. Introduction

To date, a variety of robots have been used in automated production lines for object manipulation tasks in factories, and these robotic technologies have contributed significantly to the development of modern industry. Recently, with the fourth industrial revolution, robots have been introduced to provide various services not only in factories but also in human living environments. Robots used in factories perform simple repetitive operations such as pick-and-place using specially designed grippers; however, in an environment such as cafes and restaurants, human-level object manipulation ability is required. Therefore, to provide a wider range of services in a human living environment, robots must have a higher level of object manipulation similar to that of humans.

In addition to manipulating objects using grasping, humans can manipulate objects freely by appropriately utilizing nonprehensile manipulations [1] without grasping, such as throwing, batting, rolling, pushing, and sliding. Most robots are still limited to prehensile manipulation using grasping. A robot that performs nonprehensile manipulation is rare owing to the difficulty of control using nonprehensile manipulations. When nonprehensile manipulation is performed, the object moves during manipulation. Since the moving object is not fixed to the robot, it is necessary to plan and control the future behavior of the moving object based on the state of the robot and the moving object. Despite these difficulties, nonprehensile manipulation has advantages such as increased object manipulation methods, expansion of the robot workspace, and increased manipulation dexterity [2]. To improve the ability of robots to manipulate objects, the robots must have nonprehensile manipulation capabilities with these advantages.

Fabio et al. conducted a survey on nonprehensile manipulation of robots to address the trends and open issues of nonprehensile manipulation studies [3]. According to [3], the nonprehensile manipulation task is complex and difficult. Because of the complexity and difficulty of the task, most studies divide the nonprehensile manipulation task into simpler subtasks called nonprehensile manipulation primitives. The representative nonprehensile manipulation primitives include throwing [4], catching [5], batting [6], pushing [7], sliding [8], and rolling [9]. Each of these nonprehensile manipulation primitives is selectively operated by a high-level supervisor depending on the task [10]. Among the nonprehensile manipulation primitives, batting is challenging because it requires precise and fast manipulation of the moving objects, and studies on this topic are lacking. In sporting events such as table tennis, tennis, and baseball, players have a batting ability to send the moving objects (balls) to the desired position with sophisticated and quick manipulation, while the batting performance of robots is insufficient. Therefore, this study aims to contribute to the improvement of the dexterity of robots by conducting research on batting, which requires more precise and quicker nonprehensile manipulation of moving objects than do the above-mentioned primitives.

For a robot to perform a batting task, at least four methods are required. First, image processing is required to recognize a moving ball. Second, estimation of the future trajectory of the ball is required. Third, the motion of the robot arm must be controlled to affect the ball direction. Lastly, calibration is required to convert the coordinates between the robot arm's coordinate system and the vision sensor's coordinate system. This implies that the performance of each method affects the performance of the batting primitive; studies that can improve the performance of each method are discussed.

Chen et al. [11] developed a vision module for humanoid robotic table tennis. The vision module contains two stereo vision sensors with a 200 fps and an algorithm for predicting the rebound trajectory of a table tennis ball. Nakabo et al. [12] developed a high-speed vision system capable of image acquisition and image processing at 1 kHz for moving ball recognition. A parallel computation architecture was used to reduce image transfer and processing time, and an active vision system with moving cameras was developed to track the moving objects. Tesheng et al. [13] used an aerodynamic model of a ball to account for the trajectory of the ball before and after collision to improve the performance of ball direction control. Serra et al. conducted the study of hitting a table tennis ball to the desired position [14]. To accurately control the arrival position of the ball after hitting, a more accurate aerodynamic model that that in [13] was applied for the trajectory estimation of the ball. Although the algorithms were tested via simulations, the implementation of the algorithms on an actual hardware system was left to be covered under future work. For accurate ball direction control, a batting algorithm considering impact dynamics was proposed [15,16]; however, the problem of extending the 2D algorithm to the 3D algorithm remains due to the computation time required for impact dynamics in 3D.

Schüthe et al. [17] introduced the optimal control with state and soft constraints for a simulated ball batting task. By utilizing the soft constraints, a motion utilizing the redundant degree of freedom (DOF) is automatically generated, but there is a limitation in that a motion exceeding the range of motion of the joint is generated. Pekarovskiy et al. [18] proposed a motion generation method that can adapt to rapidly changing target points in consideration of the feasibility and computation time of the motion trajectory. This method was applied to 2D planar volleyball batting. Kober et al. [19] proposed a method to generate the trajectory of the robot arm through learning. If the system is changed, the process of collecting and processing data is required again, and the ball direction control is not considered. Mori et al. [20] developed a fast and lightweight robotic arm for badminton.

The use of pneumatic actuators made the robotic arm lighter, enabling high-speed batting. The position of the shuttlecock was measured using a high-speed motion capture system of 240 Hz, but there was a limitation in that control of the direction of the ball was generated by a pre-learned feedforward motion. In the realization of a three-dimensional ball direction control system, the amount of computation in object recognition and trajectory estimation and the realizable three-dimensional motion generation are important.

Senoo et al. [21] developed a high-speed robot batting algorithm using a high-speed active vision system developed by Nakabo et al. [12]. The algorithm was extended to control the direction of the baseball in three-dimensional space [22]. For fast batting and ball direction control, the authors proposed a hybrid trajectory generator comprising a part that generates a high-speed batting motion and a part that modifies the motion through visual feedback. A high-speed image processing system (1 kHz) specially developed for object recognition was used, and a simple least square method was applied to estimate the object trajectory because the sampling rate was fast. Without this specially designed sensor system, it is difficult for other researchers to implement a batting system using this algorithm.

In this study, we propose a batting framework capable of controlling the three dimensional direction of a moving ball by using an off-the-shelf vision sensor with a relatively low fps (50 Hz). The proposed batting framework includes object recognition, object trajectory estimation, robot arm motion control, and calibration. In a condition where the image sampling rate is low, since the importance of one data point is relatively large, the performance of algorithms for object recognition, trajectory estimation, and motion control becomes more important. Therefore, we propose ways to improve the performance of each algorithm at a low sampling rate. Since the proposed methods were developed based on an off-the-shelf vision sensor, other researchers can easily implement these algorithms.

The specific contributions of this study are as follows. First, in terms of ball recognition, this study proposes an image processing method for improving ball recognition accuracy in a more general environment. Previous studies covered noise filtering after applying a difference image or color segmentation. In the real world, however, various lights affect the ball recognition accuracy. Further, we applied the method using the geometrical properties of the ball to improve the recognition accuracy under these lighting conditions. Second, by applying weighted squares regression considering the positional accuracy according to the distance of the stereo vision sensor, we improve the trajectory estimation accuracy of the target object even with a low sampling rate. Third, we analyze the factors that affect the performance of ball-direction control and propose additional constant posture control methods to reduce the influence of those factors. Finally, in actual implementation, calibration between the camera and the robot coordinate system is essential. A simple but accurate calibration method is introduced.

2. Robot Batting System

We built a robot betting system as shown in Figure 1. The hardware of the batting system consists of a stereo vision sensor for recognizing the red ball and obtaining positional information in three-dimensional space, and a six DOF robotic arm for batting the ball to the target position (Net in Figure 1). The stereo vision sensor (Bumblebee2, Point Gray Research Inc.) provides 640 × 480 pixels color images at up to 48 frames per second (fps). The Triclops library included in Bumblebee2 (Point Gray Research Inc.) provides 3D position coordinates. Bumblebee is shipped with precision calibration at the production stage, and because the two cameras are structurally fixed, there is no need to perform additional calibration between the two cameras. A slightly modified version of the arm of a humanoid robot, Hubo [23], designed by the HUBO LAB of the Korea Advanced Institute of Science and Technology, was used for batting. Link lengths were changed to extend the workspace, and the robotic hand end device was replaced with a 0.09 m diameter round aluminum plate to perform the batting task.

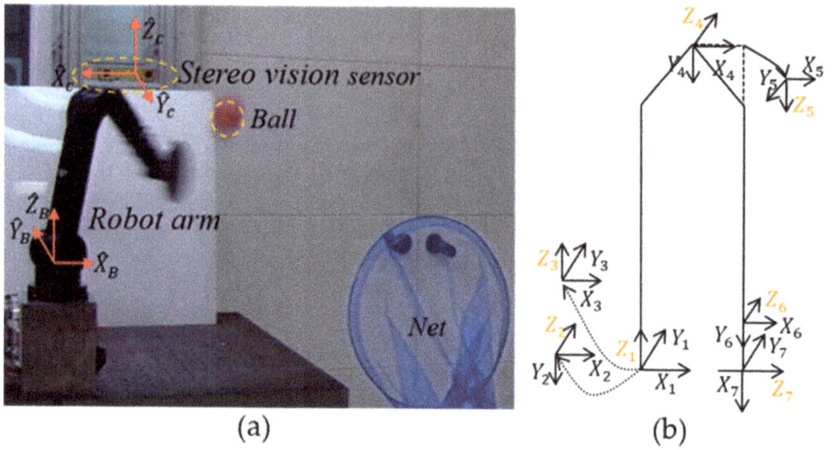

Figure 1. (**a**) Robotic batting system. The hardware of the batting system consists of a stereo vision sensor and a six degree of freedom robotic arm. (**b**) Robot arm configuration. The base coordinate system of the robot arm is the first coordinate system.

3. Method

We applied a color segmentation method [24] to recognize the red ball. To improve the recognition accuracy of the ball under the conditions of illumination, a circle fitting method [25] using the geometrical characteristics of the ball is employed.

3.1. Ball Recognition

3.1.1. Color Segmentation

The color separation method is used to find pixels with specific color values in the image. Since we used a red ball in the batting experiment, only red is segmented in the image by comparing the red component (I_c) and the threshold value ($I_{Threshold}$) of the pixel as shown in Equation (1).

$$I_b(u,v) = \begin{cases} 1, \ if \ I_c(u,v) \geq I_{threshold} \\ 0, \ if \ I_c(u,v) < I_{threshold} \end{cases} \tag{1}$$

For improved visualization, the color image is converted to a binary image comprising pixels (I_b) with only two values. The pixels corresponding to the threshold value or more are represented by "1" (white), and the other pixels are represented to "0" (black); thus, the red ball becomes white. However, since the RGB value changes according to the change in illuminance, not only the ball but also noise are binarized. To remove this noise, the morphology method [26] is applied. This method is effective for removing the salt-and-pepper noise or impulse noise between objects and the background. Figure 2 shows the results of finding the location of the ball on the desk in the lab when color segmentation and morphology methods are applied. The left side of the figure is a color image, the right side is a binary image, and only the red ball appears white in the binary image.

Figure 2. Color image (**left**) and binary image (**right**). Green circle of color image represents the position of the red ball.

3.1.2. Circle Fitting

Figure 3a shows the batting experiment environment captured using a stereo vision camera. Because the ceiling light creates a shadow on the lower part of the ball, the RGB value of the lower part of the ball changes. Figure 3b shows a binarized image with color segmentation applied to the image in Figure 3a. In the binary image of Figure 3b, the shape of the ball appears as a semicircle rather than a circle because the lower part of the color change is not properly color segmented. Incorrect color segmentation causes errors in the center position measurement of the ball. For example, if only the upper part of the ball is color-segmented, the position error in the vertical direction increases. Threshold adjustment is limited because the difference between the color value of the lower part of the ball and the color value of the upper part is large.

To overcome the limitations of color segmentation, we additionally applied a circle fitting method that uses the geometric characteristics of a circular ball. Figure 3c shows only the edge of the semicircle shown in Figure 3b. By applying a circle fitting to the edge of the semicircle, we can estimate the original circle ball shape, as shown in Figure 3d. Figure 3e shows the center position of the ball before applying the circle fitting (lime green point) and the center position of the ball after application (blue point). The positions of the lime green and blue points are (X, Y, Z) = (−0.0416, 1.3299, −0.0888) and (X, Y, Z) = (−0.0413, 1.3207, −0.0955), respectively. The true position of the ball is (X, Y, Z) = (−0.0414, 1.3187, −0.0985). Prior to the circle fitting method, the error in the z direction was 0.0097 m, which is approximately 20% of the diameter of the ball at 0.05 m. After the circle fitting method was applied, the center position of the ball moved by about 0.007 m further down the Z-axis, and the position error in the Z direction was reduced from 0.0097 m to 0.0030 m.

(a) (b) (c) (d) (e)

Figure 3. Circle fitting process. (**a**) Batting experiment environment captured using a stereo vision camera, (**b**) A binary image with color segmentation, (**c**) The edge image of the semicircle, (**d**) An image with circle fitting applied to the edge of a semicircle, (**e**) Center position before (green) and after (blue) circle fitting.

3.1.3. Calibration

Since the position of the ball measured in the vision coordinate system needs to be converted to the robot coordinate system, a coordinate transformation between the two coordinate systems is required. For coordinate transformation between two coordinates, we attached the red marker shown in Figure 4 to the end-effector of the robot arm and measured its position in the vision coordinate system and the robot coordinate system. Coordinate transformation between two coordinates is possible using a homogeneous transformation matrix representing the relationship between the camera coordinate system and the robot base coordinate system, as shown in the following equation:

$$\begin{bmatrix} {}^{B}P \\ 1 \end{bmatrix} = H \begin{bmatrix} {}^{C}P \\ 1 \end{bmatrix} \quad (2)$$

where $H = \begin{bmatrix} {}^{B}_{C}R & {}^{B}P_c \\ 0\ 0\ 0 & 1 \end{bmatrix}$.

Subscripts B and C represent the robot arm base coordinate and camera coordinate systems, ${}^{B}P$ and ${}^{C}P$ represent the positions of the markers (Figure 4) measured in the robot and vision coordinate systems, respectively, H denotes the homogeneous transformation matrix, and ${}^{B}_{C}R$ and ${}^{B}P_c$ represent the rotation matrix and the distance vector between the robot coordinate system and the camera coordinate system, respectively. The homogeneous transformation matrix can be calculated by the least square method as

$$H = EV^T\left(VV^T\right)^{-1},$$

$$\text{where } E = \left[\begin{bmatrix} {}^{B}P_1 \\ 1 \end{bmatrix} \begin{bmatrix} {}^{B}P_2 \\ 1 \end{bmatrix} \cdots \begin{bmatrix} {}^{B}P_N \\ 1 \end{bmatrix}\right], V = \left[\begin{bmatrix} {}^{C}P_1 \\ 1 \end{bmatrix} \begin{bmatrix} {}^{C}P_2 \\ 1 \end{bmatrix} \cdots \begin{bmatrix} {}^{C}P_N \\ 1 \end{bmatrix}\right]. \quad (3)$$

E and V consist of the position vectors of the markers measured in the robot coordinate system and the vision coordinate system, respectively, and N is the number of measured position vectors. From the base coordinate system of the robot, the position data set ${}^{B}P_1 \ldots {}^{B}P_N$ is calculated by forward kinematics. The position data set ${}^{C}P_1 \ldots {}^{C}P_N$ is measured using the stereo vision sensor.

The accuracy of coordinate transformation is improved by calculating a homogeneous matrix from data measured at various locations as shown in Figure 5. From the measured data, the homogeneous transformation matrix is calculated as

$$H = \begin{bmatrix} -0.9981 & -0.0066 & 0.0054 & 0.1935 \\ 0.0184 & -0.9754 & 0.0026 & 0.8037 \\ -0.0114 & 0.0055 & 0.9906 & 0.3792 \\ 0 & 0 & 0 & 1 \end{bmatrix}. \quad (4)$$

From ${}^{B}_{C}R$ and ${}^{B}P_c$ of the homogeneous transformation matrix, the camera coordinate system is at a distance of 0.1935 m in the X-axis direction, 0.8037 m in the Y-axis direction, and 0.3792 m in the Z-axis direction from the robot coordinate system. In the robot coordinate system, the camera coordinate system is rotated by 0.3188° for the X axis, 0.6539° for the Y axis, and 178.9466° for the Z axis. To evaluate the accuracy of the calculated homogeneous transformation matrix, the positions of the markers obtained from the homogeneous transformation of the marker positions measured in the vision are compared with the positions of the markers measured in the robot coordinate system as shown in Figure 6. The error is expressed as a norm value for the X, Y, and Z axes. The mean error is 0.0024 m and the standard deviation is 0.0012 m. Considering that the error is less than 0.005 m and the diameter of marker is 0.01 m, the calculated homogeneous matrix is sufficiently accurate.

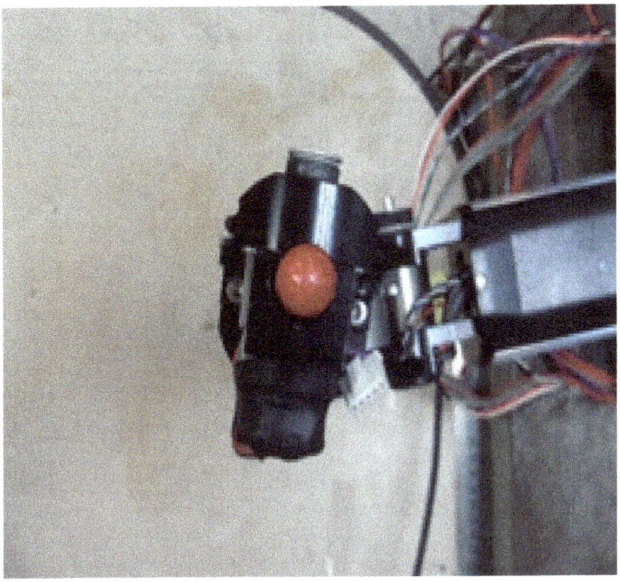

Figure 4. Red marker attached to the end-effector of the robotic arm to calculate the homogeneous matrix.

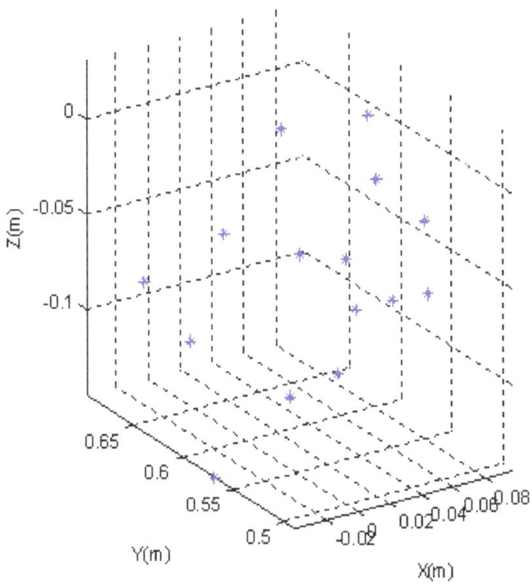

Figure 5. Red marker positions measured from the vision sensor. The positions of the markers measured at various positions are used to accurately calculate a homogeneous matrix.

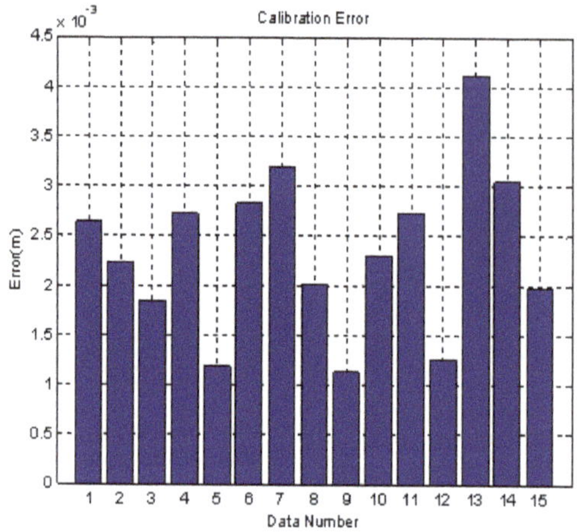

Figure 6. Histogram showing the calibration error for 15 marker positions.

3.2. *Ball Trajectory Estimation*

Considering the limitations of the joint speed, the robot arm should be able to move in advance by estimating the trajectory of the ball. To estimate the trajectory of the ball, we used weighted least-squares regression [27]. Each time a new ball position and time is acquired by the vision sensor, the trajectory of the ball is updated using weighted least-squares regression.

3.2.1. Least Square Regression

The x and y planes of the robot coordinate system are horizontal planes and the z direction is perpendicular to the x-y plane. A linear function was used to estimate the horizontal trajectory of the ball, and a quadratic function was used to estimate the vertical trajectory. Because the ball moves in a parabolic trajectory, the z-axis trajectory of the ball is fitted as a quadratic function, unlike other axes. The trajectory function for each of the x, y, and z axes of the ball is given by

$$\begin{cases} x = a_1 t + b_1 \\ y = a_2 t + b_2 \\ z = a_3 t^2 + b_3 t + c \end{cases} \tag{5}$$

Whenever three-dimensional position data are newly measured in the stereo vision sensor, the coefficient of the trajectory function (5) is newly calculated through least square regression [28]. The position data sampling interval of the ball is 20 ms, which is the image acquisition period of the stereo vision sensor. Since the flight time of the ball is 0.5 s on average, approximately 15 to 20 ball position data points are measured. Figure 7 represents the estimated trajectory of the ball according to the number of data points using least squares regression. As the number of data points increases, it can be seen that the estimated trajectory converges to one trajectory.

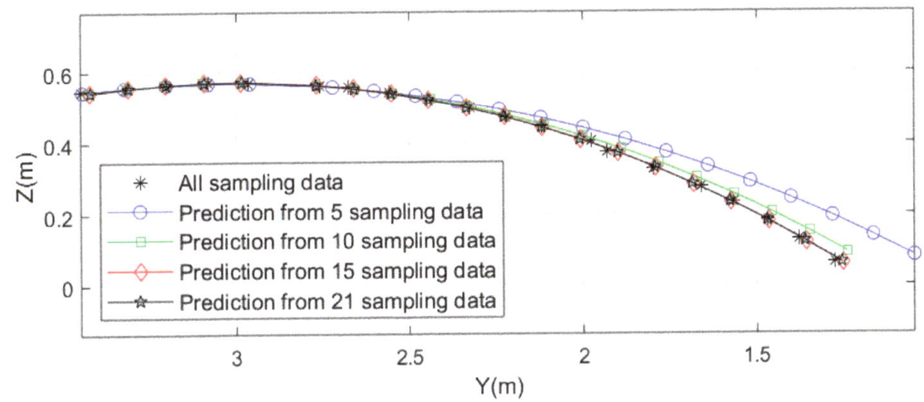

Figure 7. Estimated ball trajectory according to the number of sampled data points. As the number of sampled data points increases, the trajectory estimation becomes more accurate.

3.2.2. Weighted Least Square Regression

Least square regression was applied to the ball trajectory estimation regression described in Section 3.2.1, based on the assumption that the data measured by the stereo vision sensor are accurate. However, the accuracy of the position data measured from the stereo vision sensor is inversely proportional to the distance between the stereo vision sensor and the object. This is because the closer the distance, the greater is the number of pixels representing the object. Figure 8 shows a graph of position accuracy versus distance between Bumblebee2 and the object. Therefore, it is more appropriate to apply the weighted least-squares regression than the least-squares regression considering the accuracy of the measured value according to the distance.

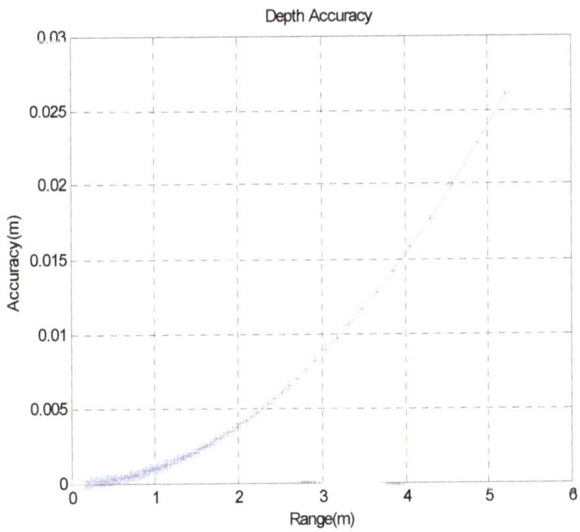

Figure 8. Depth error vs. distance. Depth accuracy of stereo vision sensor according to distance.

Least squares regression minimizes the sum of the squares of the residuals, and the weighted least-squares regression minimizes the sums (Q) of the squares of the residuals $(y_i - bx_i)$ multiplied by the weight (w_i); it is given by

$$Q = \sum_{i=1}^{n} w_i (y_i - bx_i)^2, \text{ where } w_i = w_{i+1}\lambda \tag{6}$$

The weights are calculated as a geometric sequence based on the last weight. In several tests, 0.8 was used as λ of the weighted least square. Figure 9 is a graph showing the RMS errors for the X, Y, and Z components obtained by applying the weighted least square and least square regressions. Although there is almost no difference in the RMS error for X and Y with and without the weighted least square, the RMS error decreases rapidly with weighted least square for the Z axis. This result is expected because the change in the trajectory of the ball on the z axis is larger than that on the x and y axes.

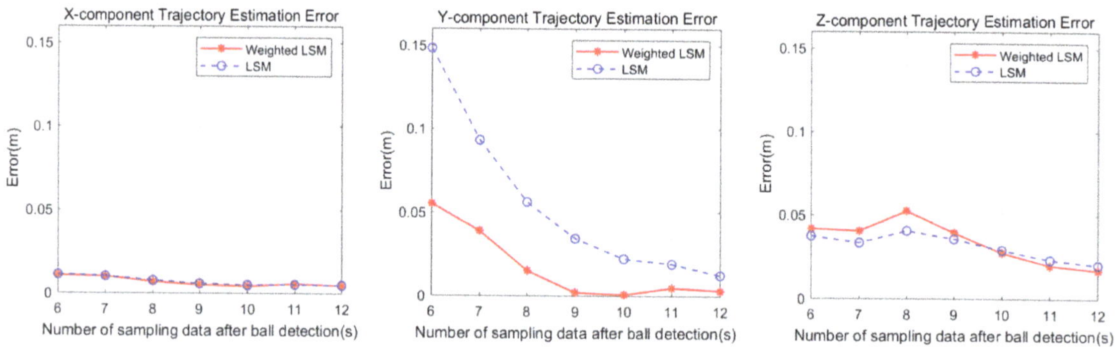

Figure 9. RMS error vs. sampling data number about the *x* axis. RMS error of the estimated ball trajectory with and without weighted least squares method. RMS error for each axis is displayed in the order of X axis, Y axis, and Z axis from the left.

3.3. *Motion Control of the Robot Arm*

The trajectory of the robot arm consists of two stages: before and after batting. The speed of the end-effector of the robot arm increases before batting, and after batting, a trajectory is generated that slows down the increased speed of the end-effector. The robotic arm used in the experiment has six joints, and, therefore, we need to generate the trajectories of the six joints. For convenience, the joint trajectory generation method is described by the notation of one joint, and the same method is applied to the remaining joints. The trajectory generation method before batting is described first, followed by the trajectory generation method after batting.

The cubic spline interpolation function for generating the motion trajectory of a joint is:

$$f(t) = \sum_{i=0}^{3} a_i t^i, \tag{7}$$

where a_i is the coefficient of the trajectory function and t is the time. Since coefficients a_i in Equation (7) are four, four constraints are needed to obtain each coefficient. The constraints of the trajectory function before hitting the ball are as follow:

$$f(0) = \theta_0,\ f(t_h) = \theta_h,\ \dot{f}(0) = 0,\ \dot{f}(t_h) = u_h \tag{8}$$

3.3.1. Trajectory Interpolation

The trajectory constraints of Equation (8) represent the initial joint position $f(0)$, joint position $f(t_h)$ to hit the ball, initial joint velocity $\dot{f}(0)$, and velocity $\dot{f}(t_h)$ to hit the ball. θ_0 and θ_h represent the initial position of the joint and the joint position at the moment of hitting the ball, t_h represents the time of the moment of hitting the ball, and u_h represents the angular velocity vector at the moment of hitting the ball. Four simultaneous equations are constructed according to four constraints, and each coefficient can be obtained by calculating the simultaneous equations.

After hitting the ball, a joint trajectory has to be generated to reduce the speed of the robot arm. By calculating the simultaneous equations with the following four constraints, a joint trajectory is generated to decrease the velocity of robot arm.

$$f(t_h) = \theta_h,\ f\left(t_f\right) = \theta_f,\ \dot{f}(t_h) = u_h,\ \dot{f}\left(t_f\right) = 0 \tag{9}$$

The initial joint position is the angle to hit the ball, the final joint position is the angle in the configuration where the robot arm stops, the initial velocity is u_h at the time of the hit, the final velocity is 0, and t_f is the time to stop.

3.3.2. Robot Batting Trajectory

The angular and velocity trajectories of joints 1 and 3 for when the batting task is performed as generated by the cubic spline function and constraints are shown in Figure 10. Through cubic spline interpolation, the trajectories of the joints were smoothly interpolated from the initial position before batting to the stop position after batting. The batting time is 0.274 s and the time it takes for the robot arm to stop completely after it starts moving is 0.914 s. The maximum angular velocity is given only to joints 1 and 3 because the maximum angular velocity of the other joints is limited and has little effect on the speed of the end-effector. After hitting, the robot arm stops smoothly by imposing a constraint on the angular velocity of joints 1 and 3 as 0. The velocity of the end-effector is calculated as

$$V_R = J\dot{\Theta} \tag{10}$$

where $J = \begin{bmatrix} \frac{\partial P_X}{\partial \theta_1} & \cdots & \frac{\partial P_X}{\partial \theta_6} \\ \frac{\partial P_Y}{\partial \theta_1} & \cdots & \frac{\partial P_Y}{\partial \theta_6} \\ \frac{\partial P_Z}{\partial \theta_1} & \cdots & \frac{\partial P_Z}{\partial \theta_6} \end{bmatrix}$, $\dot{\Theta} = \begin{bmatrix} \dot{\theta}_1 \\ \vdots \\ \dot{\theta}_6 \end{bmatrix}$

J is the Jacobian matrix of the robot arm, and $\dot{\Theta}$ is a vector consisting of the angular velocities of the joints of the robot arm. The calculated velocity of the end-effector is 6.2 m/s.

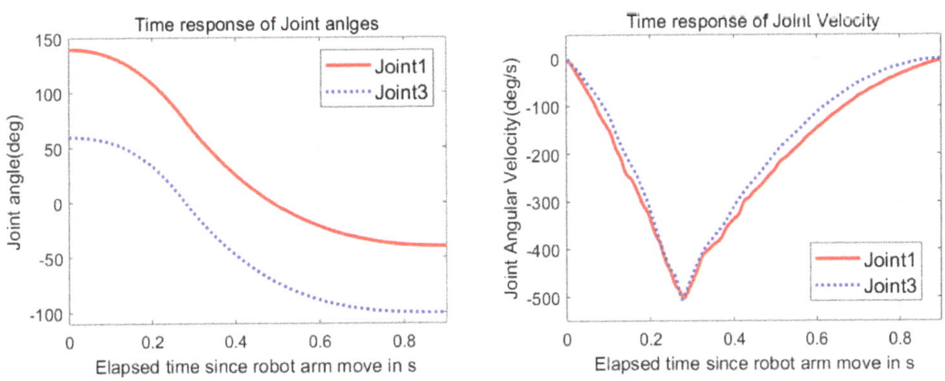

Figure 10. The angular and velocity trajectories of joints 1 and 3. When the robot arm hits the ball, the joint angular velocity of joint 1 and joint 3 was constrained to $-500°/s$, and the rest of the joints to zero.

3.3.3. Ball Direction Control

The ultimate goal of the batting task is to move the object to the desired target position in a single collision. The orientation of the end-effector of the robot must be controlled to move the ball to the desired target position. That is, the direction of the ball after the collision can be controlled by controlling the normal vector perpendicular to the circular plate of the end-effector. The normal vector can be calculated from the coefficient of restitution between the ball and the end-effector, the velocity of the ball and the end effector before and after the collision, and some assumptions. Figure 11 shows the velocity vector of the ball before the batting ($V_{B1} \in R^3$), the velocity vector of the ball after the batting ($V_{B2} \in R^3$), and the velocity vector of the end-effector of the robot ($V_R \in R^3$) and the normal vector ($n \in R^3$) perpendicular to the plane of the end-effector.

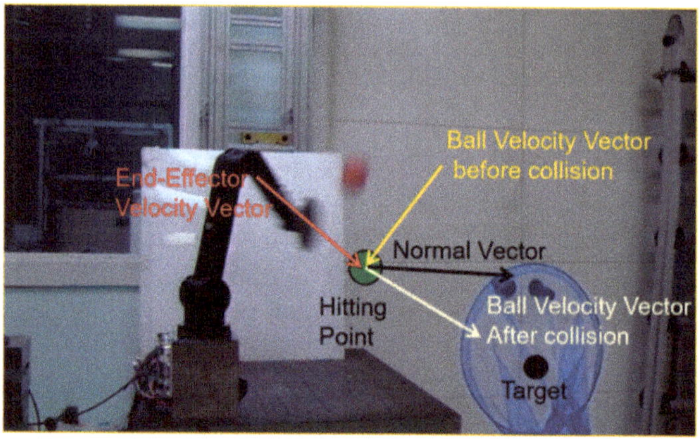

Figure 11. Vector arrangements representing end-effector velocity (V_R), ball velocity before collision (V_{B1}), ball velocity after collision (V_{B2}), and normal vector (n).

The vector shown in Figure 11 can be expressed as shown in Figure 12 under the assumption that there is no change in the velocity of the end-effector before and after the collision, and that there is only energy loss caused by the deformation of the ball in the collision.

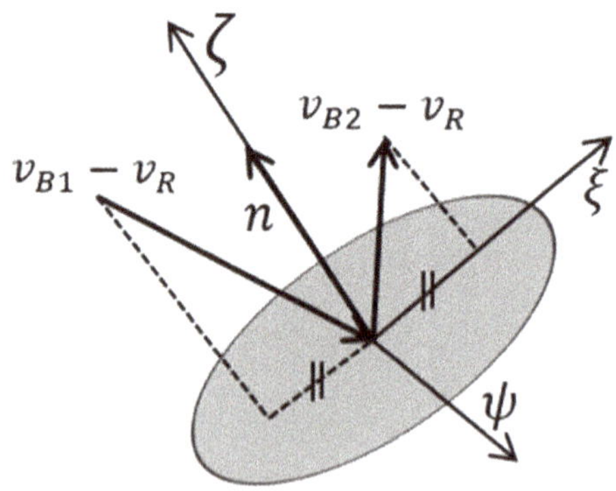

Figure 12. Vector analysis before and after collision.

From Figure 12, the relationship between the velocity vectors and the normal vector can be defined by the collision coefficient equation

$$e = \frac{(v_{B2} - v_R) \cdot n}{(v_{B1} - v_R) \cdot n} \quad (11)$$

The collision coefficient (e) is the speed ratio before and after the collision of the ball, and the material of the ball and the end device has a major influence on the collision coefficient. Considering the inherent properties between the ball and the end-effector, we obtained the collision coefficient through collision experiments. The coefficient was obtained by dropping the ball perpendicular to the end-effector to measure the speed before and after the impact. The average collision coefficient obtained from 20 experiments is 0.6506 and the standard deviation is 0.0448. Based on the assumptions above, the velocity of the horizontal component before and after the collision of the ball is preserved and can be expressed as

$$v_{B1} - v_R - \{(v_{B1} - v_R) \cdot n\}n = v_{B2} - v_R - \{(v_{B2} - v_R)n\}n \tag{12}$$

The above equation can be rearranged as

$$n = \frac{v_{B2} - v_{B1}}{||v_{B2} - v_{B1}||} \tag{13}$$

Note that the normal vector is in the same direction as that of the difference in the velocity vector before and after the collision of the ball. Therefore, the information of v_{B1} and v_{B2} is needed to calculate n. The v_{B1} is calculated by differentiating Equation (5), and the direction of v_{B2} is calculated from the estimated trajectory of the ball and the target position. Therefore, to calculate the normal vector, it is necessary to calculate the magnitude of the v_{B2}.

The direction of the velocity vector after the collision is derived from the vector geometry. The velocity vector after the collision in Figure 12 can be set to one side of the triangle as shown in Figure 13. ρ is the direction vector of v_2, and σ represents the horizontal relative velocity vector that is preserved before and after the collision of the ball. The Pythagorean theorem applies to the two triangles $\triangle abc$ and $\triangle acd$ and is represented by the following two equations:

$$||v_{B1} - v_R||_2 = ||(v_{B1} - v_R) \cdot \sigma||_2 + ||(v_{B1} - v_R) \cdot n||_2, \tag{14}$$

$$||v_{B2} - v_R||_2 = ||(v_{B1} - v_R) \cdot \sigma||_2 + ||(v_{B1} - v_R) \cdot n||_2 e^2 \tag{15}$$

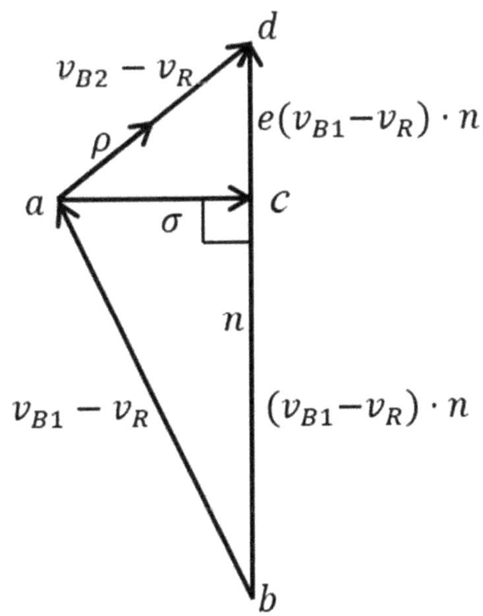

Figure 13. Vector analysis.

The difference between the two equations is

$$||v_{B1} - v_R||_2 - ||v_{B2} - v_R||_2 = ||(v_{B1} - v_R) \cdot n||_2\left(1 - e^2\right) \quad (16)$$

The cosine law is applied to $\triangle abd$ as

$$||(v_{B1} - v_R) \cdot n||_2(e+1)^2 = ||v_{B1} - v_R||_2 + ||v_{B2} - v_R||_2 - 2||v_{B1} - v_R||||v_{B2} - v_R|| \cos\theta, \quad (17)$$

$$\text{where } \theta = \cos^{-1} \frac{(v_{B1} - v_R) \cdot \rho}{||v_{B1} - v_R||||\rho||}$$

From Equations (16) and (17), we can calculate the relative velocity of the ball after the collision as

$$||v_{B2} - v_R|| = \frac{||v_{B1} - v_R||(1-e)\cos\theta + \sqrt{(||v_{B1} - v_R||(1-e)\cos\theta)^2 + 4||v_{B1} - v_R||_2 e}}{2} \quad (18)$$

From Equations (13) and (18), the normal vector perpendicular to the circular plate of the terminal end-effector can be expressed as

$$n = \frac{v_{B2} - v_{B1}}{||v_{B2} - v_1||} = \frac{(v_{B2} - v_R) - (v_{B1} - v_R)}{||(v_{B2} - v_R) - (v_{B1} - v_R)||} \quad (19)$$

From the calculated normal vector, the orientation of the ball after the collision is adjusted by adjusting the orientation of the robot end-effector.

4. Experiment

4.1. Batting Experiment

The batting task was performed based on the algorithms proposed in the previous sections. Snapshots of the robot arm hitting the oncoming ball to the target are shown in Figure 14. The target position for the ball is a blue net that is 0.57 m in the x-axis, −0.48 m in the y-axis, and −0.09 m in the z-axis from the robot base coordinate. We performed the experiment in the same way by changing the position of the target to 0.36 m for the x-axis and 1.06 m for the y-axis to −0.02 m for the z-axis. We conducted 30 batting experiments each, the success rate converged to approximately 40%.

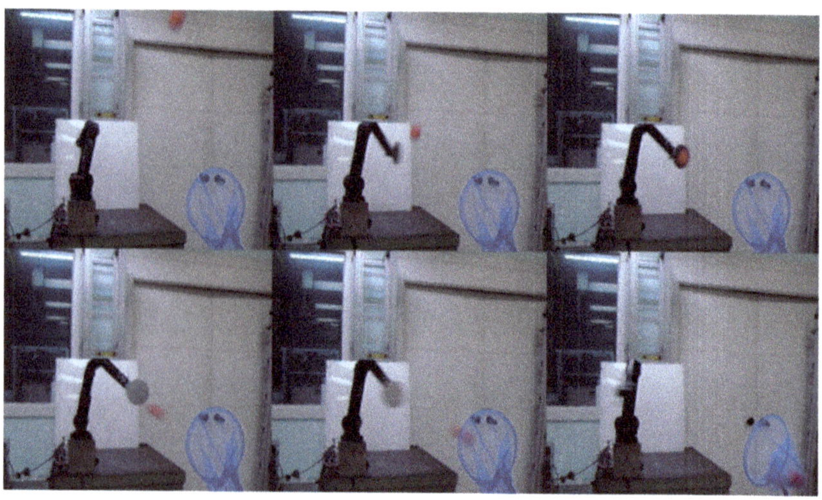

Figure 14. Successive snapshots of batting task (sequence is horizontal to vertical).

We conducted two additional experiments to analyze the factors that influence the success rate. Since the trajectory prediction function of the ball is a function of position versus time (Equation (5)), the arrival position estimation error of the ball and arrival time estimation error of the ball are analyzed. If there is a large error in the estimated ball position or time, this error will affect the success rate of the batting task.

4.2. Experiment to Analyze the Predicted Ball Position Accuracy

We stopped the robotic arm at the predicted ball position without swinging to analyze the estimated ball position accuracy. We will call this experiment a bunt experiment, similar the action of the same name in a baseball game. The effect of the time error was separated by placing the end-effector of the robot arm at the position where the ball arrived. Thirty experiments were carried out, and the robotic arm was able to touch all the thrown balls (success rate is 100%). This experiment shows that the estimated ball position is accurate for ball batting. Figure 15 shows the positions of the batted balls in 30 experiments.

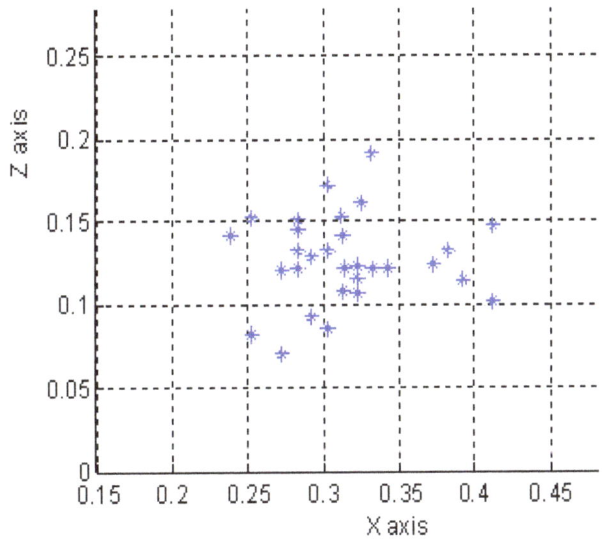

Figure 15. Shot group of the batted ball. In this bunt experiment, the robotic arm hit the ball with a 100% success rate.

4.3. Experiment to Analyze the Accuracy of the Predicted Ball Arrival Times

Data acquired from the stereo vision sensor includes time information together with image data. The time information is measured at the moment the shutter of the camera is closed, and this time information is called a time stamp. The timer cycle of the camera is 8 kHz, and a time stamp is calculated based on this cycle.

A red marker (Figure 4) was attached to the end-effector of the robot arm, and the position and time of the marker was measured with a vision sensor. To measure the time difference between each image, the robot arm drew a circle at a constant velocity of 17 degree/s with respect to the Y-Z plane. While the robot arm moved in a constant velocity circular motion, time information was measured along with the position of the marker. Figure 16 shows the marker positions of the end-effector when the robot arm moved at constant speed. A histogram comparing the time difference between two theoretical positions caused by constant circular motion and the time difference measured using a vision sensor is shown in Figure 17. The mean time difference error was −0.00057 s and the standard deviation was 0.0048 s. If the error pattern is biased, the time difference error can be compensated; however, in the case of such a random error, it is difficult to compensate for the time difference error. Therefore, the following section describes the proposed method to compensate for the time error.

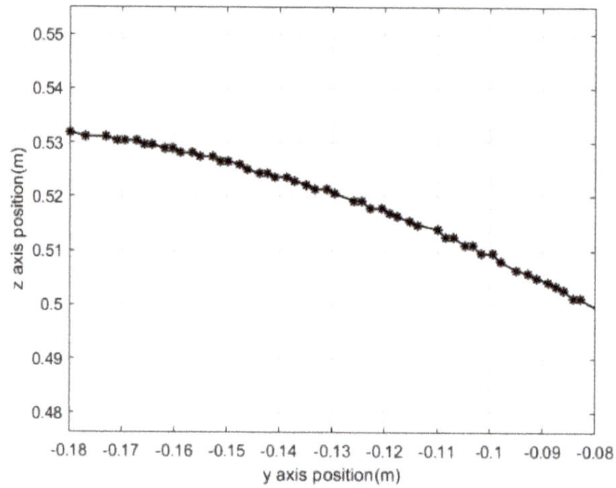

Figure 16. Marker position from vision sensor.

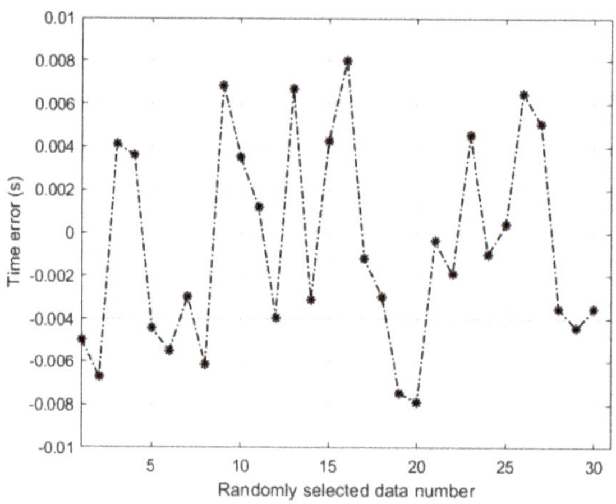

Figure 17. Time error for randomly selected data. Time error shows a random error pattern.

4.4. Constant Posture Control Method

Time errors change the timing of the hitting. Accordingly, the success rate of ball direction control is decreased. To compensate for this time error, we propose a method of maintaining the posture of the end-effector for a predetermined time before and after the ball hitting, and this method is called the constant posture control (CPC) method. By maintaining the posture of the end-effector for a predetermined time before and after the hitting timing, the posture change of the end device due to time error is prevented. We set the time that the posture of the end-effector should be maintained to a total of 0.3 s with 0.015 s before and after the hit.

We conducted a ball direction control experiment to verify the effectiveness of the CPC method. Figure 18 shows the posture change of the end-effector with and without the CPC method while the robot arm swings. Postures are expressed in Euler angles. The solid line is a change in posture considering the CPC method, and the dotted line is a change in posture without considering the CPC method. Since the robot arm strikes the ball while rotating about the Z axis (refer to the Z_1 axis in Figure 1), the Euler angle change about the Z axis is very large. For a clear comparison, the experimental results are described based on the Z axis Euler angle. The interval in which the posture should be kept is between 0.222 s and 0.252 s. The red solid Z-axis Euler angle was in the range of $-76.79°$ to $-77.63°$. This is a negligible error compared to the Z axis Euler angle of $-77.16°$ to be maintained. On the other hand, the blue dotted line without considering the CPC method shows that the Z-axis Euler angle was changed from $-73.24°$ to $-82.81°$. Finally, after 15 experiments, the success rate was 53.0% when considering the CPC method, and the success rate was increased by 13% when the CPC method was not considered. The experimental video is shown in [29].

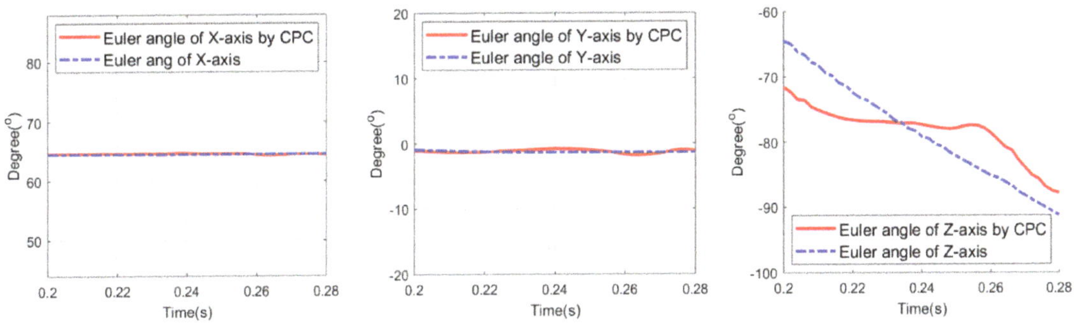

Figure 18. Posture change of end-effector with and without constant posture control (CPC) method.

5. Discussion

In this paper, we designed a ball batting system to perform batting tasks and proposed algorithms to improve the success rate of ball direction control under low sampling rate conditions. In a condition where the image sampling rate is low, since the importance of one data point is relatively large, the performance of algorithms such as object recognition and trajectory estimation becomes more important. In terms of object recognition, we applied a circular fitting method that is robust to the influence of illuminance and improved the 3D position accuracy of the ball by about 60% compared to the conventional image processing method with color segmentation and noise filtering applied. In terms of trajectory estimation, while the conventional method uses complex models, we proposed a weighted least square trajectory estimation method based on a simple model. As shown in Figure 9, it is possible to accurately estimate the trajectory with a small number of data points by considering the weight according to the distance. In terms of motion control of the robot arm, through analysis of the factors affecting the success rate of ball direction control, a time error, which is a random error, was found as shown in Figure 17, and a constant posture control method was proposed to overcome this. Through this method, the ball direction control success rate was improved by about 13%.

The methods proposed in this study can be applied to other research fields. The ball recognition method using geometric features can be used to improve recognition accuracy of other objects. Since the ball trajectory estimation method using the weighted least squares method can overcome the problem of computation time in the conventional complex model-based trajectory estimation method, this method can be applied not only at low-speed sampling times but also at high-speed sampling times. In addition to the trajectory estimation field, the algorithm considering the accuracy of the stereo vision sensor according to the distance can be easily applied to other stereo vision applications, and thus can be used as an effective performance improvement method.

Although the current ball direction control success rate is similar to that of research using high-speed vision sensors, there is a possibility that the success rate will be further improved if the algorithms proposed in this study are applied to a batting system equipped with a high-speed vision sensor. In addition, if the currently used ball model is improved, the accuracy of ball trajectory estimation is expected to be further improved. We plan to study the trajectory estimation method that combines the improved ball model and the weighted least squares method.

6. Conclusions

We conducted research on batting, one of the primitives of nonprehensile manipulation. We designed a ball batting system to perform batting tasks and proposed algorithms to improve the success rate of ball direction control under low sampling rate conditions. The recognition accuracy of the ball was improved by applying the color segmentation method and the circle fitting method to the recognition of the ball. In consideration of the measurement accuracy according to the distance of the stereo vision sensor, the estimated trajectory of the ball was predicted more accurately and in a faster manner by applying weighted least square regression to the ball trajectory estimation. A method of controlling the posture of the end-effector of the robot arm to control the direction of the ball was presented, and a smooth robot arm trajectory was generated while satisfying the constraints and adapting to the target trajectory. Furthermore, we analyzed the factors affecting the ball direction control and proposed a method of maintaining the end position of the robot arm to compensate for the time uncertainty. Through this, the posture of the end-effector was kept constant before and after hitting, and the success rate of the ball direction control was increased by about 13%.

Author Contributions: Conceptualization, H.-M.J.; Formal analysis, H.-M.J. and J.-H.O.; Funding acquisition, J.-H.O.; Investigation, H.-M.J.; Methodology, H.-M.J. and J.-H.O.; Resources, J.L.; Software, H.-M.J. and J.L.; Supervision, J.-H.O.; Validation, H.-M.J.; Visualization, H.-M.J.; Writing–original draft, H.-M.J. and J.-H.O.; Writing–review & editing, H.-M.J. and J.L. All authors have read and agreed to the published version of the manuscript.

Funding: This work was supported by the National Research Foundation of Korea (NRF) grant funded by the Korea government (Ministry of Science and ICT) (No. NRF-2020R1G1A1101767 and No. NRF-2020R1C1C1008520).

Conflicts of Interest: The authors declare no conflict of interest.

References

1. Lynch, K.M.; Mason, M.T. Dynamic nonprehensile manipulation: Controllability, planning, and experiments. *Int. J. Robot. Res.* **1999**, *18*, 64–92. [CrossRef]
2. Ruggiero, F.; Petit, A.; Serra, D.; Satici, A.; Cacace, J.; Donaire, A.; Ficuciello, F.; Buonocore, L.; Fontanelli, G.; Lippiello, V.; et al. Nonprehensile manipulation of deformable objects: Achievements and perspectives from the robotic dynamic manipulation project. *IEEE Robot. Autom. Mag.* **2018**, *25*, 83–92. [CrossRef]
3. Ruggiero, F.; Lippiello, V.; Siciliano, B. Nonprehensile dynamic manipulation: A survey. *IEEE Robot. Autom. Lett.* **2018**, *3*, 1711–1718. [CrossRef]
4. Satici, A.C.; Ruggiero, F.; Lippiello, V.; Siciliano, B. A Coordinate-Free Framework for Robotic Pizza Tossing and Catching. In Proceedings of the 2016 IEEE International Conference on Robotics and Automation (ICRA), Stockholm, Sweden, 16–21 May 2016; pp. 3932–3939.
5. Bätz, G.; Yaqub, A.; Wu, H.; Kühnlenz, K.; Wollherr, D.; Buss, M. Dynamic Manipulation: Nonprehensile Ball Catching. In Proceedings of the 18th Mediterranean Conference on Control and Automation, MED'10, Marrakech, Morocco, 23–25 June 2010; pp. 365–370.
6. Serra, D.; Satici, A.C.; Ruggiero, F.; Lippiello, V.; Siciliano, B. An Optimal Trajectory Planner for a Robotic Batting Task: The Table Tennis Example. In Proceedings of the 13th International Conference on Informatics in Control, Automation and Robotics, Lisbon, Portugal, 29–31 July 2016; pp. 90–101.
7. Nieuwenhuisen, D.; van der Stappen, A.F.; Overmars, M.H. Pushing a disk using compliance. *IEEE Trans. Robot.* **2007**, *23*, 431–442. [CrossRef]
8. Higashimori, M.; Utsumi, K.; Omoto, Y.; Kaneko, M. Dynamic manipulation inspired by the handling of a pizza peel. *IEEE Trans. Robot.* **2009**, *25*, 829–838. [CrossRef]
9. Donaire, A.; Ruggiero, F.; Buonocore, L.R.; Lippiello, V.; Siciliano, B. Passivity-based control for a rolling-balancing system: The nonprehensile disk-on-disk. *IEEE Trans. Control Syst. Technol.* **2016**, *25*, 2135–2142. [CrossRef]

10. Woodruff, J.Z.; Lynch, K.M. Planning and Control for Dynamic, Nonprehensile, and Hybrid Manipulation Tasks. In Proceedings of the 2017 IEEE International Conference on Robotics and Automation (ICRA), Singapore, 29 May–3 June 2017; pp. 4066–4073.
11. Chen, X.; Huang, Q.; Wan, W.; Zhou, M.; Yu, Z.; Zhang, W.; Yasin, A.; Bao, H.; Meng, F. A robust vision module for humanoid robotic ping-pong game. *Int. J. Adv. Robot. Syst.* **2015**, *12*, 35. [CrossRef]
12. Nakabo, Y.; Ishikawa, M.; Toyoda, H.; Mizuno, S. 1 ms Column Parallel Vision System and Its Application of High Speed Target Tracking. In Proceedings of the ICRA, Millennium Conference, IEEE International Conference on Robotics and Automation, Symposia Proceedings, San Francisco, CA, USA, 24–28 April 2000; pp. 650–655.
13. Hsiao, T.; Yang, C.M.; Lee, I.H.; Hsiao, C.C. Design and implementation of a ball-batting robot with optimal batting decision making ability. In Proceedings of the 2014 IEEE International Conference on Automation Science and Engineering (CASE), New Taipei, Taiwan, 18–22 August 2014; pp. 1026–1031.
14. Serra, D.; Ruggiero, F.; Satici, A.C.; Lippiello, V.; Siciliano, B. Time-optimal paths for a robotic batting task. In *Informatics in Control, Automation and Robotics*; Springer: Cham, Switzerland, 2018; pp. 256–276.
15. Gardner, M.; Jia, Y.B.; Lin, H. Batting Flying Objects to the Target in 2D. In Proceedings of the 2016 IEEE/RSJ International Conference on Intelligent Robots and Systems (IROS), Daejeon, Korea, 9–14 October 2016; pp. 3225–3232.
16. Jia, Y.-B.; Gardner, M.; Mu, X. Batting an in-flight object to the target. *Int. J. Robot. Res.* **2019**, *38*, 451–485. [CrossRef]
17. Schüthe, D.; Frese, U. Optimal Control with State and Command Limits for a Simulated Ball Batting Task. In Proceedings of the 2015 IEEE/RSJ International Conference on Intelligent Robots and Systems (IROS), Hamburg, Germany, 28 September–2 October 2015; pp. 3988–3994.
18. Pekarovskiy, A.; Nierhoff, T.; Hirche, S.; Buss, M. Dynamically consistent online adaptation of fast motions for robotic manipulators. *IEEE Trans. Robot.* **2017**, *34*, 166–182. [CrossRef]
19. Kober, J.; Mülling, K.; Krömer, O.; Lampert, C.H.; Schölkopf, B.; Peters, J. Movement Templates for Learning of Hitting and Batting. In Proceedings of the 2010 IEEE International Conference on Robotics and Automation, Anchorage, AK, USA, 3–7 May 2010; pp. 853–858.
20. Mori, S.; Tanaka, K.; Nishikawa, S.; Niiyama, R.; Kuniyoshi, Y. High-speed and lightweight humanoid robot arm for a skillful badminton robot. *IEEE Robot. Autom. Lett.* **2018**, *3*, 1727–1734. [CrossRef]
21. Senoo, T.; Namiki, A.; Ishikawa, M. High-Speed Batting Using a Multi-Jointed Manipulator. In Proceedings of the IEEE International Conference on Robotics and Automation, New Orleans, LA, USA, 26 April–1 May 2004; pp. 1191–1196.
22. Senoo, T.; Namiki, A.; Ishikawa, M. Ball Control in High-Speed Batting Motion Using Hybrid Trajectory Generator. In Proceedings of the 2006 IEEE International Conference on Robotics and Automation, Orlando, FL, USA, 15–19 May 2006; pp. 1762–1767.
23. Park, I.W.; Kim, J.Y.; Lee, J.; Oh, J.H. Mechanical design of the humanoid robot platform, HUBO. *Adv. Robot.* **2007**, *21*, 1305–1322. [CrossRef]
24. Plataniotis, K.N.; Venetsanopoulos, A.N. *Color Image Processing and Applications*; Springer Science & Business Media: Berlin, Germany, 2013.
25. Bradski, G.; Kaehler, A. *Learning OpenCV: Computer Vision with the OpenCV Library*; O'Reilly Media, Inc.: Sebastopol, CA, USA, 2008.
26. Gonzalez, R.C.; Woods, R.E. *Digital Image Processing*; Addison-Wesley Publishing Company: Boston, MA, USA, 1992.
27. Strutz, T. *Data Fitting and Uncertainty. A Practical Introduction to Weighted Least Squares and Beyond*; Vieweg + Teubner: Wiesbaden, Germany, 2010.
28. Lawson, C.L.; Hanson, R.J. *Solving Least Squares Problems*; Society for Industrial and Applied Mathematics: Philadelphia, PA, USA, 1995.
29. Nonprehensile Dynamic Manipulation: High Speed Ball Batting Using Visual Feedback. Available online: https://www.youtube.com/watch?v=BPGVibfnOXg (accessed on 19 January 2018).

applied sciences

MDPI

Article

Augmented Reality-Based Interface for Bimanual Robot Teleoperation

Alberto García, J. Ernesto Solanes *, Adolfo Muñoz, Luis Gracia and Josep Tornero

Instituto de Diseño y Fabricación, Universitat Politècnica de València, 46022 Valencia, Spain; agf25394@gmail.com (A.G.); amunyoz@upvnet.upv.es (A.M.); luigraca@isa.upv.es (L.G.); jtornero@isa.upv.es (J.T.)
* Correspondence: esolanes@idf.upv.es

Abstract: Teleoperation of bimanual robots is being used to carry out complex tasks such as surgeries in medicine. Despite the technological advances, current interfaces are not natural to the users, who spend long periods of time in learning how to use these interfaces. In order to mitigate this issue, this work proposes a novel augmented reality-based interface for teleoperating bimanual robots. The proposed interface is more natural to the user and reduces the interface learning process. A full description of the proposed interface is detailed in the paper, whereas its effectiveness is shown experimentally using two industrial robot manipulators. Moreover, the drawbacks and limitations of the classic teleoperation interface using joysticks are analyzed in order to highlight the benefits of the proposed augmented reality-based interface approach.

Keywords: robot system; augmented reality; robot teleoperation

Citation: García, A.; Solanes, J.E.; Muñoz, A.; Gracia, L.; Tornero, J. Augmented Reality-Based Interface for Bimanual Robot Teleoperation. *Appl. Sci.* **2022**, *12*, 4379. https://doi.org/10.3390/app12094379

Academic Editors: DaeEun Kim and Alessandro Gasparetto

Received: 21 March 2022
Accepted: 24 April 2022
Published: 26 April 2022

1. Introduction

1.1. Motivation

There is currently a proliferation of applications based on robotic systems that require interaction or cooperation with users [1–3]. This is due, on the one hand, to the fact that the complete automation of such applications is not yet well resolved due to their complexity, need for adaptability to changes, decision making, etc., and, on the other hand, to the benefits of human and robot cooperation for certain applications [4].

While many approaches can be found in the literature proposing solutions for complex applications, where human and robot have to cooperate, many of these solutions do not take into account that human–robot interaction has to be natural and intuitive for the human [5–8]. Otherwise, the benefits that such cooperation may bring a priori will be negatively affected, and the cooperative solution adopted may be rejected.

Based on the complexity presented by the interaction of users with the so-called bimanual robotic systems, this work develops a novel methodology for the design of interfaces based on augmented reality so that this interaction is natural and intuitive for the user.

1.2. Previous Research

1.2.1. Bimanual Robotics

Dual-arm robotic systems are being used in a wide range of domestic, industrial, and healthcare tasks. The main reason for this is their flexibility and manipulability. In addition, they have a behavior quite similar to that of the human, which makes it possible for humans to relate to their movements more intuitively [9–11].

More specifically, *bimanual robotics* consists of the coordination of two robotic arms that interact physically in order to achieve a common goal [9]. Many applications of bimanual robotics can be found in the literature—for instance, the handling of deformable objects [12–14], objects with unknown shape [15,16], or objects whose geometry requires two grasping points [17–19]; the emulation of human bimanual tasks [20–22]; assistive robotics [23,24]; assembly operations [25–27]; surgery tasks [28]; and simultaneous manipulation and cutting [29], manipulation and fastening [10], or manipulation and surface treatment [30,31], which is the case considered in this paper.

The vast majority of contributions in bimanual robotics present fully automated tasks based, for instance, on artificial intelligence techniques [21,23], motion planing techniques [14,15,26,32,33], or other low-level control approaches [19,24,25,27,28].

However, the presence of the human interacting with the bimanual robotic system is very interesting due to the possibility of exploiting the human's natural knowledge of bimanual configurations and motions in order to improve the task performance [31,34,35]. For this reason, human–robot interaction (HRI), which is the main focus in this work, is nowadays a trending research topic in bimanual robotics.

Some interesting approaches can be found related to HRI in bimanual robotics. For instance, the authors in [34] proposed to improve the transportation of a large workpiece, typically performed by two users, by using a bimanual robot attached to a mobile platform. In this approach, the mobile platform moved through a pre-defined trajectory, while the user was able to arbitrarily adapt this trajectory by means of an impedance control. The authors in [35] proposed a multi-layered prioritized shared controller to maintain the robot hands' orientation and contact with the manipulated surface, while the user was able to teleoperate the bimanual robot hands on a plane. The authors in [31] presented a similar approach based on the task priority and sliding mode control techniques to perform surface treatment tasks using a bimanual robotic system. In this case, the user was able to teleoperate all six Degrees of Freedom (DoF) of one robotic arm that held the workpiece, whose movement was limited in the 3D workspace, and to teleoperate two DoF of the other robotic arm, which held the surface treatment, maintaining the appropriate tool orientation and pressure. Due to the relevance of this application for this work, more details can be found in Section 2.

1.2.2. Assisted Robot Teleoperation

The remote control or *teleoperation* of robots by users has been studied for many years [36] and still represents a relevant research field in robotics. Robot teleoperation is required for a wide variety of reasons: when the working environment is dangerous to humans (e.g., in space [37], radioactive zones [29,38], aerial zones [18,39], or underwater areas [35,40]); when performing rescue operations [41], and when precision surgeries need to be performed [42–45], among others.

Nowadays, there are sophisticated artificial intelligence (AI) techniques that allow the automation of complex tasks that not so long ago had to be performed by means of human teleoperation. However, despite current advances in AI, there are still many tasks that cannot be fully automated due to their complexity or subjectivity. However, these tasks can be partially automated, allowing the cooperation between human and robot, introducing shared-control architectures [46]. Hence, many recent contributions have focused on human–robot interaction and, more specifically, on advanced robot teleoperation [16,30,31,47–51], which is also the case of this paper.

Telepresence [36] allows the user to perform the robot teleoperation task by means of an interface, achieving a result less dependent on their skills. Telepresence is currently a trending research topic thanks to the introduction of new technologies, such as augmented and virtual reality [51], visual interfaces [42], haptic devices [52], or a combination of them [16,30,43], to perform direct control teleoperation. For instance, the authors in [53] proposed a low-cost telerobotic system based on virtual reality technology and the homunculus model of mind. In this case, the user was able to move both robotic arms according to the dynamic mapping between the user and the robot developed. In addition, the user was able to see the real workspace in the virtual environment using feedback from a camera. Similarly, the authors in [54] proposed a virtual reality interface based on the three-dimensional coordinates of the shoulder, elbow, wrist, and hand captured by a

Kinect camera to model the geometry of the human arms and perform the mapping with the robot arms. As in [53], the user receives visual feedback from a camera placed on the robot. In both cases, robot manipulation tasks were performed. However, for more complex tasks (e.g., surface treatment tasks), interfaces developed with virtual reality techniques can increase the time of completion of the task and worsen the quality of the surface finishing, compared to that obtained by the human operator using direct teleoperation. This is due to two facts: on the one hand, when using virtual reality, it is difficult to incorporate all the necessary information of the task in the virtual world and in real time, and, on the other hand, the user already has a real notion of the robotic system and, hence, is able to guide it naturally and intuitively using direct teleoperation. For this reason, in order to obtain the best of both worlds (i.e., direct teleoperation and teleoperation based on virtual reality), the present work proposes to use interfaces based on augmented reality to provide a solution to a greater number of industrial tasks carried out with bimanual robots.

Other approaches try to ease the teleoperation of bimanual robotic systems, such as in [48], where the authors developed a bimanual robot application in which a robot arm is teleoperated to grasp the workpiece, whilst the other robotic arm is automatically controlled using visual servoing in order to keep the workpiece visible for the camera.

Since the performance of robot teleoperation may rely on the user's skills, some approaches are focused on incorporating restrictions that prevent the user from commanding the robot into failure situations. For example, the authors in [44] incorporate Virtual Fixtures (i.e, virtual barriers) so that the references provided by the user are automatically adapted to the allowed region. The authors in [38,52] proposed the use of haptic devices in order to prevent the user from commanding references beyond the allowed region.

Despite all the above, robot teleoperation by means of interfaces and virtual barriers is still a subject of study due to the drawbacks it presents, mainly due to direct control performed by the user [47]. In this sense, this work presents a new methodology based on augmented reality devices to improve the current assisted teleoperation interfaces for bimanual robotics.

1.2.3. Augmented Reality-Based Interfaces

Human–machine interfaces are devices that allow the interaction between a human and a machine [55,56]. If the interface is placed inside the brain or body of the human, it is known as an invasive or implanted interface [57]. On the contrary, if the interface is external to the human body, it is known as a non-invasive or wearable interface [58–60]. This work is focused on non-invasive interfaces and on how to develop this kind of interface for complex robotic applications.

Technological advances in the creation of holograms have nowadays made it possible to have devices and software tools that allow augmented reality (AR) applications in industrial sectors [61–64]. In short, augmented reality projects holograms into physical space, allowing for a more intuitive and natural interaction between human and machine [65].

Some previous works used AR interfaces to improve robot teleoperation for industrial tasks. For example, the authors in [66] proposed a new AR interface to control a robot manipulator in order to facilitate the interaction between the user and the robot. The authors in [66] proposed a mixed reality system in order to move the end-effector of the robot system. The authors in [67] proposed a mixed reality system to allow the user to visualize the intended teleoperation command prior to the real robot motion. A similar approach was developed in [68], where a mixed reality head-mounted display enabled the user to create and edit robot motions using waypoints. The authors in [69] proposed a multimodal AR interface coined as *Sixth Sense* that allowed the user to interact with information that was projected onto physical objects through hand gestures, arm movements, and, in some cases, blinking. The authors in [70] proposed a method for using hand gestures and speech inputs for AR multimodal interaction with industrial manipulators.

Note that most of the AR approaches mentioned above developed solutions for robot–object manipulation tasks. Thus, to the best of the authors' knowledge, this is the first work that proposes a new AR interface for industrial, complex tasks, such as surface treatment tasks, involving a bimanual robot system.

In addition, the interaction with the robotic system needs to be natural and intuitive not only from the point of view of the visual feedback produced by the AR but also from the point of view of the means of sending the robot commands. All AR headsets have interaction elements based on hand tracking. As demonstrated in [5–8], such prolonged interaction over time can be annoying and not ergonomic enough. This is why, similarly to [8], this work proposes the use of gamepads, which are devices ergonomically designed to be used for long periods of time.

To the best of the authors' knowledge, this is the first work proposing an AR interface together with a gamepad for bimanual robot teleoperation.

1.3. Proposed Approach

This paper develops an original augmented reality-based interface for teleoperating bimanual robots. The proposed interface is more natural to the user, which reduces the interface learning process. A full description of the proposed interface is detailed in the paper, whereas its effectiveness is shown experimentally using two industrial robot manipulators. Moreover, the drawbacks and limitations of the classic teleoperation interface using joysticks are analyzed in order to illustrate the benefits of the proposed augmented reality-based interface approach.

The content of the article is as follows. Section 2 presents a brief description of the advanced bimanual robot teleoperation application considered in this work. Then, Section 3 provides a methodology to develop AR interfaces for bimanual robot teleoperation tasks and, subsequently, develops the specific AR-based interface proposed for the application at hand. Moreover, the interface functionalities are illustrated through several experiments. Furthermore, Section 4 shows the performance and effectiveness of the proposed AR-based interface by means of real experimentation. Finally, Section 5 presents the conclusions.

2. Previous Work

Without loss of generality, this work uses the robotic application developed by the authors in [31] to demonstrate the benefits of the proposed AR-based interface with respect to conventional PC-based interfaces. It consists in a surface treatment application carried out through the cooperation of a bimanual robotic system and a user, who is able to partially command both robots at distance, i.e., by means of robot teleoperation. Moreover, both robots are partially automatically controlled to fulfill some 2D and 3D constraints, as well as to keep constant the force exerted to the workpiece by the tool and the orientation of the tool at any time during the task.

Next, a description of this application, as well as the problems of using conventional PC-based interfaces, is detailed.

2.1. Description of the Advanced Bimanual Robot Teleoperation Application

The advanced bimanual robot teleoperation is based on the *task-priority strategy* [71,72] and *conventional* and *non-conventional Sliding Mode Controllers* (SMCs) [73,74]. As commented before, the goal of this bimanual robotic application is to perform a human–robot cooperative control loop so that the user operator partially teleoperates two robotic arms to perform a surface treatment operation, whilst the robots automatically ensure the appropriate tool force and orientation; see Figure 1. Thus, the so-called *workpiece robot* (WR), which consists of a 7R collaborative robot with a workpiece of flat methacrylate fixed to the end-effector using a self-made piece (see Figure 1a), is in charge of holding the workpiece. Meanwhile, the so-called *surface treatment robot* (STR), which consists of a 6R robotic arm with a Force/Torque (F/T) sensor and a cylinder-shaped tool with a piece of cloth (see Figure 1a), operates with the surface treatment tool on the workpiece. Thus, the user controls the workpiece position and orientation and, simultaneously, controls the 2D tool motion on the workpiece surface using an interface, which consists of a gamepad to command the robots and a visual feedback screen to show the user the robots and the user reference states; see Figure 1a.

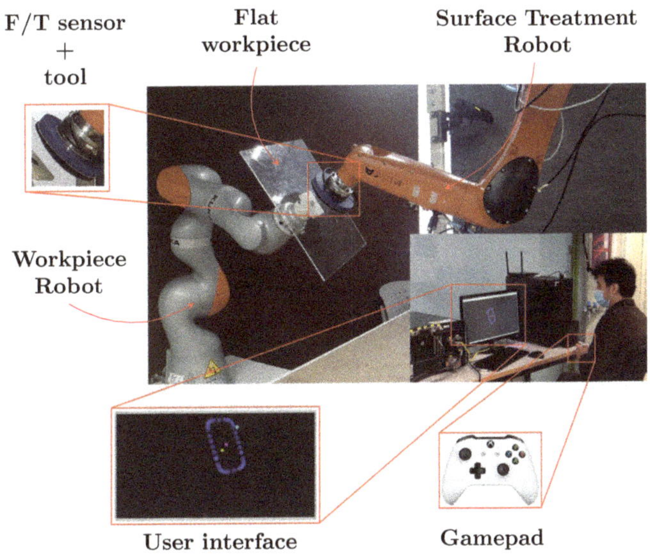

(**a**)

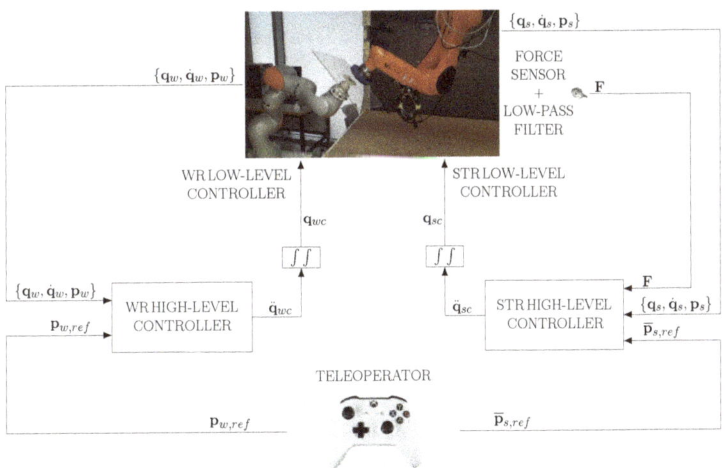

(**b**)

Figure 1. Bimanual application setup and block diagram (for further details, refer to [31]). (**a**) Previous setup used for the real experimentation. (**b**) Block control diagram for both robots (WR and STR).

Figure 1b shows the block control diagram for both robots, where subscript s stands for the STR; subscript w stands for the WR; subscript *ref* stands for the user reference; subscript c stands for the commanded control action; $\mathbf{p} = \begin{bmatrix} x & y & z & \alpha & \beta & \gamma \end{bmatrix}^T$ is the robot pose, i.e., the linear positions $\{x, y, z\}$ plus orientation angles $\{\alpha, \beta, \gamma\}$; $\overline{\mathbf{p}}_s = \begin{bmatrix} x & y \end{bmatrix}^T$ is the 2D position of the STR tool on the workpiece surface, i.e., the linear positions $\{x, y\}$ relative to this surface; $\mathbf{q} = \begin{bmatrix} q_1 & \cdots & q_n \end{bmatrix}^T$ is the robot configuration, with n the number of robot joints; and $\mathbf{F}$ is the vector containing the measured forces and torques. Thus, using the gamepad joysticks, the user is able to send the reference to the WR pose $\mathbf{p}_w$ and, simultaneously, the reference to the 2D position $\overline{\mathbf{p}}_s$ of the STR tool on the workpiece surface. Thus, the high-level controllers of both robots compute the corresponding joint commands $\mathbf{q}_c$ from the user references, the state $\{\mathbf{q}, \dot{\mathbf{q}}, \mathbf{p}\}$ of both robots, and the force sensor data $\mathbf{F}$. These joint commands are then sent to the low-level controllers of both robots, as shown in Figure 1b, in order to complete the teleoperation task. See [31] for further details on the high-level controllers of both robots and the related signals.

In addition, some constraints are considered for both robots in order to increase the safety of the task: (1) the WR is automatically controlled to maintain the workpiece center inside the allowed region that is modeled as a superellipsoid, which is similar to a rectangular prism with smooth corners; (2) the STR is automatically controlled to keep the center of the treatment tool within the allowed region on the workpiece, which is modeled as a superellipse, i.e., a rectangle with smooth edges.

2.2. Description of the Controllers

The control architecture developed in [31] for each robotic manipulator of the application is as follows.

The control of the WR is given by 4 prioritized tasks:

(W1) The highest-priority task is used to keep the workpiece center inside the aforementioned superellipsoid and is accomplished using non-conventional SMC.
(W2) The medium–high-priority task is used to keep the angular position of the workpiece within a certain range and is accomplished using non-conventional SMC.
(W3) The medium–low-priority task is used to allow the user to command the WR and is accomplished using a hybrid SMC.
(W4) The lowest-priority task is utilized to "push" the WR configuration towards a *home* (only applies for a *redundant* WR).

The control of the STR is given by 3 prioritized tasks:

(S1) The highest-priority task is used to keep the center of the treatment tool within the aforementioned modified superellipse and is accomplished using non-conventional SMC.
(S2) The medium-priority task is used to exert the appropriate pressure with the tool on the workpiece, as well as to maintain the orientation of the tool perpendicular to the surface of the workpiece. It is accomplished using conventional SMC.
(S3) The lowest-priority task is used to allow the user to command the STR and is accomplished using a hybrid SMC.

2.3. Description of the Conventional PC-Based Interface

The authors in [31] proposed a conventional PC-based interface, which shows a 3D interface on a screen, that is composed of the following visual elements (see Figure 2):

- STR reference consisting of a yellow sphere. The position of this element is controlled by the user using the gamepad input.
- STR current tool position consisting of a red sphere.
- STR boundary and WR workpiece orientation consisting of several blue spheres positioned along the curved define by the modified superellipse. When the tool collides with the boundary, the color of these spheres changes from blue to green.
- WR reference consisting of a cyan sphere. The position of this element is controlled by the user using the gamepad input.
- WR current workpiece position consisting of a pink sphere.

Note that the user commands both robots by means of the gamepad.

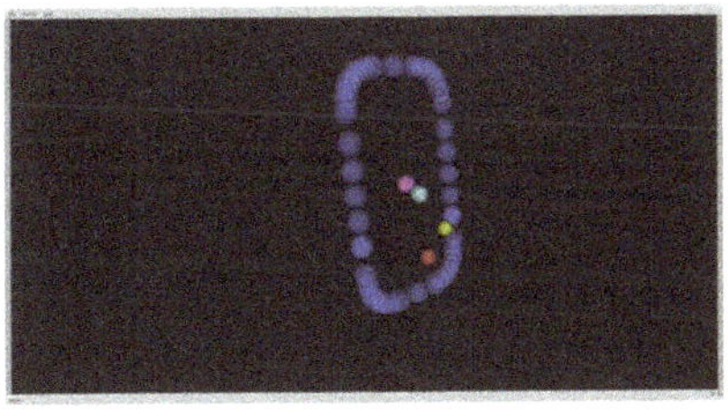

(a)

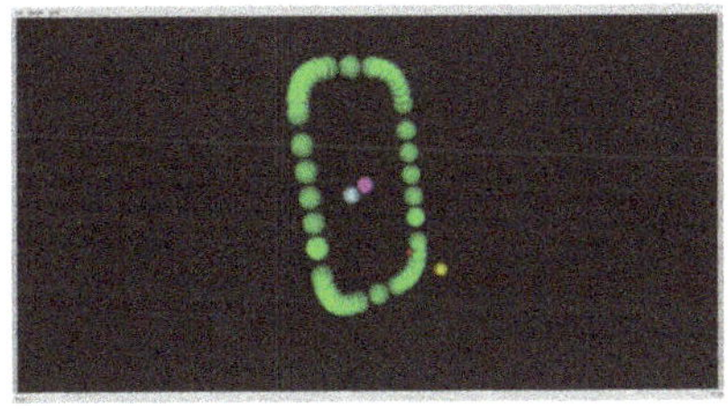

(b)

Figure 2. Conventional PC-based user interface: visual references and effects. (**a**) Video: 0 m 20 s. (**b**) Video: 0 m 23 s.

Figure 3 shows several frames for the performance of the described application, focusing on the interface; see the video at https://media.upv.es/player/?id=15ffabe0-a733-11eb-a0b0-2fbcb59aaef7 (accessed on 26 April 2022) [75].

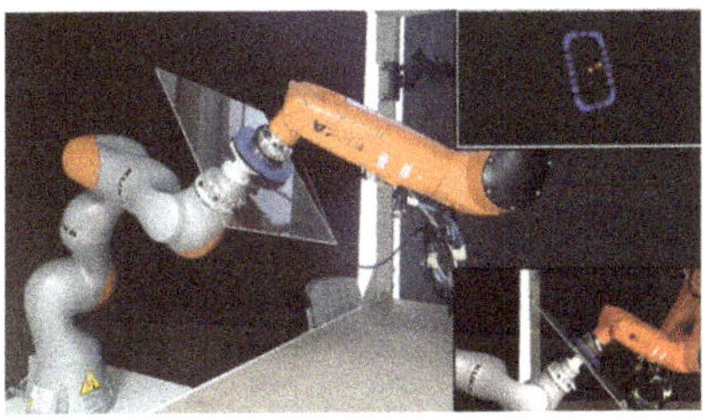

(a)

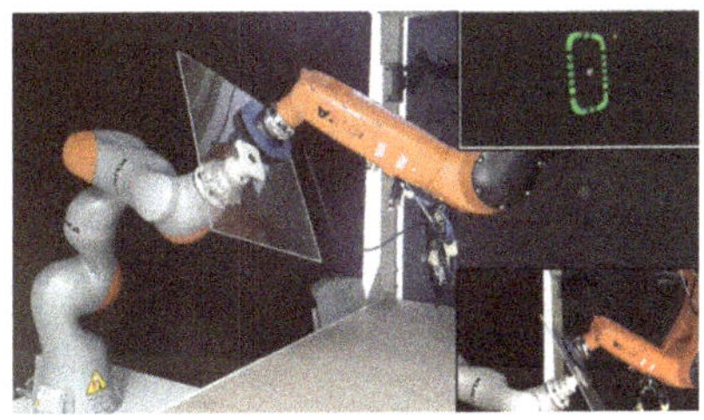

(b)

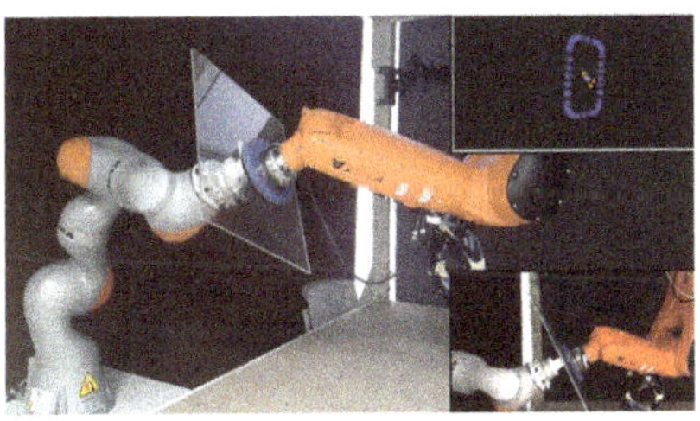

(c)

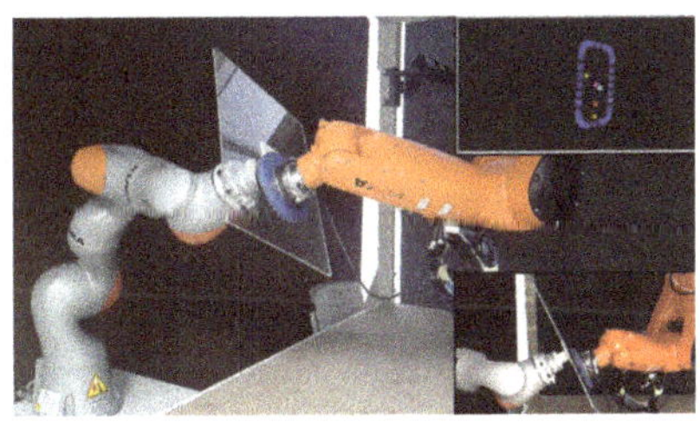

(d)

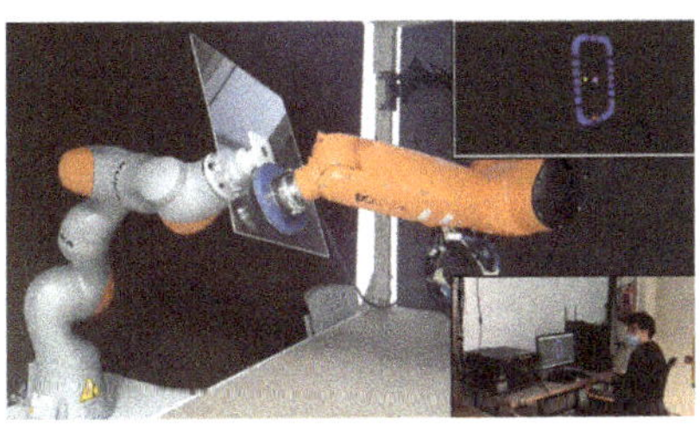

(e)

Figure 3. Frames of the video showing the functionalities of the conventional user interface. See the video at https://media.upv.es/player/?id=15ffabe0-a733-11eb-a0b0-2fbcb59aaef7 (accessed on 26 April 2022) [75]. (**a**) Video: 0 m 20 s. (**b**) Video: 1 m 00 s. (**c**) Video: 1 m 24 s. (**d**) Video: 1 m 33 s. (**e**) Video: 1 m 43 s.

2.4. Discussion of Human–Robot Interaction Using Conventional Interfaces

The conventional PC-based user interface presents several problems that directly affect the task performance. Next, the three most relevant problems, which were identified from questions asked to several users that tested the application described above, are discussed.

The first significant problem reported by the users is that their interaction with the virtual environment was not natural. In particular, the robotic system is teleoperated in the 3D space and, hence, it requires changing the screen view to properly track the task. To do this, the user has to stop the robot teleoperation and accommodate the interface, affecting the total amount of time needed to complete the task.

The second significant problem reported by the users is that it was difficult for them to see the real system at any time. In this sense, Figure 3e shows the user looking at the real system instead of the interface while performing the task. The reason given by several users, who exhibited the same behavior, was that they needed to see what the real system was doing because they did not know if the task was being done correctly or not. This means that this type of interface does not properly help the user to conduct the real task.

The third significant problem reported by several users is that it was difficult for them to move the references in the virtual 3D space, wasting a lot of time before resuming the robotic task.

All these issues and problems show the difficulties of using conventional interfaces and make evident the need to develop new interfaces allowing a more intuitive user interaction, especially when working with complex systems such as the bimanual robotic system considered in this work.

3. Proposed Augmented Reality-Based User Interface

In order to overcome the aforementioned problems of the conventional PC-based interface, this work proposes the use of AR technology to improve the user ergonomics and task performance. In particular, the conventional PC-based interface used in the previous setup (see Figure 1a) is replaced by an AR headset in the new setup—see Figure 4—allowing the user to see the relevant information in the form of holograms while still seeing at all times the real elements involved in the task: robots, workpiece, tool, etc. Note that the remaining elements of the new setup (see Figure 4) are the same as in the previous setup (see Figure 1a): an STR with an F/T sensor and a cylinder-shaped tool with a piece of cloth; a WR with a flat workpiece of methacrylate attached to the end-effector using a self-made piece; and a gamepad to command both robots.

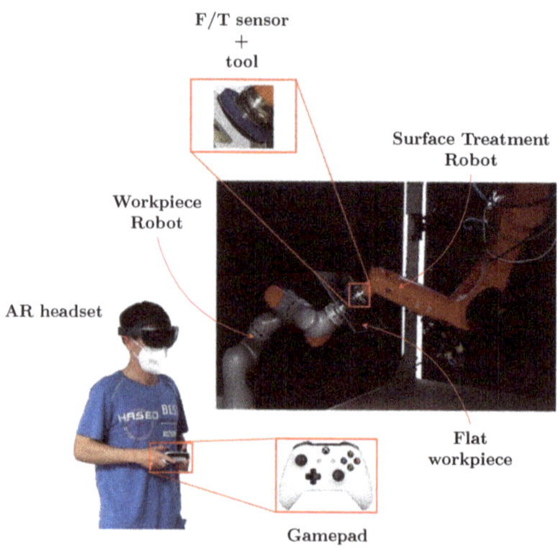

Figure 4. New setup used for the real experimentation.

Figure 5 shows the methodology considered in this work to develop and validate the proposed AR-based interface. Although this methodology is used below to design the AR interface for the specific bimanual robot teleoperation task at hand, it is generic and, in general, it can be applied to design AR interfaces for other types of applications.

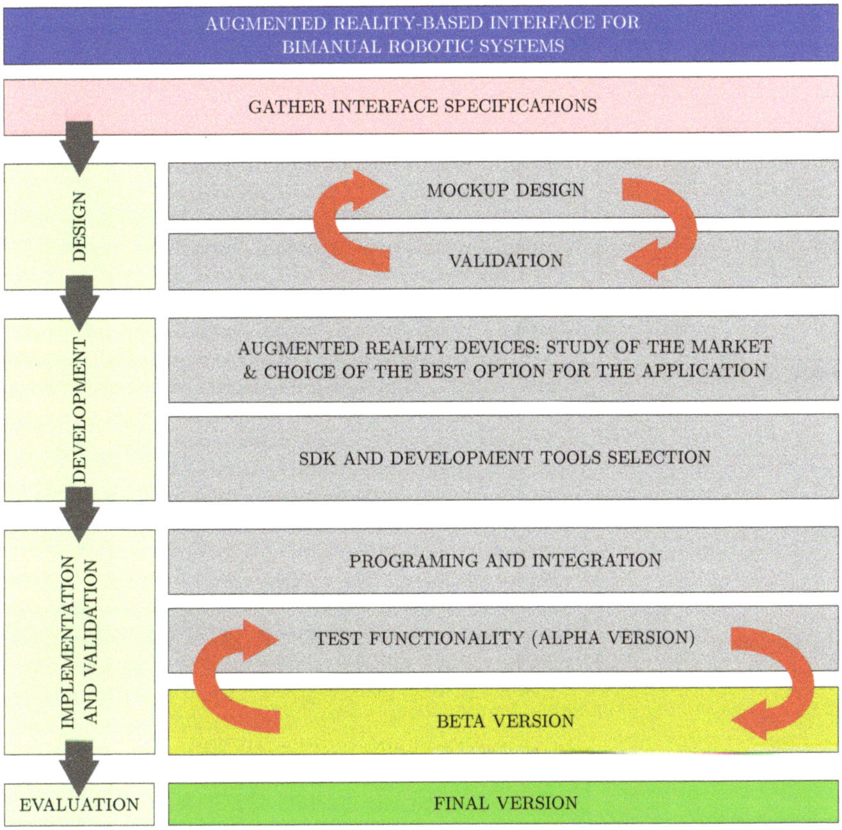

Figure 5. Flowchart of the methodology proposed in this work for designing the AR-based interface.

Firstly, the requirements of the applications were established based on the opinions of several users who previously tested the conventional PC-based interface. These requirements are summarized in Table 1.

Table 1. Application requirements.

The user should have the option to see the full boundaries when required
The part of the boundary activated should be indicated (e.g., visually, sound, etc.)
STR tool reference direction should be indicated
WR rotation reference direction should be indicated
The new interface should use a similar interaction device to that of the previous PC-based interface (i.e., gamepad, joystick, or similar)
Alarm sounds should be used to indicate boundary activation
The user should have the option to remove all holograms
Holograms should not disturb the user visibility during the task
The user should have the option to configure, activate, and deactivate the alarm sounds

A mockup design was developed taking into account this information. The designed AR-based interface has, from a functionality perspective, two kinds of virtual objects: firstly, those representing the STR and WR reference indicators; and, secondly, those corresponding to the boundaries information. In order to develop both kinds of virtual objects, several tools and strategies related to the mockup design were used. These preliminary designs were validated by some users before their implementation.

Once the preliminary design was finished, the following step was to study the best option of AR headset to be used for the application at hand. Several considerations were taken into account, especially the following: first, the capability of the device to be used in industrial environments; second, the stability of the holograms, which is important when working in this kind of application; third, the computational power of the device; fourth, the sound capabilities; and fifth, the communication capability (i.e., Bluetooth and WiFi). Note that most AR headsets in the market accomplish the aforementioned requirements. However, among all of them, Microsoft *HoloLens glasses* [76] were chosen because the second generation of this device offers several services that could be added to the final version of the interface according to the company needs [77].

Once the AR headset was selected, the interface was developed. Using a PC workstation, the proposed virtual objects were created and assembled in a virtual space using Blender 2.7 [78] and *Unity* [79], respectively. This was an iterative design process, where the main characteristics of the virtual objects (e.g., size, color, shape, etc.) and their interactions were verified and modified, connecting the workstation with the AR headset in a remote mode from the Unity editor (note that the perception of the holograms is different when showing them in a PC screen compared to when projecting them in the real world through the AR headset), until the result was satisfactory.

Figure 6 shows the holograms designed for the robot references. In the case of the WR, the user can command the robot through the 3D workspace and modify the end-effector orientation. For this reason, two different holograms were designed. The translation reference hologram was modeled by a 3D orange cube; see Figure 6a. This hologram appears when the user teleoperates the WR translation reference. To reduce the number of holograms present at any moment, this hologram disappears 3 s after the user has stopped moving the WR translation reference. The orientation reference hologram was modeled by an animated arrowed yellow circle; see Figure 6b. This hologram appears when the user teleoperates the WR rotation reference, and disappears 3 s after the user has stopped moving the WR rotation reference. It should be noted that, in both cases, the movement of the references is relative to the position of the user, i.e., the AR headset, making their use more intuitive and natural. The STR translation reference was modeled by a yellow arrow attached to a green sphere; see Figure 6b. Note that this hologram is constrained to the plane of the workpiece surface, allowing a 2D movement. This hologram disappears 3 s after the user has stopped moving the STR translation reference.

Figure 7 shows the holograms designed for the 2D and 3D boundaries.

The 3D boundary is modeled by a superellipsoid—see Figure 7a—which is defined as:

$$\left|\frac{x}{W}\right|^m + \left|\frac{y}{H}\right|^m + \left|\frac{z}{M}\right|^m = 1, \tag{1}$$

where $\{W, H, M\}$ are the superellipsoid axes and m represents the smoothing parameter of the superellipsoid, i.e., it is equivalent to an ellipsoid for $m = 2$, whereas it tends to a cuboid as m tends to infinity. For the bimanual robot application at hand, it has been chosen $m = 4$.

The 2D boundary is modeled by a modified superellipse—see Figure 7c—which is defined as:

$$\left|\frac{x}{W}\right|^m + \left(\frac{\max(|y| - (H - W), 0)}{W}\right)^m = 1, \tag{2}$$

where it is implicitly assumed that the value of axis H is greater than that of axis W (the expression is easily modified for the analogous case $H < W$). This equation represents a rectangle with smooth corners, with $2H$ for its long side and $2W$ for its short side, by joining a $2W \times 2(H - W)$ rectangle to two offset halves of an even-sided $2W \times 2W$ superellipse.

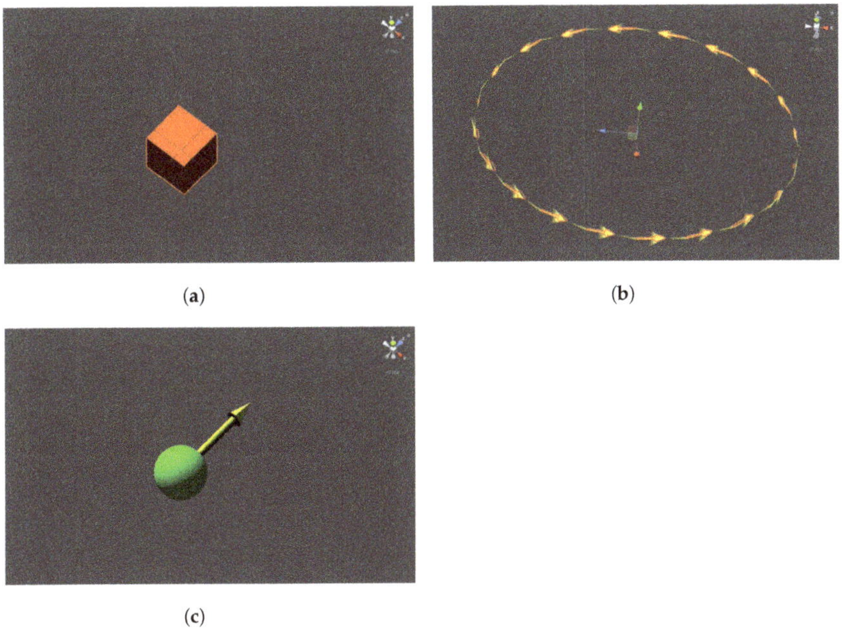

Figure 6. Proposed holograms for the robot references. (**a**) WR: translation reference hologram. (**b**) WR: rotation reference hologram. (**c**) STR: translation reference hologram.

Note that if the proposed boundary holograms were permanently shown, they could occlude some real elements from the user's view, affecting the task performance. For this reason, a new material shader [80] was designed; see Figure 8. This shader computes the minimum distance between the robot end-effector and the 3D boundary, for the case of the WR, or the closest point of the robot tool to the 2D boundary, for the case of the STR. Thus, the shader only displays the affected part of the boundary hologram. That is, as the WR end-effector and/or the STR tool approach to the 3D and 2D boundaries, respectively, the part of the boundary hologram affected is progressively displayed; see Figure 7b,d.

In addition to this, and according to the user requirements, two warning sounds were included in the interface: the first one to indicate that the STR tool is close to the 2D boundary; and the second one to indicate that the WR end-effector is close to the 3D boundary. Moreover, the user is able to deactivate this warning sound at any time.

Once the main holograms and sound elements were implemented, some communication protocols were used and programmed. Bluetooth communication between the Microsoft *HoloLens glasses* and the gamepad was established to allow the user to provide commands to the interface. Moreover, in order to avoid non-desired interactions with the interface, voice and gesture commands were deactivated by default. In addition, the AR interface and the robot controller communicate via WiFi with Protocol TCP/UDP at 10 Hz.

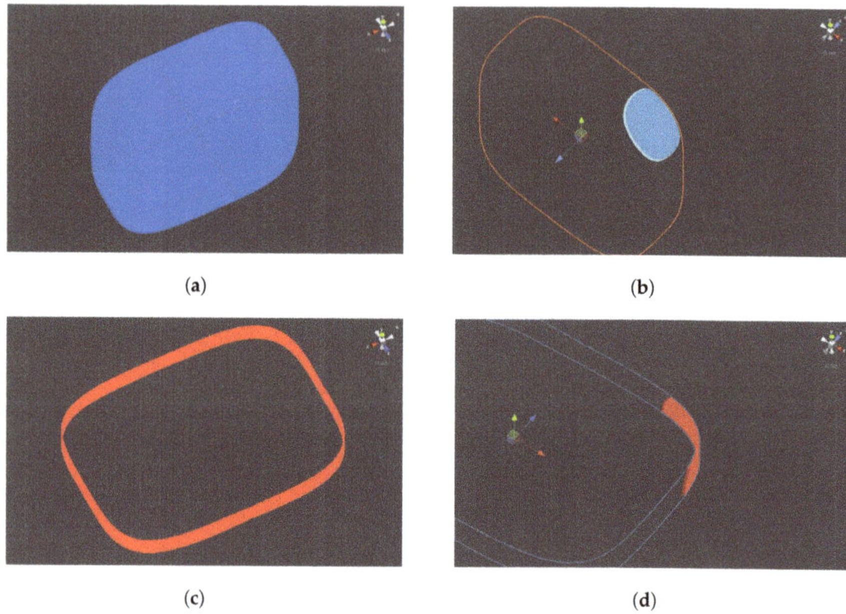

Figure 7. Proposed holograms for the robot 3D and 2D boundaries. (**a**) 3D boundary hologram (full). (**b**) 3D boundary hologram (local). (**c**) 2D boundary hologram (full). (**d**) 2D boundary hologram (local).

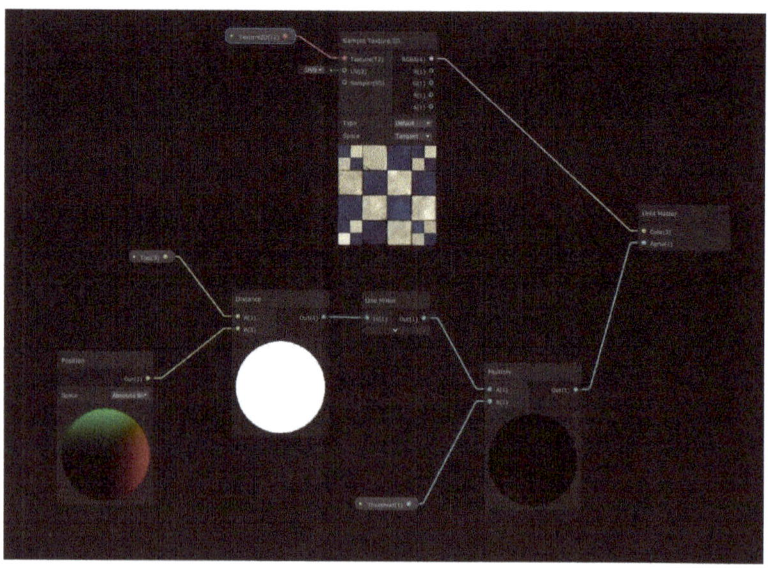

Figure 8. Material shader designed for controlling the visibility of the 3D and 2D boundaries depending on the proximity of the WR end-effector and STR tool, respectively.

4. Results

This section presents four experiments to show the main functionalities of the developed AR-based interface; the performance of the 2D boundary and the STR reference hologram; the performance of the 3D boundary and the WR reference hologram; and the performance of the overall system when the user commands simultaneously both robots using the proposed AR-based interface.

Figure 9 depicts several frames of the *first experiment*, which shows the main functionalities of the AR interface implemented in the Microsoft *HoloLens glasses*; see the video at https://media.upv.es/player/?id=a64014f0-8a5a-11ec-ac0a-b3aa330d3dad (accessed on 26 April 2022) [81]. Figure 9a shows the full 3D boundary hologram, whilst Figure 9b shows the full 2D boundary hologram. Note that both holograms are hidden by default. Figure 9c shows the WR end-effector translation reference hologram, whilst Figure 9e,f show the WR end-effector rotation reference hologram. Note that, in the case of the rotation, the animated arrows indicate the direction of the commanded angle while the yellow circle indicates the rotation in the roll, pitch, and yaw angles, or a combination of them. Figure 9d shows the STR reference hologram.

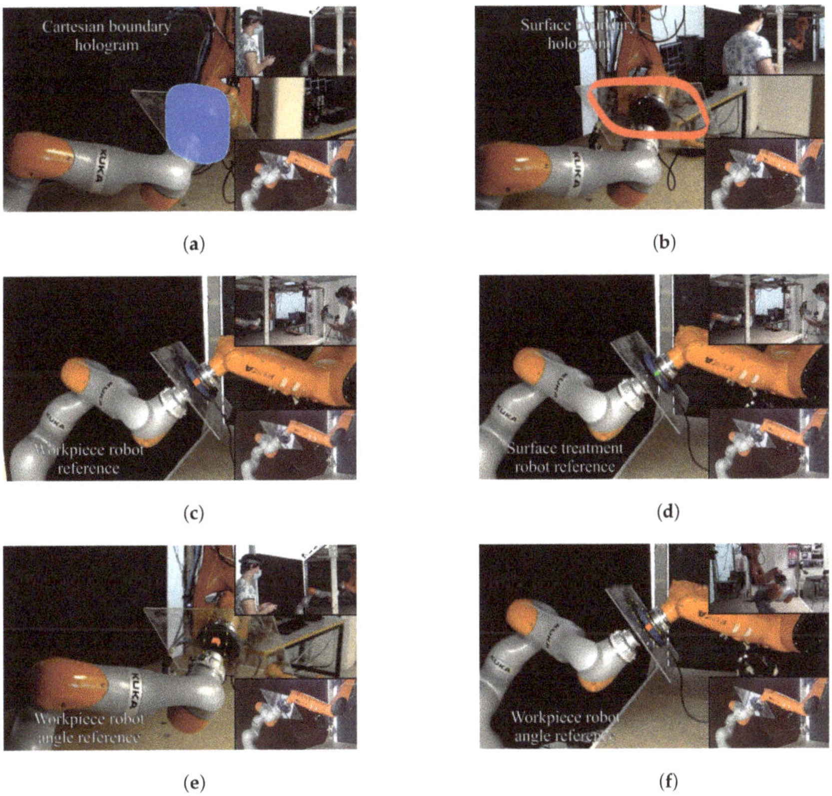

Figure 9. First experiment: frames of the video showing the functionalities of the proposed AR-based interface. See the video at https://media.upv.es/player/?id=a64014f0-8a5a-11ec-ac0a-b3aa330d3dad (accessed on 26 April 2022) [81]. (**a**) Video: 0 m 20 s. (**b**) Video: 0 m 23 s. (**c**) Video: 0 m 30 s. (**d**) Video: 0 m 36 s. (**e**) Video: 0 m 43 s. (**f**) Video: 0 m 57 s.

Figure 10 depicts several frames of the *second experiment*, which shows the performance of the 2D boundary and the STR reference hologram; see the video at https://media.upv.es/player/?id=9504e6f0-8a61-11ec-b7c7-7d27dda7c5d5 (accessed on 26 April 2022) [82]. Figure 10a shows how the user is commanding the STR tool towards one side of the workpiece and, when the tool approaches the 2D boundary, the boundary region closest to the STR tool is shown in red and the warning sound is activated; see Figure 10b,c. Note that, when the user reference exceeds the 2D boundary, the tool is automatically kept within the allowed region. More details about this aspect can be further analyzed in Figure 11, which shows the allowed region on the workpiece surface, the trajectory followed by the user reference, and the trajectory followed by the STR tool. Figure 10d shows how the 2D boundary hologram automatically disappears when the STR tool is far from the 2D boundary.

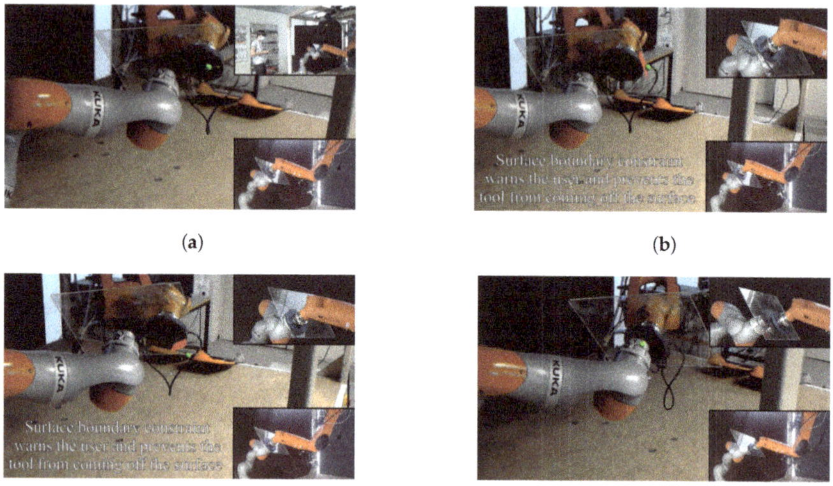

(**a**) (**b**)

(**c**) (**d**)

Figure 10. Second experiment: frames of the video showing the performance of the 2D boundary and the STR reference hologram. See the video at https://media.upv.es/player/?id=9504e6f0-8a61-11ec-b7c7-7d27dda7c5d5 (accessed on 26 April 2022) [82]. (**a**) Video: 0 m 31 s. (**b**) Video: 0 m 41 s. (**c**) Video: 1 m 5 s. (**d**) Video: 1 m 46 s.

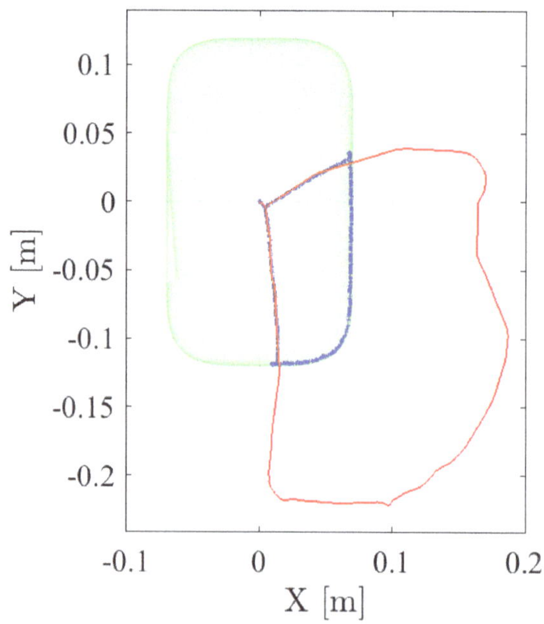

Figure 11. The 2D trajectory performance for the second experiment, showing the 2D boundary and the STR reference hologram (see the video at https://media.upv.es/player/?id=9504e6f0-8a61-11ec-b7c7-7d27dda7c5d5 (accessed on 26 April 2022) [82]): 2D allowed workpiece region in green; trajectory followed by the user reference in thin red line; and trajectory followed by the STR tool in thick blue line.

Figure 12 shows the position followed by the STR tool on the workpiece surface, which is due to the STR teleoperation, together with the reference values provided by the user. In particular, it can be appreciated that the trajectory described by the STR tool corresponds closely to the user reference values, except obviously when the 2D boundary constraint is active; see the bottom graph in Figure 12. In fact, the maximum deviation of the actual STR position values compared to the user reference values, when the 2D boundary constraint was not active, was around 3.2 cm, with a standard deviation of around 0.8 cm; see Table 2. Note that these teleoperation error values include all the potential sources of error: communication delays, high-level and low-level robot control, the accuracy of the workpiece location, teleoperation system, etc. Therefore, it can be concluded that the accuracy of the proposed AR-based teleoperation of the STR is sufficient for the task at hand.

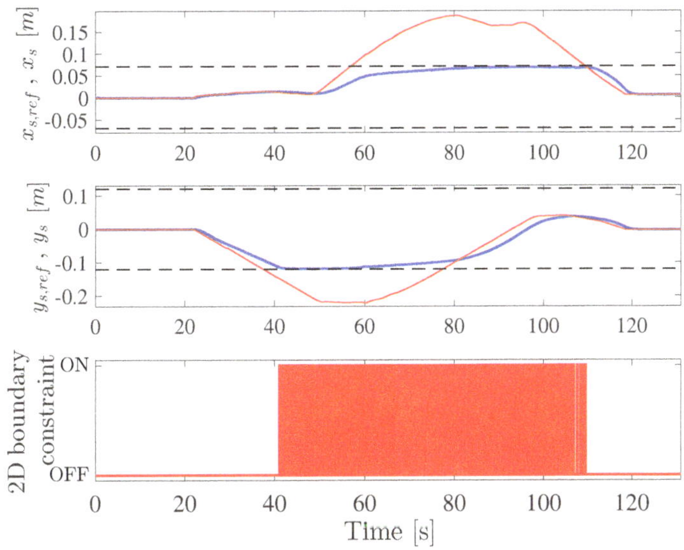

Figure 12. Performance of the STR position teleoperation for the second experiment. First two graphs: user position references in thin red line, actual position values of the STR tool on the workpiece surface (coordinates relative to the surface) in thick blue line, and position limits given by the 2D boundary constraint in dashed lines. Bottom graph: activation of the 2D boundary constraint for the position of the STR tool on the workpiece surface.

Table 2. Teleoperation errors for the 2D position $\overline{\mathbf{p}}_s$ of the STR tool on the workpiece surface.

	Position (cm)	
	x	y
Maximum deviation	1.8	3.2
Standard deviation	0.5	0.8

Figure 13 shows several frames of the *third experiment*, which shows the performance of the 3D boundary and the WR reference hologram; see the video at https://media.upv.es/player/?id=17d88200-8f0b-11ec-be22-d786eca82090 (accessed on 26 April 2022) [83]. Figure 13a shows how the user is commanding the WR and, when the WR end-effector approaches the 3D boundary, the boundary region closest to the WR end-effector is shown in blue and the warning sound is activated; see Figure 13b–d. Note that, when the user reference exceeds the 3D boundary, the WR end-effector is automatically kept within the allowed region. More details about this aspect can be further analyzed in Figure 14, which shows the allowed 3D region, the trajectory followed by the user reference, and the trajectory followed by the WR end-effector.

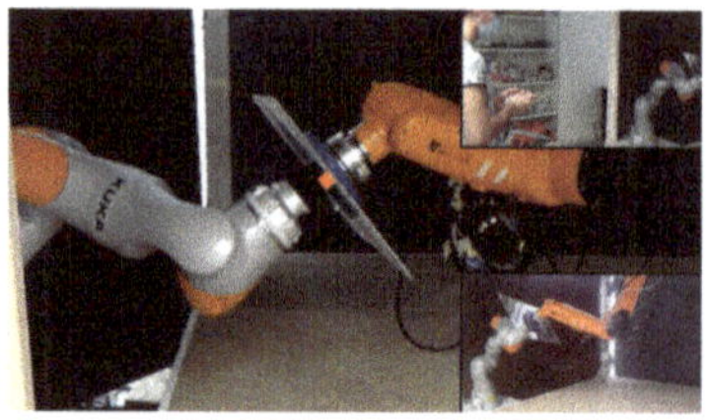

(**a**)

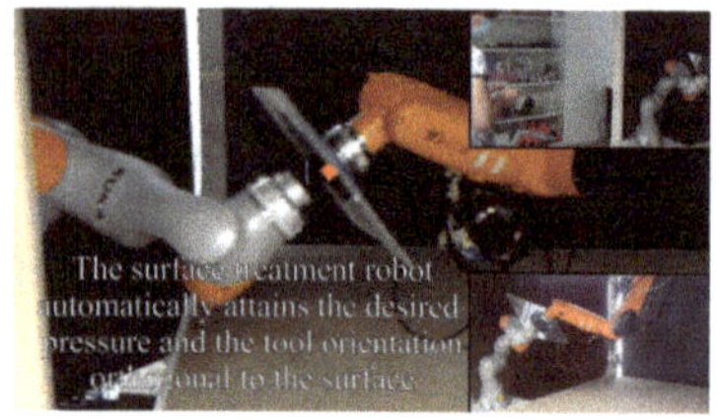

(**b**)

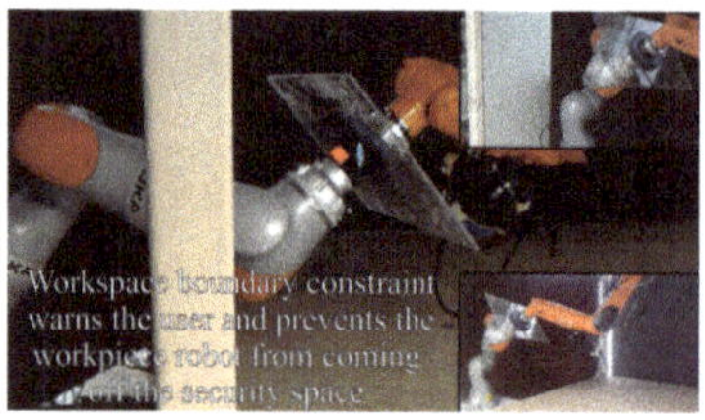

(**c**)

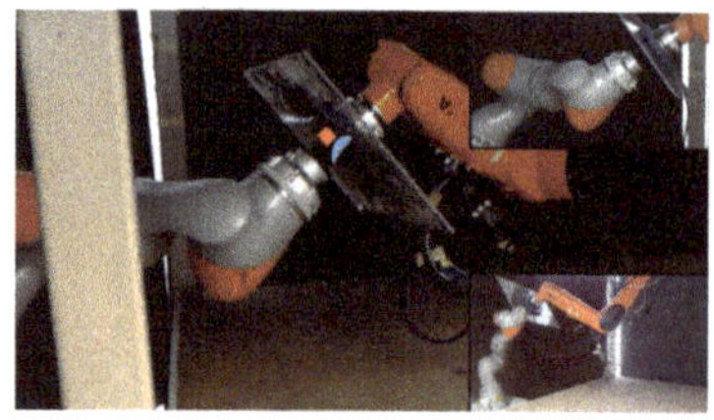

(**d**)

Figure 13. Third experiment: frames of the video showing the performance of the 3D boundary and the WR reference hologram. See the video at https://media.upv.es/player/?id=17d88200-8f0b-11ec-be22-d786eca82090 (accessed on 26 April 2022) [83]. (**a**) Video: 0 m 22 s. (**b**) Video: 0 m 24 s. (**c**) Video: 1 m 04 s. (**d**) Video: 1 m 46 s.

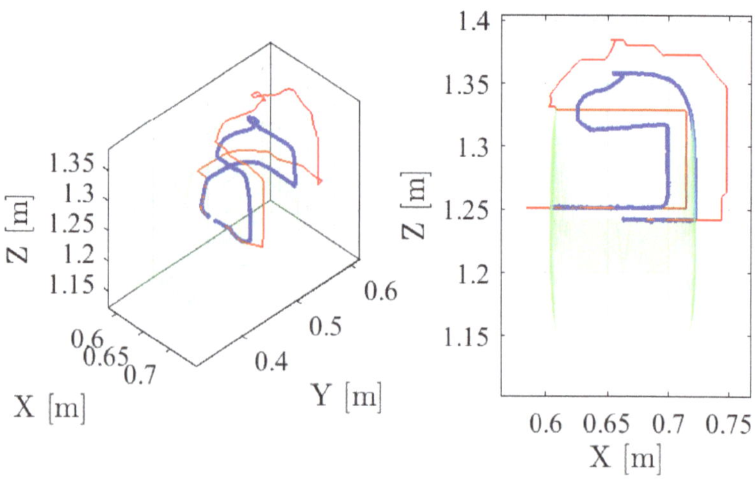

Figure 14. The 3D trajectory performance for the third experiment, showing the 3D boundary and the WR reference hologram (see the video at https://media.upv.es/player/?id=17d88200-8f0b-11ec-be22-d786eca82090 (accessed on 26 April 2022) [83]): 3D allowed region in green; trajectory followed by the user reference in thin red line; and trajectory followed by the WR end-effector in thick blue line.

Figures 15 and 16 show the position and orientation, respectively, followed by the workpiece, which are due to the WR teleoperation, together with the reference values provided by the user. In particular, it can be appreciated that the trajectory described by the workpiece corresponds closely to the user reference values, except obviously when the 3D boundary constraint is active; see the bottom graph in Figure 15. In fact, the maximum deviation of the actual workpiece position values compared to the user reference values, when the 3D boundary constraint was not active, was around 1.2 cm, with a standard deviation of around 0.4 cm; see Table 3. Moreover, the maximum deviation of the actual workpiece orientation values compared to the user reference values was around 1.7°, with a standard deviation of around 0.3°; see Table 3. Note that these teleoperation error values include all the potential sources of error: communication delays, high-level and low-level robot control, teleoperation system, etc. Therefore, it can be concluded that the accuracy of the proposed AR-based teleoperation of the WR is sufficient for the task at hand.

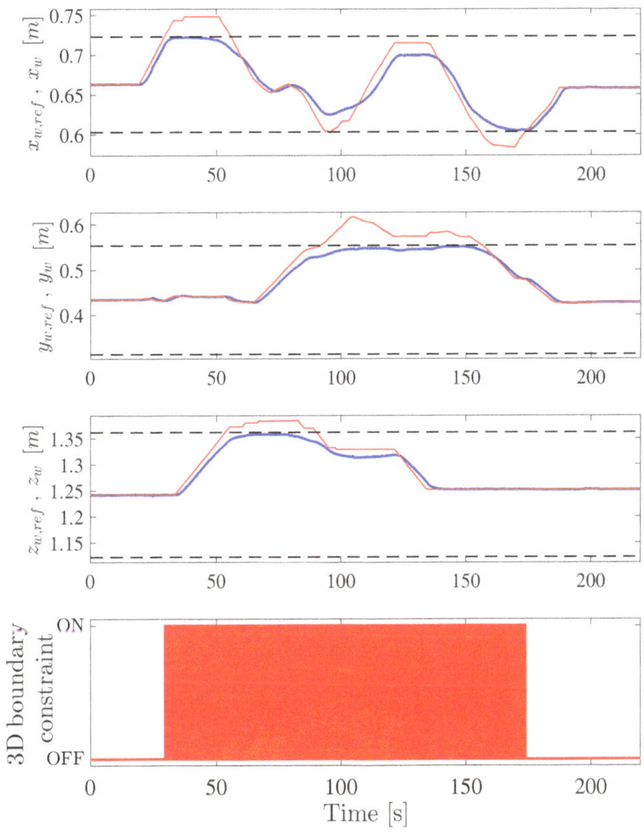

Figure 15. Performance of the WR position teleoperation for the third experiment. First three graphs: user position references in thin red line, actual position values of the workpiece in thick blue line, and position limits given by the 3D boundary constraint in dashed lines. Bottom graph: activation of the 3D boundary constraint for the workpiece position.

Table 3. Teleoperation errors for the pose $\mathbf{p}_w$ (i.e., position and orientation) of the WR.

	Position (cm)			**Orientation (deg)**		
	x	y	z	α	β	γ
Maximum deviation	1.2	0.9	0.1	1.7	1.5	1.7
Standard deviation	0.4	0.3	0.03	0.3	0.2	0.2

Figure 17 depicts several frames of the *fourth experiment*, which shows the performance of the overall system when the user commands simultaneously both robots using the proposed AR-based interface; see the video at https://media.upv.es/player/?id=29330720

-8a8b-11ec-97cd-ab744f931636 (accessed on 26 April 2022) [84]. Figure 17a–d show how the user modifies the orientation of the WR while, at the same time, commanding the STR tool towards one side of the workpiece. Note that, in this situation, when the WR end-effector is close to one side of the 3D boundary, it is partially shown by the corresponding blue hologram. Furthermore, Figure 17e shows how the user simultaneously commands both robots to reach both 2D and 3D boundaries, which are partially shown by the red and blue holograms, respectively. It is worth noting that, in addition to the mentioned holograms, the user hears different warning sounds. Figure 17f,g show how the user modifies again the orientation of the WR while, at the same time, commanding the STR tool towards the other side of the workpiece. Finally, Figure 17h shows how the STR tool reaches the 2D boundary while the user is also commanding the WR end-effector.

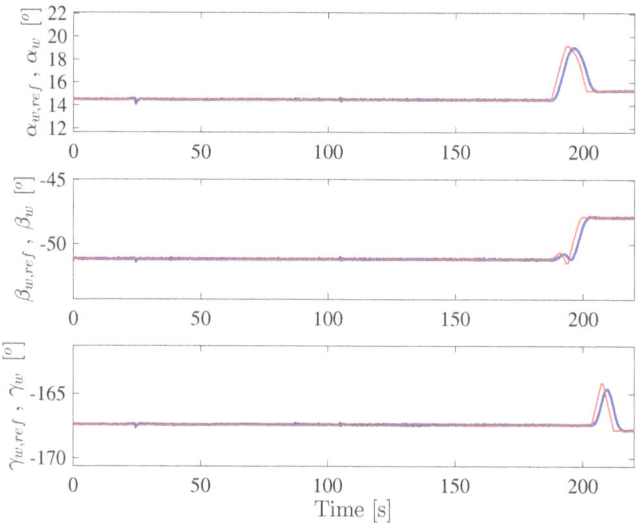

Figure 16. Performance of the WR angle teleoperation for the third experiment: user angular references in thin red line and actual angular values of the workpiece in thick blue line.

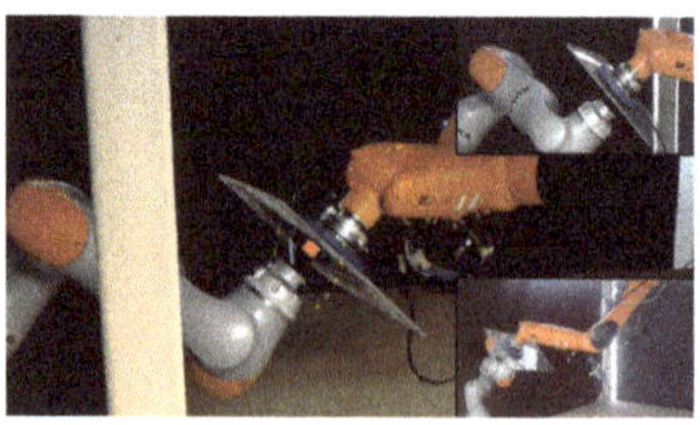

(a)

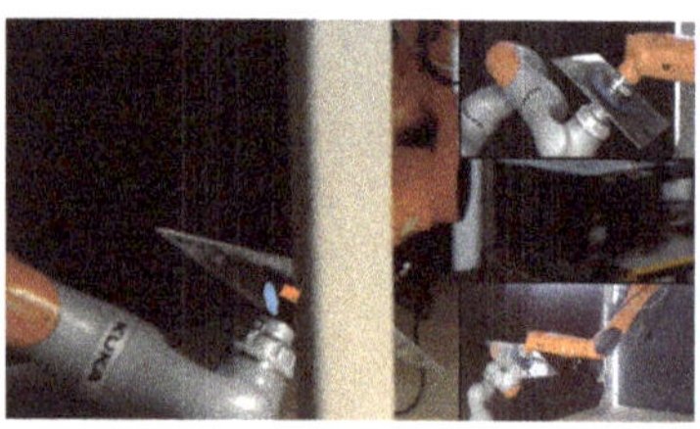

(b)

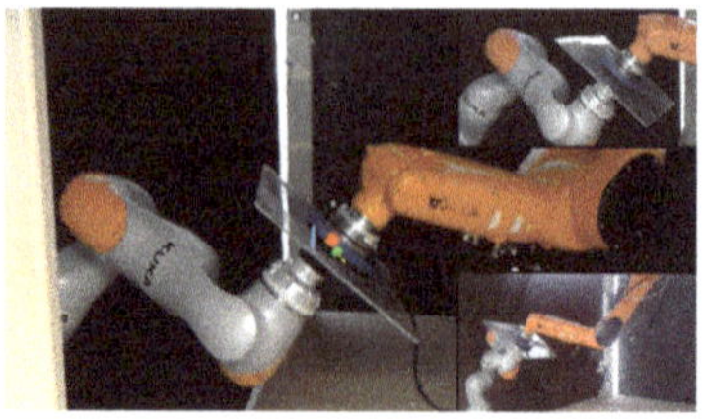

(c)

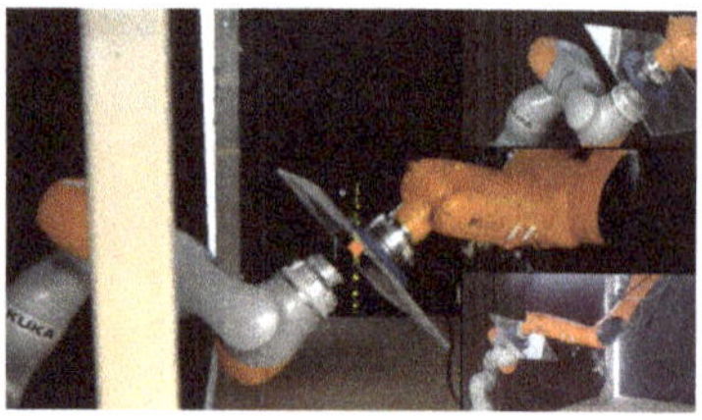

(d)

Figure 17. *Cont.*

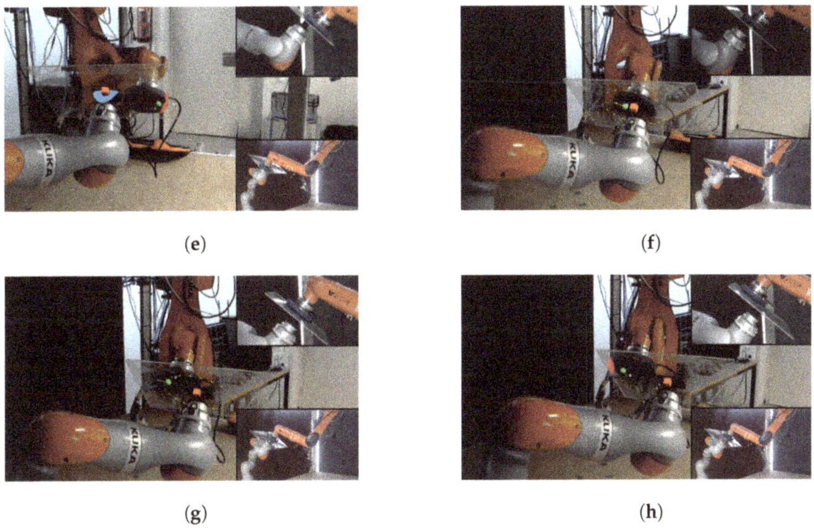

Figure 17. Fourth experiment: frames of the video showing the simultaneous teleoperation of both robots with the proposed AR-based interface. See the video at https://media.upv.es/player/?id=29330720-8a8b-11ec-97cd-ab744f931636 (accessed on 26 April 2022) [84]. (**a**) Video: 1 m 19 s. (**b**) Video: 1 m 44 s. (**c**) Video: 1 m 54 s. (**d**) Video: 1 m 55 s. (**e**) Video: 2 m 13 s. (**f**) Video: 2 m 42 s. (**g**) Video: 2 m 52 s. (**h**) Video: 3 m 5 s.

For the fourth experiment, Figure 18 shows the complete 2D trajectories followed by the user STR reference and the STR tool, whilst Figure 19 shows the complete 3D trajectories followed by the user WR reference and the WR end-effector. In both cases, as in the second and third experiments, the STR tool and the WR end-effector are automatically kept within the allowed regions despite the fact that, at some point, the user references exceed the 2D and 3D boundaries, respectively.

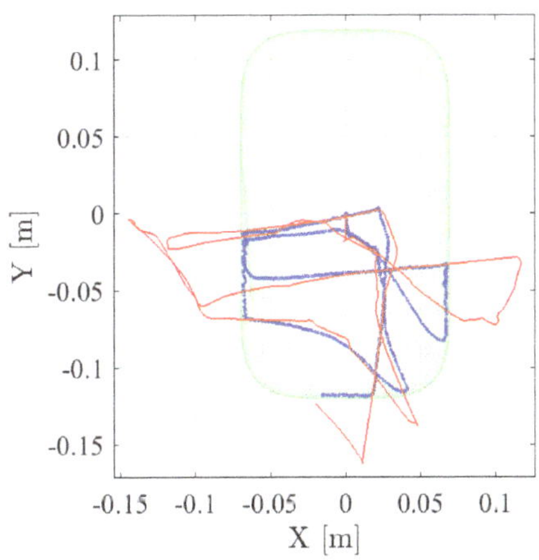

Figure 18. The 2D trajectory performance for the fourth experiment, showing the simultaneous teleoperation of both robots (see the video at https://media.upv.es/player/?id=29330720-8a8b-11ec-97cd-ab744f931636 (accessed on 26 April 2022) [84]): 2D allowed workpiece region in green; trajectory followed by the user reference in thin red line; and trajectory followed by the STR tool in thick blue line.

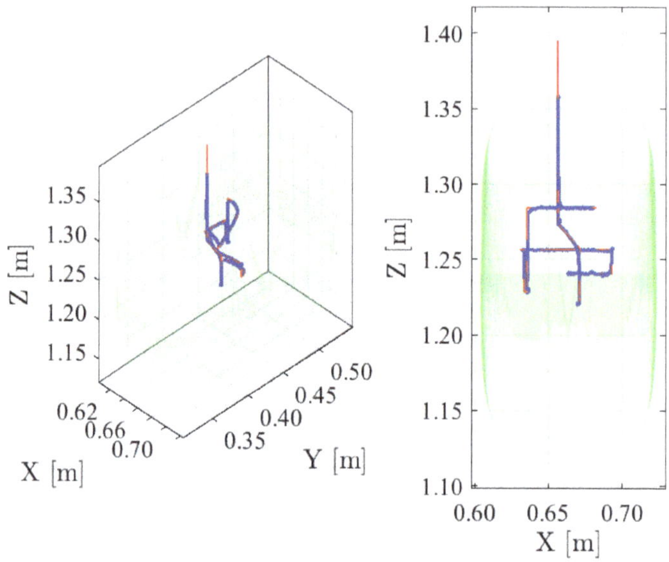

Figure 19. The 3D trajectory performance for the fourth experiment, showing the simultaneous teleoperation of both robots (see the video at https://media.upv.es/player/?id=29330720-8a8b-11ec-97cd-ab744f931636 (accessed on 26 April 2022) [84]): 3D allowed region in green; trajectory followed by the user reference in thin red line; and trajectory followed by the WR end-effector in thick blue line.

The teleoperation errors for the fourth experiment, in which the user commands simultaneously both robots using the proposed AR-based interface, are similar to those shown above for the second experiment (STR teleoperation) and third experiment (WR teleoperation): approximately 0.8 cm standard deviation for the position of the STR tool—see Table 2—and approximately 0.4 cm and 0.3° standard deviation for the WR position and orientation, respectively—see Table 3. As mentioned above, these teleoperation error values include all the potential sources of error: communication delays, high-level and low-level control of both robots, teleoperation system, etc. Therefore, it is concluded that the accuracy achieved by the proposed AR-based approach for teleoperating the bimanual robot system is satisfactory.

5. Conclusions

A solution to improve the assisted bimanual robot teleoperation has been developed in this work using augmented reality (AR) technology and tools. In particular, a new AR interface using the Microsoft *HoloLens glasses* has been proposed to mitigate the problems in terms of user ergonomics and task performance (i.e., completion time and finishing quality) raised from the use of conventional PC-based user interfaces. In addition, this work has proposed and followed a new methodology to design and develop AR interfaces for bimanual robotic systems.

The effectiveness and applicability of the proposed AR interface were shown by means of real experimentation with an advanced bimanual robot application consisting of two robotic arms: a 7R cobot and a 6R industrial manipulator.

It is worth noting that several users, who tested both the conventional PC-based interface and the proposed AR interface, found the latter more intuitive and were able to conduct the robot teleoperation task faster. Note that when the users teleoperated the bimanual robot system using the conventional PC-based interface, most of them complained about the difficulty of checking whether the robots were performing the task correctly or not. In addition, the users indicated that with the conventional PC-based interface, it was not easy for them to command both robots simultaneously because they could not pay attention to so many reference signals shown. These facts negatively affected the performance of the users in terms of the time required to complete the task. Thus, the mentioned issues were mitigated with the proposed AR interface, significantly improving the user performance in the teleoperation task.

Another relevant remark is that the users also indicated that the warning sounds helped them in the early stages of the teleoperation task but, as the time of use of the interface increased, these sounds were annoying and they preferred only the visual warnings.

Author Contributions: Conceptualization, J.E.S. and L.G.; Funding acquisition, J.T.; Investigation, J.E.S., A.M. and L.G.; Methodology, J.E.S. and L.G.; Resources, J.T.; Software, A.G. and A.M.; Supervision, L.G. and J.T.; Writing—original draft, J.E.S.; Writing—review & editing, L.G. All authors have read and agreed to the published version of the manuscript.

Funding: This research was funded by the Spanish Government (Grant PID2020-117421RB-C21 funded by MCIN/AEI/10.13039/501100011033) and by the Generalitat Valenciana (Grants GV/2021/181 and ACIF/2019/007).

Conflicts of Interest: The authors declare no conflict of interest.

References

1. Wang, W.; Chen, Y.; Li, R.; Zhang, Z.; Krovi, V.; Jia, Y. Human-robot collaboration for advanced manufacturing by learning from multi-modal human demonstrations. In *Manufacturing In The Era Of 4th Industrial Revolution*; World Scientific Publishing Co.: Singapore, 2021; pp. 87–116.
2. Admoni, B.A.; Aronson, R.M.; Srinivasa, S.S.; Kitani, K.; Admoni, H. HARMONIC: A multimodal dataset of assistive human–robot collaboration. *Int. J. Robot. Res.* **2022**, *41*, 3–11. [CrossRef]
3. Kim, S. Working With Robots: Human Resource Development Considerations in Human–Robot Interaction. *Hum. Resour. Dev. Rev.* **2022**, *21*, 48–74. [CrossRef]
4. Gracia, L.; Solanes, J.E.; Muñoz-Benavent, P.; Valls Miro, J.; Perez-Vidal, C.; Tornero, J. Adaptive Sliding Mode Control for Robotic Surface Treatment Using Force Feedback. *Mechatronics* **2018**, *52*, 102–118. [CrossRef]
5. Cardoso, J.C.S. Comparison of Gesture, Gamepad, and Gaze-Based Locomotion for VR Worlds. In *Proceedings of the 22nd ACM Conference on Virtual Reality Software and Technology*; Association for Computing Machinery: New York, NY, USA, 2016; pp. 319–320.
6. Kitson, A.; Hashemian, A.M.; Stepanova, E.R.; Kruijff, E.; Riecke, B.E. Comparing leaning-based motion cueing interfaces for virtual reality locomotion. In Proceedings of the 2017 IEEE Symposium on 3D User Interfaces (3DUI), Los Angeles, CA, USA, 18–19 March 2017; pp. 73–82.
7. Zhao, J.; Allison, R.S. Comparing head gesture, hand gesture and gamepad interfaces for answering Yes/No questions in virtual environments. *Virtual Real.* **2019**, *24*, 515–524. [CrossRef]
8. Solanes, J.E.; Muñoz, A.; Gracia, L.; Martí, A.; Girbés-Juan, V.; Tornero, J. Teleoperation of industrial robot manipulators based on augmented reality. *Int. J. Adv. Manuf. Technol.* **2020**, *111*, 1077–1097. [CrossRef]
9. Smith, C.; Karayiannidis, Y.; Nalpantidis, L.; Gratal, X.; Qi, P.; Dimarogonas, D.V.; Kragic, D. Dual arm manipulation—A survey. *Robot. Auton. Syst.* **2012**, *60*, 1340–1353. [CrossRef]
10. Makris, S.; Tsarouchi, P.; Matthaiakis, A.S.; Athanasatos, A.; Chatzigeorgiou, X.; Stefos, M.; Giavridis, K.; Aivaliotis, S. Dual arm robot in cooperation with humans for flexible assembly. *CIRP Ann.* **2017**, *66*, 13–16. [CrossRef]
11. Sepúlveda, D.; Fernández, R.; Navas, E.; Armada, M.; González-De-Santos, P. Robotic Aubergine Harvesting Using Dual-Arm Manipulation. *IEEE Access* **2020**, *8*, 121889–121904. [CrossRef]
12. Chatzilygeroudis, K.; Fichera, B.; Lauzana, I.; Bu, F.; Yao, K.; Khadivar, F.; Billard, A. Benchmark for Bimanual Robotic Manipulation of Semi-Deformable Objects. *IEEE Robot. Autom. Lett.* **2020**, *5*, 2443–2450. [CrossRef]
13. Garcia-Camacho, I.; Lippi, M.; Welle, M.C.; Yin, H.; Antonova, R.; Varava, A.; Borras, J.; Torras, C.; Marino, A.; Alenyá, G.; et al. Benchmarking Bimanual Cloth Manipulation. *IEEE Robot. Autom. Lett.* **2020**, *5*, 1111–1118. [CrossRef]
14. Sintov, A.; Macenski, S.; Borum, A.; Bretl, T. Motion Planning for Dual-Arm Manipulation of Elastic Rods. *IEEE Robot. Autom. Lett.* **2020**, *5*, 6065–6072. [CrossRef]
15. Mitash, C.; Shome, R.; Wen, B.; Boularias, A.; Bekris, K. Task-Driven Perception and Manipulation for Constrained Placement of Unknown Objects. *IEEE Robot. Autom. Lett.* **2020**, *5*, 5605–5612. [CrossRef]
16. Clark, J.P.; Lentini, G.; Barontini, F.; Catalano, M.G.; Bianchi, M.; O'Malley, M.K. On the role of wearable haptics for force feedback in teleimpedance control for dual-arm robotic teleoperation. In Proceedings of the 2019 International Conference on Robotics and Automation (ICRA), Montreal, QC, Canada, 20–24 May 2019; pp. 5187–5193.
17. Salehian, S.S.M.; Figueroa, N.; Billard, A. A unified framework for coordinated multi-arm motion planning. *Int. J. Robot. Res.* **2018**, *37*, 1205–1232. [CrossRef]

18. Suarez, A.; Real, F.; Vega, V.M.; Heredia, G.; Rodriguez-Castaño, A.; Ollero, A. Compliant Bimanual Aerial Manipulation: Standard and Long Reach Configurations. *IEEE Access* **2020**, *8*, 88844–88865. [CrossRef]
19. Wu, Q.; Li, M.; Qi, X.; Hu, Y.; Li, B.; Zhang, J. Coordinated control of a dual-arm robot for surgical instrument sorting tasks. *Robot. Auton. Syst.* **2019**, *112*, 1–12. [CrossRef]
20. Rakita, D.; Mutlu, B.; Gleicher, M.; Hiatt, L.M. Shared control–based bimanual robot manipulation. *Sci. Robot.* **2019**, *4*, eaaw0955. [CrossRef] [PubMed]
21. García, N.; Rosell, J.; Suárez, R. Motion Planning by Demonstration With Human-Likeness Evaluation for Dual-Arm Robots. *IEEE Trans. Syst. Man Cybern. Syst.* **2019**, *49*, 2298–2307. [CrossRef]
22. Qu, J.; Zhang, F.; Wang, Y.; Fu, Y. Human-like coordination motion learning for a redundant dual-arm robot. *Robot. Comput.-Integr. Manuf.* **2019**, *57*, 379–390. [CrossRef]
23. Joshi, R.P.; Tarapure, J.P.; Shibata, T. Electric Wheelchair-Humanoid Robot Collaboration for Clothing Assistance of the Elderly. In Proceedings of the 2020 13th International Conference on Human System Interaction (HSI), Tokyo, Japan, 6–8 June 2020; pp. 300–306.
24. Li, Y.; Guo, S.; Mukai, T. Position Adjustment Control of A Nursing-care Robot Holding A Patient in Its Arms. In Proceedings of the 2019 IEEE International Conference on Mechatronics and Automation (ICMA), Tianjin, China, 4–7 August 2019; pp. 976–981.
25. Chen, H.; Li, J.; Wan, W.; Huang, Z.; Harada, K. Integrating combined task and motion planning with compliant control. *Int. J. Intell. Robot. Appl.* **2020**, *4*, 149–163. [CrossRef]
26. Zimmermann, S.; Hakimifard, G.; Zamora, M.; Poranne, R.; Coros, S. A Multi-Level Optimization Framework for Simultaneous Grasping and Motion Planning. *IEEE Robot. Autom. Lett.* **2020**, *5*, 2966–2972. [CrossRef]
27. Parigi Polverini, M.; Zanchettin, A.M.; Rocco, P. A constraint-based programming approach for robotic assembly skills implementation. *Robot. Comput.-Integr. Manuf.* **2019**, *59*, 69–81. [CrossRef]
28. Zhong, F.; Wang, Y.; Wang, Z.; Liu, Y. Dual-Arm Robotic Needle Insertion With Active Tissue Deformation for Autonomous Suturing. *IEEE Robot. Autom. Lett.* **2019**, *4*, 2669–2676. [CrossRef]
29. Bandala, M.; West, C.; Monk, S.; Montazeri, A.; Taylor, C.J. Vision-Based Assisted Tele-Operation of a Dual-Arm Hydraulically Actuated Robot for Pipe Cutting and Grasping in Nuclear Environments. *Robotics* **2019**, *8*, 42. [CrossRef]
30. Girbés-Juan, V.; Schettino, V.; Demiris, Y.; Tornero, J. Haptic and Visual Feedback Assistance for Dual-Arm Robot Teleoperation in Surface Conditioning Tasks. *IEEE Trans. Haptics* **2021**, *14*, 44–56. [CrossRef]
31. García, A.; Solanes, J.E.; Gracia, L.; Muñoz-Benavent, P.; Girbés-Juan, V.; Tornero, J. Bimanual robot control for surface treatment tasks. *Int. J. Syst. Sci.* **2022**, *53*, 74–107. [CrossRef]
32. McConachie, D.; Dobson, A.; Ruan, M.; Berenson, D. Manipulating deformable objects by interleaving prediction, planning, and control. *Int. J. Robot. Res.* **2020**, *39*, 957–982. [CrossRef]
33. Liang, J.; Xu, Z.; Zhou, X.; Li, S.; Ye, G. Recurrent Neural Networks-Based Collision-Free Motion Planning for Dual Manipulators Under Multiple Constraints. *IEEE Access* **2020**, *8*, 54225–54236. [CrossRef]
34. Ibarguren, A.; Eimontaite, I.; Outón, J.L.; Fletcher, S. Dual Arm Co-Manipulation Architecture with Enhanced Human–Robot Communication for Large Part Manipulation. *Sensors* **2020**, *20*, 6151. [CrossRef]
35. Brantner, G.; Khatib, O. Controlling Ocean One: Human–robot collaboration for deep-sea manipulation. *J. Field Robot.* **2021**, *38*, 28–51. [CrossRef]
36. Niemeyer, G.; Preusche, C.; Stramigioli, S.; Lee, D., Telerobotics. In *Springer Handbook of Robotics*; Siciliano, B., Khatib, O., Eds.; Springer International Publishing: Cham, Switzerland, 2016; pp. 1085–1108.
37. Chen, H.; Huang, P.; Liu, Z. Mode Switching-Based Symmetric Predictive Control Mechanism for Networked Teleoperation Space Robot System. *IEEE/ASME Trans. Mechatron.* **2019**, *24*, 2706–2717. [CrossRef]
38. Abi-Farraj, F.; Pacchierotti, C.; Arenz, O.; Neumann, G.; Giordano, P.R. A Haptic Shared-Control Architecture for Guided Multi-Target Robotic Grasping. *IEEE Trans. Haptics* **2020**, *13*, 270–285. [CrossRef] [PubMed]
39. Isop, W.A.; Gebhardt, C.; Nägeli, T.; Fraundorfer, F.; Hilliges, O.; Schmalstieg, D. High-Level Teleoperation System for Aerial Exploration of Indoor Environments. *Front. Robot. AI* **2019**, *6*, 95. [CrossRef] [PubMed]
40. Sivčev, S.; Coleman, J.; Omerdić, E.; Dooly, G.; Toal, D. Underwater manipulators: A review. *Ocean Eng.* **2018**, *163*, 431–450. [CrossRef]
41. Kono, H.; Mori, T.; Ji, Y.; Fujii, H.; Suzuki, T. Development of Perilous Environment Estimation System Using a Teleoperated Rescue Robot with On-board LiDAR. In Proceedings of the 2019 IEEE/SICE International Symposium on System Integration (SII), Paris, France, 14–16 January 2019; pp. 7–10.
42. Yoon, H.; Jeong, J.H.; Yi, B. Image-Guided Dual Master–Slave Robotic System for Maxillary Sinus Surgery. *IEEE Trans. Robot.* **2018**, *34*, 1098–1111. [CrossRef]
43. Saracino, A.; Oude-Vrielink, T.J.C.; Menciassi, A.; Sinibaldi, E.; Mylonas, G.P. Haptic Intracorporeal Palpation Using a Cable-Driven Parallel Robot: A User Study. *IEEE Trans. Biomed. Eng.* **2020**, *67*, 3452–3463. [CrossRef]
44. Chen, Y.; Zhang, S.; Wu, Z.; Yang, B.; Luo, Q.; Xu, K. Review of surgical robotic systems for keyhole and endoscopic procedures: State of the art and perspectives. *Front. Med.* **2020**, *14*, 382–403. [CrossRef]
45. Kapoor, A.; Li, M.; Taylor, R.H. Spatial Motion Constraints for Robot Assisted Suturing Using Virtual Fixtures. In *Medical Image Computing and Computer-Assisted Intervention–MICCAI 2005*; Duncan, J.S., Gerig, G., Eds.; Springer: Berlin/Heidelberg, Germany, 2005; pp. 89–96.

46. Johnson, M.; Vera, A. No AI Is an Island: The Case for Teaming Intelligence. *AI Mag.* **2019**, *40*, 16–28. [CrossRef]
47. Selvaggio, M.; Abi-Farraj, F.; Pacchierotti, C.; Giordano, P.R.; Siciliano, B. Haptic-Based Shared-Control Methods for a Dual-Arm System. *IEEE Robot. Autom. Lett.* **2018**, *3*, 4249–4256. [CrossRef]
48. Nicolis, D.; Palumbo, M.; Zanchettin, A.M.; Rocco, P. Occlusion-Free Visual Servoing for the Shared Autonomy Teleoperation of Dual-Arm Robots. *IEEE Robot. Autom. Lett.* **2018**, *3*, 796–803. [CrossRef]
49. Lu, Z.; Huang, P.; Liu, Z. Predictive Approach for Sensorless Bimanual Teleoperation Under Random Time Delays With Adaptive Fuzzy Control. *IEEE Trans. Ind. Electron.* **2018**, *65*, 2439–2448. [CrossRef]
50. Girbés-Juan, V.; Schettino, V.; Gracia, L.; Solanes, J.E.; Demeris, Y.; Tornero, J. Combining haptics and inertial motion capture to enhance remote control of a dual-arm robot. *J. Multimodal User Interfaces* **2022**, 1–20. *in press*. [CrossRef]
51. Gorjup, G.; Dwivedi, A.; Elangovan, N.; Liarokapis, M. An Intuitive, Affordances Oriented Telemanipulation Framework for a Dual Robot Arm Hand System: On the Execution of Bimanual Tasks. In Proceedings of the 2019 IEEE/RSJ International Conference on Intelligent Robots and Systems (IROS), Macau, China, 3–8 November 2019; pp. 3611–3616.
52. Selvaggio, M.; Ghalamzan, A.; Moccia, R.; Ficuciello, F.; Siciliano, B. Haptic-guided shared control for needle grasping optimization in minimally invasive robotic surgery. In Proceedings of the IEEE/RSJ International Conference on Intelligent Robots and Systems, Macau, China, 3–8 November 2019.
53. Lipton, J.I.; Fay, A.J.; Rus, D. Baxter's Homunculus: Virtual Reality Spaces for Teleoperation in Manufacturing. *IEEE Robot. Autom. Lett.* **2018**, *3*, 179–186. [CrossRef]
54. Bian, F.; Li, R.; Zhao, L.; Liu, Y.; Liang, P. Interface Design of a Human-Robot Interaction System for Dual-Manipulators Teleoperation Based on Virtual Reality. In Proceedings of the 2018 IEEE International Conference on Information and Automation (ICIA), Wuyishan, China, 11–13 August 2018; pp. 1361–1366. [CrossRef]
55. Tonin, L.; Millan, J.D.R. Noninvasive Brain-Machine Interfaces for Robotic Devices. In *Annual Review of Control, Robotics, and Autonomous Systems*; Leonard, N., Ed.; Annual Reviews: Palo Alto, CA, USA, 2021; Volume 4, pp. 191–214. [CrossRef]
56. Tang, G.; Shi, Q.; Zhang, Z.; He, T.; Sun, Z.; Lee, C. Hybridized wearable patch as a multi-parameter and multi-functional human-machine interface. *Nano Energy* **2021**, *81*, 105582. [CrossRef]
57. Jin, Y.; Chen, J.; Zhang, S.; Chen, W.; Zheng, X. Invasive Brain Machine Interface System. In *Neural Interface: Frontiers and Applications*; Zheng, X., Ed.; Springer: Singapore, 2019; pp. 67–89._3. [CrossRef]
58. Dumitrescu, C.; Costea, I.M.; Semenescu, A. Using Brain-Computer Interface to Control a Virtual Drone Using Non-Invasive Motor Imagery and Machine Learning. *Appl. Sci.* **2021**, *11*, 1876. [CrossRef]
59. Xu, B.; Li, W.; Liu, D.; Zhang, K.; Miao, M.; Xu, G.; Song, A. Continuous Hybrid BCI Control for Robotic Arm Using Noninvasive Electroencephalogram, Computer Vision, and Eye Tracking. *Mathematics* **2022**, *10*, 618. [CrossRef]
60. Cao, L.; Li, G.; Xu, Y.; Zhang, H.; Shu, X.; Zhang, D. A brain-actuated robotic arm system using non-invasive hybrid brain-computer interface and shared control strategy. *J. Neural Eng.* **2021**, *18*, 046045. [CrossRef]
61. Muñoz, A.; Mahiques, X.; Solanes, J.E.; Martí, A.; Gracia, L.; Tornero, J. Mixed reality-based user interface for quality control inspection of car body surfaces. *J. Manuf. Syst.* **2019**, *53*, 75–92. [CrossRef]
62. Muñoz, A.; Martí, A.; Mahiques, X.; Gracia, L.; Solanes, J.E.; Tornero, J. Camera 3D positioning mixed reality-based interface to improve worker safety, ergonomics and productivity. *CIRP J. Manuf. Sci. Technol.* **2020**, *28*, 24–37. [CrossRef]
63. Chu, Y.B.; Chang, C.W. A Mobile Augmented Reality Interface on Additive Manufacturing. In *Advances in Electrical and Electronic Engineering and Computer Science*; Zakaria, Z., Emamian, S.S., Eds.; Springer: Singapore, 2021; pp. 1–12.
64. Tedesco, A.; Dallet, D.; Arpaia, P. Augmented Reality (AR) and Brain-Computer Interface (BCI): Two Enabling Technologies for Empowering the Fruition of Sensor Data in the 4.0 Era. In *Sensors and Microsystems*; Di Francia, G., Di Natale, C., Eds.; Springer International Publishing: Cham, Switzerland, 2021; pp. 85–91.
65. Craig, A.B. Chapter 2-Augmented Reality Concepts. In *Understanding Augmented Reality*; Craig, A.B., Ed.; Morgan Kaufmann: Boston, MA, USA, 2013; pp. 39–67.
66. Li, C.; Fahmy, A.; Sienz, J. An Augmented Reality Based Human-Robot Interaction Interface Using Kalman Filter Sensor Fusion. *Sensors* **2019**, *19*, 4586. [CrossRef]
67. Rosen, E.; Whitney, D.; Phillips, E.; Chien, G.; Tompkin, J.; Konidaris, G.; Tellex, S. Communicating and controlling robot arm motion intent through mixed-reality head-mounted displays. *Int. J. Robot. Res.* **2019**, *38*, 1513–1526. [CrossRef]
68. Gadre, S.Y.; Rosen, E.; Chien, G.; Phillips, E.; Tellex, S.; Konidaris, G. End-User Robot Programming Using Mixed Reality. In Proceedings of the 2019 International Conference on Robotics and Automation (ICRA), Montreal, QC, Canada, 20–24 May 2019; pp. 2707–2713.
69. Mistry, P.; Maes, P. SixthSense: A Wearable Gestural Interface. In *ACM Siggraph Asia 2009 Sketches*; SIGGRAPH ASIA '09; Association for Computing Machinery: New York, NY, USA, 2009. [CrossRef]
70. Ismail, A.W.; Billinghurst, M.; Sunar, M.S.; Yusof, C.S. Designing an Augmented Reality Multimodal Interface for 6DOF Manipulation Techniques. In *Intelligent Systems and Applications*; Arai, K., Kapoor, S., Bhatia, R., Eds.; Springer International Publishing: Cham, Switzerland, 2019; pp. 309–322.
71. Nakamura, Y.; Hanafusa, H.; Yoshikawa, T. Task-Priority Based Redundancy Control of Robot Manipulators. *Int. J. Robot. Res.* **1987**, *6*, 3–15. [CrossRef]
72. Nenchev, D. Tracking manipulator trajectories with ordinary singularities: A null space-based approach. *Int. J. Robot. Res.* **1995**, *14*, 399–404. [CrossRef]

73. Garelli, F.; Gracia, L.; Sala, A.; Albertos, P. Sliding mode speed auto-regulation technique for robotic tracking. *Robot. Auton. Syst.* **2011**, *59*, 519–529. [CrossRef]
74. Garelli, F.; Mantz, R.; De Battista, H. Sliding mode reference conditioning to preserve decoupling of stable systems. *Chem. Eng. Sci.* **2007**, *62*, 4705–4716. [CrossRef]
75. Video of the PC-Based Interface. 2022. Available online: https://media.upv.es/player/?id=15ffabe0-a733-11eb-a0b0-2fbcb59aaef7 (accessed on 26 April 2022).
76. Microsoft Hololens. HoloLens (1st Gen) Hardware Details. Available online: https://docs.microsoft.com/en-us/windows/mixed-reality/hololens-hardware-details (accessed on 26 April 2022).
77. Microsoft Hololens (2nd Gen) Hardware Details. Available online: https://www.microsoft.com/en-us/hololens/hardware (accessed on 26 April 2022).
78. Hess, R. *Blender Foundations: The Essential Guide to Learning Blender 2.6*; Focal Press, Elsevier: Oxford, UK, 2010. Available online: https://www.sciencedirect.com/book/9780240814308/blender-foundations (accessed on 26 April 2022).
79. Jackson, S. *Unity 3D UI Essentials*; Packt Publishing: Birmingham, UK, 2015. Available online: https://dl.acm.org/citation.cfm?id=2789365 (accessed on 26 April 2022).
80. Unity. Shaders Core Concepts. Available online: https://docs.unity3d.com/Manual/Shaders.html (accessed on 26 April 2022).
81. First Experiment Video. 2022. Available online: https://media.upv.es/player/?id=a64014f0-8a5a-11ec-ac0a-b3aa330d3dad (accessed on 26 April 2022).
82. Second Experiment Video. 2022. Available online: https://media.upv.es/player/?id=9504e6f0-8a61-11ec-b7c7-7d27dda7c5d5 (accessed on 26 April 2022).
83. Third Experiment Video. 2022. Available online: https://media.upv.es/player/?id=17d88200-8f0b-11ec-be22-d786eca82090 (accessed on 26 April 2022).
84. Fourth Experiment Video. 2022. Available online: https://media.upv.es/player/?id=29330720-8a8b-11ec-97cd-ab744f931636 (accessed on 26 April 2022).

applied sciences

MDPI

Article

Evaluation of RGB-D Multi-Camera Pose Estimation for 3D Reconstruction

Ian de Medeiros Esper [1,*], Oleh Smolkin [2,3], Maksym Manko [2,4], Anton Popov [2,3,4] and Pål Johan From [1] and Alex Mason [1,5,*]

1 Faculty of Science and Technology, Norwegian Univiersity of Life Sciences, Universitetstunet 3, 1430 Ås, Norway; pal.johan.from@nmbu.no
2 Ciklum Data & Analytics, 03680 Kyiv, Ukraine; osmo@ciklum.com (O.S.); mman@ciklum.com (M.M.); poant@ciklum.com (A.P.)
3 Faculty of Applied Sciences, Ukrainian Catholic University, 79000 Lviv, Ukraine
4 Electronic Engineering Department, Igor Sikorsky Kyiv Polytechnic Institute, 03056 Kyiv, Ukraine
5 Animalia, Norwegian Meat Research Institute, 0513 Oslo, Norway
* Correspondence: ian.esper@nmbu.no (I.d.M.E.); alex.mason@nmbu.no (A.M.)

Abstract: Advances in visual sensor devices and computing power are revolutionising the interaction of robots with their environment. Cameras that capture depth information along with a common colour image play a significant role. These devices are cheap, small, and fairly precise. The information provided, particularly point clouds, can be generated in a virtual computing environment, providing complete 3D information for applications. However, off-the-shelf cameras often have a limited field of view, both on the horizontal and vertical axis. In larger environments, it is therefore often necessary to combine information from several cameras or positions. To concatenate multiple point clouds and generate the complete environment information, the pose of each camera must be known in the outer scene, i.e., they must reference a common coordinate system. To achieve this, a coordinate system must be defined, and then every device must be positioned according to this coordinate system. For cameras, a calibration can be performed to find its pose in relation to this coordinate system. Several calibration methods have been proposed to solve this challenge, ranging from structured objects such as chessboards to features in the environment. In this study, we investigate how three different pose estimation methods for multi-camera perspectives perform when reconstructing a scene in 3D. We evaluate the usage of a charuco cube, a double-sided charuco board, and a robot's tool centre point (TCP) position in a real usage case, where precision is a key point for the system. We define a methodology to identify the points in the 3D space and measure the root-mean-square error (RMSE) based on the Euclidean distance of the actual point to a generated ground-truth point. The reconstruction carried out using the robot's TCP position produced the best result, followed by the charuco cuboid; the double-sided angled charuco board exhibited the worst performance.

Keywords: pose estimation; robotics; 3D reconstruction; charuco cuboid

Citation: de Medeiros Esper, I.; Smolkin, O.; Manko, M.; Popov, A.; From, P.J.; Mason, A. Evaluation of RGB-D Multi-Camera Pose Estimation for 3D Reconstruction. *Appl. Sci.* **2022**, *12*, 4134. https://doi.org/10.3390/app12094134

Academic Editors: Luis Gracia and Carlos Perez-Vidal

Received: 15 March 2022
Accepted: 17 April 2022
Published: 20 April 2022

Publisher's Note: MDPI stays neutral with regard to jurisdictional claims in published maps and institutional affiliations.

1. Introduction

The idea and use of 3D imaging dates back to the 19th century, and laser scanning to the 1960s [1], but only recently has it been capable of revolutionising the interaction between robots, the environment and humans. Many advances in computational power, sensor precision and affordability have made this possible [2–4].

The recent development of RGB-D cameras has provided visual sensor devices capable of generating pixel-wise depth information, together with a colour image. The technology behind these cameras has been constantly improving, with developers working to reduce noise and increase precision, e.g., Microsoft Kinect Azure and Intel RealSense L515 and D455 [3].

The depth information from those devices can be used to generate a three-dimensional projection of the captured object. Understanding the well-known pinhole camera model [5] is important to understand how the reprojection works, and how it is affected by noise in the depth data.

The model describes the transformation from the 3D world to the 2D image plane, as shown in Figure 1 and in the Equation (1) [6]. It can also be used to calculate the inverse path for reprojecting from 2D to 3D.

Equation (1) has two matrices: one for the intrinsic parameters and another for the extrinsic parameters. The first contains the camera's internal parameters, which are constant for each camera. The second describes where the camera is in the world, i.e., the pose of the camera in relation to an origin coordinate system.

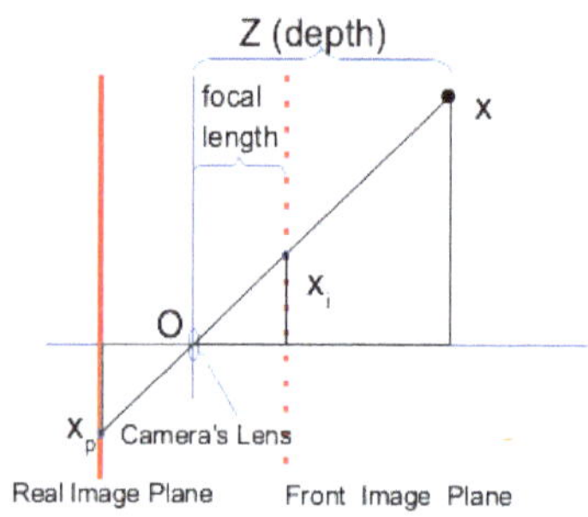

Figure 1. Pinhole camera's projective geometry.

The direction vector of the ray from the camera projection centre can be found using these parameters, but the length of the vector cannot. This information is lost in the conversion from 3D to 2D. However, when using an RGB-D camera this information is saved, as a depth that determines where the point lies in the world. The set of points reprojected from these data is called a point cloud.

Other parameters that are important for reprojection are the distortion coefficients which are used to correct the radial and tangential distortions of the lens [7]. In this work, the FRAMOS Industrial Depth Camera D415e, which was built with Intel® RealSense™ technology, was used. The Intel® RealSense module claims to have no lens distortions [8,9].

$$s\begin{bmatrix}u\\v\\1\end{bmatrix} = \underbrace{\begin{bmatrix}f_x & 0 & c_x\\0 & f_y & c_y\\0 & 0 & 1\end{bmatrix}}_{\text{intrinsic}}\underbrace{\begin{bmatrix}r_{11} & r_{12} & r_{13} & t_1\\r_{21} & r_{22} & r_{23} & t_2\\r_{31} & r_{32} & r_{33} & t_3\end{bmatrix}}_{\text{extrinsic}}\begin{bmatrix}X\\Y\\Z\\1\end{bmatrix} \tag{1}$$

The intrinsic parameters of a camera are normally represented as a 3×3 matrix, as shown in Equation (2).

$$K = \begin{bmatrix}f_x & 0 & c_x\\0 & f_y & c_y\\0 & 0 & 1\end{bmatrix} \tag{2}$$

where f_x and f_y are the focal lengths in the x and y directions, respectively. Furthermore, c_x and c_y are the optical centres of the image plane, shown as a solid red line in Figure 1.

As illustrated in Figure 1, the focal length is the distance from the camera lens to the image plane; since the pixel is not necessarily square, the focal length is divided by the pixel size in x and y, resulting in the variables f_x and f_y, respectively, expressing the values in pixels.

The extrinsic parameters are typically represented by a homogeneous transformation matrix, shown in Equation (3), and this was well explained by Briot and Khalil (2015) [10]. This contains the rotation matrix $R_{3\times3}$ and the translation vector $T_{3\times1}$, representing the camera's transformation in relation to the origin of the reference coordinate system in the desired scene.

$$D = \begin{bmatrix} R_{3\times3} & T_{3\times1} \\ 0_{1\times3} & 1 \end{bmatrix} \quad (3)$$

The distortion coefficients are the parameters used to describe the radial and the tangential distortion. They are represented as k_n and p_n, respectively. The most notable distortion model is the Brown–Conrady model [11].

The term calibration normally refers to methods of estimating the intrinsic parameters, distortion coefficients and extrinsic parameters.

Quaternions are another way to express rotations in the three-dimensional vector space. This method has a compact representation and has some mathematical advantages [12,13]. It is commonly used by the robotics industry because it is more mathematically stable and avoids the gimbal lock phenomenon, where two axes align and prevent the rotation in one dimension.

The robotic arm used in this work, the ABB IRB 4600, uses quaternions for the orientation of its TCP. Equation (4) [14] shows the conversion method from quaternions to Euler angles used in the homogeneous transformation matrix which was used in this work.

$$\begin{bmatrix} \phi \\ \theta \\ \psi \end{bmatrix} = \begin{bmatrix} \text{atan2}(2(q_0q_1 + q_2q_3), 1 - 2(q_1^2 + q_2^2)) \\ \text{asin}(2(q_0q_2 - q_3q_1)) \\ \text{atan2}(2(q_0q_3 + q_1q_2), 1 - 2(q_2^2 + q_3^2)) \end{bmatrix} \quad (4)$$

1.1. Point Cloud

Reprojecting the points in 3D using the intrinsic parameters in a pinhole model, as shown in Equation (1), with no distortion is carried out according to Equations (5)–(7). For an RGB-D device, Z is the depth information extracted from the depth frame, x is the column index and y is the row index. In these equations, a point is a 3D structure with (x, y, z) data representing a point in the camera's coordinate frame.

$$point.z = Z; \quad (5)$$

$$point.x = ((X - cx)/fx) * Z; \quad (6)$$

$$point.y = ((X - cy)/fy) * Z; \quad (7)$$

The origin of the coordinate system for each point cloud is one sensor of the camera, in this case, the left infrared sensor. To reconstruct the scene from multiple cameras or perspectives, the camera's positionin the scene, i.e., the global coordinate system, must be known.

The global coordinate system's origin is chosen according to the task. It can be the optical centre of one of the camera's sensors, for example. In this work, the chosen origin was the base coordinate system of an ABB IRB 4600 robot [15].

1.2. Related Work

The problem of scene reconstruction has been well studied in different scenarios. To solve the challenge of registering two or more point clouds together, the six degrees of freedom (DoF) transformation between them has to be found. This can be calculated using the point clouds by either selecting the matching points manually or using algorithms to find possible matching points. Another method is to calculate a known position for the sensors.

Chen and Medioni [16] and Besl and McKay [17] proposed one of the most widely used algorithms for the registration of 3D shapes, the iterative closest point (ICP). It tries to find the best match between the point clouds by finding the closest points from one point cloud to the other, then it determines the transformation matrix that minimises the distance between the points, and finally, it iterates until it converges. Due to fact that it relies on the data association of the points between point clouds, a good overlap is necessary for convergence.

Wang and Solomon [18] proposed a replacement of ICP called the deep closest point (DCP) method, which uses a three-step approach, first embedding the point clouds into a common space, then capturing co-contextual information in an attention-based module and finally using singular value decomposition to find the transformation matrix.

Aoki et al. [19] used the Pointnet [20] method with the Lukas and Kanade (LK) [21] algorithm to solve the registration problem.

Other methods include that of Stoyanov et al. [22], who used a three-dimensional normal distribution transform representation of the distance between the models, followed by a Newton optimisation, and that of Zhan et al. [23], who proposed an algorithm based on normal vector and particle swarm optimisation. These methods all rely on having a sufficient overlap between the point clouds to solve the problem. Moreover, these methods tend to be slow.

Performing an estimation of the position of the sensorprovides a more reliable transformation for the point clouds. Initially, work on camera calibration was focused on finding the intrinsic parameters of a single camera. Zhang [24] was the first to propose the solution to this challenge using a chessboard pattern, i.e., a planar target, through least-square approximation.

With the intrinsic parameters, in theory, it would be possible to calculate the pose of the camera with four coplanar points that are not collinear, but Schweighofer and Axel [25], discussing pose ambiguities, proved that two local minima exist, and proposed an algorithm to solve this problem.

Fiducial markers became popular for camera pose estimations, and several markers have been proposed, including point-based [26], circle-based [27], and square-based [28,29] markers, which can determine the pose using the four corners and have the ID in the middle of the marker.

Garrido-Jurado et al. [28] proposed a system with configurable marker dictionaries, specially designed for camera localization. The authors developed a marker generator, as well as an automatic detection algorithm. These ArUco markers form the basis of the charuco board used in this work.

Other publications have proposed ways to solve the problem of pose estimation using an arbitrary scene with texture [30–32]. These are not relevant to this project, since it was designed in a contained environment.

In this work, we propose and carry out a novel evaluation method for multiple camera perspectives. We used a charuco to identify and label the points on which the metrics were based, and calculated the errors of three different methods in order to carry out the transformation of the cameras to the global coordinate system.

2. Materials and Methods

This investigation was part of two projects called Meatable [33] and RoButcher (https://robutcher.eu, accessed on 15 February 2022). The projects involved researching the design and robotisation of a cell to process pig carcasses, called the Meat Factory Cell, and proposing a significant deviation from existing meat processing practices. The process was briefly described by Alvseike et al. (2018) [34,35]. The projects defined constraints on the data captured due to the specific data configuration of the scene and the application requirements.

Due to the inherent characteristics of the projects, precision was a key aspect of the system; otherwise, the cutting and manipulation of carcasses would be unsatisfactory or ineffective (e.g., the robot could cut too deep into the carcass, or not cut it at all).

For these projects, a bespoke machine called a carcass handling unit (CHU) was used, as shown in Figure 2. Its function was to grip a pig's carcass and present it to the robot, which would perform automated cutting to segment or dissect the parts.

Figure 2. Carcass handling unit in the RoButcher laboratory.

The robot needed to have a complete understanding of the environment in the 3D space. However, the camera had a limited angle of view. To solve this constraint, multiple camera perspectives had to be transformed to obtain a complete view of the scene.

The robotic arm had a camera attached to the tool centre point (TCP), as illustrated in Figure 3, to capture the data. It cycled the camera to six positions around the CHU, with two on the right, two on top, and two on the left side. With this configuration, almost 360° of the scene could be captured.

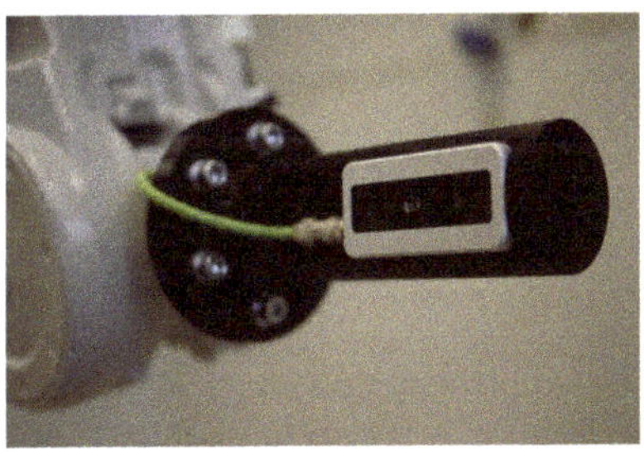

Figure 3. FRAMOS D415e Camera and its bespoke support

The cameras are referred to as left/right/up and front/back. The left/right is defined in relation to the CHU, not to the pig, which can be rotated once grabbed by the CHU. TThe left, right, front, and back sides of the CHU were defined as shown in Figure 4.

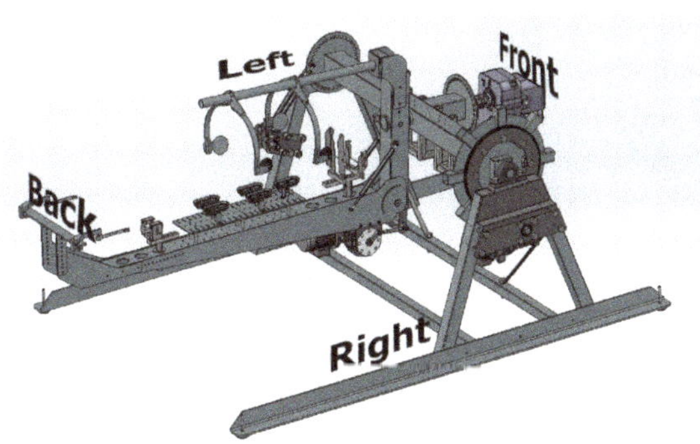

Figure 4. Carcass handling unit (CHU) with sides labelled.

The FRAMOS D415e has 3 image sensors (2 infrared (IR) and 1 RGB) and one IR laser projector. It calculates depth based on the disparity between the two IR sensors. Furthermore, it is stated by the documentation that the infrared cameras have no distortion; hence, all the distortion coefficients are 0.0 [9]. Furthermore, the cameras come from the factory calibrated with intrinsic parameters recorded in the memory and can be easily read using the appropriate Intel RealSense SDK library.

The Intel RealSense D415 was evaluated in regard to its the accuracy, performance, and precision by Lourenco and Araujo [36], with an RMSE accuracy distance for an analysis of 100 images of 0.07927, an average failed points ratio of 0.5414%, and average outliers (±10 cm) of 0.0667%, measured at 1 m. The repetitive pattern of the board also prejudices the depth calculation, and consequentially the point cloud, as illustrated in Figure 5.

Figure 5. Charuco board with noise and artefacts due to the repetitive pattern.

The data are recorded in a bag file (http://wiki.ros.org/Bags, accessed on 1 February 2022) using the SDK. The bagfile is a container file that contains the image and the depth frame, the intrinsic parameters for both sensors, and metadata information. These data are used to extract the parameters needed to execute the calibration.

The file-naming conventions used in the work used camera positions numbered from 0 to 5, and the sequential take number, with the following relation to the CHU:

- Left/Front: Cam 0
- Left/Back: Cam 1
- Up/Back: Cam 2
- Up/Front: Cam 3
- Right/Front: Cam 4
- Right/Back: Cam 5

2.1. Evaluation Metrics

To quantify the quality of the reconstruction in each method, the root-mean-square error (RMSE), shown in Equation (8), was used. The error used in the RMSE calculation was the Euclidean distance in three-dimensional space $\mathbb{R}^3$, and it was calculated after reprojecting the points in the point cloud, as shown in Equation (9).

$$rmse = \sqrt{\frac{\sum_{i=0}^{n} error^2}{n}} \tag{8}$$

$$error = \sqrt{(x_o - x_r)^2 + (y_o - y_r)^2 + (z_o - z_r)^2} \tag{9}$$

The depth information received from the camera was able to have an error of 2%, thus increasing the RMSE. However, as this affected all methods similarly, it did not interfere with the final comparison.

To measure the distance between the correspondents points, these points must be known, i.e, which point in one point cloud should match which point in the other point cloud. Since the point cloud from the RGB-D device did not contain information enabling us to identify which point was which, a method was proposed to label some points.

In this work we used OpenCV [37], as it is a stable and robust library for computer vision. ArUco markers were used as fiducial markers and a charuco board was used as the pattern board for the pose estimation, as proposed by Garrido et al. (2014) [28].

The ArUco module in OpenCV has all the necessary functions for pose estimation using the implemented charuco board [38].

2.2. Labelling Points

The method used to identify and label the points in order to enable error calculation was based on the unique ID of each ArUco marker. Each ID and the corners of the marker could be found on the board during the pose estimation process, as shown in Figure 6.

Figure 6. Charuco Board showing the identified ArUco markers.

Equation (10) shows the calculation of the point label values at points on the charuco board. Each corner receives a label to identify the specific point in the point cloud.

$$label = (i * 4) + j \tag{10}$$

where i is the ArUco ID and it is multiplied by 4 because every marker gives four corner points. j is the index of the ArUco in the vector where it is stored.

2.3. Methodology

The methodology used to measure the reconstruction error was based on a charuco cuboid with 3 faces, which was used to identify the points and calculate the RMSE. The chosen charuco boards were 300 mm by 200 mm, with a checker size of 20 mm and a marker size of 15.56 mm with 14 columns and 9 rows. This was manufactured using precision technology in aluminium by Calib.io (https://calib.io, accessed on 13 January 2022).

A charuco rectangular cuboid was built, as shown in Figure 7, using three boards.

2.4. Pre-Processing RGB-D Data

The first step was to improve the data by minimising the noise and the artefacts in the point cloud. To perform this task, some post-processing was applied to the captured frames.

In this work, the chosen post-processing methods were: alignment of colour and depth frame, sharpening of the colour frame and temporal and spatial filters applied to the frameset. Although the alignment and sharpening were always applied, the results were tested with both temporal and spatial filters enabled and disabled.

Figure 7. Charuco rectangular cuboid.

2.4.1. Alignment of Frames

The FRAMOS D415e camera had its origin reference in the left IR imager sensor and the RGB imager on the right side, as shown in Figure 8. As the sensors are different, an alignment between the colour image and the information was performed.

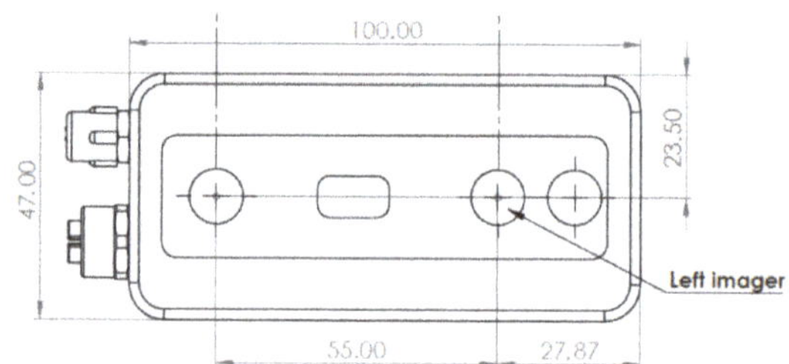

Figure 8. FRAMOS D415e origin reference system.

The alignment was performed in relation to the colour imager, i.e., the depth information was aligned to the colour frame. Since the alignment changes the relation of the pixels in the depth frame to match the colour sensor, the intrinsic matrix used to reproject the point cloud was also derived from the colour imager.

2.4.2. Sharpening Filter

Besides the alignment, a sharpening filter was applied to the colour image to improve the ArUco marker identification. The filter kernel used is shown in Equation (11).

$$\begin{vmatrix} -1 & -1 & -1 \\ -1 & 9 & -1 \\ -1 & -1 & -1 \end{vmatrix} \quad (11)$$

Figure 9a shows the charuco board before the sharpening filter was applied and the identified ArUco markers. Figure 9b shows the image after filtering, showing a significant improvement of the corners and edges. The sharpening filter improved the edges between the white and the black pixels of the board, increasing the number of identified ArUco markers.

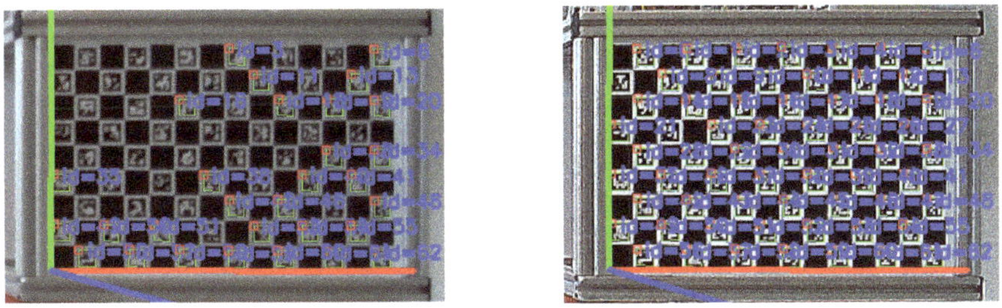

(a) (b)

Figure 9. Use of a sharpening filter to improve ArUco recognition. (a) No filter applied. (b) Filter applied.

2.4.3. Temporal and Spatial Filters

To diminish the noise artefacts of the data, two post-processing techniques were applied to the captured frames, a temporal filter and a spatial filter. Both filters were implemented in RealSense SDK and these have been well explained by Grunnet-Jepsen and Tong [39].

The temporal filter performs a time average of the frames to improve the depth calculation. The filter implements three parameters, those being the alpha, delta, and persistence filter. The alpha parameters represent the temporal history of the frames. The delta is a threshold to preserve edges, and the persistence filter tries to minimise holes by keeping the last known value for a pixel.

The spatial filter implemented in the SDK is based on the work of Gastal and Oliveira [40]. It preserves edges while smoothing the data. It takes four parameters, alpha and delta, which have the same function as in the temporal filter. This method also involves the filter magnitude, which defines the number of iterations, and hole filling parameters, which improve artefacts.

2.4.4. The Ground-Truth

After post-processing, the next step for the evaluation of the reconstruction is to generate a cuboid of reference points, in relation to the robot's base, to be used as ground-truth data. This was performed based on the first camera data. The charuco board was identified and the markers' corner was extracted from the image, as shown in Figure 6. Based on the cuboid geometry, the top reference was generated by rotating $-90°$ around the x-axis, followed by a rotation of 180° around the y-axis. Then, a translation of 280 mm, 187 mm and -190 mm in the x, y and z directions, respectively, was carried out, as shown in Equation (12). The backside was generated by rotating 180° around the y-axis and translating 280 mm and -200 mm in the x and z directions, respectively, as shown in Equation (13).

$$_{top}T^{b} = T^{0} *_{cub} T^{cam0} * trans_{a} * rot_{x^{-90}} * rot_{y^{180}} * (_{cub}T^{cam0})^{-1} \quad (12)$$

where $_{top}T^{b}$ is the transformation from the top of the charuco origin to the robot's base, T^{0} is the transformation from TCP to the base, $_{cub}T^{cam0}$ is the transformation from the first camera to the charuco, $trans_{a}$ is the translation, $rot_{x^{-90}}$ is the rotation in the x-axis, $rot_{y^{180}}$ is the rotation in the y-axis, and $(_{cub}T^{cam0})^{-1}$ is the inverse transformation from the first camera to the charuco.

$$_{r}T^{b} = T^{0} *_{cub} T^{cam0} * trans_{b} * rot_{y^{180}} * (_{cub}T^{cam0})^{-1} \quad (13)$$

where $_{top}T^{b}$ is the transformation from the top of the charuco origin to the robot's base, and the other terms have been explained in the previous paragraph.

The final reference cuboid can be seen in Figure 10 with the coordinate system of every side shown as a visual aid.

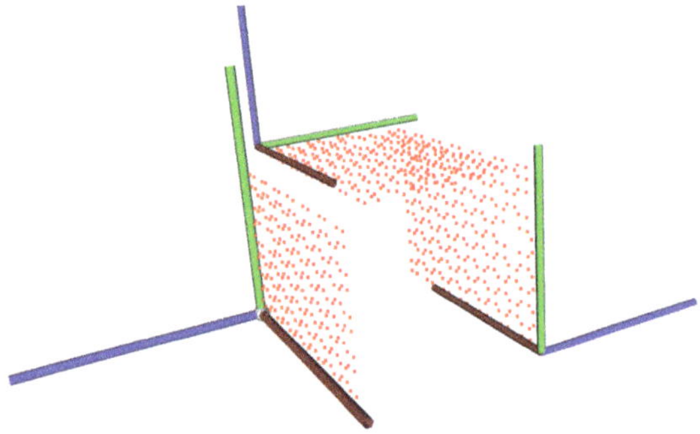

Figure 10. Cuboid with reference points and the coordinate frame of each board.

The camera positions make it impossible to see only one charuco board at a time. To solve this challenge, the board can be rotated when it can be seen in at least two camera positions, or two (or more) boards can be used with known geometries.

In this work, two reconstructions were performed using the charuco board and one using a robotic arm's position.

2.5. Charuco Cuboid

When using the charuco cuboid, each side is seen in two camera positions. Consequently, the pose estimation is based on the visible board. The defined origin for the reconstruction was set to be the first camera position. To obtain an optimal reconstruction, one must perform a rotation and translation on the data from camera positions, where the camera is facing a different board, based on the cuboid geometry, as explained in Section 2.4.4.

Notably, to calibrate all camera views (i.e., in all six possible positions), the charuco cuboid can remain stationary within the environment (i.e., it does not require repositioning for each camera).

2.6. Charuco Double-Sided Angled Board

Using a double-sided board follows the same principle as the cuboid but is more simple since the camera can see the same side in 4 positions, and the back of the board in the other two, as shown in Figure 11a–c.

To perform the transformation, the left and the upper cameras were transformed to the inverse matrix of charuco to camera transformation (${}_{cha}T^{cam}$), to make the charuco board origin the same for these cameras. Then, the right cameras had to perform a 180° rotation around the y-axis and a translation of 280 mm and 1.2 mm in the x and z directions, respectively, according to the board geometry.

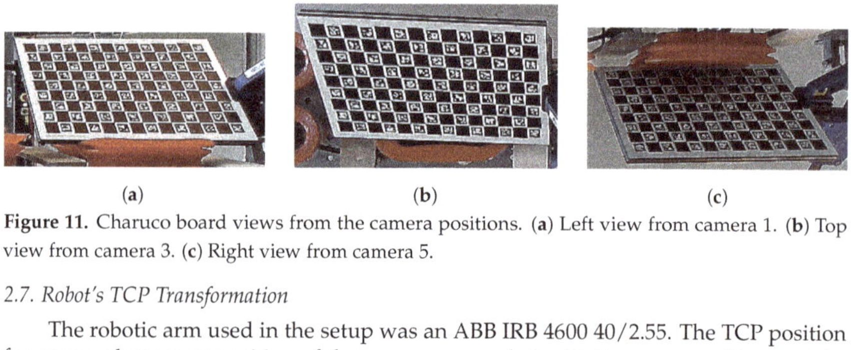

(**a**) (**b**) (**c**)

Figure 11. Charuco board views from the camera positions. (**a**) Left view from camera 1. (**b**) Top view from camera 3. (**c**) Right view from camera 5.

2.7. Robot's TCP Transformation

The robotic arm used in the setup was an ABB IRB 4600 40/2.55. The TCP position for every robot target position of the arm was recorded with the x, y, z position and the rotation $q1, q2, q3, q4$ in quaternions. A transformation from the camera's origin to the TCP was applied.

This transformation was calculated using the hand-eye calibration method proposed by Horaud and Dornaike [41]. A set of 30 different positions were defined for the calibration. The result was compared with the holder geometry, as seen in Figure 12.

The transformation found during calibration was the translation of -0.050 mm in the x direction, -0.0442 in the y direction, and 0.0780 in the z direction, and rotations of $90.9179°$around the x-axis, $1.1774°$around the y-axis, and $0.4859°$around the z-axis.

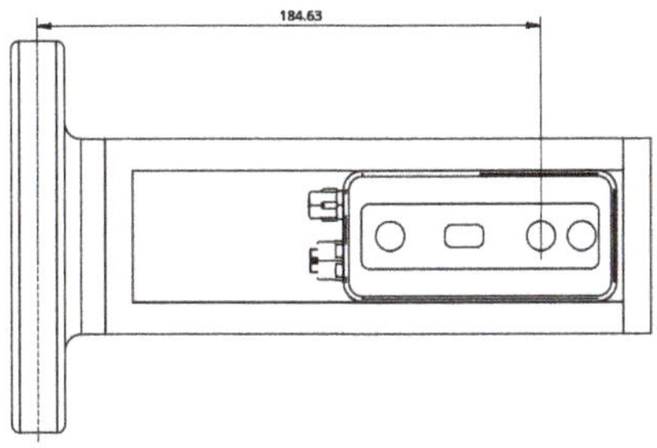

Figure 12. Camera holder CAD model with dimensions in TCP's z-axis.

Pair Matching

With the reference points ready, the transformation for each camera pose was solved based on the chosen methods explained above. The correspondent marker points were identified, and the metrics were calculated according to the explanation given in Section 2.1. Figure 13 shows arrows between the reference point and the board points, showing the correct identification of the pairs in which the error was calculated.

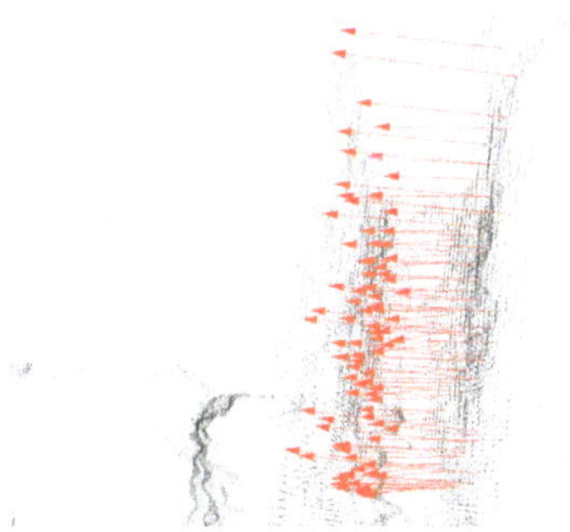

Figure 13. Arrows showing the correct correspondence between measured pair pointsbased on the labelling method.

3. Results

The results of the reconstruction follow the metrics explained in Section 2.1. The first step was to identify the two most accurate methods for the 3D reconstruction of the scene. For this, each system used 1188 points to calculate the reconstruction error for the six camera views.

Table 1 shows the mean absolute error (MAE), the mean square error (MSE), and the root-mean-square error (RMSE) of each system, expressed in millimetres. It is interesting to note that the MAE and the RMSE exhibit a large difference, meaning that there was not a high discrepancy between the errors.

Table 1. Summary results with MAE, MSE, and RMSE for one set of data.

Method	MAE	MSE	RMSE
Charuco cuboid	7.5812	5.7475×10^{-2}	6.8093
Double-sided Angled	10.12	0.1156	10.75
Robotic Arm	3.9888	1.5886×10^{-2}	3.7121

Based on these results, a program was written to run 20 sets of data for each method, with each set calculating the error between approximately 1188 points, depending on how many markers were found by the charuco algorithm.

3.1. Charuco Double-Sided Angled Board

The double-sided board with an angle (to enable four cameras to see it simultaneously) exhibited a poorer performance when compared with the other two methods. The angle applied to the board made it harder to identify the markers, as shown in Figure 14a,b, possibly making it less accurate than the other methods.

Visually, it was able to calculate the pose well in the 2D image, but there was a degradation in the RMSE, which in this case was 10.12 mm.

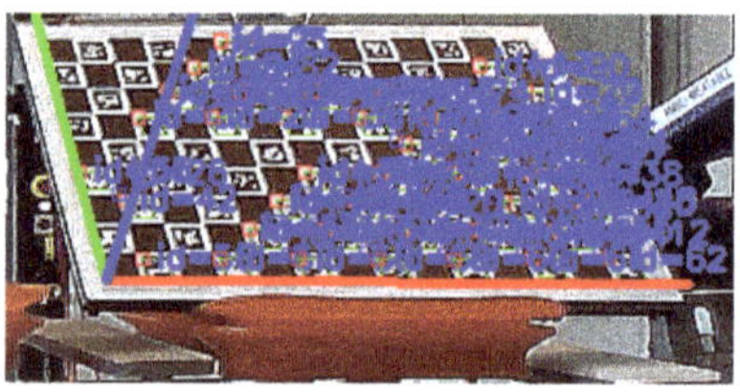
(**a**)

(**b**)

Figure 14. Charuco double-sided board marker detection for pose estimation. (**a**) Camera 1. (**b**) Camera 2.

Figure 15 shows a detail of the reconstructed image with the double-sided charuco board, where small gaps between the top and lateral panel can be observed.

Figure 15. Reconstruction view using charuco double-sided angled board.

Due to this method's less accurate 3D reconstruction compared to the other methods, we focused more on the other methods discussed below.

3.2. Cuboid vs. TCP Reconstruction Accuracy

The program was written in C++ using the Qt Library (https://www.qt.io, accessed on 8 February 2022), and it is available at GitHub (Please see "Data Availability Statement"), together with the dataset captured.

The program iterated over a directory with the files and calculated the RMSE, the mean squared error (MSE), and the mean absolute error (MAE) for both chosen methods. In addition, it assisted with the visualization of the point cloud.

The program was executed with both filtered and unfiltered data, as explained in Section 2.4.

3.2.1. Unfiltered Data

After running the program for the 20 sets, a total of 23,117 paired points were used to calculate the error values. Figure 16 shows a screenshot of the program after the calculation with the reconstruction of the point cloud for both methods, with the TCP method on the left and the cuboid on the right side.

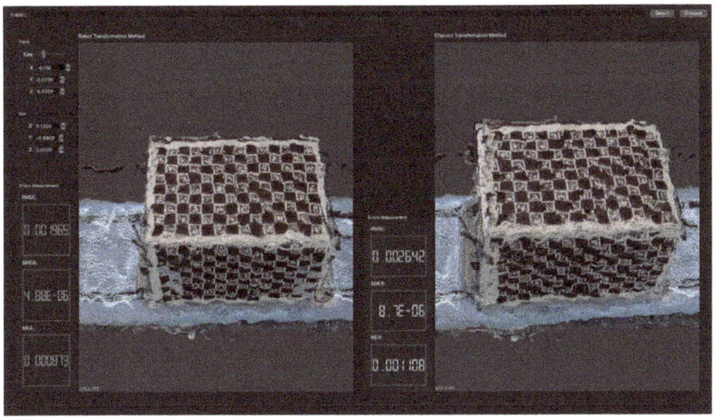

Figure 16. GUI showing the resulting 3D reconstruction using unfiltered data.

Table 2 summarises the results, showing a smaller error for the TCP method.

Table 2. Summary results for cuboid and TCP reconstruction with MAE, MSE, and RMSE for unfiltered data.

Method	MAE	MSE	RMSE
Charuco cuboid	1.10809	8.6994×10^{-3}	2.6417
Robotic Arm	0.8734	4.8808×10^{-3}	1.9651

3.2.2. Filtered Data

For the filtered data, a total of 23,603 paired points were used to calculate the error values. Figure 17 shows a screenshot of the program after the calculation.

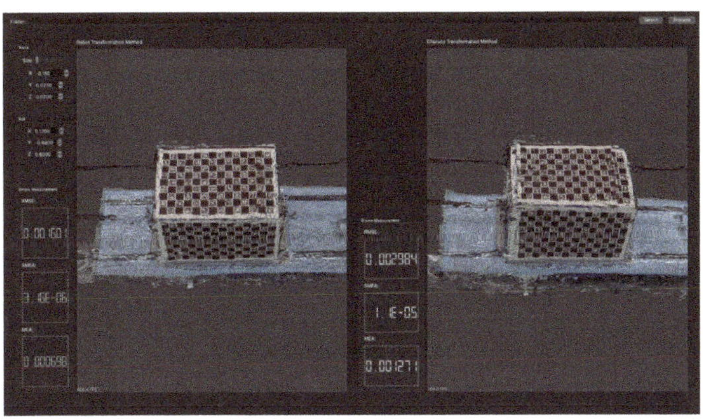

Figure 17. GUI showing the final result of the error measurement of the 3D reconstruction using filtered data.

Table 3 summarises the results obtained, again showing a smaller error for the TCP method.

Table 3. Summary results for cuboid and TCP reconstruction with MAE, MSE, and RMSE for filtered data.

Method	MAE	MSE	RMSE
Charuco cuboid	1.2708	1.0987×10^{-2}	2.9837
Robotic Arm	0.6983	3.1597×10^{-3}	1.7432

3.3. Charuco Cuboid

The reconstruction using the charuco cuboid had a low RMSE of 2.9837 mm for the filtered set, and an RMSE of 2.6417 mm for the unfiltered set, showing a better result for the unfiltered dataset.

3.4. Robot's TCP Transformation

The robotic arm ABB IRB 4600 had a payload of 40 kg and a reach of 2.55 m. It had a position repeatability of 0.06 mm, a path repeatability of 0.28 mm, and an accuracy of 1 mm [15]. With this precision, the reconstruction showed the best performance with an RMSE of 1.7432 mm for the filtered dataset and an RMSE of 1.9651 mm for the unfiltered. In this method, the filtered dataset showed an improved result in relation to the unfiltered dataset.

4. Discussion

The accuracy of the reconstruction is vital to robotics tasks when using depth information to generate trajectories for a robot in a large scene. The usage of depth stream to reconstruct the scene, as in Newcombe et al. [42], may not be appropriate if the system is time-sensitive. It is faster to move the robotic arm to positions at a high speed and then capture a frame than it is to cycle through at a slower speed to capture the data.

The algorithms currently available for point cloud registration, such as iterative closest point (ICP) [16,43] and its variants, as well as other methods that try to find the transformation between point clouds, use overlapping, which in this case is minimal due to the fact that the observed scene is large. Moreover, they are too slow to be used in a time-sensitive system.

The ABB IRB 4600 has a preset maximum speed of 7000 mm/s, and the axis speed varies from 175°/s (axis 1) to 500 °/s (axis 6). With these high speeds, the data acquisition and reconstruction can be sped up through a straightforward mathematical approach, such as the TCP transformation method. The idea is to have a fast and reliable transformation for the camera positions.

Further investigations could involve the study of how the trajectory generated for the TCP is affected by the errors in the reconstructions. Understanding this impact could help in developing a faster method without having repercussions on the desired output.

This work will assist in developing new designs and understanding one's options when responding to reconstruction challenges using multi-camera views and a robotic arm. It casts light on two different approaches and how to evaluate the reconstruction of the 3D scene. However, it is not an exhaustive study on the topic.

5. Conclusions

In this study, we evaluated three pose estimation methods for RGB-D (3D) cameras in the context of 3D point cloud reconstruction. We proposed a method to identify the point pairs, and performed error measurements. Using a charuco cuboid, we identified and labeled the corner of the ArUco markers in order to create a pairwise set of points to measure the error values.

Two of our developed methods used the charuco board to estimate the pose of the cameras, whereas the third method used the robots' TCP position to calculate and transform the point clouds to reconstruct the scene. The data were captured with a fixed position and at a distance of around 1 m from the board in each direction.

This distance reduces the resolution of the charuco board and consequently reduces the camera's ability to recognise markers on it. To improve the recognition of markers, a sharpening filter was applied. Further filters were used to mitigate artefacts in the point cloud, due to the stereo vision system used by the camera.

Two methods, the cuboid and TCP methods, provided a good reconstruction with low RMSE values, taking into account the often noisy nature of point clouds derived from RGB-D cameras. The double-sided angled board had the highest RMSE, and it was excluded from further assessments. The increased error was assumed to be a result of the combination of the angle and distance from the camera. The robot's TCP position had a high accuracy, exhibiting the best overall performance.

Author Contributions: Conceptualisation, I.d.M.E., M.M., O.S. and A.P.; methodology, I.d.M.E.; software, I.d.M.E.; validation, A.M.; formal analysis, I.d.M.E.; investigation, I.d.M.E.; writing—original draft preparation, I.d.M.E.; writing—review and editing, A.M.; visualisation, I.d.M.E.; supervision, A.M. and P.J.F.; project administration, A.M.; funding acquisition, A.M. All authors have read and agreed to the published version of the manuscript.

Funding: This work was partially supported by the Research Council of Norway through the funding to the "MeaTable—Robotised cells to obtain efficient meat production for the Norwegian meat industry" project no. 281234. The work is also, in part, supported by the EC H2020 project "RoBUTCHER" grant agreement no. 871631.

Data Availability Statement: The source code can be downloaded from https://github.com/imesper/Calibration.git (accessed on 10 March 2022). For the data please contact the author at ian.esper@nmbu.no.

Acknowledgments: We wish to acknowledge the help provided by Tommy Jonsson and Haris Hadzic from Byte Motion, and Håvard Bakke from RobotNorge for providing us with very valuable knowledge on the matter.

Conflicts of Interest: The authors declare no conflict of interest. The funders had no role in the design of the study; in the collection, analyses, or interpretation of data; in the writing of the manuscript, or in the decision to publish the results.

Abbreviations

The following abbreviations are used in this manuscript:

TCP	Tool centre point
RMSE	Root mean squared error
MAE	Mean absolute error
SME	Squared mean error
CHU	Carcass handling unit

References

1. Edl, M.; Mizerák, M.; Trojan, J. 3D laser scanners: History and applications. *Acta Simul.-Int. Sci. J. Simul.* **2018**, *4*, 4–5. [CrossRef]
2. Munoz, F.I.I.I.; Israel, F.; Munoz, I. Global Pose Estimation and Tracking for RGB-D Localization and 3D Mapping. Ph.D. Thesis, Université Côte d'Azur, Nice, France, 2018.
3. Carfagni, M.; Furferi, R.; Governi, L.; Santarelli, C.; Servi, M.; Uccheddu, F.; Volpe, Y. Metrological and critical characterization of the intel D415 stereo depth camera. *Sensors* **2019**, *19*, 489. [CrossRef]
4. Zollhöfer, M.; Stotko, P.; Görlitz, A.; Theobalt, C.; Nießner, M.; Klein, R.; Kolb, A. State of the Art on 3D Reconstruction with RGB-D Cameras. In Proceedings of the EUROGRAPHICS 2018, Delft, The Netherlands, 16–20 April 2018; Volume 37.
5. Hartley, R.; Zisserman, A. *Multiple View Geometry in Computer Vision*; Cambridge University Press: Cambridge, UK, 2004. [CrossRef]
6. Ferrara, P.; Piva, A.; Argenti, F.; Kusuno, J.; Niccolini, M.; Ragaglia, M.; Uccheddu, F. Wide-angle and long-range real time pose estimation: A comparison between monocular and stereo vision systems. *J. Visual Commun. Image Represent.* **2017**, *48*, 159–168. [CrossRef]

7. Harwin, S.; Lucieer, A.; Osborn, J. The Impact of the Calibration Method on the Accuracy of Point Clouds Derived Using Unmanned Aerial Vehicle Multi-View Stereopsis. *Remote Sens.* **2015**, *7*, 11933–11953. [CrossRef]
8. Intel. Intel ®RealSense™ D400 Series Product Family Datasheet. 2019. Available online: https://dev.intelrealsense.com/docs/intel-realsense-d400-series-product-family-datasheet (accessed on 1 March 2022).
9. Keselman, L.; Woodfill, J.I.; Grunnet-Jepsen, A.; Bhowmik, A. Intel RealSense Stereoscopic Depth Cameras. In Proceedings of the IEEE Computer Society Conference on Computer Vision and Pattern Recognition Workshops, Honolulu, HI, USA, 21–26 July 2017; pp. 1267–1276.
10. Briot, S.; Khalil, W. Homogeneous transformation matrix. *Mech. Mach. Sci.* **2015**, *35*, 19–32. [CrossRef]
11. Brown, D. Decentering Distortion of Lenses. *Photom. Eng.* **1966**, *32*, 444–462.
12. Pervin, E.; Webb, J.A. *Quaternions in Computer Vision and Robotics*; Carnegie-Mellon Univ Pittsburgh Pa Department of Computer Science: Pittsburgh, PA, USA, 1991. [CrossRef]
13. Mukundan, R. Quaternions: From Classical Mechanics to Computer Graphics, and Beyond. In Proceedings of the 7th Asian Technology Conference in Mathematics, Cyberjaya, Malaysia, 7–17 December 2002; pp. 97–105.
14. Ben-Ari, M. *A Tutorial on Euler Angles and Quaternions*; Weizmann Institute of Science: Rehovot, Israel, 2014.
15. ABB. *Product Specification—IRB 4600*; 2009. Available online: https://search.abb.com/library/Download.aspx?DocumentID=3HAC032885-001&LanguageCode=en&DocumentPartId=&Action=Launch (accessed on 1 March 2022).
16. Yang, C.; Medioni, G. Object modelling by registration of multiple range images. *Image Vis. Comput.* **1992**, *10*, 145–155. [CrossRef]
17. Besl, P.J.; McKay, N.D. A Method for Registration of 3-D Shapes. *IEEE Trans. Pattern Anal. Mach. Intell.* **1992**, *14*, 239–256. [CrossRef]
18. Wang, Y.; Solomon, J. Deep Closest Point: Learning Representations for Point Cloud Registration. In Proceedings of the 2019 IEEE/CVF International Conference On Computer Vision (ICCV), Seoul, Korea, 27 October–2 November 2019; pp. 3522–3531.
19. Aoki, Y.; Goforth, H.; Srivatsan, R.A.; Lucey, S. PointNetLK: Robust & Efficient Point Cloud Registration using PointNet. In Proceedings of the IEEE Computer Society Conference on Computer Vision and Pattern Recognition, Long Beach, CA, USA, 15–20 June 2019; IEEE Computer Society: Washington, DC, USA, 2019; pp. 7156–7165.
20. Qi, C.R.; Su, H.; Mo, K.; Guibas, L.J. PointNet: Deep learning on point sets for 3D classification and segmentation. In Proceedings of the 30th IEEE Conference on Computer Vision and Pattern Recognition, CVPR 2017, Honolulu, HI, USA, 21–26 July 2017; Institute of Electrical and Electronics Engineers Inc.: New York, NY, USA, 2017; pp. 77–85. [CrossRef]
21. Lucas, B.D.; Kanade, T. An iterative image registration technique with an application to stereo vision. In Proceedings of the IN IJCAI81, Vancouver, BC, Canada, 24–28 August 1981; pp. 674–679.
22. Stoyanov, T.; Magnusson, M.; Lilienthal, A.J. Point set registration through minimization of the L2 distance between 3D-NDT models. In Proceedings of the IEEE International Conference on Robotics and Automation, Guangzhou, China, 11–14 December 2012; pp. 5196–5201. [CrossRef]
23. Zhan, X.; Cai, Y.; Li, H.; Li, Y.; He, P. A point cloud registration algorithm based on normal vector and particle swarm optimization. *Meas. Control* **2019**, *53*, 265–275. [CrossRef]
24. Zhang, Z. A flexible new technique for camera calibration. *IEEE Trans. Pattern Anal. Mach. Intell.* **2000**, *22*, 1330–1334. [CrossRef]
25. Schweighofer, G.; Pinz, A. Robust Pose Estimation from a Planar Target. *IEEE Trans. Pattern Anal. Mach. Intell.* **2006**, *28*, 2024–2030. [CrossRef] [PubMed]
26. Dorfmüller, K.; Wirth, H. Real-Time Hand and Head Tracking for Virtual Environments Using Infrared Beacons. In Proceedings of the Modelling and Motion Capture Techniques for Virtual Environments, Geneva, Switzerland, 26–27 November 1998; Springer: Berlin/Heidelberg, Germany, 1998; Volume 1537, pp. 113–127. [CrossRef]
27. Naimark, L.; Foxlin, E. Circular data matrix fiducial system and robust image processing for a wearable vision-inertial self-tracker. In Proceedings of the International Symposium on Mixed and Augmented Reality, ISMAR 2002, Darmstadt, Germany, 30 September–1 October 2002; pp. 27–36. [CrossRef]
28. Garrido-Jurado, S.; Muñoz-Salinas, R.; Madrid-Cuevas, F.J.; Marín-Jiménez, M.J. Automatic generation and detection of highly reliable fiducial markers under occlusion. *Pattern Recognit.* **2014**, *47*, 2280–2292. [CrossRef]
29. Atcheson, B.; Heide, F.; Heidrich, W. CALTag: High Precision Fiducial Markers for Camera Calibration. In *Vision, Modeling, and Visualization*; The Eurographics Association: Reims, France, 2010.
30. Barazzetti, L.; Mussio, L.; Remondino, F.; Scaioni, M. Targetless Camera Calibration. *Remote Sens. Spat. Inf. Sci.* **2011**, *38*, 8. [CrossRef]
31. Cavallari, T.; Golodetz, S.; Lord, N.A.; Valentin, J.; Prisacariu, V.A.; Di Stefano, L.; Torr, P.H.S. Real-Time RGB-D Camera Pose Estimation in Novel Scenes using a Relocalisation Cascade. *IEEE Trans. Pattern Anal. Mach. Intell.* **2019**, *42*, 2465–2477. [CrossRef] [PubMed]
32. Wang, C.; Guo, X. Feature-based RGB-D camera pose optimization for real-time 3D reconstruction. *Comput. Vis. Media* **2017**, *3*, 95–106. [CrossRef]
33. Mason, A. *Robotised Cells to Obtain an Efficient Meat Production for the Norwegian Meat Industy*; Animalia/NMBU. 2018. Available online: https://prosjektbanken.forskningsradet.no/project/FORISS/281234 (accessed on 2 February 2022).
34. Alvseike, O.; Prieto, M.; Torkveen, K.; Ruud, C.; Nesbakken, T. Meat inspection and hygiene in a Meat Factory Cell—An alternative concept. *Food Control* **2018**, *90*, 32–39. [CrossRef]

35. Alvseike, O.; Prieto, M.; Bjørnstad, P.H.; Mason, A. Intact gastro-intestinal tract removal from pig carcasses in a novel Meat Factory Cell approach. *Acta Vet. Scand.* **2020**, *62*, 1–5. [CrossRef] [PubMed]
36. Lourenco, F.; Araujo, H. Intel realsense SR305, D415 and L515: Experimental evaluation and comparison of depth estimation. In Proceedings of the16th International Joint Conference on Computer Vision, Imaging and Computer Graphics Theory and Applications (VISIGRAPP 2021), Online, 8–10 February 2021. [CrossRef]
37. Bradski, G. The OpenCV Library. *Dr. Dobb's J. Softw. Tools* **2000**, *25*, 120–123.
38. OpenCV. OpenCV: Detection of ChArUco Corners. Available online: https://docs.opencv.org/4.x/df/d4a/tutorial_charuco_detection.html (accessed on 15 January 2022).
39. Grunnet-Jepsen, A.; Tong, D. *Depth Post-Processing for Intel® RealSense™ D400 Depth Cameras*; New Technologies Group, Intel Corporation. 2021. Available online: https://dev.intelrealsense.com/docs/depth-post-processing (accessed on 15 January 2022).
40. Gastal, E.S.; Oliveira, M.M. Domain transform for edge-aware image and video processing. In Proceedings of the ACM SIGGRAPH 2011, Vancouver, BC, Canada, 7–11 August 2011; Volume 30. [CrossRef]
41. Horaud, R.; Dornaika, F. Hand-eye Calibration. *Int. J. Robot. Res.* **1995**, *14*, 195–210. [CrossRef]
42. Newcombe, R.A.; Hilliges, O.; Molyneaux, D.; Kim, D.; Davison, A.J.; Kohli, P.; Shotton, J.; Hodges, S.; Fitzgibbon, A. KinectFusion: Real-Time Dense Surface Mapping and Tracking. In Proceedings of the 2011 10th IEEE International Symposium on Mixed and Augmented Reality, Basel, Switzerland, 26–29 October 2011.
43. Segal, A.V.; Haehnel, D.; Thrun, S. Generalized-ICP. In *Robotics: Science and Systems*; University of Washington, Seattle, WA, USA, 2009; Volume 2.

applied sciences

MDPI

Article

Towards a Multi-Perspective Time of Flight Laser Ranging Device Based on Mirrors and Prisms

Luka Pogačnik * and Marko Munih

Faculty of Electrical Engineering, University of Ljubljana, 1000 Ljubljana, Slovenia; marko.munih@fe.uni-lj.si
* Correspondence: luka.pogacnik@fe.uni-lj.si

Abstract: This paper investigates the feasibility of redirecting the field of view (FOV) of a light-based time-of-flight (ToF) ranging device, commonly known as a pulsed lidar, using fixed mirrors and prisms for possible future use in robotics. The emphasis is on configurations where the FOV redirection element is positioned beyond the ranging device's dead zone. A custom made direct ToF ranging device with time-over-threshold (TOT)-based walk error compensation was used to evaluate the effects of the FOV redirecting optics on range measurement accuracy and precision. The tests include redirecting the FOV with a clean prism with anti-reflective (AR) coating on its legs, as well as with a regular and a first surface mirror in both a clean and dusted state. The study finds the prism to be unsuitable due to parasitic reflections, which ruin the ranging data. The clean mirrors were found to have no noticeable effect on ranging accuracy. When they are dusty, mirrors introduce a negative measurement error. This effect is the most pronounced when a mirror is positioned toward the end of the partial dead zone of the ToF rangefinder, but loses influence as the mirror is moved farther away. The error is attributed to the parasitic reflection off dust on the mirror, which reduces the time of detection of the pulse reflected off the real target, and interferes with the walk error compensation by widening the detected pulse.

Keywords: time-of-flight; lidar; field of view redirection; passive redirection; clean mirror; dusted mirror; prism

Citation: Pogačnik, L.; Munih, M. Towards a Multi-Perspective Time of Flight Laser Ranging Device Based on Mirrors and Prisms. *Appl. Sci.* **2022**, *12*, 7121. https://doi.org/10.3390/app12147121

Academic Editors: Luis Gracia and Carlos Perez-Vidal

Received: 1 June 2022
Accepted: 12 July 2022
Published: 14 July 2022

1. Introduction

In recent years, we have observed a change in the way robots are used in the industry. The workspaces of robots and humans are beginning to overlap, which is allowed by the introduction of collaborative robots that are designed so that they cannot harm humans. Inherent safety mostly stems from monitoring the forces acting on the robot and stopping the motion when the measured values exceed the calculated expected values. In addition to using robots designed not to harm humans, speed and separation monitoring (SSM) can also be used [1]. This approach is becoming increasingly common. In its simplest form, SSM is implemented using an industrial 2D lidar scanner to monitor the robot's surroundings. Further improvements can be achieved by monitoring the robot's environment with a depth camera [2,3]. This way, it is possible to determine not only where the operator is standing, but also whether the operator is reaching in the direction of the robot, allowing for even more nuanced SSM. A problem with observing the robot's surroundings form a single field of view (FOV) is occlusions. To obtain a better representation of the robot's surroundings, multiple depth cameras, placed in different locations, can be used [4,5]. Another approach that may be used to solve occlusion is to mount a complementary depth camera or laser time of flight (ToF) proximity sensors onto a robot [6,7]. A similar safety system can be implemented by distributing individual proximity sensors over the robot without using an external depth camera [8–10]. To reap the full benefits of such a configuration, the position and orientation of the sensors must be considered to achieve maximum data throughput while minimizing blind spots [11], as sensors may interfere with one another's operation when their FOVs overlap. Readings from the sensors can be used to implement a continuous SSM response-slow down or stop when an obstacle is detected-or to implement an advanced collision avoidance algorithm [12,13].

All of the cited work using ToF proximity sensors relied on mounting them all over the robot or introduced massive dead zones around the robot where targets could not be detected. We believe that imitating the effect of distributing a plurality of individual ToF sensors across the robot is a promising approach. This could be achieved by mounting a centralized range finder with multiple FOVs on the end of the robot's segment and distributing its FOVs across the joint by the means of mirrors or prisms. The use of a centralized multi-perspective ranging device could allow us to achieve better performance or lower system cost, since some common components could be reused for all channels.

The redirecting of lidar's FOV with a mirror, located within its dead zone, is a common practice in modern lidar scanners [14]. Much less research has been conducted for configurations where the redirection element may be located anywhere within the ranging device's FOV. Notable exceptions are researches where mirrors were used to expand the FOV of a scanning lidar [15–17] and the study that achieved a simulated multiple depth camera setup for 3D object capture by reflecting parts of the depth camera's FOV in [18]. The listed studies proved that mirrors can be used for redirecting light-based ToF ranging device's FOV but did not go into detail about how the mirror affects measurement characteristics. In order to pursue the direction of equipping a robot with a multi-perspective light ToF ranging device with redirected FOVs, this paper analyzes the effects of mirrors and prisms in expected real conditions at different distances from the ranging device on its ranging performance. The direct ToF measurement approach with time over threshold (TOT)-based walk error compensation was selected as it does not require a long integration time, and therefore allows a higher measurement rate and introduces only negligible motion blur [19].

This paper is organized as follows: after Section 1, the introduction, Section 2 presents the underlying principles of ToF measurement. It explains how ToF measurements work, the difference between direct and redirected measurement, the construction of the ranging device used in the experiments, and the measurements setup, as well how the data were gathered and processed. Section 3 presents the experimental results in the form of graphs for different measurement configurations. That includes direct and reflected ranging with the ranging device at different distances from the FOV redirection optics. Results are interpreted in Section 4. Conclusions and ideas for further work based on said interpretations are gathered in Section 5.

2. Materials and Methods

In this section, basic functioning and measurement procedures are discussed. This includes the explanation of the difference between laser ToF-based ranging in direct and redirected configurations, how the ranging device that was used in later experiments works, a description of the measurement setup, and the description of the measurement procedure.

2.1. Direct and Redirected Time of Flight Measurement

The oath of light from a laser ToF ranging device may be direct or redirected before reaching the target surface. A direct measurement is common in laser rangefinders, where only one dimensional ranging is required. Redirected measurements are common in lidar scanners, where the FOV of a single device is redirected in multiple directions. The distance between the device and the reflecting surface that is used for redirection of light in this configuration is constant and never changes. Redirection may be carried out in one or two dimensions. The former results in 2D measurements of the surroundings, while the later generates a point cloud on the surface of observed objects. Both approaches have been thoroughly explored and are standard in industrial environments. Even though redirection using mirrors is the most common, beam steering using prisms can be used instead, such as those detailed in [20,21].

The comparison of how a direct or redirected laser ToF ranging works is depicted in Figure 1. Figure 1a shows direct operation, Figure 1b shows how light FOV can be redirected by using a reflective surface within the devices dead zone, which is commonly used in industrial lidar scanners, and lastly, Figure 1c shows the operation where the reflecting surface lies beyond the device's dead zone. Boxes marked Rx and Tx represent light transmitters (lasers) and receivers (photodiodes), respectively. The system's dead zone ranges from the transmitter/receiver to the dashed line, and partially dead zone extends up to the solid line. While operation between the first two configurations should not be critically different, the third one may pose problems, as any imperfections on the reflecting surface may cause reflections that will interfere with the ranging device's operation.

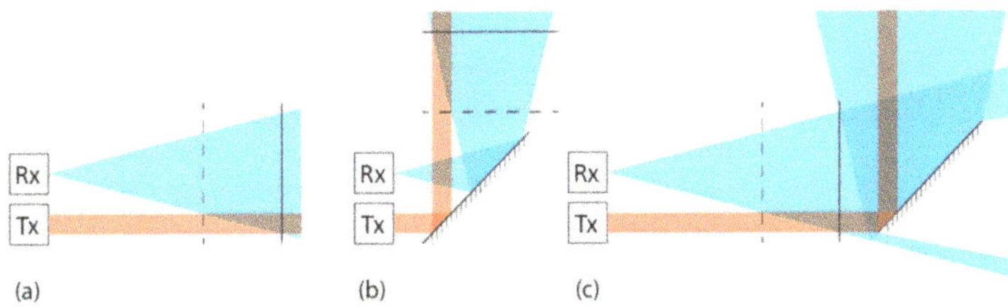

Figure 1. Diagrams of different areas of operation for light-based ToF distance measuring devices are shown for (**a**) direct method, (**b**) redirected method, where reflecting surface is entirely within the ranging device's dead zone, and (**c**) for when the reflecting surface is located beyond the ranging device's dead zone. Boxes marked Rx and Tx represent light transmitters (lasers) and receivers (photodiodes), respectively. The system's dead zone ranges from the transmitter/receiver to the dashed line, and the partially dead zone extends up to the solid line.

This article addresses a similar approach to redirecting the FOV as described previously, with the major difference that the device used to redirect the FOV is stationary and immovable, with respect to the ranging device. The idea is to transmit a laser beam along the length of the robot's joint and redirect it perpendicularly outwards when it hits the redirection device. The light, reflected from a target, then travels essentially the same path back to reach the receiver. Multiple optical paths are established along the robot's segment, leading to a multi-perspective view of the surroundings. A multi-perspective view might provide a significant advantage in monitoring the robot's surroundings, enabling enhanced safety features. The principle of operation is depicted in Figure 2, where orange bands represent individual light paths.

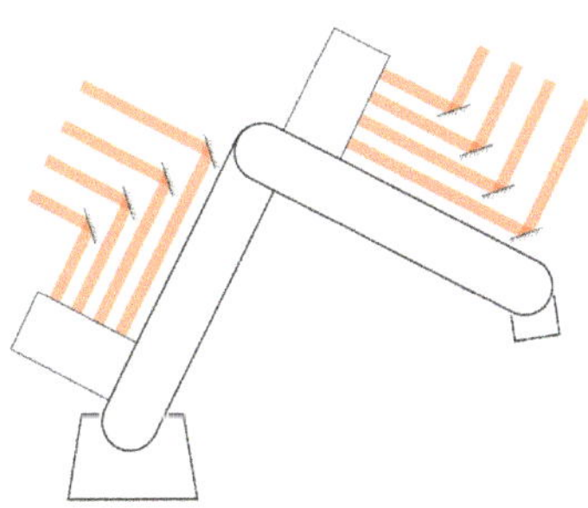

Figure 2. The principle of operation of a robot proximity monitoring system based on one centralized ranging device per robot's segment and individual fields of view redirected by mirrors. The robot segment lengths are usually from 0.2 m to 0.8 m, and the detection range from the robot's segment to the potential obstacle should be in compliance with the SSM mode of operation, as specified in ISO/TS 15066:2016. In a real application, the field of view redirection optics would be located as close to the robot's surface as possible.

2.2. Laser ToF Ranging Device Device

Many commercial ToF-based ranging devices are available on the market as either off the shelf products or as development kits. Those offerings, however, do not offer much insight into what is going on with the measurement. For that reason, we opted to use a custom direct time of flight ranging device, which not only allows us to measure the distance to the target, but also to indirectly observe the amplitude of the return signal. This section discusses the laser ToF ranging device, its construction, and principles of operation.

The laser ToF ranging device that was used has a 905 nm laser diode as a light source and an avalanche photodiode (APD) connected to a transimpedance amplifier (TIA), with the gain of 25 kV/A on the receiving end. The receiver APD is also masked with an optical low pass filter to eliminate much of the ambient illumination. The model used is the Optolite IR filter that cuts off light with wavelengths below approximately 750 nm. For outdoor use, a band pass filter would be more appropriate but is unnecessary for indoor use, as modern lights produce little to no light in the IR spectrum [22]. Both the transmitters and the receiver are equipped with focusing optics to achieve a narrow FOV of 5°. Time of flight measurement is taken care of by a pair of TDC7200 time to digital converters (TDC). This model has a resolution of 55 ps (8.8 mm) and a standard deviation of 35 ps (5.2 mm). This is improved by using a running average of eight samples. Both TDCs start the measurement on the positive front of the signal to emit a light pulse but their stop signals are different. One TDC stops the measurement when the current through the laser diode, as measured on a shunt resistor, exceeds the threshold. The other TDC measures both the time of detecting the reflected pulse and its width. This is achieved by passing the signal from the TIA through a comparator with a selected threshold voltage, and then passing the resulting waveform through a pulse generator for both the falling and rising front from the first comparator. The described setup eliminates measurement error that comes from the temperature dependent delay between triggering the transmitted pulse and the laser actually turning on. The system can select between three different illumination channels, all of which trigger the same stop signal for the reference TDC. Ranging is controlled with an external microcontroller. A simplified schematic is shown in Figure 3. The dashed line encircles components of each of the three equal transmit channels and the dotted line encircles components on the laser ranging device's printed circuit board (PCB), which comprise the analog front end. Throughout the experiment, only one channel 1, which is located 2 cm from the receiver, was used. The remaining two channels will allow the same ranging device to be used in later research.

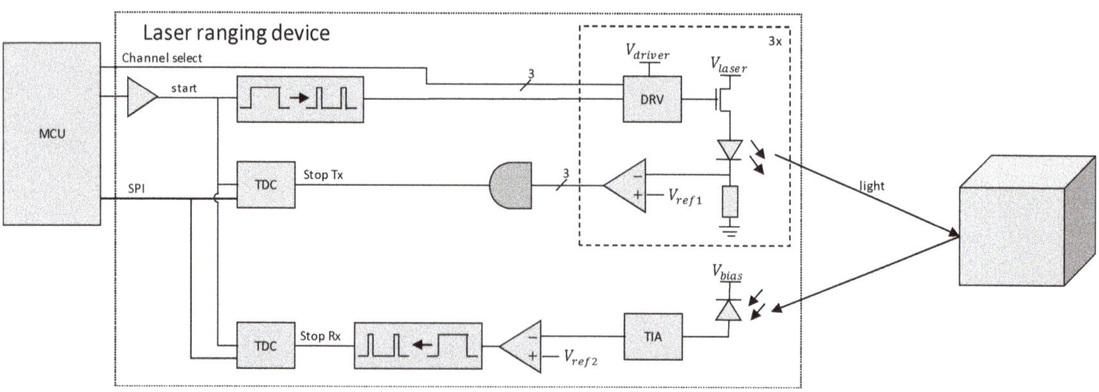

Figure 3. Block diagram of laser ToF ranging device. Dotted line encircles the crucial components that are mounted on the same PCB. The dashed line encircles the components of one of three transmit channels, constructed in the same way.

The microcontroller, which controlled the ranging, was an ATmega328P on an Arduino Nano development board. It was used for generating appropriate transmit pulses and controlling TDCs, which includes initialization, arming, and reading measurement data. The later was sent over to the PC with minimal additional processing carried out on the microcontroller.

2.3. Walk Error

When transmitted light is focused in a parallel beam, the amount of the returned light follows an inverse square law with increasing distance to the target. Total received power for diffuse targets may be estimated using a modified radar Equation (1) [23], where P_{Rx} is received power, G_{Rx} is receiver gain, G_{Tx} transmitter gain, P_{Tx} power of the transmitted signal, r is the radius of receiving lens (πr^2 is its area), and d the distance between measuring device and the target. 2π in the equation represents the solid angle, at which the light is reflected. In this case it is assumed that the target behaves as a perfect point source that reflects light evenly in the hemisphere towards the light source.

$$P_{Rx} = \frac{G_{Tx} G_{Rx} P_{Tx} \pi r^2}{2\pi d^2} \tag{1}$$

The equation does not take into account the target's albedo, which provides information on what portion of light is reflected off of the target. This factor could be added to the equation as a multiplicative factor in the fraction's numerator. The possibility for the target to be non-Lambertian is omitted as well. In such a case, another orientation dependent factor that addresses the material's physical reflective properties should be added to the equation. Since real materials rarely have a uniform reflection pattern, Equation (1) is only good to obtain a rough estimate for diffuse targets, which remains valid so long as the target's orientation in regard to the illuminator and the observer is fixed. To summarize the main takeaways, the further the target or lower its albedo, the less signal is captured by the receiving diode. For real signals with finite rise and fall times, the reduction in amplitude effectively shrinks the pulse width after the comparator. Figure 4a shows the return signal, captured after TIA with the target at different distances. It should be noted that TIA is saturated for signal responses from targets at 80 cm or closer. To make the effect more pronounced, a dark target with low albedo was selected when gathering data for Figure 4a. Traces of different colors represent return signals for measurements with targets at different distances. Which color corresponds to which distance can be read from the legend. For strong return signals, the pulse width can exceed 150 ns, which is notably wider than the transmitted pulse, set at 100 ns.

The returned signal's amplitude plays a huge role in when and if the signal will cross the detection threshold, which introduces the walk error. Ideally, neighboring traces in Figure 4a would cross the threshold, set at 1.68 V, 667 ps apart, the same time it takes the light to traverse the roundabout distance between subsequent measurements. In our case, spread between traces was greater than this and varied by signal strength, as observed in Figure 4a. Walk error is a systemic error, inherent to ToF based ranging devices [24]. Some laser ToF ranging devices are constructed so that they do not need additional walk error compensation. More often than not, that is due to having an analog compensation built into the receiver or having a design that is inherently less sensitive to walk error. When that is not the case, walk error compensation has to be implemented to obtain accurate measurements. Common methods for measurement correction for systems with PIN photodiode or APD receivers are based on the signal's time over threshold (TOT) or rise time. The TOT-based compensation function relies on detected pulse width [25], and the later one is based on signal's rise time, measured by the time it takes for the received signal to rise from first threshold to the next [26]. The ranging device used in this article used TOT-based walk error compensation.

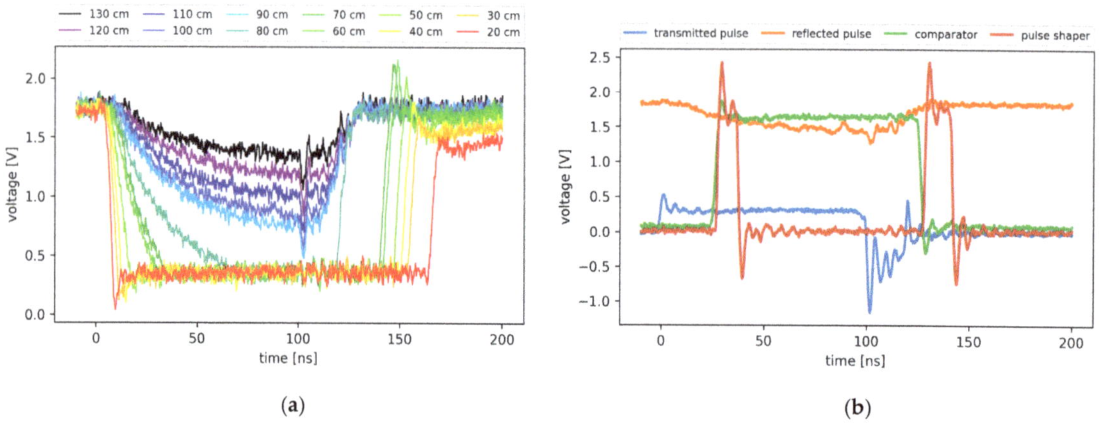

Figure 4. (**a**) Reflected pulse after TIA at different target distances, measured after receiving TIA. All pulses are synchronized to the rising edge of the transmitted pulse. (**b**) Pulse shape of the transmitted pulse, as measured on the shunt resistor (blue), reflected pulse, as measured after TIA (orange), pulse shape after the first comparator (green, and the pulse pair after the pulse shaper (red). The comparator's and the pulse shaper's waveforms were attenuated by 50%.

The measured pulse width is a systemic parameter that only depends on the target reflectivity and distance to the target, assuming that the ranging device's operating parameters, such as supply voltage and temperature, are fixed. TDC7200, the TDC used on the selected ranging device, has the ability to measure time from the start pulse to multiple stop pulses if a sufficient amount of time passes between individual pulses. However, all stop pulses must have the same direction (rising or falling edge). By transmitting a 100 ns light pulse and converting the returned signal into two 10 ns pulses, one for rising and one for the falling edge of the return signal after passing through a comparator, we ensure that the condition for measuring time to multiple stop pulses is met. Figure 4b shows the pulse shape of the transmitted pulse, as measured on the shunt resistor (blue), reflected pulse, as measured after TIA (orange), pulse shape after the first comparator (green, and the pulse pair after the pulse shaper (red). The comparator's and the pulse shaper's waveforms were attenuated by 50% to make the graph more comprehendible.

To obtain the compensation function, multiple measurements were taken off of a spinning color wheel with varying albedo, set at multiple different known distances. Both the time to the first stop signal and TOT were recorded for all samples. The difference between measured distance and the expected value (error) versus TOT was plotted, and a piecewise linear (PWL) function was fitted to the resulting data cluster, as shown in Figure 5. The yellow points correspond to single measurements, the green points are for the data captured with running averages on eight samples, and the solid blue line represents the manual PWL fit. Beyond calibration range, linear continuation with the same coefficient as the last segment within the calibration range is assumed.

After the time from the transmitted to received pulse and the corresponding TOT are measured, walk error compensation is applied. As observed in Figure 5, a wider pulse means that less time is subtracted from the first stop signal and vice versa. Walk error compensation has to be applied either immediately on the microcontroller or afterwards on the master computer. The later was chosen for our experiments, as we wanted to obtain raw as well as compensated measurements.

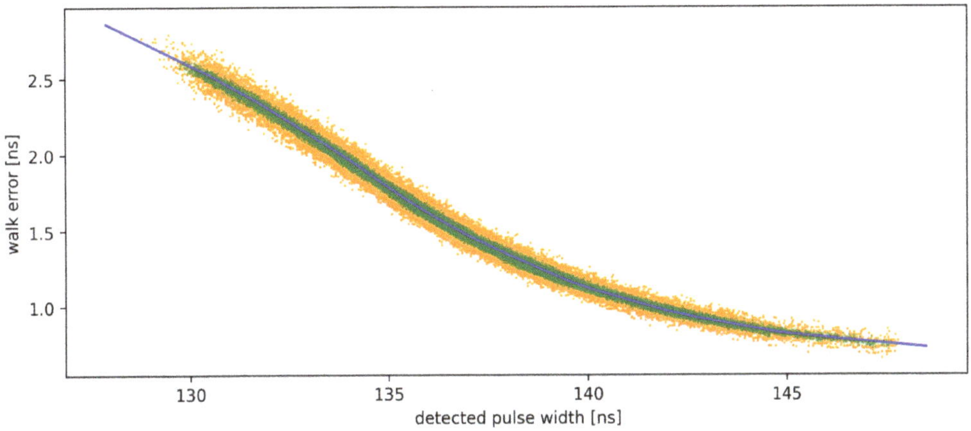

Figure 5. Walk error vs. pulse width for raw data (yellow), average of eight samples (green) and piecewise linear function fit (blue).

2.4. Measurement Setup

This section discusses the tools and procedures used in experiments for this article. It is split into three subsections. In them, measurement hardware, measurement software, and measurement procedure are discussed respectively.

2.4.1. Measurement Hardware

Ranging was carried out using a direct ToF laser ranging device, which was thoroughly discussed in the previous sections. The device was validated in advance to achieve sub centimeter accuracy on the range of zero to eighty centimeters for a range of targets with different albedos. The ToF ranging device was mounted on a stand that allowed for some minor height and angle adjustments.

For ranging through an FOV redirection device (mirrors and prisms), a special jig was devised. The jig comprises a linear actuator, and an acrylic snap-in plate, fixedly mounted on a solid particle board baseplate. They were mounted in a way that their axis intersected at a right angle. The FOV redirection device was mounted on a pedestal at the desired height, angle and position. Mirrors were angled at 45° to both the linear rail and the optical axis of the ranging device. This was the same with the prism, when observing the hypotenuse. Minor adjustments could be made before fixing the stand in place to ensure the desired orientation. That means that experiments were repeatable so long as the redirection device was not changed. The measurement setup is shown in Figure 6.

Optics for redirecting FOV were Thorlabs' first surface protected aluminum mirror (ME2S-G01), a generic aluminum back coated mirror from Duratool (S5003), and Knight Optical's right angle prisms made of N-BK7 glass with AR coated legs (PTK4001). It was known from our previous experiments that using an uncoated prism does not work, as too much light is reflected off of the first surface back into the receiver diode; thus, we did not test that configuration. Mirrors and prism were large enough such that the entire laser beam was safely within the bounds of mirrors or the prism. The target was covered with multiple layers of white office paper. During the aligning process, we made sure that the entire transmitted pulse was hitting the target.

To control for errors rising from variable system parameters, such as temperature and varying voltage levels, all power supplies were left to warm up and then left powered for the entire duration of experiments. The ranging device was left ranging for at least one day before the experimental data were gathered. The walk error compensation curve was recorded on a warmed up system, which was not powered off until all data were collected.

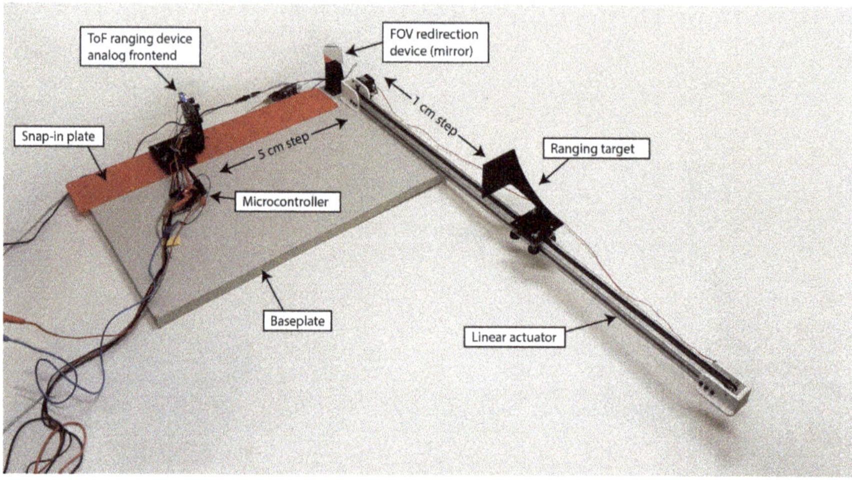

Figure 6. Measurement setup as used in the experiments. Photographed configuration depicts a state during testing of effects of a clean first surface mirror on ranging where distance between the ranging device and the mirror was 40 cm.

2.4.2. Measurement Software

The ranging device was set in autonomous ranging mode and was continuously sending measurement data to the master PC through a USB serial data link. The microcontroller, connected to the ranging device, was only used to initiate the measurement, gather time of flight data, calculate the running average, and send data over to the master PC with minimal additional processing. Walk error compensation and time to distance conversions were carried out on the PC.

The linear actuator waited for commands to go to the home position, detected by a microswitch, or to move to a desired location. Once the command was executed, the PC was informed that linear actuator is finished with moving.

The master PC was running a python script that took care of setting the linear actuator and saving the ranging data. Data were captured on a span of 800 mm in 10 mm increments. For each set distance, 128 raw samples were taken and saved to a file.

2.4.3. Measurement Procedure

For each FOV redirection optic and its configuration, we conducted the same experimental procedure. The ranging device was mounted on the furthermost position (55 cm) and the target on the linear actuator was set to its furthermost distance as well. Optics, mounted onto a stand, were placed into the ranging device's FOV such that the laser spot was redirected onto a correct spot on the target. It was ensured that all the transmitted light was hitting either the mirror or the prism, and that all reflected light was hitting the target. Following that, autonomous ranging commenced.

The linear actuator was instructed to move to the home position (5 cm from the optics) and the system started ranging. The target moved one centimeter further away each time, for the total of 80 set distances with 128 raw samples collected at each one.

When data for the entire range were collected, the ToF ranging device was manually moved one position (5 cm) closer to the redirection optics, and the procedure was repeated for a total capture of eleven datasets per redirection optics configuration.

For the experiments with clean optics, they were thoroughly cleaned using nonwoven cotton wipes (Webril Handi-Pads) damped in isopropanol. When the solvent had evaporated, an optical inspection was conducted. If any streaking or imperfections were observed, the cleaning process was repeated.

2.5. Measurement Data Processing

Data for each optics configuration and each distance between them and the ranging device were saved in individual files. After measurements were completed, data for optics configuration were pooled together and individual datasets were plotted with Python and Matplotlib.

For error plots, all 128 raw samples for each set distance were averaged and the expected measurement was subtracted. Standard deviation was calculated for each distance individually as well. Raw and compensated measurement data were plotted against non-discarded measurement number. Plots for same optics configuration but different distances between them and the optics were overlaid.

3. Results

This section contains measurement results for different configurations of ranging. First, measurement data for a direct measurement method are provided for reference. This is followed by the measurement data for setups where light was reflected by a mirror, and the data for ranging by redirecting FOV by a prism.

3.1. Reference Measurements

The characteristics of the ranging device without FOV redirection were first measured. Results of those measurements are shown in Figure 7. Figure 7a shows uncompensated and walk error compensated measurements with orange and blue lines, respectively. Figure 7b shows uncompensated errors plotted against distance, and Figure 7c shows TOT measurement in regard to distance to the target. Lastly, Figure 7d shows the measurement errors between expected and measured distance in blue, and calculated standard deviations in orange.

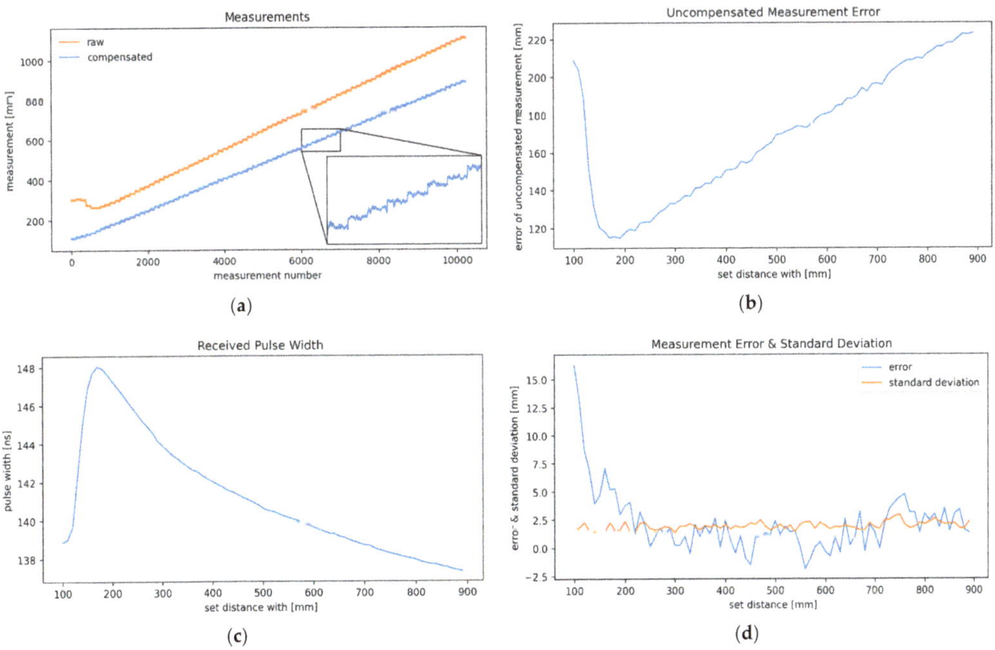

Figure 7. (**a**) shows raw and compensated measurements, (**b**) depicts errors of uncompensated measurements, (**c**) reflected pulse width, and (**d**) errors and standard deviations of compensated measurements in blue and orange, respectively.

3.2. Reflecting Off of Mirrors

Measurement errors for the configurations where light is reflected off of a mirror are shown in Figure 8. Figure 8a shows data for a clean first surface mirror, Figure 8b for the dusted first surface mirror, Figure 8c for clean regular mirror, and Figure 8d for a dusted regular mirror. Different traces show measurement errors for configurations with different distances between the ranging device and mirror, ranging from 5 cm, shown in blue, to 55 cm, shown in magenta, and other configurations in 5 cm increments in between.

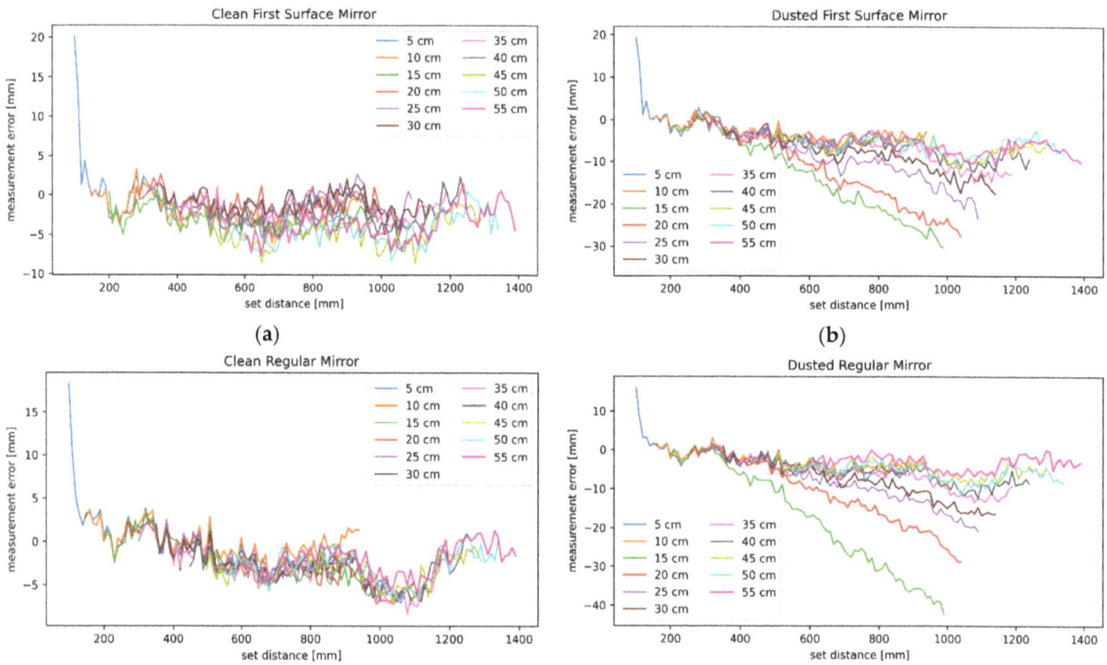

(c) (d)

Figure 8. Measurement error measured on a clean (**a**) and dusted (**b**) protected first surface mirror, and clean (**c**) and dusted (**d**) regular mirror. Different traces correspond to configurations with different distances between ranging device and mirror's reflecting surface. Blue trace corresponds to the configuration where this distance is 5 cm, and magenta trace for when it is 55 cm all other distances are displayed in the legend key.

The blue (5 cm) traces in all graphs in Figure 8 show a significant positive measurement error in the first few data points. Even larger errors appear in setups with a dusted mirror at 15 cm or further from the ranging device. There the measurement error grows increasingly negative as the target is moved away. The effect is less noticeable when the mirror is positioned further from the ToF ranging device and becomes unnoticeable beyond 40 cm.

To help explain the disparity between the error plot for clean and dusted mirror, Figure 9 shows the shape of the reflected pulse on a system with a dusted mirror without a detectable target in sight (blue) and with a weak reflection from a target (orange). The signal is sampled after the TIA. One can notice that some signal is present even when there should be none (ranging at an infinitely distant target).

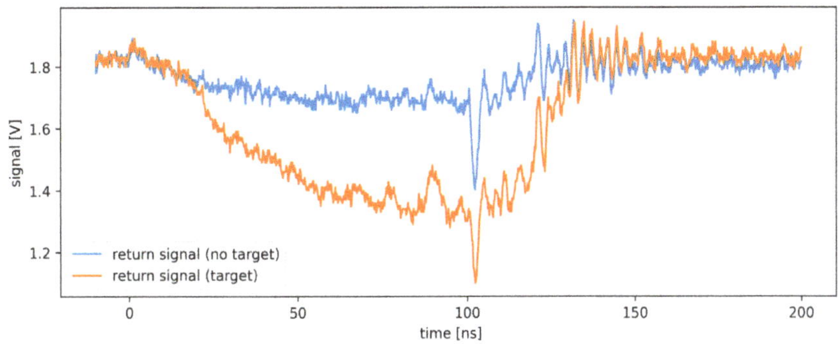

Figure 9. Shape of return signal for the configuration with a dusted protected silver first surface mirror. Return signal with no target in the detection range of the ranging device is shown in blue. Return signal with a poorly reflective target in sight is shown in orange. The root cause of the presence of the unwanted signal, and the consequent waveform's changes are presented in the discussion section, where Figure 9 is referenced.

3.3. *Reflecting from a Prism*

Measurements with the AR-coated prism behaved quite differently than the mirrors. Figure 10a shows the compensated measurement error for the entire suite of measurements, from 5 cm to 55 cm distance between the ranging device and the prism. Figure 10b shows the compensated measurement data for the same dataset. Measurements for all but the first configuration (blue, 5 cm) plateau at some distance.

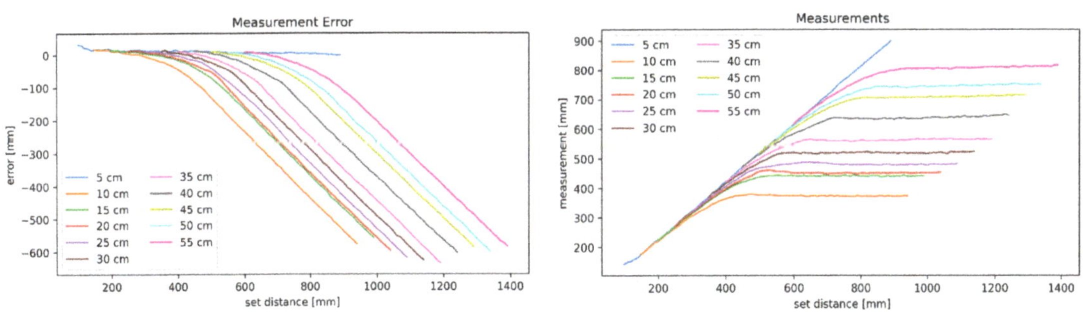

(**a**) (**b**)

Figure 10. (**a**) Measurement error for the entire suite of measurements, from 5 cm to 55 cm distance between the ranging device and prism. (**b**) Compensated measurement data for the same datasets where x-axis is the measurement number. Which line corresponds to which configuration can be read from the legend. The explanation for the presented data is in the discussion section, where Figure 10 is referenced.

4. Discussion

This section discusses the meaning of measured data. Firstly, the direct measurement method is discussed where the importance of walk error compensation is illustrated. Following that, redirecting FOV with mirrors and reasons for the trend of measurement becoming increasingly negative for only some distances between the ranging device and dusted mirror is explained. Our results are also compared to similar data from preexisting studies. Lastly, we take a look at what measurement results for redirecting light with a prism suggest.

The importance of walk error compensation is observed in Figure 7a. It shows uncompensated and compensated measurements with orange and blue lines, respectively. It can be observed that the uncompensated measurements show a target at significantly greater distance than expected. Beyond the partially dead zone, the error increases with distance, as best illustrated in Figure 7b. It shows the uncompensated measurement error versus the set distance. Figure 7c shows the reflected pulse's width, which decreases with distance beyond the partially dead zone, extending to approximately 17 cm beyond the sensor. In this zone, not all transmitted light is in the receiver's FOV, as illustrated in Figure 1. Figure 7d provides a reference for further ranging methods, as it shows the measurement error and the standard deviation of measurements for the direct ranging. A trend observed here and in all remaining measurements is that standard deviation increases slightly with distance but does not vary greatly between different configurations, as it can be observed in Table 1. It provides an overview of the maximal measurement standard deviations. The values are gathered only from data points where the total distance between the ranging device and target is greater than 13 cm, as to eliminate the data points just beyond the ranging device's dead zone, which shows a large deviation from the expected values even in a direct measurement configuration.

Table 1. Maximal measurement standard deviations in various configurations.

Distance to Mirror (cm)	Maximum Measurement Standard Deviation (mm)				
	First Surface Mirror (Clean)	First Surface Mirror (Dusted)	Regular Mirror (Clean)	Regular Mirror (Dusted)	without Mirror
/	/	/	/	/	3.0
5	2.8	2.7	2.7	3.7	/
10	2.7	2.7	2.6	2.9	/
15	3.0	3.4	2.9	3.7	/
20	2.8	4.5	2.8	3.3	/
25	3.0	3.5	3.3	2.9	/
30	2.8	3.6	3.2	3.0	/
35	3.8	3.5	3.6	3.4	/
40	3.3	3.4	3.2	3.5	/
45	3.7	3.5	3.8	3.7	/
50	3.5	3.9	3.8	3.9	/
55	3.4	4.1	3.6	3.9	/
min	2.7	2.7	2.6	2.9	/
max	3.8	4.5	3.8	3.9	/
avg	3.2	3.5	3.2	3.4	/

A more or less direct comparison between direct measurements and the ones reflected by a mirror at 5 cm can be made. This is because the minimal distance between the ranging device and target in both cases was 10 cm in both cases. By comparing Figure 8a to Figure 8c (FOV redirected by a clean first surface and a regular mirror), any differences are within the margin of measurement error and show outstanding similarity in shape. When Figure 7d (direct measurement) is taken into consideration as well, some minor differences can be observed among the three datasets, but those are once again below the designed accuracy threshold. The rest of the measurements with FOV redirected by a clean mirror were still very similar to the ones obtained directly in the overlapping measurement range.

Worse results were expected for the configurations with dusted mirrors. When comparing measurement results for clean mirrors to their dusted counterparts in Figure 8b and Figure 8d for the dusted first surface and regular mirror, respectively, essentially no differences can be observed up to the 300 mm mark. From that point on, traces for configurations with the mirror set closer than 150 mm from the ranging device continue the same way as in Figure 8a or Figure 8b, but the remaining traces trend downwards. The effect diminishes as the distance between the mirror and the ranging device increases. Measurements with the mirror 40 cm or further from the ranging device seem effectively unaffected by the phenomenon. An explanation for the effect can be found by examining the reflected pulse's shape, as shown in Figure 9. When ranging at an infinity, no return signal should be observed, but when ranging at an infinity while redirecting the FOV with a dusted mirror, some weak return signal can be detected (blue trace in Figure 9). This is a reflection off the dust on the mirror's surface. It is below the detection threshold on its own, but when ranging on a target is carried out with such a setup, reflections off the target and dust combine into a single pulse (orange trace in Figure 9), resulting in a measurement error. It comprises two effects, moving the ranging result towards lower measured time, and pulse widening. In a direct time of flight ranging system with TOT-based walk error compensation, those effects work in opposing directions. The first lowers the compensated ranging result, and the other increases it. This study finds that moving the measurement towards lower time has more influence on the ranging, which results in a negative measurement error. As the target moves further away, the desired reflection grows dimmer, but the reflection off the dust remains unchanged. This means that the error will grow in magnitude, which can be observed in error plots in Figure 8b,d. The effect will be the most pronounced when the mirror is positioned around the end of the partially dead zone, as there the most parasitic reflection will be coupled into the receiver. As observed in Figure 8b,d, the amount of light from the parasitic reflection is negligible when the mirror is positioned at 10 cm from the ranging device or closer. With increasing distance, the parasitic reflection's intensity decreases alongside its influence on the measurement. Therefore, ranging with a dusted mirror 40 cm from the target or further is hardly affected by the parasitic reflection.

The data about the effects of mirrors on ranging characteristics, presented in Figures 7 and 8, cannot be directly compared to other researches. In spite of that, some relatable data are found by taking multiple separate researches into account simultaneously. It has been found in [18] that redirecting Microsoft Kinect 2.0's FOV with mirrors results in an RMS error ranging between 4 mm and 8 mm for raw data, which was reduced to roughly 2.5 mm to 7 mm after the enhancement algorithm was applied. The data have to be read from the provided graph, as no numerical data were provided. It can be compared to the results of [27], where a camera's raw performance was evaluated. The mean measurement error has been found to be within 10 mm on the entire advertised measurement range of the Microsoft Kinect 2.0 (4.5 m). Even though the data cannot be compared directly because the distance to the object in the first study was not disclosed, it can be observed that the performance of the camera with reflected FOV was close to the study with direct measurements. This confirms our findings that clean mirrors do not have a large influence on the camera's performance. Another study that can be used for comparison used a movable mirror to redirect a portion of VLP-16's FOV [16]. It obtained measurement errors that ranged from comparable to significantly worse than the reference measurements obtained by direct measurements in [28]. Five of the twelve provided error measurements where FOV was redirected by a mirror that exceeded the advertised measurement tolerance of the lidar, which was in itself twice the expected amount according to [28]. A large contributor to this discrepancy may have been the tolerances for mirror positioning, as we found measurement results for direct ranging and ranging through a clean mirror to be effectively indistinguishable from one another. The alternative explanation is that the mirror used in [16] was not clean in at least some parts of the experiments.

In our tests, the use of prisms has proven to be more problematic than mirrors. When a prism is placed within the ranging device's dead zone, its influence is rather small, although still observable. This can be observed in Figure 10a, where the blue trace (prism 5 cm from the ranging device) trends towards increasingly negative measurement errors as the target moves away. Since the prism's first leg was in the ranging device's dead zone, this reflection likely did not contribute much. Therefore, internal reflections are the likely culprit.

When the prism is positioned 10 cm from the ranging device or further in our setup, both of its legs rest within the ranging device's partially dead zone. This means that reflections off of both surfaces may be coupled straight back into the receiving photodiode. When the target is nearby, the signal's strength is strong enough to dominate the parasitic reflection; thus, the measurements are accurate. As the target moves further away, its influence slowly diminishes until eventually, the parasitic reflections take over entirely When this happens, the measurements plateau, which can be clearly observed in Figure 10b

The discussed results were obtained with a clean prism with AR coated legs, which is the best-case scenario for redirecting FOV of a light ToF ranging device with a prism. With that in mind, it can be concluded that prisms are inappropriate for redirecting the FOV when the reflecting surfaces are beyond the ranging device's dead zone.

5. Conclusions

In this article, we explored the influence of first surfaces and regular mirrors, as well as AR-coated prisms, on the performance of a light ToF ranging device with TOT-based walk error compensation. This was carried out for eleven distances to mirrors or prisms, while the target was moved in 1 cm increments. It has been determined that there is no significant difference between direct measurements and the ones where the FOV was with clean mirrors. Furthermore, there was no significant difference between a first surface or a regular mirror. When the mirror is dusted, the measurements are essentially the same as with direct measurements, so long as the reflecting surface is within the partially dead zone or closer. Beyond that, light reflecting off dust introduces a negative measurement error that is more pronounced when the target is further away. This effect is more noticeable when the mirror is closer to the ranging device. Beyond a certain distance, between the redirection optics and ranging device, dust's influence on the measurements becomes undetectable. A method for detecting and compensating for the reflection off lint on a mirror might be explored in the future to determine whether redirecting FOVs with mirrors shows potential for use in industrial environments.

This study has also examined the usability of prisms for redirecting FOV of a direct ToF ranging device. As long as the entire prism is well within the ranging device's partially dead zone or closer, a strong correlation between the set and measured distance can be observed, even though parasitic reflections have some influence on measurements. When the prism is placed further, parasitic reflections quickly exceed and overpower the strength of reflections off the target when the target moves away, at which point the measurements plateau. This was observed in the best-case scenario for the prisms—a clean prism with AR-coated legs. The results of this study show that the prisms in general are not a feasible option for achieving the set goals, unless transmitted and received light paths could be separated entirely.

Author Contributions: Conceptualization, M.M. and L.P.; methodology, M.M. and L.P.; software, L.P.; validation, M.M. and L.P.; formal analysis, M.M. and L.P.; investigation, L.P.; resources, M.M. and L.P.; data curation, L.P.; writing—original draft preparation, L.P.; writing—review and editing, M.M.; visualization, L.P.; supervision, M.M.; project administration, M.M.; funding acquisition, M.M. All authors have read and agreed to the published version of the manuscript.

Funding: The authors acknowledge the financial support from the Slovenian Research Agency (research core founding No. P2-0228).

Institutional Review Board Statement: Not applicable.

Informed Consent Statement: Not applicable.

Data Availability Statement: All data are an integral part of the corresponding author's PhD thesis.

Conflicts of Interest: The authors declare no conflict of interest.

References

1. *ISO/TS 15066:2016*; Robots and Robotic Devices—Collaborative Robots. International Organization for Standardization (ISO): Geneva, Switzerland, 2016.
2. Rosenstrauch, M.J.; Pannen, T.J.; Krüger, J. Human Robot Collaboration—Using Kinect v2 for ISO/TS 15066 Speed and Separation Monitoring. *Procedia CIRP* **2018**, *76*, 183–186. [CrossRef]
3. Chemweno, P.; Torn, R.-J. Innovative Safety Zoning for Collaborative Robots Utilizing Kinect and LiDAR Sensory Approaches. *Procedia CIRP* **2022**, *106*, 209–214. [CrossRef]
4. Fischer, M.; Henrich, D. 3D Collision Detection for Industrial Robots and Unknown Obstacles Using Multiple Depth Images. In *Advances in Robotics Research: Theory, Implementation, Application*; Springer Science & Business Media: Berlin, Germany, 2009; pp. 111–122, ISBN 978-3-642-01212-9.
5. Pasinetti, S.; Nuzzi, C.; Lancini, M.; Sansoni, G.; Docchio, F.; Fornaser, A. Development and Characterization of a Safety System for Robotic Cells Based on Multiple Time of Flight (TOF) Cameras and Point Cloud Analysis. In Proceedings of the 2018 Workshop on Metrology for Industry 4.0 and IoT, Brescia, Italy, 16–18 April 2018; pp. 1–6.
6. Strong, M.; Escobedo, C.; Roncone, A. Volumetric Data Fusion of External Depth and Onboard Proximity Data For Occluded Space Reduction. *arXiv* **2021**, arXiv:2110.11512.
7. Rybski, P.; Anderson-Sprecher, P.; Huber, D.; Niessl, C.; Simmons, R. Sensor Fusion for Human Safety in Industrial Workcells. In Proceedings of the 2012 IEEE/RSJ International Conference on Intelligent Robots and Systems, Vilamoura, Portugal, 7–12 October 2012; pp. 3612–3619.
8. Tsuji, S.; Kohama, T. Proximity Skin Sensor Using Time-of-Flight Sensor for Human Collaborative Robot. *IEEE Sens. J.* **2019**, *19*, 5859–5864. [CrossRef]
9. Cheng, G.; Dean-Leon, E.; Bergner, F.; Rogelio Guadarrama Olvera, J.; Leboutet, Q.; Mittendorfer, P. A Comprehensive Realization of Robot Skin: Sensors, Sensing, Control, and Applications. *Proc. IEEE* **2019**, *107*, 2034–2051. [CrossRef]
10. Tsuji, S.; Kohama, T. Proximity and Contact Sensor for Human Cooperative Robot by Combining Time-of-Flight and Self-Capacitance Sensors. *IEEE Sens. J.* **2020**, *20*, 5519–5526. [CrossRef]
11. Adamides, O.A.; Modur, A.S.; Kumar, S.; Sahin, F. A Time-of-Flight On-Robot Proximity Sensing System to Achieve Human Detection for Collaborative Robots. In Proceedings of the 2019 IEEE 15th International Conference on Automation Science and Engineering (CASE), Vancouver, BC, Canada, 22–26 August 2019; pp. 1230–1236.
12. Escobedo, C.; Strong, M.; West, M.; Aramburu, A.; Roncone, A. Contact Anticipation for Physical Human–Robot Interaction with Robotic Manipulators Using Onboard Proximity Sensors. In Proceedings of the 2021 IEEE/RSJ International Conference on Intelligent Robots and Systems (IROS), Prague, Czech Republic, 27 September–1 October 2021; IEEE: Piscataway, NJ, USA, 2021; pp. 7255–7262.
13. Ding, Y.; Wilhelm, F.; Faulhammer, L.; Thomas, U. With Proximity Servoing towards Safe Human-Robot-Interaction. In Proceedings of the 2019 IEEE/RSJ International Conference on Intelligent Robots and Systems (IROS), Macau, China, 3–8 November 2019; pp. 4907–4912.
14. Royo, S.; Ballesta-Garcia, M. An Overview of Lidar Imaging Systems for Autonomous Vehicles. *Appl. Sci.* **2019**, *9*, 4093. [CrossRef]
15. Matsubara, K.; Nagatani, K. Improvement in Measurement Area of Three-Dimensional LiDAR Using Mirrors Mounted on Mobile Robots. In *Field and Service Robotics*; Ishigami, G., Yoshida, K., Eds.; Springer: Singapore, 2021; pp. 85–99.
16. Matsubara, K.; Nagatani, K.; Hirata, Y. Improvement in Measurement Area of 3D LiDAR for a Mobile Robot Using a Mirror Mounted on a Manipulator. *IEEE Robot. Autom. Lett.* **2020**, *5*, 6350–6356. [CrossRef]
17. Kümmerle, R.; Ruhnke, M.; Steder, B.; Stachniss, C.; Burgard, W. Autonomous Robot Navigation in Highly Populated Pedestrian Zones. *J. Field Robot.* **2014**, *32*, 565–589. [CrossRef]
18. Nguyen, T.-N.; Huynh, H.-H.; Meunier, J. 3D Reconstruction With Time-of-Flight Depth Camera and Multiple Mirrors. *IEEE Access* **2018**, *6*, 38106–38114. [CrossRef]
19. Horaud, R.; Hansard, M.; Evangelidis, G.; Ménier, C. An Overview of Depth Cameras and Range Scanners Based on Time-of-Flight Technologies. *Mach. Vis. Appl.* **2016**, *27*, 1005–1020. [CrossRef]
20. Nelson, D. CAROTHERS Compact Chip Scale LIDAR Solution. U.S. Patent 20170146639, 20 November 2015.
21. Xiaoping, H.; Huai, H.; Jiebin, X. Lidar Sensing System with Small Form Factor. U.S. Patent 201816167264, 21 October 2018.
22. Kim, S.; Jahandar, M.; Jeong, J.; Lim, D. Recent Progress in Solar Cell Technology for Low-Light Indoor Applications. *Curr. Altern. Energy* **2019**, *3*, 3–17. [CrossRef]
23. Chevrier, M.; Campanella, G. *LIDAR Pulsed Time of Flight Reference Design*; Texas Instruments: Dallas, TX, USA, 2018.
24. Willimas George, M., Jr. *Range-Walk Correction Uing Time Over Threshold*; Allegro MicroSystems: Manchester, NH, USA, 2021.
25. Kurtti, S.; Kostamovaara, J. Pulse Width Time Walk Compensation Method for a Pulsed Time-of-Flight Laser Rangefinder. In Proceedings of the 2009 IEEE Instrumentation and Measurement Technology Conference, Singapore, 5–7 May 2009; pp. 1059–1062.
26. Yang, J.; Liu, X.; Gu, G.; Qian, W.; Ren, K.; Zhang, J.; Guo, P.; Gao, Q.; Chen, Q. A Double Threshold Correction Method for Walk Error in Pulsed Laser Ranging System. *Infrared Phys. Technol.* **2019**, *100*, 28–36. [CrossRef]
27. Lachat, E.; Macher, H.; Landes, T.; Grussenmeyer, P. Assessment and Calibration of a RGB-D Camera (Kinect v2 Sensor) Towards a Potential Use for Close-Range 3D Modeling. *Remote Sens.* **2015**, *7*, 13070–13097. [CrossRef]
28. Kidd, J. Performance Evaluation of the Velodyne VLP-16 System for Surface Feature Surveying. Ph.D. Thesis, University of New Hampshire, Durham, NH, USA, 2017.

applied sciences

MDPI

Review

6D Pose Estimation of Objects: Recent Technologies and Challenges

Zaixing He [1], **Wuxi Feng** [1], **Xinyue Zhao** [1,*] **and Yongfeng Lv** [2]

[1] School of Mechanical Engineering, the State Key Lab of Fluid Power & Mechatronic Systems, Zhejiang University, Hangzhou 310027, China; zaixinghe@zju.edu.cn (Z.H.); feng_wx@zju.edu.cn (W.F.)
[2] Zhejiang Institute of Mechanical and Electrical Engineering, Hangzhou 310053, China; luyongfeng@zime.edu.cn
* Correspondence: zhaoxinyue@zju.edu.cn

Abstract: 6D pose estimation is a common and important task in industry. Obtaining the 6D pose of objects is the basis for many other functions such as bin picking, autopilot, etc. Therefore, many corresponding studies have been made in order to improve the accuracy and enlarge the range of application of various approaches. After several years of development, the methods of 6D pose estimation have been enriched and improved. Although some predecessors have analyzed the methods and summarized them in detailed, there have been many new breakthroughs in recent years. To understand 6D pose estimation better, this paper will make a new and more detailed review of 6D pose estimation. We divided these methods into two approaches: Learning-based approaches and non-learning-based approaches, including 2D-information-based approach and 3D-information-based approach. Additionally, we introduce the challenges that exist in 6D pose estimation. Finally, we compare the performance of different methods qualitatively and discuss the future development trends of the 6D pose estimation.

Keywords: 6D pose estimation; learning-based approach; 2D-information-based approach; 3D-information-based approach; textureless and reflective objects; foreground occlusion; background clutter

Citation: He, Z.; Feng, W.; Zhao, X.; Lv, Y. 6D Pose Estimation of Objects: Recent Technologies and Challenges. *Appl. Sci.* **2021**, *11*, 228. https://doi.org/10.3390/app11010228

Received: 11 November 2020
Accepted: 24 December 2020
Published: 29 December 2020

1. Introduction

1.1. Overview

6D pose refers to the posture of an object, specifically on the basis of a translation vector and a rotation vector. 6D pose estimation is an important step in many industrial fields highly related to another challenge—problem tracking [1]—such as bin picking [2–6], autonomous driving [7–9], augmented reality [10–12], SLAM (Simultaneous Localization and Mapping) [13–15] and so on (Figure 1). There have been an increasing number of applications of pose estimation developed in recent years. Autonomous vehicles use the technology of 6D pose estimation to recognize roads and obstacles. In the factory, the robots use the technology of 6D pose to recognize and grab objects. In the field of augmented reality, 6D pose estimation is used to measure the pose of objects in the real environment and add the virtual objects onto them in a correct pose. Some previous approaches could only detect the object and ensure its position, as is the case for GPS (Global Positioning System) [16] and radar detection [17]. These methods cannot measure the 6D pose of objects accurately. In industrial developments, higher demands are made for new application scenarios. Therefore, 6D pose estimation has become a hot topic in industry in recent years. 6D pose estimation uses a number of kinds of information to solve problems. It obtains texture information, geometric information, and color information to measure the 6D pose of objects. Due to the development of hardware in recent years, depth information is also used frequently in 6D pose estimation. However, 6D pose estimation is faced with many challenges, such as background clutter and inadequate information. Many methods have been proposed to improve the performance and enlarge the range of applications of 6D pose estimation, and many new methods have been proposed in recent years. Additionally there are many challenges in the 6D pose estimation field. A deeper comprehension of these challenges will also help arrive at more practical methods. To understand these methods and challenges more deeply, more detailed classification and performance evaluation need to be carried out. This review summarizes some relevant studies published in recent years and divides these methods into three categories. At the same time, we also analyze the advantages and disadvantages of these categories and challenges in 6D pose estimation.

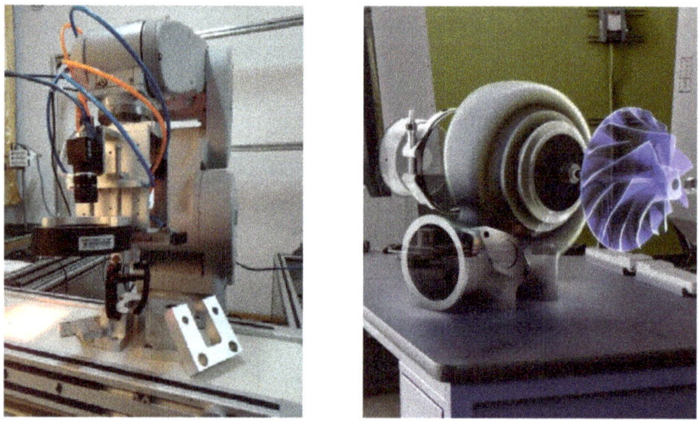

Figure 1. 6D pose estimation applied in bin picking and augmented reality.

1.2. Classification

In this paper, the 6D pose estimation approaches are divided into two categories: 1. Learning-based approach, and 2. Non-learning-based approach. The non-learning-based approach is divided into two categories: 1. 2D-information-based approach, and 2. 3D-information-based approach. The classification in this paper is mainly based on the core principle and the input information of the various methods: Learning-based approaches mainly use CNN, regression or some other methods based on deep learning to train a learning model with adequate training data and then obtain the 6D pose estimation result on the basis of these models. Approaches that do not use deep learning belong to the following two categories. 2D-information-based approaches mainly use the 2D information of the scene, such as RGB images. 3D-information-based approaches mainly use the 3D information of the scene, such as point clouds and RGB-D images. Both 2D-information-based approaches and 3D-information-based approaches convert the 6D pose estimation into image retrieval. The two types of approach both calculate the key points or key features and match the input image with the most similar image in the dataset according to the key points or key features. However, they also have some obvious differences, which will be covered in following sections.

The main purpose of learning-based approaches is to train a proper model to measure the 6D pose of an unknown situation according to the training data. Many kinds of model can be used to measure the 6D pose, such as regression models and CNN models. There are many classification methods for learning-based approaches, and those widely accepted among them are introduced in this paper. Keypoints-based approaches adopt a two-step category to measure the 6D pose, which is easier to implement than other approaches. Meanwhile, the aim of holistic approaches is to train an end-to-end network to measure the 6D pose of an object. It sees the image as a whole and tries to predict the location and orientation of the object in a single step and discretize the 6D space, converting the pose estimation task into a classification task. However, holistic approaches are more complex and time-consuming than keypoints-based approaches.

As for 2D-information-based approaches, the main purpose is to find the correlation between the input image and one of the template images through the 2D information contained in the image. Actually, 2D-information-based approaches converts the pose estimation into an image-matching problem. The matching results have a great influence on the results of the pose estimation. 2D-information-based approaches can be divided into real-image-based approaches and CAD-image-based approaches according to the kind of template used. When the approach uses real images as a template, it belongs to the real-image-based approaches. If the approach uses images generated by CAD model, it can be regarded as a CAD-image-based approach. In general, CAD-image-based approaches are more accurate than real-image-based approaches because the images generated by CAD models contain little noise. However, sometimes CAD models cannot be easily obtained, so real images are used as a template in such situations.

3D-information-based approaches also focus on the matching between the input and the dataset; however, they use the 3D information of the object, such as point clouds and RGB-D images. 3D-information-based approaches can be divided into two categories. The main idea of matching-based approaches is to match the input image and the template directly and to take the 6D pose of the matched template as the pose estimation result of the input image. Local descriptor-based approaches measure the 6D pose using the correspondence between the descriptor of input images and templates. Matching-based approaches require large storage to save enough templates to ensure the accuracy of pose estimation, and the more templates it has, the more accurate the pose estimation result will be.

1.3. Challenge

Although great progress has been made in the research of 6D pose estimation in recent years, there are still some challenging problems to be solved in practical application. When the background is messy, the viewpoint and illumination change greatly, or the scene has less texture, the accuracy and the robustness of 6D pose estimation needs to be improved. Learning-based approaches are relatively robust in these conditions. However, 2D-information-based approaches and 3D-information-based approaches do not perform well. However, for learning-based approaches, adequate training data and training time are required, and challenges related to the requirement of offline training and practicality are also problems to be solved.

1.4. Structure Layout

As shown in Figure 2, the rest of this paper can be divided into four parts: First, we introduce some research contributions in recent years in the three approaches and describe the advantages and challenges they are faced with in detail. We introduce learning-based approaches in the second section, 2D-information-based approaches and 3D-information-based approaches in the third section. Next, we focus on the challenges of 6D pose estimation. Then we compare the three approaches qualitatively. Finally, we give some views on the future development of the 6D pose estimation.

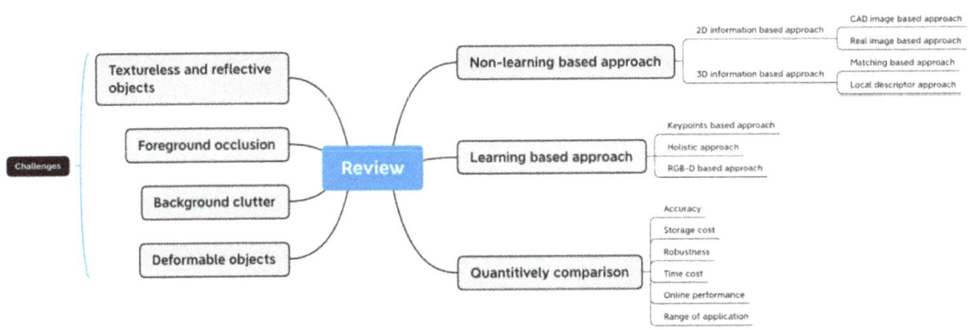

Figure 2. Structure layout of the review.

2. Learning-Based Approaches

In recent years, machine learning-based algorithms have become a hot topic due to new concepts' emergence, such as deep learning and neural networks. Many scholars have applied machine learning-based methods to 6D pose estimation and achieved good results. Posenet [18] is a monocular 6D relocalization system that trains a convolutional neural network (CNN) to regress 6D poses. The network transformed the problem of 6D pose estimation into a regression problem for which the input is a single RGB image and the output is the camera's 6D pose by using an end-to-end approach. To address the limitation of the lack of training data, a method was proposed that could generate large regression datasets of camera pose automatedly based on structure from motion. Crivellaro [19] also trained a CNN (Figure 3) to predict the 6D poses of objects that are partially visible. The key idea of the method was to predict the 3D-2D projections of feature points on each part of the object. When the test image is partially visible, the method could measure the 6D pose according to the feature points of visible part. When the test image is fully visible, the method is able to achieve more accurate results by combining all the feature points of the part. Particle Swarm Optimization (PSO)-based methods [20,21] demonstrate superior performance compared to Iterative Closest Point (ICP) algorithms. Hoang et al. [22] combined CNN with Simultaneous Localization and Mapping (SLAM), which improved the method [23] by adding a 6D object pose detector and measuring the 6D pose from different viewpoints to achieve a robust object detector system. However, there is a special problem for 6D pose estimation methods based on deep learning.

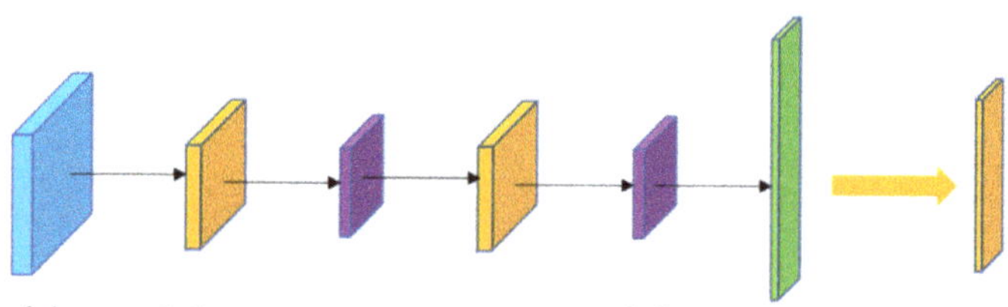

Figure 3. Architecture of the CNN predicting the projections of the control points.

Symmetrical objects' 6D poses (Figure 4) are difficult to measure correctly using normal deep learning methods [24], because the 6D pose of an object does not change from a fixed point of view when it is rotated 180 degrees. However, their actual ground truth is obviously different. For instance, if a network is trained to predict the pose using the squared loss between the ground truth poses and the predicted poses, it will converge to a model predicting the average of the possible poses for an input image, which is of course meaningless. Pitteri et al. [25] proposed an efficient method combined with Faster-RCNN, which relies on the normalization of the pose rotation. Manhardt et al. [26] proposed a method that was able to detect the rotation ambiguities and characterization of the uncertainty in the problem without further annotation or supervision. Zhang et al. [27] used the rigid transformation-invariant point-wise features of the point clouds as input features and used a hierarchical neural network that combined global point cloud information with the local patches to predict the key point coordinates.

2.1. Keypoint-Based Approaches

Keypoint-based approaches establish 2D-3D correspondences between images and then measure the pose according to these correspondences [28–30]. The procedures for keypoint-based approaches can be divided into two steps: 1. extract the 2D feature points in the input image; and 2. regress the 6D pose results using a PnP algorithm. BB8 [31] leveraged CNN to predict the 2D projections of eight vertices of the 3D bounding box of the object (Figure 5). To solve the challenges presented by textureless symmetrical objects,

BB8 restricted the range of the rotation angle of the training data and used a classifier to predict the rotation angle during the estimation step. However, when the object is partially invisible, BB8 may not obtain the correct 3D bounding box, which would have an adverse influence on PnP. To solve this problem, Hu [32] et al. proposed a method that segmented the image into several patches and made them predict both to which object they belonged and where the 2D projections were. Then, all the patches belonging to the same objects would be combined to measure the 6D pose based on PnP. Because each patch of the object is used to measure their respective local pose, this method was able to perform well when faced with occlusion. PVNet [33] predicted the direction of each pixel to each keypoint; thus, the spatial probability distribution of 2D key points can be obtained in a manner like RANSAC. According to the distribution, uncertainty-driven PnP could be used to measure the 6D pose. Predicting the direction of pixels and keypoints makes the local features more prominent. Even if one feature point is invisible, it can be positioned by means of another visible part. Jeon et al. [34] proposed a method involving learning orientation-induced primitives, rather than employing 3D bounding boxes, and calculated the rotation and translation vector in different modules. The methods mentioned above are all two-step-based; however, Hu [35] revealed the weakness of this kind of method. First, it is not an end-to-end system. Additionally, the loss function of the neural network cannot represent the accuracy of 6D pose estimation. Therefore, Hu et al. proposed a single-stage 6D pose estimation method that could directly regress the 6D pose on the basis of groups of 3D-to-2D correspondences associated with each 3D object keypoint.

Figure 4. Symmetrical objects.

Figure 5. The red bounding boxes for the pose estimation results using BB8.

2.2. Holistic Approaches

Unlike keypoint-based approaches, holistic approaches are an end-to-end architecture that can be faster than keypoint-based approaches. Kendall et al. [36] proposed PoseNet, which firstly applies the CNN architecture to 6D pose estimation, and found that it was able to adapt well to the environment. Liu et al. [37] proposed SSD, which was the first method to associate bounding box priors with the feature maps of different spatial resolutions in the network that was able to detect objects in images using a single deep neural network. This method improved accuracy and retained a low time cost. Kehl et al. [38] proposed SSD-6D, which extended the SSD method to 6D pose estimation and allowed for easy training and handling of symmetries. Do et al. [39] proposed the deep-6DPose network, the detection and segmentation in which leverage the Region Proposal Network (RPN) [40] based on Mask R-CNN [41]. In pose estimation, it decouples the parameters into translation and rotation so that the rotation can be regressed via a Lie algebra representation. However, because the network uses the ROIs from RPN as inputs and predicts the 6D pose of the object in ROIs, the network was not able to work well when measuring the 6D pose of small or symmetrical objects. To overcome this problem, Xiang et al. [42] proposed a new network PoseCNN. This method calculated the translation vector by ensuring the center of the objects in the image and estimating the distance between the center and the camera. Then it calculated the rotation matrix by regressing to a quaternion representation. Additionally, it especially employed a novel loss function for symmetric objects. The method was able to handle occlusion and symmetric objects in cluttered scenes with RGB or RGB-D images as input.

2.3. RGB-D-Based Appraoches

Compared with the learning-based approaches mentioned above, which only use the RGB information, RGB-D-based learning approaches combine color information and depth information to measure the pose of objects, and are able to solve the problem of insufficient information in approaches that only use color information. Additionally, due to the emergence of lots of RGB-D datasets, an increasing number of studies on RGB-D-based learning approaches are being performed. In [43], Wang et al. proposed a novel method named DenseFusion that provided a two-stage method for measuring the 6D pose. In the first stage, DenseFusion uses a heterogeneous network to deal with the RGB data and point cloud data, and to save their original structure. In the second stage, a full convolutional network is used to map each pixel in RGB crop to colored feature space and uses a network based on PointNet to map each point in the point cloud to geometrical feature space. Then it merges the feature points in the colored feature space and the geometrical feature space and outputs a 6D pose estimation result. In addition, it finally refines the result by loop learning. In [44], Chen et al. proposed a 6D pose estimation framework named G2L-Net. Firstly, it extracts the coarse point cloud from RGB-D images. Then, the point cloud is added into the network to achieve 3D segmentation and object translation predictions. Finally, the fine point cloud is transferred into a local canonical coordinate to estimate initial object rotation. In [45], PVN3D was proposed, in which the method based on 2D key points was extended to 3D key points, making full use of geometric constraint information of rigid objects and improving the accuracy of the 6D estimation significantly. In [46], a method named CosyPose was proposed, which used multiple cameras to estimate 6D pose. Firstly, it estimates the 6D pose of objects in each image and then matches the individual 6D object pose hypotheses across different input images in order to jointly estimate the camera viewpoints and 6D poses of all objects in a single consistent scene.

2.4. Conclusions

In this section, we introduced learning-based approaches and classified the appraches into three categories: keypoints-based approaches, holistic approaches, and RGB-D-based approaches. Keypoints-based approaches are two-step approaches that extract 2D-3D point pairs and then use PnP to calculate the 6D pose of the object. Holistic approaches use an end-to-end structure to measure the 6D pose, which is faster and more robust than keypoints-based approached. RGB-D-based learning approaches combine color information and depth information to achieve a more accurate and robust 6D pose result.

The comprehensive performance of learning-based approaches is better, as can be seen in the Amazon Robotic Challenge. The first Amazon Robotic Challenge was held in 2015 in the USA [47]. The competition presents a challenging problem that integrates many fields, including pose estimation. Many teams use learning-based approach and obtain robust results. In [48], the researchers presented a method for multi-class segmentation from RGB-D data. Objects were segmented by computing the possibility of each pixel belonging to each object by using a network. However, learning-based approaches still have some weaknesses. They require plenty of storage for storign training data and enough time to train the model. In conclusion, there are three challenges facing learning-based approaches. Firstly, the approaches require plenty of training data to train the model. Secondly, they require plenty of time to train the model before online use. For some complex objects that need much training data, this may take serval hours or even several days to train the model, thus restricting the possible applications of such approaches. Finally, these approaches cannot perform well when measuring poses that do not exist in the dataset.

3. Non-Learning-Based Approaches

3.1. 2D-Information-Based Approaches

Compared with 3D information, 2D information can be obtained more easily by simpler devices such as Charge-Coupled Device (CCD) cameras, Complementary Metal Oxide Semiconductor (CMOS) cameras, or even color cameras that can obtain the color information of objects. There is much 2D information that can be used to measure the 6D pose of objects, for instance, geometric information, texture information, color information, and so on. Scale-invariant feature transform (SIFT) features [49] and speeded up robust features (SURF) [50] are the early and classical features for pose estimation based on the texture of objects. SIFT and SURF features are both reliable and can achieve precise matching; however, they rely on the texture information of objects. Therefore, they cannot be used to solve the pose estimation of textureless objects. Zhang et al. [51] used geometric information, which does not rely on texture, to solve the pose estimation problem of textureless objects. E. Miyake et al. [52] combined the color information in the 6D pose estimation to improve accuracy and robustness. 2D information is extracted for the matching of the template. According to the dimensions of the template, approaches based on 2D information can be divided into two categories: 1. CAD image-based methods; and 2. real image-based methods.

3.1.1. CAD Image-Based Approaches

In contrast to real image-based approaches, CAD image-based approaches are more suitable for industrial products. The virtual images (used as templates) generated by CAD models are more accurate than real images, as the render process is not affected by illumination or blur. Moreover, CAD models can be obtained in industrial applications. Therefore, many methods based on 3D CAD model have been proposed. A hierarchical model [53] has been proposed (Figure 6), combining a coarse-to-fine search with similarity scores [54] calculated between a template and a real image or between templates. In [55], a perspective cumulated orientation feature (PCOF) was proposed based on the orientation histograms extracted from randomly generated 2D projection images using CAD models. Muñoz et al. [56] proposed the use of edge correspondences to estimate poses, with a similarity measure encoded using a pre-computed linear regression matrix. The Fine pose Parts-based Model (FPM) [57] was introduced to localize objects in an image, and to estimate their fine pose using the given CAD models. Pei et al. [58] proposed a robust method that only used one pair of vanishing points and one structural line to estimate the relative pose between image pairs. Peng et al. [59] proposed a method which used several cameras to detect geometrical features and then combined their results to obtain the final result.

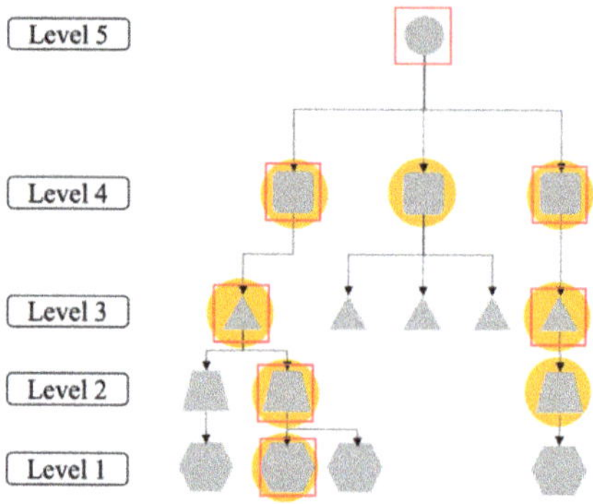

Figure 6. Object recognition using the hierarchy of views.

In [60], the Epipolar Geometry method and direct estimation method were used to estimate the 3D parameters, which were then used to construct the transformation matrix. In [61], 6D pose estimation was used in augmented reality. A local moving edge tracker was used to provide real-time tracking of points normal to the object contours. In addition, an M-estimator was used, integrated with a robust control law, to obtain good robustness. Straight lines and curves were both used in this method to complement the virtual visual servoing. In [62], 6D pose estimation was used for end effector tracking in a scanning electron microscope to aid in enabling more precise automated manipulations and measurements. Visible line features were also used to update the pose results. Kemal et al. [6] proposed a CAD model-based tracking method for visually guided microassembly. They used multiple cameras to track objects and find feature points along the edges of objects. Then, the 3D-3D for each feature point was built, and the 6D pose was calculated.

Whether real images or CAD models are being used as the template, the performance of the matching step determines the accuracy of pose estimation. Improving the efficiency and accuracy of template matching has become an important problem to ensure the results of pose estimation.

Additionally, the number of templates influences the accuracy of the pose estimation. The more templates there are, the more accurate the pose estimation will be. However, a large number of templates requires lots of storage and search time. In [63], a part-based efficient template matching method was proposed which was able to accelerate the matching step and improve the accuracy of pose estimation. Each of the templates leveraged a different forest and independently encoded similarity function.

3.1.2. Real Image-Based Approaches

Although it is possible to achieve precise results when using 3D CAD models, sometimes accurate 3D CAD models cannot be obtained. Therefore, real images are used as the template under these conditions. The histogram of gradients (HoG) [64] is a popular method that is computed on a dense grid with uniform intervals for better performance. Guo et al. [65] used multi-cooperative logos to measure 6D pose. Hinterstoisser [66] proposed a method including a novel image representation for template matching designed to be robust to small image transformations. It used the gradient orientation of the edges of objects to create templates. The method was able to be extended if 3D information was available. However, because obtaining adequate real image templates is time-consuming and challenging, and generating the images by CAD model is becoming easier, there is not much research on real image-based approaches.

3.1.3. Conclusions

In this part, we introduced 2D-information-based approaches and classified the approaches into two categories: CAD image-based approaches and real image-based approaches. The difference between them is which type of template is used. CAD image-based approaches require a CAD model of the object, but templates can be generated conveniently. Real image-based approaches use real images as templates, but if there is clutter in the real images, the information may be extracted incorrectly.

Compared with 3D-information-based approaches, 2D-information-based approaches are less robust, because these approaches use 2D information, which has less data than 3D information. Additionally, complex scenes have a bad influence on the performance of these approaches. The biggest weakness for 2D-information-based approaches is that they are not able to adapt to some special scenes, such as scenes with strong changes in illumination, large numbers of repeated structures, textureless scenes, and so on.

3.2. *3D-Information-Based Approaches*

Although there are many kinds of 2D information, it is difficult to use 2D information to measure the 6D pose under some special conditions, or the method requires complex algorithms to obtain precise results. With the development of hardware, more and more devices that can record 3D scene information are appearing, such as depth cameras [67] and 3D scanners [68]. Compared with 2D information, 3D information preserves the original appearance of the object, which is more useful for measuring 6D pose [44,69]. Combined with 3D information, the method is more robust and can obtain more accurate results. Mainstream methods can be broadly divided into the following two categories.

3.2.1. Matching-Based Approaches

The aim of matching-based approaches is to search for the most similar template in the dataset and return the 6D pose of the template. Because the 3D original data is always too large, processing these data directly can be computationally expensive. Therefore, many preprocessing methods have been proposed to reduce the complexity of the task. Zhang et al. [70] proposed two methods for solving this problem. One was to use a 2D/2.5D object detector for scene point clouds. YOLO was used to segment the scene point cloud with 2D bounding boxes due to their lower time consumption. The other preprocessing method was to extract the keypoints in the template point clouds. Points with more information, such as the points on edges, were preserved and points on surfaces were removed in order to compress the point cloud. Konishi et al. [71] combined PCOF-MOD (multimodal PCOF), balanced pose tree (BPT), and optimum memory rearrangement into 6D pose estimation to optimize data storage structure and lookup speed. To improve the accuracy of matching, Park [72] et al. proposed a novel multi-task template matching (MTTM) framework that finds the nearest template of a target object from an image while predicting segmentation masks and a pose transformation between the template and a detected object in the scene using the same feature map of the object region. In [73], research was carried out on tracking and control for micro-electro-mechanical system (MEMS) microassembly. The correlation between the real-time 3D vision tracking method and the control law based on 3D vision was demonstrated, and a pose-based visual servoing approach was used to to enable a precise regulation toward zero of 3D error.

3.2.2. Local Descriptor Approaches

Approaches based on the local descriptor define and calculate a global descriptor on the model offline. The global descriptor should be invariant with respect to rotation and translation. Then, the local descriptor is calculated and matched with the global descriptor online. The iterative closest point (ICP) [74,75] algorithm is a classical one that is able to calculate the pose relation between two coordinates according to two sets of point clouds. 6D pose can be measured by the correspondence between them or as the result of voting. Guo et al. [76] proposed a global method named Super Key 4-Points Congruent Sets (SK-4PCS), combined with invariant local features of 3D shapes, thus reducing the amount of processed data. Akizuki et al. [77] proposed a method aimed at everyday tools that does not rely on the 3D model of the measured object. It assumed that the same kinds of tools possess the same part-affordance. Therefore, using the 3D models for similar kinds of object, the method was able to measure the pose of objects belonging to this kind on the basis of the spatial relationships of part-affordance. Yu et al. [78] used the improved Oriented Fast and Rotated Brief (ORB) [79] feature and rBRIEF descriptor (a descriptor developed based on binary robust independent elementary features (BRIEF) [80]) to obtain a coarse match, and then culled mismatches to retain more correct matches. They also proposed a hybrid reprojection errors optimization model (HREOM) to improve the accuracy of the result by minimizing 3D-3D and 3D-2D reprojection errors. To measure the 6D pose of large-scale objects that are partially visible, David et al. [81] proposed a method based on semi-global descriptors. They used semi-global descriptors for scene segments and model views in combination with up-sampling and segment label merging techniques and obtained more reliable results than with other descriptors.

3.2.3. Conclusions

In this part, we introduced 3D-information-based approaches and classified the approaches into two categories: matching-based approaches and local descriptor approaches. Matching-based approaches are more computationally expensive; however, local descriptor approaches require some preprocessing oofline.

The biggest weakness for approaches based on 3D information is that they do not perform well when the object is reflective, which is due to how the approaches work. When measuring the 6D pose of reflective objects, these approaches are not able to obtain the depth information (or accurate point clouds) of the object. Additionally, another weakness is that the efficiency of this approach is relatively low, because point clouds and depth images include a lot of data, causing computational burden.

4. Comparison

In this section, we will compare these approaches in detail according to their performance in Table 1. Specifically, the accuracy, the storage cost, the robustness, the time cost, online performance and range of application will be discussed. The mentioned indicators are compared qualitatively in Table 1 according to the information in these papers, where A represents the best performance and C represents the worst performance.

Table 1. Comparison of three kinds of approaches.

		Accuracy	Storage Cost	Robustness	Time Cost	Online Performance	Range of Application
Learning-based approaches	Keypoint-based approaches	B	B	B	C	C	B
	Holistic approaches	C	B	B	B	B	B
	RGB-D-based approaches	A	C	A	C	C	C
Non-learning-based approaches	2D-information-based approaches	B	A	C	A	A	A
	3D-information-based approaches	A	B	B	B	B	C

Accuracy: The accuracy of holistic approaches is worse than that of 2D-information-based approaches and 3D-information-based approaches. This is because holistic approaches transform the problem of 6D pose estimation into a problem of classification. The accuracy of classification decides the accuracy of 6D pose estimation. However, RGB-D-based approaches use more comprehensive information. The depth information is able to provide the overall morphology of a rigid body, while the color information is able to describe the position of keypoints. Therefore, RGB-D-based approaches that combine depth information and color information are able to improve the accuracy. 2D-information-based approaches and 3D-information-based approaches convert the 6D pose estimation problem into a template matching or coordinate transformation problem. Both of them are able to measure the

6D pose more specifically. 3D-information-based approaches use more information than 2D-information-based approaches, so they perform better on accuracy.

Robustness: Robustness in this paper mainly refers to the anti-interference performance of approaches to noise and environmental changes. Plenty of data is used to train a network in learning-based approaches. The model adequately considers the information and situation in the input scene, so learning-based approaches have better robustness than the other two approaches. Additionally, using only color information may lead to missing keypoints, while depth information provides some global pose information and is complementary to color information. After a long period of training, learning models are able to distinguish object features and environmental noise. Although the other two approaches are also able to distinguish information from the environment efficiently, some wrong information may be taken into consideration under some complex situations where background clutter and foreground occlusion are present.

Storage cost: Learning-based approaches need the most storage because the training process of the learning models needs plenty of data. The other two approaches need to store templates. However, the number of templates is lower than the training data required for learning-based approaches. In particular, for coarse-to-fine methods, which just need to match a basically similar template with the input image, much fewer templates are needed. In terms of a comparison between 2D-information-based approaches and 3D-information-based approaches, the former needs less storage, because 2D information is smaller than 3D information.

Time cost and real-time performance: Because the time an approach takes determines the real-time performance of the approach, the two indicators are discussed together in this part. Learning-based approaches can be divided into two steps: offline training and online measuring. The offline step is time-consuming because it needs to train a model using plenty of data, which means that repeated calculations are necessary, although a GPU could accelerate the training process. However, in the online step, the 6D pose can be measured directly by the trained model, which costs very little time. 2D-information-based approaches and 3D-information-based approaches an also be divided into an offline step and an online step. In the offline step, many templates are generated from different angles (however, there are also some methods that do not need a lot of templates). Meanwhile, in the online step, the proper template needs to be retrieved from template dataset, which is a little time-consuming. Therefore, in the online step, the real-time performance of learning-based approaches is the best. 2D-information-based approaches cost less time than 3D-information-based approaches because the retrieval of 3D information is more time-consuming.

Range of application: Due to their principle, 3D-information-based approaches are not able to handle the problem of 6D pose estimation of reflective objects. However, such objects are common in industry, such as metal parts. Learning-based approaches need plenty of time to train a network, which may not satisfy the requirements of real-time performance.

In this section, we compared the approaches with respect to six different aspects. In general, learning-based approaches have the best robustness, but these approaches need lots of storage to save training data and plenty of time to train the model. 3D-information-based approaches achieve the most accurate results, but they cannot be used on reflective objects. The range of application of 2D-based approaches is the widest; however, 2D information is easily affected by the environment. Therefore, the three approaches all have their advantages. In different situations, different approaches are needed.

5. Challenges

5.1. *Textureless and Reflective Objects*

As a special case in pose estimation, the pose estimation of textureless objects is challenging. Due to the lack of reliable texture information on the surface of such objects, it is difficult to extract feature points on them. Therefore, many methods based on the surface texture information of objects cannot effectively measure the 6D pose of such objects.

However, textureless objects are common in industry. Therefore, the pose estimation of textureless objects is very important.

Zhang et al. [82] transformed sliding windows to scale-invariant RGB-D patches and applied a hash voting-based hypothesis generation scheme to compute a rough 6D pose hypothesis and then employed particle swarm optimization to improve the result, which achieved high precision and good performance on three datasets. Pan et al. [83] coarsely segmented the object in the point cloud and precisely measured the pose results in the gray image using a view-based matching method.

The methods mentioned above used a depth camera or 3D scanning device to obtain the depth image or the point cloud of objects. However, for objects with reflective surfaces, such as metal parts, depth information is hard to acquire. Therefore, some other kinds of information are used in pose estimation. Geometrical information is a reliable kind of information for textureless objects. Based on the above perspective, He et al. [84] proposed a new method (Figure 7) making full use of the geometrical information of objects. It used geometric features, such as straight lines, to generate descriptors of the objects, and proposed the GIIBOLD algorithm for matching the input image and the template image. Furthermore, accurate 2D-3D point pairs were acquired on the basis of the matched geometric features. Finally, the 6D pose was measured using the PnP-RANSAC algorithm. This method leveraged simple geometrical information, achieving fast matching and accurate measurement without the requirement for plenty of storage and time. Zhang et al. [85] first detected the object in the RGB image using a 2D bounding box and then measured the pose result in the edge image. Pan et al. [86] used multiple appearance features including color, size and aspect ratio to distinguish objects from environmental clutter and measured the 6D pose. Zhang et al. [51] proposed a novel method for measuring textureless objects on the basis of RGB images. It followed a coarse-to-fine procedure, using only the shape and contour information of the input image. Several template images with poses similar to the input image were selected to match with the input. Then the contour and shape information, specifically the ORB features, were used to establish 2D-3D correspondence and finally to calculate the 6D pose. On the basis of the studies above, without reliable texture information, there are many kinds of information can be leveraged, such as contour, color, shape, and so on. Accurate results can be obtained through the proper use of this information.

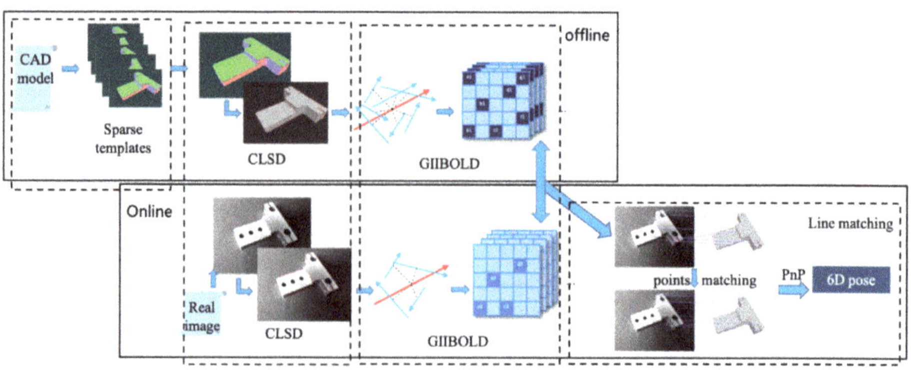

Figure 7. The overall workflow chart.

5.2. Foreground Occlusion

In complex industrial scenes, the condition of object occlusion appears frequently. Because the target is obscured by other objects, the recognition of the target's features is disturbed. In addition, due to part of the object information being missing, it is difficult to calculate accurate pose results. As a common condition, object occlusion has been studied by many researchers.

Crivellaro et al. [19] proposed a method for representing the pose of each part on the basis of the 2D reprojections of a small set of 3D control points. It was able to predict the 6D pose of the object by predicting the pose of the visible part through the reprojections of 2D control points. Even if the object was only partially visible, the method was able to calculate the 6D pose of the object. If the object had several visible parts, the method was able to combine all the information and obtain more accurate results. Distinct from the method above, Dong et al. [87] used 3D information as input and chose an end-to-end strategy. They proposed a novel network named Point-wise Pose Regression Network (PPR-Net). For each of the points in the point cloud, the network regressed a 6D pose of the object instance that the point belonged to. In the pose space, a clustering method was adopted in order to segment multiple instances from the clutter point cloud, and an instance's pose can be computed by averaging each subsidiary point's pose prediction. Essentially, the method used the information of visible parts to predict the pose of the object. The more parts of an object that could be seen, the more reliable the results obtained. Chen et al. [88] proposed a network which took the point cloud as input and regressed the point-wise unit vectors pointing to the 3D keypoints. Then the vectors were used to generate keypoint hypotheses from which 6D object pose hypotheses were computed. Tekin et al. [28] predicted the 2D image locations of the projected vertices of the object's 3D bounding box and used a PnP algorithm to estimate the object's 6D pose. Taking RGB-D images as input, Zhang et al. [89] combined holistic and local patches to measure the 6D object pose and obtained high precision and good performance under conditions of foreground occlusion and background clutter.

In conclusion, the principle of the method for dealing with object occlusion is to use the information of any visible parts to predict the 6D pose of the whole object. Generally, the number of methods to achieve this based on 3D information is greater than the number of those based on 2D information. In addition, among the methods based on 2D information, it is more appropriate to use outer information such as bounding boxes and contours. Because inner information, such as the texture, is occluded by other objects, the extraction of the information is affected to a great extent.

5.3. Background Clutter

Background clutter is also a challenge in 6D pose estimation. Because the target is surrounded by much useless information, it is difficult to measure the 6D pose directly. However, due to the complexity of practical scenarios, there are many conditions under which it is necessary to measure the 6D pose of objects in clutter.

He et al. [41] proposed Mask R-CNN, which efficiently detected objects in an image while simultaneously generating a high-quality segmentation mask for each instance. It extended Faster R-CNN by adding a branch for predicting an object mask in parallel with the existing branch for bounding box recognition. Mitash et al. proposed a method for measuring the 6D pose of objects in clutter. A global optimization process was employed to improve candidate poses by taking into account scene-level physical interactions between objects. Then, the combinations of candidate object poses were searched using a Monte Carlo Tree Search (MCTS) process that used the similarity between the observed depth image of the scene and the rendering of the scene given the hypothesized pose as a score, guiding the search procedure. Li et al. [90] proposed a two-step method for measuring the 6D pose. The first step was few-shot instance segmentation to segment the known objects from RGB images. Chen et al. [91] proposed a method based on point clouds that detected objects in an end-to-end manner. They introduced a point cloud-based 6D target object detection method that used segmented object point cloud patches to predict object 6D poses and identity. It used a point cloud segmentation procedure that was easier to visualize and tune in order to overcome the problem caused by background clutter.

5.4. Deformable Objects

The 6D pose estimation of deformable objects is a huge challenge in the field, because the posture of the objects is unpredictable and there are many ways for the objects to deform. Thus, lots of conditions need to be taken into consideration, resulting in pressure on the algorithm and calculation.

Li et al. [92] proposed a novel method that was able to classify and estimate the categories and poses of deformable objects. They simulated the deformable objects and obtained the depth images from different viewpoints in different postures as a dataset. By extracting features and using deep learning, they set up a codebook for the object, which could be used in training process. This method uses learning-based approaches and sets up a two-layer framework to solve the problem of 6D pose estimation. Lots of storage space is needed, but the method would be robust in such complex situations. In [93], a predictive and model-driven approach was proposed to solve this problem. A dataset was built up by using the picked-up garments in multiple poses under gravity. The dataset was orginized in an efficient way, increasing the speed of the searching process. The proposed method constructed a fully featured 3D model of the garment in real time and used volumetric features to obtain the most similar model in the database in order to predict the object category and pose. Accurate model simulation could also be used to optimize the trajectories of the manipulation of deformable objects. Caporali et al. [94] proposed a four-step method to solve this problem of grasping clothes by using a point cloud. Firstly, the instance segmentation was performed, and then a wrinkledness measure was implemented to robustly detect the graspable regions of the cloth. Next, the identification of each individual wrinkle was accomplished by fitting a piecewise curve, and finally a pose for each detected wrinkle was estimated.

5.5. Conclusions

In this section, four challenges were discussed. For reflective objects, 3D-information-based approaches cannot accurately measure their pose. For textureless objects, only the geometrical features can be extracted for calculating the 6D pose of objects. Faced with foreground occlusion, many methods use the visible part to describe the invisible part of objects and measure their 6D pose. For background clutter, instance segmentation is used to separate the object, and then the 6D pose is measured. For deformable objects, images of the objects are captured in different poses in different views and a dataset is set up. The best matching is found in the dataset when measuring the pose of objects.

6. Conclusions

This paper divided solutions of 6D pose estimation into three kinds of approaches and made some detailed introductions to their advantages and disadvantages. Then, this paper focused on the challenges in 6D pose estimations, introducing the difficulties of these problems and summarizing some feasible solutions. Finally, some approaches were qualitatively compared with respect to several different indicators. Learning-based approaches achieved the best robustness. However, they are time-consuming and require lots of storage. 2D-information-based approaches are easy to implement and can be applied online. 3D-information-based approaches can achieve higher accuracy than 2D-information-based approaches, but they require more information (depth information) to be collected and dealt with, and they cannot measure the 6D pose of reflective objects. Generally, the methods presented have already satisfied the requirements of industrial application for the 6D pose detection of general objects. However, these methods cannot maintain their excellent performance under some challenging conditions. In future research, learning-based approaches should be further developed. On one hand, their robust performance should be retained. On the other hand, the offline training time should be decreased. These approaches can also be combined with 2D-information-based and 3D-information-based approaches to obtain more accurate results.

Author Contributions: Conceptualization, Z.H. and X.Z.; methodology, Z.H.; software, W.F.; validation, X.Z. and Y.L.; writing—original draft preparation, W.F.; writing—review and editing, Z.H. and X.Z. All authors have read and agreed to the published version of the manuscript.

Funding: This research was supported by the National Natural Science Foundation of China (51775498, 51775497) and Zhejiang Province Public Welfare Technology Application Research Project (LGG19E050019).

Data Availability Statement: The data presented in this study is contained in the article itself.

Conflicts of Interest: The authors declare no conflict of interest.

References

1. Drummond, T.; Cipolla, R. Real-time visual tracking of complex structures. *IEEE Trans. Pattern Anal. Mach. Intell.* **2002**, *24*, 932–946. [CrossRef]
2. Peng, L.; Zhao, Y.; Qu, S.; Zhang, Y.; Weng, F. Real Time and Robust 6D Pose Estimation of RGBD Data for Robotic Bin Picking. In Proceedings of the 2019 Chinese Automation Congress (CAC), Hangzhou, China, 22–24 November 2019; pp. 5283–5288.
3. Yan, W.; Xu, Z.; Zhou, X.; Su, Q.; Li, S.; Wu, H. Fast Object Pose Estimation Using Adaptive Threshold for Bin-Picking. *IEEE Access* **2020**, *8*, 63055–63064. [CrossRef]
4. Xu, J.; Pu, S.; Zeng, G.; Zha, H. 3D pose estimation for bin-picking task using convex hull. In Proceedings of the 2012 IEEE International Conference on Mechatronics and Automation, Chengdu, China, 5–8 August 2012; pp. 1381–1385.
5. Buchholz, D. *Bin-Picking: New Approaches for a Classical Problem*; Springer: Berlin/Heidelberg, Germany, 2016.
6. Yang, J.; Zeng, G.; Wang, W.; Zuo, Y.; Yang, B.; Zhang, Y. Vehicle Pose Estimation Based on Edge Distance Using Lidar Point Clouds (Poster). In Proceedings of the 2019 22th International Conference on Information Fusion (FUSION), Ottawa, ON, Canada, 2–5 July 2019; pp. 1–6.
7. Gu, R.; Wang, G.; Hwang, J. Efficient Multi-person Hierarchical 3D Pose Estimation for Autonomous Driving. In Proceedings of the 2019 IEEE Conference on Multimedia Information Processing and Retrieval (MIPR), San Jose, CA, USA, 28–30 March 2019; pp. 163–168.
8. Kothari, N.; Gupta, M.; Vachhani, L.; Arya, H. Pose estimation for an autonomous vehicle using monocular vision. In Proceedings of the 2017 Indian Control Conference (ICC), Guwahati, India, 4–6 January 2017; pp. 424–431.
9. Zhang, S.; Song, C.; Radkowski, R. Setforge-Synthetic RGB-D Training Data Generation to Support CNN-Based Pose Estimation for Augmented Reality. In Proceedings of the 2019 IEEE International Symposium on Mixed and Augmented Reality Adjunct (ISMAR-Adjunct), Beijing, China, 10–18 October 2019; pp. 237–242.
10. Lu, Y.; Kourian, S.; Salvaggio, C.; Xu, C.; Lu, G. Single Image 3D Vehicle Pose Estimation for Augmented Reality. In Proceedings of the 2019 IEEE Global Conference on Signal and Information Processing (GlobalSIP), Ottawa, ON, Canada, 11–14 November 2019; pp. 1–5.
11. Hachiuma, R.; Saito, H. Recognition and pose estimation of primitive shapes from depth images for spatial augmented reality. In Proceedings of the 2016 IEEE 2nd Workshop on Everyday Virtual Reality (WEVR), Greenville, SC, USA, 20–20 March 2016; pp. 32–35.
12. Li, X.; Ling, H. Hybrid Camera Pose Estimation with Online Partitioning for SLAM. *IEEE Robot. Autom. Lett.* **2020**, *5*, 1453–1460. [CrossRef]
13. Ruan, X.; Wang, F.; Huang, J. Relative Pose Estimation of Visual SLAM Based on Convolutional Neural Networks. In Proceedings of the 2019 Chinese Control Conference (CCC), Guangzhou, China, 27–30 July 2019; pp. 8827–8832.
14. Xiao, Z.; Wang, X.; Wang, J.; Wu, Z. Monocular ORB SLAM based on initialization by marker pose estimation. In Proceedings of the 2017 IEEE International Conference on Information and Automation (ICIA), Macau, China, 18–20 July 2017; pp. 678–682.
15. Malyavej, V.; Torteeka, P.; Wongkharn, S.; Wiangtong, T. Pose estimation of unmanned ground vehicle based on dead-reckoning/GPS sensor fusion by unscented Kalman filter. In Proceedings of the 2009 6th International Conference on Electrical Engineering/Electronics, Computer, Telecommunications and Information Technology, Pattaya, Thailand, 6–9 May 2009; pp. 395–398.
16. Zhaoyang, N.; Jianyun, Z.; Zhidong, Z. Angle Estimation for Bi-static MIMO Radar Based on Tri-iterative Algorithm. In Proceedings of the 2010 First International Conference on Pervasive Computing, Signal Processing and Applications, Harbin, China, 17–19 September 2010; pp. 1264–1267.
17. Kendall, A.; Grimes, M.; Cipolla, R. PoseNet: A Convolutional Network for Real-Time 6-DOF Camera Relocalization. In Proceedings of the 2015 IEEE International Conference on Computer Vision (ICCV), Santiago, Chile, 7–13 December 2015; pp. 2938–2946.
18. Crivellaro, A.; Rad, M.; Verdie, Y.; Yi, K.M.; Fua, P.; Lepetit, V. A Novel Representation of Parts for Accurate 3D Object Detection and Tracking in Monocular Images. In Proceedings of the 2015 IEEE International Conference on Computer Vision (ICCV), Santiago, Chile, 7–13 December 2015; pp. 4391–4399.
19. Zabulis, X.; Lourakis, M.; Koutlemanis, P. 3D Object Pose Refinement in Range Images. In *International conference on Computer Vision Systems*; Springer: Cham, Switzerland, 2015; pp. 263–274.
20. Zabulis, X.; Lourakis, M.I.; Koutlemanis, P. Correspondence-free pose estimation for 3D objects from noisy depth data. *Vis. Comput.* **2018**, *34*, 193–211. [CrossRef]
21. Hoang, D.; Stoyanov, T.; Lilienthal, A.J. Object-RPE: Dense 3D Reconstruction and Pose Estimation with Convolutional Neural Networks for Warehouse Robots. In Proceedings of the 2019 European Conference on Mobile Robots (ECMR), Prague, Czech Republic, 4–6 September 2019; pp. 1–6.

22. Hoang, D.-C.; Stoyanov, T.; Lilienthal, A. High-Quality Instance-Aware Semantic 3D Map Using RGB-D Camera. *arXiv* **2019**, arXiv:1903.10782.
23. Wen, Y.; Pan, H.; Yang, L.; Wang, W. Edge Enhanced Implicit Orientation Learning With Geometric Prior for 6D Pose Estimation. *IEEE Robot. Autom. Lett.* **2020**, *5*, 4931–4938. [CrossRef]
24. Pitteri, G.; Ramamonjisoa, M.; Ilic, S.; Lepetit, V. On Object Symmetries and 6D Pose Estimation from Images. In Proceedings of the 2019 International Conference on 3D Vision (3DV), Québec City, QC, Canada, 16–19 September 2019; pp. 614–622.
25. Manhardt, F.; Arroyo, D.M.; Rupprecht, C.; Busam, B.; Birdal, T.; Navab, N.; Tombari, F. Explaining the Ambiguity of Object Detection and 6D Pose From Visual Data. In Proceedings of the 2019 IEEE/CVF International Conference on Computer Vision (ICCV), Seoul, Korea, 27 October–2 November 2019; pp. 6840–6849.
26. Zhang, W.; Chenkun, Q.I. Pose Estimation by Key Points Registration in Point Cloud. In Proceedings of the 2019 3rd International Symposium on Autonomous Systems (ISAS), Shanghai, China, 29–31 May 2019; pp. 65–68.
27. Tekin, B.; Sinha, S.N.; Fua, P. Real-Time Seamless Single Shot 6D Object Pose Prediction. In Proceedings of the 2018 IEEE/CVF Conference on Computer Vision and Pattern Recognition, Salt Lake City, UT, USA, 18–23 June 2018; pp. 292–301.
28. Li, Z.; Wang, G.; Ji, X. CDPN: Coordinates-Based Disentangled Pose Network for Real-Time RGB-Based 6-DoF Object Pose Estimation. In Proceedings of the 2019 IEEE/CVF International Conference on Computer Vision (ICCV), Seoul, Korea, 27 October–2 November 2019; pp. 7677–7686.
29. Oberweger, M.; Rad, M.; Lepetit, V. Making Deep Heatmaps Robust to Partial Occlusions for 3D Object Pose Estimation. In Proceedings of the European Conference on Computer Vision (ECCV), Munich, Germany, 8–14 September 2018.
30. Rad, M.; Lepetit, V. BB8: A Scalable, Accurate, Robust to Partial Occlusion Method for Predicting the 3D Poses of Challenging Objects without Using Depth. In Proceedings of the 2017 IEEE International Conference on Computer Vision (ICCV), Venice, Italy, 22–29 October 2017; pp. 3848–3856.
31. Hu, Y.; Hugonot, J.; Fua, P.; Salzmann, M. Segmentation-Driven 6D Object Pose Estimation. In Proceedings of the 2019 IEEE/CVF Conference on Computer Vision and Pattern Recognition (CVPR), Long Beach, CA, USA, 15–20 June 2019; pp. 3380–3389.
32. Peng, S.; Liu, Y.; Huang, Q.; Bao, H.; Zhou, X. PVNet: Pixel-wise Voting Network for 6DoF Pose Estimation. In Proceedings of the IEEE Conference on Computer Vision and Pattern Recognition, Salt Lake City, UT, USA, 18–23 June 2018.
33. Jeon, M.H.; Kim, A. PrimA6D: Rotational Primitive Reconstruction for Enhanced and Robust 6D Pose Estimation. *IEEE Robot. Autom. Lett.* **2020**, *5*, 4955–4962. [CrossRef]
34. Hu, Y.; Fua, P.; Wang, W.; Salzmann, M. Single-Stage 6D Object Pose Estimation. In Proceedings of the 2020 IEEE/CVF Conference on Computer Vision and Pattern Recognition (CVPR), Seattle, WA, USA, 13–19 June 2020; pp. 2927–2936.
35. Kendall, A.; Grimes, M.; Cipolla, R. PoseNet: A Convolutional Network for Real-Time 6-DOF Camera Relocalization. *Educ. Inf.* **2015**, *31*, 2938–2946.
36. Liu, W.; Anguelov, D.; Erhan, D.; Szegedy, C.; Reed, S. SSD: Single Shot MultiBox Detector. In *European Conference on Computer Vision*; Springer: Cham, Switzerland, 2016.
37. Kehl, W.; Manhardt, F.; Tombari, F.; Ilic, S.; Navab, N. SSD-6D: Making RGB-based 3D detection and 6D pose estimation great again. In Proceedings of the IEEE International Conference on Computer Vision, Venice, Italy, 22–29 October 2017.
38. Do, T.-T.; Cai, M.; Pham, T.; Reid, I. Deep-6DPose: Recovering 6D Object Pose from a Single RGB Image. *arXiv* **2018**, arXiv:1802.10367.
39. Ren, S.; He, K.; Girshick, R.; Sun, J. Faster R-CNN: Towards Real-Time Object Detection with Region Proposal Networks. *IEEE Trans. Pattern Anal. Mach. Intell.* **2015**, *39*, 1137–1149. [CrossRef]
40. He, K.; Gkioxari, G.; Dollár, P.; Girshick, R. Mask R-CNN. In Proceedings of the 2017 IEEE International Conference on Computer Vision (ICCV), Venice, Italy, 22–29 October 2017; pp. 2980–2988. [CrossRef]
41. Xiang, Y.; Schmidt, T.; Narayanan, V.; Fox, D. PoseCNN: A Convolutional Neural Network for 6D Object Pose Estimation in Cluttered Scenes. *arXiv* **2017**, arXiv:1711.00199.
42. Wang, C.; Xu, D.; Zhu, Y.; Martín-Martín, R.; Lu, C.; Fei-Fei, L.; Savarese, S. DenseFusion: 6D Object Pose Estimation by Iterative Dense Fusion. In Proceedings of the 2019 IEEE/CVF Conference on Computer Vision and Pattern Recognition (CVPR), Long Beach, CA, USA, 15–20 June 2019; pp. 3338–3347.
43. Chen, W.; Jia, X.; Chang, H.J.; Duan, J.; Leonardis, A. G2L-Net: Global to Local Network for Real-Time 6D Pose Estimation With Embedding Vector Features. In Proceedings of the 2020 IEEE/CVF Conference on Computer Vision and Pattern Recognition (CVPR), Seattle, WA, USA, 13–19 June 2020; pp. 4232–4241.
44. He, Y.; Sun, W.; Huang, H.; Liu, J.; Fan, H.; Sun, J. PVN3D: A Deep Point-Wise 3D Keypoints Voting Network for 6DoF Pose Estimation. In Proceedings of the 2020 IEEE/CVF Conference on Computer Vision and Pattern Recognition (CVPR), Seattle, WA, USA, 13–19 June 2020; pp. 11629–11638.
45. Labbé, Y.; Carpentier, J.; Aubry, M.; Sivic, J. CosyPose: Consistent Multi-view Multi-object 6D Pose Estimation. In *European Conference on Computer Vision*; Springer: Cham, Switzerland, 2020; pp. 574–591.
46. Correll, N.; Bekris, K.E.; Berenson, D.; Brock, O.; Causo, A.; Hauser, K.; Okada, K.; Rodriguez, A.; Romano, J.M.; Wurman, P.R. Analysis and Observations From the First Amazon Picking Challenge. *IEEE Trans. Autom. Sci. Eng.* **2018**, *15*, 172–188. [CrossRef]
47. Jonschkowski, R.; Eppner, C.; Höfer, S.; Martín-Martín, R.; Brock, O. Probabilistic multi-class segmentation for the Amazon Picking Challenge. In Proceedings of the 2016 IEEE/RSJ International Conference on Intelligent Robots and Systems (IROS), Daejeon, Korea, 9–14 October 2016; pp. 1–7.
48. Lowe, D.G. Distinctive Image Features from Scale-Invariant Keypoints. *Int. J. Comput. Vis.* **2004**, *60*, 91–110. [CrossRef]

49. Bay, H.; Ess, A.; Tuytelaars, T.; Gool, L.V. Speeded-Up Robust Features (SURF). *Comput. Vis. Image Underst.* **2008**, *110*, 346–359. [CrossRef]
50. Zhang, X.; Jiang, Z.; Zhang, H.; Wei, Q. Vision-Based Pose Estimation for Textureless Space Objects by Contour Points Matching. *IEEE Trans. Aerosp. Electron. Syst.* **2018**, *54*, 2342–2355. [CrossRef]
51. Miyake, E.; Takubo, T.; Ueno, A. 3D Pose Estimation for the Object with Knowing Color Symbol by Using Correspondence Grouping Algorithm. In Proceedings of the 2020 IEEE/SICE International Symposium on System Integration (SII), Honolulu, HI, USA, 12–15 January 2020; pp. 960–965.
52. Ulrich, M.; Wiedemann, C.; Steger, C. Combining Scale-Space and Similarity-Based Aspect Graphs for Fast 3D Object Recognition. *IEEE Trans. Pattern Anal. Mach. Intell.* **2012**, *34*, 1902–1914. [CrossRef]
53. Steger, C. Occlusion, clutter, and illumination invariant object recognition. *Int. Arch. Photogramm. Remote Sens.* **2003**, *34*, 345–350.
54. Konishi, Y.; Hanzawa, Y.; Kawade, M.; Hashimoto, M. Fast 6D Pose Estimation from a Monocular Image Using Hierarchical Pose Trees. In *European Conference on Computer Vision*; Springer: Cham, Switzerland, 2016; Volume 9905, pp. 398–413.
55. Muñoz, E.; Konishi, Y.; Murino, V.; Bue, A.D. Fast 6D pose estimation for texture-less objects from a single RGB image. In Proceedings of the 2016 IEEE International Conference on Robotics and Automation (ICRA), Stockholm, Sweden, 16–21 May 2016; pp. 5623–5630.
56. Khosla, A.; Torralba, A.; Lim, J. FPM: Fine Pose Parts-Based Model with 3D CAD Models. In *European Conference on Computer Vision*; Springer: Cham, Switzerland, 2014. [CrossRef]
57. Pei, L.; Liu, K.; Zou, D.; Li, T.; Wu, Q.; Zhu, Y.; Li, Y.; He, Z.; Chen, Y.; Sartori, D. IVPR: An Instant Visual Place Recognition Approach Based on Structural Lines in Manhattan World. *IEEE Trans. Instrum. Meas.* **2020**, *69*, 4173–4187. [CrossRef]
58. Peng, J.; Xu, W.; Liang, B.; Wu, A. Virtual Stereovision Pose Measurement of Noncooperative Space Targets for a Dual-Arm Space Robot. *IEEE Trans. Instrum. Meas.* **2020**, *69*, 76–88. [CrossRef]
59. Chaumette, F.; Hutchinson, S. Visual servo control. II. Advanced approaches [Tutorial]. *IEEE Robot. Autom. Mag.* **2007**, *14*, 109–118. [CrossRef]
60. Comport, A.I.; Marchand, E.; Pressigout, M.; Chaumette, F. Real-time markerless tracking for augmented reality: The virtual visual servoing framework. *IEEE Trans. Vis. Comput. Graph.* **2006**, *12*, 615–628. [CrossRef] [PubMed]
61. Kratochvil, B.E.; Dong, L.; Nelson, B.J. Real-time rigid-body visual tracking in a scanning electron microscope. In Proceedings of the 2007 7th IEEE Conference on Nanotechnology (IEEE NANO), Hong Kong, China, 2–5 August 2007; pp. 442–447. [CrossRef]
62. Yesin, K.B. A CAD model based tracking for visually guided three-dimensional microassembly. *Robotica* **2005**, *23*, 409–418. [CrossRef]
63. Muñoz, E.; Konishi, Y.; Beltran, C.; Murino, V.; Bue, A.D. Fast 6D pose from a single RGB image using Cascaded Forests Templates. In Proceedings of the 2016 IEEE/RSJ International Conference on Intelligent Robots and Systems (IROS), Daejeon, Korea, 9–14 October 2016; pp. 4062–4069.
64. Dalal, N.; Triggs, B. Histograms of oriented gradients for human detection. In Proceedings of the 2005 IEEE Computer Society Conference on Computer Vision and Pattern Recognition (CVPR'05), San Diego, CA, USA, 20–25 June 2005; Volume 881, pp. 886–893.
65. Guo, J.; Wu, P.; Wang, W. A precision pose measurement technique based on multi-cooperative logo. *J. Phys. Conf. Ser.* **2020**, *1607*, 012047. [CrossRef]
66. Hinterstoisser, S.; Cagniart, C.; Ilic, S.; Sturm, P.; Navab, N.; Fua, P.; Lepetit, V. Gradient Response Maps for Real-Time Detection of Textureless Objects. *IEEE Trans. Pattern Anal. Mach. Intell.* **2012**, *34*, 876–888. [CrossRef] [PubMed]
67. Li, X.; Wang, H.; Yi, L.; Guibas, L.J.; Abbott, A.L.; Song, S. Category-Level Articulated Object Pose Estimation. In Proceedings of the 2020 IEEE/CVF Conference on Computer Vision and Pattern Recognition (CVPR), Seattle, WA, USA, 13–19 June 2020; pp. 3703–3712.
68. Wang, K.; Xie, J.; Zhang, G.; Liu, L.; Yang, J. Sequential 3D Human Pose and Shape Estimation From Point Clouds. In Proceedings of the 2020 IEEE/CVF Conference on Computer Vision and Pattern Recognition (CVPR), Seattle, WA, USA, 13–19 June 2020; pp. 7273–7282.
69. Zhang, Z.; Hu, L.; Deng, X.; Xia, S. Weakly Supervised Adversarial Learning for 3D Human Pose Estimation from Point Clouds. *IEEE Trans. Vis. Comput. Graph.* **2020**, *26*, 1851–1859. [CrossRef] [PubMed]
70. Zhang, Y.; Zhang, C.; Rosenberger, M.; Notni, G. 6D Object Pose Estimation Algorithm Using Preprocessing of Segmentation and Keypoint Extraction. In Proceedings of the 2020 IEEE International Instrumentation and Measurement Technology Conference (I2MTC), Dubrovnik, Croatia, 25–28 May 2020; pp. 1–6.
71. Konishi, Y.; Hattori, K.; Hashimoto, M. Real-Time 6D Object Pose Estimation on CPU. In Proceedings of the 2019 IEEE/RSJ International Conference on Intelligent Robots and Systems (IROS), Macau, China, 3–8 November 2019; pp. 3451–3458.
72. Park, K.; Patten, T.; Prankl, J.; Vincze, M. Multi-Task Template Matching for Object Detection, Segmentation and Pose Estimation Using Depth Images. In Proceedings of the 2019 International Conference on Robotics and Automation (ICRA), Montreal, QC, Canada, 20–24 May 2019; pp. 7207–7213.
73. Tamadazte, B.; Marchand, E.; Dembélé, S.; Le Fort-Piat, N. CAD Model-based Tracking and 3D Visual-based Control for MEMS Microassembly. *Int. J. Robot. Res.* **2010**, *29*, 1416–1434. [CrossRef]
74. Chen, Y.; Medioni, G. Object modelling by registration of multiple range images. *Image Vis. Comput.* **1992**, *10*, 145–155. [CrossRef]
75. Besl, P.J.; McKay, N.D. A Method for Registration of 3-D Shapes. *IEEE Trans. Pattern Anal. Mach. Intell.* **1992**, *14*, 239–256. [CrossRef]

76. Guo, Z.; Chai, Z.; Liu, C.; Xiong, Z. A Fast Global Method Combined with Local Features for 6D Object Pose Estimation. In Proceedings of the 2019 IEEE/ASME International Conference on Advanced Intelligent Mechatronics (AIM), Hong Kong, China, 8–12 July 2019; pp. 1–6.
77. Akizuki, S.; Aoki, Y. Pose alignment for different objects using affordance cues. In Proceedings of the 2018 International Workshop on Advanced Image Technology (IWAIT), Chiang Mai, Thailand, 7–9 January 2018; pp. 1–3.
78. Yu, H.; Fu, Q.; Yang, Z.; Tan, L.; Sun, W.; Sun, M. Robust Robot Pose Estimation for Challenging Scenes With an RGB-D Camera. *IEEE Sens. J.* **2019**, *19*, 2217–2229. [CrossRef]
79. Rublee, E.; Rabaud, V.; Konolige, K.; Bradski, G. ORB: An efficient alternative to SIFT or SURF. In Proceedings of the 2011 International Conference on Computer Vision, Barcelona, Spain, 6–13 November 2011; pp. 2564–2571.
80. Calonder, M.; Lepetit, V.; Strecha, C.; Fua, P. BRIEF: Binary robust independent elementary features. In Proceedings of the Proceedings of the 11th European Conference on Computer Vision: Part IV, Heraklion, Crete, Greece, 22 December 2011; pp. 778–792.
81. Nospes, D.; Safronov, K.; Gillet, S.; Brillowski, K.; Zimmermann, U.E. Recognition and 6D Pose Estimation of Large-scale Objects using 3D Semi-Global Descriptors. In Proceedings of the 2019 16th International Conference on Machine Vision Applications (MVA), Tokyo, Japan, 27–31 May 2019.
82. Zhang, H.; Cao, Q. Texture-less object detection and 6D pose estimation in RGB-D images. *Robot. Auton. Syst.* **2017**, *95*, 64–79. [CrossRef]
83. Pan, W.; Zhu, F.; Hao, Y.; Zhang, L. Fast and precise 6D pose estimation of textureless objects using the point cloud and gray image. *Appl. Opt.* **2018**, *57*, 8154–8165. [CrossRef] [PubMed]
84. He, Z.; Jiang, Z.; Zhao, X.; Zhang, S.; Wu, C. Sparse Template-Based 6-D Pose Estimation of Metal Parts Using a Monocular Camera. *IEEE Trans. Ind. Electron.* **2020**, *67*, 390–401. [CrossRef]
85. Zhang, H.; Cao, Q. Detect in RGB, Optimize in Edge: Accurate 6D Pose Estimation for Texture-less Industrial Parts. In Proceedings of the 2019 International Conference on Robotics and Automation (ICRA), Montreal, QC, Canada, 20–24 May 2019; pp. 3486–3492.
86. Pan, W.; Zhu, F.; Hao, Y.; Zhang, L. 6D Pose Estimation Based on Multiple Appearance Features from Single Color Image. In Proceedings of the 2017 IEEE 7th Annual International Conference on CYBER Technology in Automation, Control, and Intelligent Systems (CYBER), Honolulu, HI, USA, 31 July–4 August 2017; pp. 406–411.
87. Dong, Z.; Liu, S.; Zhou, T.; Cheng, H.; Zeng, L.; Yu, X.; Liu, H. PPR-Net:Point-wise Pose Regression Network for Instance Segmentation and 6D Pose Estimation in Bin-picking Scenarios. In Proceedings of the 2019 IEEE/RSJ International Conference on Intelligent Robots and Systems (IROS), Macau, China, 3–8 November 2019; pp. 1773–1780.
88. Chen, W.; Duan, J.; Basevi, H.; Chang, H.J.; Leonardis, A. PointPoseNet: Point Pose Network for Robust 6D Object Pose Estimation. In Proceedings of the 2020 IEEE Winter Conference on Applications of Computer Vision (WACV), Snowmass Village, CO, USA, 1–5 March 2020; pp. 2813–2822.
89. Cao, Q.; Zhang, H. Combined Holistic and Local Patches for Recovering 6D Object Pose. In Proceedings of the 2017 IEEE International Conference on Computer Vision Workshops (ICCVW), Venice, Italy, 22–29 October 2017; pp. 2219–2227.
90. Li, W.; Sun, J.; Luo, Y.; Wang, P. 6D Object Pose Estimation using Few-Shot Instance Segmentation and 3D Matching. In Proceedings of the 2019 IEEE Symposium Series on Computational Intelligence (SSCI), Xiamen, China, 6–9 December 2019; pp. 1071–1077.
91. Chen, X.; Chen, Y.; You, B.; Xie, J.; Najjaran, H. Detecting 6D Poses of Target Objects From Cluttered Scenes by Learning to Align the Point Cloud Patches with the CAD Models. *IEEE Access* **2020**, *8*, 210640–210650. [CrossRef]
92. Li, Y.; Chen, C.; Allen, P.K. Recognition of deformable object category and pose. In Proceedings of the 2014 IEEE International Conference on Robotics and Automation (ICRA), Hong Kong, China, 31 May–7 June 2014; pp. 5558–5564.
93. Li, Y.; Wang, Y.; Yue, Y.; Xu, D.; Case, M.; Chang, S.; Grinspun, E.; Allen, P.K. Model-Driven Feedforward Prediction for Manipulation of Deformable Objects. *IEEE Trans. Autom. Sci. Eng.* **2018**, *15*, 1621–1638. [CrossRef]
94. Caporali, A.; Palli, G. Pointcloud-based Identification of Optimal Grasping Poses for Cloth-like Deformable Objects. In Proceedings of the 2020 25th IEEE International Conference on Emerging Technologies and Factory Automation (ETFA), Vienna, Austria, 8–11 September 2020; pp. 581–586.

applied sciences

MDPI

Article

Using Full Pose Measurement for Serial Robot Calibration

Marek Franaszek * and **Jeremy A. Marvel**

Intelligent Systems Division, National Institute of Standards and Technology, Gaithersburg, MD 20899, USA; jeremy.marvel@nist.gov
* Correspondence: marek.franaszek@nist.gov

Abstract: To ensure smooth robot operations, parameters of its kinematic model and a registration transformation between robot base and world coordinate frame must be determined. Both tasks require data acquired by external sensors that can measure either 3D locations or full 6D poses. We show that use of full pose measurements leads to much smaller robot orientation errors when compared with the outcome of calibration and registration procedures based on 3D data only. Robot position errors are comparable for both types of data. The conclusion is based on extensive simulations of 7 degrees of freedom robot arm and different levels of pseudo-noise perturbing both positional and rotational components of pose.

Keywords: robot calibration; robot remastering; calibration uncertainty; part probing; uncertainty reduction; sensor feedback

Citation: Franaszek, M.; Marvel, J.A. Using Full Pose Measurement for Serial Robot Calibration. *Appl. Sci.* **2022**, *12*, 3680. https://doi.org/10.3390/app12073680

Academic Editor: Luis Gracia and Carlos Perez-Vidal

Received: 4 March 2022
Accepted: 31 March 2022
Published: 6 April 2022

1. Introduction

The topic of robot calibration is well-established, yet it is still a significant factor identified by end-users as being negatively impactful for robot usability and utility [1]. Calibration is followed by registration of robot frame to world frame so the accurate encoder angles can be obtained from inverse kinematic and fed to the robot's controller. Both procedures have a profound impact on robot performance and, as pointed out in [2], "it is impossible to distinguish the end-effector error contributed either by" incorrect model parameters or by inaccurate registration transformation.

Various methods of calibrating a robot's kinematic chain have been developed (e.g., [3–5]). Many of these methods rely on intrinsic kinematic models (e.g., [6–9]), which minimize complicated, nonlinear error functions (unless only linearized error models are considered, which may exchange uncertainty for mathematical simplicity) in at least N-dimensional space, where N is the number of controllable joints in the serial kinematic chain. Calibrations based on extended modeling (i.e., beyond rigid kinematics) include compensating for thermal effects [10], and elastostatic [11] and higher order errors [12]. Likewise, examples of non-kinematics-based calibrations can be seen in [13,14]. There are also compensation techniques that can handle both kinematic and non-kinematic errors, but they require steady calculations and application of corrections during on-line operations [15–17], or dynamically selected pre-calculated, hand–eye calibrations from a table [18].

Robot calibration procedures depend on theoretical models of the mechanical system's forward kinematic. For a serial open chain robot, the Product of Exponentials (POE)—based on screw theory—is thought to be one of the most versatile models that can handle singularities in the popular Denavit–Hartenberg (DH) parameter model [19]. For robots with revolute joints only, each joint is parametrized by a three-dimensional (3D) unit vector indicating axis of rotation, and a 3D vector of any point on the axis line. Calibration procedures for such models rely on Circle Point Analysis (CPA) applied to 3D data acquired with laser tracker or other sensor: positioning the robot into a zero-reference configuration (i.e., where all joint angles are set to zero), and then rotating each joint one by one while keeping all other joints fixed at zero [20,21]. Unfortunately, POE-based models do not explicitly include zero offsets of encoder angles. Accurate estimating of zero offsets is critical because the largest contribution to the robot positioning error (97%) comes from incorrect zero offsets [22]. Performing the zero offsets calibration in CPA causes that errors in registration transformation and in the individual offsets accumulate. This may lead to inconsistent calibration results. For some poses, the calibration process reduced robot pose error seven-fold; for others, it actually increased error twofold [23].

A desired outcome of calibration is the error reduction in full pose of robot end-effector, i.e., in its position and orientation. However, both components belong to two different spaces: position is a vector in 3D space and its components have length units, like millimeters, while orientation matrix is parameterized by three angles in degrees. This causes a fundamental scaling problem when a full pose error is minimized (as discussed in [24], ad hoc introduced scaling factors put more weight either on linear or angular part of pose error and push optimizer towards different solutions). This may become a problem in commercial applications where not only position but also orientation of end-effector is important. For example, in automated drilling, a parallelism between the spindle axis and the normal axis of the drilling plate surface should be below 0.2° [25]. Small orientation error 0.05° required for automated riveting, drilling and spot welding was demonstrated by applying online pose corrections in [26]. Automated fiber placement is another example of industrial application where the orientation of robot end-effector is important [27].

The approach that we introduce in this paper avoids the pitfall of minimization of unbalanced 6D error. First, link twists are determined in the CPA-like procedure from 3D data. Then, using full 6D poses measured by sensors, encoder zero offsets are determined in a separate minimization. The error function used in this minimization does not depend on linear DH parameters (link lengths and linear offsets) nor on the position components of noisy 6D poses acquired by sensors. Once twists and zero offsets are known, they are inserted into another error function, which depends only on position components of sensor data. The remaining linear DH parameters are determined by minimizing this second error function. For comparison, robot calibration based on only 3D sensor data is also performed. Obtained results clearly show that orientation errors of end-effector are smaller when orientation part of 6D data is used. At the same time, the position errors are comparable for both methods.

2. Background

The frame $\boldsymbol{F}_{TCP}$ associated with the robot's Tool Center Point (TCP) coordinate system can be expressed as a 4×4 homogeneous transformation consisting of a 3×3 rotation matrix $\boldsymbol{R}$ and a 3×1 translation vector $\boldsymbol{t}$:

$$\boldsymbol{F}_{TCP} = \begin{bmatrix} \boldsymbol{R}_{3\times3} & \boldsymbol{t}_{3\times1} \\ \boldsymbol{0}_{1\times3} & 1 \end{bmatrix} \tag{1}$$

For a serial, open-chain collaborative robot arm with N revolute joints, the frame $\boldsymbol{F}_{TCP}$ in the robot's base coordinate system can be determined using a forward kinematic model:

$$\boldsymbol{F}_{TCP}(\boldsymbol{\theta}_k) = \boldsymbol{F}_1\boldsymbol{F}_2\ldots\boldsymbol{F}_N\boldsymbol{F}_T \tag{2}$$

where

$$\boldsymbol{\theta}_k = [\theta_{1,k}, \theta_{2,k}, \ldots, \theta_{n,k}, \ldots, \theta_{N,k}] \tag{3}$$

and $\theta_{n,k}$ is the encoder angle of the n-th revolute joint for the k-th robot configuration, $\boldsymbol{F}_n$ is the homogeneous transformation associated with the n-th joint, $n = 1, \ldots, N$, and $\boldsymbol{F}_T$ is a transformation from the robot's flange frame to the TCP. Using the nominal DH parameters, the rotation component of each $\boldsymbol{F}_n$ can be written as

$$\boldsymbol{R}_{n,k}(\theta_{n,k}, \epsilon_n, \alpha_n) = \boldsymbol{R}_z(\theta_{n,k} + \epsilon_n)\boldsymbol{R}_x(\alpha_n) \tag{4}$$

where R_z and R_x are rotations around z and x axis, respectively. Two angular DH parameters in (4) are α_n (the link twist) and ϵ_n (the zero offset angle for the n-th encoder). The translation component of F_n can be expressed as

$$t_n = \left[\begin{array}{ccc} r_n \cos(\theta_{n,k} + \epsilon_n) & r_n \sin(\theta_{n,k} + \epsilon_n) & d_n \end{array} \right]^T, \quad (5)$$

where r_n and d_n are two linear DH parameters (link length and offset), and $[\ldots]^T$ denotes the vector transpose.

From (2) and (4) it can be seen that the rotation part R_{TCP} of the F_{TCP} frame depends only on the rotation components

$$R_{TCP}(\theta_k) = R_{1,k} R_{2,k} \ldots R_{N,k} R_T. \quad (6)$$

This is a general property of serial chain manipulators with revolute joints, and is not dependent on a particular kinematic model (here, we use the DH model for illustration purposes only). In the remainder of this paper, we use the notation

$$R_k = R_{TCP}(\theta_k, \alpha, \epsilon) \quad (7)$$

where $\alpha = [\alpha_1, \alpha_2, \ldots, \alpha_N]$ and $\epsilon = [\epsilon_1, \epsilon_2, \ldots, \epsilon_N]$ are the vectors of the DH angular parameters. Note that the positional component t_k of the $F_{TCP}(\theta_k)$ frame depends on joint angles and all four vectors of the DH parameters:

$$t_k = t_{TCP}(\theta_k, \alpha, \epsilon, r, d). \quad (8)$$

3. Determination of Link Twist

To ensure that forward kinematics correctly predict the tool pose in the robot coordinate frame, the DH parameters must be determined first during the robot calibration process. Once calibrated, they remain fixed during robot on-line operations. Calibration may be performed by installing a spherically-mounted retro-reflector (SMR) at the robot's TCP, and tracking it with a laser tracker. From four vectors of DH parameters (α, ϵ, r, d), the twist angles α can be determined independently of other DH parameters by using 3D data acquired for CPA procedure. The twist angle α_n is defined as the angle between two consecutive joint axes of rotation, u_n and u_{n+1} (the last twist α_N is, by definition, set to zero). If $C_{n,K}$ denotes a set of K 3D points t_k calculated in (8) and acquired for n-th joint in CPA procedure, then these points are distributed along an arch (section of a circle) on a plane in 3D space. Thus, for each joint $n = 1, \ldots, N$, a unit vector c_n normal to the fitted plane can be calculated. While the exact locations of $C_{n,K}$ points depend on all DH parameters (ϵ, α, r, d), vector c_n is parallel to the axis of rotation u_n and, therefore, α_n can be determined from the scalar product of two consecutive axes, $\alpha_n = \arccos(c_n \cdot c_{n+1})$. If $B_{n,K}$ is the set of 3D points measured by laser tracker which correspond to $C_{n,K}$, then a unit vector b_n normal to the plane fitted to $B_{n,K}$ can be calculated and the angle between two consecutive b_n and b_{n+1} is used as the estimate of α_n.

To get correctly estimated twist angles α_n, two important steps must be followed. First, since arccos () is an even function, a sign of estimated angle must be equal to the sign of the default (i.e., theoretical) twist angle, $sign(\alpha_{0,n})$. Second, plane fitting procedure provides only a normal to the plane, its particular direction (up or down) depends on a bounding box containing the points. To remove this ambiguity, fitted normal $\tilde{b}_n$ must obey the right hand rule together with acquired 3D points $B_{n,K}$, which are located on a section of a circle. Thus, the estimated corrected twist angle $\tilde{\alpha}_n$ is determined as

$$\tilde{\alpha}_n = sign(\alpha_{0,n}) \arccos\left(\tilde{b}_n \cdot \tilde{b}_{n+1}\right). \quad (9)$$

4. Robot Calibration Based on 3D Measurements

Once the twist angles $\tilde{\boldsymbol{\alpha}}$ are estimated, they can be inserted in (4) and the remaining DH parameters $(\boldsymbol{\epsilon}, \boldsymbol{r}, \boldsymbol{d})$ can be found in traditional calibration procedure using 3D data. Given K configurations of the arm (i.e., "poses") defined by $\boldsymbol{\theta}_k$ $(k = 1, \ldots, K)$, the SMR is moved to K positions $\boldsymbol{g}_k = [x_k, y_k, z_k]$ in 3D Cartesian space. Since $\boldsymbol{g}_k$ and $\boldsymbol{t}_k$ in (8) are determined in different coordinate frames, the error function Err_{pos} in the calibration process is based on relative distances between two 3D points to avoid a dependence on a registration. For convenience, the whole set of K points can be divided in two halves and then

$$Err_{pos}(\tilde{\boldsymbol{\alpha}}, \boldsymbol{\epsilon}, \boldsymbol{r}, \boldsymbol{d}) = \frac{1}{K/2} \sum_{k=1}^{K/2} (L_k(\tilde{\boldsymbol{\alpha}}, \boldsymbol{\epsilon}, \boldsymbol{r}, \boldsymbol{d}) - h_k)^2 \tag{10}$$

where

$$L_k(\tilde{\boldsymbol{\alpha}}, \boldsymbol{\epsilon}, \boldsymbol{r}, \boldsymbol{d}) = \|\boldsymbol{t}_k - \boldsymbol{t}_{k+K/2}\| \tag{11}$$

and $\boldsymbol{t}_k, \boldsymbol{t}_{k+K/2}$ are determined in (8), $\|\ldots\|$ is the Euclidean norm and $h_k = \|\boldsymbol{g}_k - \boldsymbol{g}_{k+K/2}\|$ is a distance between two points measured by laser tracker.

Thus, the fitted DH parameters $(\tilde{\boldsymbol{\epsilon}}, \tilde{\boldsymbol{r}}, \tilde{\boldsymbol{d}})$ can be estimated by minimizing Err_{pos}

$$(\tilde{\boldsymbol{\epsilon}}, \tilde{\boldsymbol{r}}, \tilde{\boldsymbol{d}}) = \min_{\epsilon, r, d} Err_{pos}(\tilde{\boldsymbol{\alpha}}, \boldsymbol{\epsilon}, \boldsymbol{r}, \boldsymbol{d}). \tag{12}$$

providing the vector of link twist angles $\tilde{\boldsymbol{\alpha}}$ is known. The actual dimension of search space is $3N - 2$ since the distance between two points in (11) does not depend on d_1 and ϵ_1 (the two parameters may have arbitrary values which only affect the registration transformation between robot and sensor). In the remainder of this paper, we call this procedure Method 1.

5. Calibration Based on 6D Measurements

Such data were used for robot calibration using different procedures [14,28,29]. The approach we propose calculates zero offsets $\tilde{\boldsymbol{\epsilon}}$ in a separate minimization based on orientation components of 6D poses and determined earlier twist angles $\tilde{\boldsymbol{\alpha}}$.

In the remainder of this paper, we assume that, for each robot configuration defined by $\boldsymbol{\theta}_k$, there is a corresponding 3×3 rotation matrix $\boldsymbol{G}_k$ provided by an external sensor. Both $\boldsymbol{G}_k$ and $\boldsymbol{R}_k$ in (7) are determined in different coordinate frames. If $\boldsymbol{\Omega}$ denotes the rotation component of registration matrix then, for each k-th robot orientation $\boldsymbol{R}_k$ in (7) and the corresponding $\boldsymbol{G}_k$ measured with the external sensor, the following relation holds:

$$\boldsymbol{G}_k = \mathcal{N}_k \, \boldsymbol{\Omega} \, \boldsymbol{R}_k \tag{13}$$

where $\mathcal{N}_k$ is a small, random rotation accounting for noise in the orientation part of 6D data acquired by sensors. For a pair of orientations $\boldsymbol{G}_k$ and $\boldsymbol{G}_{k'}$ (where $k' = k + K/2$), matrix $\boldsymbol{D}_k$ can be defined as

$$\boldsymbol{D}_k = \boldsymbol{G}_k \boldsymbol{R}_k^{-1} \boldsymbol{R}_{k'} \boldsymbol{G}_{k'}^{-1} = \mathcal{N}_k \mathcal{N}_{k'}^{-1} \tag{14}$$

and its angle of rotation $\psi_k \in [0^\circ \; 180^\circ]$ is calculated as

$$\psi_k = \arccos\left(\frac{1}{2}(trace(\boldsymbol{D}_k) - 1)\right). \tag{15}$$

Matrix $\boldsymbol{D}_k$ and its angle ψ_k depend on the measured joint angle vectors $\boldsymbol{\theta}_k$ and $\boldsymbol{\theta}_{k'}$, the twist angles $\tilde{\boldsymbol{\alpha}}$ estimated earlier, and all zero offsets ϵ_n for $n = 2, \ldots, N$, which can be obtained by minimizing the error function Err_{rot}

$$(\tilde{\epsilon}_2, \ldots, \tilde{\epsilon}_N) = \min_{\epsilon_2, \ldots, \epsilon_N} Err_{rot}(\tilde{\boldsymbol{\alpha}}, \epsilon_2, \ldots, \epsilon_N), \tag{16}$$

where

$$Err_{rot}(\tilde{\boldsymbol{\alpha}}, \epsilon_2, \ldots, \epsilon_N) = \frac{1}{K/2} \sum_{k=1}^{K/2} \psi_k(\tilde{\boldsymbol{\alpha}}, \epsilon_2, \ldots, \epsilon_N). \tag{17}$$

Once the zero offsets $\tilde{\epsilon}$ are estimated, they can be inserted in (12) and the linear DH parameters $\boldsymbol{r}$ and $(d_2, \ldots, d_N)$ can be found by minimizing $Err_{pos}(\tilde{\boldsymbol{\alpha}}, \tilde{\boldsymbol{\epsilon}}, \boldsymbol{r}, \boldsymbol{d})$ in (10). In the remainder of this paper, we call this procedure Method 2.

To show a scaling problem when both position and rotation errors are simultaneously minimized, robot calibration was attempted by minimizing the following error function:

$$Err_{full}(\tilde{\boldsymbol{\alpha}}, \boldsymbol{\epsilon}, \boldsymbol{r}, \boldsymbol{d}) = Err_{pos} + wErr_{rot}, \quad (18)$$

where Err_{pos} is defined in (10), Err_{rot} in (17) and positive scaling factor w ensures correct dimensionality of Err_{full}. In the remainder of this paper, we call this procedure Method 3.

6. Registering Robot Frame

When all robot model parameters are known, i.e., estimated $\tilde{\epsilon}_2, \ldots, \tilde{\epsilon}_N, \tilde{\boldsymbol{\alpha}}, \tilde{\boldsymbol{r}}, \tilde{d}_2, \ldots, \tilde{d}_N$ and arbitrary values are assigned to d_1 and ϵ_1, then a registration transformation (rotation $\boldsymbol{\Omega}$ and translation $\boldsymbol{\tau}$) between the coordinate systems of robot and laser tracker can be determined.

There are many registration techniques, one of the commonly used was developed in [30] and is based on 3D data. For calibration Method 1 described in Section 4, where only 3D data acquired by sensor are available, there is only one possible registration transformation $(\boldsymbol{\Omega}, \boldsymbol{\tau})$. When 6D data are available, the registration transformation can be calculated in two ways. In the first (which we name Registration (1)) $\boldsymbol{\Omega}_1$ is calculated using only the 3D positional parts of full poses, as in [30]. In the second (named hereafter as Registration (2)), the rotation matrix $\boldsymbol{\Omega}_2$ is calculated as the mean rotation $\overline{\boldsymbol{\Omega}}$ calculated properly [31] from the individual matrices $\boldsymbol{G}_k\boldsymbol{R}_k^{-1}$ in (13). Once $\boldsymbol{\Omega}_{1,2}$ are known, the translation vectors $\boldsymbol{\tau}_{1,2}$ can be determined as

$$\boldsymbol{\tau}_i = \overline{\boldsymbol{g}}_s - \boldsymbol{\Omega}_i \, \overline{\boldsymbol{t}}_r \, , \; i = 1,2 \quad (19)$$

where $\overline{\boldsymbol{t}}_r$ and $\overline{\boldsymbol{g}}_s$ are the centroids of the collected 3D positions in the robot and the external sensor frame, respectively.

7. Simulation

All calculations were performed in Matlab. Built-in nonlinear least-square (NLS) optimizer *lsqnonlin* with default input parameters was used to minimize the error function Err_{pos} in (10), Err_{rot} in (17) and Err_{full} in (18). As a starting point for all optimizations, default DH parameters were used.

To test the proposed calibration method, a kinematic model of a 7 degrees-of-freedom (DoF) industrial robot arm KUKA LWR 4+ was used. The robot's default DH parameters $(\boldsymbol{\epsilon}_0, \boldsymbol{\alpha}_0, \boldsymbol{r}_0, \boldsymbol{d}_0)$ are provided in Table 1 (all angular parameters are in degrees and all linear in millimeters). Ground truth (GT) parameters used in simulations were defined as a sum of the defaults and deviations, for example $\boldsymbol{\epsilon}_{GT} = \boldsymbol{\epsilon}_0 + \Delta\boldsymbol{\epsilon}$. Deviations from the default DH parameters are provided in Table 2. Two sets of arbitrarily chosen deviations were used in simulations: small deviations $(\Delta\boldsymbol{\alpha}_1, \Delta\boldsymbol{r}_1, \Delta\boldsymbol{d}_1)$ and large deviations $(\Delta\boldsymbol{\alpha}_2, \Delta\boldsymbol{r}_2, \Delta\boldsymbol{d}_2)$. GT parameters were used to generate noisy sensor data $\boldsymbol{G}_k$ from (7) and $\boldsymbol{g}_k$ from (8)

$$\boldsymbol{G}_k = \mathcal{N}_k \, \boldsymbol{\Omega} \, \boldsymbol{R}_k(\boldsymbol{\theta}_k, \boldsymbol{\alpha}_{GT}, \boldsymbol{\epsilon}_{GT}) \quad (20)$$

and

$$\boldsymbol{g}_k = \boldsymbol{\Omega}\boldsymbol{t}_k(\boldsymbol{\theta}_k, \boldsymbol{\alpha}_{GT}, \boldsymbol{\epsilon}_{GT}, \boldsymbol{r}_{GT}, \boldsymbol{d}_{GT}) + \boldsymbol{\tau} + \boldsymbol{\zeta}_k \quad (21)$$

where $(\boldsymbol{\Omega}, \boldsymbol{\tau})$ is arbitrarily selected transformation between robot and sensor frame, $\boldsymbol{\zeta}_k$ is 3D positional Gaussian noise with standard deviation σ_p, and $\boldsymbol{\xi}_k$ is 3D angular Gaussian noise with standard deviation σ_a, which was used to generate small random rotations.

$$\mathcal{N}_k = \boldsymbol{R}_{ZYX}(\boldsymbol{\xi}_k). \quad (22)$$

Table 1. Default DH parameters.

	1	2	3	4	5	6	7
ϵ_0	180	90	0	0	0	0	0
α_0	−90	−90	−90	90	90	−90	0
r_0	0	0	0	0	0	0	0
d_0	310.5	0	400	0	390	0	78

Table 2. Deviations of DH parameters.

	1	2	3	4	5	6	7
$\Delta\epsilon$	0	−1.4	0.68	0.24	0.54	1.37	0.85
$\Delta\alpha_1$	0.15	−0.1	0.07	−0.04	0.02	−0.06	0
$\Delta\alpha_2$	3.35	−4.1	2.7	−3.4	4.2	−3.6	0
Δr_1	0.25	0.4	0.09	0.3	0.28	0.17	0.06
Δr_2	0.8	1.3	0.65	1.4	0.86	0.38	0.55
Δd_1	0	0.06	−0.14	0.12	0.27	0.08	0.3
Δd_2	0	0.27	−1.45	0.4	1.26	0.3	0.35

In Figure 1a, examples of histograms for x component of vectors ξ_k are shown (histograms for y and z components look similar). In Figure 1b, histograms of corresponding angles of rotation β of small random rotations $\mathcal{N}_k$ are plotted. Note that histograms of ξ_k are well approximated by a Gaussian distribution while non-symmetric histograms of β are well approximated by a Fisher–Bingham–Kent (FBK) distribution [32]. Similar histograms of angles were observed for experimental data acquired with a marker-based pose measuring system, see Figures 1 and 3 in [33].

Tool transformation $\boldsymbol{F}_T$ needed in (7) and (8) was arbitrarily chosen with the caveat that the TCP center is not located on the last axis of rotation so that 3D data acquired for CPA procedure are located on a circle.

For each n-th join, $K_\alpha = 40$ vectors of encoder angles $\boldsymbol{\theta}_{n,k}$ were created such that their components were all zero except $\theta_{n,k}$

$$\theta_{n,k} = \theta_{min,n} + k\delta_\theta, \quad k = 1, \ldots, K_\alpha \tag{23}$$

where $\theta_{min,n}$ and δ_θ were such that all $\theta_{n,k}$ were within a valid range of n-th encoder angles. These angles were then inserted in (21) to generate 3D sensor data from which the twist angles $\tilde{\boldsymbol{\alpha}}$ were estimated as described in Section 3. In order to estimate the remaining DH parameters $(\tilde{\boldsymbol{\epsilon}}, \tilde{\boldsymbol{r}}, \tilde{\boldsymbol{d}})$ and calculate registration transformation $(\boldsymbol{\Omega}, \boldsymbol{\tau})$, another set of $K = 100$ joint angle vectors $\boldsymbol{\theta}_k$ was selected in such a way that corresponding poses $\boldsymbol{F}_{TCP}(\boldsymbol{\theta}_k)$ in (1) were randomly scattered in the workspace that is accessible to the robot arm. In computer simulations, this is the only restriction for selection of tool poses, but additional limitations may arise in lab experiments due to a use of a line-of-sight sensor for pose acquisition.

In addition, a separate batch of $J = 50$ joint angles $\boldsymbol{\theta}_j$ was selected for evaluation of calibration and registration procedures. These test poses were used neither in calibration nor registration. To test the performance of all three procedures, the robot kinematic model $\boldsymbol{F}_{TCP}(\boldsymbol{\theta}_j)$ in (1) was used with the parameters $(\tilde{\boldsymbol{\epsilon}}, \tilde{\boldsymbol{\alpha}}, \tilde{\boldsymbol{r}}, \tilde{\boldsymbol{d}})$ estimated by Method 1, 2 and 3. For Method 1, the registration transformation $(\boldsymbol{\Omega}, \boldsymbol{\tau})$ was calculated using 3D data. For Method 2 and 3, both registrations $(\boldsymbol{\Omega}_1, \boldsymbol{\tau}_1)$ and $(\boldsymbol{\Omega}_2, \boldsymbol{\tau}_2)$ were calculated, as described in Section 6. For each tested arm configuration $\boldsymbol{\theta}_j$ and selected m-th noise levels $(\sigma_a, \sigma_p)_m$, the corresponding rotation $\boldsymbol{G}_j$ and position $\boldsymbol{g}_j$ were calculated in (20) and (21) to simulate noisy 6D measurements acquired by sensor. Then, the mean of J angles $\langle \eta_j \rangle$ of rotations $\boldsymbol{Q}_j$ and the mean of J relative distances $\langle q_j \rangle$ were calculated, where

$$\boldsymbol{Q}_j(\eta_j) = \boldsymbol{G}_j^{-1}\boldsymbol{\Omega}\boldsymbol{R}_j, \; q_j = \|\boldsymbol{\Omega}\boldsymbol{t}_j + \boldsymbol{\tau} - \boldsymbol{g}_j\|, \tag{24}$$

and the transformation $(\boldsymbol{\Omega}, \boldsymbol{\tau})$ was appropriate for each of the three calibration procedures and (for Method 2 and 3) the appropriate Registration 1 or 2. Both calculated means $\langle \eta_j \rangle$ and $\langle q_j \rangle$ were used as metrics to gauge a performance of tested procedures.

These steps were repeated for each of the selected noise level $m = 1, \ldots, M_n$ and both sets of GT parameters corresponding to two deviation vectors: small $(\Delta\boldsymbol{\alpha}_1, \Delta\boldsymbol{r}_1, \Delta\boldsymbol{d}_1)$ and large $(\Delta\boldsymbol{\alpha}_2, \Delta\boldsymbol{r}_2, \Delta\boldsymbol{d}_2)$, as shown in Table 2. $M_n = 16$ noise levels were equally spaced between zero and 0.15 (degrees for σ_a and millimeters for σ_p). In order to estimate a variability of the calculated metrics, all the above calculations were repeated for $N_{rep} = 25$ different realizations of noise (different sequences of pseudo-numbers). Thus, for each i-th instance of noise and each m-th pair of noise levels $(\sigma_a, \sigma_p)_m$, the end-effector errors were calculated: $v_{m,i} = \langle q_j \rangle$ for positional error and $\rho_{m,i} = \langle \eta_j \rangle$ for angular error. As the final results, the averages and standard deviations from all repeats N_{rep} were stored for each m-th noise level:

$$\bar{\rho}_m = \frac{1}{N_{rep}} \sum_{i=1}^{Nrep} \rho_{m,i}, \; \delta\rho_m^2 = \frac{1}{N_{rep}} \sum_{i=1}^{Nrep} (\rho_{m,i} - \bar{\rho}_m)^2 \tag{25}$$

and similarly for positional errors $\bar{v}_m$ and δv_m^2 .

To test a performance of the three error functions Err_{pos}, Err_{rot} and Err_{full} used in calibration, for a few randomly selected noise repeats and strengths, minimization was restarted from 300 randomly scattered initial points (i.e., starting DH parameters) and the final optimized parameters were analyzed. In addition, for Method 3, minimization of Err_{full} was repeated for a few scaling factors w in (18).

In all simulations performed in this study, the distal variant of DH parameters was used [34]. Alternatively, the proximal variant could be used, which would affect derived from it homogeneous matrix $\boldsymbol{F}_{TCP}$. However, not every kinematic model is suitable for describing any robot: a well-known example is a robot with two consecutive joint axes that are parallel to each other. In such a case, the DH model is not continuous and must be replaced by another model, e.g., POE [20], and parameters specific for a given model must be determined. Whichever kinematic model is selected, it is important to consistently use it in a calibration process along with other basic definitions (like use of a right-hand or left-hand coordinate system). With all procedural steps clearly defined and consistently followed, there is no ambiguity in the calibration process.

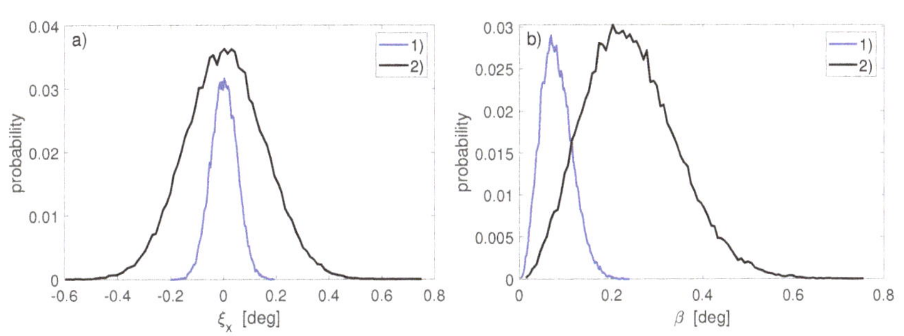

Figure 1. Characteristics of simulated small random rotations $\mathcal{N}_k$ in (22): (**a**) histograms of x component of angle vectors $\boldsymbol{\xi}_k$; (**b**) histograms of angle of rotation β of rotation matrix $\mathcal{N}_k$. Blue lines correspond to weak noise with $\sigma_a = 0.05°$ and black lines correspond to strong noise with $\sigma_a = 0.15°$.

8. Results

Fitted DH parameters revealed different amounts of variations for different simulated conditions. The twist angles $\tilde{\boldsymbol{\alpha}}$ estimated from 3D data generated for the CPA procedure showed moderate variations. The largest absolute deviation $\delta\alpha_{max}$ from the GT value over all N joints and all simulated conditions (M_n noise levels, N_{rep} repeats and both deviations $\Delta\boldsymbol{\alpha}_{1,2}$ from the default values $\boldsymbol{\alpha}_0$) was 0.3°. Zero offsets $\tilde{\boldsymbol{\epsilon}}$ revealed larger deviations: the largest absolute deviation $\delta\epsilon_{max} = 1.18°$. The largest link length deviation was $\delta d_{max} = 4.7$ mm and the largest link offset deviation was $\delta r_{max} = 4.2$ mm. Such large differences between the fitted and the GT parameters were observed mostly for large noise levels σ_p and σ_a.

Figure 2 shows an example of robot end-effector errors at $J = 50$ test poses. Position errors q_j and orientation errors η_j were calculated in (24) for robot DH parameters calibrated with Method 1 and Method 2. Presented errors were calculated for simulated sensor poses perturbed by $i = 14$ noise realization (selected arbitrary from N_{rep} repeats) and $m = 7$ noise levels $(\sigma_a, \sigma_p)_m$. These $(q_j, \eta_j)_{m,i}$ errors were then used to calculate $(v_{m,i}, \rho_{m,i})$ and then, mean errors $\overline{v}_m$ and $\overline{\rho}_m$ in (25) and the corresponding standard deviations δv_m and $\delta\rho_m$ for each m-th noise level. These means and standard deviations were then used to create the plots in the remaining Figures 3–6.

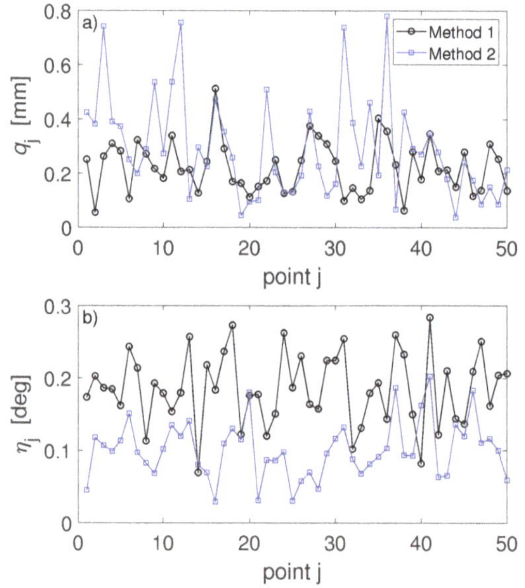

Figure 2. Robot end-effector errors calculated at $J = 50$ test poses for fixed sensor noise $\sigma_p = 0.055$ mm, $\sigma_a = 0.055°$ and one, arbitrary selected noise realization: (**a**) positional errors q_j; (**b**) orientation errors η_j. Robot was calibrated with Method 1 (black lines) and Method 2 (blue lines).

Figure 3 shows the outcomes of two registration transformations $(\boldsymbol{\Omega}_1, \boldsymbol{\tau}_1)$ and $(\boldsymbol{\Omega}_2, \boldsymbol{\tau}_2)$ described in Section 6. In both cases, robot was calibrated with Method 2. GT parameters used in simulation of 6D data, i.e., end-effector poses and noisy poses as measured by sensor, were obtained by modifying the default DH parameters with deviations shown in Table 2. For both registrations, mean errors were calculated at the same values of sensor noise (σ_p in Figure 3a,c and σ_a in Figure 3b,d). In each subplot, two graphs are slightly shifted horizontally only for better visualization. Error bars δv_m in Figure 3a,c and $\delta\rho_m$ in Figure 3b,d are the corresponding standard deviations calculated in (25) from N_{rep} repeated simulations of noisy sensor data.

Figure 4 shows the outcomes of two registration procedures applied after robot was calibrated using Method 3 and the error function Err_{full} defined in (18) with the scaling factor $w = 1$. Presented results were obtained for 6D data generated with GT values of DH parameters deviating from their default values by $(\Delta\boldsymbol{\alpha}_2, \Delta\boldsymbol{r}_2, \Delta\boldsymbol{d}_2)$ shown in Table 2.

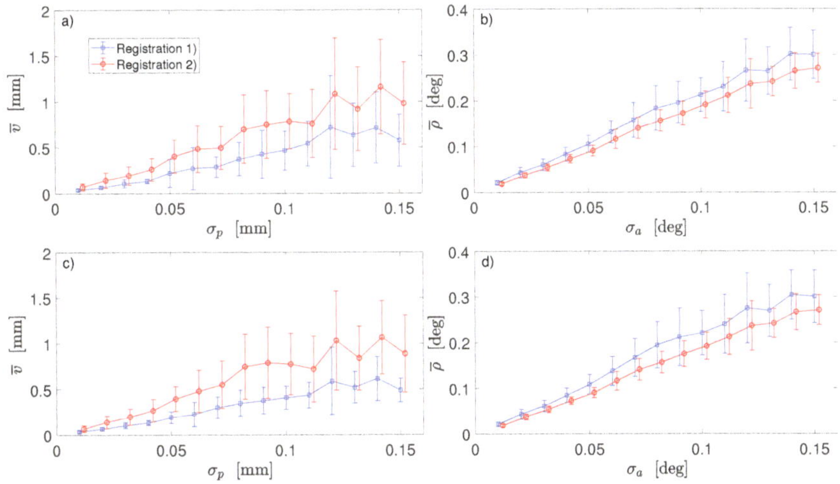

Figure 3. Comparison of two registration procedures for robot calibrated with Method 2 and data generated using: (**a**,**b**)—small deviations from the default DH parameters $(\Delta\boldsymbol{\alpha}_1, \Delta\boldsymbol{r}_1, \Delta\boldsymbol{d}_1)$; (**c**,**d**)—large deviations $(\Delta\boldsymbol{\alpha}_2, \Delta\boldsymbol{r}_2, \Delta\boldsymbol{d}_2)$. Dependence of the mean positional error $\overline{v}$ of robot end-effector on positional noise σ_p in sensor 6D data in (**a**,**d**); dependence of the mean orientation error $\overline{\rho}$ of robot end-effector on angular noise σ_a in sensor 6D data in (**b**,**d**).

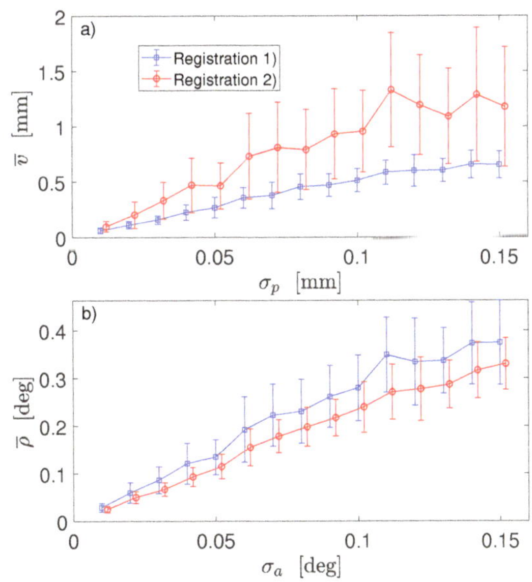

Figure 4. Comparison of two registration procedures for robot calibrated with Method 3 and data generated using large deviations from the default DH parameters: (**a**) dependence of the mean positional error $\overline{v}$ of robot end-effector on positional noise σ_p in sensor 6D data; (**b**) dependence of the mean orientation error $\overline{\rho}$ of robot end-effector on angular noise σ_a in sensor 6D data.

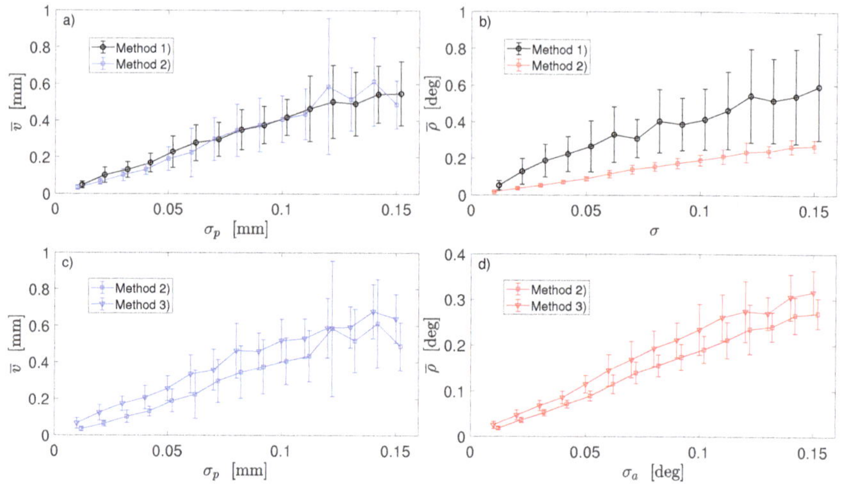

Figure 5. Comparison of three calibration methods: (**a**) dependence of the mean positional error $\overline{v}$ of robot end-effector on positional noise σ_p —Registration 1 was used in Method 2 (blue line); (**b**) dependence of the mean orientation error $\overline{\rho}$ of robot end-effector on noise ($\sigma = \sigma_p$ for Method 1 and $\sigma = \sigma_a$ for Method 2)—Registration 2 was used in Method 2 (red line); (**c**) dependence of error $\overline{v}$ on positional noise σ_p—Registration 1 was used in both methods; (**d**) dependence of error $\overline{\rho}$ on noise σ_a—Registration 2 was used in both methods.

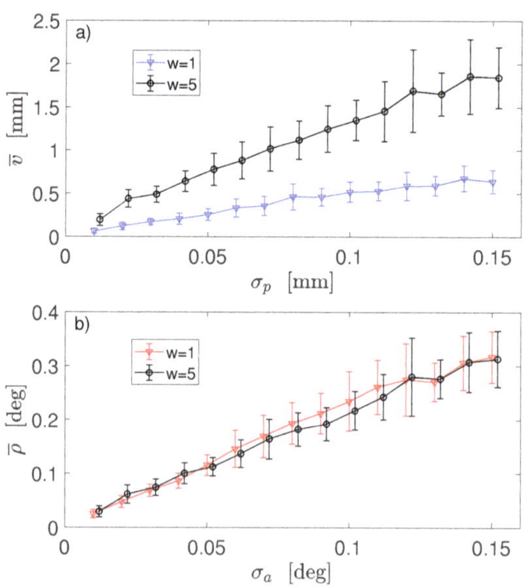

Figure 6. Comparison of robot calibrations using two different scaling factors w in error function Err_{full} in Method 3: (**a**) dependence of the mean positional error $\overline{v}$ of robot end-effector on positional noise σ_p in sensor 6D data—Registration 1 was used; (**b**) dependence of the mean orientation error $\overline{\rho}$ of robot end-effector on angular noise σ_a in sensor 6D data—Registration 2 was used. Data generated using large deviations from the default DH parameters.

Figure 5 shows the outcomes of three calibration procedures: Method 1 based on 3D sensor data, and Method 2 and 3 based on 6D sensor data (in Figure 5b,d noise $\sigma = \sigma_p$ in mm for Method 1 and $\sigma = \sigma_a$ in degrees for Method 2 and 3). Two different registration procedures were used in robot calibration with Method 2 and 3: for positional error $\overline{v}$, Registration 1 was used (blue line in Figure 5a,c, the same as in Figure 3c for Method 2 and the blue line with triangle markers in Figure 5c, the same as blue line in Figure 4a for Method 3). For angular error $\overline{\rho}$, Registration 2 was used (red line in Figure 5b,d, the same as in Figure 3d for Method 2 and the red line with triangle markers in Figure 5d, the same as red line in Figure 4b). Error bars δv_m in Figure 5a,c and $\delta \rho_m$ in Figure 5b,d are the corresponding standard deviations calculated in (25) from N_{rep} repeated simulations of noisy sensor data. On each subplot, the two graphs are slightly shifted horizontally for a visualisation effect. Robot GT parameters used in simulation of 6D data were obtained by modifying the default DH parameters with large deviations $(\Delta\boldsymbol{\alpha}_2, \Delta\boldsymbol{r}_2, \Delta\boldsymbol{d}_2)$ shown in Table 2. Similar results for $\overline{v}$ and $\overline{\rho}$ were obtained when small deviations $(\Delta\boldsymbol{\alpha}_1, \Delta\boldsymbol{r}_1, \Delta\boldsymbol{d}_1)$ were used in simulations.

Figure 6 shows outcome of robot calibration for Method 3 with two different values of the scaling factor w in Err_{full} in (18). Results for Method 3 presented in Figures 4 and 5c,d were obtained for $w = 1$.

For each of the selected cases where the minimization of the error function was repeated from 300 different starting points, all initial DH parameters led to the same solution. Fitted DH parameters depended on noise strengths, choice of error function and GT values of DH parameters.

9. Discussion

In this study, an open-chain robotic manipulator with N revolute joints was calibrated using three different methods and two different sets of data: 3D positions only, and full 6D poses. All three methods share the same strategy for determining link twists $\tilde{\boldsymbol{\alpha}}$. Then, in Method 1, the error function Err_{pos} in (10) was minimized, and the remaining DH parameters $(\tilde{\boldsymbol{\epsilon}}, \tilde{\boldsymbol{r}}, \tilde{\boldsymbol{d}})$ were found by using 3D data only. In Method 2, a search for the zero offsets $\tilde{\boldsymbol{\epsilon}}$ was performed separately by minimizing Err_{rot} in (17), which depends only on the orientation part of full 6D data. Once the zero offsets were known, the remaining DH parameters $(\tilde{\boldsymbol{r}}, \tilde{\boldsymbol{d}})$ were found by minimizing $Err_{pos}(\boldsymbol{r}, \boldsymbol{d})$ in (10) using only the positional part of 6D data. Such an approach reduces the dimensionality of the search space when compared with minimization of Err_{pos} in Method 1. In addition, by using angles ψ_k of relative rotations $\boldsymbol{D}_k$ in error function Err_{rot} in (17) and relative distances L_k between pairs of 3D points in error function Err_{pos} in (10), the proposed strategy decouples robot calibration from registration of the robot frame to the world frame. Different calibration strategies yielded different sets of fitted DH parameters which, in turn, led to different end-effector errors. This is expected, as the optimizer which uses different error functions and different sensor data usually converges to different solutions for the same kinematic model. It should be noted that both Methods 1 and 2 are equally valid and it is a matter of practicality which one is more useful.

In Method 2, two different approaches to registration were used. Rotation $\boldsymbol{\Omega}_1$ from the first approach minimizes distances between the sensor's 3D positions and robot's TCP points for K robot arm configurations [30]. Rotation $\boldsymbol{\Omega}_2$ is calculated as the mean rotation $\overline{\boldsymbol{\Omega}}$ of K relative rotations $\boldsymbol{G}_k\boldsymbol{R}_k^{-1}$ and, thus, minimizes angular distances between orientations of TCP frame and orientations provided by sensor. Therefore, one may expect that $\boldsymbol{\Omega}_2$ is better than $\boldsymbol{\Omega}_1$ in aligning robot orientations with sensor orientations. Indeed, end-effector angular errors $\overline{\rho}$ shown in Figure 3b,d are smaller for $\boldsymbol{\Omega}_2$ in Registration 2 (red line) than for $\boldsymbol{\Omega}_1$ in Registration 1 (blue line).

When it comes to the positional errors $\overline{v}$, the situation is exactly opposite. Both translation vectors $\boldsymbol{\tau}_1$ and $\boldsymbol{\tau}_2$ are calculated in (19). Since $\boldsymbol{\Omega}_2$ does not depend on positional data, the transformation $(\boldsymbol{\Omega}_2, \boldsymbol{\tau}_2,)$ does not minimize (in the least-square sense) the distances between sensor 3D positions and robot TCP points for K robot arm configurations. Transformation $(\boldsymbol{\Omega}_1, \boldsymbol{\tau}_1,)$ does minimize the distances, and therefore is expected to better align the sensor 3D positions with the robot TCP. Indeed, end-effector position errors $\overline{v}$ shown in Figure 3a,c are smaller for $\mathbf{\Omega}_1$ in Registration 1 (blue line) than for $\mathbf{\Omega}_2$ in Registration 2 (red line).

Analysis of the plots in Figure 3 suggests the optimal strategy: instead of choosing either the first $(\mathbf{\Omega}_1, \boldsymbol{\tau}_1,)$ or the second $(\mathbf{\Omega}_2, \boldsymbol{\tau}_2,)$ registration, take the best part from both. Use $\mathbf{\Omega}_2$ to transform orientations $\boldsymbol{R}_k$ of robot end-effector and use $(\mathbf{\Omega}_1, \boldsymbol{\tau}_1,)$ to transform robot TCP $\boldsymbol{t}_k$. Outcome of such strategy is displayed in Figure 5a,b: note blue line for positional errors $\overline{v}$ and red line for angular error $\overline{\rho}$ indicating a use of different registrations in Method 2.

Another advantage of using $\mathbf{\Omega}_2$ to transform the TCP orientations rather than $\mathbf{\Omega}_1$ is that there is a much smaller dispersion of orientation errors $\overline{\rho}$ for different noise realizations. The error bars in Figure 5b are much smaller for Method 2 (which leverages $\mathbf{\Omega}_2$) than for Method 1. This implies that orientations from the world coordinate system can be fed into an inverse kinematic solver more consistently and accurately.

It may appear counter intuitive that mean position errors $\overline{v}(\sigma_p)$ and mean orientation errors $\overline{\rho}(\sigma_a)$ calculated for the same m-th pair of noise strengths $(\sigma_p, \sigma_a)_m$ but different GT values of DH parameters are almost the same, as Figure 3 shows. However, it should not be a surprise since we used NLS optimizer with exact error function. Scale of deviation from the default DH parameters may become an issue when the calibration is performed using approximated, linearlized errors and the Jacobian is calculated at the default DH values.

Results of robot calibration obtained with Method 3 clearly reveal the consequences of scaling problem when simultaneous minimization of both position and orientation errors in one optimization is attempted, as demonstrated in Figure 6. While the mean orientation errors $\overline{\rho}$ are almost equal for two selected values of w, the corresponding position errors $\overline{v}$ differ substantially. This method, similarly as Method 2, uses 6D data and, therefore, two registration procedures are available. In Method 3, similarly to Method 2, smaller position errors are obtained when Registration 1 is applied to the position data and smaller orientations errors are observed when Registration 2 is applied to the orientation data, as results in Figure 4 clearly indicate. Even as both Method 2 and 3 share a possibility of using different registrations for position and rotation components of a full pose, a direct comparison between the two methods clearly points to Method 2 as a better procedure, as demonstrated by the results shown in Figure 5c,d. Thus, a use of Method 3 is discouraged.

The calibration strategy outlined in this paper was tested on a kinematic model of a serial open chain robot with revolute joints only. A question can be asked if the strategy can be applied to a more complex kinematic model when a serial chain has both revolute and prismatic joints. Acquisition of full 6D poses enables calculation of two registrations defined in (19): one of them minimizes a position error and the other minimizes an orientation error. Therefore, as long as full 6D poses are acquired, the outlined calibration strategy could in principle be used for robots with a mixture of revolute and prismatic joints. However, a presence of prismatic joints complicates the error function Err_{pos} in (10) by increasing a number of search variables and it requires further study to verify whether the strategy is beneficial also for robots with revolute and prismatic joints.

The simulation results presented in this paper raise an important, practical question about the characteristics of 6D pose measuring sensors which are used for robot calibration. Commercially available sensors allow quick acquisition of many repeated measurements, which enables the noise in recorded data to be substantially reduced by calculating mean poses. The mean position error of robot end-effector $\overline{v}$ calculated by Method 2 is increasing with sensor position noise σ_p, as Figure 5a shows. If the three sigma rule is followed and approximate relation $\overline{v} \approx 4\sigma_p$ holds, then the upper bound $\hat{\sigma}_p$ for sensor position noise should satisfy $12\hat{\sigma}_p < tol_p$, where tol_p is the acceptable robot position tolerance. For orientation data, due to the strong non-symmetric FBK-like distribution of angles β (which accounts for deviation of noisy, instantaneous rotations from the mean rotation), the three sigma rule can be replaced by calculating quantile $\hat{\beta}_{997}$ of angles β at 0.997 level. Assuming the mean orientation error of robot end-effector $\overline{\rho}$ is four times larger than sensor's orientation noise (as shown in Figure 5b for Method 2), the upper bound for sensor orientation noise should satisfy $4\hat{\beta}_{997} < tol_a$ where tol_a is the acceptable robot orientation tolerance. For different robot, the dependence of end-effector error on sensor noise may be different from that shown in Figure 5a,b. Then, the estimates for upper bounds of position noise $\hat{\sigma}_p$ and orientation noise $\hat{\beta}_{997}$ need to be updated.

The proposed calibration strategy reduces both the position and orientation errors of the robot end-effector. Recommended procedure for serial robot calibration consists of: (1) acquiring the full 6D poses; (2) getting link twists in CPA-like procedure; (3) getting encoder zero offsets using orientation data only; (4) getting link lengths and offsets using position data only. Then, use two separate registrations to transform position and orientation component of a pose from a world to the robot frame. In summary, the dilemma of having only the position or the orientation error of the robot's end-effector minimized can be avoided and a pose with both optimized components can be fed into inverse kinematic solver.

Author Contributions: Conceptualization, M.F. and J.A.M.; methodology, M.F. and J.A.M.; software, M.F.; validation, M.F. and J.A.M.; formal analysis, M.F. and J.A.M.; investigation, M.F.; writing—original draft preparation, M.F.; writing—review and editing, J.A.M.; visualization, M.F.; project administration, J.A.M. All authors have read and agreed to the published version of the manuscript.

Funding: This research received no external funding.

Conflicts of Interest: The authors declare no conflict of interest. Certain commercial equipment, instruments, or software are identified in this paper to foster understanding. Such identification does not imply recommendation or endorsement by the National Institute of Standards and Technology, nor does it imply that the equipment or software identified are necessarily the best available for the purpose.

References

1. Marvel, J.; Messina, E.; Antonishek, B.; Fronczek, L.; Wyk, K.V. *NISTIR 8093: Tools for Collaborative Robots within SME Workcells*; Technical Report; National Institute of Standards and Technology: Gaithersburg, MD, USA, 2015.
2. He, R.; Zhao, Y.; Yang, S.; Yang, S. Kinematic-parameter identification for serial-robot calibration based on POE formula. *IEEE Trans. Robot.* **2010**, *26*, 411–423.
3. Roth, Z.; Mooring, B.; Ravani, B. An overview of robot calibration. *IEEE J. Robot. Autom.* **1987**, *3*, 377–385. [CrossRef]
4. Hollerbach, J.M.; Wampler, C.W. The calibration index and taxonomy for robot kinematic calibration methods. *Int. J. Robot. Res.* **1996**, *15*, 573–591. [CrossRef]
5. Elatta, A.; Gen, L.P.; Zhi, F.L.; Daoyuan, Y.; Fei, L. An overview of robot calibration. *Inf. Technol. J.* **2004**, *3*, 74–78. [CrossRef]
6. Chen, H.; Fuhlbrigge, T.; Choi, S.; Wang, J.; Li, X. Practical industrial robot zero offset calibration. In Proceedings of the 2008 IEEE International Conference on Automation Science and Engineering, Arlington, VA, USA, 23–26 August 2008; pp. 516–521.
7. Nubiola, A.; Bonev, I.A. Absolute calibration of an ABB IRB 1600 robot using a laser tracker. *Robot. Comput.-Integr. Manuf.* **2013**, *29*, 236–245. [CrossRef]
8. Messay, T.; Ordóñez, R.; Marcil, E. Computationally efficient and robust kinematic calibration methodologies and their application to industrial robots. *Robot. Comput.-Integr. Manuf.* **2016**, *37*, 33–48. [CrossRef]
9. Yang, P.; Guo, Z.; Kong, Y. Plane kinematic calibration method for industrial robot based on dynamic measurement of double ball bar. *Precis. Eng.* **2020**, *62*, 265–272. [CrossRef]
10. Li, R.; Zhao, Y. Dynamic error compensation for industrial robot based on thermal effect model. *Measurement* **2016**, *88*, 113–120. [CrossRef]
11. Kalas, V.J.; Vissière, A.; Company, O.; Krut, S.; Noiré, P.; Roux, T.; Pierrot, F. Application-oriented selection of poses and forces for robot elastostatic calibration. *Mech. Mach. Theory* **2021**, *159*, 104176. [CrossRef]
12. Ma, L.M.; Fong, T.; Micire, M.J.; Kim, Y.K.; Feigh, K. Human-Robot teaming: Concepts and components for design. In *Field and Service Robotics*; Springer: Berlin/Heidelberg, Germany, 2018; pp. 649–663.
13. Joubair, A.; Bonev, I.A. Non-kinematic calibration of a six-axis serial robot using planar constraints. *Precis. Eng.* **2015**, *40*, 325–333. [CrossRef]
14. Chen, X.; Zhang, Q.; Sun, Y. Non-kinematic calibration of industrial robots using a rigid–flexible coupling error model and a full pose measurement method. *Robot. Comput.-Integr. Manuf.* **2019**, *57*, 46–58. [CrossRef]
15. Yang, J.; Wang, D.; Fan, B.; Dong, D.; Zhou, W. Online absolute pose compensation and steering control of industrial robot based on six degrees of freedom laser measurement. *Opt. Eng.* **2017**, *56*, 034111. [CrossRef]
16. Zeng, Y.; Tian, W.; Li, D.; He, X.; Liao, W. An error-similarity-based robot positional accuracy improvement method for a robotic drilling and riveting system. *Int. J. Adv. Manuf. Technol.* **2017**, *88*, 2745–2755. [CrossRef]

17. Chen, D.; Yuan, P.; Wang, T.; Ying, C.; Tang, H. A compensation method based on error similarity and error correlation to enhance the position accuracy of an aviation drilling robot. *Meas. Sci. Technol.* **2018**, *29*, 085011. [CrossRef]
18. Franaszek, M.; Cheok, G.S. Using locally adjustable hand-eye calibrations to reduce robot localization error. *SN Appl. Sci.* **2020**, *2*, 839. [CrossRef]
19. Park, F.C. Computational aspects of the product-of-exponentials formula for robot kinematics. *IEEE Trans. Autom. Control* **1994**, *39*, 643–647. [CrossRef]
20. Wang, H.; Shen, S.; Lu, X. A screw axis identification method for serial robot calibration based on the POE model. *Ind. Robot Int. J.* **2012**, *39*, 146–153. [CrossRef]
21. Santolaria, J.; Conte, J.; Ginés, M. Laser tracker-based kinematic parameter calibration of industrial robots by improved CPA method and active retroreflector. *Int. J. Adv. Manuf. Technol.* **2013**, *66*, 2087–2106. [CrossRef]
22. Conrad, K.L.; Shiakolas, P.S.; Yih, T. Robotic calibration issues: Accuracy, repeatability and calibration. In Proceedings of the 8th Mediterranean Conference on Control and Automation (MED2000), Rio, Patras, Greece, 17–19 July 2000; Volume 1719.
23. Choi, Y.; Cheong, J.; Kyung, J.H.; Do, H.M. Zero-offset calibration using a screw theory. In Proceedings of the 2016 13th International Conference on Ubiquitous Robots and Ambient Intelligence (URAI), Xi'an, China, 19–22 August 2016; pp. 526–528.
24. Wu, Y.; Klimchik, A.; Caro, S.; Furet, B.; Pashkevich, A. Geometric calibration of industrial robots using enhanced partial pose measurements and design of experiments. *Robot. Comput.-Integr. Manuf.* **2015**, *35*, 151–168. [CrossRef]
25. Posada, D.J.; Schneider, U.; Pidan, S.; Geravand, M.; Stelzer, P.; Verl, A. High Accurate Robotic Drilling with External Sensor and Compliance Model-Based Compensation. In Proceedings of the IEEE International Conference on Robotics and Automation, Stockholm, Sweden, 16–21 May 2016; pp. 3901–3907.
26. Gharaaty, S.; Shu, T.; Joubair, A.; Xie, W.F.; Bonev, I.A. Online pose correction of an industrial robot using an optical coordinate measure machine system. *Int. J. Adv. Robot. Syst.* **2018**, *4*, 1–16. [CrossRef]
27. Rousseau, G.; Wehbe, R.; Halbritter, J.; Harik, R. Automated Fiber Placement Path Planning: A state-of-the-art review. *Comput.-Aided Des. Appl.* **2019**, *16*, 172–203. [CrossRef]
28. Nguyen, H.N.; Zhou, J.; Kang, H.J. A new full pose measurement method for robot calibration. *Sensors* **2013**, *13*, 9132–9147. [CrossRef] [PubMed]
29. Nubiola, A.; Bonev, I.A. Absolute robot calibration with a single telescoping ballbar. *Precis. Eng.* **2014**, *38*, 472–480. [CrossRef]
30. Arun, K.S.; Huang, T.S.; Blostein, S.D. Least-Squares Fitting of Two 3-D Point Sets. *IEEE Trans. Pattern Anal. Mach. Intell.* **1987**, *9*, 698–700. [CrossRef] [PubMed]
31. Moakher, M. Means and averaging in the group of rotations. *SIAM J. Matrix Anal. Appl.* **2002**, *24*, 1–16. [CrossRef]
32. Kent, J.T. The Fisher-Bingham Distribution on the Sphere. *J. R. Stat. Soc.* **1982**, *44*, 71–80. [CrossRef]
33. Franaszek, M.; Shah, M.; Cheok, G.S.; Saidi, K.S. The axes of random infinitesimal rotations and the propagation of orientation uncertainty. *Measurement* **2015**, *72*, 68–76. [CrossRef]
34. Lipkin, H. A note on Denavit-Hartenberg notation in robotics. In Proceedings of the ASME International Design Engineering Technical Conferences & Computers and Information in Engineering Conference, Long Beach, CA, USA, 24–28 September 2005; pp. 1–6.

applied sciences

MDPI

Article

Longitudinal Mode System Identification of an Insect-like Tailless Flapping-Wing Micro Air Vehicle Using Onboard Sensors

Steven Aurecianus [1], Gi-Heon Ha [2], Hoon-Cheol Park [2] and Tae-Sam Kang [3,*]

[1] Department of Aerospace Information Engineering, Konkuk University, Seoul 05029, Korea; steven.aurecianus@gmail.com
[2] Department of Smart Vehicle Engineering, Konkuk University, Seoul 05029, Korea; hgh6302@gmail.com (G.-H.H.); hcpark@konkuk.ac.kr (H.-C.P.)
[3] School of Mechanical and Aerospace Engineering, Konkuk University, Seoul 05029, Korea
* Correspondence: tskang@konkuk.ac.kr

Featured Application: This paper presents the development of a longitudinal mode dynamic model via experiments of insect-like tailless flapping-wing micro air vehicles (FWMAVs).

Abstract: In this paper, model parameter identification results are presented for a longitudinal mode dynamic model of an insect-like tailless flapping-wing micro air vehicle (FWMAV) using angle and angular rate data from onboard sensors only. A gray box model approach with indirect method was utilized with adaptive Gauss–Newton, Levenberg–Marquardt, and gradient search identification methods. Regular and low-frequency reference commands were mainly used for identification since they gave higher fit percentages than irregular and high-frequency reference commands. Dynamic parameters obtained using three identification methods with two different datasets were similar to each other, indicating that the obtained dynamic model was sufficiently reliable. Most of the identified dynamic model parameters had similar values to the computationally obtained ones, except stability derivatives for pitching moment with forward velocity and pitching rate variations. Differences were mainly due to certain neglected body, nonlinear dynamics, and the shift of the center of gravity. Fit percentage of the identified dynamic model (~49%) was more than two-fold higher than that of the computationally obtained one (~22%). Frequency domain analysis showed that the identified model was much different from that of the computationally obtained one in the frequency range of 0.3 rad/s to 5 rad/s, which affected transient responses. Both dynamic models showed that the phase margin was very low, and that it should be increased by a feedback controller to have a robustly stable system. The stable dominant pole of the identified model had a higher magnitude which resulted in faster responses. The identified dynamic model exhibited much closer responses to experimental flight data in pitching motion than the computationally obtained dynamic model, demonstrating that the identified dynamic model could be used for the design of more effective pitch angle-stabilizing controllers.

Keywords: system identification; flapping-wing micro air vehicle; longitudinal mode; model refinement; gray box model; onboard sensors

Citation: Aurecianus, S.; Ha, G.-H.; Park, H.-C.; Kang, T.-S. Longitudinal Mode System Identification of an Insect-like Tailless Flapping-Wing Micro Air Vehicle Using Onboard Sensors. *Appl. Sci.* **2022**, *12*, 2486. https://doi.org/10.3390/app12052486

Academic Editors: Luis Gracia and Carlos Perez-Vidal

Received: 26 January 2022
Accepted: 24 February 2022
Published: 27 February 2022

1. Introduction

Nature provides investigators with abundant insights and inspirations that aid in solving engineering problems and creating new inventions [1,2]. With the advancement of science, engineering, and technology, a growing number of robots have been designed in different shapes and sizes to mimic biological systems [3,4]. These biomimetic robots mimic animals, such as four-legged animals (quadrupeds) [5], monkeys [6], snakes [7], fish [8,9], bats [10,11], and various birds and insects [12–14]. These robots have diverse potential applications, ranging from simple toys to exploration robots in dangerous and hazardous environments. They have the potential to perform tasks that conventional robots cannot accomplish, such as climbing steep surfaces [15,16], maneuvering through rubble [17], and walking on uneven terrains [5]. These unique capabilities that only animals can perform have motivated many researchers to develop biomimetic robots.

The flapping-wing micro air vehicle (FWMAV) is one such type of robot. It has been popular for more than a decade [12,18–20]. With birds and insects as the main source of inspiration for designing robots, FWMAVs are typically categorized into bird- or insect-inspired FWMAVs. Alternatively, they are differentiated into tailed or tailless FWMAVs based on the existence of the control surface on their tails, by which the differentiation agrees with the physical configurations of birds or insects, respectively [21–25]. In general, birds and insects have unique flight characteristics, which can be observed from the manner in which their wings are manipulated and modulated. Compared to birds, insects normally flap with higher flapping frequency and larger stroke amplitude to produce sufficient force to fly. The flapping stroke plane of insects is nearly horizontal, which causes similar lift force to be produced during a downstroke and an upstroke, whereas the flapping stroke plane of birds is practically vertical, which causes most of the lift force to be produced during downstroke [23,24]. These flight characteristics are mimicked by FWMAVs using various approaches and engineering designs, especially for producing wing flapping motion and other kinematics to generate force and moment [19,20,26].

Unlike tailed FWMAVs, tailless FWMAVs can exhibit steady hover and high maneuverability during a flight [27], indicating their potential applications in restricted spaces. To date, several tailless FWMAV designs have successfully performed stable flight, including the Nano Hummingbird [27], RoboBee [28], Delfly Nimble [29], KUBeetle [30–35], robotic hummingbird [36], Colibri [37], NUS-Robobird [38], and Purdue Hummingbird [39]. There are some similarities between each design. Nano Hummingbird, KUBeetle, robotic hummingbird, and Colibri use only one DC motor to generate thrust through one pair of wings and a few servos to manipulate the wing kinematics for control moment generation. They are mainly different in system integration and various transmission mechanisms, such as combination of string and crankshaft [27], combination of 4-bar linkage and pulley-string [30], and combination of slider-crank and 4-bar linkage [37]. In comparison, Delfly Nimble and NUS-Robobird use two DC motors to manipulate the flapping of two pairs of wings to generate thrust and roll moments. Servos are used to adjust the relative angle of two flapping mechanisms for pitch and yaw moment generation. A different control strategy was found in Purdue Hummingbird where each wing, one left and one right, is controlled by a DC motor to realize fully decoupled and independent wing kinematics manipulation. In this way, no servos are needed because control moments are generated by changing the wing flapping range and speed. While most tailless FWMAVs with successful flights use DC motors and servos, RoboBee uses piezoelectric actuators driven with high voltage. Stabilizing tailless FWMAVs is a significantly important objective in the development of tailless FWMAVs for flying. Owing to the inherent instability of tailless FWMAVs, feedback controllers are required to stabilize them during flight. In comparison, tailed FWMAVs, such as the Microbat [40], Delfly II [41], TL-Flowerfly [42], Konkuk University Biomimetic Ornithopter [43], H2Bird Ornithopter [44], and Robird [45], have an inherent stability [46] so that they can maintain equilibrium without any active manipulation of wing or tail kinematics. Recent findings on Hawkmoth [47] have indicated that the vibration of flapping wings potentially contributes to the improved stability of large insects and FWMAVs through a passive vibrational stabilization mechanism.

Owing to their feasibility and simplicity, proportional–integral–derivative (PID) feedback controllers and their simple variants are often implemented, particularly during the early development of tailless FWMAVs [19,30,37,38]. More complex variants of PID feedback controllers incorporating different strategies (such as neural network [48], particle swarm optimization [49], and dominant pole placement [50]) and autotuning algorithms (such as optimal approach [51,52] and fuzzy adaptive [53]) are being developed for various applications. As with many biomimetic robots, tailless FWMAVs have less complicated body structures and control mechanisms than their biological counterparts. Thus, they can be considered as simplified versions of insects. However, owing to the nonlinear, time-varying, and periodic nature of these systems, describing the mechanics of tailless FWMAVs is still a considerable challenge. In addition, as tailless FWMAVs have different wing configurations [54–57], actuators, and control mechanisms [19,27–39], they cannot be fully generalized. Many efforts have been made to model the flapping kinematics, aerodynamics, and flight dynamics of flyers and FWMAVs. They are derived computationally or experimentally with various assumptions, ranging from time-invariant linear to nonlinear time-periodic models [58–60]. One of the main purposes of obtaining a dynamic model is to design a flight controller with various control techniques and schemes, such as PID [34,61], linear–quadratic–Gaussian [62], state feedback [63,64], nonlinear [65–67], sliding mode [68,69], robust [70,71], adaptive [72], adaptive neural network [73,74], and model-free [75]. While many dynamic models and controllers have been successfully developed, only a limited number have been implemented on tailless FWMAVs because some tailless FWMAVs cannot generate sufficient lift to perform free flight for implementation and verification.

In this study, the system identification of a tailless FWMAV called KUBeetle was conducted to improve the previously derived longitudinal mode extended dynamic model [34] and obtain a more accurate dynamic model to design a controller. The fabrication of tailless FWMAVs with successful flight makes it possible for system identification based on measured input and output data to be conducted [65,76–82]. This approach can potentially identify real dynamics and address the design changes of tailless FWMAVs to obtain a dynamic model with a high degree of similarity to actual systems. In a previous study [65], three single-axis closed-loop system identifications of the robotic hummingbird were conducted using a gimbal system and a Vicon motion tracking system with step reference command as the input. A second-order system was used to describe the pitch, roll, and yaw dynamic. A theoretical linear model for the piezo actuator positioning of Robobee wings' [76] was developed using an off-board displacement laser sensor. The identified model had an error of around 5%, but it was restricted to the wing driving system. A longitudinal gray box model system identification using open-loop system identification procedures was conducted on Delfly Nimble [77]. Data were obtained using the OptiTrack motion tracking system and the onboard system with doublet input as a reference command. The obtained linear time-averaged longitudinal dynamic models for hover and forward flight up to 1 m/s showed more than 80% fit.

The majority of system identification research uses high-speed camera systems with high data rates to obtain experimental flight data. High-speed camera systems are very expensive. In addition, the camera marker balls are relatively heavy compared to the small mass of the KUBeetle (only 17.7 g), which could affect the dynamics. Moreover, there exists little difference between the angle measured by a high-speed camera system and that obtained by an onboard sensor, due to misalignment, scale factor errors, and added measurement noises. So, the model obtained using the high-speed camera system should be modified to be used in a real feedback control system, while the dynamic model obtained using onboard sensors can be used directly without modification.

Many system identification methods have been developed. They are available for open-loop and closed-loop system identification [78–82]. Because KUBeetle flight is unstable, a closed-loop system identification was chosen. Considering prior knowledge of the dynamics of the KUBeetle and the PD feedback controller, a gray box model approach with indirect methods [79,82] was utilized. The longitudinal mode extended dynamic model of KUBeetle, categorized as a linear, time-invariant, and single-input multiple-output system with state space model structure, was refined with the help of system identification tools in MATLAB®. Several variations of input and output data pairing and different identification methods on model refinement were investigated and compared. By investigating dynamic model parameters and time-domain responses, the most appropriate refined dynamic model was obtained. The main contributions of this paper are: (1) dynamic model parameters for the new KUBeetle using only low-cost onboard sensors outputs; and (2) stability and frequency domain analyses, which could be used for designing a closed-loop feedback controller.

The remainder of this paper is organized as follows. Section 2 describes the design and longitudinal mode dynamic model of the KUBeetle. Section 3 presents experimental flight, system identification setups, and the results. Finally, Section 4 provides conclusions and indicates towards future research topics.

2. KUBeetle

2.1. KUBeetle Design

Figure 1 shows the KUBeetle [30–35,83,84], a tailless FWMAV used for parameter identification in this study. The design of the KUBeetle in this study slightly differed from that of a previously developed KUBeetle [34]. They were mainly different in their flapping mechanisms and the location of the control board. One pair of wings of the KUBeetle were modulated by a DC motor using a rack and pinion mechanism to perform a high-amplitude flapping motion for thrust generation [85], whereas previous models used a pulley-string mechanism [30,31]. In the current KUBeetle, the control board was placed between the yaw servo motor and battery, which could lower the center of gravity (CG) position compared with the older version of the KUBeetle. The control board was designed to receive the reference command to control the FWMAV and transmit flight data in real time through a wireless communication RF module extension. A previous study [34] explains the detailed design and specifications of the control board. Figure 1b shows that the flight stability and control are maintained by manipulating the wing stroke plane using three servo motors to generate control moments for pitch, roll, and yaw.

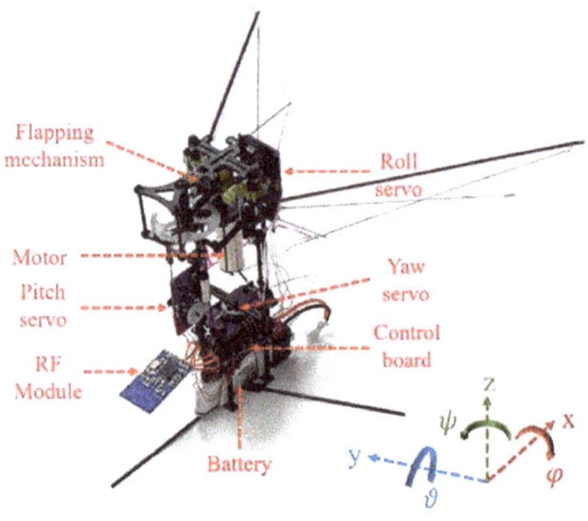

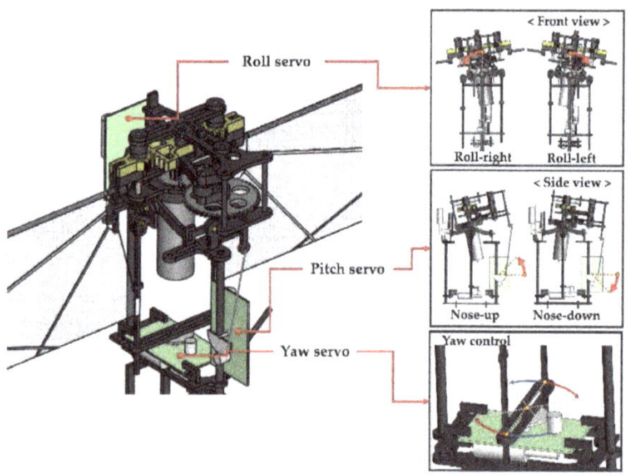

Figure 1. A flight model of the KUBeetle: (**a**) detailed construction components used in the KUBeetle; (**b**) principle of the stroke-plane-change mechanism for control moment generation.

2.2. Dynamic Model

System identification using a gray box model requires an initial dynamic model with starting parameters. The longitudinal mode dynamic model of the KUBeetle used in this study was based on Newton–Euler equations of motion at near-hover conditions [86–89]. The dynamic model was extended to include filters and servo motor dynamics [34]. Parameters (specifically, the moments of inertia and filter parameters) were updated according to the latest KUBeetle design. Table 1 shows the stability and control derivatives obtained from the computational fluid dynamics (CFD) analysis and force and moment measurements. Figure 1 shows the body frame of the KUBeetle. The extended longitudinal mode dynamic model is described below [34,86–89]:

$$\dot{x}_{ext} = A_{ext}\, x_{ext} + B_{ext}\, u_{ext}, \tag{1}$$

$$y_{ext} = C_{ext}\, x_{ext}, \tag{2}$$

$$x_{ext} = \begin{bmatrix} u \\ w \\ q \\ \vartheta \\ \hat{q}_{lpf} \\ \hat{\vartheta}_{lpf} \\ \hat{\vartheta}_{kf,angle} \\ \hat{\vartheta}_{kf,bias} \\ u_{servo} \end{bmatrix}, \tag{3}$$

where u and w are velocities along the x- and z-axis, respectively; q is the pitch rate (rad/s); ϑ is the pitch angle (rad); u_{ext} is the control input in the form of pitch servo angle; y_{ext} is the output or response of the KUBeetle; $\hat{q}_{lpf}$ is the low-pass filtered pitch rate; $\hat{\vartheta}_{lpf}$ is the low-pass filtered pitch angle; $\hat{\vartheta}_{kf,angle}$ and $\hat{\vartheta}_{kf,bias}$ are the pitch angle and bias estimation of the Kalman filter, respectively, and u_{servo} is the servo output. The system and input matrices of the extended dynamic model are given below [34,86–89]:

$$A_{ext} = \begin{bmatrix} \frac{X_u}{m} & \frac{X_w}{m} & \frac{X_q}{m} & g & 0 & 0 & 0 & 0 & \frac{X_{\gamma p}}{m} \\ \frac{Z_u}{m} & \frac{Z_w}{m} & \frac{Z_q}{m} & 0 & 0 & 0 & 0 & 0 & \frac{Z_{\gamma p}}{m} \\ \frac{M_u}{I_{yy}} & \frac{M_w}{I_{yy}} & \frac{M_q}{I_{yy}} & 0 & 0 & 0 & 0 & 0 & \frac{M_{\gamma p}}{I_{yy}} \\ 0 & 0 & 1 & 0 & 0 & 0 & 0 & 0 & 0 \\ 0 & 0 & 204.3 & 0 & -204.3 & 0 & 0 & 0 & 0 \\ 0 & 0 & 0 & 10.13 & 0 & -10.13 & 0 & 0 & 0 \\ 0 & 0 & 0 & 0 & 1 & 0.329 & -0.329 & 1 & 0 \\ 0 & 0 & 0 & 0 & 0 & -0.05 & 0.05 & 0 & 0 \\ 0 & 0 & 0 & 0 & 0 & 0 & 0 & 0 & -30.67 \end{bmatrix}, \tag{4}$$

$$B_{ext} = \begin{bmatrix} 0 \\ 0 \\ 0 \\ 0 \\ 0 \\ 0 \\ 0 \\ 0 \\ 30.91 \end{bmatrix}, \tag{5}$$

where X and Z are aerodynamic forces along the x- and z-axes, respectively, and M is the aerodynamic pitching moment about the CG. Subscripts u, w, and q indicate their aerodynamic derivatives with respect to u, w, and q. The subscript γp implies control input by the pitch servo. m is the mass of the KUBeetle (17.7×10^{-3} kg), I_{yy} is the pitching moment of inertia (12.43×10^{-6} kg m^2), and g is the gravitational acceleration (9.81 m s^{-2}). The first four rows of the system matrix consisted of parameters of the basic dynamic model [34]. In the fifth and sixth rows, parameters of low-pass filters for pitch rate and pitch angle were included, respectively. The seventh and eighth rows included Kalman filter parameters for pitch angle and bias estimation. Finally, the last row contained servo motor dynamic parameters. The output matrix of the extended dynamic model, C_{ext} (open-loop) or $C_{ext,cl}$ (closed-loop), can be adjusted according to the number of available outputs for system identification. For pitch angle and pitch rate outputs, the output matrix is given as follows:

$$C_{ext1} = C_{ext,cl1} = \begin{bmatrix} 0 & 0 & 0 & 0 & 1 & 0 & 0 & 0 & 0 \\ 0 & 0 & 0 & 0 & 0 & 0 & 1 & 0 & 0 \end{bmatrix}. \tag{6}$$

For pitch angle output only, the output matrix was given as:

$$C_{ext2} = C_{ext,cl2} = \begin{bmatrix} 0 & 0 & 0 & 0 & 0 & 0 & 1 & 0 & 0 \end{bmatrix}. \tag{7}$$

Table 1. Stability and control derivatives.

X_u [N s m^{-1}]	X_w [N s m^{-1}]	X_q [N s rad^{-1}]	Z_u [N s m^{-1}]	Z_w [N s m^{-1}]	Z_q [N s rad^{-1}]
-2.39×10^{-2}	2.79×10^{-4}	6.22×10^{-4}	1.06×10^{-3}	-1.34×10^{-2}	1.15×10^{-4}
M_u [N s]	M_w [N s]	M_q [N s m rad^{-1}]	$X_{\gamma p}$ [N rad^{-1}]	$Z_{\gamma p}$ [N rad^{-1}]	$M_{\gamma p}$ [N m rad^{-1}]
-8.3×10^{-4}	-1.6×10^{-5}	-6.2×10^{-6}	1.32×10^{-1}	-2.03×10^{-2}	1.38×10^{-2}

In general, it is not feasible or practical to identify an unstable system. Therefore, a simple PD feedback controller was added to the unstable system to obtain a stable closed-loop control system. The same PD gains in a previous study [34] were used as the starting reference. They were then slightly tuned through experimental flight. Figure 2 shows the data acquisition structure of system identification with a PD feedback controller. Pitch rate estimation was obtained after the pitch rate output of the dynamic model was filtered by a low-pass filter [34]. Considering real time implementation, a first order low-pass filter (cutoff frequency 32 Hz) was used to reduce the high-frequency noise because of its simplicity which took a small calculation time and memory of the microprocessor. Pitch angle estimation was obtained by combining low-pass filtered pitch rate output and pitch angle estimated from acceleration outputs using the Kalman filter [34]. The KUBeetle showed stable flight performance with the current filters, especially on hovering, although there were some oscillating motions, indicating that the current setup was sufficient for an experimental flight.

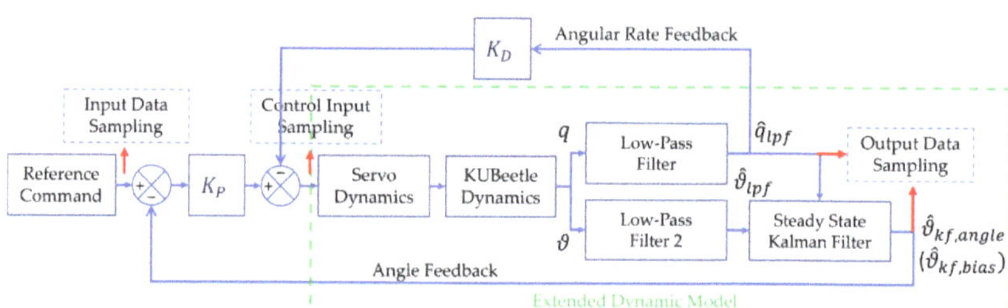

Figure 2. Data acquisition structure of system identification with a PD feedback controller.

A closed-loop system identification is more challenging than an open-loop system identification due to existing feedback. The feedback can affect system identification performance by making the closed-loop system less sensitive to parameter changes in an open-loop system [79]. In addition, the natural dynamic of the KUBeetle toward the reference command might be regarded as a disturbance that needs to be damped and minimized by the controller [77,78]. However, the feedback also allows for more input in a specific frequency range without increasing the output magnitude [79]. Two different methods [79,82], direct and indirect, were tested to identify the longitudinal mode dynamic model of KUBeetle. According to Figure 2, the direct method used sets of control input and output data to obtain the open-loop dynamic model, while the indirect method used sets of reference command and output data to obtain the closed-loop dynamic model. It is possible to use the indirect approach, because the structure of the state space of the closed-loop extended dynamic model is fully known.

A simple PD controller with reference command is given bellow [34]:

$$u_{pd} = K_P\left(\vartheta_{ref} - \hat{\vartheta}_{kf,angle}\right) - K_D\hat{q}_{lpf}, \tag{8}$$

where u_{pd} is the control input, ϑ_{ref} is the angle reference command, K_P is the proportional controller gain, and K_D is the derivative controller gain. The system matrix and input matrix of the closed-loop extended dynamic model can be expressed as follows:

$$A_{ext,cl} = \begin{bmatrix} & & & & A_{ext}(1:8,\ 1:9) & & & & \\ 0 & 0 & 0 & 0 & -30.91K_D & 0 & -30.91K_P & 0 & -30.67 \end{bmatrix}, \tag{9}$$

$$B_{ext,cl} = \begin{bmatrix} B_{ext}(1:8,\ 1) \\ 30.91K_P \end{bmatrix}. \tag{10}$$

The addition of a PD controller to the extended dynamic model only changed the last row of the state space matrix where the control input was given to the servo. Either an open-loop or a closed-loop extended dynamic model could be used as the starting reference in system identification with the gray box model approach. The open-loop extended dynamic model could be retrieved simply by setting controller gain parameters on the last row of the system matrix to zero and changing the last parameter of the input matrix to 30.91, according to Equations (4) and (5). Moreover, it should be noted that velocities along the x- and z-axes were not measured. They could not be used for system identification, which makes the system identification more challenging.

Parameters of the filters, servo motor, and PD controller were fixed, because their values were already known and verified [34]. The gravity, integrator, and zero constants were also fixed. Thus, only nine stability derivatives and three control derivates were refinable in system identification. Each parameter had different physical characteristics and an effect on the dynamic of the KUBeetle [87–89]. Based on CFD analysis [89], X_u, Z_w, and M_u should be negative and have relatively large magnitudes; X_w, Z_u, Z_q, and M_w should be close to zero; and X_q and M_q should have relatively small magnitudes. Overall, X_u, X_q, Z_w, M_u, and M_q, the five parameters of stability derivatives, and $X_{\gamma p}$, $Z_{\gamma p}$, and $M_{\gamma p}$, the three parameters of control derivatives, were mainly focused on for system identification. During system identification, these eight parameters were refined with a relatively wide range of restriction. Conversely, non-focused parameters of stability derivatives were kept close to zero with a narrow range of restriction. Restrictions were adjusted considering computational fluid dynamics (CFD) results [87–89] and current KUBeetle design specifications to maintain the integrity of the identified dynamic model in relation to the physical system. While relaxing parameter restrictions might improve the result of system identification, some refined parameters might not match the real physical system.

3. Experimental Flight and System Identification

In this study, the system identification process was performed in the following steps: (1) experimental flight setup; (2) data acquisition for system identification via experimental flight; (3) linear model refinement in system identification using the gray box model approach; and (4) dynamic model analysis and verification.

3.1. Experimental Flight Setup

Figure 3 shows the experimental flight setup to obtain flight data for system identification. This setup consisted of a remote control, the KUBeetle, and a ground station. The remote control was used for throttle adjustment for taking off, hovering, and landing the KUBeetle. It was also used for adjusting and maintaining trim conditions during the experimental flight. The flight was conducted indoors without wind disturbance. All experimental flight data were obtained and processed with an onboard control board equipped with an inertial measurement unit (MPU-9250, InvenSense, San Jose, CA, USA). Flight data were then transmitted via wireless communication (NRF24L01+, Nordic Semiconductor, Trondheim, Norway) to the ground station. Considering that the bandwidth of the longitudinal mode dynamics with the PD controller was less than 20 rad/s, the sampling rate was set to be 100 Hz. Recorded data consisted of time, pitch reference command, pitch angle, and pitch rate. In addition, the control input was calculated from the reference command and output data with known control gains.

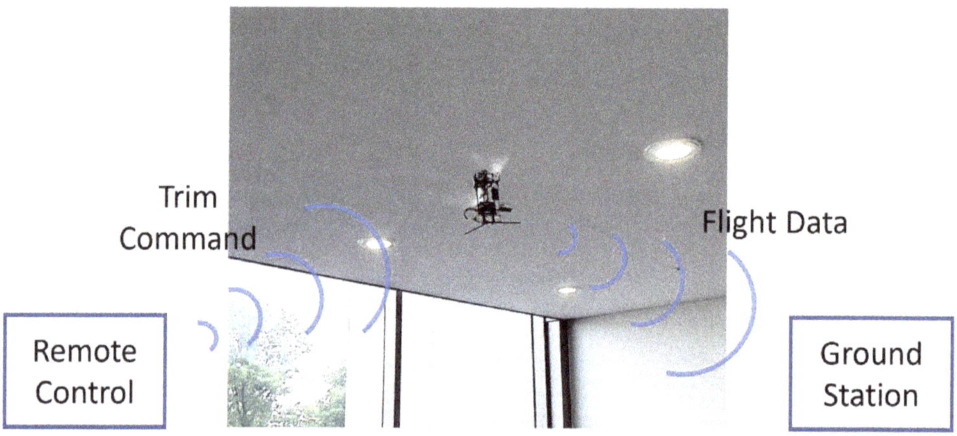

Figure 3. Experimental flight setup for data acquisition.

3.2. Data Acqusition for System Identification via Experimental Flight

Initially, the KUBeetle was manually kept in a hovering condition with minimum drift via a remote control. In hovering condition, the preprogrammed command in the control board was triggered through the remote control to automatically apply reference commands to the KUBeetle. Preprogrammed command was used to ensure the repeatability and consistency during each experimental flight. The reference command is a very important component of the system identification because it affects entire aspects of the system identification process. The reference command should be able to excite as many dynamic modes of interest and natural dynamics as possible to provide comprehensive data for system identification [77–80]. With the dynamic of the KUBeetle in consideration, two different reference commands were used in the experimental flight. The first reference command was in the form of a regular square signal with a relatively long and constant period (amplitude: $\pm 17^\circ$ (0.3 rad); period: 6 s) to analyze transient and steady state responses of the KUBeetle (Supplementary Materials Video S1). The second reference command consisted of an irregular square signal with varying periods (amplitude: $\pm 17^\circ$ (0.3 rad); period: 0.2 to 1.4 s) to cover the frequency bandwidth of interest (Supplementary Materials Video S2). The second reference command was generated using a pseudo-random binary signal, which was a periodic and deterministic signal with properties such as white noise [79]. The amplitude and period of the reference command were chosen considering the operating range of the KUBeetle. Assuming the same amplitude and frequency range, the square signal covered a similar range to that of a chirp signal, as indicated by the frequency spectrum of reference commands shown in Figure 4:

For clarity, the extended dynamic model with parameters obtained from CFD and force and moment measurements, later used as the initial dynamic model in system identification, is called the CFD dynamic model. Figure 5 shows experimental and simulation results of the KUBeetle with a regular square command for pitch. Together with the preprogrammed command, the constant trim control input from the remote control was applied to the KUBeetle. The constant trim command was subtracted to obtain the reference command from the actual command input applied. Small noises were added while the trim control input was being read from the pulse-width-modulated signal. The reference command was the summation of the preprogrammed command and the added trim command noise. Figure 5a shows that the closed-loop KUBeetle has a relatively slow response, high overshoot, and large steady state error. After the reference command was provided, the KUBeetle attempted to follow the reference command in the beginning by responding quickly according to the stroke plane change via the servo motor. However, it was gradually stabilized to a low pitching angle, which maintained the large steady state error. This phenomenon was mainly caused by the position of the CG that affected the force and the moment generated by the gravity force during flight. The CG was located at a relatively large distance below the aerodynamic center (AC). Figure 5b presents the pitch rate responses to a regular square reference command, showing that the measured pitch rate still had a large periodic and high-frequency noise due to wing flapping.

Figure 6 demonstrates experimental and simulation results of the KUBeetle with an irregular square reference command. Figure 6a shows the pitch angle responses. In some intervals, the pitch angle response misleadingly looks as if it is following the command signal well, although some phase delays exist. This is due to overshooting responses at the beginning of step inputs. However, as the frequency of the reference command increased, it could not follow the reference command. Figure 6b illustrates that the pitch rate was contaminated with large noise, as in the regular square reference command case shown in Figure 5b. The high-frequency noise was still very large on the angular rate signal, although it was low-pass filtered (cutoff frequency 32 Hz). However, it was not easy to remove it without invoking much phase delay, which degraded the closed-loop stability. Thus, the current setup was kept for the experimental flight.

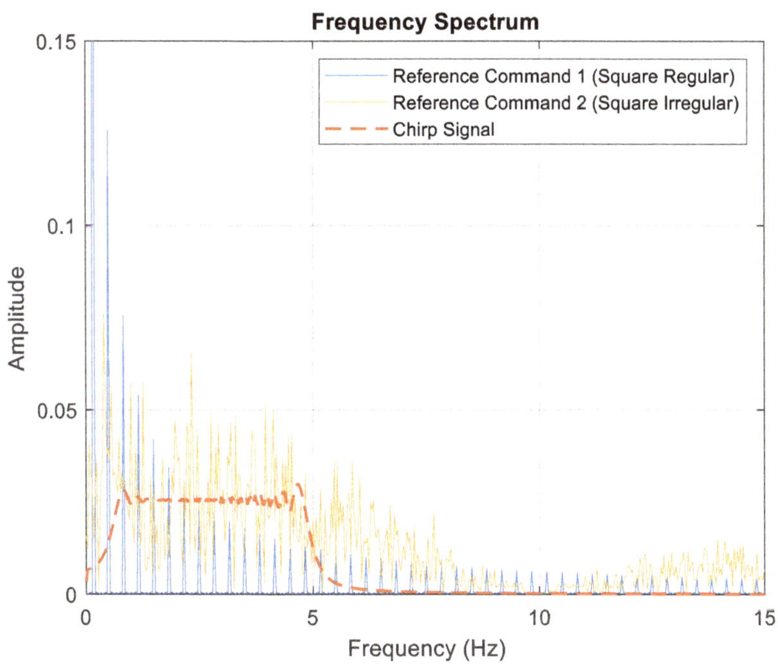

Figure 4. Frequency spectrum of reference commands.

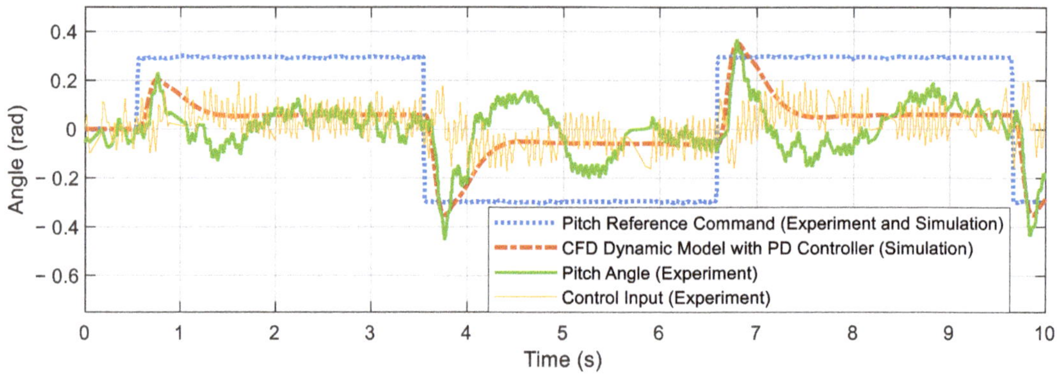

(a)

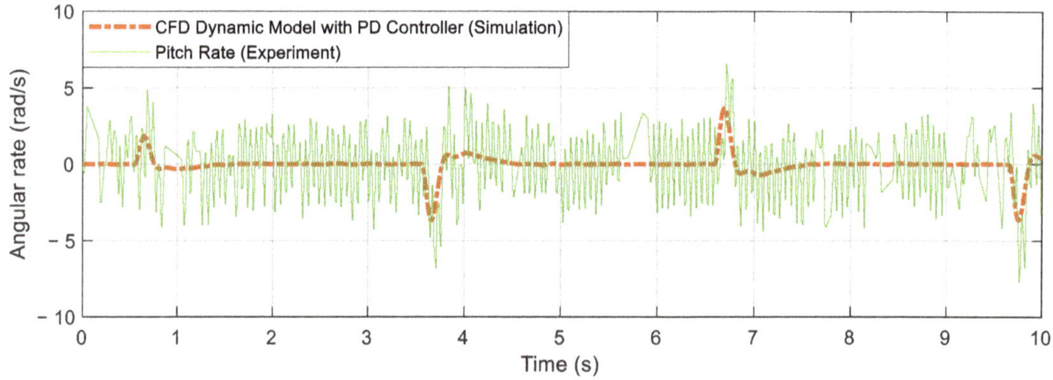

(b)

Figure 5. Experimental flight data and simulation results for system identification with regular square reference inputs: (**a**) pitch angle responses, and (**b**) pitch rate responses.

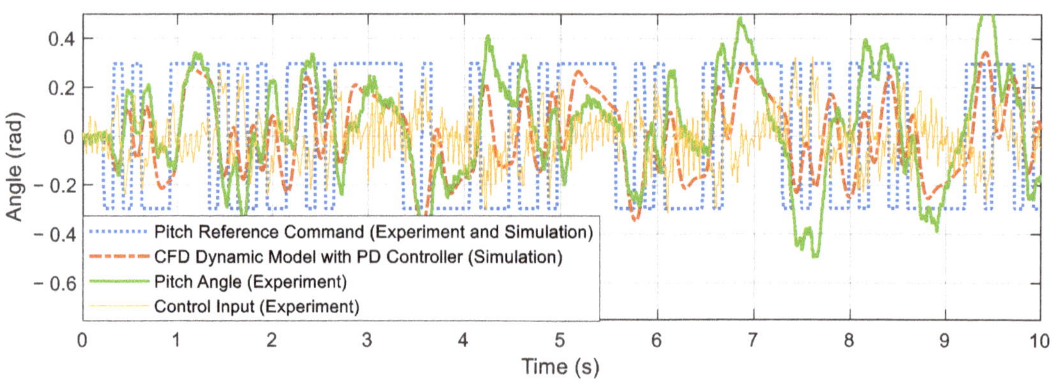

(a)

Figure 6. *Cont.*

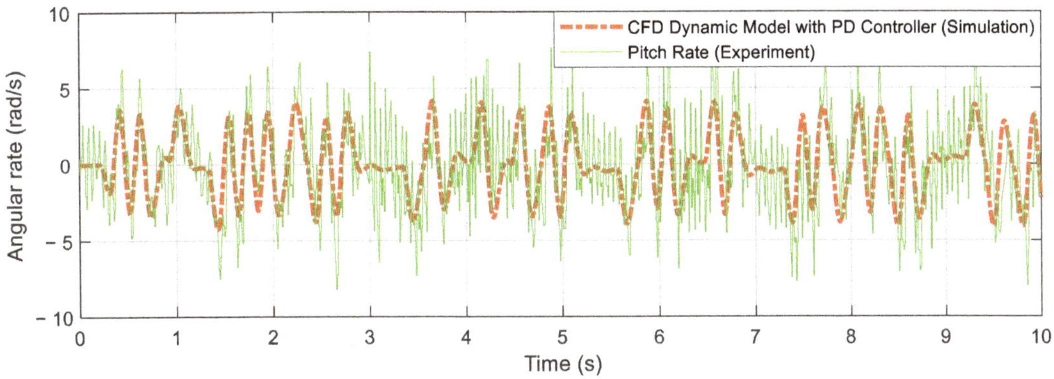

(**b**)

Figure 6. Experimental flight data and simulation results for system identification with irregular square reference inputs: (**a**) pitch angle responses, and (**b**) pitch rate responses.

3.3. Linear Model Refinement in System Identification

System identification processes were performed with the help of system identification tools in MATLAB®. The dynamic model generated from the system identification using the black box model approach was found to be very complex. It could not be matched with real physical parameters of the KUBeetle. Thus, this research only focused on the system identification using the gray box model approach. The gray box model approach with direct and indirect methods was tested to refine open- and closed-loop dynamic models. Two different pairings were considered for the direct and indirect methods, respectively. Firstly, the control input was given as input, while the pitch rate and pitch angle were given as outputs. Secondly, the reference command was given as input, while the pitch rate and pitch angle were given as outputs. Equations (4) and (5) present the structure of the system and input matrices of the dynamic model as fixed for the open-loop dynamic model, while Equations (9) and (10) present those for the closed-loop dynamic model. In addition, given values of parameters in Equations (4)–(10), obtained from computations and measurements, were used as the initial values for system identification. The pitch angle output was given higher weighting, because it had smaller noise compared to the pitch rate output which had larger high-frequency noise that could affect the accuracy of the identified dynamic model. According to the experimental flight setup and data, the obtained dynamic model mainly represented the dynamic of KUBeetle during hover and forward flights with pitch angle under 0.3 radian. Adaptive Gauss–Newton (GNA), Levenberg–Marquardt (LM), and gradient search (GRAD) methods were utilized for model refinements. Table 2 summarizes the general characteristics of each identification method [90–92]:

The total iteration for each search method was limited to 2000 iterations, as further iterations provided negligible improvement. Results from each search method were compared and evaluated. One of the indicators used was the fit percentage of the responses of the identified dynamic model to the experimental flight data. It was defined as:

$$Fit\ Percentage = 100\left(1 - \frac{\sqrt{\sum_{i=1}^{N}\left(y_{measured} - y_{model}\right)^2}}{\sqrt{\sum_{i=1}^{N}\left(y_{measured} - \overline{y_{measured}}\right)^2}}\right), \tag{11}$$

where, N was the number of samples, $y_{measured}$ was the experimental flight data measured, y_{model} was the simulated response of the dynamic model identified, and $\overline{y_{measured}}$ was the mean of experimental flight data measured.

Table 2. General characteristics of each identification method.

Identification Method	Characteristics
GNA	The adaptive subspace Gauss–Newton search method. This minimizes the objective function, which is the sum of squared errors that is approximately quadratic in the parameters near the optimal solution. Convergence is not guaranteed, and it might converge only to a local optimum, depending on the starting parameters.
LM	The Levenberg–Marquardt least square search method. This updates the parameters by adaptively combining the Gauss–Newton and gradient descent approaches, depending on the parameter errors. This is a refinement to the Gauss–Newton search method that improves the chance of local convergence and prevents divergence. This method is more robust than the Gauss–Newton and gradient descent search methods.
GRAD	The Steepest descent or Gradient descent search method. This updates the parameters towards the largest directional derivative or negative gradient direction of the objective function to reduce the sum of squared errors. When close to minimum, the iteration process becomes relatively slow.

Firstly, system identification using the gray box model approach with the direct method was conducted. Closed-loop system identification with the direct method is often used, because it works regardless of the complexity of the regulator. In addition, it does not require knowledge of the characteristic of the feedback [79]. However, the system identification result depended on the quality of the disturbance model, which should also be identified during model refinement. The open-loop CFD dynamic model in Equations (4) and (5) was used as the initial dynamic model. However, the direct method was found to be unsuitable for the system identification of the KUBeetle using the current experimental flight data as the disturbance model could not be identified well, which caused responses of the identified model not to match experimental flight data. Note that some working identification methods for open-loop system identification may fail when they are applied to a closed-loop system identification [79].

Secondly, considering that the open-loop dynamic model could easily be retrieved from the closed-loop dynamic model, system identification using the gray box model approach with the indirect method was conducted. In the indirect method, the system matrix of the closed-loop CFD dynamic model in Equations (9) and (10) was used as the initial dynamic model. Through system identification with several different methods and various restriction settings, several identified dynamic models were obtained. The criteria to select the identified dynamic model were: (1) the fit percentage should be high enough; (2) time-domain responses of the closed-loop dynamic model should have similar characteristics to the experimental flight data; and (3) the open-loop dynamic model should have similar stable and unstable poles and zeros to those of the computationally calculated dynamic model, whereas the closed-loop dynamic model with the PD feedback controller should be stable.

3.4. Dynamic Model Analysis and Verification

Single or a combination of multiple sets of experimental flight data were tested to obtain the best identified dynamic model. This was completed to evaluate the consistency of the identification results and sensitivity of the identification methods. It was found that identified parameters related to velocities along the x- and z-axes had relatively larger deviations because they were not available using onboard sensors. In addition, high-frequency noises and nonlinear body dynamics not included in the model also affected identification results. By following all the selection criteria, six identified dynamic models were considered. Table 3 shows parameters of the identified dynamic model with the indirect method. Each of the selected dynamic models was named based on the identification method used to refine it and numbers represent datasets used for the identification. Fit percentages of identified models obtained using irregular and high-frequency reference signals were much lower than those obtained using regular and low-frequency reference signals. Thus, they were discarded. The fit percentage of the dynamic model from CFD was only about 22%. This low fit percentage was mainly because only wing parts were considered to obtain the stability derivative-related parameters, instead of using the whole body of the KUBeetle. The whole body of the KUBeetle, including the wings, was considered to obtain mass and moments of inertia. The average fit percentage of the identified dynamic models was approximately 49%, which was more than double that of the CFD dynamic model. Most parameter values obtained from the three different identification methods and CFD were quite similar to each other. The CG of the new KUBeetle design, being lower than the CG of the previous KUBeetle design, increased the magnitude of M_u. This was in accordance with the shift of the CG position due to the relocation of the control board and RF module from the top of the flapping mechanism to under the motor. The increase in M_q showed that the system had higher damping than that initially given from CFD. This indicated that the body of the KUBeetle greatly affected system stability. According to system identification results, aerodynamic forces along the x- and z-axes, $X_{\gamma p}$ and $Z_{\gamma p}$, had larger magnitudes than that previously measured. When measuring the control derivative, only the wings and frames near the wing were included. Because of a mounting problem, the lower body of the KUBeetle could not be included, which could be one of the possible reasons for the difference. Other parameters maintained similar values to CFD values.

Table 3. Refined parameters of identified dynamic models using an indirect method.

Parameters	CFD (Initial)	Identified Dynamic Model						
		GNA 1	LM 1	GRAD 1	GNA 2	LM 2	GRAD 2	Average
X_u/m	−1.351	0.000	0.000	0.000	0.000	0.000	0.000	0.000
Z_u/m	0.060	0.001	0.000	0.000	0.000	0.028	0.000	0.005
M_u/I_{yy}	−66.755	−79.656	−79.870	−75.722	−87.300	−99.508	−83.732	−84.298
X_w/m	0.016	4.237	4.237	4.237	4.237	4.237	4.237	4.237
Z_w/m	−0.756	−1.464	−1.742	−1.645	−1.268	−1.113	−1.646	−1.480
M_w/I_{yy}	−1.287	−1.126	−1.126	−1.126	−1.126	−1.227	−1.126	−1.143
X_q/m	0.035	0.294	0.560	0.347	−0.044	−0.058	−0.002	0.183
Z_q/m	0.006	0.006	0.006	0.006	0.006	0.006	0.006	0.006
M_q/I_{yy}	−0.499	−7.715	−1.532	−3.622	−6.470	−7.467	−4.460	−5.211
$X_{\gamma p}/m$	7.458	16.553	20.746	21.329	12.767	9.796	18.457	16.608
$Z_{\gamma p}/m$	−1.147	−11.333	−14.579	−14.826	−8.953	−7.405	−13.671	−11.795
$M_{\gamma p}/I_{yy}$	1109.171	1015.702	1092.554	1108.319	1016.950	1058.045	1107.437	1066.501
Fit Percentage	21.982	49.090	48.640	49.920	45.110	43.180	46.520	48.300

Using the current flight data and dynamic model structure, it was not easy to obtain dynamic model parameters with high fit percentage, even using the black box model approach where there were no parameter restrictions. Moreover, it was found that the parameters obtained were a little sensitive to the datasets used for system identification. The possible reasons for this were: (1) velocities along x- and z-axes were not measured, which degraded the accuracy of the related parameters, X_u, X_w, X_q, Z_u, Z_w, and Z_q; (2) high-frequency noise also contributed to accuracy degradation; and (3) there were nonlinear coupling dynamic terms not included in the model. It has been reported that accuracy can be improved if velocities along the x- and z-axes are available and nonlinear coupling terms are included [77].

Although all dynamic model candidates showed similar fit percentages, GRAD1 was chosen for analysis and evaluation since it had the highest fit percentage. Identified parameters were evaluated in the basic form of the dynamic model [34], where filters and servo dynamics were not included. The basic dynamic model is described as follows:

$$\dot{x}_{basic} = A_{basic}\, x_{basic} + B_{basic}\, u_{basic}, \tag{12}$$

$$y_{basic} = C_{basic}\, x_{basic}, \tag{13}$$

$$x_{basic} = \begin{bmatrix} u \\ w \\ q \\ \vartheta \end{bmatrix}, \tag{14}$$

$$A_{basic} = \begin{bmatrix} \frac{X_u}{m} & \frac{X_w}{m} & \frac{X_q}{m} & g \\ \frac{Z_u}{m} & \frac{Z_w}{m} & \frac{Z_q}{m} & 0 \\ \frac{M_u}{I_{yy}} & \frac{M_w}{I_{yy}} & \frac{M_q}{I_{yy}} & 0 \\ 0 & 0 & 1 & 0 \end{bmatrix}, \tag{15}$$

$$B_{basic} = \begin{bmatrix} \frac{X_{\gamma p}}{m} \\ \frac{Z_{\gamma p}}{m} \\ \frac{M_{\gamma p}}{I_{yy}} \\ 0 \end{bmatrix}, \tag{16}$$

and C_{basic} was set as $\begin{bmatrix} 0 & 0 & 0 & 1 \end{bmatrix}$ to describe the pitch angle measurement.

Figure 7 shows the open-loop frequency responses and a Nyquist diagram of the GRAD1 and CFD dynamic models. When the frequency was under 0.01 rad/s, the low-frequency gain of the identified GRAD1 dynamic model was similar to that of the CFD dynamic model. As the frequency increased to higher than 0.3 rad/s, the gain of the GRAD1 dynamic model was smaller than that of the CFD dynamic model. Moreover, at mid-frequency, the GRAD1 dynamic model had a local minimum point. As the frequency further increased to higher than 5 rad/s, both dynamic models had similar characteristics. Poles of the GRAD1 were (2.86 ± 8.46i, −9.35, −1.64), while those of the CFD model were (3.69 ± 7.58i, −9.22, −0.76). The unstable pole locations did not move much. However, the magnitude of the stable dominant pole of the GRAD1 dynamic model was larger than that of the CFD dynamic model. Thus, it is expected that response of the GRAD1 dynamic model is faster than that of the CFD dynamic model. Both basic dynamic models encircle the (−1, 0) point two times in the Nyquist diagram, meaning that if the loop is closed, the two systems will be stable. However, Figure 7b shows that both models have very small phase margins, with less than 8 degrees.

The simulation was used to analyze and compare extended dynamic models as shown in Figure 8. Gain and phase margins of the GRAD1 dynamic model were 17.2 dB and 39.0°, respectively, while the gain and phase margins of the CFD dynamic model were 16.5 dB and 30.3°, respectively. In the Nyquist diagram, both extended dynamic models encircled the (−1, 0) point two times. Both dynamic models had small gains in the low-frequency range, which resulted in a large steady state error. To reduce the steady state error, the loop gain should be increased. However, this may decrease the stability margins.

Figure 9 shows experimental flight data and simulation results of the GRAD1 dynamic model using the PD feedback controller. As shown in Figure 9a, the pitch angle response of the GRAD1 dynamic model was closer to the pitch angle from the experimental flight data compared to that of the dynamic model from CFD. The peak and valley points predicted using the GRAD1 dynamic model were more accurate than those of the CFD dynamic model. Figure 9b also shows that the GRAD1 dynamic model was better than the CFD dynamic model in following the pitch rate from the experimental flight data, neglecting the high-frequency noise. The high-frequency noise was due to wing flapping, which degraded the performances of system identification and control.

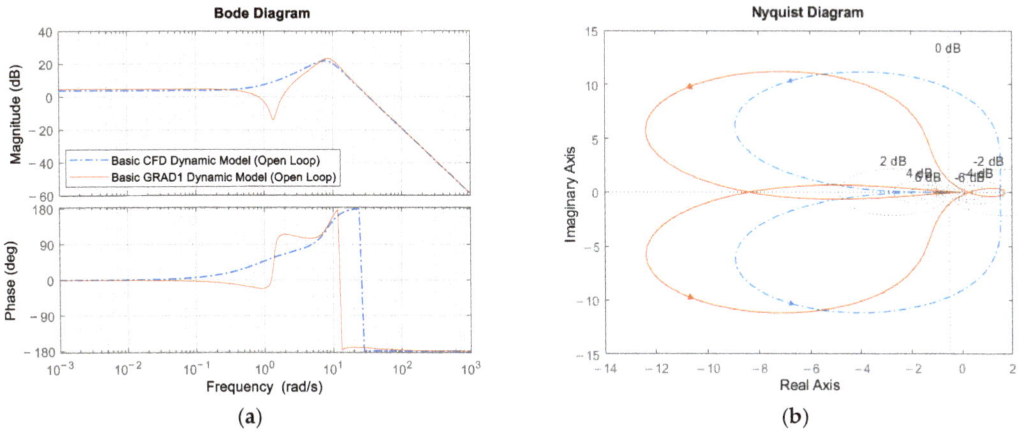

Figure 7. Frequency responses (basic dynamic models only): (**a**) Bode plot; (**b**) Nyquist diagram.

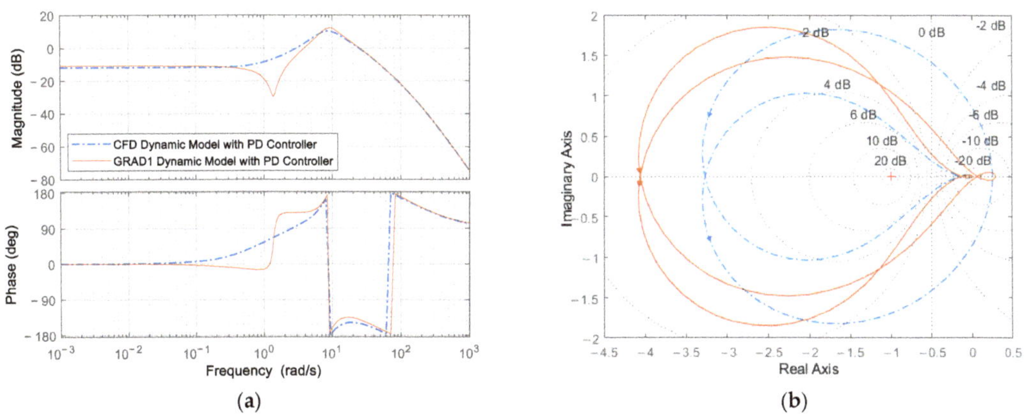

Figure 8. Frequency responses of extended dynamic models with a PD controller: (**a**) Bode plot; (**b**) Nyquist diagram.

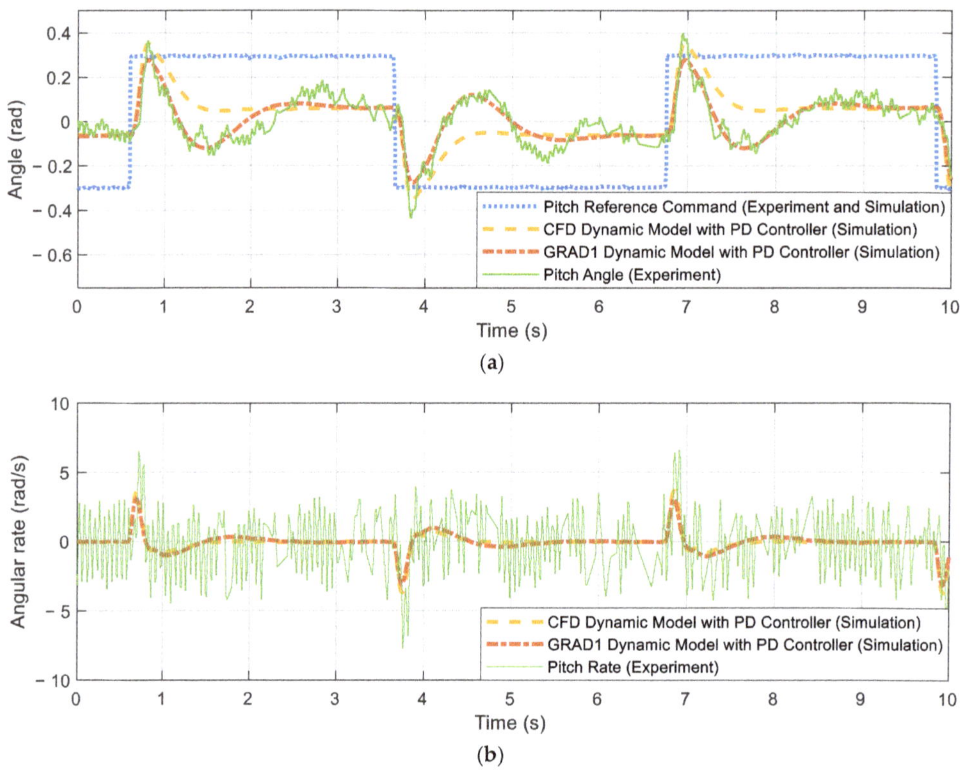

Figure 9. Experimental flight data and simulation results of the GRAD1 dynamic model using a PD feedback controller: (**a**) pitch angle, and (**b**) pitch rate for the provided reference command of the square signal with long period and constant frequency.

4. Conclusions

In this study, we identified a dynamic model of the KUBeetle with several identification approaches and methodologies using angle and angular rate data from onboard sensors only. The gray box model approach was preferred to the black box model approach, because the structure and integrity of parameters of the dynamic model could be maintained. Higher weighting was given to the angle output because it had much lower noises compared to angular rate output. It was found that dynamic models with high fit percentage were obtained using datasets which included the data obtained using regular and low-frequency reference commands. In addition, obtained parameters were slightly sensitive to the quality of the dataset. GNA, LM, and GRAD identification methods showed similar results to each other for two different datasets, while GRAD1 showed the highest fit percentage. These consistent results indicate that the dynamic model was quite reliable and could represent the main dynamic of the KUBeetle, although nonlinear dynamics were not fully identified. Compared to that of the CFD dynamic model, the fit percentage was improved from 22% to 49%. Increases in M_u complied with the lower CG of the KUBeetle. In addition, the increases in M_q indicated higher damping caused by the body dynamics of the KUBeetle that were not considered previously. The obtained dynamic model mainly works for predicting hover and forward flight responses with pitch angle reference command under 0.3 radian, which is used in an experimental flight. The obtained dynamic model described the pitch angular motion quite well, while the linear velocity-related parameters needed more improvement. This is because there was no onboard speed sensor.

From frequency domain analysis, it was found that the open-loop GRAD1 model had a small phase margin indicating that KUBeetle stability could be easily affected by the angular rate feedback gain changes. Moreover, the GRAD1 dynamic model had a faster response because its stable dominant pole had a larger magnitude than the CFD dynamic model. In addition, it showed a similar response to the experimental flight data, neglecting high-frequency noises. It is expected that if the high-frequency noise due to flapping could be reduced and the velocity could be estimated, the fit percentage could be further improved. The time-domain and frequency responses of the identified dynamic model were considerably closer to experimental flight data than those obtained by numerical computation. Further studies are needed to reduce the high-frequency noise due to flapping, to better estimate forward and vertical velocities, and to improve the accuracy of identification.

Supplementary Materials: The following are available online at https://www.mdpi.com/article/10.3390/app12052486/s1, Video S1: Experimental Flight PD controller with regular frequency command (speed 0.5×); Video S2: Experimental Flight PD controller with irregular frequency command (speed 0.5×).

Author Contributions: Conceptualization, T.-S.K.; Data curation, S.A. and G.-H.H.; Formal analysis, S.A.; Investigation, S.A. and G.-H.H.; Resources, H.-C.P.; Supervision, T.-S.K.; Validation, S.A. and H.-C.P.; Writing—original draft, S.A.; Writing—review & editing, H.-C.P. and T.-S.K. All authors have read and agreed to the published version of the manuscript.

Funding: This work was partially supported by a grant (No. 2020R1F1A1077026) of the National Research Foundation (NRF) funded by MSIT and a grant (No. 2020R1A6A1A03046811) from the Ministry of Education, Korea.

Institutional Review Board Statement: Not applicable.

Informed Consent Statement: Not applicable.

Data Availability Statement: Data sharing is not applicable to this article.

Acknowledgments: The authors also wish to express their gratitude to Hoang Vu Phan, Thi Kim Loan Au, and Khanh Nguyen for their support in this study.

Conflicts of Interest: The authors declare no conflict of interest.

References

1. Sanchez, C.; Arribart, H.; Giraud Guille, M.M. Biomimetism and Bioinspiration as Tools for the Design of Innovative Materials and Systems. *Nat. Mater.* **2005**, *4*, 277–288. [CrossRef] [PubMed]
2. Tan, R.; Liu, W.; Cao, G.; Shi, Y. Creative Design Inspired by Biological Knowledge: Technologies and Methods. *Front. Mech. Eng.* **2019**, *14*, 1–14. [CrossRef]
3. Fukuda, T.; Chen, F.; Shi, Q. Special Feature on Bio-Inspired Robotics. *Appl. Sci.* **2018**, *8*, 817. [CrossRef]
4. Wang, J.; Chen, W.; Xiao, X.; Xu, Y.; Li, C.; Jia, X.; Meng, M.Q.-H. A Survey of the Development of Biomimetic Intelligence and Robotics. *Biomim. Intell. Robot.* **2021**, *1*, 100001. [CrossRef]
5. Biswal, P.; Mohanty, P.K. Development of Quadruped Walking Robots: A Review. *Ain Shams Eng. J.* **2021**, *12*, 2017–2031. [CrossRef]
6. Zhu, H.; Gu, S.; He, L.; Guan, Y.; Zhang, H. Transition Analysis and Its Application to Global Path Determination for a Biped Climbing Robot. *Appl. Sci.* **2018**, *8*, 122. [CrossRef]
7. Liljebäck, P.; Pettersen, K.Y.; Stavdahl, Ø.; Gravdahl, J.T. A Review on Modelling, Implementation, and Control of Snake Robots. *Robot. Auton. Syst.* **2012**, *60*, 29–40. [CrossRef]
8. Razif, M.; Mohd Faudzi, A.A.; Mohd Nordin, I.N.A.; Natarajan, E.; Yaakob, O. A Review on Development of Robotic Fish. *J. Transp. Syst. Eng.* **2014**, *1*, 12–22.
9. Duraisamy, P.; Kumar Sidharthan, R.; Nagarajan Santhanakrishnan, M. Design, Modeling, and Control of Biomimetic Fish Robot: A Review. *J. Bionic Eng.* **2019**, *16*, 967–993. [CrossRef]
10. A Robot That Flies like a Bat. *Nature* **2017**, *542*, 140. [CrossRef]
11. Bie, D.; Li, D.; Xiang, J.; Li, H.; Kan, Z.; Sun, Y. Design, Aerodynamic Analysis and Test Flight of a Bat-Inspired Tailless Flapping Wing Unmanned Aerial Vehicle. *Aerosp. Sci. Technol.* **2021**, *112*, 106557. [CrossRef]
12. Han, J.-H.; Lee, J.-S.; Kim, D.-K. Bio-Inspired Flapping UAV Design: A University Perspective. In *Health Monitoring of Structural and Biological Systems 2009*; Kundu, T., Ed.; SPIE: Bellingham, WA, USA, 2009; pp. 466–477.

13. Tan, X.; Zhang, W.; Ke, X.; Chen, W.; Zou, C.; Liu, W.; Cui, F.; Wu, X.; Li, H. Development of Flapping-Wing Micro Air Vehicle in Asia. In Proceedings of the 10th IEEE World Congress on Intelligent Control and Automation, Beijing, China, 6–8 July 2012; IEEE: Piscataway, NJ, USA, 2012; pp. 3939–3942.
14. Mwongera, V.M. A Review of Flapping Wing MAV Modelling. *Int. J. Aeronaut. Aerosp. Res.* **2015**, *2*, 27–36. [CrossRef]
15. Unver, O.; Uneri, A.; Aydemir, A.; Sitti, M. Geckobot: A Gecko Inspired Climbing Robot Using Elastomer Adhesives. In Proceedings of the 2006 IEEE International Conference on Robotics and Automation, ICRA 2006, Orlando, FL, USA, 15–19 May 2006; IEEE: Piscataway, NJ, USA, 2006; pp. 2329–2335.
16. Nansai, S.; Mohan, R. A Survey of Wall Climbing Robots: Recent Advances and Challenges. *Robotics* **2016**, *5*, 14. [CrossRef]
17. Sanfilippo, F.; Azpiazu, J.; Marafioti, G.; Transeth, A.; Stavdahl, Ø.; Liljebäck, P. Perception-Driven Obstacle-Aided Locomotion for Snake Robots: The State of the Art, Challenges and Possibilities. *Appl. Sci.* **2017**, *7*, 336. [CrossRef]
18. Ward, T.A.; Fearday, C.J.; Salami, E.; Binti Soin, N. A Bibliometric Review of Progress in Micro Air Vehicle Research. *Int. J. Micro Air Veh.* **2017**, *9*, 146–165. [CrossRef]
19. Phan, H.V.; Park, H.C. Insect-Inspired, Tailless, Hover-Capable Flapping-Wing Robots: Recent Progress, Challenges, and Future Directions. *Prog. Aerosp. Sci.* **2019**, *111*, 100573. [CrossRef]
20. Park, H.C.; Vu, P.H.; Lee, J.H. Tailless Insect-Mimicking Flapping-Wing Micro Air Vehicle: A Review and Perspective. *J. ICROS* **2019**, *25*, 960–965. [CrossRef]
21. Gerdes, J.W.; Gupta, S.K.; Wilkerson, S.A. A Review of Bird-Inspired Flapping Wing Miniature Air Vehicle Designs. *J. Mech. Robot.* **2012**, *4*, 021003. [CrossRef]
22. Taylor, G.K. Mechanics and Aerodynamics of Insect Flight Control. *Biol. Rev.* **2001**, *76*, 449–471. [CrossRef]
23. Dudley, R. *The Biomechanics of Insect Flight: Form, Function, Evolution*; Princeton Paperbacks; Princeton University Press: Princeton, NJ, USA, 2002; ISBN 978-0-691-09491-5.
24. Alexander, D.E. *Nature's Flyers: Birds, Insects, and the Biomechanics of Flight*; Johns Hopkins University Press: Baltimore, MD, USA, 2004; ISBN 978-0-8018-8059-9.
25. Sun, M. Insect Flight Dynamics: Stability and Control. *Rev. Mod. Phys.* **2014**, *86*, 615–646. [CrossRef]
26. Zhang, C.; Rossi, C. A Review of Compliant Transmission Mechanisms for Bio-Inspired Flapping-Wing Micro Air Vehicles. *Bioinspir. Biomim.* **2017**, *12*, 025005. [CrossRef] [PubMed]
27. Keennon, M.; Klingebiel, K.; Won, H. Development of the Nano Hummingbird: A Tailless Flapping Wing Micro Air Vehicle. In Proceedings of the 50th AIAA Aerospace Sciences Meeting including the New Horizons Forum and Aerospace Exposition, Nashville, TN, USA, 9–12 January 2012; American Institute of Aeronautics and Astronautics: Nashville, TN, USA, 2012.
28. Ma, K.Y.; Chirarattananon, P.; Fuller, S.B.; Wood, R.J. Controlled Flight of a Biologically Inspired, Insect-Scale Robot. *Science* **2013**, *340*, 603–607. [CrossRef] [PubMed]
29. Karásek, M.; Muijres, F.T.; De Wagter, C.; Remes, B.D.W.; de Croon, G.C.H.E. A Tailless Aerial Robotic Flapper Reveals That Flies Use Torque Coupling in Rapid Banked Turns. *Science* **2018**, *361*, 1089–1094. [CrossRef] [PubMed]
30. Phan, H.V.; Kang, T.; Park, H.C. Design and Stable Flight of a 21 g Insect-like Tailless Flapping Wing Micro Air Vehicle with Angular Rates Feedback Control. *Bioinspir. Biomim.* **2017**, *12*, 036006. [CrossRef]
31. Phan, H.V.; Aurecianus, S.; Kang, T.; Park, H.C. KUBeetle-S: An Insect-like, Tailless, Hover-Capable Robot That Can Fly with a Low-Torque Control Mechanism. *Int. J. Micro Air Veh.* **2019**, *11*, 175682931986137. [CrossRef]
32. Aurecianus, S.; Na, Y.S.; Vu, P.H.; Park, H.C.; Kang, T. Study of Gyroscope and Accelerometer Dynamic Characteristics on Flapping Wing Micro Aerial Vehicle. *J. ICROS* **2019**, *25*, 981–990. [CrossRef]
33. Phan, H.V.; Park, H.C. Mechanisms of Collision Recovery in Flying Beetles and Flapping-Wing Robots. *Science* **2020**, *370*, 1214–1219. [CrossRef]
34. Aurecianus, S.; Phan, H.V.; Kang, T.; Park, H.C. Longitudinal Mode Model-Based Controller Design for Tailless Flapping Wing Robot with Loop Shaping Compensator. *Bioinspir. Biomim.* **2020**, *15*, 056004. [CrossRef]
35. Aurecianus, S.; Phan, H.V.; Park, J.; Park, H.C.; Kang, T. Lateral Mode Controller Design for Insect-like Tailless Flapping-Wing Micro Air Vehicle. *J. ICROS* **2021**, *27*, 1–10. [CrossRef]
36. Coleman, D.A.; Benedict, M.; Amp, T.A.; Hrishikeshavan, V.; Chopra, I. Design, Development and Flight-Testing of a Robotic Hummingbird. In Proceedings of the AHS 71st Annual Forum, Virginia Beach, VA, USA, 5–7 May 2015.
37. Roshanbin, A.; Altartouri, H.; Karásek, M.; Preumont, A. COLIBRI: A Hovering Flapping Twin-Wing Robot. *Int. J. Micro Air Veh.* **2017**, *9*, 270–282. [CrossRef]
38. Nguyen, Q.-V.; Chan, W.L. Development and Flight Performance of a Biologically-Inspired Tailless Flapping-Wing Micro Air Vehicle with Wing Stroke Plane Modulation. *Bioinspir. Biomim.* **2018**, *14*, 016015. [CrossRef] [PubMed]
39. Tu, Z.; Fei, F.; Deng, X. Untethered Flight of an At-Scale Dual-Motor Hummingbird Robot with Bio-Inspired Decoupled Wings. *IEEE Robot. Autom. Lett.* **2020**, *5*, 4194–4201. [CrossRef]
40. Keennon, M.; Grasmeyer, J. Development of Two MAVs and Vision of the Future of MAV Design. In Proceedings of the AIAA International Air and Space Symposium and Exposition: The Next 100 Years, Dayton, OH, USA, 14–17 July 2003; American Institute of Aeronautics and Astronautics: Dayton, OH, USA, 2003.
41. Karásek, M.; Percin, M.; Cunis, T.; van Oudheusden, B.W.; De Wagter, C.; Remes, B.D.; de Croon, G.C. Accurate Position Control of a Flapping-Wing Robot Enabling Free-Flight Flow Visualisation in a Wind Tunnel. *Int. J. Micro Air Veh.* **2019**, *11*, 175682931983368. [CrossRef]

42. Nguyen, Q.; Chan, W.-L.; Debiasi, M. Performance Tests of a Hovering Flapping Wing Micro Air Vehicle with Double Wing Clap-and-Fling Mechanism. In Proceedings of the International Micro Air Vehicles Conference and Flight Competition, Aachen, Germany, 15–18 September 2015.
43. Park, J.H.; Yoon, K.-J. Designing a Biomimetic Ornithopter Capable of Sustained and Controlled Flight. *J. Bionic Eng.* **2008**, *5*, 39–47. [CrossRef]
44. Rose, C.; Fearing, R.S. Comparison of Ornithopter Wind Tunnel Force Measurements with Free Flight. In Proceedings of the 2014 IEEE International Conference on Robotics and Automation (ICRA), Hong Kong, China, 31 May–7 June 2014; IEEE: Piscataway, NJ, USA, 2014; pp. 1816–1821.
45. Folkertsma, G.A.; Straatman, W.; Nijenhuis, N.; Venner, C.H.; Stramigioli, S. Robird: A Robotic Bird of Prey. *IEEE Robot. Automat. Mag.* **2017**, *24*, 22–29. [CrossRef]
46. Taylor, G.K.; Thomas, A.L.R. Animal Flight Dynamics II. Longitudinal Stability in Flapping Flight. *J. Theor. Biol.* **2002**, *214*, 351–370. [CrossRef]
47. Taha, H.E.; Kiani, M.; Hedrick, T.L.; Greeter, J.S.M. Vibrational Control: A Hidden Stabilization Mechanism in Insect Flight. *Sci. Robot.* **2020**, *5*, eabb1502. [CrossRef]
48. Benrabah, M.; Kara, K.; AitSahed, O.; Hadjili, M.L. Adaptive Fourier Series Neural Network PID Controller. *Int. J. Control Autom. Syst.* **2021**, *19*, 3388–3399. [CrossRef]
49. Freire, H.; Moura Oliveira, P.B.; Solteiro Pires, E.J. From Single to Many-Objective PID Controller Design Using Particle Swarm Optimization. *Int. J. Control Autom. Syst.* **2017**, *15*, 918–932. [CrossRef]
50. Du, H.; Hu, X.; Ma, C. Dominant Pole Placement with Modified PID Controllers. *Int. J. Control Autom. Syst.* **2019**, *17*, 2833–2838. [CrossRef]
51. Memon, F.; Shao, C. An Optimal Approach to Online Tuning Method for PID Type Iterative Learning Control. *Int. J. Control Autom. Syst.* **2020**, *18*, 1926–1935. [CrossRef]
52. Pongfai, J.; Angeli, C.; Shi, P.; Su, X.; Assawinchaichote, W. Optimal PID Controller Autotuning Design for MIMO Nonlinear Systems Based on the Adaptive SLP Algorithm. *Int. J. Control Autom. Syst.* **2021**, *19*, 392–403. [CrossRef]
53. Fadaei, A.; Salahshoor, K. A Novel Real-Time Fuzzy Adaptive Auto-Tuning Scheme for Cascade PID Controllers. *Int. J. Control Autom. Syst.* **2011**, *9*, 823–833. [CrossRef]
54. Nguyen, A.T.; Han, J.-H. Wing Flexibility Effects on the Flight Performance of an Insect-like Flapping-Wing Micro-Air Vehicle. *Aerosp. Sci. Technol.* **2018**, *79*, 468–481. [CrossRef]
55. Nguyen, K.; Au, L.T.K.; Phan, H.-V.; Park, S.H.; Park, H.C. Effects of Wing Kinematics, Corrugation, and Clap-and-Fling on Aerodynamic Efficiency of a Hovering Insect-Inspired Flapping-Wing Micro Air Vehicle. *Aerosp. Sci. Technol.* **2021**, *118*, 106990. [CrossRef]
56. Nan, Y.; Peng, B.; Chen, Y.; Feng, Z.; McGlinchey, D. Can Scalable Design of Wings for Flapping Wing Micro Air Vehicle Be Inspired by Natural Flyers? *Int. J. Aerosp. Eng.* **2018**, *2018*, 9538328. [CrossRef]
57. Nan, Y.; Karásek, M.; Lalami, M.E.; Preumont, A. Experimental Optimization of Wing Shape for a Hummingbird-like Flapping Wing Micro Air Vehicle. *Bioinspir. Biomim.* **2017**, *12*, 026010. [CrossRef] [PubMed]
58. Taha, H.E.; Hajj, M.R.; Nayfeh, A.H. Flight Dynamics and Control of Flapping-Wing MAVs: A Review. *Nonlinear Dyn.* **2012**, *70*, 907–939. [CrossRef]
59. Orlowski, C.T.; Girard, A.R. Dynamics, Stability, and Control Analyses of Flapping Wing Micro-Air Vehicles. *Prog. Aerosp. Sci.* **2012**, *51*, 18–30. [CrossRef]
60. Khan, Q.; Akmeliawati, R. Review on System Identification and Mathematical Modeling of Flapping Wing Micro-Aerial Vehicles. *Appl. Sci.* **2021**, *11*, 1546. [CrossRef]
61. Kajak, K.M.; Karásek, M.; Chu, Q.P.; de Croon, G.C.H.E. A Minimal Longitudinal Dynamic Model of a Tailless Flapping Wing Robot for Control Design. *Bioinspir. Biomim.* **2019**, *14*, 046008. [CrossRef] [PubMed]
62. Deng, X.; Schenato, L.; Sastry, S.S. Flapping Flight for Biomimetic Robotic Insects: Part II-Flight Control Design. *IEEE Trans. Robot.* **2006**, *22*, 789–803. [CrossRef]
63. Rifaï, H.; Marchand, N.; Poulin-Vittrant, G. Bounded Control of an Underactuated Biomimetic Aerial Vehicle—Validation with Robustness Tests. *Robot. Auton. Syst.* **2012**, *60*, 1165–1178. [CrossRef]
64. Tahmasian, S.; Woolsey, C.A.; Taha, H.E. Longitudinal Flight Control of Flapping Wing Micro Air Vehicles. In Proceedings of the AIAA Guidance, Navigation, and Control Conference, National Harbor, MD, USA, 13–17 January 2014; American Institute of Aeronautics and Astronautics: National Harbor, MD, USA, 2014.
65. Zhang, J.; Tu, Z.; Fei, F.; Deng, X. Geometric flight control of a hovering robotic hummingbird. In Proceedings of the 2017 IEEE International Conference on Robotics and Automation (ICRA), Singapore, 29 May–3 June 2017; IEEE: Piscataway, NJ, USA, 2017; pp. 5415–5421. [CrossRef]
66. Rakotomamonjy, T.; Ouladsine, M.; Moing, T.L. Longitudinal Modelling and Control of a Flapping-Wing Micro Aerial Vehicle. *Control Eng. Pract.* **2010**, *18*, 679–690. [CrossRef]
67. Serrani, A. Robust Nonlinear Control Design for a Minimally-Actuated Flapping-Wing MAV in the Longitudinal Plane. In Proceedings of the 50th IEEE Conference on Decision and Control and European Control Conference, Orlando, FL, USA, 12–15 December 2011; IEEE: Piscataway, NJ, USA, 2011; pp. 7464–7469.

68. Tran, X.-T.; Oh, H.; Kim, I.-R.; Kim, S. Attitude Stabilization of Flapping Micro-Air Vehicles via an Observer-Based Sliding Mode Control Method. *Aerosp. Sci. Technol.* **2018**, *76*, 386–393. [CrossRef]
69. Banazadeh, A.; Taymourtash, N. Adaptive Attitude and Position Control of an Insect-like Flapping Wing Air Vehicle. *Nonlinear Dyn.* **2016**, *85*, 47–66. [CrossRef]
70. Alkitbi, M.; Serrani, A. Robust Control of a Flapping-Wing MAV by Differentiable Wingbeat Modulation. *IFAC-PapersOnLine* **2016**, *49*, 290–295. [CrossRef]
71. Lee, J.; Ryu, S.; Kim, H.J. Stable Flight of a Flapping-Wing Micro Air Vehicle Under Wind Disturbance. *IEEE Robot. Autom. Lett.* **2020**, *5*, 5685–5692. [CrossRef]
72. He, W.; Mu, X.; Zhang, L.; Zou, Y. Modeling and Trajectory Tracking Control for Flapping-Wing Micro Aerial Vehicles. *IEEE/CAA J. Autom. Sin.* **2021**, *8*, 148–156. [CrossRef]
73. He, W.; Yan, Z.; Sun, C.; Chen, Y. Adaptive Neural Network Control of a Flapping Wing Micro Aerial Vehicle With Disturbance Observer. *IEEE Trans. Cybern.* **2017**, *47*, 3452–3465. [CrossRef]
74. Cheng, B.; Deng, X. A Neural Adaptive Controller in Flapping Flight. *J. Robot. Mechatron.* **2012**, *24*, 602–611. [CrossRef]
75. Khosravi, M.; Novinzadeh, A.B. A multi-body control approach for flapping wing micro aerial vehicles. *Aerosp. Sci. Technol.* **2021**, *112*, 106525. [CrossRef]
76. Finio, B.M.; Pérez-Arancibia, N.O.; Wood, R.J. System Identification and Linear Time-Invariant Modeling of an Insect-Sized Flapping-Wing Micro Air Vehicle. In Proceedings of the 2011 IEEE/RSJ International Conference on Intelligent Robots and Systems, San Francisco, CA, USA, 25–30 September 2011; IEEE: Piscataway, NJ, USA, 2011; pp. 1107–1114. [CrossRef]
77. Nijboer, J.; Armanini, S.F.; Karasek, M.; de Visser, C.C. Longitudinal Grey-Box Model Identification of a Tailless Flapping-Wing MAV Based on Free-Flight Data. In Proceedings of the AIAA Scitech 2020 Forum, Orlando, FL, USA, 6 January 2020; American Institute of Aeronautics and Astronautics: Orlando, FL, USA, 2020.
78. Klein, V.; Morelli, E.A. *Aircraft System Identification: Theory and Practice*; AIAA education series; American Institute of Aeronautics and Astronautics: Reston, VA, USA, 2006; ISBN 1-56347-832-3.
79. Ljung, L. *System Identification: Theory for the User*, 2nd ed.; Prentice Hall Information and System Sciences Series; Prentice Hall PTR: Upper Saddle River, NJ, USA, 1999; ISBN 978-0-13-656695-3.
80. Hoffer, N.V.; Coopmans, C.; Jensen, A.M.; Chen, Y. A Survey and Categorization of Small Low-Cost Unmanned Aerial Vehicle System Identification. *J. Intell. Robot. Syst.* **2014**, *74*, 129–145. [CrossRef]
81. Landau, I.D. Identification in Closed Loop: A Powerful Design Tool (Better Design Models, Simpler Controllers). *Control Eng. Pract.* **2001**, *9*, 51–65. [CrossRef]
82. Gustavsson, I.; Ljung, L.; Söderström, T. Identification of Processes in Closed Loop—Identifiability and Accuracy Aspects. *Automatica* **1977**, *13*, 59–75. [CrossRef]
83. Phan, H.V.; Aurecianus, S.; Au, T.K.L.; Kang, T.; Park, H.C. Towards the Long-Endurance Flight of an Insect-Inspired, Tailless, Two-Winged, Flapping-Wing Flying Robot. *IEEE Robot. Autom. Lett.* **2020**, *5*, 5059–5066. [CrossRef]
84. Phan, H.; Aurecianus, S.; Kang, T.; Park, H.C. Attitude Control Mechanism in an Insect-like Tailless Two-Winged Flying Robot by Simultaneous Modulation of Stroke Plane and Wing Twist. In Proceedings of the International Micro Air Vehicle Conference and Competition, Melbourne, Australia, 17–23 November 2018.
85. Nguyen, T.A.; Vu Phan, H.; Au, T.K.L.; Park, H.C. Experimental Study on Thrust and Power of Flapping-Wing System Based on Rack-Pinion Mechanism. *Bioinspir. Biomim.* **2016**, *11*, 046001. [CrossRef]
86. Karásek, M.; Preumont, A. Simulation of Flight Control of a Hummingbird Like Robot Near Hover. In Proceedings of the 18th International Conference Engineering Mechanics, Svratka, Czech Republic, 14–17 May 2012.
87. Au, L.T.K.; Phan, V.H.; Park, H.C. Longitudinal Flight Dynamic Analysis on Vertical Takeoff of a Tailless Flapping-Wing Micro Air Vehicle. *J. Bionic Eng.* **2018**, *15*, 283–297. [CrossRef]
88. Au, L.T.K.; Park, H.C. Influence of Center of Gravity Location on Flight Dynamic Stability in a Hovering Tailless FW-MAV: Longitudinal Motion. *J. Bionic Eng.* **2019**, *16*, 130–144. [CrossRef]
89. Nguyen, K.; Au, L.T.K.; Phan, H.-V.; Park, H.C. Comparative Dynamic Flight Stability of Insect-Inspired Flapping-Wing Micro Air Vehicles in Hover: Longitudinal and Lateral Motions. *Aerosp. Sci. Technol.* **2021**, *119*, 107085. [CrossRef]
90. Björck, Å. *Numerical Methods for Least Squares Problems*; Society for Industrial and Applied Mathematics: Philadelphia, PA, USA, 1996; ISBN 978-0-89871-360-2.
91. Lourakis, M.I.A. A Brief Description of the Levenberg-Marquardt Algorithm Implemented by Levmar. *Found. Res. Technol.* **2005**, *4*, 1–6.
92. Gavin, H.P. *The Levenberg-Marquardt Method for Nonlinear Least Squares Curve-Fitting Problems*; Duke University: Durham, NC, USA, 2020.

applied sciences

MDPI

Article

Automated Parking Space Allocation during Transition with both Human-Operated and Autonomous Vehicles

Mingkang Wu [1], Haobin Jiang [1,*] and Chin-An Tan [2]

[1] School of Automotive and Traffic Engineering, Jiangsu University, Zhenjiang 212013, China; 2221804039@stmail.ujs.edu.cn
[2] Department of Mechanical Engineering, Wayne State University, 5050 Anthony Wayne Drive, Detroit, MI 48202, USA; tan@wayne.edu
* Correspondence: jianghb@ujs.edu.cn; Tel.: +86-13805283226

Abstract: As fully automated valet parking systems are being developed, there is a transition period during which both human-operated vehicles (HVs) and autonomous vehicles (AVs) are present in the same parking infrastructure. This paper addresses the problem of allocation of a parking space to an AV without conflicting with the parking space chosen by the driver of a HV. A comprehensive assessment of the key factors that affect the preference and choice of a driver for a parking space is established by the fuzzy comprehensive method. The algorithm then generates a ranking order of the available parking spaces to first predict the driver's choice of parking space and then allocate a space for the AV. The Floyd algorithm of shortest distance is used to determine the route for the AV to reach its parking space. The proposed allocation and search algorithm is applied to the examples of a parking lot with three designed scenarios. It is shown that parking space can be reasonably allocated for AVs.

Keywords: automated parking system; fuzzy comprehension evaluation; Floyd algorithm; human-operated vehicle; autonomous vehicle

Citation: Wu, M.; Jiang, H.; Tan, C.-A. Automated Parking Space Allocation during Transition with both Human-Operated and Autonomous Vehicles. *Appl. Sci.* **2021**, *11*, 855. https://doi.org/10.3390/app11020855

Received: 11 December 2020
Accepted: 12 January 2021
Published: 18 January 2021

1. Introduction

According to the International Parking Institute (IPI), the number of vehicles on the road will reach 2.5 billion in 2050 [1]. With this projected increase in the volume of vehicles, parking has become an emerging issue that affects not only drivers looking for parking spaces, but also city governments in their planning, particularly in urban areas where land resources are limited and constrained. It has been reported that about 30% of traffic backup in a typical downtown area is caused by drivers searching for parking spaces [2]. The expected increase in the number of vehicles likely implies more new drivers and drivers who are unskilled in parking, thus leading to more road congestion and increased waste of valuable manpower-hours and resources. In recent years, with the continuous advancement and development in computer and control technologies, automated parking has become feasible and various strategies have been proposed to help alleviate the unskilled parking problem [3,4]. Advances in V2X technology have also led related researchers to develop a more robust system of automated parking, namely the Automated Valet Parking System [5].

Compared with the automated parking system, the concept of automated valet parking system is based on V2X communication technology, which enables self-driving vehicles to interact and collaborate with an intelligent parking infrastructure during the entire parking space search process, from the entrance to the self-parking space [6]. Existing intelligent parking administration systems can detect the status of the parking spaces in real time through camera recognition, infrared sensing, and other technologies [7,8]. According to SAE (Society of Automotive Engineers) classification for autonomous driving levels, the automated valet parking system is classified as L4 autonomous driving, that is, under certain scenarios, an equipped vehicle can complete all the driving tasks autonomously without the participation of a driver. Although self-driving cars are gaining popularity, related research and testing are far from the level of completion required to bring automated valet parking to fruition. Even if L4 level self-driving cars were to enter the market in the short term, there is likely going to be a long transition period during which both human-operated vehicles (HVs), i.e., with drivers, and autonomous vehicles (AVs), i.e., self-driving or driverless, co-exist, and augmented search strategies need to be developed for the AVs. Self-driving vehicles execute well-defined algorithms based on sensor information, while drivers make cognitive decisions based on perception and surveying of the surrounding environment. The objective of this paper is to investigate parking space allocation and route selection (i.e., path planning to the allocated space) for self-driving vehicles during such a transition period in which significant interactions between the two types of vehicles are expected.

1.1. Motivation for Research

In this paper, as shown in Figure 1, we define an intelligent parking infrastructure (which refers to either a parking lot or a parking structure) as one that is equipped with a central command station that receives information from all sensors and interacts with autonomous vehicles in real time. In an automated valet parking system, the central command station is capable of determining the number of available parking spaces through geomagnetic sensors and obtaining the locations of these parking spaces through a pre-stored layout of the parking infrastructure. After an AV enters the parking infrastructure, the central command station utilizes information from multiple cameras to provide a road map to the allocated parking space, enabling the self-parking to be efficient. However, as development and implementation of automated parking systems move forward, there will be a transition period to fully automated valet parking unless the city government phases out HVs and mandates purchase of AVs, which is very unlikely, or designates parking spaces only for AVs. Thus, in an intelligent parking infrastructure, there will generally be both self-driving and human-driving vehicles. Self-driving vehicles can interact with the parking infrastructure in real time based on V2X communication. Human-driving vehicles are limited by non-intelligent devices that disengage any interaction with the central command station.

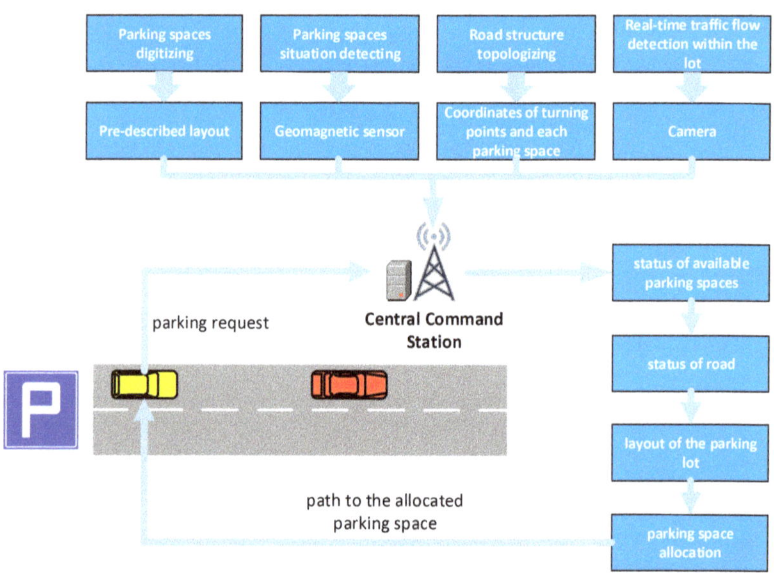

Figure 1. Flowchart of the entire process of parking space allocation and route selection for an autonomous vehicle.

Consider the scenario when both a HV and an AV are looking for parking spaces. In order for the central command station not to allocate the same parking space for the AV as the driver would choose, the station must be able to predict with high probability the choice of parking space that the driver would make. Liang et al. proposed four preference factors that could affect the choice of parking spaces for an individual driver [9]. Logit models, based on fuzzy logic theory, were established to describe driver preference selection, which were then used to improve parking experience and alleviate difficulties in parking [9,10]. Although fuzzy evaluation methods have been widely employed with proven success to determine factors, which are otherwise difficult to quantify, and make optimal choices incorporating expert opinions, there are only few published studies related to parking choice preference. In this paper, by adopting the fuzzy evaluation method, a preference ranking method of parking spaces in an intelligent parking infrastructure is established based on four representative factors that would affect the parking space selection of drivers.

In an automated valet parking system, route selection is also a key enabling technology [11,12]. Since the AV might not know a priori the overall layout of the parking infrastructure, it is necessary for the central command station to guide the self-driving vehicle to the allocated parking space according to the current conditions of the intended route. In general, in searching for a parking space, the shorter the path distance, the better. There are several methods to find the shortest path, such as the Dijkstra and Floyd algorithms [13,14]. Shi et al. [15] studied the shortest path planning problem of mobile robots based on the Floyd algorithm, focusing on the node selection problem for mobile robot path planning and determination of the weighting factor of each passable road. Experiments have also illustrated that the Floyd algorithm has the advantage of providing the shortest path selection for mobile robots [15]. Based on a known layout of the environment, Dijkstra algorithm can also efficiently find the shortest path between two points. Although the Floyd algorithm is slightly more time-consuming than the Dijkstra algorithm, it is a dynamic programming algorithm aimed to solve the shortest path problem between multiple source points. In a complex environment such as parking, the Floyd algorithm appears to be more suitable for our current problem of interest.

1.2. Organization and Contributions of This Paper

In this paper, state-of-the-art research of automated parking space allocation and route selection is reviewed in Section 2. Section 3 describes the four factors that affect the choice of parking space for drivers. In Section 4, a fuzzy comprehensive evaluation method is proposed to evaluate and score (i.e., rank) the available parking spaces in an intelligent parking infrastructure [16]. The fuzzy algorithm is based on first predicting which space the driver would select and then allocating one of the remaining spaces to the autonomous vehicle. In Section 5, the Floyd algorithm is introduced to provide path navigation for autonomous vehicles [14] according to the nodes of available parking spaces and road information in the parking infrastructure. Section 6 provides three examples of a parking lot on the campus of Jiangsu University to illustrate the step-by-step implementation of the proposed algorithm via Python. Concluding remarks of this work are given in Section 7.

The contributions of this paper are listed as follows:

1. In this paper, a fuzzy comprehensive evaluation method is used to first predict the parking preference of a driver based on four main factors that influence their choice of parking spaces. This procedure prevents conflict of allocation of parking space to the AV with that chosen by the HV.
2. Based on the node information in a parking infrastructure, the Floyd algorithm is used to provide path navigation for the AVs.
3. The proposed methodology, combining fuzzy theory and the Floyd algorithm, provides a novel scheme for a central command station to assign parking spaces for AVs in the presence of HVs. The merit of this work thus provides a foundation for future work to investigate the problem of automated parking for multiple vehicles.

2. Review of Related Work

In 2007, the DARPA (Defense Advanced Research Projects Agency) Urban Challenge (DUC) initiated numerous research projects to address challenges of autonomous driving. Since then, a great deal of research has been conducted in related topics. In addition to highway and urban driving, automated valet parking is also of major interest to relieve drivers from the stress of parking [17].

In 2012, a parking information acquisition and release system was designed by Dou et al. based on the dynamic allocation algorithm of parking spaces in the parking lot [18]. The main idea is to optimize among the distribution of parking spaces arranged in a tabulated format and driving paths/distances. However, the algorithm can obtain the availability of parking spaces only after the human-operated vehicles have completed their parking. Thus, if there are more than two vehicles, at least one HV and one AV, choosing the same parking space and converging near that space, this scenario could lead to traffic congestion and the wasting of time and fuel consumption.

In 2012, Audi Corporation developed a parking guidance system that could assist and orchestrate the entire parking process [19]. The approach of the system includes three steps: scanning, positioning, and arranging parking [20]. The distance to the obstacle is detected by eight ultrasonic sensors installed in the bumper and the parking space is detected by 10 ultrasonic sensors. This parking system can only be executed if the speed of the vehicle is below 30 km/h. While this is an excellent demonstration of an automated valet parking system, during the transition period to fully autonomous driving, HVs will also be represent; the guidance system cannot allocate parking spaces to the HVs with no communication devices equipped and hence cannot complete the space allocation task for the AVs.

In 2016, Kotb et al. [21] proposed a system to reduce the time wasted in looking for parking spaces by offering guaranteed parking reservations with the lowest possible cost and searching time for drivers and the highest revenue and resource utilization for parking managers. However, they did not include other important factors that influence how drivers choose parking spaces, such as the safe locations of the parking spaces and the distance to reach the parking space.

In 2018, Tcheumadjeu et al. [22] presented an architecture for Automated Valet Parking (AVP) connected to cloud-based IoT services and mobile user interfaces. Autonomous vehicles can share information and data via the phones of their occupants. Moreover, under this communication architecture, some functions of the AVs, such as drop-off and pick-up, can be activated by phones. However, the paper only considers the interactions between users and AVs. In an actual scenario, many vehicles will be present at the same time, making efficient planning of traffic flow in the parking lot not possible by just integrating the shared information provided by the users.

In 2019, a new path planning system based on the Dijkstra algorithm was designed by Yu et al. [23]. The road occupancy factor is calculated and added to the path weight of the Dijkstra algorithm, and a shortest path is selected by comparing the weight of each path. However, the traffic situation in the parking lot is complex, and each section of the path may be occupied at any time. Hence, the optimal path planning in the parking lot requires a dynamic planning.

In 2020, a parking guidance system based on the multi-objective point A* algorithm was designed by Xiao et al. [24], which combines the distance factor between the entrance and the parking spaces and path planning using a heuristic function of the A* algorithm to quickly generate multiple driving paths and select the most efficient path. However, the heuristic function of the A* algorithm cannot take into consideration traffic congestion. If congestion happens, the time loss on queuing will reduce the efficiency of the system and subsequently increase the fuel consumption.

Therefore, in an intelligent parking infrastructure with both HVs and AVs involved, it is important to have a central command station to schedule the overall traffic flow and interface with autonomous vehicles. Moreover, it is necessary to consider in detail the factors that would affect parking space selection by the drivers and efficiently guide/allocate potentially available parking spaces for the AVs.

3. Factors for Choosing Optimal Parking Spaces

In an automated valet parking system, AVs are assumed to be able to interface with the central command station in an intelligent parking infrastructure. However, under the scenario in which there are both HVs and AVs in a parking infrastructure, no matter how intelligent the parking infrastructure is, the central command station cannot allocate parking spaces for the HVs which are not equipped with any communication devices. Based on this scenario, we consider the problem of a HV looking for a parking space in an intelligent parking infrastructure when an AV enters at the same time. The central command station, before allocating a parking space for the AV, must predict which parking space that the driver of the HV would choose and then allocate one of the remaining parking spaces to the autonomous vehicle. When a driver enters a parking infrastructure, his choice of parking space is often affected by his perception of various factors in the parking environment. Chen et al. [25] proposed six major factors that influence drivers to choose parking spaces: the walking distance from the parking spaces to the exit of the parking facility, the distance to the parking space from the entrance, the status of the path to the parking spaces, the status of the available parking spaces, parking safety, and shade under sunlight when parked outdoors. Since parking safety is somewhat implied in the status of the path for the AVs and the test case considered in this paper (see Section 6) is an outdoor parking lot with no shaded area, the last two factors will not be discussed in this paper. Thus, this paper focuses on the following four factors.

- *Walking distance*

Walking distance is the distance between the parking space and the exit of the parking infrastructure. For drivers, they likely prefer to arrive at their destination as quickly as possible after parking. Therefore, the shorter the walking distance, the more likely the parking space will be chosen.

- *Distance to parking*

Distance to parking is the distance between the entrance of the parking infrastructure and the parking space. For drivers and occupants, they generally prefer to arrive at their vehicles as quickly as possible, particularly when they forget something (e.g., phones, wallets) or need to place something back in the vehicle. Therefore, the shorter this distance, the more likely the parking space will be chosen.

- *Status of the path to the parking spaces*

Vehicles that break down (e.g., dead batteries) may block the path to some available spaces. Generally, lanes in parking infrastructures are narrow and often only one vehicle can pass through the lane. Lane occupancy increases traffic congestion and reduces traffic efficiency in the parking infrastructure. Thus, drivers are concerned with the status of the path leading to the available parking spaces and will likely choose those parking spaces that have clear paths.

- *Status of available parking spaces*

Ma [26] hypothesized that, in general, particularly new drivers are more inclined to choose available parking spaces with spaces on both sides unoccupied. The order of the priority of choice is followed by those spaces with one adjacent space unoccupied, those adjacent to the road, and finally those with spaces on both sides occupied.

Characteristics of parking spaces can be grouped into two main types: the cost type and the benefit type. For the four factors mentioned above, walking distance and distance to parking are cost types. Status of the path lane to parking spaces and status of available parking spaces are benefit types. The lower the value of the cost type factor, the higher the priority of the parking space would have. Contrarily, the higher the value of the benefit type factor, the higher priority of the parking space. Walking distance and distance to parking can be described by measurable values based on the layout of the entire parking infrastructure. Fuzzy factors such as the status of lanes and available parking spaces can also be represented by quantitative values. Following [27], a value of three is assigned when the lane is occupied; nine when the lane to the parking space is clear; eight to available parking space whose both sides are unoccupied; seven to an available parking space with one adjacent space unoccupied; six to an available parking space adjacent to the road; five to an available parking space with spaces on both sides occupied. Note that these numbers are relative and their absolute values are irrelevant.

4. Optimal Parking Space Selection Model

For a driver, parking space selection is a decision based on his perception and assessment of the parking environment. As it is difficult to quantify these fuzzy concepts for humans, fuzzy theory is introduced to solve these types of problems. The concept of fuzzy theory was put forward by Professor Zadeh in 1965, aiming to quantify the uncertainty of issues [28]. Based on the fuzzy theory, the four factors affecting driver's choice of parking space selection, discussed in Section 3, are quantified. In this section, we will introduce how to assign and sort the weights of available parking spaces based on the Fuzzy Comprehensive Evaluation (FCE) method [28]. As a result, the order of parking space preference will be displayed with different weights.

Procedure for Model Construction Based on FCE:

(1) Establish a factor vector $U = (u_1, u_2, \ldots, u_m)$ where u_i is the $i-$th factor that affects the parking selection. These factors usually have varying degrees of fuzziness. In this paper, the factor vector for evaluating parking preference is $U = (u_1, u_2, u_3, u_4)$, where u_1 represents "walking distance", u_2 "distance to parking", u_3 "status of lane to the parking spaces" and u_4 "status of available parking spaces".

(2) Establish an evaluation matrix A whose i-th row, $(a_{i1}, a_{i2}, \ldots, a_{im})$, is the evaluation vector for the i-th available parking space, and a_{ij} is the evaluation value for the $j-$th factor in U. In this paper, the evaluation vector for evaluating each factor in the vector U is $(a_{i1}, a_{i2}, a_{i3}, a_{i4})$ for the i-th parking space, where a_{i1}, a_{i2} are defined as the actual values of "walking distance" and "distance to parking" for evaluating u_1 and u_2, respectively; a_{i3} and a_{i4} are the scores under the evaluation standard discussed in Section 3 for evaluating the "status of lane to the parking spaces" u_3 and "status of available parking spaces" u_4, respectively.

(3) Establish the fuzzy comprehensive evaluation matrix $R = (r_{ij})_{n\times m}$ by normalizing A, where n is the total number of targeted parking spaces that need to be evaluated and m is the number of factors.

$$R = \begin{pmatrix} r_{11} & r_{12} & \cdots & r_{1m} \\ r_{21} & r_{21} & \cdots & r_{2m} \\ \vdots & \vdots & \ddots & \vdots \\ r_{n1} & r_{n2} & \cdots & r_{nm} \end{pmatrix} \tag{1}$$

The purpose of this normalization is to eliminate the impact caused by the differences in the orders of numbers of the physical measurements in the decision making process. For example, distance of "walking distance" maybe several meters while there is no particular unit for "status of parking spaces"; the orders of these numeric values are very different. The normalizations are:

$$r_{ij} = \frac{\min(a_{ij})}{a_{ij}},\ i \in (1,2, \ldots, n),\ j \in I_1 \tag{2}$$

$$r_{ij} = \frac{a_{ij}}{max(a_{ij})},\ i \in (1,2, \ldots, n),\ j \in I_2 \tag{3}$$

where, I_1 and I_2 represent the cost type factors and benefit factors, respectively [25].

(1) According to [29], to eliminate a potentially large number of combinatorial comparisons due to a large dimension of the factor vector, it is necessary to evaluate the independent relationship between every two factors by the pair-wise comparison matrix $B = \left(b_{ij}\right)_{n\times m}$ which is defined by drivers with different driving experiences, where $b_{ii} = 0.5$, $b_{ij} + b_{ji} = 1$, and $b_{ij} \geq 0$. b_{ij} represent the value of factor i compared to the preference to factor j, b_{ii} represents the value of a factor compared to the preference to itself. In this paper, we have collected opinions from twenty drivers to form the pair-wise comparison matrix. Our sample included five males and five females, with ages varying from 20 to 55 years old and with driving experiences ranging from less than one month to more than 35 years. This set of samples represents sufficiently broad variability.

(2) Define the weight vector of the factors: $w = (w_1, w_2, \ldots, w_n)$, where w_i can be calculated by the least variance method (LVM) for further ranking priority of the factors based on the pair-wise comparison matrix. [30]:

$$w_i = \frac{1}{n}\left(\sum_{j=1}^{n} b_{ij} + 1 - \frac{n}{2}\right),\ i = (1, 2, \ldots, n) \quad (4)$$

(3) According to Equations (1)–(3) we established a priority vector $z_{i(w)}$, which can be described as follows:

$$z_{i(w)} = WAA_w(r_{i1}, r_{i2}, \ldots, r_{in}) = \sum_{j=1}^{n} R_{n\times m} w_i \quad (5)$$

where, $WAA_w(r_{i1}, r_{i2}, \ldots, r_{in})$ is the weighted arithmetic average operator. The higher the weight value of a certain parking space in $z_{i(w)}$, the more likely it would be selected by the driver [31].

5. Optimal Route Selection

Shortest path planning is very important in path navigation. Traffic efficiency can be improved by planning the shortest path based on the distance information between locations. Floyd algorithm is a classic dynamic programming algorithm that uses dynamic programming to find the shortest path between multiple source points in a given weighted graph. The algorithm aims to find the shortest path from one point to another [32].

Assume that the shortest path from a node i to another node j has no more than two possibilities, one is directly from i to j, and the other is from i through node k to j. Hence, the dynamic transfer function of this algorithm is: $dis(i,j) = min\left(D_{i,j},\ D_{i,k} + D_{k,j}\right)$, where $dis(i,j)$ represents the shortest distance between node i and node j. The specifics of our proposed algorithm is shown as follows:

(1) Number each parking location in the map as a node;

(2) Initialize an adjacent matrix $D_{p\times q}$, where $D_{i,j}$ represents the distance between node i and j; if i and j are not adjacent, $D_{i,j}$ will be assigned with ∞; if $i = j$, the value of $D_{i,j}$ will be 0.

$$D_{p\times q} = \begin{pmatrix} 0 & D_{1,2} & D_{1,3} & \cdots & D_{1,\,q-1} & D_{1,q} \\ D_{2,1} & 0 & D_{2,3} & \cdots & D_{2,q-1} & D_{2,q} \\ D_{3,1} & D_{3,2} & 0 & \cdots & D_{3,q-1} & D_{3,q} \\ \vdots & \vdots & \vdots & \ddots & \vdots & \vdots \\ D_{p-1,1} & D_{p-1,2} & D_{p-1,3} & \cdots & 0 & D_{p-1,q} \\ D_{p,1} & D_{p,2} & D_{p,3} & \cdots & D_{p,q-1} & 0 \end{pmatrix} \quad (6)$$

(3) Update the transfer function. For example, consider three nodes i, j, and k in the map, where k is an intermediate node of i and j, and i and j represents the start

point and the end point, respectively. The transfer function $dis(i,j)$ is updated by $dis(i,j) = \min\left(D_{i,j},\ D_{i,k} + D_{k,j}\right)$ to select the smaller of the two distances from i to j and from i to k to j. The Floyd algorithm finds an intermediate node k to determine whether there is a shorter distance through this node k.

6. Example: Results and Discussion

The example we consider is a single-entrance parking lot, see Figure 2. The layout of the parking lot has been redesigned according to the Engineering Construction Industry Standard JGJ100-98 and the layout of the example parking lot is shown in Figure 3. Since the entrance is located in the lower right of the entire layout, we define the coordinate system with the origin at the lower right corner, as shown in Figure 4.

Figure 2. Aerial view of the parking lot of our example problem.

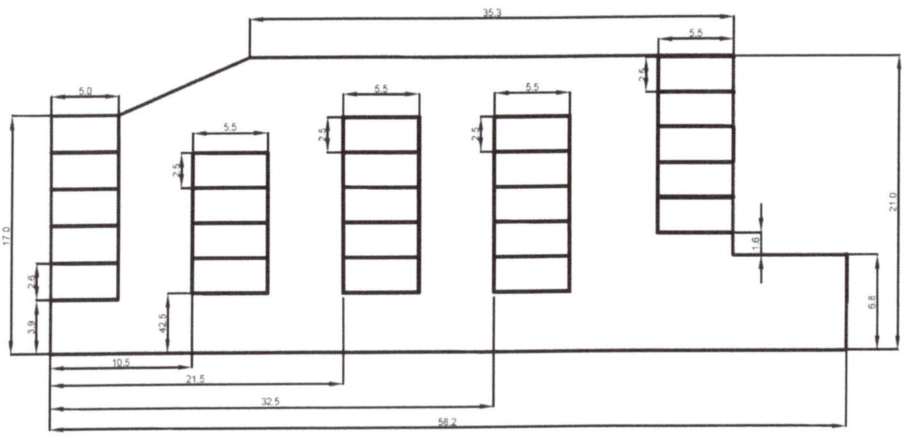

Figure 3. Layout of the parking lot (distance unit in meter).

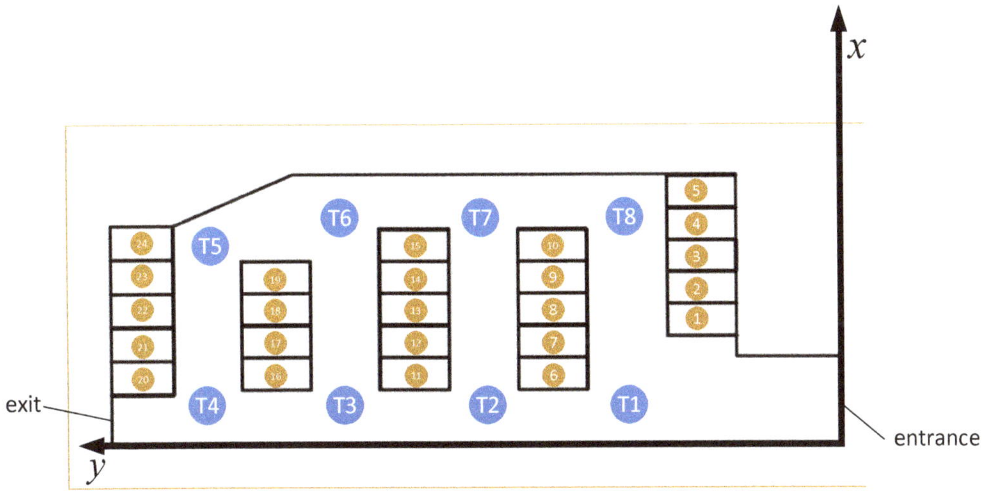

Figure 4. Layout of nodes in the parking lot and reference to a coordinate system.

As shown in Figure 4, in order to design the route selection, we denote each parking space and each turning intersection as a node (1–24) and a number (T1–T8). Additionally, considering the differences in sizes of vehicles entering the parking lot, we define the coordinates of each of the nodes and intersection numbers as the centered position of the parking space and the central line of the route at the turning. The coordinates of all nodes are listed in Table 1.

Table 1. Coordinates in metric units of all nodes in test graph

*T*1	(2.1, 17.0)	1	(9.7, 17.0)	9	(13.0, 17.0)	17	(8.0, 39.5)
*T*2	(2.1, 28.5)	2	(12.2, 17.0)	10	(15.5, 17.0)	18	(10.5, 39.5)
*T*3	(2.1, 39.5)	3	(14.7, 17.0)	11	(5.5, 28.5)	19	(13.0, 39.5)
*T*4	(2.1, 50.4)	4	(17.2, 17.0)	12	(8.0,28.5)	20	(5.2, 50.5)
*T*5	(15.7, 50.4)	5	(19.7, 17.0)	13	(10.5, 28.5)	21	(7.8, 50.5)
*T*6	(19.7, 39.5)	6	(5.5, 17.0)	14	(13.0, 28.5)	22	(10.4, 50.5)
*T*7	(19.7, 28.5)	7	(8.0, 17.0)	15	(15.5, 28.5)	23	(13.1, 50.5)
*T*8	(19.7, 17.0)	8	(10.5, 17.0)	16	(5.5, 39.5)	24	(15.7, 50.5)

In this section, we have designed three specific parking scenarios to test our proposed algorithm via Python. In the following, we will take scenario 1 as an example to explain the application of our method in detail. As shown in Figure 5, red nodes represent the parking spaces that are occupied, brown ones are those that are currently available, and the light blue node (node 11) represents the parking space that is being parked by a vehicle. A blue vehicle, which is an AV, is just entering into the intelligent parking lot and sending a parking request to the central command station. Meanwhile, an orange vehicle, which is a HV and cannot communicate with the central command station, is seeking an available parking space. Under this situation, the problem is to allocate an available parking space for the blue autonomous vehicle by the central command station.

In order to solve this problem, we need to first determine which parking space that the driver of the orange vehicle would choose and offer the shortest path between the entrance and the allocated parking space. The overview scheme is described as a flowchart (Figure 6). The specific steps in the parking spaces selection and the shortest path planning are shown as follows:

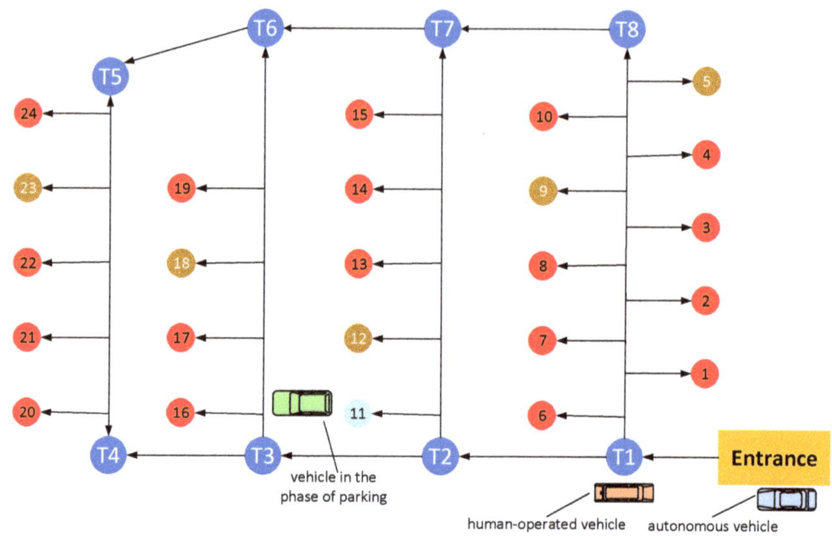

Figure 5. Example of a parking scenario 1.

Figure 6. Flowchart for parking space selection and the shortest path planning.

Step 1: Set the properties table of all available parking spaces currently in the example parking lot. As shown in Table 2.

Table 2. Properties of the available parking spaces in the parking lot of scenario 1.

Parking Space	Walking Distance (m)	Distance to Parking (m)	Status of Parking	Status of Available Parking Spaces
5	51.0	36.7	clear	adjacent to the road
9	44.3	30.0	clear	both sides are occupied
12	27.8	36.5	occupied	both sides are occupied
18	41.8	27.5	clear	both sides are occupied
23	11.1	63.6	clear	both sides are occupied

For the walking distance and distance to parking, distances in metric units are used in evaluating the values in matrix A, as defined in Section 4. For the status of lane and status of available parking spaces, fuzzy evaluation is assigned with values proposed in Section 3. Based on the values from Table 2, the evaluation matrix A is:

$$A = \begin{pmatrix} 51.0 & 36.7 & 9.0 & 6.0 \\ 44.3 & 30.0 & 9.0 & 5.0 \\ 27.8 & 36.5 & 3.0 & 5.0 \\ 41.8 & 27.5 & 9.0 & 5.0 \\ 11.1 & 63.6 & 9.0 & 5.0 \end{pmatrix}$$

Step 2: Normalize the evaluation matrix A to form the fuzzy comprehensive evaluation matrix $R = (r_{ij})_{n\times m}$ according to Equations (2) and (3):

$$R = \begin{pmatrix} 0.217 & 0.749 & 1.000 & 1.000 \\ 0.250 & 0.917 & 1.000 & 0.833 \\ 0.399 & 0.753 & 0.333 & 0.833 \\ 0.266 & 1.000 & 1.000 & 0.833 \\ 1.000 & 0.432 & 1.000 & 0.833 \end{pmatrix}$$

Step 3: Establish the pair-wise comparison matrix by the comparison method, based on the opinions of our sample set of twenty drivers with a range of driving experiences, as described in Section 4. The following shows the response from one of the drivers:

$$B_i = \begin{pmatrix} 0.5 & 0.8 & 0.9 & 0.4 \\ 0.2 & 0.5 & 0.1 & 0.4 \\ 0.1 & 0.9 & 0.5 & 0.9 \\ 0.6 & 0.6 & 0.1 & 0.5 \end{pmatrix}$$

where, $i = 1, \ldots, 10$ for our data set. As shown in B_i, all diagonal values are 0.5 as each factor complements itself. The values of $b_{12} = 0.8$ and $b_{21} = 0.2$ indicate that in the view of this driver, the factor "walking distance" is more important than "distance to parking". "distance to parking" is still a factor that the driver would consider when choosing a parking space, but it is just less important.

Step 4: Obtain the weight vector for the pair-wise comparison matrix in Step 3 according to Equation (4):

$$w_i = (0.40,\ 0.05,\ 0.35,\ 0.20)$$

Applying Equation (4), the pair-wise comparison matrix from the twenty drivers is converted into the following weight matrix:

$$w = \begin{pmatrix} 0.400 & 0.050 & 0.350 & 0.200 \\ 0.200 & 0.325 & 0.450 & 0.275 \\ 0.500 & 0.125 & 0.175 & 0.175 \\ 0.175 & -0.050 & 0.625 & 0.750 \\ 0.150 & 0.375 & 0.250 & 0.225 \\ 0.225 & 0.125 & 0.250 & 0.400 \\ 0.125 & 0.550 & 0.350 & 0.350 \\ 0.375 & 0.100 & 0.350 & 0.175 \\ 0.125 & 0.225 & 0.375 & 0.250 \\ 0.225 & 0.200 & 0.400 & 0.225 \\ 0.250 & 0.075 & 0.475 & 0.200 \\ 0.425 & 0.075 & 0.475 & 0.200 \\ 0.425 & 0.075 & 0.325 & 0.175 \\ 0.125 & 0.125 & 0.425 & 0.325 \\ 0.050 & 0.275 & 0.225 & 0.450 \\ 0.175 & 0.250 & 0.275 & 0.300 \\ 0.300 & 0.125 & 0.250 & 0.325 \\ 0.325 & 0.175 & 0.100 & 0.425 \\ 0.300 & 0.000 & 0.450 & 0.250 \\ 0.150 & 0.150 & 0.425 & 0.325 \\ 0.100 & 0.300 & 0.325 & 0.275 \end{pmatrix}$$

To make the above matrix representative of the preference of most drivers, we remove the largest and smallest values of each factor in the above matrix and average the remaining values to obtain an average weight vector:

$$\widetilde{w} = (0.233,\ 0.170,\ 0.336,\ 0.286)$$

Step 5: With the average weight vector and Equation (5), the priority vector that contains the weights of every available parking space in the given situation can be obtained:

$$z_i(w) = (0.799,\ 0.788,\ 0.571,\ 0.806,\ 0.881)$$

Step 6: Ranking the priorities of the available parking spaces in this example parking lot according to $z_i(w)$ gives: $z_{23}(w) > z_{18}(w) > z_5(w) > z_9(w) > z_{12}(w)$.

Based on step 6, parking space 23 has the highest priority for the driver of the orange vehicle. Thus, parking space 18 will be allocated to the blue autonomous vehicle by the central command station. By analyzing the preferences of the human driver, the algorithm is able to provide the optimal parking space for the autonomous vehicle. Since an intelligent parking lot system cannot influence the choice of the driver, a suboptimal parking space, which is space 18, is allocated to autonomous vehicle in this situation.

Step 7: Establish the adjacent matrix with all the road intersection nodes and the node of parking space 18 for route selection based on the adjacent matrix assignment rule according to Equation (6) in Section 5:

$$D = \begin{pmatrix} 0 & 11.5 & \infty & \infty & \infty & \infty & \infty & 17.6 & \infty \\ \infty & 0 & 11.0 & \infty & \infty & \infty & 17.6 & \infty & \infty \\ \infty & \infty & 0 & 10.9 & \infty & 17.6 & \infty & \infty & 8.4 \\ \infty & \infty & \infty & 0 & 13.6 & \infty & \infty & \infty & \infty \\ \infty & \infty & \infty & 13.6 & 0 & \infty & \infty & \infty & \infty \\ \infty & \infty & \infty & \infty & 11.6 & 0 & \infty & \infty & 9.2 \\ \infty & \infty & \infty & \infty & \infty & 11.0 & 0 & \infty & \infty \\ \infty & \infty & \infty & \infty & \infty & \infty & 11.5 & 0 & \infty \\ \infty & \infty & \infty & \infty & \infty & \infty & \infty & 0 & 0 \end{pmatrix}$$

As shown in matrix D, there is only one passable path from T1 to parking space 18: starting from T1, go through T2, subsequently arrive at T3, then turn right to parking space 18, which is also the shortest path, with a total distance of 30.9 m. Through this method, under the scenario of the presence of one HV and one AV in this example, we simultaneously predict the parking space preference of the driver and allocate a parking space to the AV through V2X communication based on the driver preferences. Moreover, a parking lot may be constrained with traffic rules (e.g., one way road) and limited space, the adjacent matrix D can, in addition to providing the shortest route navigation, also reflect the constraints of the traffic rules and other conditions.

Using the same approach as used in scenario one, we can also solve the similar issues, such as in the following two scenarios:

Scenario 2:

Step 1: Set the properties table of all available parking spaces currently in the example parking scenario (Figure 7). As shown in Table 3.

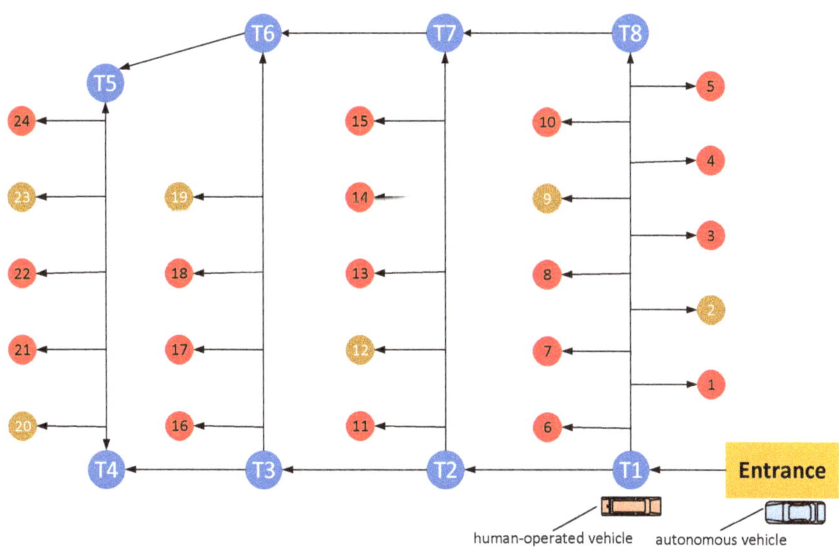

Figure 7. Example of a parking scenario 2.

Table 3. Properties of the available parking spaces in the parking lot of scenario two.

Parking Space	Walking Distance (m)	Distance to Parking (m)	Status of Parking	Status of Available Parking Spaces
2	43.5	29.2	clear	both sides are occupied
9	44.3	30.0	clear	both sides are occupied
12	27.8	36.5	clear	both sides are occupied
19	21.8	42.5	clear	adjacent to the road
20	3.2	55.7	clear	adjacent to the road
23	11.1	63.6	clear	both sides are occupied

Based on the values from Table 3 and assignment rule in Section 3, the evaluation matrix A is:

$$A = \begin{pmatrix} 43.5 & 29.2 & 9.0 & 5.0 \\ 44.3 & 30.0 & 9.0 & 5.0 \\ 27.8 & 36.5 & 9.0 & 5.0 \\ 21.8 & 42.5 & 9.0 & 6.0 \\ 3.2 & 55.7 & 9.0 & 6.0 \\ 11.1 & 63.6 & 9.0 & 5.0 \end{pmatrix}$$

Step 2: Normalize the evaluation matrix A to form the fuzzy comprehensive evaluation matrix $R = (r_{ij})_{n \times m}$ according to Equations (2) and (3):

$$R = \begin{pmatrix} 0.074 & 1.000 & 1.000 & 0.833 \\ 0.072 & 0.973 & 1.000 & 0.833 \\ 0.115 & 0.799 & 1.000 & 0.833 \\ 0.147 & 0.687 & 1.000 & 1.000 \\ 1.000 & 0.524 & 1.000 & 1.000 \\ 0.288 & 0.459 & 1.000 & 0.833 \end{pmatrix}$$

Step 3: With the average weight vector calculated in scenario one and Equation (5), the priority vector that contains the weights of every available parking space in scenario two can be obtained:

$$z_i(w) = (0.761,\ 0.756,\ 0.737,\ 0.773,\ 0.944,\ 0.719)$$

Step 4: Ranking the priorities of the available parking spaces in this example parking lot according to $z_i(w)$ gives: $z_{20}(w) > z_{19}(w) > z_2(w) > z_9(w) > z_{12}(w) > z_{23}(w)$.

Based on Step 4, parking space 20 has the highest priority for the driver of the orange vehicle based on scenario two. Thus, parking space 19 will be allocated to the blue autonomous vehicle by the central command station.

Scenario 3:

Step 1: Set the properties table of all available parking spaces currently in the example parking scenario (Figure 8). As shown in Table 4.

Based on the values from Table 4 and assignment rule in Section 3, the evaluation matrix A is:

$$A = \begin{pmatrix} 51.0 & 36.7 & 9.0 & 6.0 \\ 41.8 & 27.5 & 9.0 & 5.0 \\ 25.3 & 34.0 & 9.0 & 6.0 \\ 27.8 & 36.5 & 9.0 & 7.0 \\ 21.8 & 42.5 & 9.0 & 6.0 \\ 3.2 & 55.7 & 9.0 & 6.0 \\ 11.1 & 63.6 & 9.0 & 5.0 \end{pmatrix}$$

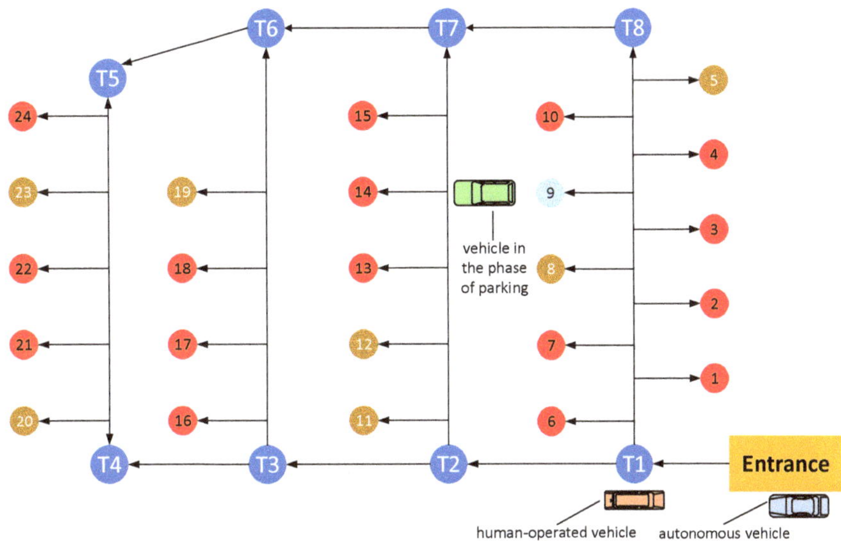

Figure 8. Example of a parking scenario three.

Table 4. Properties of the available parking spaces in the parking lot of scenario two.

Parking Space	Walking Distance (m)	Distance to Parking (m)	Status of Parking	Status of Available Parking Spaces
5	51.0	36.7	clear	adjacent to the road
8	41.8	27.5	clear	both sides are occupied
11	25.3	34.0	clear	adjacent to the road
12	27.8	36.5	clear	one adjacent space unoccupied
19	21.8	42.5	clear	adjacent to the road
20	3.2	55.7	clear	adjacent to the road
23	11.1	63.6	clear	both sides are occupied

Step 2: Normalize the evaluation matrix A to form the fuzzy comprehensive evaluation matrix $R = (r_{ij})_{n \times m}$ according to Equations (2) and (3):

$$R = \begin{pmatrix} 0.063 & 0.749 & 1.000 & 0.857 \\ 0.077 & 1.000 & 1.000 & 0.714 \\ 0.126 & 0.809 & 1.000 & 0.857 \\ 0.115 & 0.753 & 1.000 & 1.000 \\ 0.147 & 0.647 & 1.000 & 0.857 \\ 1.000 & 0.494 & 1.000 & 0.857 \\ 0.288 & 0.432 & 1.000 & 0.714 \end{pmatrix}$$

Step 3: With the average weight vector calculated in scenario one and Equation (5), the priority vector that contains the weights of every available parking space in scenario three can be obtained:

$$z_i(w) = (0.723,\ 0.728,\ 0.748,\ 0.777,\ 0.725,\ 0.898,\ 0.681)$$

Step 4: Ranking the priorities of the available parking spaces in this example parking lot according to $z_i(w)$ gives: $z_{20}(w) > z_{12}(w) > z_{11}(w) > z_8(w) > z_{19}(w) > z_5(w) > z_{23}(w)$.

Based on step 4, parking space 20 has the highest priority for the driver of the orange vehicle based on scenario two. Thus, parking space 12 will be allocated to the blue autonomous vehicle by the central command station.

As can be seen from the above three test examples, people are more inclined to choose the available parking space near the exit of the parking lot.

In this paper, to validate of the proposed algorithm, we have collected opinions from fifty drivers on their most preferred parking spaces to check against the data from the twenty drivers mentioned in Step 3. Figure 9 shows that, in addition to the preferred choice of space 23, 20 and 20 in designed scenarios one, two and three, respectively, there are only four, six and four other choices among the fifty drivers who are surveyed. The accuracy of the prediction of the proposed algorithm is thus, 92%, 88% and 92%, respec-tively, in the three test cases.

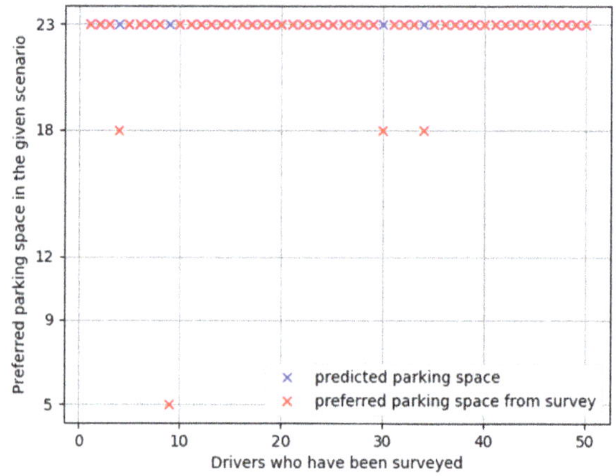

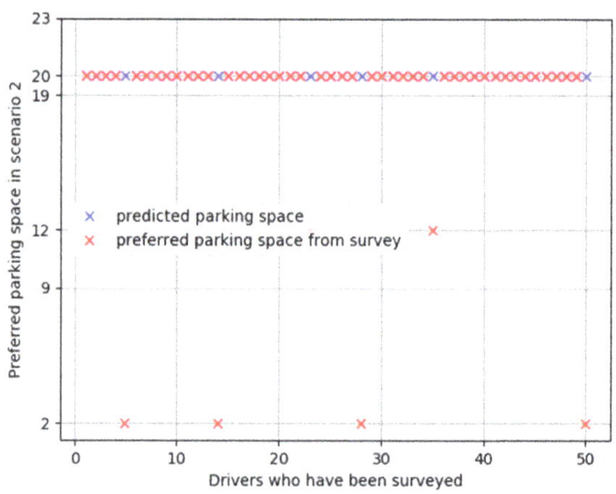

Figure 9. *Cont.*

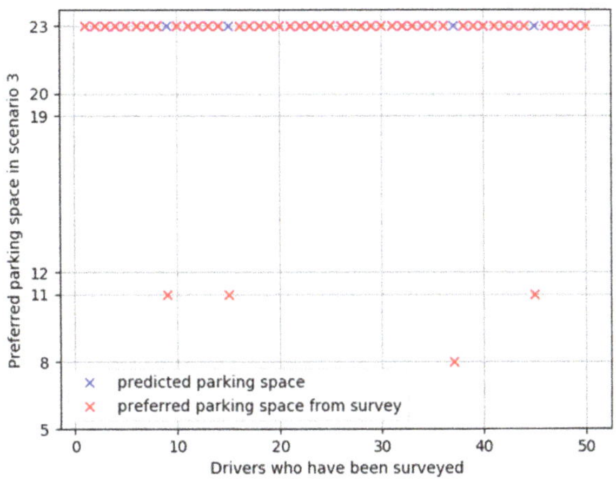

Figure 9. Comparison between the predicted parking space preference and the opinions of the most preferred parking space from fifty drivers in three example scenarios.

7. Conclusions and Future Work

In this paper, we studied the problem of automated parking space allocation during the transition period when both human-operated vehicles (HVs) and autonomous (driverless) vehicles (AVs) are present simultaneously in an intelligent parking infrastructure. Since not all vehicles will be autonomous in a near foreseeable future and fully automated parking systems are still being developed, the problem of interest, thus, has important and relevant applications. Based on four key factors that affect the choice of parking spaces for drivers, the fuzzy comprehensive evaluation (FCE) method is used to first predict the driver's choice of parking space, since a HV cannot communicate with the central command station. This prediction then allows an intelligent parking system to allocate the remaining available parking spaces to the AVs without conflict arising with the HVs. Furthermore, by incorporating information of the parking spaces and road map of the parking facility, an adjacent matrix in the Floyd algorithm is then developed to determine the shortest feasible paths to arrive at the parking spaces safely. The proposed algorithms are applied to three parking lot scenarios in which both a HV and an AV are entering. It is shown that a parking space can be reasonably allocated to the AV after a prediction of the driver's choice of parking space.

This work lays the foundation for future investigation into the automated parking problem with multiple HVs and AVs. Moreover, since the test parking lot discussed in this paper is not very large, it is likely that drivers can observe the entire layout and the situation from the entry point. More complexities of the parking infrastructure, such as multiple entrances, exits, and levels, can be considered in the future. A faster and more robust dynamic planning method needs to be developed.

Author Contributions: M.W. and H.J. designed the algorithm; C.-A.T. verified the rationality of the algorithm; M.W. performed the software simulation; H.J. and C.-A.T. analyzed the results of the simulation; M.W. wrote the paper with help of H.J. and C.-A.T.; C.-A.T. completed the final editing. All authors have read and agreed to the published version of the manuscript.

Funding: This work is financially supported by The Major University Nature Science Research Project of Jiangsu Province (No.16KJA580001).

Institutional Review Board Statement: Not applicable.

Informed Consent Statement: Not applicable.

Data Availability Statement: Not applicable.

Conflicts of Interest: The authors declare no conflict of interest.

References

1. Kim, O.T.T.; Tran, N.H.; Pham, C.; Leanh, T.; Thai, M.T.; Hong, C.S. Parking Assignment: Minimizing Parking Expenses and Balancing Parking Demand among Multiple Parking Lots. *IEEE Trans. Autom. Sci. Eng.* **2020**, *17*, 1320–1331.
2. Nawaz, S.; Efstratiou, C.; Mascolo, C. ParkSense: A Smartphone Based Sensing System for On-Street Parking. In Proceedings of the 19th International Conference on Mobile Computing & Networking, Miami, FL, USA, 30 Sepetember–4 October 2013.
3. Jang, C.; Kim, C.; Lee, S.; Kim, S.; Lee, S.; Sunwoo, M. Re-Plannable Automated Parking System with a Standalone around View Monitor for Narrow Parking Lots. *IEEE Trans. Intell. Transp. Syst.* **2020**, *21*, 777–790. [CrossRef]
4. Li, C.X. Research on Automatic Vertical Parking System Based on Intelligent Recognition Technology of Parking Space Scene. Ph.D. Thesis, Jiangsu University, Zhenjiang, China, 2016.
5. Schwarz, S.; Zochmann, E.; Muller, M.; Guan, K. Dependability of Directional Millimeter Wave Vehicle-to-Infrastructure Communications. *IEEE Access* **2020**, *8*, 53162–53171. [CrossRef]
6. Sedighi, S.; Nguyen, D.V.; Kuhnert, K.D. Guided Hybrid A-star Path Planning Algorithm for Valet Parking Applications. In Proceedings of the 5th IEEE International Conference on Control, Automation and Robotics (ICCAR), Beijing, China, 19–22 April 2019.
7. Shih, S.; Tsai, W. A Convenient Vision-Based System for Automatic Detection of Parking Spaces in Indoor Parking Lots Using Wide-Angle Cameras. *IEEE Trans. Veh. Technol.* **2014**, *63*, 2521–2532. [CrossRef]
8. Mohandes, M.; Deriche, M.; Abuelma'atti, M.T.; Tasadduq, N. Preference-based smart parking system in a university campus. *IET Intell. Transp. Syst.* **2019**, *13*, 417–423. [CrossRef]
9. Liang, W.; Hu, J.M.; Zhang, Y.; Zhang, Z.; Wang, Z.W. Modeling and empirical analyses of parking space choices based on individual preferences. *Tsinghua Univ. (Sci. Technol.)* **2017**, *57*, 100–106.
10. Leephakpreeda, T. Car-parking guidance with fuzzy knowledge-based decision making. *Build. Environ.* **2007**, *42*, 803–809. [CrossRef]
11. Marina, D.; Vladislav, L.; Inna, S.; Igor, K. Parking Guide Service for Large Urban Areas. In Proceedings of the 10th IEEE International Conference on Intelligent Data Acquisition and Advanced Computing Systems: Technology and Applications (IDAACS), Metz, France, 18–21 September 2019.
12. Nesic, Z.; Ljubic, L.; Radojicic, M.; Vasovic, J.V. Some considerations on improvement of parking service information system. In Proceedings of the 24th Telecommunications Forum (TELFOR), Belgrade, Serbia, 22–23 November 2016.
13. Dijkstra, E.W. A Note on Two Problems in Connexion with Graphs. *Numer. Math.* **1959**, *1*, 269–271. [CrossRef]
14. Floyd, R.W. Algorithm 97: Shortest path. *Commun. ACM* **1962**, *5*, 345. [CrossRef]
15. Shi, W.; Wang, K. Floyd algorithm for the shortest path planning of mobile robot. *Chin. J. Sci. Instrum.* **2009**, *10*, 2088–2092.
16. Ji, Y.J.; Wang, W.; Deng, W. Micro-level parking behaviors within parking facilities and optimal parking space choice model. *J. Southeast Univ. (Nat. Sci. Ed.)* **2009**, *2*, 399–403.
17. Banzhaf, H.; Nienhuser, D.; Knoop, S.; Zollner, J.M. The future of parking: A survey on automated valet parking with an outlook on high density parking. Proceeding of the IEEE Intelligent Vehicles Symposium (IV), Los Angeles, CA, USA, 11–14 June 2017.
18. Dou, Y.X.; Zhang, M.M.; Zhang, J.; Fan, X.Y. Study on Dynamic Distribution Algorithm in Parking. *J. Zhenzhou Univ. Light Ind.* **2012**, *6*, 24–27.
19. Audi Industrial Group. Audi Technology Portal. Available online: https://www.audi-technology-portal.de/en/electrics-electronics/driver-assistant-systems/parking-assist (accessed on 10 September 2020).
20. Alfatihi, S.; Chihab, S.; Alj, Y.S. Intelligent Parking System for Car Parking Guidance and Damage Notification. In Proceedings of the International Conference on Intelligent Systems Modelling & Simulation, Bangkok, Thailand, 29–31 January 2013.
21. Kotb, A.O.; Shen, Y.C.; Zhu, X.; Huang, Y. iParker—A New Smart Car-Parking System Based on Dynamic Resource Allocation and Pricing. *IEEE Trans. Intell. Transp. Syst.* **2016**, *17*, 2637–2647. [CrossRef]
22. Tcheumadjeu, L.C.T.; Andert, F.; Tang, Q.; Sohr, A.; Stürzl, W. Integration of an Automated Valet Parking Service into an Internet of Things Platform. In Proceedings of the 21st International Conference on Intelligent Transportation Systems (ITSC), Maui, HI, USA, 4–7 November 2018.
23. Yu, L.Y.; Jiang, H.B.; Hua, L. Anti-Congestion Route Planning Scheme Based on Dijkstra Algorithm for Automatic Valet Parking System. *Appl. Sci.* **2019**, *9*, 5016. [CrossRef]
24. Xiao, W.; Zhang, L.; Qiu, Z.H.; Zhong, Y.; Zhang, T.N. Design of Parking Lot Path Guidance System Based on Multi-objective Point A* Algorithm. *Comput. Mod.* **2020**, *6*, 40–45.
25. Chen, M.; Hu, C.; Chang, T. The Research on Optimal Parking Space Choice Model in Parking Lots. In Proceedings of the 3th International Conference on Computer Research and Development (ICCRD), Shanghai, China, 11–13 March 2011.
26. Ma, J. Design and Implementation of Parking Space Recommendation System Considering User Preference. Master's Thesis, Henan University, Henan, China, 2018.
27. Chen, J. Study on Urban Parking Facilities' Planning Methods. Ph.D. Thesis, Southeast University, Nanjing, China, 2000.
28. Zadeh, L.A. Fuzzy sets. *Inf. Control* **1965**, *8*, 338–353. [CrossRef]
29. Xu, Z.S. *Uncertain Multiple Attribute Decision Making: Methods and Applications*; Tsinghua University Press: Beijing, China, 2004.

30. Xu, Z.S. The Least Variance Priority Method (LVM) for Fuzzy Complementary Judgement Matrix. *Syst. Eng. Theory Pract.* **2001**, *21*, 93–96.
31. Waerden, P.; Borgers, A.; Timmermans, H. Travelers Micro-Behavior at Parking Lots: A Model of Parking Choice Behavior. In Proceedings of the 82nd Annual Meeting of the Transportation Research Board, Washington, DC, USA, 12–16 January 2003.
32. Sathyaraj, B.M.; Jain, L.C.; Finn, A.; Drake, S. Multiple UAVs path planning algorithms: A comparative study. *Fuzzy Optim. Decis. Mak.* **2008**, *7*, 257–267. [CrossRef]

applied sciences

MDPI

Article

Leader-Follower Formation Control of Wheeled Mobile Robots without Attitude Measurements

Jonathan Hirata-Acosta [1], Javier Pliego-Jiménez [1], César Cruz-Hernádez [1,*] and Rigoberto Martínez-Clark [2]

[1] Electronics and Telecommunication Department, Scientific Research and Advanced Studies Center of Ensenada (CICESE), Ensenada 22860, Mexico; jonathanhi.ac@gmail.com (J.H.-A.); jpliego@cicese.mx (J.P.-J.)
[2] Faculty of Engineering, Administrative and Social Science, Universidad Autónoma de Baja Calfifornia, FCIAS, Tecate 21100, Mexico; rigoberto.martinez@uabc.edu.mx
* Correspondence: ccruz@cicese.mx

Abstract: The problem of leader-follower formation of a platoon of differential-drive wheeled mobile robots without using attitude measurements is addressed in this paper. Contrary to the position-distance approaches existing in the literature, the formation and collision avoidance is achieved by introducing a state-dependent delay in the desired trajectory. The delay is obtained as the output of a dynamical system and its magnitude will decrease/increase depending on the distance between the robots. To guarantee trajectory tracking and to overcome the lack of orientation measurements, an output feedback control and attitude observer are proposed based on the kinematic model of the robots. The attitude observer is designed directly on the special orthogonal group $SO(2)$ and it can be used in open-loop schemes. The proposed control-observer scheme ensures asymptotic convergence of the tracking and observer errors. Finally, experimental results are presented to show the performance of the proposed approach.

Keywords: mobile robot; formation; motion coordination; attitude observer; trajectory tracking

Citation: Hirata-Acosta, J.; Pliego-Jiménez, J.; Cruz-Hernández, C.; Martínez-Clark, R. Leader-Follower Formation Control of Wheeled Mobile Robots without Attitude Measurements. *Appl. Sci.* **2021**, *11*, 5639. https://doi.org/10.3390/app11125639

Academic Editor: Luis Gracia

Received: 28 May 2021
Accepted: 7 June 2021
Published: 18 June 2021

1. Introduction

In recent years, autonomous mobile vehicles have attracted interest from the scientific community, mainly due to the wide range of applications in which they can be implemented; ranging from searching, surveillance and exploration applications to cargo transportation and cooperative manipulation [1,2]. Cooperative formation control focuses more on the efficiency and fault tolerance that a single mobile robot could not provide [3]. A particular problem of multi-robot coordination that has received much attention in the last decades is formation control. The objective of formation control of multiple mobile robots is to achieve a desired formation pattern while guaranteeing that the multiple robots as a group accomplish a given task cooperatively [4].

The formation control approach has been implemented in different types of vehicles, this in order to perform the tasks, with greater ease and robustness. For example, in [5] this approach was used for underwater vehicles, where the follower tracks a reference trajectory based on the leader position and predetermined formation without the need for leader's velocity and dynamics. This is desirable in marine robotics due to weak underwater communication and low bandwidth. In order to tackle the harsh conditions of underwater environment, in [6] the authors drive unmanned underwater vehicle using a deterministic artificial intelligence approach. This technique is based on self-awareness of the robot and relies on the dynamic of the vehicle and linear regression instead of stochastic or traditional control theory methods. Another application of this technique is presented in [7], where the authors address formation control for a team of quadrotor UAVs in which the robots follow a specified group trajectory while safely changing the shape of the formation according to the specifications of the task. On the other hand, in [8] the formation control of a group of unicycle-type wheeled mobile robots at the dynamics level with a little amount of inter-robot communication is investigated. Another interesting approach arises from bio-inspired control techniques. The robotic swarm control is a new paradigm of multi-robot control system aiming to achieve task in collective way using low level interactions between the members of the swarm. For instance, in [9] the authors proposed to employ a variation of the particle swarm optimization (PSO) algorithm to achieve formation in a swarm of agents while tracking a dynamic target. This techniques is inspired on the pheromone based communication of the ant colonies.

Several formation control approaches have been proposed in the literature, and they are mainly divided in three categories: behavior-based methods, leader–follower and virtual structure methods [10,11]. The behavior-based approach is inspired by the emerging behaviors in nature such as flock of birds, random walks of ants and school of fish [12]. In this case, a group behavior (or mission) comprises some low-level actions (or sub-tasks) and is constructed to achieve the global objective, where the individual robot needs to perform low-level actions to accomplish the group behavior [13]. In the leader-follower formation [14,15], one robot is chosen as the leader which decides the whole formation group's moving trajectory, the other ones are the followers which are tasked to follow the leader, and the desired relative separations and bearings are expected to be maintained [16]. This strategy is easily implemented by using two controllers only and is suitable to describe the formation of robots, but it is hard to take into account the functioning capabilities of different robots, i.e., ability gap of a robot [17]. Finally, in the virtual structure formation, robots behave like particles embedded in a rigid virtual structure [18].

Some very interesting works where the leader-follower formation is used are, for example [19], in which the formation problem is converted to a trajectory tracking problem, where each follower robot tracks its corresponding generated reference trajectory such that the whole group forms and maintains the desired shape. In this work, some experiments were successfully conducted and reported using a group of four TURTLEBOTs. In [20], the authors tackle the leader-follower formation control problem of non-holonomic mobile robots. In this case, the trajectory tracking control for a single non-holonomic mobile robot is extended to the formation control for two non-holonomic robots in which one is the leader and the second is the follower. The controllers proposed by the authors are based on the PI control technique. Simulation results are presented to demonstrate the good performance of the proposed controller.

In this work, we addressed the leader-follower formation control problem for a group of nonholonomic wheeled mobile robots (WMRs). We do not follow the common distance-based approaches where the follower tracks the trajectory generated by the leader with an offset to avoid a collisions. In our approach, the followers track the delayed desired trajectory of the leader. The time delay is not arbitrary, on the contrary, it is obtained as the output of a dynamical system whose inputs are the position of the robots. The aforementioned dynamical system is designed in such way that, when the distance between the robots increases, the magnitude of the delay decreases and vice versa. The proposed methodology allows to achieve a convoy formation or platooning without collisions. Another advantage of the proposed approach is that the followers do not deviate from the leader's path during cornering [21] like the distance-based approaches. Moreover, the distance between robots can be modified by simple tuning the parameters of the delay's dynamical system. To track the desired trajectory, a novel control is proposed that exploits the cascade structure of the robot's kinematic model.

On the other hand, one of the most common problems during the implementation of controllers is the lack of state measurements such as velocity, acceleration, orientation, to name a few. This absence of information could be treated by using different sensors to mediate it. However, this would make the system more complex and above all more expensive. Another factor, for example, is that accurate velocity measurements can be difficult, and actually, may be contaminated by the noises in real environments, which can deteriorate the control performance [22].

One of the alternatives to solve the problem of lack of information, is the design of state observers to estimate the measures that the controller needs. There are many works that use observers to estimate information of a system. In [23], the Cartesian position and the kinematic model is employed to design nonlinear observers to estimate the orientation angle and the linear velocity of a mobile robot. On the other hand, in [24] a state-feedback controller for the non-linear error dynamics of the robot is combined with an observer that estimates the orientation error based on available trajectory information and measurement of the position coordinates. Furthermore, in [25] kinematic and dynamic models of the WMR are described, and an output feedback controller is proposed using adaptive sliding mode controller and a high gain observer is designed for velocity estimation to obtain WMR trajectory tracking.

In this paper, we assume that only the Cartesian position and its time derivative are available from measurements. To overcome the lack of attitude measurements a nonlinear observer is proposed based on the kinematic model of the robot. The orientation observer is designed directly on $SO(2)$ and can be used in either open and closed loop schemes. The stability analysis is carried out by means of Lyapunov theory.

The rest of the paper is organized as follows: In Section 2 the kinematic model of a unicycle mobile robot is presented. The design of the attitude observer and its stability analysis is presented in Section 3, and in Section 4, the control algorithm for Leader-follower formation is described. In Section 5 the stability analysis for the complete closed-loop system is presented. Experimental results with a group of three wheeled mobile robots are presented in Section 6. Finally, conclusions and future research directions are stated in Section 6.

2. Kinematic Model

Consider a group of differential-drive mobile robots as it is shown in the Figure 1. According to the figure, the Cartesian position and orientation of each robot are denoted by $\boldsymbol{x}_i = \mathrm{col}(x_i, y_i) \in \Re^2$ and $\theta_i \in \Re$, respectively. The posture of each robot can be described by the unicycle kinematic model as follows

$$\dot{\boldsymbol{x}}_\mathrm{i} = \boldsymbol{\Theta}_i \nu_\mathrm{i}, \tag{1a}$$

$$\dot{\theta}_i = \omega_i, \quad i = \ell, \mathrm{f}_j, \quad j = 1, \dots, n, \tag{1b}$$

where $\boldsymbol{\Theta}_i \triangleq \mathrm{col}(\cos(\theta_i), \sin(\theta_i)) \in \Re^2$ represents the direction of the Cartesian velocity and n is the number of slave robots. Finally, ν_i and ω_i are the linear and angular velocities, respectively. The orientation angle and the Cartesian velocity satisfy the nonholonomic constraint

$$\dot{y}_\mathrm{i} \cos(\theta_\mathrm{i}) - \dot{x}_\mathrm{i} \sin(\theta_\mathrm{i}) = 0 \tag{2}$$

or equivalently,

$$\tan(\theta_\mathrm{i}) = \frac{\dot{y}_\mathrm{i}}{\dot{x}_\mathrm{i}}. \tag{3}$$

The nonholonomic constraint (3) implies that the velocity in the direction of wheel axis is zero, i.e., the robot cannot move in the lateral directions.

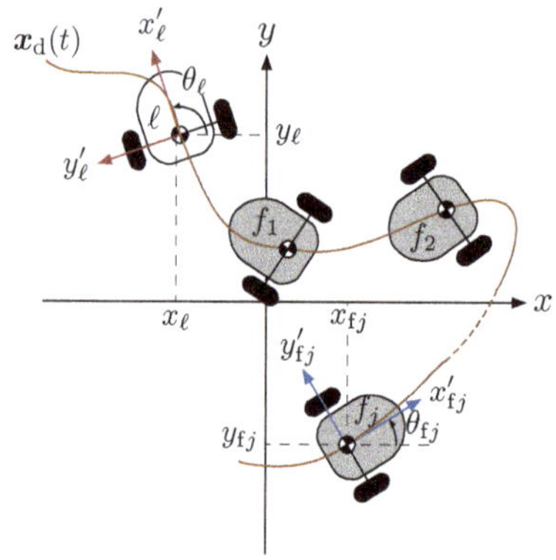

Figure 1. A group of nonholonomic mobile robots with one leader and n followers.

3. Attitude Observer

In order to develop the attitude observer, first notice that an equivalent representation of the kinematic model (1) is the following

$$\dot{\boldsymbol{x}}_i = v_i \boldsymbol{R}_i \boldsymbol{e}_1, \quad \boldsymbol{e}_1 = \text{col}(1, 0) \tag{4a}$$
$$\dot{\boldsymbol{R}} = \omega_i \boldsymbol{S} \boldsymbol{R}_i \tag{4b}$$

where $\boldsymbol{R}_i = \begin{bmatrix} \boldsymbol{\Theta}_i & \boldsymbol{S}\boldsymbol{\Theta}_i \end{bmatrix} \in \mathcal{SO}(2) = \{\boldsymbol{R} \in \Re^{2\times 2} \,|\, \boldsymbol{R}^\top \boldsymbol{R} = \boldsymbol{I},\ \text{de}(\boldsymbol{R}) = +1\}$ is the rotation matrix and $\boldsymbol{S} \in \Re^{2\times 2}$ is a skew-symmetric matrix given by

$$\boldsymbol{S} = \begin{bmatrix} 0 & -1 \\ 1 & 0 \end{bmatrix}. \tag{5}$$

For the case of $SO(2)$ the rotation matrix $\boldsymbol{R}_i$ and skew-symmetric matrix $\boldsymbol{S}$ commute, i.e., $\boldsymbol{S}\boldsymbol{R}_i = \boldsymbol{R}_i\boldsymbol{S}$. On the other hand, from (1a) the rotation matrix $\boldsymbol{R}_i$ can be reconstructed in an algebraic way as follows

$$\boldsymbol{R} = \begin{bmatrix} \boldsymbol{v}_i & \boldsymbol{S}\boldsymbol{v}_i \end{bmatrix}, \quad \boldsymbol{v}_i \triangleq \frac{\dot{\boldsymbol{x}}_i}{\|\dot{\boldsymbol{x}}_i\|} \tag{6}$$

as long as $\|\dot{x}_i(t)\| \nrightarrow 0$ for all $t \geq 0$. Based on the foregoing equation, the attitude observation error is defined as

$$\widetilde{\boldsymbol{R}}_i = \widehat{\boldsymbol{R}}_i^\top \boldsymbol{R}_i \tag{7}$$

where $\widehat{\boldsymbol{R}}_i = \begin{bmatrix} \widehat{\boldsymbol{\Theta}}_i & \boldsymbol{S}\widehat{\boldsymbol{\Theta}}_i \end{bmatrix} \in \mathcal{SO}(2)$ and $\widehat{\boldsymbol{\Theta}}_i$ are estimates of $\boldsymbol{R}_i$ and $\boldsymbol{\Theta}_i$, respectively. With the previous definition the objective is to design an attitude observer such that $\widetilde{\boldsymbol{R}}_i(t) \to \boldsymbol{I}$ as $t \to \infty$ where $\boldsymbol{I} \in \Re^{2\times 2}$ is the identity matrix. Motivated by the work reported in [26] the following attitude observer is proposed

$$\dot{\widehat{\boldsymbol{R}}}_i = \widehat{\omega}_i \boldsymbol{S} \widehat{\boldsymbol{R}}_i \tag{8a}$$
$$\widehat{\omega}_i = \omega_i + k_{ai} \boldsymbol{v}_i^\top \boldsymbol{S} \widehat{\boldsymbol{\Theta}}_i \tag{8b}$$

where $k_{ai} \in \Re$ is a positive constant and $\boldsymbol{v}_i \in \Re^2$ is given in (6). Notice that the proposed attitude observer can be used in open-loop schemes. Now we can establish the first result of the paper.

Theorem 1. *Assume that the angular velocity ω_i and the Cartesian velocity $\dot{\boldsymbol{x}}_i$ are available. Moreover, assume that the robot's motion satisfies $\|\dot{\boldsymbol{x}}_i(t)\| \nrightarrow 0 \; \forall \; t \geq 0$. Then, the attitude observer given in (8) guarantees that $\widetilde{\boldsymbol{R}}_i \to \boldsymbol{I}$ as $t \to \infty$.*

Proof. By taking into account (4b) and (8) the time derivative of the attitude observation error (7) is given by

$$\begin{aligned}\dot{\widetilde{\boldsymbol{R}}}_i =& \dot{\widehat{\boldsymbol{R}}}_i^\top \boldsymbol{R}_i + \widehat{\boldsymbol{R}}_i^\top \dot{\boldsymbol{R}}_i \\ =& (\omega_i - \widehat{\omega}_i)\widehat{\boldsymbol{R}}_i^\top \boldsymbol{S}\boldsymbol{R}_i \\ =& (\omega_i - \widehat{\omega}_i)\widetilde{\boldsymbol{R}}_i\boldsymbol{S}. \end{aligned} \tag{9}$$

where $\boldsymbol{R}_i\boldsymbol{S} = \boldsymbol{S}\boldsymbol{R}_i$ has been used. Now consider the positive scalar function

$$V_{\mathrm{R}i} = \frac{1}{2}\mathrm{tr}(\boldsymbol{I} - \widetilde{\boldsymbol{R}}_i) \tag{10}$$

whose time derivative along (9) is given by

$$\begin{aligned}\dot{V}_{\mathrm{R}i} &= -\frac{1}{2}(\omega_i - \widehat{\omega}_i)\mathrm{tr}(\widetilde{\boldsymbol{R}}_i\boldsymbol{S}) \\ &= \frac{1}{2}k_{\mathrm{a}_i}(\boldsymbol{v}_i^\top \boldsymbol{S}\widehat{\boldsymbol{\Theta}}_i)\mathrm{tr}(\widetilde{\boldsymbol{R}}_i\boldsymbol{S}).\end{aligned} \tag{11}$$

By taking into account (6), the elements of $\widetilde{\boldsymbol{R}}_i$ can be expressed as

$$\widetilde{\boldsymbol{R}}_i = \begin{bmatrix} \boldsymbol{v}_i^\top \widehat{\boldsymbol{\Theta}}_i & -\boldsymbol{v}_i^\top \boldsymbol{S}\widehat{\boldsymbol{\Theta}}_i \\ \boldsymbol{v}_i^\top \boldsymbol{S}\widehat{\boldsymbol{\Theta}}_i & \boldsymbol{v}_i^\top \widehat{\boldsymbol{\Theta}}_i \end{bmatrix}. \tag{12}$$

Therefore, the trace of the matrix $\widetilde{\boldsymbol{R}}_i\boldsymbol{S}$ is given by

$$\mathrm{tr}(\widetilde{\boldsymbol{R}}_i\boldsymbol{S}) = -2\boldsymbol{v}_i^\top \boldsymbol{S}\widehat{\boldsymbol{\Theta}}_i. \tag{13}$$

Substituting the previous result in (11) yields

$$\dot{V}_{\mathrm{R}i} = -k_{\mathrm{a}_i}(\boldsymbol{v}_i^\top \boldsymbol{S}\widehat{\boldsymbol{\Theta}}_i)^2 < 0 \tag{14}$$

Therefore $V_{\mathrm{R}i}$ converges asymptotically to zero, this in turn implies that $\widetilde{\boldsymbol{R}}_i \to \boldsymbol{I}$ as $t \to \infty$. This completes the proof. □

4. Formation Control Algorithm

The kinematic model (1) is an underactuated nonlinear system. To overcome this problem, consider the auxiliary control input

$$\boldsymbol{u}_i = \nu_i \boldsymbol{R}_{\mathrm{d}} \boldsymbol{e}_1 = \nu_i \boldsymbol{\Theta}_{\mathrm{d}i} \tag{15}$$

where $\boldsymbol{R}_{\mathrm{d}i} = \begin{bmatrix} \boldsymbol{\Theta}_{\mathrm{d}i} & \boldsymbol{S}\boldsymbol{\Theta}_{\mathrm{d}i} \end{bmatrix} \in SO(2)$ represents the desired orientation with $\boldsymbol{\Theta}_{\mathrm{d}i} = \mathrm{col}(\cos(\theta_{\mathrm{d}i}), \sin(\theta_{\mathrm{d}i}))$. Therefore, the kinematic model can be written as

$$\dot{\boldsymbol{x}}_i = \boldsymbol{u}_i + \nu_i(\boldsymbol{\Theta}_i - \boldsymbol{\Theta}_{\mathrm{d}i}) \tag{16a}$$

$$\dot{\theta}_i = \omega_i. \tag{16b}$$

In this case, the translational subsystem given by (16a) can be analyzed as a completely actuated system perturbed by the coupling term $\nu_i(\boldsymbol{\Theta}_i - \boldsymbol{\Theta}_{\mathrm{d}i})$ which relates the translational subsystem with the attitude subsystem (16b). On the other hand, given the control input $\boldsymbol{u}_i$ the desired vector $\boldsymbol{\Theta}_{di}$ and ν_i can be computed as

$$\boldsymbol{\Theta}_{di} = \frac{\boldsymbol{u}_i}{\|\boldsymbol{u}_i\|}, \quad \nu_i = \|\boldsymbol{u}_i\|. \tag{17}$$

The proposed formation control strategy is based on the leader-follower approach and delayed reference signals. Contrary to the conventional distance-based leader-follower approach, where the follower robot follows the trajectory generated by the leader, in our proposed approach, the followers track the delayed desired trajectory of the leader robot. To avoid collisions, the time delay depends on the distance between the robots. The time delay becomes larger when the follower is closed to the leader and vice versa. The control objective can be stated as follows: design the control inputs $\boldsymbol{u}_i$ and ω_i such that the position and attitude tracking errors defined as

$$\widetilde{\boldsymbol{x}}_\ell = \boldsymbol{x}_d(t) - \boldsymbol{x}_\ell, \qquad \widetilde{\theta}_\ell = \theta_{d\ell}(t) - \theta_\ell \tag{18a}$$

$$\widetilde{\boldsymbol{x}}_{fi} = \boldsymbol{x}_d(t-\tau_i) - \boldsymbol{x}_{fi}, \quad \widetilde{\theta}_{fi} = \theta_{dfi}(t-\tau_i) - \theta_{fi} \tag{18b}$$

converges asymptotically to zero without using attitude measurements. In (18), the subscripts ℓ and f denote the leader and follower robots and $\boldsymbol{x}_d(t) \in \Re^2$ and $\theta_{di}(t)$ are the desired Cartesian trajectory and desired orientation, respectively. Finally, τ_i denotes the time delay ($i = 1, \ldots, n$) which is obtained as the output of the system

$$\dot{s}_i = -\gamma_i s_i + a_i(1 - \exp(b_i/\|\widetilde{\boldsymbol{\rho}}_i\|)), \quad s_i(0) = s_{i0} \tag{19a}$$

$$\tau_i = s_i + s_{i-1}, \tag{19b}$$

where a_i, b_i and γ_i are positive parameters and $s_0 = 0$. For the first leader we have $\widetilde{\boldsymbol{\rho}}_1 = \boldsymbol{x}_\ell - \boldsymbol{x}_f$, and for $i > 1$ we have $\widetilde{\boldsymbol{\rho}}_i = \boldsymbol{x}_{fi-1} - \boldsymbol{x}_{fi}$. The second term in (19a) increases or decreases the magnitude of the time delay depending on the distance between the robots.

4.1. Leader Robot Controller

Before presenting the leader's controller, let us introduce the following auxiliary error variable

$$\boldsymbol{\eta}_\ell = \widetilde{\boldsymbol{x}}_\ell + \boldsymbol{\beta}_\ell \tag{20}$$

where $\boldsymbol{\beta}_\ell \in \Re^2$ is the state of the following auxiliary linear system

$$\dot{\boldsymbol{\beta}}_\ell = -\boldsymbol{K}_{\beta\ell}\boldsymbol{\beta}_\ell - \boldsymbol{K}_{\eta\ell}\boldsymbol{\eta}_\ell \tag{21}$$

where $\boldsymbol{K}_{\eta\ell} = \boldsymbol{K}_{\eta\ell}^{\mathrm{T}} > \boldsymbol{O} \in \Re^{2\times 2}$ and $\boldsymbol{K}_{\beta\ell} = \boldsymbol{K}_{\beta\ell}^{\mathrm{T}} > \boldsymbol{O} \in \Re^{2\times 2}$. Based on (20) and (21) the proposed leader's controller is given by

$$\boldsymbol{u}_\ell = \dot{\boldsymbol{x}}_d(t) + \boldsymbol{K}_{\beta\ell}\boldsymbol{\beta}_\ell \tag{22a}$$

$$\omega_\ell = \omega_{d\ell} + k_{o\ell}\boldsymbol{\Theta}_{d\ell}^{\mathrm{T}}\boldsymbol{S}\widehat{\boldsymbol{\Theta}}_\ell \tag{22b}$$

where $k_{o\ell} \in \Re$ is a positive gain, $\widehat{\boldsymbol{\Theta}}_\ell$ is extracted from $\widehat{\boldsymbol{R}}_\ell$ and $\boldsymbol{\Theta}_{d\ell}$ is computed according to (17). Regarding $\omega_{d\ell}$ can be computed as

$$\omega_{d\ell} = \boldsymbol{\Theta}_{d\ell}^{\mathrm{T}}\boldsymbol{S}^{-1}\dot{\boldsymbol{\Theta}}_{d\ell}. \tag{23}$$

4.2. Follower Robot Controller

The next step is to design the tracking controller for the follower. The proposed position and attitude control laws for the followers have a similar structure to the leader's controllers and are given by

$$u_{fj} = \dot{x}_d(t-\tau_i) + (1-\dot{\tau}_i)x_d(t-\tau_i) + K_{\beta fi}\beta_{fi} \tag{24a}$$

$$\omega_{fi} = \omega_{dfi} + k_{ofi}\Theta_{dfi}^\top S\widehat{\Theta}_{fi} \tag{24b}$$

where $K_{\beta fi} = K_{\beta fi}^\top > O$, $k_{ofi} > 0$ are the control gains and $\beta_{fi} \in \Re^2$ is obtained as the solution of

$$\dot{\beta}_{fi} = -K_{\beta fi}\beta_{fi} - K_{\eta fi}\eta_{fi} \tag{25}$$

with $K_{\eta fj} = K_{\eta fj}^\top > O$ and $\eta_{fj} = \widetilde{x}_{fj} + \beta_{fj}$.

To avoid complex calculations, the time-derivative of Θ_{di} can be approximated by a low-pass filter, $\dot{\Theta}_{di} = \frac{s}{\lambda s + 1}\Theta_{di}$ with $\lambda > 0$ is the cutoff frequency. It is important to point out that the attitude control laws (22b) and (24) does not explicitly use the orientation error $\widetilde{\theta}_i$.

5. Stability Analysis

Once the leader and follower control laws and attitude observers have been presented we can state the following theorem which summarizes the main result of the paper.

Theorem 2. *Consider a group of nonholonomic mobile robots described by (1) in closed loop with the control laws (22) and (24) in combination with the attitude observer (8). Furthermore, assume that $|\widetilde{\theta}_i(0)| < \pi$ and $\|\dot{x}_i(t)\| \nrightarrow 0$ for all $t \geq 0$. Then, the leader-follower formation is achieved, i.e.,*

$$\lim_{t\to\infty} \widetilde{x}_i(t) = \mathbf{0}, \quad \lim_{t\to\infty} \widetilde{\theta}_i(t) = 0$$

with $i = \ell, f_j$.

Proof. First, we develop the closed loop dynamics of the position and attitude tracking errors. By taking into account (16a), (18), (20)–(22), (24) and (25), the position error dynamics is given by

$$\dot{\beta}_i = -K_{\beta i}\beta_i - K_{\eta i}\eta_i \tag{26}$$

$$\dot{\eta}_i = -K_{\eta i}\eta_i - \nu_i(\Theta_i - \Theta_{di}). \tag{27}$$

By using trigonometric identities, the term $\nu_i(\Theta_i - \Theta_{di})$ can be written as

$$\nu_i(\Theta_i - \Theta_{di}) = \psi_i(t, \widetilde{\theta}_i)\widetilde{\theta}_i = \nu_i R_i \begin{bmatrix} (1-\cos(\widetilde{\theta}_i))/\widetilde{\theta}_i \\ -\sin(\widetilde{\theta}_i)/\widetilde{\theta}_i \end{bmatrix} \widetilde{\theta}_i. \tag{28}$$

On the other hand, by taking into account the attitude observer (8) and attitude control laws (22b) and (24b), the attitude error dynamics is given by

$$\dot{\widetilde{\theta}}_i = -k_{oi}\sin(\widetilde{\theta}_i) + k_{oi}\Theta_{di}^\top S R_i(I - \widetilde{R}_i^\top)e_1 \tag{29}$$

$$\dot{\widetilde{R}}_i = (\omega_i - \widehat{\omega}_i)\widetilde{R}_i S \tag{30}$$

where the equalities $\Theta_{di}^\top S\Theta_i = \sin(\widetilde{\theta}_i)$ and $\Theta_i - \widehat{\Theta}_i = R_i(I - \widetilde{R}_i^\top)e_1$ have been used.

Finally, by defining $z_i = \begin{bmatrix} \beta_i^\top & \eta_i^\top \end{bmatrix}^\top \in \Re^4$ the whole closed-loop dynamics can be written as follows

$$\Sigma_1 : \left\{ \dot{z}_i = A_i z_i + \Upsilon_i(t, \tilde{\theta}_i)\tilde{\theta}_i \right. \tag{31a}$$

$$\Sigma_2 : \left\{ \begin{aligned} \dot{\tilde{\theta}}_i &= -k_{oi}\sin(\tilde{\theta}_i) + k_{oi}\Theta_{di}^\top S R_i (I - \tilde{R}_i^\top) e_1 \\ \dot{\tilde{R}}_i &= (\omega_i - \widehat{\omega}_i)\tilde{R}_i S \end{aligned} \right. \tag{31b}$$

where

$$A_i = \begin{bmatrix} -K_{\beta i} & -K_{\eta i} \\ O & -K_{\eta i} \end{bmatrix}, \quad \Upsilon_i(t, \tilde{\theta}_i) = \begin{bmatrix} 0 \\ \psi_i(t, \tilde{\theta}_i) \end{bmatrix}.$$

Clearly, the complete closed-loop system presents a cascade structure between the position and attitude error dynamics. If $\tilde{R}_i = I$, then, $\widehat{R}_i = R_i$ which implies that $\widehat{\Theta}_i = \Theta_i$. In this case, the term $\omega_i - \widehat{\omega}_i$ becomes

$$\omega_i - \widehat{\omega}_i = k_{ai} v_i^\top S\Theta_i = \frac{k_{ai}}{\|\dot{x}_i\|}\dot{x}_i^\top S\Theta_i = 0 \implies \dot{\tilde{R}} = O$$

The result follows from the nonholonomic constrain (2). The previous result shows that the equilibrium point of (31) is $(z_i^\star, \tilde{\theta}_i^\star, R_i^\star) = (0, k\pi, I)$ with $k = 0, 1, 2, \ldots$.

Now let us analyze the subsystem Σ_2 which is independent of the state z_i and has a cascade structure with $k_{oi}\Theta_{di}^\top S R_i (I - \tilde{R}_i^\top) e_1$ as an interconnection term. This term is bounded and according to Theorem 1 vanishes since the attitude observer error $\tilde{R}_i \to I$ as $t \to \infty$. Moreover, the equilibrium point $\tilde{\theta}_i = 0$ of the unperturbed system

$$\dot{\tilde{\theta}}_i = -k_{oi}\sin(\tilde{\theta}_i) \tag{32}$$

is locally asymptotically stable with Lyapunov function $V_{\theta i} = 1 - \cos(\tilde{\theta}_i)$ $\forall \tilde{\theta}_i \in (-\pi, \pi)$. Furthermore, $\tilde{\theta}_i = 0$ is locally exponentially stable since the linear approximation of (32) is given by $\dot{\tilde{\theta}}_i = -k_{oi}\tilde{\theta}_i$. Therefore, according with Lemma A1 (see Appendix A) the subsystem $\dot{\tilde{\theta}}_i = -k_{oi}\sin(\tilde{\theta}_i) + k_{oi}\Theta_{di}^\top S R_i (I - \tilde{R}_i^\top) e_1$ is Input-to-State Stable (ISS) with input $(I - \tilde{R}_i^\top) e_1$. Clearly, the attitude subsystem Σ_2 satisfies the condition of Theorem A1 given in Appendix A, thus, it is concluded that the equilibrium point $(\tilde{\theta}_i, \tilde{R}_i) = (0, I)$ is uniformly asymptotically stable.

The position subsystem Σ_1 can be analyzed using similar arguments. It is straightforward to show that the matrix A_i is Hurwitz. Therefore, the equilibrium point $z_i = 0$ of the unforced subsystem $\dot{z}_i = A_i z_i$ (with $\tilde{\theta}_i = 0$) is exponentially stable. This implies that the subsystem Σ_1 is ISS with input $\tilde{\theta}_i$. The closed loop system Σ_1 and Σ_2 satisfy the conditions of Theorem A1. As a result, it is concluded that $(z_i, \tilde{\theta}_i, \tilde{R}_i) \to (0, 0, I)$ as $t \to \infty$. The convergence of z_i to zero implies that η_i and β_i also converge asymptotically to zero. Then, it follows that the position tracking error $\tilde{x}_i = \eta_i + \beta_i \to 0$ as $t \to \infty$. This completes the proof. □

6. Experimental Results

In this section, experimental results are presented to validate the performance of the attitude observer and control laws developed in Sections 3 and 4. The testbed is composed of three Khepera III mobile robots from *K-Team* and six infrared *Optitrack* cameras which measure the Cartesian position of the robots (see Figure 2). Although the infrared cameras can also measure the orientation of the robot, This measurement is used only for comparison purposes and do not influence the behavior of the controller. The control laws and the observer were programmed in Matlab with a sample time of 20 [ms]. The control signals were sent to the robots via WIFI communication channel. Table 1 summarizes the parameter values of the control law and the orientation observer employed in the experiments. It is worth to notice, that the initial postures of the robots are selected arbitrarily in every case.

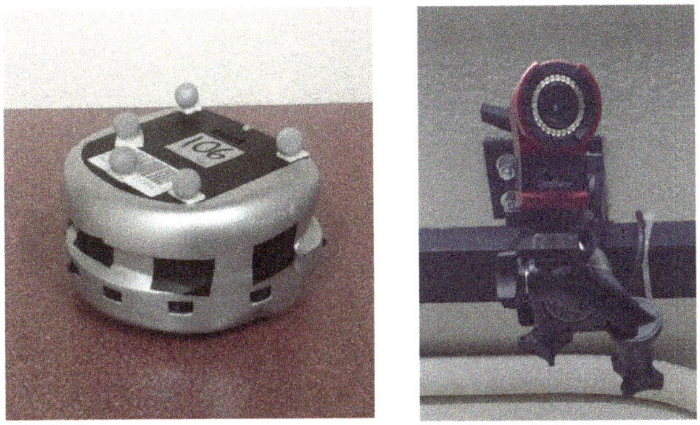

(a) (b)

Figure 2. Experimental testbed: (**a**) Differential drive Khepera III mobile robot, and (**b**) Optitrack infrared camera.

Table 1. Observer, controller gains and delay parameters.

Robot	k_{ai}	$K_{\eta i}$	$K_{\beta i}$	b_i	γ_i	λ_i	κ_i	ζ_i
Leader	3	$4I$	$3.5I$	-	-	10	$20I$	$5I$
Follower 1	3	$4I$	$3.5I$	1	1	10	$20I$	$5I$
Follower 2	3	$4I$	$3.5I$	1	1	10	$20I$	$5I$

The control algorithms together with the attitude observer were tested using two desired trajectories, a circular path and a lemniscate curve. The parametric equations of both desired trajectories are shown below

$$x_{\mathrm{dm}_1} = \begin{bmatrix} 0.3\cos\left(\frac{\pi}{15}t\right) \\ 0.3\sin\left(\frac{\pi}{15}t\right) \end{bmatrix} [\mathrm{m}], \quad x_{\mathrm{dm}_2} = \begin{bmatrix} 0.35\sin\left(\frac{2\pi}{45}t\right) \\ 0.35\sin\left(\frac{4\pi}{45}t\right) \end{bmatrix} [\mathrm{m}] \tag{33}$$

The Cartesian velocity $\dot{x}_i$ can be computed by means of numerical differentiation. However, we obtained better results with the following velocity observer [27]

$$\begin{aligned} \dot{\widehat{x}}_i &= \kappa_i(x_i - \widehat{x}_i) + \mu_i \\ \dot{\mu}_i &= \zeta_i(x_i - \widehat{x}_i) \end{aligned}$$

where $\kappa_i, \zeta_i \in \Re^{2\times 2}$ are positive definite matrices and $\widehat{x}_i \in \Re^2$ is an estimate of the Cartesian position x_i.

The observer, controller gains and the parameters of the delay dynamic equation are shown in Table 1. Regarding the parameter a_j, for the first trajectory was set to $a_1 = 3$ and for the second one we have $a_2 = 5$. All other parameters were the same for both trajectories.

The trajectories of the robots obtained during the two experiments with the desired Cartesian trajectories are shown in Figures 3 and 4, respectively. The figures also shown the robots' positions at the time instants $t = 0$ [s], $t = 13$ [s] for the first experiment and $t = 0$ [s] and $t = 20$ [s] for the second experiment. In both cases, after the transient response the robots successfully achieve the convoy formation.

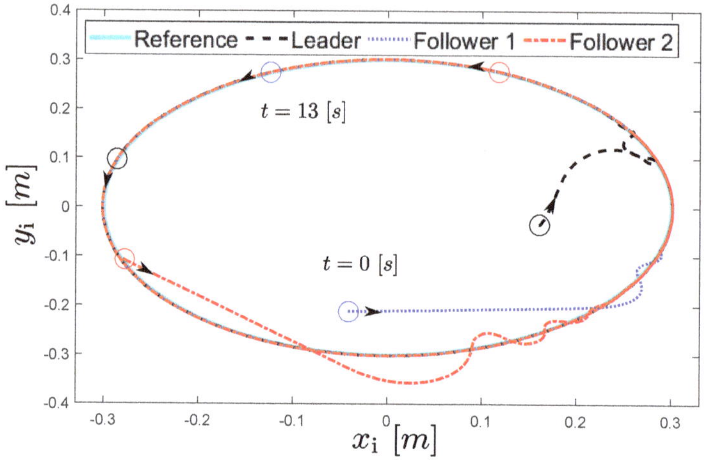

Figure 3. Trajectory of the robot group following the circular path: Robots at the time instants $t =$ (and $t = 15$ s.

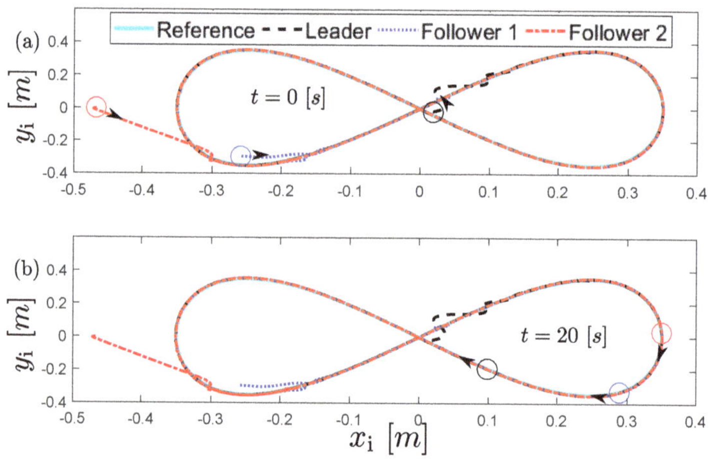

Figure 4. Trajectory of the robot group following the lemniscate curve: (**a**) Robots at $t = 0$ s, and (**b**) Robots at $t = 30$ s.

In order to assess the performance of the attitude observer and control laws we compute the orientation errors $\widehat{\theta}_i = \theta_i - \theta_{\mathrm{d}i}$ and $\bar{\theta}_i = \theta_i - \widehat{\theta}_i$ where θ_i is the angle measured by the cameras and the estimated angle $\widehat{\theta}_i$ is extracted from $\widehat{\mathbf{\Theta}}_i$ as follows

$$\widehat{\theta}_i = \mathrm{atan2}(\widehat{\Theta}_{2i}, \widehat{\Theta}_{1i})$$

where atan2$(\cdot,\cdot)$ is the two argument arctangent function and $\widehat{\mathbf{\Theta}}_i = \mathrm{col}(\widehat{\Theta}_{1i}, \widehat{\Theta}_{2i})$. The time evolution of the position and attitude tracking errors obtained in each experiment are shown in Figures 5 and 6. It is observed in the Figures that despite the unmodeled dynamics and discretization of the control laws, a good tracking was achieved. The time evolution of the time delays are shown in Figures 7 and 8. Notice that for the circular path the delays converge to a constant value while for the second trajectory the time delays change slowly while their magnitude increases at the points of the curve with greater curvature (see Figure 4b). This behavior was expected since at this points the robots come closer to each other.

In order to assess the performance of the proposed algorithm, the RMS error is computed for the distance and orientation of each robot to its desired trajectory, additionally to the RMS error of the estimation of the orientation observer is presented. Table 2 collects the results for the circular desired trajectory while Table 3 shows the results corresponding to the lemniscate curve desired trajectory experiment. On both cases it is observed that the RMS distance error is bellow 0.04 m, while RMS orientation error is under 0.155 radians.

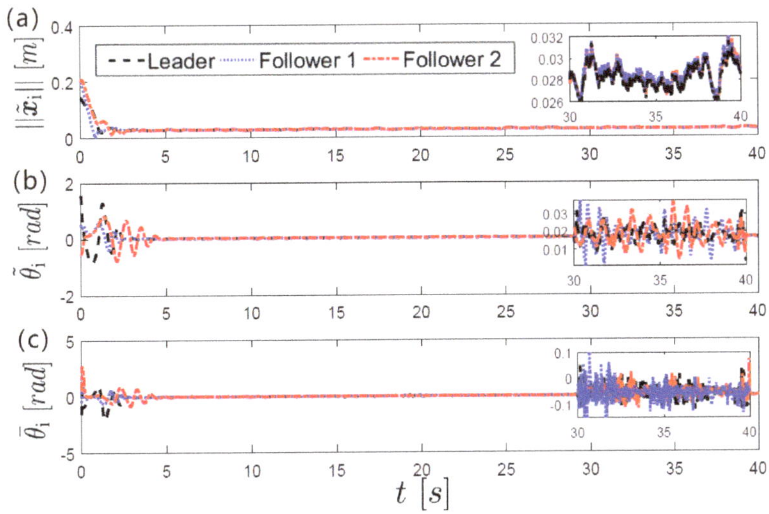

Figure 5. Time evolution of the position and attitude errors in the first experiment (circular path): (**a**) Norm of the position error $\tilde{x}_i$, (**b**) orientation error $\tilde{\theta}_i = \theta_{di} - \theta_i$, (**c**) observation error $\bar{\theta}_i = \theta_i - \hat{\theta}_i$.

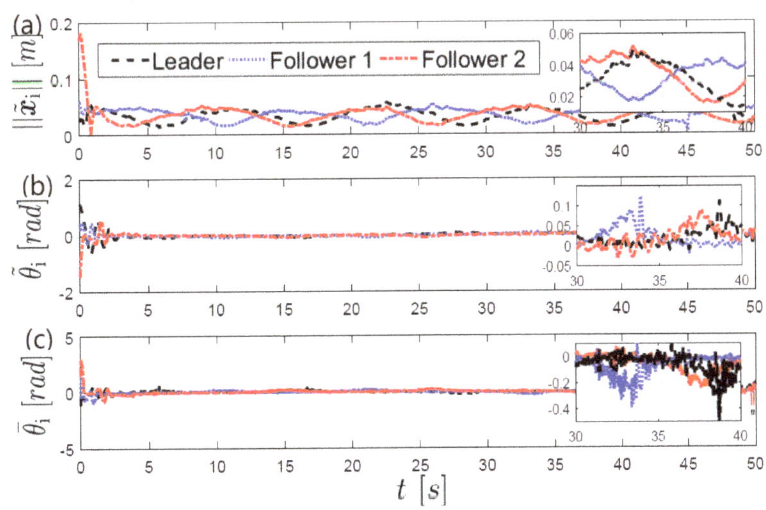

Figure 6. Time evolution of the position and attitude errors in the second experiment (lemniscate curve): (**a**) Norm of the position error $\tilde{x}_i$, (**b**) orientation error $\tilde{\theta}_i = \theta_{di} - \theta_i$, (**c**) observation error $\bar{\theta}_i = \theta_i - \hat{\theta}_i$.

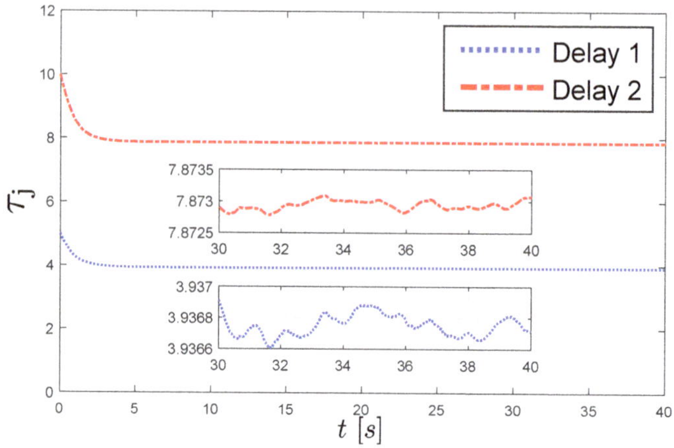

Figure 7. Time evolution of the delay for the first trajectory.

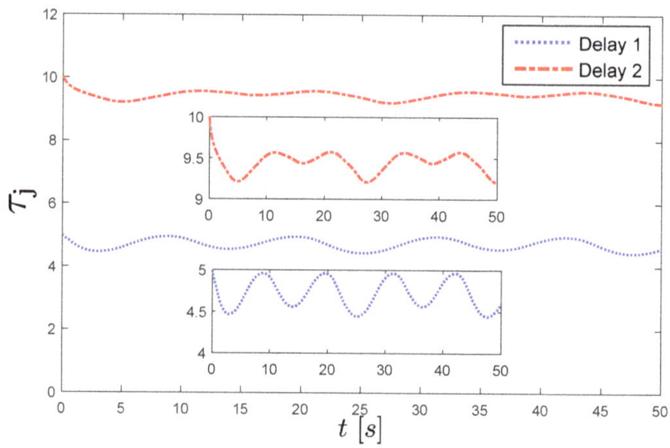

Figure 8. Time evolution of the delay for the second trajectory.

Table 2. RMS errors of distance, orientation and observed orientation of the robots in the circular trajectory experiment.

Robot	Distance [m]	Orientation [rad]	Observation [rad]
Leader	0.0330	0.1533	0.2115
Follower 1	0.0336	0.0806	0.1217
Follower 2	0.0367	0.1416	0.1949

Table 3. RMS errors of distance, orientation and observed orientation of the robots in the lemniscate trajectory experiment.

Robot	Distance [m]	Orientation [rad]	Observation [rad]
Leader	0.0360	0.1221	0.1814
Follower 1	0.0365	0.0654	0.1862
Follower 2	0.0397	0.0775	0.2060

Finally, Figures 9 and 10 show the control inputs. Notice that $\nu_i(t) \nrightarrow 0$ for all $t \geq 0$ this implies that the assumption $\|\dot{x}_i(t)\| \nrightarrow 0$ is satisfied in both experiments.

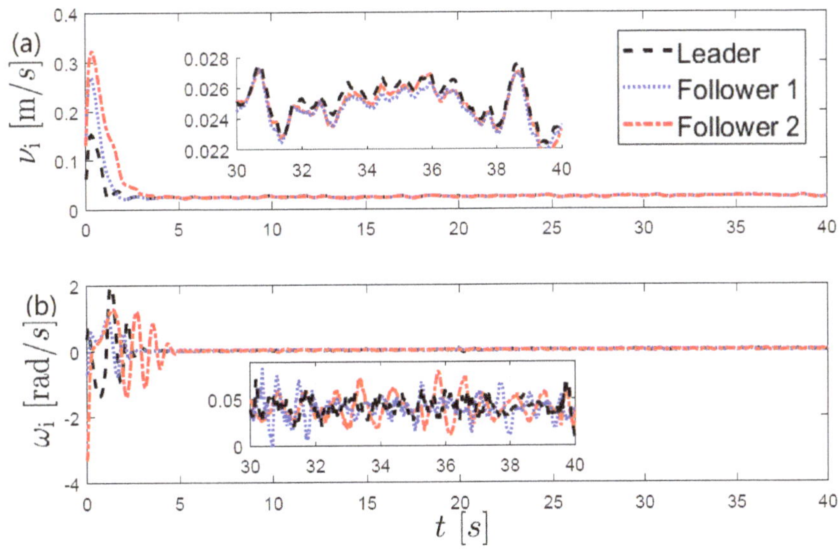

Figure 9. Control inputs obtained in the first experiment: (**a**) ν_i, (**b**) ω_i.

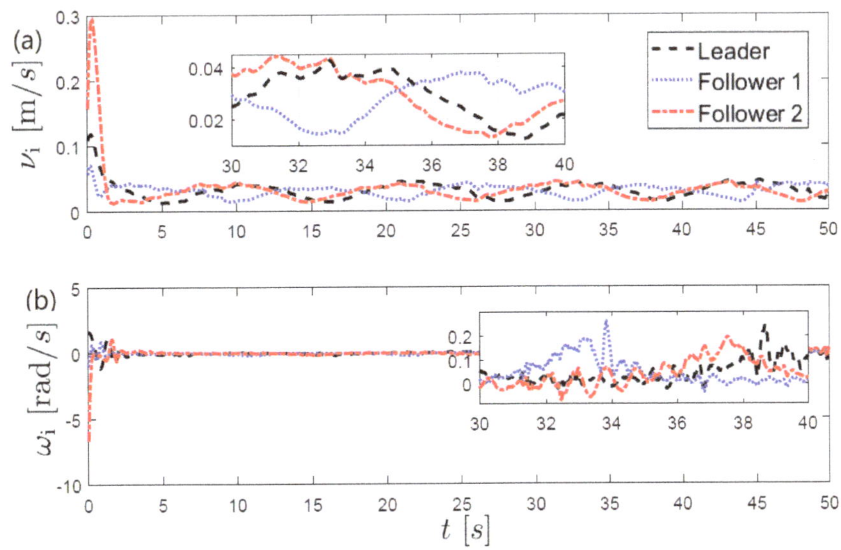

Figure 10. Control inputs obtained in the second experiment: (**a**) ν_i, (**b**) ω_i.

7. Conclusions

In this paper, we proposed kinematic control laws in combination with an attitude observer that solve the problem of convoy formation for a group of nonholonomic mobile robots without using attitude measurements. The proposed control approach is based on the leader-follower scheme but contrary to other works, we used delayed reference signals for the follower robots. The time delays depend on the distance between the robots and are obtained as the outputs of a dynamical system that couples the leader and follower dynamics. As result, collisions between the members of the group are avoided. The kinematic control laws were designed by exploiting the cascade structure of the robots' kinematic model. The proposed controllers are decentralized since only require its own position and the position of the nearest leader. On the other hand, the attitude observer was designed directly on $SO(2)$ and it can be used in open and closed loop schemes. Finally, real-time experiments are presented to show the effectiveness of the proposed control-observer approach.

Author Contributions: Conceptualization, J.H.-A. and J.P.-J.; Formal analysis, J.H.-A. and J.P.-J.; Funding acquisition, C.C.-H.; Investigation, J.H.-A.; Methodology, R.M.-C.; Project administration, C.C.-H.; Software, R.M.-C.; Supervision, C.C.-H.; Validation, J.H.-A.; Writing—original draft, J.H.-A and J.P.-J.; Writing—review & editing, C.C.-H. and R.M.-C. All authors have read and agreed to the published version of the manuscript.

Funding: This research was funded by Consejo Nacional de Ciencia y Tecnología, grant number 166654 and CICESE, grant number F0F156.

Institutional Review Board Statement: Not applicable.

Informed Consent Statement: Not applicable.

Data Availability Statement: Not applicable.

Acknowledgments: This work was supported by the CONACYT research project Synchronization of complex systems with applications under grant 166654 (A1-S-31628) and CICESE internal project F0F156 Generation of collective behaviors and their applications.

Conflicts of Interest: The authors declare no conflict of interest.

Appendix A

In this Appendix A, we recall some results regarding the stability of interconnected systems and Input-to-State stability (ISS) [28].

Lemma A1. *Consider the system* $\dot{\boldsymbol{x}} = \boldsymbol{f}(t, \boldsymbol{x}, \boldsymbol{u})$ *where* $\boldsymbol{f}(t, \boldsymbol{x}, \boldsymbol{u})$ *is locally Lipschitz in* $(\boldsymbol{x},\boldsymbol{u})$ *and uniformly in t. If the unforced system* $\dot{\boldsymbol{x}} = \boldsymbol{f}(t, \boldsymbol{x}, \mathbf{0})$ *has a uniformly asymptotically stable stable equilibrium point at* $\boldsymbol{x} = \mathbf{0}$*, then the system is locally ISS.*

Theorem A1. *Consider the interconnected system*

$$\dot{\boldsymbol{x}} = \boldsymbol{f}(t, \boldsymbol{x}, \boldsymbol{y}) \tag{A1a}$$

$$\dot{\boldsymbol{y}} = \boldsymbol{g}(t, \boldsymbol{y}) \tag{A1b}$$

if the subsystem (A1a) with $\boldsymbol{y}$ *as input is ISS and* $\boldsymbol{y} = \mathbf{0}$ *is a uniformly asymptotically stable equilibrium point of the subsystem (A1b), then, the origin* $(\boldsymbol{x}, \boldsymbol{y})=(\mathbf{0}, \mathbf{0})$ *of the interconnected system (A1a) and (A1b) is uniformly asymptotically stable.*

References

1. Spletzer, J.; Das, A.K.; Fierro, R.; Taylor, C.J.; Kumar, V.; Ostrowski, J.P. Cooperative localization and control for multi-robot manipulation. In Proceedings of the 2001 IEEE/RSJ International Conference on Intelligent Robots and Systems. Expanding the Societal Role of Robotics in the the Next Millennium (Cat. No. 01CH37180), Maui, HI, USA, 29 October–3 November 2001; IEEE: New York, NY, USA, 2001; Volume 2, pp. 631–636.
2. Hosseinzadeh, M. UAV geofencing: Navigation of UVAs in constrained environments. In *Unmanned Aerial Systems*; Elsevier: Amsterdam, The Netherlands, 2021; pp. 567–594.
3. Parker, L.E. Current state of the art in distributed autonomous mobile robotics. In *Distributed Autonomous Robotic Systems 4*; Springer: Berlin/Heidelberg, Germany, 2000; pp. 3–12.
4. Dai, S.L.; He, S.; Chen, X.; Jin, X. Adaptive Leader-Follower Formation Control of Nonholonomic Mobile Robots With Prescribed Transient and Steady-State Performance. *IEEE Trans. Ind. Inform.* **2019**, *16*, 3662–3671 [CrossRef]
5. Cui, R.; Ge, S.S.; How, B.V.E.; Choo, Y.S. Leader—Follower formation control of underactuated autonomous underwater vehicles. *Ocean Eng.* **2010**, *37*, 1491–1502. [CrossRef]
6. Sands, T. Development of Deterministic Artificial Intelligence for Unmanned Underwater Vehicles (UUV). *J. Mar. Sci. Eng.* **2020**, *8*, 578. [CrossRef]

7. Turpin, M.; Michael, N.; Kumar, V. Decentralized formation control with variable shapes for aerial robots. In Proceedings of the 2012 IEEE International Conference on ROBOTICS and Automation, Saint Paul, MN, USA, 14–18 May 2012; IEEE: New York, NY, USA, 2012; pp. 23–30.
8. Do, K.; Pan, J. Nonlinear formation control of unicycle-type mobile robots. *Robot. Auton. Syst.* **2007**, *55*, 191–204. [CrossRef]
9. Coquet, C.; Arnold, A.; Bouvet, P.J. Control of a Robotic Swarm Formation to Track a Dynamic Target with Communication Constraints: Analysis and Simulation. *Appl. Sci.* **2021**, *11*, 3179. [CrossRef]
10. Beard, R.W.; Lawton, J.; Hadaegh, F.Y. A coordination architecture for spacecraft formation control. *IEEE Trans. Control Syst. Technol.* **2001**, *9*, 777–790. [CrossRef]
11. Chen, J.; Sun, D.; Yang, J.; Chen, H. Leader-follower formation control of multiple non-holonomic mobile robots incorporating a receding-horizon scheme. *Int. J. Robot. Res.* **2010**, *29*, 727–747. [CrossRef]
12. Martínez-Clark, R.; Cruz-Hernández, C.; Pliego-Jimenez, J.; Arellano-Delgado, A. Control algorithms for the emergence of self-organized behaviours in swarms of differential-traction wheeled mobile robots. *Int. J. Adv. Robot. Syst.* **2018**, *15*, 1729881418806435. [CrossRef]
13. Balch, T.; Arkin, R.C. Behavior-based formation control for multirobot teams. *IEEE Trans. Robot. Autom.* **1998**, *14*, 926–939. [CrossRef]
14. Wang, X.; Li, S.; Shi, P. Distributed finite-time containment control for double-integrator multiagent systems. *IEEE Trans. Cybern.* **2013**, *44*, 1518–1528. [CrossRef] [PubMed]
15. Yang, Z.; Zhu, S.; Chen, C.; Feng, G.; Guan, X. Leader-follower formation control of nonholonomic mobile robots with bearing-only measurements. *J. Frankl. Inst.* **2020**, *357*, 1628–1643. [CrossRef]
16. Xiao, H.; Li, Z.; Chen, C.P. Formation control of leader–follower mobile robots' systems using model predictive control based on neural-dynamic optimization. *IEEE Trans. Ind. Electron.* **2016**, *63*, 5752–5762. [CrossRef]
17. Sun, D.; Wang, C.; Shang, W.; Feng, G. A synchronization approach to trajectory tracking of multiple mobile robots while maintaining time-varying formations. *IEEE Trans. Robot.* **2009**, *25*, 1074–1086.
18. Lewis, M.A.; Tan, K.H. High precision formation control of mobile robots using virtual structures. *Auton. Robot.* **1997**, *4*, 387–403. [CrossRef]
19. Besseghieur, K.L.; Trkebiski, R.; Kaczmarek, W.; Panasiuk, J. Leader-follower formation control for a group of ROS-enabled mobile robots. In Proceedings of the 2019 6th International Conference on Control, Decision and Information Technologies (CoDIT), Paris, France, 23–26 April 2019; IEEE: New York, NY, USA, 2019; pp. 1556–1561.
20. Halima, M.; Abdelouahab, H.; Yaakoub, T.M. Leader-follower formation control using PI controller. In Proceedings of the 2018 International Conference on Electrical Sciences and Technologies in Maghreb (CISTEM), Algiers, Algeria, 28–31 October 2018; IEEE: New York, NY, USA, 2018; pp. 1–4.
21. Bayuwindra, A.; Lefeber, E.; Ploeg, J.; Nijmeijer, H. Extended Look-Ahead Tracking Controller With Orientation-Error Observer for Vehicle Platooning. *IEEE Trans. Intell. Transp. Syst.* **2019**, *21*, 4808–4821. [CrossRef]
22. Liang, X.; Wang, H.; Liu, Y.H.; Chen, W.; Liu, T. Formation control of nonholonomic mobile robots without position and velocity measurements. *IEEE Trans. Robot.* **2017**, *34*, 434–446. [CrossRef]
23. Pliego-Jiménez, J.; Martínez-Clark, R.; Cruz-Hernández, C. Orientation and velocity observers for unicycle mobile robots. In Proceedings of the 2019 18th European Control Conference (ECC), Naples, Italy, 25–28 June 2019; IEEE: New York, NY, USA, 2019; pp. 2134–2139.
24. Noijen, S.P.M.; Lambrechts, P.F.; Nijmeijer, H. An observer-controller combination for a unicycle mobile robot. *Int. J. Control* **2005**, *78*, 81–87. [CrossRef]
25. Asif, M.; Memon, A.Y.; Junaid Khan, M. Output feedback control for trajectory tracking of wheeled mobile robot. *Intell. Autom. Soft Comput.* **2016**, *22*, 75–87. [CrossRef]
26. Mahony, R.; Hamel, T.; Pflimlin, J.M. Nonlinear complementary filters on the special orthogonal group. *IEEE Trans. Autom. Control* **2008**, *53*, 1203–1218. [CrossRef]
27. Pliego-Jiménez, J.; Arteaga-Pérez, M. Telemanipulation of cooperative robots: A case of study. *Int. J. Control* **2018**, *91*, 1284–1299. [CrossRef]
28. Khalil, H. *Nonlinear Systems*; Prentice Hall: Hoboken, NJ, USA, 1996.

applied sciences

MDPI

Article

Autonomous Mobile Robot Navigation in Sparse LiDAR Feature Environments

Phuc Thanh-Thien Nguyen, Shao-Wei Yan, Jia-Fu Liao and Chung-Hsien Kuo *

Department of Electrical Engineering, National Taiwan University of Science and Technology, Taipei City 106335, Taiwan; d10907813@mail.ntust.edu.tw (P.T.-T.N.); m10807301@mail.ntust.edu.tw (S.-W.Y.); m10707306@mail.ntust.edu.tw (J.-F.L.)

* Correspondence: chkuo@mail.ntust.edu.tw; Tel.: +886-2-2737-6679

Abstract: In the industrial environment, Autonomous Guided Vehicles (AGVs) generally run on a planned route. Among trajectory-tracking algorithms for unmanned vehicles, the Pure Pursuit (PP) algorithm is prevalent in many real-world applications because of its simple and easy implementation. However, it is challenging to decelerate the AGV's moving speed when turning on a large curve path. Moreover, this paper addresses the kidnapped-robot problem occurring in spare LiDAR environments. This paper proposes an improved Pure Pursuit algorithm so that the AGV can predict the trajectory and decelerate for turning, thus increasing the accuracy of the path tracking. To solve the kidnapped-robot problem, we use a learning-based classifier to detect the repetitive pattern scenario (e.g., long corridor) regarding 2D LiDAR features for switching the localization system between Simultaneous Localization And Mapping (SLAM) method and Odometer method. As experimental results in practice, the improved Pure Pursuit algorithm can reduce the tracking error while performing more efficiently. Moreover, the learning-based localization selection strategy helps the robot navigation task achieve stable performance, with 36.25% in completion rate more than only using SLAM. The results demonstrate that the proposed method is feasible and reliable in actual conditions.

Keywords: path planning; pure pursuit controller; trajectory tracking; deep learning; robot kidnapping detection

Citation: Nguyen, P.T.-T.; Yan, S.-W.; Liao, J.-F.; Kuo, C.-H. Autonomous Mobile Robot Navigation in Sparse LiDAR Feature Environments. *Appl. Sci.* **2021**, *11*, 5963. https://doi.org/10.3390/app11135963

Academic Editor: Luis Gracia

Received: 26 May 2021
Accepted: 24 June 2021
Published: 26 June 2021

1. Introduction

In recent years, due to the dramatic development and evolution of technology, an increasing number of industries have turned to automation. The AGV plays a significant role in the automation and is widely used in various other fields.

1.1. Path Planning and Trajectory-Tracking Algorithms

In unmanned vehicle navigation, path planning is essential to search for an optimal path from one point to another point in the environment. Researchers have adopted different methods to solve the problem of AGV path planning, two of which are grid search-based methods and intelligent-based methods. Grid Search-based methods include the A* algorithm and its variants. Chang et al. [1] proposed an improved A* path planning algorithm based on a compressed map to reveal actual narrow areas the robot cannot reach although this approach produces some precision loss, leading the path to be conservative. To reduce the redundant points in A* algorithm pathfinding process, Zeng et al. [2] used Jump Point Search to obtain jump points in the raster map and speed up the A* algorithm based on obtained jump points, though the search time fluctuates in different practical scenarios. For intelligent-based methods, Huang et al. [3] proposed an improved genetic algorithm under a global static environment, which improved the slow convergence and precocity problems. Meanwhile, Zhang et al. [4] refined inertia weights and acceleration factors in Particle Swarm Optimization to prevent local minimum value falling and increase convergence speed.

Following path planning process, trajectory tracking is required so that the AGV can track the movement according to a set trajectory path. With the development of technology, various trajectory-tracking methods have been proposed. Wu et al. [5] introduced a local linear Model Predictive Control (MPC) to track the nonlinear vehicle model velocity and path simultaneously. In [6], a reference trajectory is predefined using a sigmoid function. Then the trajectory is adjusted dynamically by a nonlinear MPC when an obstacle appears in the predictive horizon. Besides MPC, Yang et al. [7] proposed a Fixed-Time Control method and a Fixed-Time Sliding Mode Controller to trajectory-tracking control while meeting the predetermined performance and disturbance suppression. Furthermore, an adaptive trajectory-following strategy was proposed in [8] that constructs a knowledge database through the Particle Swarm Optimization (PSO) algorithm to optimize the controller parameters set according to various vehicle speed and heading error combinations. Meanwhile, Yan et al. [9] proposed a hybrid visual trajectory strategy in which a 2.5D visual servo framework was used to enhance trajectory-tracking behavior.

Although non-geometric controllers such as MPC can be applied to linear or nonlinear models with multiple constraints, their limitations are heavy computation and an inability to provide a closed-form solution when the model is sophisticated. On the other hand, the Pure Pursuit (PP) algorithm is a popular trajectory-tracking algorithm because of its simplicity, efficiency, and low computational requirements, even in limited resource conditions. It computes angular velocity to move the robot from its current position to some look-ahead point in front of the robot. However, the tracking performance is poor due to improper selection of the look-ahead distance. Chen et al. [10] combined the PP algorithm with Proportional Integral (PI) Controller to smooth the final output steering angle through a low-pass filter and verify its feasibility through simulation experiments. By analyzing the vehicle speed and the shortest distance between the GPS trajectory and the current vehicle position, Wang et al. [11] proposed an algorithm that can reduce the lateral error when the vehicle tracks the ideal path. Meanwhile, a Pure Pursuit algorithm based on the optimized look-ahead distance (OLDPPA) [12] introduced an adaptive random motion mechanism of particles in the Salp Swarm Algorithm to improve mining and exploration capabilities.

1.2. AGV Localization Algorithms

To navigate autonomously and safely, the AGV needs to be able to locate its position in its environment. Consequently, the localization problem has been studied and various techniques are proposed to solve the localization problem [13]. The simple form for localization is to use odometry methods, which provide the current position from odometry information estimated by velocity and rotation of wheels (wheel odometry), inertial measurement units (IMU odometry), laser source (laser odometry) or images (visual odometry), etc. For instance, a free-sensor LiDAR-based odometry method [14] integrated the LiDAR-only odometry (LOAM) algorithm to estimate odometry then segment the local map by Convolutional Neural Network (CNN) before using a two-stage RANSAC for verifying the position matches in the local map. Moreover, Zhao et al. [15] proposed a multi-model sensor fusion framework that uses different tightly coupled and loosely coupled optimization methods around the primary IMU odometry factors and can work in several challenging environments.

In contrast, the Simultaneous Localization And Mapping (SLAM) technology consists of the map building process and the localization process. In [16], the authors enhanced the localization method using least square-based geometric matching to compensate for the predicted position. Using 2D LiDAR scan, Millance et al. [17] use a Determinant of Hessian-based detector to find points of high curvature on the Signed Distance Function (SDF) for place recognition. Although the LiDAR-based SLAM method provides helpful information to determine free-space regions and characterizes places for localization, it seems inefficient in structure-less environments, e.g., long corridors, tunnels, dusty or foggy areas, etc. On the other hand, the Sensor-based odometry method proves their accuracy and robustness in various scenarios, even in challenging environments.

Currently, modern image classification systems based on deep neural networks, including Inception V3 [18] and YOLO V3 [19], are more accurate than traditional machine learning classification methods. In general, mobile robots are usually equipped with LiDAR for robot localization due to its accuracy, speed, and 3D reconstruction ability. Therefore, a deep neural network can extract the features of the LiDAR point-cloud data. For example, Chen et al. [20] extracted 2D LiDAR features and used SVM to recognize front pedestrians and track them. However, in the repetitive pattern environment, e.g., in the long corridor, where LiDAR point clouds are sparse to collect. As the result, it is challenging to localize precisely the AGV position, leading to mislocalization or the kidnapped-robot problem. When a mobile robot fails to localize itself due to sparse LiDAR point-cloud, some methods are developed to relocate AGV's position. In the SLAM localization system, it localizes AGV's position through Monte Carlo Localization (MCL) method, which takes a long time and is not helpful in broad-space scenarios. Therefore, Wi-Fi fingerprinting was proposed to solve the problem of robot kidnapping [21], and MCL was integrated with the Fast Library for Approximate Nearest Neighbors (FLANN) machine learning technology to solve this problem [22].

1.3. Contributions

Motivated by discussion above, this paper focused on control movement ability of the AGV on curve path and localizing the AGV on the localization system in the structure-less environment (long corridor). The main contributions of our work are as follows:

- For trajectory tracking, we adopt the PP algorithm and improves it. The traditional PP algorithm often causes errors when it encounters a turn because it is overdue to decelerate speed. Therefore, an improved PP algorithm is proposed that incorporates turning prediction-based deceleration to reduce the impact caused by late attempts at deceleration.
- To solve the kidnapped-robot problem, we combine 2D LiDAR point-cloud features with a deep convolutional network-based classifier to distinguish the current situation for selecting SLAM or odometry localization system. Thus, if the AGV is in a situation where SLAM fails to determine robot position, the task can still be continued.
- In addition, practical experiments in the long corridor terrain are carried out to verify the feasibility of the proposed system.

The remain of this paper is organized as follows. In Section 2, the hardware platform and vehicle kinematic of robot system are described. In Section 3, the improved Pure Pursuit algorithm using turning prediction-based speed adjustment is introduced. the deep learning-based selection strategy using 2D LiDAR point-cloud features for localization task is discussed Section 4. In Section 5, the practical experimental results and verification of the proposed method is reported. Finally, in Section 6, the conclusions are presented.

2. Robot System

The mobile robot has four differential wheels that use two motors on both left and right sides. In addition, the hardware platform is equipped with two LiDAR systems that can obtain 360-degree point-cloud information in the front and back of the robot for SLAM [23]. The schematic diagram of our mobile robot hardware platform is shown in Figure 1.

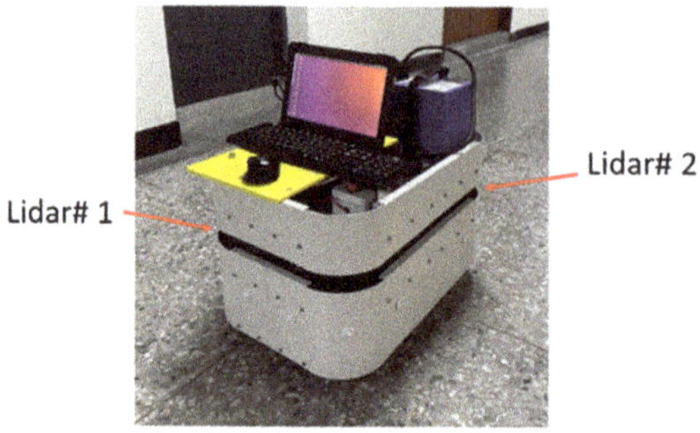

Figure 1. The schematic diagram of mobile robot hardware platform.

3. Design of Trajectory-Tracking System

The mobile robot in this paper is driven in a differential-wheel mode. The left and right wheels are related to the overall velocity and angular velocity of the mobile robot. The coordinate system of the mobile robot is shown in Figure 2, where (x, y) is the location of the mobile robot, L is the distance between the left and right wheels, θ is the angle between the mobile robot and the X-axis, v_R is the velocity of the right wheel, v_L is the velocity of the left wheel, v is the velocity of the mobile robot and ω is the angular velocity of the mobile robot. The kinematic model of the differential wheel is as follows:

$$x = v\cos\theta \tag{1}$$

$$y = v\cos\theta \tag{2}$$

$$\omega = \frac{v_R - v_L}{L} \tag{3}$$

$$v = \frac{v_R + v_L}{2} \tag{4}$$

$$v_L = v - \frac{L\omega}{2} \tag{5}$$

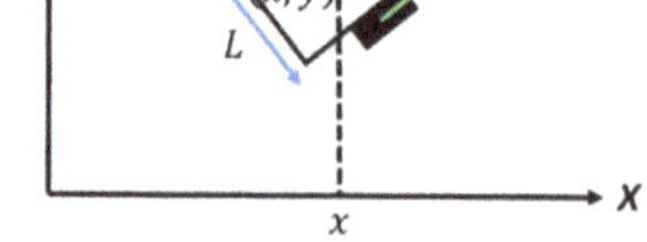

Figure 2. The schematic diagram of the differential-wheel model.

3.1. Introduction of Pure Pursuit (PP) Algorithm

In the PP algorithm, the target point g in tracking path is the target point between the forward-looking distance of the center of the vehicle body and the path. As shown in Figure 3, the target point g belongs to one of the points along the entire travel path. The forward-looking distance L_f is calculated using Equation (6):

$$L_f = k_f * v + L_{fm} \tag{6}$$

where k_f is the custom speed weight, v is the linear velocity of the mobile robot and L_{fm} is the minimum forward-looking distance limit.

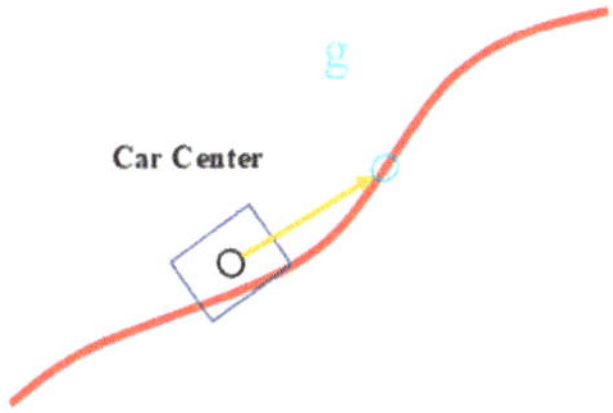

Figure 3. The definition of the target point g (blue circle) in Pure Pursuit algorithm.

The algorithm uses the PD controller to follow the path calculates the angle deviation according to the current position of the robot and the forward-looking distance point g, and then keeps the robot moving on the trajectory through the PD controller. The control structure block diagram of the PD controller is shown in Figure 4. In the PP algorithm, the forward-looking distance can impact to path tracking accuracy and may cause the mobile robot to oscillate, shown in Figure 5.

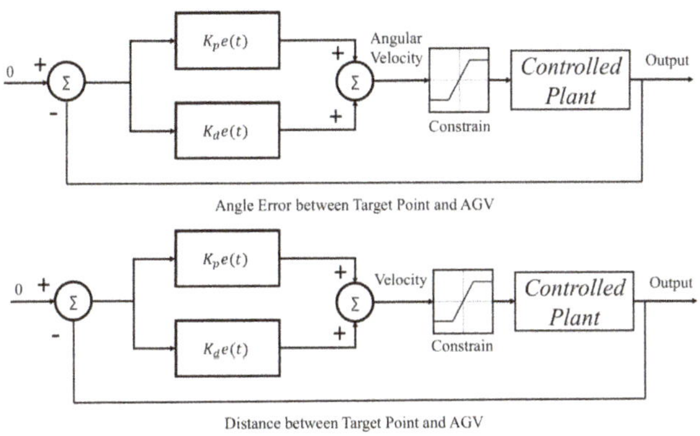

Figure 4. The control structure block diagram of PD controller.

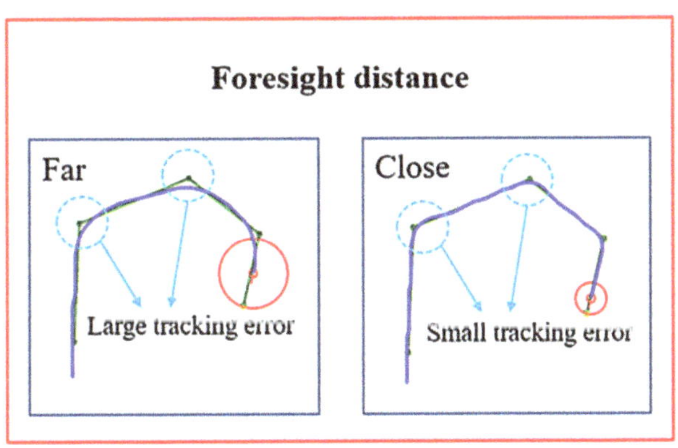

Figure 5. The impact of forward-looking (foresight) distance on generating the tracking error in the Pure Pursuit algorithm. A longer forward-looking distance represents smoother path tracking, and a shorter forward-looking distance give accurate tracking, but the PD controller is more challenging to adjust.

3.2. Improved Pure Pursuit Algorithm

In the original Pure Pursuit algorithm, the shorter the forward-looking distance, the higher the trajectory-tracking accuracy. As Equation (6) is adopted, the forward-looking distance L_f is longer when the velocity v is high; by contrast, L_f is shorter when v is slow. This leads that the turning time can be predicted when the angular velocity ω is large. However, the long forward-looking distance decreases trajectory-tracking accuracy, making it impossible to slow down the turn in time (Figure 6).

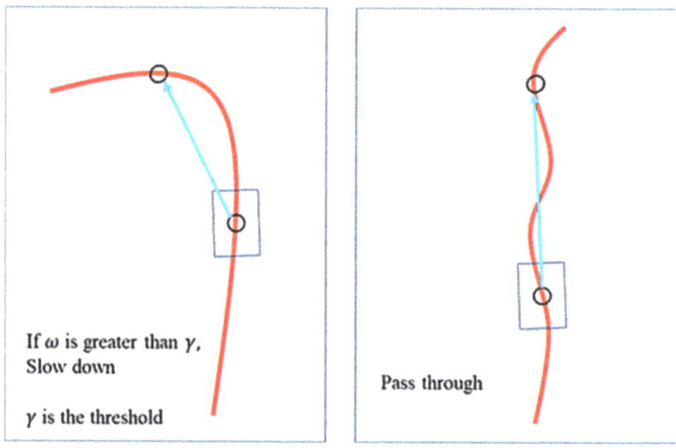

Figure 6. The impact forward-looking distance in trajectory-tracking.

To improve the PP algorithm, this paper adopts a fixed short forward-looking distance to increase the trajectory-tracking ability at any time. Then the proposed method judges the current turn to decelerate it and keeps the angular velocity ω as a deceleration basis at a certain level so that the mobile robot can better track the path when the path is more rugged, shown in Figure 7.

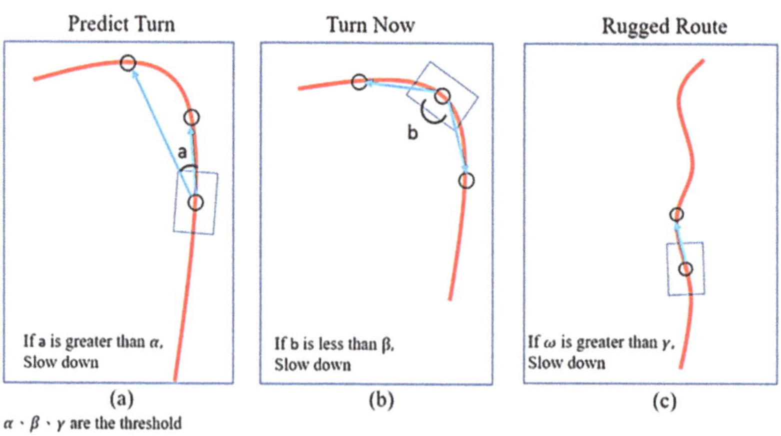

Figure 7. The turning decision strategy of the proposed improved forward-looking distance tracking. (**a**) Predict Turn: Predicting a front turn or not. (**b**) Turn Now: Making the turn action. (**c**) Rugged Route: Making a turn in the case of rugged route.

We separate the turning decision strategy into three steps. First, we predict the turning distance using an angle a between two vectors created from the current robot center to the two next forward-look points (shown in Figure 7a). Assume that two vectors $\left|\vec{V1}\right|, \left|\vec{V2}\right|$ of angle a in the ideal situation (shown in Figure 8) as the following definition:

$$\left|\vec{V2}\right| = 2 \times \left|\vec{V1}\right| \quad (7)$$

where $\left|\vec{V1}\right|$ is the predicted distance, and the ideal angle α is 60°. When the ideal situation is encountered, the AGV will decelerate. However, when the route has a radius of gyration R, the angle a will never reach α value and the AGV cannot decelerate. Without considering R, the maximum value of α can be obtained from $\left|\vec{V1}\right|$ and $\left|\vec{V2}\right|$ as below:

$$\alpha \leq \cos^{-1}\left(\frac{\left|\vec{V2}\right|}{\left|\vec{V1}\right|}\right) \quad (8)$$

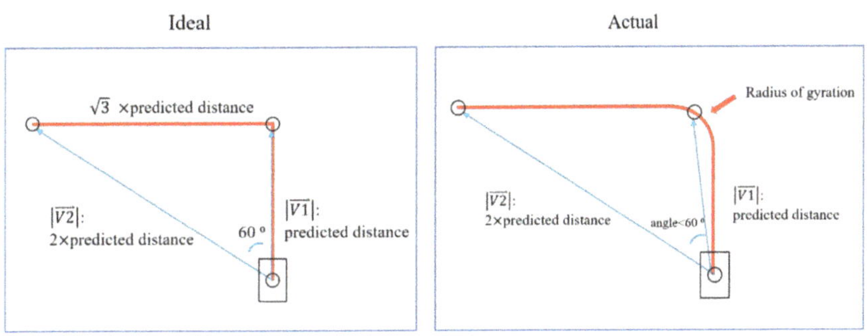

Figure 8. The curve prediction in the ideal case and the actual case.

From Equation (8), we can clearly define the predicted turning distance with the corresponding radius of gyration R:

$$\cos^{-1}\left(\frac{\left|\vec{V2}\right|}{\left|\vec{V1}\right|}\right) > \tan^{-1}\left(\frac{R}{\left|\vec{V1}\right|}\right) \quad (9)$$

$$\alpha \leq \cos^{-1}\left(\frac{\left|\vec{V2}\right|}{\left|\vec{V1}\right|}\right) - \tan^{-1}\left(\frac{R}{\left|\vec{V1}\right|}\right) \quad (10)$$

In the following step (shown in Figure 7b), to make a turn action, we define a current turning angle b between two vectors created from the current robot center to the next and the previous forward-looking points and ensure $b \leq \beta$ value, where $\beta = 135^{\circ}$ in ideal case. When the vehicle is traveling on rugged terrain (shown in Figure 7c), this paper uses $\gamma = 0.1$ (rad/s) as the threshold to indicate that the angular velocity ω is too high if exceeding γ, helping the mobile robot can increase the tracking accuracy.

4. The Localization Switching Method in the Structure-Less Environment

4.1. Two-Dimensional (2D) LiDAR SLAM

2D LiDAR SLAM technology uses LiDAR sensors to collect point-cloud data and scan matches. The SLAM technology then uses algorithms to optimize and loop closure detection for map building (Figure 9) and localization. However, in a structure-less environment as a long corridor, SLAM cannot determine its position on the map, leading to cause unexpected accidents easily. To avoid the mislocalization problem, we use the characteristics of 2D LiDAR data to recognize where the mobile robot is lost.

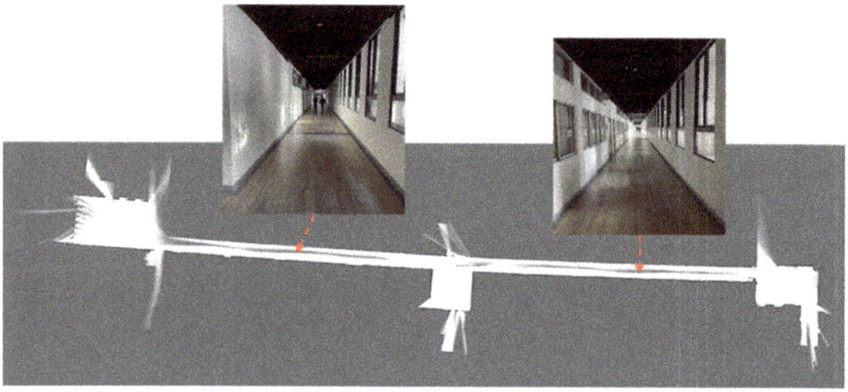

Figure 9. The environment map constructed by 2D LiDAR; black indicates obstacles (e.g., walls); white indicates no obstacles and gray represents unknown areas.

In scan matching, SLAM will match the point clouds with the map features. If the localization is successful, the point clouds are superimposed on the black edge of the SLAM map. At this time, we can use all point clouds and the point clouds superimposed on the black edge of the map to determine whether the AGV is mislocalized as follows:

$$m_r = 1 - \frac{p_{match}}{p_{all}} \tag{11}$$

where m_r represents the missing rate (ranged from 0 to 1). If m_r is greater than 50%, most of the point clouds are not superimposed on the map features. In this case, it can be judged that the mobile robot is getting lost; otherwise, it represents localization success. p_{all} represents the number of all point clouds in a frame, and p_{match} represents the number of point clouds in a frame superimposed on the map features.

However, this judgment method cannot detect the localization status under all conditions. when the surrounding environment has s repetitive pattern, the map features will be too consistent despite the point-cloud is superimposed on the map features. This makes SLAM localization impossible to confirm. For example, in the corridor part of the map (Figure 9), the point clouds extracted from the walls on both sides are too sparse, making it impossible to determine where the mobile robot is in the corridor.

4.2. Deep Learning-Based Corridor Recognition for Switching Localization Systems

To avoid the corridor effect, this paper proposes to use deep learning to identify where is corridor area to switch the localization system. Because we need to know whether the current environment belong to corridor area or not, we will define the corridor recognition problem as the binary classification problem. The process will be following as below:

1. First, to collect images that represent the current area, we need to convert the LiDAR point-cloud data into 2D images by the following formula:

$$p_{pic} = \underbrace{r\begin{bmatrix} \cos pic_\theta & -\sin pic_\theta \\ \sin pic_\theta & \cos pic_\theta \end{bmatrix}}_{R_{pic}} p_{lidar} + \underbrace{\begin{bmatrix} pic_x \\ pic_y \end{bmatrix}}_{t_{pic}} \tag{12}$$

where p_{pic} is the position of the point-cloud on the picture, p_{lidar} is the position of the point-cloud on real world, R_{pic} is the transfer matrix from the LiDAR point-cloud position to the image point-cloud position and t_{pic} is the offset of the LiDAR point-cloud position from the image point-cloud position. To convert the real scale to image pixels, and we set a pixel equal to 0.05 m with r is the image resolution. The point-cloud range is set within a square of 10 m $\times$ 10 m with the center of the mobile robot as the base, as shown in Figure 10a. Finally, the point-cloud information is drawn on the two-dimensional picture with the map coordinates (100, 200) as the center of the mobile robot through a conversion matrix, as shown in Figure 10b.

2. When putting the 2D point-cloud image into the deep neural network for recognition, it is found that if there are people in the image, this will cause noise, and the recognition performance of the corridor is poor. Therefore, image edge detection is used to empirically determine the Region of Interests (ROI) of $x \geq 100$ and $90 \leq y \leq 110$ in the range of the image. The ROI content then is filtered noise, as shown in Figure 11. After image preprocessing, it is put into a deep neural network to determine corridor area.
3. For the corridor recognition network, we use 2 different InceptionV3 [18] and LeNet-5 [24] architectures. Despite having impressive performance in classification tasks, most deep neural networks require powerful hardware support for their heavy computation, leading to difficulties deploying the deep learning method into edge devices such as AGV. In this paper, we choose the lightweight deep neural networks, which have a small number of parameters but still give a good performance, to implement on our system. In the Inception V3-based corridor classification model, we apply the fine-tuning approach to adopt ImageNet for speeding up the training phase and the model accuracy. Moreover, we also define a lightweight model, inspiring by LeNet. The proposed LeNet-inspired model, shown in Table 1, has fewer parameters than the InceptionV3-based model but keeps a good classification performance.
4. When a long corridor area is detected by the trained deep neural networks, the AGV avoids the mislocalization problem by switching the SLAM localization system into the IMU-based Odometer localization system.

Table 1. The LeNet-inspired architecture for corridor recognition.

Layer	Kernel Size	Input Size
Conv	5 × 5	128 × 128 × 1
Batch norm	-	124 × 124 × 8
Avg Pooling	2 × 2	124 × 124 × 8
Conv	5 × 5	62 × 62 × 8
Batch norm	-	58 × 58 × 16
Avg Pooling	2 × 2	58 × 58 × 16
Conv	5 × 5	29 × 29 × 16
Batch norm	-	25 × 25 × 32
Avg Pooling	2 × 2	25 × 25 × 32
Linear	-	4608 × 1
Linear	-	256

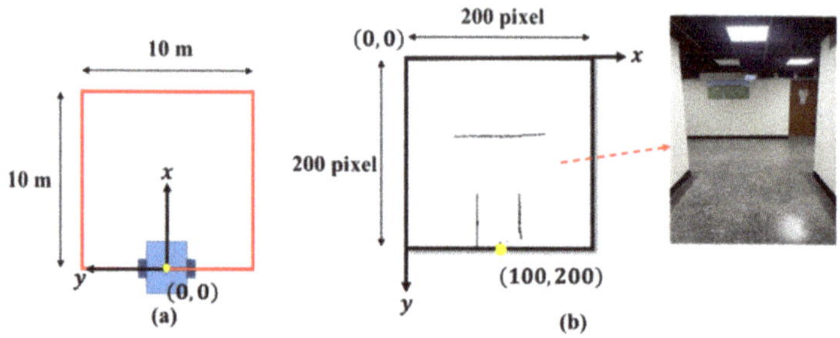

Figure 10. (**a**) Mobile robot receives point-cloud range. (**b**) Point-cloud is drawn on picture.

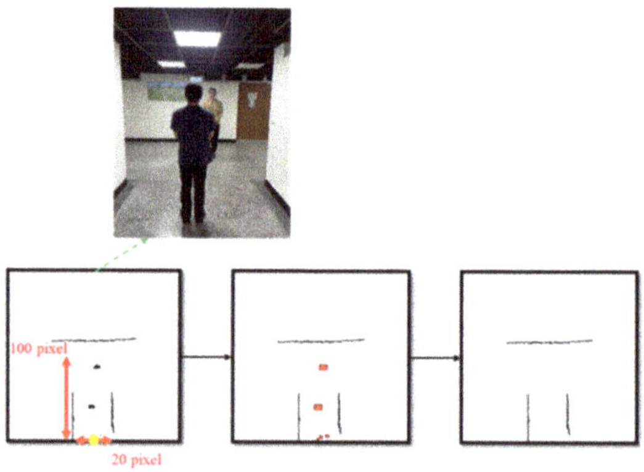

Figure 11. The preprocessing for 2D LiDAR images.

5. Experimental Results

This paper includes three main experiments to verify the performance of the improved Pure Pursuit algorithm and the effectiveness of the LiDAR point-cloud feature-based deep learning classifier for switching localization systems. The first part is a trajectory-tracking accuracy experiment. The second part is a trajectory-tracking speed experiment. The third part verifies the deep learning-based classifier to recognize long corridor terrain using the LiDAR point-cloud feature for switching localization systems.

5.1. Trajectory-Tracking Accuracy Experiment

This experiment will verify the trajectory-tracking accuracy of the proposed method in this paper. The experimental method sets two preset paths. The first is the Double-L-shaped path, as shown in Figure 12a, and the second is the S-shaped path, as shown in Figure 13a. The coordinates reached by the mobile robot during navigation and the trajectory errors of the preset paths are recorded. The experiment is repeated 10 times on each path from the same starting point. The Model Predictive Control (MPC) and the original Pure Pursuit (PP) are used to compare in this paper, as shown in Table 2. Because the starting point is joystick migration, there is a slight artificial error at the starting point, and the error data are calculated after 5 s. The results verified that the maximum error of the improved Pure Pursuit is within 45 mm, with a 77% improvement rate compared to the original Pure Pursuit, while our method has a similar error rate as the MPC method.

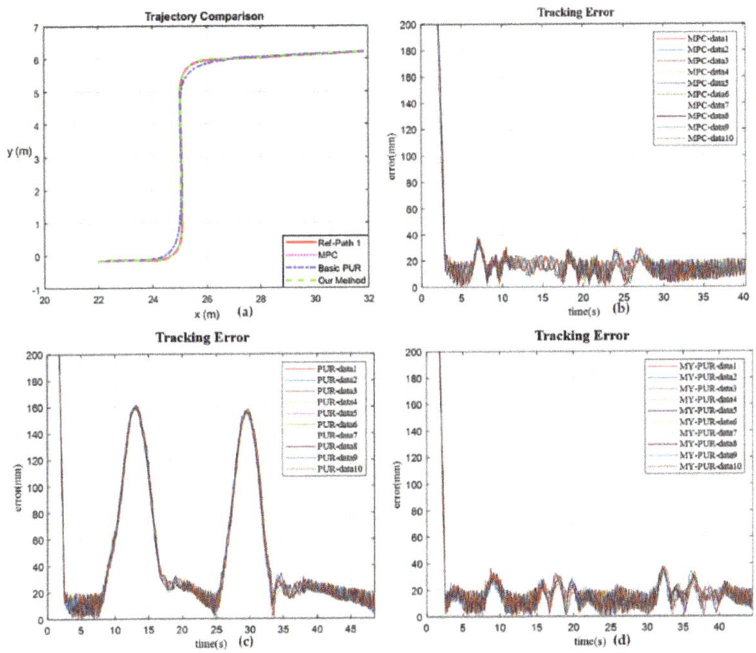

Figure 12. The path error comparison between MPC (purple), original PP (blue) and proposed improved PP (green) methods in Double-L-shaped path (red): (**a**) Trajectory comparison chart. (**b**) MPC trajectory error path. (**c**) Original PP trajectory error graph. (**d**) Improved PP trajectory error graph.

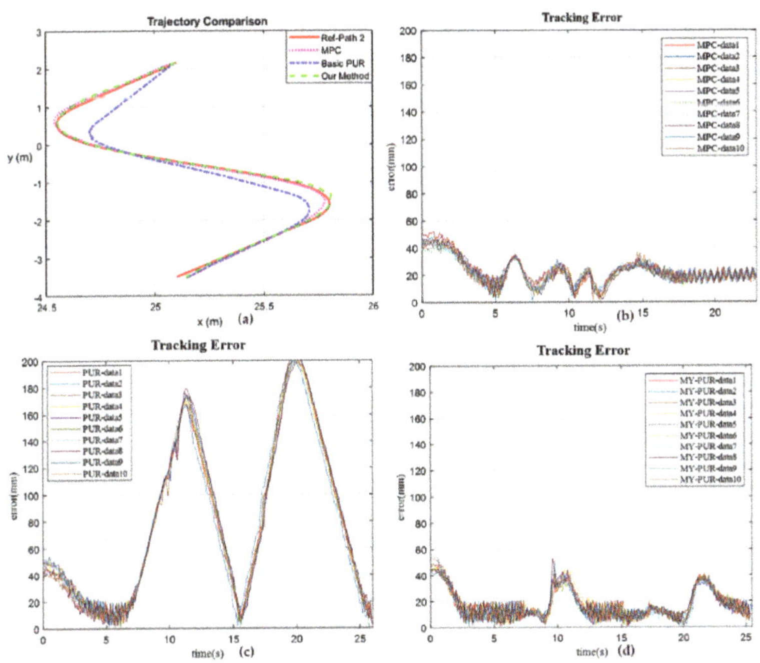

Figure 13. The path error comparison between MPC (purple), original PP (blue) and proposed improved PP (green) methods in S-shaped path (red): (**a**) Trajectory comparison chart. (**b**) MPC trajectory error path. (**c**) Original PP trajectory error graph. (**d**) Improved PP trajectory error graph.

Table 2. Results of Trajectory-Tracking Accuracy Experiment. MPC stands for Model Predictive Control method, PP stands for Pure Pursuit method.

Trajectory-Tracking Algorithm	Maximum Error (mm)	Average Error (mm)	Standard Deviation of Error (mm)
	Double-L-shaped path (14.9 m)		
MPC	35.959	14.644	±0.131
PP	160.215	48.158	±0.289
Our improved PP	**35.967**	**14.892**	**±0.223**
	S-shaped path (8.2 m)		
MPC	34.282	19.329	±0.449
PP	202.026	91.625	±0.885
Our improved PP	**44.609**	**15.742**	**±0.330**

5.2. Trajectory-Tracking Speed Experiment

Besides accuracy, speed is also an essential factor. Thus, the verification experiment was conducted. According to Figures 14 and 15, and Table 3, the average speed, task time and speed standard deviation of the improved PP are better than those of the original PP. The speed performance of the Double-L-shaped path increases by 11.2% and the speed of the S-shaped path increases by 5.6%. The performance of the improved PP is similar MCP method. This experiment proves that the improved PP performs tasks more efficiently.

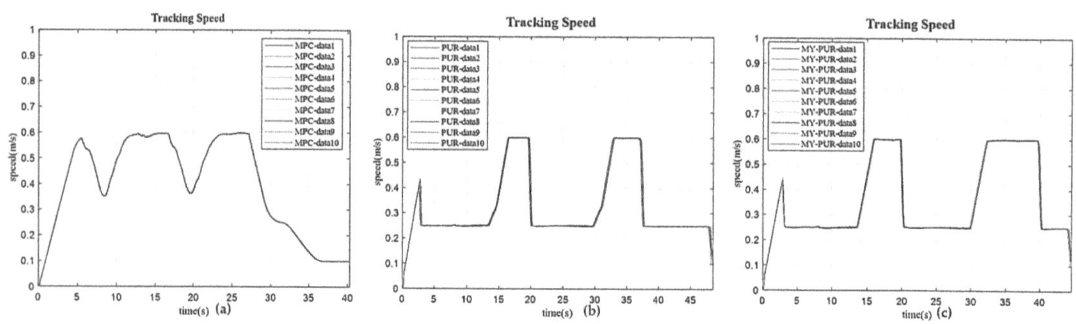

Figure 14. The speed comparison between MPC, original PP and proposed improved PP methods in Double-L-shaped path: (**a**) MPC speed curve. (**b**) Original PP speed curve. (**c**) Improved PP speed curve.

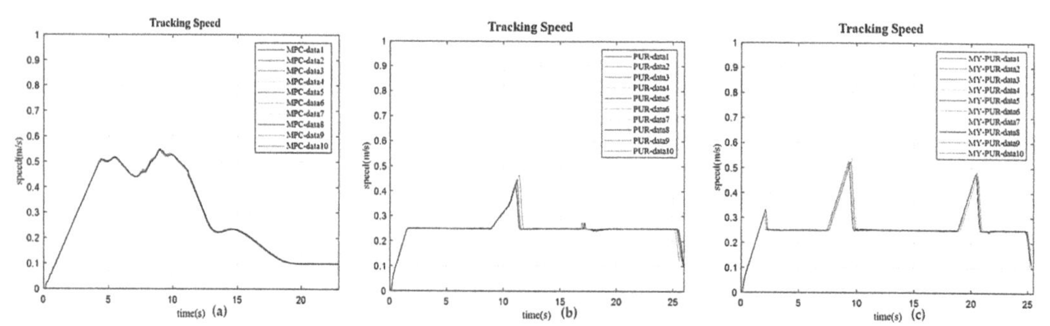

Figure 15. The speed comparison between MPC, original PP and proposed improved PP methods in S-shaped path: (**a**) MPC speed curve. (**b**) Original PP speed curve. (**c**) Improved PP speed curve.

Table 3. Results of Trajectory-Tracking Speed Experiment. MPC stands for Model Predictive Control method, PP stands for Pure Pursuit method.

Trajectory-Tracking Algorithm	Average Speed (m/s)	Task Time (s)	Standard Deviation of Speed (s)
	Double-L-shaped path (14.9 m)		
MPC	0.396	40.31	±0.070
PP	0.322	48.42	±0.125
Our improved PP	**0.358**	**44.63**	**±0.078**
	S-shaped path (8.2 m)		
MPC	0.287	22.92	±0.060
PP	0.248	25.93	±0.118
Our improved PP	**0.262**	**25.43**	**±0.064**

5.3. *Verifying the Deep Learning-Based Localization Switching Method to Solve Corridor Effect*

The location of the experiment is a corridor at the National Taiwan University of Science and Technology, as shown in Figure 16. The red line is the ground truth of the experiment. Marks are spaced every 5 m, and the total length is 88 m.

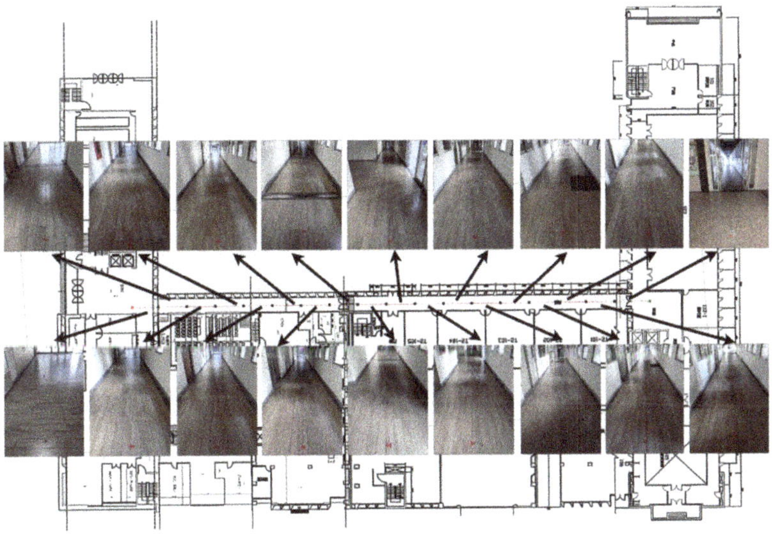

Figure 16. The practical corridor environment for experiments.

5.3.1. Evaluation of the Deep Learning-Based Corridor Recognition Method

Because we consider the corridor recognition problem as the binary classification problem, we collect two types of point-cloud data: the data of corridor area and the data of non-corridor data. Then we preprocess the collected 2D LiDAR images as mentioned in Section 4.2 and split the training/test dataset in a ratio of 9:1. Furthermore, due to small amount of 2D LiDAR data, we apply some data augmentation operations, such as flip and rotation, to enrich training data. As shown in Table 4, compared to the traditional Support Vector Machine (SVM) [24] classifier, the deep learning-based.

Table 4. Results of corridor recognition models.

Models	Accuracy (%)	Number of Parameters
SVM [25]	80%	-
InceptionV3-based model	**100%**	~22 million
LeNet-based model	**100%**	**~1.1 million**

Models have better accuracy results in the test dataset. The LeNet-based model has only about 1.2 million parameters for the model size comparison, while its accuracy is similar to the bigger InceptionV3-based model. This guarantees that our proposed deep-learning-based model can be deployed on AGV and give a reliable and effective performance.

5.3.2. Verification of the Localization Switching Method in Practice

On the experimental corridor, we first manually move the mobile robot along the ground truth to record the trajectory of SLAM localization. Simultaneously, the deep learning-based classifier also is used to detect the long corridor regions. As the experimental results (shown in Figures 17 and 18 and Table 5), it is impossible to complete the trajectory tracking and localization task if using SLAM only. Otherwise, by switching between SLAM and odometry localization system using our proposed method, the AGV can complete the trajectory tracking even in sparse LiDAR feature environment. Our experimental results proved the effectiveness of the deep learning-based localization switching method that involve improved Pure Pursuit robustness and feasibility.

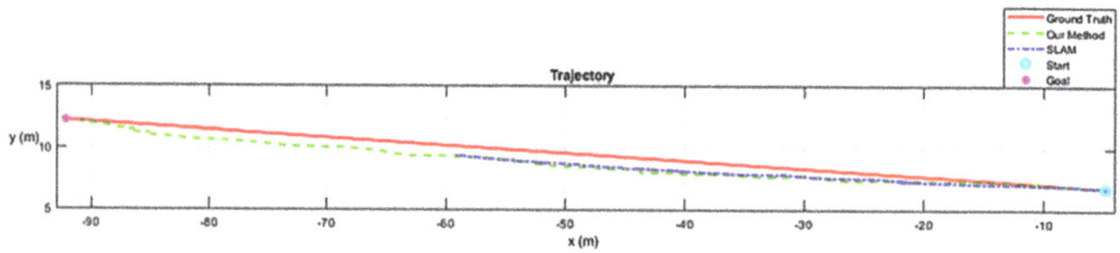

Figure 17. The tracked trajectory comparison. The red line is the ground truth, The blue line is the SLAM method, and the green line is our method's trajectory method.

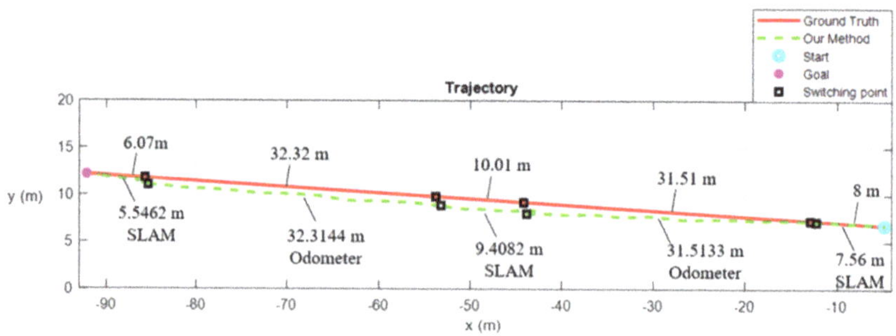

Figure 18. The complete trajectory tracking of our proposed method in practice.

Table 5. Results of corridor recognition models.

Experiment	Track Length (m)
Ground Truth	88
SLAM	54.4
Our Method	**86.3**

6. Conclusions

To improve the trajectory-tracking accuracy of the original Pure Pursuit algorithm when following the turning path, we propose an improved Pure Pursuit algorithm that adds the functions of predicting the next turn and adjusting speed in the current turn. In structure-less environment AGV localization, this paper introduces a deep-learning-based corridor area classifier using 2D LiDAR data to select a suitable localization system to solve the corridor effect. The practical experimental results verified that the maximum error of the modified Pure Pursuit is within 45 mm, with a 77% improvement rate compared to the original Pure Pursuit. The improved Pure Pursuit algorithm also increased the speed by more than 5.6%. Moreover, the proposed localization switching method using deep learning helps to increase 36.25% of completion rate higher than that only using SLAM localization, prove the robust effectiveness of the proposed method in practice.

Author Contributions: P.T.-T.N. is responsible for revising papers, design deep-learning based localization switching method, verify experiments; S.-W.Y. is responsible for designing SLAM and Odometry localization, design deep-learning based localization switching method; J.-F.L. is responsible for path planning part and improve PurePursuit trajectory tracking method; C.-H.K. is the advisor who orients research direction for the paper, gives comments and advices to do this research. All authors have read and agreed to the published version of the manuscript.

Funding: Ministry of Science and Technology, Taiwan under Grant MOST 109-2221-E-011-112-MY3 and the "Center for Cyber-physical System Innovation" from The Featured Areas Research Center Program within the framework of the Higher Education Sprout Project by the Ministry of Education (MOE) in Taiwan.

Institutional Review Board Statement: Not applicable.

Informed Consent Statement: Not applicable.

Data Availability Statement: Not applicable.

Conflicts of Interest: The authors declare no conflict of interest.

References

1. Chang, L.; Shan, L.; Li, J.; Dai, Y. The Path Planning of Mobile Robots Based on an Improved A* Algorithm. In Proceedings of the 2019 IEEE 16th International Conference on Networking, Sensing and Control (ICNSC), Banff, AB, Canada, 9–11 May 2019; pp. 257–262.
2. Zeng, Z.; Sun, W.; Wu, W.; Xue, M.; Qian, L. An Efficient Path Planning Algorithm for Mobile Robots. In Proceedings of the 2019 IEEE 15th International Conference on Control and Automation (ICCA), Edinburgh, UK, 16–19 July 2019; pp. 487–493.
3. Li, Y.; Huang, Z.; Xie, Y. Path planning of mobile robot based on improved genetic algorithm. In Proceedings of the 2020 3rd International Conference on Electron Device and Mechanical Engineering (ICEDME), Suzhou, China, 1–3 May 2020; pp. 691–695.
4. Zhang, L.; Zhang, Y.; Li, Y. Mobile Robot Path Planning Based on Improved Localized Particle Swarm Optimization. *IEEE Sens. J.* **2020**, *21*, 6962–6972. [CrossRef]
5. Wu, H.; Si, Z.; Li, Z. Trajectory Tracking Control for Four-Wheel Independent Drive Intelligent Vehicle Based on Model Predictive Control. *IEEE Access* **2020**, *8*, 73071–73081. [CrossRef]
6. Li, S.; Li, Z.; Yu, Z.; Zhang, B.; Zhang, N. Dynamic trajectory planning and tracking for autonomous vehicle with obstacle avoidance based on model predictive control. *IEEE Access* **2019**, *7*, 132074–132086. [CrossRef]
7. Yang, T.; Li, Z.; Yu, J. Trajectory Tracking Control of Surface Vehicles: A Prescribed Performance Fixed-Time Control Approach. *IEEE Access* **2020**, *8*, 209441–209451. [CrossRef]
8. Amer, N.H.; Hudha, K.; Zamzuri, H.; Aparow, V.R.; Abidin, A.F.Z.; Kadir, Z.A.; Murrad, M. Adaptive Trajectory Tracking Controller for an Armoured Vehicle: Hardware-in-the-loop Simulation. In Proceedings of the 2018 57th Annual Conference of the Society of Instrument and Control Engineers of Japan (SICE), Nara, Japan, 11–14 September 2018; pp. 462–467.
9. Yan, F.; Li, B.; Shi, W.; Wang, D. Hybrid visual servo trajectory tracking of wheeled mobile robots. *IEEE Access* **2018**, *6*, 24291–24298. [CrossRef]
10. Chen, Y.; Shan, Y.; Chen, L.; Huang, K.; Cao, D. Optimization of Pure Pursuit Controller based on PID Controller and Low-pass Filter. In Proceedings of the 2018 21st International Conference on Intelligent Transportation Systems (ITSC), Maui, HI, USA, 4–7 November 2018; pp. 3294–3299.
11. Wang, W.-J.; Hsu, T.-M.; Wu, T.-S. The improved pure pursuit algorithm for autonomous driving advanced system. In Proceedings of the 2017 IEEE 10th International Workshop on Computational Intelligence and Applications (IWCIA), Hiroshima, Japan, 11–12 November 2017; pp. 33–38.
12. Wang, R.; Li, Y.; Fan, J.; Wang, T.; Chen, X. A Novel Pure Pursuit Algorithm for Autonomous Vehicles Based on Salp Swarm Algorithm and Velocity Controller. *IEEE Access* **2020**, *8*, 166525–166540. [CrossRef]
13. Yousif, K.; Bab-Hadiashar, A.; Hoseinnezhad, R. An Overview to Visual Odometry and Visual SLAM: Applications to Mobile Robotics. *Intell. Ind. Syst.* **2015**, *1*, 289–311. [CrossRef]
14. Rozenberszki, D.; Majdik, A.L. LOL: Lidar-only Odometry and Localization in 3D point cloud maps. In Proceedings of the 2020 IEEE International Conference on Robotics and Automation (ICRA), Paris, France, 31 May–31 August 2020; pp. 4379–4385.

15. Zhao, S.; Zhang, H.; Wang, P.; Nogueira, L.; Scherer, S. Super Odometry: IMU-centric LiDAR-Visual-Inertial Estimator for Challenging Environments. *arXiv* **2021**, arXiv:2104.14938.
16. Cho, H.; Kim, E.K.; Kim, S. Indoor SLAM application using geometric and ICP matching methods based on line features. *Robot. Auton. Syst.* **2018**, *100*, 206–224. [CrossRef]
17. Millane, A.; Oleynikova, H.; Nieto, J.; Siegwart, R.; Cadena, C. Free-Space Features: Global Localization in 2D Laser SLAM Using Distance Function Maps. In Proceedings of the 2019 IEEE/RSJ International Conference on Intelligent Robots and Systems (IROS), Macau, China, 3–8 November 2019; pp. 1271–1277.
18. Szegedy, C.; Vanhoucke, V.; Ioffe, S.; Shlens, J.; Wojna, Z. Rethinking the Inception Architecture for Computer Vision. In Proceedings of the 2016 IEEE Conference on Computer Vision and Pattern Recognition (CVPR), Las Vegas, NV, USA, 27–30 June 2016; pp. 2818–2826.
19. Redmon, J.; Farhadi, A. YOLOv3: An Incremental Improvement. *arXiv* **2018**, arXiv:1804.02767.
20. Chen, J.; Ye, P.; Sun, Z. Pedestrian Detection and Tracking Based on 2D Lidar. In Proceedings of the 2019 6th International Conference on Systems and Informatics (ICSAI), Shanghai, China, 2–4 November 2019; pp. 421–426.
21. Luo, R.C.; Hsiao, T.J. Kidnapping and Re-Localizing Solutions for Autonomous Service Robotics. In Proceedings of the IECON 2018—44th Annual Conference of the IEEE Industrial Electronics Society, Washington, DC, USA, 21–23 October 2018; pp. 2552–2557.
22. Luo, R.C.; Yeh, K.C.; Huang, K.H. Resume navigation and re-localization of an autonomous mobile robot after being kidnapped. In Proceedings of the 2013 IEEE International Symposium on Robotic and Sensors Environments (ROSE), Washington, DC, USA, 21–23 October 2013; pp. 7–12.
23. Hess, W.; Kohler, D.; Rapp, H.; Andor, D. Real-time loop closure in 2D LIDAR SLAM. In Proceedings of the 2016 IEEE International Conference on Robotics and Automation (ICRA), Stockholm, Sweden, 16–21 May 2016; pp. 1271–1278.
24. LeCun, Y.; Bottou, L.; Bengio, Y.; Haffner, P. Gradient-based learning applied to document recognition. *Proc. IEEE* **1998**, *86*, 2278–2324. [CrossRef]
25. Cortes, C.; Vapnik, V. Support-vector networks. *Mach. Learn.* **1995**, *20*, 273–297. [CrossRef]

applied sciences

MDPI

Article

Topological Navigation and Localization in Tunnels—Application to Autonomous Load-Haul-Dump Vehicles Operating in Underground Mines

Mauricio Mascaró, Isao Parra-Tsunekawa, Carlos Tampier and Javier Ruiz-del-Solar *

Advanced Mining Technology Center, Department of Electrical Engineering, Universidad de Chile, Santiago 8370451, Chile; mauricio.mascaro@amtc.cl (M.M.); isao.parra@amtc.cl (I.P.-T.); carlos.tampier@amtc.cl (C.T.)
* Correspondence: jruizd@ing.uchile.cl

Abstract: Mobile robots are no longer used exclusively in research laboratories and indoor controlled environments, but are now also used in dynamic industrial environments, including outdoor sites. Mining is one industry where robots and autonomous vehicles are increasingly used to increase the safety of the workers, as well as to augment the productivity, efficiency, and predictability of the processes. Since autonomous vehicles navigate inside tunnels in underground mines, this kind of navigation has different precision requirements than navigating in an open environment. When driving inside tunnels, it is not relevant to have accurate self-localization, but it is necessary for autonomous vehicles to be able to move safely through the tunnel and to make appropriate decisions at its intersections and access points in the tunnel. To address these needs, a topological navigation system for mining vehicles operating in tunnels is proposed and validated in this paper. This system was specially designed to be used by Load-Haul-Dump (LHD) vehicles, also known as scoop trams, operating in underground mines. In addition, a localization system, specifically designed to be used with the topological navigation system and its associated topological map, is also proposed. The proposed topological navigation and localization systems were validated using a commercial LHD during several months at a copper sub-level stoping mine located in the Coquimbo Region in the northern part of Chile. An important aspect to be addressed when working with heavy-duty machinery, such as LHDs, is the way in which automation systems are developed and tested. For this reason, the development and testing methodology, which includes the use of simulators, scale-models of LHDs, validation, and testing using a commercial LHD in test-fields, and its final validation in a mine, are described.

Keywords: autonomous navigation; topological mapping; field robotics; mining automation

Citation: Mascaró, M.; Parra-Tsunekawa, I.; Tampier, C.; Ruiz-del-Solar, J. Topological Navigation and Localization in Tunnels—Application to Autonomous Load-Haul-Dump Vehicles Operating in Underground Mines. *Appl. Sci.* **2021**, *11*, 6547. https://doi.org/10.3390/app11146547

Academic Editors: Luis Gracia and Carlos Perez-Vidal

Received: 1 June 2021
Accepted: 13 July 2021
Published: 16 July 2021

1. Introduction

The development of robotic applications has increased significantly in the last decade, and currently, robotic systems are being utilized for many purposes, in various environments. Mobile robots are no longer used exclusively in research laboratories and indoor controlled environments, but are now also used in dynamic industrial environments and outdoor sites. Moreover, the efforts for developing autonomous cars and drones have had the effect of strengthening the development and use of other autonomous machines and vehicles in various industries.

Mining is one industry in which autonomous vehicles have been in use for at least 13 years. Industrial use of autonomous hauling trucks started in 2008 in the Gabriela Mistral copper open-pit mine, located in the north of Chile. Currently, the use of autonomous mining equipment, mainly vehicles, is an important requirement in the whole mining industry. This is because mining operations need to increase the safety of the workers, as well as to augment the productivity, efficiency, and predictability of the processes. Safety is, without doubt, a key factor, and has been the top priority of mining companies in recent years. This is true, especially, in underground mining operations with their hazardous environments in which workers are constantly exposed to the risks of rock falls, rock bursts, and mud rushes, and where the presence of dust in the air can result in a number of associated occupational diseases in the workers [1].

In an underground mine, autonomous vehicles navigate inside tunnels. This kind of navigation has different precision requirements from those where they are navigating in an open environment. First, tunnels are GNSS-denied environments and thus vehicles cannot use any GNSSs (Global Navigation Satellite Systems) to self-localize. Secondly, when driving inside tunnels, it is not relevant to have an accurate localization system, but it is essential to be able to move safely through the tunnel, and to make appropriate decisions at its intersections and access points. This is completely different from most robotic applications, where safe navigation requires an accurate determination of the robot's position and orientation at every moment.

To address this need, a topological navigation system for mining vehicles operating in tunnels is proposed and validated in this paper. This system was specially designed to be used by Load-Haul-Dump (LHD) vehicles, also known as scoop trams, operating in underground mines. An LHD is a four-wheeled, center-articulated vehicle with a frontal bucket used to load and transport ore on the production levels of an underground mine (See Figure 1). These machines are a key component in the extraction of ore from underground mines because the ore extraction rate from the mine depends directly on the efficiency of the LHD. The proposed system permits a commercial LHD, which is a very large vehicle, to navigate inside tunnels that are just a couple of meters wider than the LHD. In addition, it allows bi-directional navigation and so-called *inversion* maneuvers, both required in standard LHD operations inside productive sectors of underground mines. (See an example in Figure 2). In addition, a localization system, specifically designed to be used with the topological navigation system and its associated topological map, is also proposed.

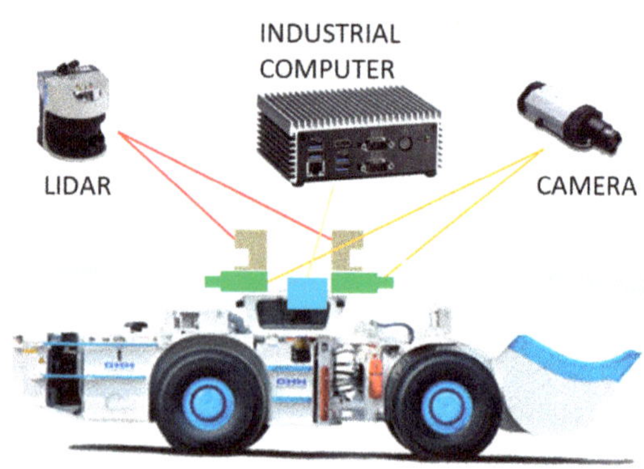

Figure 1. LHD and sensors for the Autonomous Navigation System.

An important aspect to be addressed when working with heavy-duty machinery, such as the LHDs, is the way in which automation systems are developed and tested. In order to address this important issue, we use a development process for autonomous systems for mining equipment, which is comprised of the following four-stages: (i) development using specific simulation tools, which are fed with real data from mining environments, (ii) development using scale-models of the actual machines, (iii) validation and testing using real machines in test-fields, and (iv) validation and testing using real machines in actual mine operations. This development process was used for achieving autonomous navigation of LHDs. The proposed topological navigation and location systems were validated using a commercial LHD, during several months at a medium-scale sub-level stoping mine, located in the Coquimbo Region of Chile.

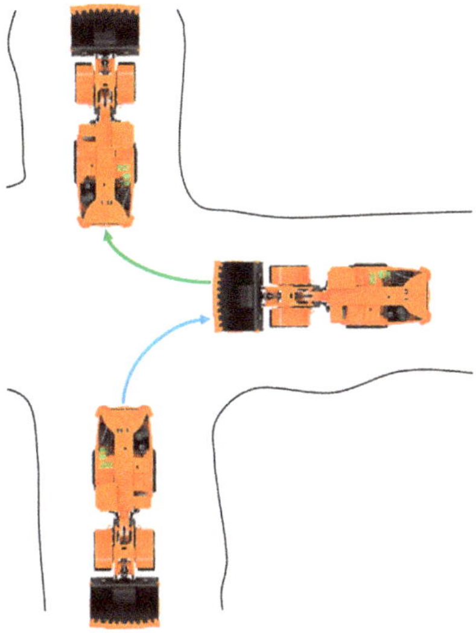

Figure 2. Example of *inversion* maneuver.

The main contributions of this paper are the following:

- A topological navigation system for mining vehicles operating in tunnels, which can be used in any tunnel network by any articulated vehicle.
- A novel localization system specifically designed for tunnel-like environments, which estimates the vehicle's global and local pose using the topological map.
- A development and testing methodology that can be used in the development process of heavy-duty machinery.
- Full-scale navigation experiments using a commercial LHD on a productive level of a sublevel stoping mine.

This paper is organized as follows: First, related work on autonomous LHD navigation is presented in Section 2. The proposed topological navigation system for LHD is described in Section 3. Then, in Section 4, the development and testing methodology is described, and in Section 5, validation results achieved in an actual mine are presented. Finally, conclusions of this work are drawn in Section 6.

2. Related Work

Autonomous navigation of LHDs has been a subject of scientific research since the late 1990s [2–4]. The main objectives have been to improve productivity and increase safety for the personnel, but simultaneously to benefit from reduced machine maintenance due to less wear of components. The theoretical development, and experimental evaluation, of a navigation system for an autonomous articulated vehicle is described in [4]. This system is based on the results obtained during extensive in-situ field trials and showed the relevance of wheel-slip for the navigation of center-articulated machines. In [5], one of the earlier industrial automation implementations is reviewed. It is mentioned there, that the use of automation in day-to-day operations offers flexibility and convenience for the operators. The development of what would become the first commercially available solutions for autonomous navigation of LHD machines followed shortly thereafter.

The work of Mäkelä [6] set the basis for the AutoMine software of the LHD manufacturer Sandvik, while the work of Duff, Roberts, and Corke [7–9] set a similar precedent for the software MINEGEM of Caterpillar. Only a few years later, the work by Marshall, Barfoot, and Larsson [10–12] configured what would become Scooptram Automation of the company Atlas Copco (now Epiroc). These companies have applied their automation solutions directly to their articulated vehicles [13]. Because of their commercial application, only the initial work on the development of the autonomous systems from LHD manufacturers is available in the literature. For instance, autonomous navigation from the manufacturer Sandvik is based on an absolute localization paradigm, i.e., it relies on odometry and detection of natural markers of the tunnel network [6]. Localization is achieved by taking a profile of the tunnel in a 5 [m] long section and comparing it to a known map. Caterpillar, on the other hand, initially based its system on mainly reactive techniques (wall following), in conjunction with a topological map with information about loading points, dump points, intersections, and other markers that are used in an opportunistic location scheme [9]. Finally, Epiroc made use of a hybrid navigation paradigm [11]. A set of behaviors was programmed under a fuzzy logic scheme to form the reactive part, while a higher level (deliberative) planner was used at intersections or open spaces.

Despite its success in delivering a product to market, research in autonomous navigation for underground tunnels continues to be a relevant topic. In [14], a review is given on the performance of automated LHDs in mining operations. It also mentions some issues and challenges that remain. The dynamic and highly variable nature of mining operations underlines the need for flexible and quickly deployable systems, features that earlier commercial solutions lacked [15,16]. These shortcomings are also known to LHD manufacturers, who continue to improve their systems [17].

Automation in mining is, most certainly, a widespread trend that has already shown corporate benefits, and it will continue to drive the modernization of the mining industry [18]. Cost, productivity, and safety are still the driving forces for investments in automated systems. Recent publications in the field also suggest the increasing interest of China in the application of automation technologies [19–21]. Particular attention has been paid to modeling, and control techniques.

The autonomous navigation system presented in this paper is based on topological navigation, and model predictive control (MPC). Underground mining environments have been shown to be suited for the extraction of features needed to build topological maps [22], and a mixture of topological and metric maps has been used successfully to map and navigate in large environments [23]. MPC has also been proven to perform well in the high-speed control of vehicles with nonlinear kinematics [24–26].

3. Topological Navigation and Localization for LHD

3.1. General System Overview

LHD are large vehicles used in underground mining. In this work a LHD model LF-11H, from the GHH Fahrzeuge manufacturer, is used. The vehicle's size is 9.71 [m] in length, 2.45 [m] in width, and 2.45 [m] in height. The LHD's navigation is based on the data provided by two laser scanners (2D LIDARs), one pointing towards the front of the machine, and the other pointing towards the back. For supervision and occasional teleoperation, two cameras give the front and back images to the control station [27]. All on-board processing is done on an industrial computer running Linux OS, while the direct machine control is handled by an internal PLC unit. The LHD and the hardware components mentioned above can be seen in Figure 1.

In order to navigate autonomously inside the network of tunnels of an underground mine a proper representation of the mine is required, in this case a navigation map that includes the topological structure of the mine, tunnels, and intersection, as well as LIDAR measurements. The LHD, therefore, builds a map of the operation area during the system setup, using measurements from LIDAR sensors, as it navigates the tunnels. This map is then linked to a topological representation, thus giving names to all the relevant locations, henceforth referred to as "nodes". The map, once built, allows for route planning and self-localization of the machine. The latter is carried out by means of scan matching between current LIDAR measurements and the LIDAR landmarks stored in the map. The scan matching is implemented using the ICP (Iterative Closest Point) algorithm, and the final pose estimation is the result of applying a Kalman Filter.

Every time a new mission in which the LHD is requested to move to a target node in the mine is executed, a topological route composed of nodes is defined, and then target poses inside each node are calculated. Afterwards, the path required to reach each target pose is computed, and a reactive control algorithm is put in charge of following the path, keeping the vehicle away from colliding with the walls of the tunnel. A model predictive control strategy is implemented because of the complexity of following the path with this articulated machine, which normally moves at a speed of between 12 and 24 km per hour inside the mine.

3.2. Topological Map and Physical Representation

The main component of the representation of the mine is the "Topological Map" (TM). Two types of nodes are defined on this map: (i) tunnel nodes and (ii) intersection nodes. A tunnel node is a section of the mine that can be traversed back and forth, i.e., it corresponds to a single path or trajectory between two physical points. Topologically, a tunnel node has two edges, connecting it with two intersection nodes. An intersection node may have multiple edges that connect it to other nodes, and it can also traverse back and forth. An example would be a fork in the tunnel, where the vehicle must choose one of the possible path alternatives.

Each topological node contains a number of access points (APs) and waypoints (WPs), which are represented by a 2D pose within the node's coordinate system. APs connect different topological nodes; they signal the transitions between map nodes. WPs, on the other hand, correspond to specific poses within the topological node, each one containing relevant information for the navigation system, such as maximum driving speed. Examples of waypoints can be a location for dumping ore, a place to park the vehicle, or a location the vehicle must go through when traversing the node.

As an example, Figure 3 shows a portion of a real mine, where a TM has been created. In this scenario, four main tunnels, T6, T7, T9, and T10, and one intersection, I0 (in green), define the map. It can also be seen that T9 contains a number of waypoints throughout its path, shown by arrows.

Figure 3. Example of a topological map of a real mine.

Besides their associated 2D pose, APs and WPs have a defined heading. APs always face the outer side of the node they are in, but WPs always face in the predefined direction of the node. Because of this, APs and WPs can be connected in 4 different ways: front-to-front, front-to-back, back-to-front, and back-to-back (see Figure 4, the orange arrow is the default direction of the node). In Figure 5, a TM with 2 tunnel nodes, 1 intersection node, and its AP, WP, and the connections between them, is shown.

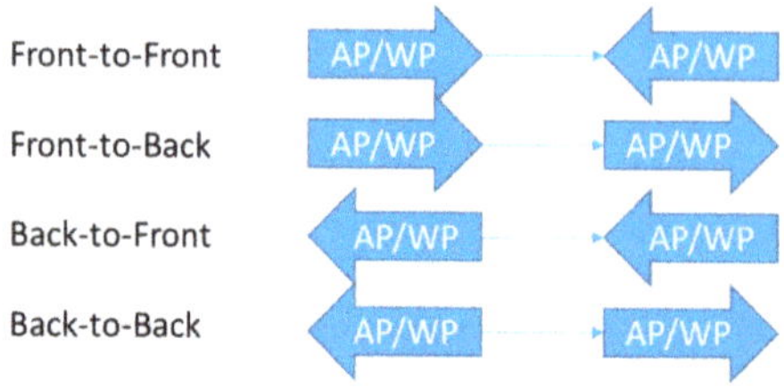

Figure 4. Types of connection between AP and WP.

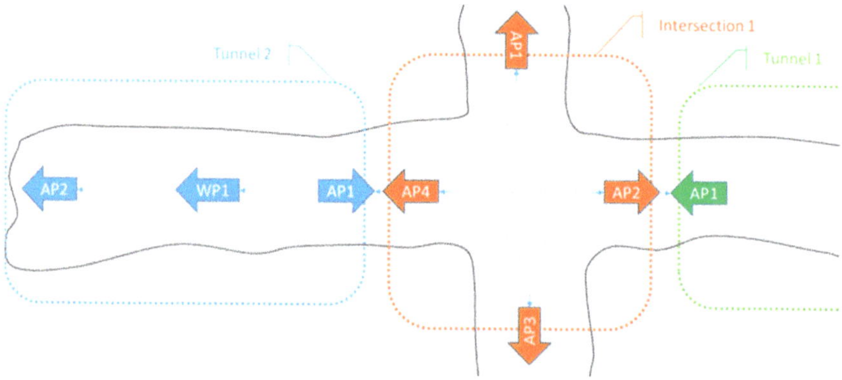

Figure 5. TM example showing one intersection node (Intersection 1), two tunnel nodes (Tunnel 1 and Tunnel 2), and the corresponding APs, WPs and connections between them.

One of the problems that articulated vehicles face while traversing an underground mine is that they sometimes need to change their direction of movement, using a maneuver known as *inversion*. (See Figure 2). To tackle this problem, a Topological Movement's Map (TMM) is built automatically from the TM. To do this, for each AP and each WP in the TM, 6 nodes are created in the TMM. These 6 nodes represent how the vehicle would move through an AP or WP in the TM:

(i) Vehicle going towards the AP/WP moving forward (bucket facing the AP/WP).
(ii) Vehicle going towards the AP/WP moving backwards (rear bumper facing the AP/WP).
(iii) Vehicle going away from the AP/WP moving forward.
(iv) Vehicle going away from the AP/WP moving backwards.
(v) Vehicle standing on the AP/WP facing in its same direction.
(vi) Vehicle standing on the AP/WP facing in its opposite direction.

These new nodes in the TMM are then connected using the information on the topological map and a predefined set of rules that reflect how a vehicle would move between two different pairs of nodes in the TM. These rules are presented on Tables 1 and 2. Rules are slightly different when considering the special case of connecting APs from different TM nodes (see Table 2), because as these APs connect different TM nodes, they have the same physical location, and then connections between "stopped" nodes must be allowed.

Table 1. TMM's connection rules for connections within the same TM node.

Origin Node's Type	Destiny Node's Type of Connection for Different AP/WP Connection Types			
	Front-to-Front	Front-to-Back	Back-to-Front	Back-to-Back
(i)	-	-	(i) & (vi)	(iii) & (v)
(ii)	-	-	(ii) & (v)	(iv) & (vi)
(iii)	(i) & (vi)	(iii) & (v)	-	-
(iv)	(ii) & (v)	(iv) & (vi)	-	-
(v)	(i)	(iii)	(i)	(iv)
(vi)	(ii)	(iv)	(ii)	(iii)

Table 2. TMM's connection rules for connections between different TM nodes.

Origin Node's Type	Destiny Node's Type of Connection for Different AP/WP Connection Types			
	Front-to-Front	Front-to-Back	Back-to-Front	Back-to-Back
(i)	-	-	(i) & (vi)	(iii) & (v)
(ii)	-	-	(ii) & (v)	(iv) & (vi)
(iii)	(i) & (vi)	(iii) & (v)	-	-
(iv)	(ii) & (v)	(iv) & (vi)	-	-
(v)	(i) & (vi)	(iii) & (v)	(i) & (vi)	(iv) & (vi)
(vi)	(ii) & (v)	(iv) & (vi)	(ii) & (v)	(iii) & (v)

It is important to note that these connections are directional, and it is required to go through a "stopped" node ((v) or (vi)) to change the vehicle's direction of movement. Therefore, these rules must be applied in both directions of each connection. Nodes (v) and (vi), with the same origin AP/WP, are always connected. An example of the TMM creation from the TM of Tunnel 1 (in Figure 5) is shown in Figures 6 and 7. For clarity of purpose, only half of the connections are shown in each figure. In Figure 6, only connections in the direction of the tunnel are shown, while in Figure 7, only connections against the direction of the tunnel are shown. Each node in the TMM is represented by a blue arrow. Each node has an associated LHD heading and movement direction. The heading is represented by the LHD image and the movement direction by the orange arrow next to the LHD.

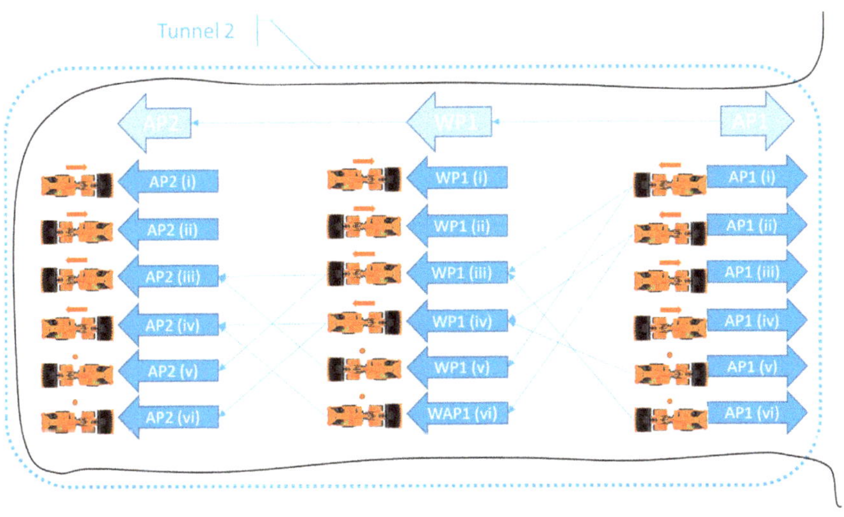

Figure 6. TMM construction example. Connections heading in the same direction of the Tunnel are shown.

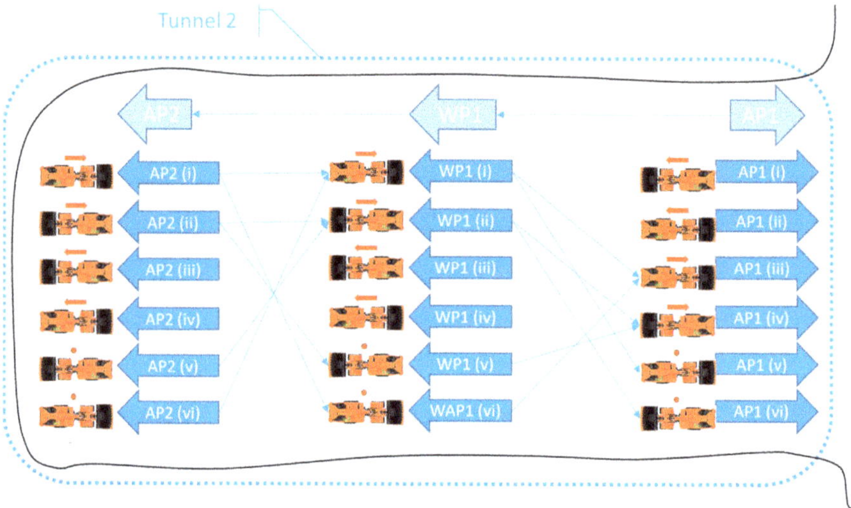

Figure 7. TMM construction example. Connections heading in the opposite direction of the Tunnel are shown.

An example is shown in Figure 8 for the inversion maneuver. Relevant TMM nodes involved in the inversion movement are highlighted.

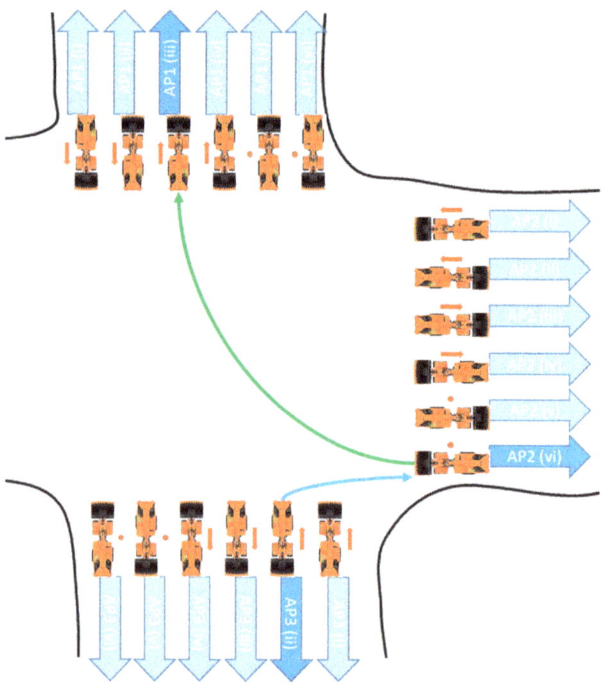

Figure 8. Representation of an *inversion* maneuver in the TMM.

Once the TMM is set up, topological paths can be calculated upon demand. When a new request is received, the vehicle's current state is pushed to the TMM as a starting point, and then the optimal route to the desired destination is calculated using Dijkstra's algorithm [28]. Traveling costs between TMM nodes are set up initially based on the distance between AP/WP, but can be fined-tuned with additional considerations such as: the change in the vehicle's movement direction, the feasibility of the movement (sometimes a curve is too sharp for the vehicle to turn in a certain direction), and unwanted paths for other reasons such as road repairs, narrow spaces, or uneven tracks.

Path planning using the TMM is extremely fast. Its execution in a very large mine layout was simulated in a layout similar to that of the new Chuquicamata underground mine, with eight parallel streets, each one having 19 intersections. Under these conditions, the TMM contained 5378 nodes and 8760 vertices, and path planning using the TMM took less than 5 [ms] on an i7 Intel processor [27].

Each node contains so-called LIDAR-landmarks that are stored when the map is built, and are used for the LHD self-localization. In regard to tunnel nodes, the LIDAR-landmarks correspond to LIDAR point clouds acquired at certain selected positions of the tunnel. In the case of intersection nodes, the LIDAR-landmark corresponds to an integration of the point clouds acquired inside the intersection.

Map Building—Determination of LIDAR-Landmarks

For map building the LHD needs to visit all the tunnels of the mine in one direction. In order to achieve this, the LHD is teleoperated. At the end of this process the LIDAR-landmarks need to be determined.

The following procedure is followed for the tunnel nodes:

1. LIDAR point clouds are stored at fixed positions inside each node (e.g., every 2 [m] in our current implementation). Each of these point clouds are built by joining the LIDAR scans acquired using the front-facing and the rear-facing LIDARs. Then, a filtering process is applied: points colliding with the LHD's body are removed, points that are located too far (>10 [m]) to the center of the LHD are also removed, and finally, the density of the point cloud is normalized. This final point clouds will be the LIDAR-landmarks.
2. The similarity among all pairs of stored LIDAR-landmarks is computed using the ICP algorithm, generating a similarity matrix. The ICP's fitness value is used as a measurement of the geometric similarity between two LIDAR-landmarks (a low fitness value means that ICP managed to fit both landmarks correctly).
3. Using the similarity matrix, representative/key LIDAR-landmarks are chosen. The selected representative LIDAR-landmarks have two important characteristics: (i) They can be matched with their local neighbors using the ICP algorithm (a radius is used for determining the local neighbors); (ii) they are different from other representative LIDAR-landmarks, guaranteeing that two different representative LIDAR-landmarks cannot be mistaken. This selection process is done by the following algorithm:
    - For each LIDAR-landmark:
        - The values of similarity with other LIDAR-landmarks are ordered from lowest to highest.
        - A delta threshold and a radius are defined:
            - The last LIDAR-landmark for which the next highest similarity value does not belong to one of the current LIDAR-landmark's local neighbors is selected (searching from lowest to highest similarity value).
            - This similarity value between the current LIDAR-landmark and the selected LIDAR-landmark is stored as the threshold of this LIDAR-landmark. This value represents the minimum ICP's fitness value that must be met for a valid geometric fit.
            - The difference between the threshold and the next similarity value is stored as the delta threshold of this LIDAR-landmark. This value represents how different the current LIDAR-landmark is from the other LIDAR-landmarks that are not its neighbors.

- The distance to the selected LIDAR-landmark is stored as the radius index of the current LIDAR-landmark. This value represents the distance (in LIDAR-landmark measurement/sample distance units) from which a valid geometric fit can be achieved.

- An empty list is created for the LIDAR-key-landmarks.
- The LIDAR-landmark with the largest radius among the first 10 LIDAR-landmarks is added to the list.
- The LIDAR-landmark with the largest radius among the last 10 LIDAR-landmarks is added to the list.
- The remaining LIDAR-landmarks are ordered by highest to lowest radius, and highest to lowest delta threshold (if they have the same radius).
- For each of these sorted LIDAR-landmarks with radius greater than 2 LIDAR-landmark measurement distances (4 [m] in this case):

If it is not near an existing LIDAR-key-landmark (farther than 5 LIDAR-landmark measurement distance) then it is added to the LIDAR-key-landmark list.

In the case of an intersection node, the LIDAR point clouds obtained in the intersection are integrated and stored as a single cloud point. This cloud point is the LIDAR-landmark of the intersection.

3.3. *Self-Localization*

Self-localization is composed of two modules: a global topological localization estimation, and a local node localization estimation. The topological localization estimates the location of the LHD inside the TMM (and also the TM, because each node in the TMM has an equivalent in the TM), as well as the distance between its current location and the closest AP/WP. The intra-node localization estimates the localization inside the current tunnel or intersection node.

3.3.1. Localization Inside Tunnel Nodes

Inside tunnels the LHD's pose is defined as the one-dimensional distance or trajectory ρ, measured from the beginning of the tunnel. This distance is updated incrementally as:

$$\rho_{t+1} = \rho_t + v_t \cdot dir_t \cdot \Delta t \tag{1}$$

where v_t is the current linear speed of the LHD, and dir_t is the current direction of movement, with the value 1 meaning that the LHD is oriented to the tunnel orientation and -1 if it is not. The values of ρ_t, v_t, and dir_t are estimated using a standard Kalman filter. The filter is updated using two different observations sources: first, using the LHD wheel's odometry obtained directly from the LHD encoders, and second, using the LIDAR point cloud. In the latter case, the ICP algorithm is used to match the current LIDAR point cloud with the LIDAR-landmarks that are near the current LHD position and the key LIDAR-landmarks within 50 [m] of distance. The result of the ICP matching is the distance to the most similar LIDAR-landmark, which is then used to estimate the relative position of the LHD in the node's coordinates, and its orientation. Both the wheel's odometry, and the ICP-estimated position and orientation, are used in the corrective stage of the Kalman filter asynchronously, i.e., every time they arrive.

It is important to mention that the ICP matching is local, not global, meaning that the matching is made only with LIDAR-landmarks that are near the current position of the LHD. The main reasons for this decision are time efficiency and to avoid wrong matching due to the similarity that different LIDAR-landmarks may have. (Mining tunnels have regular surfaces and LIDAR-landmarks computed in different tunnels may be similar). This also prevents the self-localization from changing drastically in its localization estimation.

3.3.2. Localization Inside Intersection Nodes

Inside intersections, the LHD's pose is defined as its 2D position and orientation. Considering that intersections are small, of just a few meters, the wheel's odometry, and the results of the ICP matching between the LIDAR point cloud and the LIDAR-landmark that describes the intersection node are used directly to update the LHD pose.

3.3.3. Global Localization

This module knows in each moment the node in which the LHD is located. The module also receives the current LHD's pose in the current node's coordinate frame. Using this information it evaluates if the LHD is still in the current node, or if it has moved to a neighbor node. If this is the case, then the estimation of the LHD's pose is transferred into the neighbor node's coordinate system. Each node has the geometric transformation to adjacent nodes so the transition is smooth.

3.4. Navigation

The navigation scheme is composed of two layers of nodes that enable the path planning and autonomous hauling through the underground tunnels of the mine: the high level and the low level. The high level encompasses two modules: "Navigation Control" and "Deliberative Path Planning", while the low level is comprised of another two modules: "Guidance" and "Command Executor" (see Figure 9).

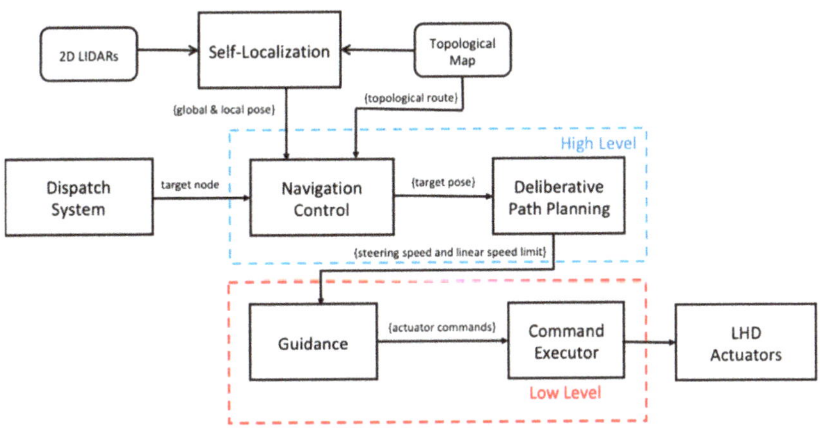

Figure 9. Diagram of the navigation system.

3.4.1. Navigation Control

This module receives a target for the LHD navigation, which could be a relevant location within the mine, such as an extraction/draw point or a dumping point. This target is usually defined by a dispatch system, or in some cases by a human operator. It consists of a destination node and a topological route, which is computed using the TMM as is explained in Section 3.2. This path is represented as a sequence of TMM nodes, each one containing information about the position, orientation, heading direction of the vehicle, and an indication of whether the vehicle must go through the node or come to a full stop on it.

Given that the topological route is composed of TMM nodes that are originated from tunnel and intersection nodes, this module has two navigation modes: *tunnel tramming*, used inside tunnels, and *path-following*, used inside intersections. In tunnel tramming mode, the navigation modules follow the path of the tunnel's walls, while in path-following mode, the navigation modules follow a trajectory in an area were more than a single path can be taken. In addition, inside tunnel nodes, intermediate sub-goals may be generated depending on the defined waypoints (See Section 3.2).

To achieve a smooth navigation across TMM nodes, transitions between different goals must be seamless. A naive implementation to identify when a goal has been reached would be to check when the global localization estimation equals the current goal, but this often makes the movement of the vehicle not continuous and clumsy. A better approach requires that *Navigation Control* anticipates when the LHD is going to reach a certain goal. In order to do this, a set of conditions are applied in addition to monitoring the global localization estimation:

$$P\left(\vec{x}_{LHD} = \vec{x}_T\right) = \frac{1}{2\pi\sqrt{|\sigma|}} e^{-\frac{1}{2}\left[\vec{x}_{LHD} - \vec{x}_T\right]^T \sigma^{-1}\left[\vec{x}_{LHD} - \vec{x}_T\right]_{1,1}} > P\left(\vec{x}_{LHD} = \vec{x}_T\right)_{MIN} \tag{2}$$

$$\dot{d}\left(\vec{x}_{LHD}, \vec{x}_T\right) > \dot{d}_{MIN} \tag{3}$$

$$|\theta_{LHD} - \theta_T| < \Delta\theta_{MAX} \tag{4}$$

where:
$\vec{x}_{LHD}$ = 2D position estimation of the LHD.
$\vec{x}_T$ = 2D position of the current navigation target.
σ = 2D Self-localization estimation variance (without orientation estimation).
$P\left(\vec{x}_{LHD} = \vec{x}_T\right)_{MIN}$ = Minimum 2D target reached likelihood threshold.
$\dot{d}$ = Euclidean distance function derivative with respect to time.
$\dot{d}_{MIN}$ = Minimum Euclidean distance function derivative threshold.
θ_{LHD} = LHD orientation (heading) estimation.
θ_T = Current target orientation (heading).
$\Delta\theta_{MAX}$ = Maximum orientation difference threshold.

The condition (2) is the probabilistic estimation of actually reaching the desired target position. Condition (3) measures if the LHD is actually getting closer to the target and condition (4) measures the difference between the LHD's orientation and the current target's orientation. When navigating in *tunnel tramming* mode, only condition (2) is used, but when navigating in *path-following* mode, conditions (2)–(4) must be met. This way, lower values for $P\left(\vec{x}_{LHD} = \vec{x}_T\right)_{MIN}$ on condition (2) can be used (which helps to anticipate transitions and obtain a smooth movement), because conditions (3) and (4) indicate that the vehicle is going to the target goal (often a tunnel entrance) in an intersection.

In *tunnel tramming* node, Navigation Control also checks that the LHD does not miss the tunnel end, checking the following conditions:

$$d_{ODOM}(t) - d_{TUNNEL} > e_{MAX}^{ODOM} \tag{5}$$

$$\frac{d_{ODOM}(t)}{d_{TUNNEL}} > e_{MAX\,\%}^{ODOM} \tag{6}$$

where:

$d_{ODOM}(t)$ = Accumulated linear odometry of the current tunnel.
d_{TUNNEL} = Total length of the tunnel.
e_{MAX}^{ODOM} = Maximum odometry error magnitude threshold.
$e_{MAX\,\%}^{ODOM}$ = Maximum odometry error percentage threshold.

If both of these conditions are true, *Navigation Control* stops the vehicle and asks for assistance to the operator/supervisor of the system. Both conditions are required because, for short tunnels, condition (6) can trigger false alarms, while for long tunnels, condition (5) can trigger false alarms.

Other important information stored in the TMM is the maximum speed at which a node should be transited, and an indicator forcing the vehicle to drive closer to one of the walls of the road (instead of trying to remain in the center of the road). Both of these parameters can be manually tuned to optimize the way the vehicle approaches certain curves or traverses through the mine.

3.4.2. Deliberative Path Planning

This module receives the next target position, which needs to be reached with a certain speed, as a relative pose from *Navigation Control*. Then, it calculates the path to be followed between the current pose and the desired destination, as a spline $S(t) = \left[S_x(t), S_y(t)\right]$. The desired steering speed ($\overline{\omega}$) and speed limit (v_{MAX}) are then computed, and sent to *Guidance* (See Figure 9).

In order to calculate the spline's coefficients, the following border conditions are used:

$$S\,(t=0) = X_0;\; S\left(t = t^* = \frac{\tilde{d}}{v}\right) = X_1 \tag{7}$$

$$\dot{X}_0 = (v\cos\gamma, v\sin\gamma);\; \dot{X}_1 = (v\cos\theta, v\sin\theta) \tag{8}$$

where:

X_0, $\dot{X}_0$ = Position and speed of the front bumper of the vehicle (See Figure 10).

X_1, $\dot{X}_1$ = Position and speed at the desired target destination (See Figure 10).

$\tilde{d}$ = Estimated distance between X_0 and X_1.

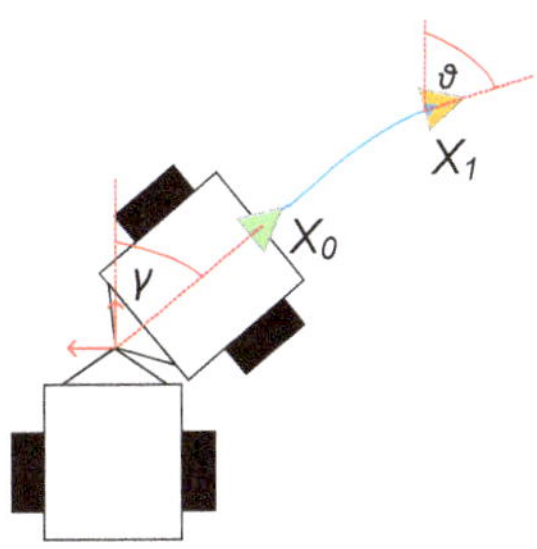

Figure 10. Graphic representation of the kinematic variables of the vehicle.

Using the calculated derivative of the spline, and the vehicle's kinematic model, given by (10) and (11), the desired steering rate γ can be calculated as:

$$\dot{\gamma} = \frac{\left(L_f\cos\gamma + L_r\right)\dot{\alpha}(t=0) - v_i\sin\gamma}{L_r} \tag{9}$$

with $\dot{\alpha}(t=0)$ the calculated angle derivative of the spline, evaluated in $t = 0$; L_r the length from the LHD's pivot to the rear wheel axis; L_f the length from the LHD's pivot to the front wheel axis; v_i the linear speed of the LHD; γ the steering angle in the LHD's pivot; $\omega = \dot{\gamma}$ the steering speed in the LHD's pivot.

3.4.3. Guidance

This module performs the task of selecting the appropriate commands for the machine's actuators, given the high-level general directives of the expected motion and, at the same time, ensuring that the LHD will not hit any obstacles or mine infrastructure. For that purpose, a model-based predictive control (MPC) scheme was implemented using the vehicle's kinematic equations and a cost function that simultaneously considers the following: the high level reference commands, the distance to the walls of the tunnel, and the smooth variation of the actuator commands over time.

The kinematic model of a center-articulated vehicle has been presented in a number of previous publications, such as in [11]. Equations (10) and (11) show an incremental model for the machine's pose.

$$\Delta[x, y, \theta] = \Delta t \cdot \left[v \cos(\theta), v \sin(\theta), (v \sin(\gamma) + L_r\omega) / \left(L_f \cos(\gamma) + L_r\right)\right] \quad (10)$$

$$\Delta\gamma = \Delta t \cdot \omega \quad (11)$$

where:

$[x, y, \theta]$ = pose of the LHD (2D position and angle).
Δt = sampling time of the discrete model.
v = linear speed of the LHD.
γ = steering angle in the LHD's pivot.
$\omega = \dot{\gamma}$ = steering speed in the LHD's pivot.
L_r = length from the LHD's pivot to the rear wheel axis.
L_f = length from the LHD's pivot to the front wheel axis.

The previous model is used in the MPC to predict the trajectory of the machine over a predefined timespan. Then, the optimization process is carried out, in which the best actuator command ($u = [u_v, u_\omega]$) for each time step is selected to minimize the following cost function:

$$Q = Q_{map} + Q_{steering} + Q_{smooth} \quad (12)$$

This equation shows that the cost function is composed of three parts: one for keeping the vehicle away from the tunnel walls (Q_{map}), another ($Q_{steering}$) for following the high-level reference commands, and a final one to smooth the optimization result over time (Q_{smooth}).

In Equation (13), it can be seen that the cost associated with keeping the machine away from the walls relies on maximizing the distance between certain key points of the vehicle and the closest data point in the registered point cloud of the environment. These key points are the corners of the front and rear vehicle bodies.

$$Q_{map} = \sum_{i=1}^{n} R_F \frac{|D_{FL,i} - D_{FR,i}|}{D_{FL,i}^2\, D_{FR,i}^2} + R_M \frac{|D_{ML,i} - D_{MR,i}|}{D_{ML,i}^2\, D_{MR,i}^2} + R_R \frac{|D_{RL,i} - D_{RR,i}|}{D_{RL,i}^2\, D_{RR,i}^2} \quad (13)$$

With $D_{FL,i}$ the distance between the front left corner of the machine and the closest point of the tunnel walls, predicted at time step i of the optimization process. Similarly, D_{FR}, D_{ML}, D_{MR}, D_{RL}, and D_{RR}, refer to the distances from the front right, middle left, middle right, rear left, and rear right corners of the vehicle, respectively. The cost function weights, R_F, R_M, and R_R, are selected to obtain proper behavior.

Equation (14) details the cost related to following the command directives issued from the high-level software modules. Here, only the reference for the steering speed ($\overline{\omega}$) is considered, since the reference for the machine's maximum speed (v_{MAX}) is directly set as an upper bound restriction for the optimization function. Again, the cost function weight R_ω is selected to obtain proper behavior.

$$Q_{steering} = \sum_{i=1}^{n} R_\omega\, |\omega_i - \overline{\omega}| \quad (14)$$

Finally, the smoothing component of the cost function (Q_{smooth}), is intended to ensure that the command has a controlled variation (i.e., limits the change in the command between time steps), and that a newly computed optimal command vector has some degree of continuity after the time span for which it was selected. Namely, the Q_{smooth} component comprises, in turn, two other terms, as stated above.

$$Q_{smooth} = Q_{acc} + Q_{proj} \quad (15)$$

The first term assigns an additional cost to commands that cause a linear or steering acceleration above predefined limits, as stated in Equation (16), while the second term, shown in Equation (18), rewards commands that, when maintained past their time horizon, for up to twice as long as originally intended, will not cause a collision with a tunnel wall.

$$Q_{acc} = \sum_{i=2}^{n} R_{\delta\omega} \cdot f(\omega_i - \omega_{i-1}, \Delta\omega_m, \Delta\omega_M) + R_{\delta v} \cdot f(v_i - v_{i-1}, \Delta v_m, \Delta v_M) \tag{16}$$

$$f(x, x_{min}, x_{MAX}) = \begin{cases} x - x_{MAX} & if\ x_{MAX} < x \\ 0 & if\ x_{min} < x < x_{MAX} \\ x_{min} - \mathrm{x} & if\ x < x_{min} \end{cases} \tag{17}$$

$$Q_{proj} = \sum_{j=1}^{m} R_p \frac{\Delta[x_j, y_j, \theta_j | v_n, \omega_n] \cdot crash(x_j, y_j)}{\Delta[x_j, y_j, \theta_j | v_n, \omega_n]} \tag{18}$$

$$crash(x, y) = \begin{cases} 1 & if\ position\ (x, y)\ is\ in\ collision \\ 0 & if\ position\ (x, y)\ is\ not\ in\ collision \end{cases} \tag{19}$$

where $R_{\delta\omega}$, $R_{\delta v}$, and R_p are the cost weights, selected for proper behavior; $\Delta\omega_m$, $\Delta\omega_M$, Δv_m and Δv_M are the parameters for the minimum and maximum steering acceleration and linear acceleration, respectively; $\Delta[x_j, y_j, \theta_j | v_n, \omega_n]$ are the displacement caused by the kinematic model of the machine, at time step j when the last optimization command of the previous process is applied.

The outcome of the former process is a command vector for every time step in the selected timespan ($\vec{u} = [u_{t_0}, \ldots, u_{t_f}]$), in which each element ($u_{t_i} = [u_v, u_\omega, t_i]$) represents a speed and steering command pair, alongside the timestamp on which this command is to be executed.

3.4.4. Command Executor

In opposition to the traditional philosophy of an MPC, the result of the *Guidance* module is not directly fed to the machine's actuators. It is first filtered and merged with previous results of the optimization process in order to always keep a consistent queue of commands that will sustain the operation of the vehicle for a short period of time. This filtering is carried out by the *Command Executor* module. The goal of this module is to ensure that the signals sent to the actuators will be appropriate, both for avoiding long-term damage of the devices involved and also for keeping the operation running as expected.

The *Command Executor's* input is a "trajectory" of commands to be executed at specific times. Each command of the trajectory is inserted in a command queue. The queue insertion process entails finding the time at which the current command is to be inserted, erasing any command previously queued from that moment onwards. Then, the new command is appended at the end of the queue, effectively overriding outdated directives.

Before the *Command Executor* issues a new command to the machine actuators, the upcoming command is filtered. The velocity command u_v is limited to a maximum value $u_{v,max}$ and a "dead zone" is applied to the steering command u_ω, namely:

$$u_v = \begin{cases} u_v & if\ u_v < u_{v,max} \\ u_{v,max} & if\ u_{v,max} < u_v \end{cases} \tag{20}$$

$$u_\omega = \begin{cases} u_\omega & if\ u_\omega < -u_{\omega,\ min}\ or\ u_{\omega,\ min} < u_\omega \\ 0 & if - u_{\omega,\ min} < u_\omega < u_{\omega,\ min} \end{cases} \tag{21}$$

where $u_{\omega,\ min}$ is a predefined constant value for the steering command "dead zone" and $u_{v,max}$ is a value computed, so that if the machine were to be commanded to stop at the present time, it would effectively stop before the last queued command. That is, given a command queue with a total duration of Q_{dt} seconds and a machine deceleration of $\overline{D}_v$ meters per second squared, then: $u_{v,max} = \overline{D}_v \cdot Q_{dt}$, where $\overline{D}_v$ is the mean deceleration of the machine when a full brake is applied, a parameter that can be determined experimentally.

A diagram of the described process is shown in Figure 11 for a single command of the input command trajectory. As mentioned, the same steps are executed for all elements.

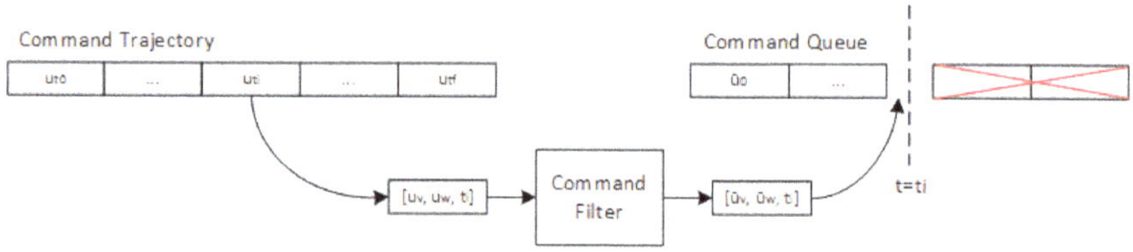

Figure 11. Command management process of the command executor node.

4. Development and Testing Methodology

The methodology used for the development and testing of the proposed navigation system consists of four steps. The first is the development of the automation system in a simulated environment, which is a safe and cost efficient platform for that purpose. The second is the use of scale models to verify the behaviors that are too complex or impractical to be tested on a simulated environment. The third is validation and testing in real equipment, using a safe location intended for that purpose. Fourth is the validation and testing in a real operation environment under controlled conditions before moving on to production. Details are discussed further in the following section.

4.1. Development in a Simulated Environment

The system was initially developed and tested in a simulated environment using Gazebo [29] and integrated with ROS [30]. At first, an underground scenario with wide tunnels and perfect self-localization, using the real position from the simulator, was used as a testing environment. When the system could perform reasonably well, the wide tunnels were substituted by realistic tunnels, using laser scans acquired in a real underground mine. The realistic tunnels were much narrower and had irregular shapes. Finally, when the challenges of the new scenario were solved, the system was tested with a functional self-localization module, and with other factors that added complexity, such as a simulation of the LHD's controller, in order to validate all low-level communication and security schemes.

4.2. Development Using Scale Models

Not all of the functions of the system can be tested in a simulated environment, either because of the complexity of the problem, which makes the simulation approach impractical, or because not enough data is available to simulate certain interactions between the equipment and the environment. To address this issue, a scale model can be built in order to validate some of the design assumptions before implementing the solution on a commercial vehicle. The scale models need to have a certain similarity in the aspects related to the phenomena that needs to be validated. In the case described here, a 1:5 scale model was built based on a commercial 5 [yd^3] LHD, shown in Figure 12, with an electric power train and hydraulic actuation for the steering and bucket movements, mimicking real equipment. A scaled-down ore extraction point was built, including ore from an actual mine. The scale model was used to perform navigation in the laboratory before installing the control system in the commercial LHD.

Figure 12. 1:5 scaled LHD built for testing and validation.

4.3. Validation in Test-Fields

The LHD automation navigation was installed in a GHH's LF11H LHD (GHH Fahrzeuge). The installation required mechanical and electrical modifications of the equipment to place the system's sensors, processing units, and wireless communication equipment. All the automation software runs in an industrial fan-less computer equipped with an Intel i7 processor and with 4 logical cores. The interface between the automation and the machine was implemented on the machine controller (IFM mobile mini controller), based on GHH's factory program (implemented in Codesys).

After the system was installed and all basic control, communications, and safety functions were thoroughly tested, the equipment was moved to a test-field nearby the OEM (Original Equipment Manufacturer) facilities in Santiago, Chile. The test-field emulated an underground tunnel by using light material to mimic the walls of the mine (Figure 13), which was enough to trick the system. It also had a dummy loading point, a dumping point, and a truck loading point. These tests were used to calibrate the system controller's parameters and the kinematic characteristics of the vehicle's model, such as the acceleration, steering, and breaking response to different operation inputs.

Figure 13. Test-field close to GHH facilities in Santiago de Chile.

Because the test-field was located just outside the city where our development team is based (Santiago, Chile), it was possible to make a short trip to test new versions of the software, which contained bug fixes or improvements to the system. Usually a team of developers would go to the test-field two or three times a week to try out different modifications in the algorithms of the automation system.

The last milestone of this stage was to validate the reactive navigation algorithm. In order to do this in a safe manner, a hybrid operation mode was used, in which the speed of the LHD was remotely controlled by an operator while the autonomous navigation system handled the steering of the vehicle in an assisted tele-operation mode. To ensure safety precautions, an onboard operator who could shut down or override the automation system commands was on the vehicle for every test.

4.4. Validation in a Real Mining Operation

In order to carry out the final stages of development, the equipment had to be tested in its real operation environment, where the last design assumptions and algorithms needed to be validated, and the presence of personnel from the operation site were required.

In the case of the automation system described here, after test site validation, the LHD was transported to a real sublevel stopping mine in the north of Chile, where the development team, the OEM, and the mine personnel coordinated the final system validation and tests.

5. Results and Discussion

The validation in the test-field was executed from March to June of 2017, requiring approximately 300 h of work. On-site tests were carried out in a medium-scale sublevel stoping mining operation called the *Mina 21 de Mayo* (21st of May Mine), the property of *Compañía Minera San Gerónimo*, located in the north of Chile. The tests were comprised of two phases:

- During the first batch of tests in 2017, approximately 2300 work-hours were needed to test the system, of which about 800 were on-site; 600 were for remote support and system troubleshooting; 800 were with OEM remote support on-site; 100 h were spent traveling from the nearest city, La Serena, Chile, to the mine. The first batch included 66 days of testing, including installation of hardware on the LHD, network infrastructure on the tunnel, tele-operation station, and CCTV cameras. First underground tele-operation tests were done on day 32, which were followed by assisted tele-operation and self-localization tests.
- During the second batch of tests, performed in 2018, about 2900 work-hours were required, including 1000 on-site, 1000 with OEM support onsite, 750 with remote support, and 150 spent traveling from the nearest city to the mine. This second batch of tests lasted for 77 days. First, autonomous navigation tests took place on day 32. Further tests included approximately 150 h of autonomous navigation. It is important to note that the LHD was also used to test an autonomous loading system, so not all on site test were for the autonomous navigation system.

Between the first and second phases, several upgrades were made to the system in order to improve its robustness, consistency, and performance. The most important improvement was on the self-localization system, because the first batch of tests proved that the initial method (not described here) could not maintain the self-localization estimation along the test tunnel. Assisted tele-operation tests during the first phase were mainly used to tune the parameters of the *Guidance* module for the tunnel and intersection navigation modes (Equations (13), (14), (16), and (18)). Once the autonomous navigation was operating properly, further adjustments were made to all system's parameters, including the parameters for the *Command Executor* module (Equations (20) and (21)). Parameters of the map were tuned, such as the maximum speed for certain segments of the tunnel, 2D poses of APs/WPs, and navigation modes for different parts of the tunnel.

On-site, at the mine, two validations were carried out: surface level tests, and underground tests. Surface level tests were done to test all the modules before entering the mine, and to visualize any problem that the LHD or the implemented automation system could have. After the arrival of the machine at the mine site, all sensors, antennas, and communication modules were re-installed and tested. The first teleoperation tests were carried out on the surface, on one of the dump sites of the mine, to verify that the operation of the LHD was correct.

The second validation was done inside the mine in a production tunnel. The system was tested incrementally from teleoperation to full autonomous operation. A network infrastructure was installed inside the test tunnel, and an operating station, consisting of a computer, screens, and controls, was installed inside the mine. Communication tests were carried out between the LHD inside the mine and the computer in the operation center. Teleoperation and assisted teleoperation modes were the first functionalities tested. In the first mode, the operator drives the equipment just as would be done aboard, and in the second mode the operator mainly indicates the direction of movement and the system keeps the LHD away from the walls keeping it from colliding with them. The system was successful in avoiding collisions between the equipment and the inner walls of the tunnel, and the general performance of the operation was similar to manual navigation.

The autonomous navigation tests showed that the system allowed tramming along a 180 [m] tunnel from its entrance to the loading point. The LHD took approximately 2 [min] to go from one point to another, which is comparable to the performance achieved by an experienced human operator. Some of the difficulties that were found included the tunnel being too narrow for the LHD (sized according to the manufacturer's specifications), and the floor having a large number of irregularities, pot holes, and varying inclinations. Of these factors, only the narrowness of the tunnel was included in the simulated environment. A view of the operator's control interface is shown in Figure 14.

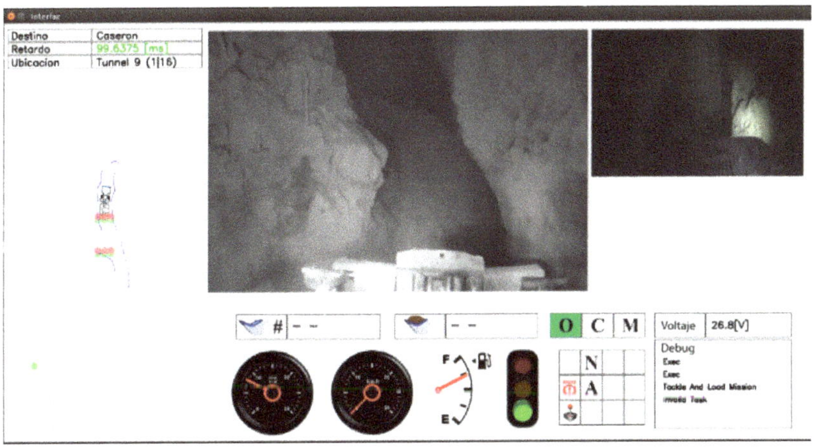

Figure 14. Operator's graphical interface of the navigation system.

Because of time constraints in the mine, testing, development and parameter tuning were done simultaneously. Because of this, most datasets of the tests in the mine are from a work-in-progress version of the navigation system. Results presented in this section are from 14 datasets (labeled 1, 2, 3, etc.), taken on a single afternoon two weeks before the end of on-site tests. 4 manual operation datasets (labeled M1, M2, M3, and M4) are also presented to have a reference for the performance of an experienced human LHD operator. These manual operation datasets were compiled a week later than the autonomous navigation datasets.

An important problem during tests was roaming between different Wi-Fi access points inside the tunnel. For safety reasons, the system stops accelerating the LHD if communication with the operation station becomes unstable, generating an emergency stop if the loss of communications is longer than a few seconds. Because of this, and a wireless network that did not have fast roaming capabilities, the system often stopped when switching from one access point to another. This can be seen in Figure 15, where stops produced by roaming, and by unstable communications, are shown. The Figure 15 also shows the instant speed of the vehicle (in km/h), the operation mode (with a value of 10 for autonomous navigation, 0 for idle, and −10 for tele-operation), distance traveled (in decameters), the Wi-Fi channel of the access point, at which the LHD is connected (different channels are used for faster roaming), and, finally, the RSSI and Noise values reported by the wireless modem of the vehicle. All the scales have been selected to fit in a single figure, to show the relation better between these variables.

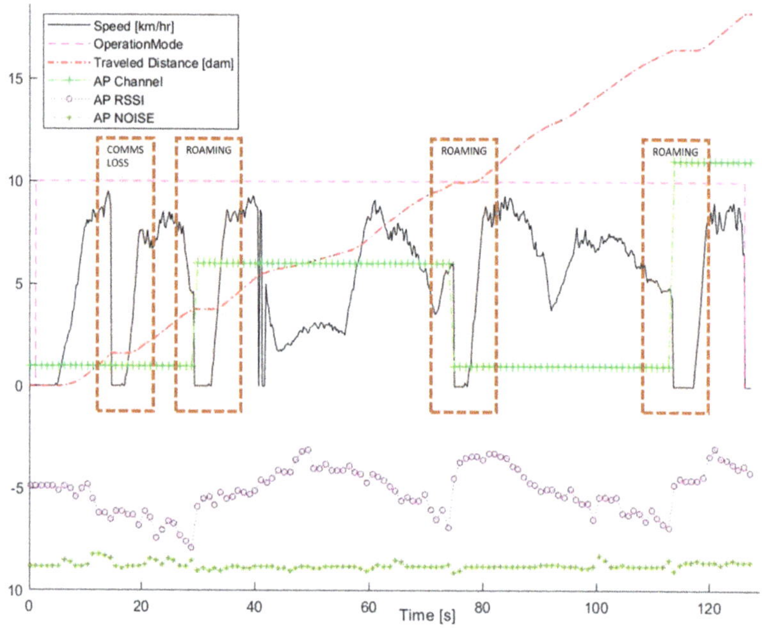

Figure 15. Instant speed, operation mode, distance traveled, wireless communication channel, RSSI, and noise for dataset 1. It can be seen in the selected areas that the LHD comes to a stop when switching between different access points, or when the communication network becomes unstable.

Another consideration for these datasets, since the navigation map was still being tuned, is the intervention of the operator through tele-operation (or assisted teleoperation) to help the LHD go through some narrow passages, or to get back and try again to pass autonomously through a given part of the tunnel. This is shown in Figure 16, as the vehicle needed to stop, then go back a couple of meters (with teleoperation assistance), to later reengage in the autonomous navigation mode, this time getting to the desired destination without further intervention.

Taking these factors (stops because of communication problems and tele-operation) into account, a series of performance indicators were computed for all the datasets. The mean and max speed of the LHD are presented in Table 3. When analyzing the results, it is important to consider that some datasets were compiled with the LHD having a fully loaded bucket, and others with an empty bucket. In some of these datasets, the LHD is moving forward, towards the draw point of the tunnel, and in others, it is moving backward, towards the dump point of the tunnel. To better understand the performance of the system, and the effects of roaming and tele-operation, other indicators are presented, such as the length of the dataset, the total distance of the movement, and the total distance that the LHD was driven by the autonomous system. With an empty bucket, the seasoned operator drove through the tunnel at an average speed of 6.4 [km/h], while the autonomous system did the same at 5.8 [km/h], thus slightly underperforming. The maximum speed achieved by the autonomous system was 11.3 [km/h] with a loaded bucket, while the seasoned operator achieved a maximum speed of 10.6 [km/h] with an empty bucket.

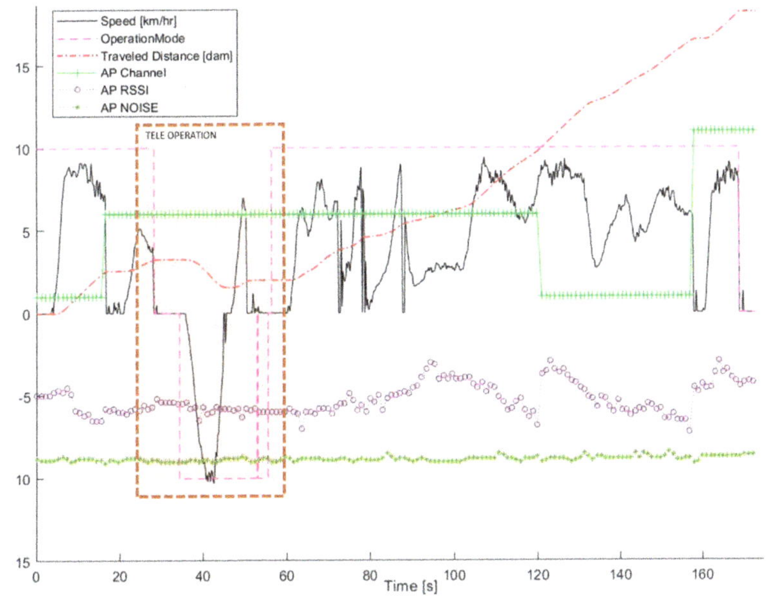

Figure 16. Instant speed, operation mode, distance traveled, and wireless communication channel for dataset 5.

Table 3. Mean Speed, Maximum Speed, Navigation Time, and Navigation Distance for autonomous navigation and manual operation datasets. ID: Dataset Identifier. T.Op Time: Tele-operation Time. TD: Total Distance. TAD: Total Autonomous Distance. E/L B: Empty or Loaded Bucket. H F/B: Heading Forward or Backwards.

ID	Mean Speed [km/h]	Max Speed [km/h]	Length [s]	Nav. Time [s]	Stop Time [s]	T.OpTime [s]	TD [m]	TAD [m]	E/L B	H F/B
1	6.01	9.50	127.4	109.3	18.1	0.0	182.6	182.6	E	F
2	4.80	10.80	189.7	139.4	38.1	12.2	184.4	185.8	L	B
3	6.06	9.80	215.6	112.1	21.6	81.8	183.2	188.7	E	F
4	4.23	9.60	300.8	154.2	117.1	29.5	187.2	181.7	L	B
5	5.91	9.90	150.8	109.5	41.3	0.0	179.6	179.6	E	F
6	4.58	9.00	169.0	105.6	63.4	0.0	134.4	134.4	L	B
7	5.98	9.70	143.6	111.4	25.3	6.9	191.0	185.4	E	F
8	4.15	10.50	336.0	192.2	69.0	74.8	174.8	221.8	L	B
9	5.54	9.40	172.9	126.1	25.7	21.1	182.1	194.6	E	F
10	4.94	10.20	205.6	134.5	46.6	24.5	172.8	184.6	L	B
11	5.33	9.70	323.9	159.0	84.7	80.2	174.3	235.1	E	F
12	4.44	10.80	165.9	147.5	18.4	0.0	181.8	181.7	L	B
13	6.04	9.50	122.9	109.2	13.7	0.0	183.7	183.3	E	F
14	4.93	11.30	147.6	133.3	14.3	0.0	182.9	182.9	L	B
M1	5.84	7.80	124.8	124.6	0.2	0.0	200.6	-	L	F
M2	6.39	8.70	116.7	116.5	0.2	0.0	205.7	-	E	B
M3	6.37	9.20	120.9	119.8	1.0	0.0	208.0	-	E	F
M4	6.44	10.60	109.8	109.6	0.2	0.0	195.0	-	E	B
μ(1–14) w/E	5.84	9.64	179.6	119.5	32.9	27.1	182.3	192.8	E	F
σ(1–14) w/E	0.29	0.18	71.0	18.4	24.4	37.5	5.0	19.3	E	F
μ(1–14) w/L	4.58	10.31	216.4	143.8	52.4	20.1	174.0	181.8	L	B
σ(1–14) w/L	0.32	0.79	72.8	26.3	35.2	27.0	18.2	25.4	L	B
μ(M2–M4)	6.10	9.50	115.8	115.3	0.5	0.0	202.9	-	E	-
σ(M2–M4)	0.04	0.98	5.6	5.2	0.5	0.0	7.0	-	E	-

The Navigation time is either an autonomous navigation time or a manual operation time, depending on the dataset. Stop time is the time the machine was stopped, which includes the time at the start and the end of each dataset. Tele-operation time is the amount of time spent tele-operating the LHD so that it is able to resume autonomous tele-operation, usually because the autonomous navigation system didn't approach a curve appropriately, and reached a point where it didn't know how to proceed. The LHD was able to go through the tunnel without remote assistance in only 6 datasets, but it is important to remember that these were done during development, and small tweaks and adjustments, some of them that worked and some of them did not, were made in between.

To further assess the driving abilities of the autonomous system, the smoothness of its operation is considered. For this, two different metrics are used. The first measures the change between two consecutive command inputs (propel and steering), as is shown in equation (22). The second, in a similar way, measures the difference between two consecutive measures of the dynamic state of the LHD, namely, its speed and the angle of its articulation, as is shown in Equation (23).

$$E_u(k) = (u_v(k+1) - u_v(k))^2 \Delta t + (u_\omega(k+1) - u_\omega(k))^2 \Delta t \quad (22)$$

$$E_{LHD}(k) = (v(k+1) - v(k))^2 \Delta t + (\gamma(k+1) - \gamma(k))^2 \Delta t \quad (23)$$

The average and maximum values of both metrics for all datasets are presented in Table 4. Again, the human operator shows a better performance than the autonomous system. The consistency of the human operator is quite remarkable, and it shows its expertise and knowledge of the machine and the tunnel. The autonomous system is also quite consistent on these metrics, but that is usually expected of an automation system In order to have a better idea of the difference between them, Figures 17–20 show the machine inputs (propel and steering) as well as the instant speed and steering angle of the LHD. Figures 17 and 19 show dataset 14, while Figures 18 and 20 show dataset M1 For clarity, Steering command and steering angle have been plotted separately from and propel command and LHD's speed. Both were made with the LHD having a fully loaded bucket, and with the vehicle moving backward, towards the dump point of the tunnel Straight lines can be seen in Figure 16 on the propel command line, showing a constant output by the autonomous system. Looking at both figures, it can be seen that the human operator uses fewer steering commands, perhaps showing a better understanding of the LHD kinematics, and, therefore, greater abilities to predict the behavior of the vehicle.

Table 4. LHD input difference, state difference and distance to the walls for autonomous navigation and manual operation. ID: Dataset Identifier. ACD: Average Command Difference (E_u). MCD: Max Command Difference (E_u). ASD: Average State Difference (E_{LHD}). MSD: Max State Difference (E_{LHD}). H F/B: Heading Forward or Backwards. ADLW: Average Distance to Left Wall. MDLW: Minimum Distance to Left Wall. ADRW: Average Distance to Right Wall. MDRW: Minimum Distance to Right Wall.

ID	ACD	MCD	ASD	MSD	ADLW [m]	MDLW [m]	ADRW [m]	MDRW [m]
1	0.0060	0.2776	0.0451	5.0400	0.62	0.19	0.67	0.16
2	0.0053	0.3941	0.0358	3.4820	0.84	0.10	0.47	0.10
3	0.0066	0.2510	0.0380	4.7157	0.62	0.12	0.64	0.19
4	0.0043	0.3777	0.0347	6.8018	0.71	0.10	0.43	0.10
5	0.0068	0.2184	0.0484	5.4908	0.62	0.20	0.61	0.23
6	0.0058	0.2607	0.0323	1.9258	0.55	0.10	0.46	0.10
7	0.0059	0.3070	0.0309	5.3291	0.61	0.10	0.64	0.12
8	0.0058	0.2110	0.0558	5.8182	0.68	0.10	0.41	0.10
9	0.0071	0.3096	0.0514	7.8898	0.63	0.17	0.65	0.11
10	0.0056	0.2268	0.0717	6.1329	0.71	0.11	0.50	0.10
11	0.0059	0.2157	0.0583	5.7393	0.56	0.10	0.65	0.17
12	0.0056	0.2526	0.0361	3.4673	0.73	0.10	0.51	0.10
13	0.0072	0.3440	0.0359	4.3288	0.64	0.16	0.63	0.11
14	0.0048	0.1749	0.0383	4.5030	0.78	0.13	0.49	0.10
M1	0.0018	0.0406	0.0076	0.1779	0.60	0.10	0.56	0.10
M2	0.0016	0.0339	0.0090	0.1044	0.47	0.10	0.63	0.11
M3	0.0016	0.0452	0.0091	0.2017	0.48	0.10	0.69	0.10
M4	0.0013	0.0260	0.0070	0.1617	0.44	0.10	0.70	0.12
μ(1–14)	0.0053	0.2711	0.0435	4.5901	0.66	0.13	0.55	0.13
σ(1–14)	0.0006	0.0834	0.0147	1.7500	0.08	0.04	0.10	0.04
μ(M1–M4)	0.0016	0.0364	0.0082	0.1614	0.50	0.10	0.65	0.11
σ(M1–M4)	0.0002	0.0083	0.0010	0.0414	0.07	0.00	0.06	0.01

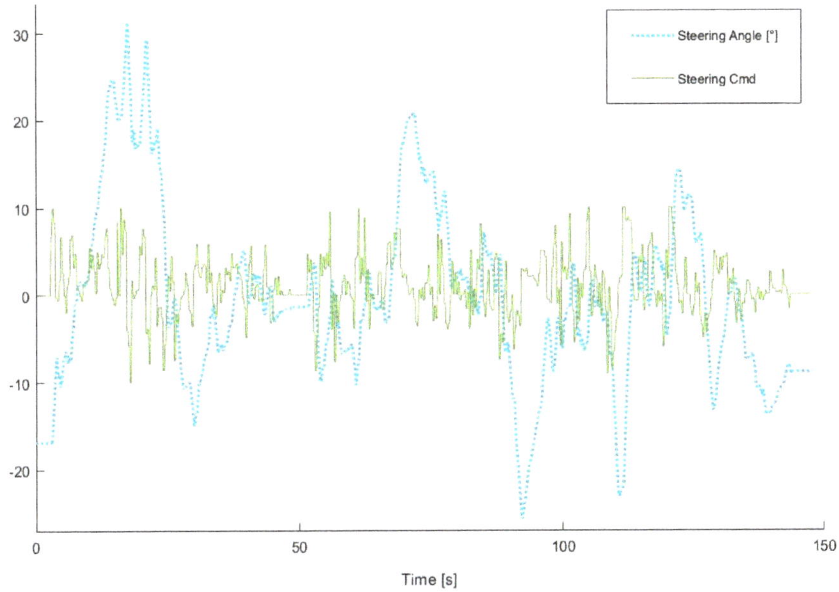

Figure 17. Autonomous System steering commands and LHD steering angle on dataset 14.

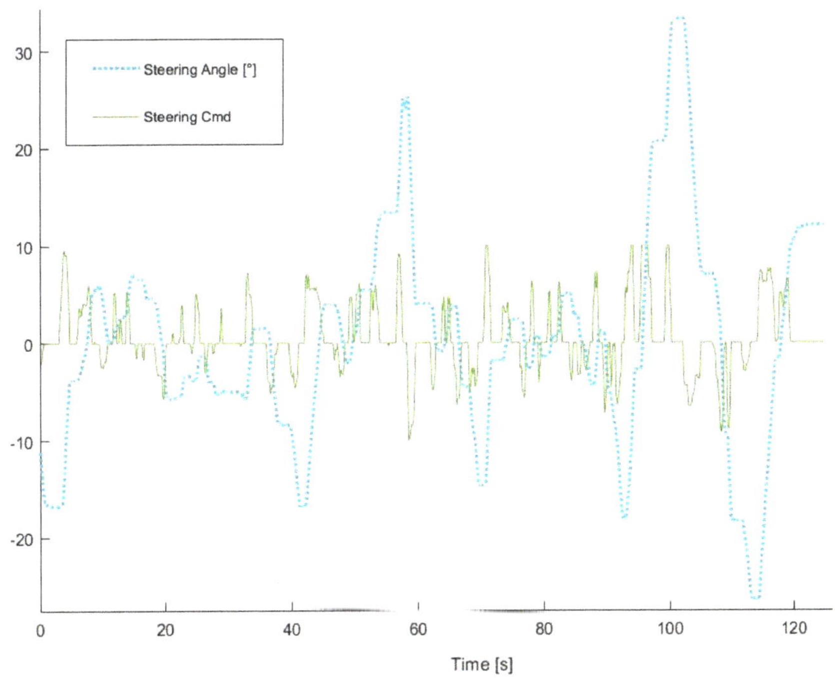

Figure 18. Human operator steering commands and LHD steering angle on dataset M1.

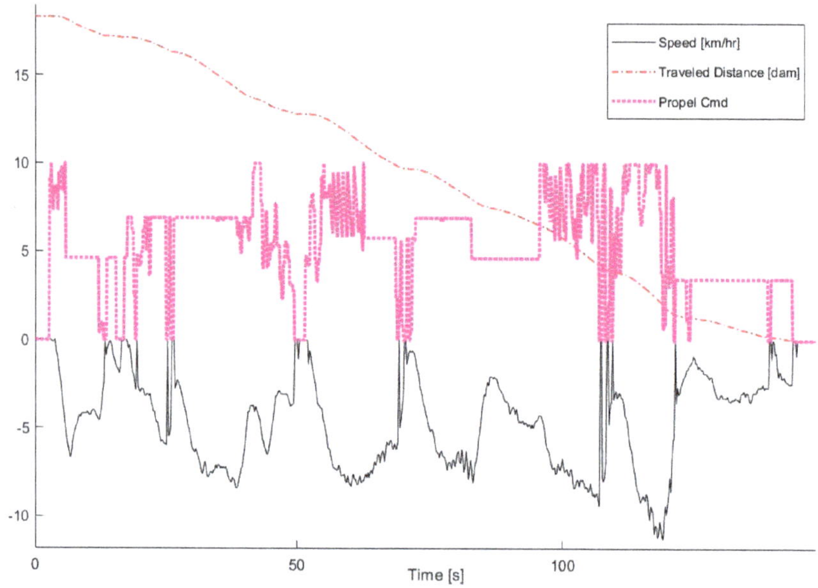

Figure 19. Autonomous System propel commands and LHD speed on dataset 14.

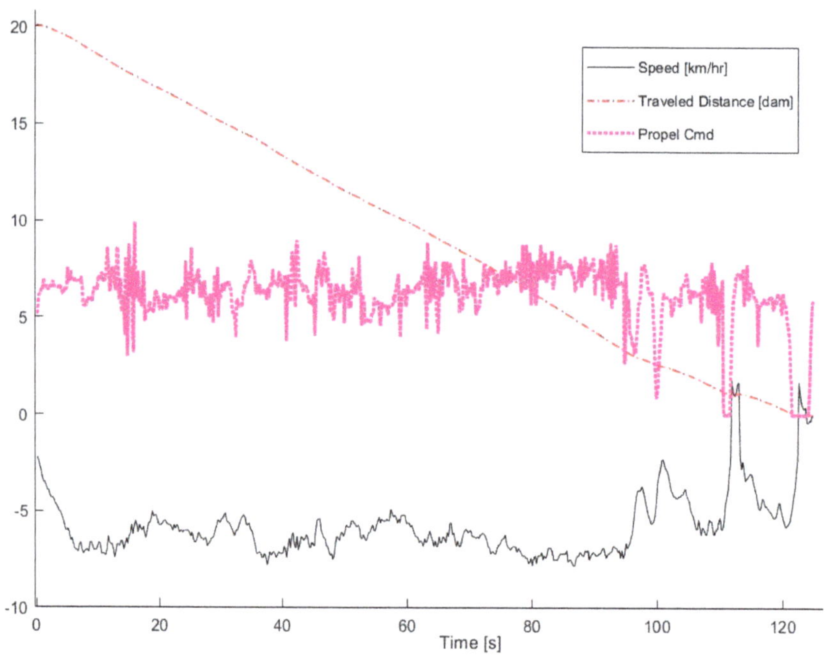

Figure 20. Human operator propel commands and LHD speed on dataset M1.

The average and minimum distances to both tunnel walls are also shown on Table 4. In this regard, the system and the human operator have similar performance, with the human operator preferring to be slightly closer to the left wall (since the cabin is on that side, therefore the operator has better visibility on that side), while the autonomous system is usually closer to the right side of the tunnel.

6. Conclusions

The proposed topological navigation and localization system for LHDs was developed and tested in simulation, field trials, and finally, in a production tunnel of a copper, underground, sublevel stoping mine. Using this system, the LHD was able to navigate safely inside the mine, maintaining a safe distance between the LHD and the tunnel's walls at all times.

Parameterization of the navigation conditions for each individual TM node was crucial for achieving the desired behavior on the underground industrial tests. The software modularization allowed the development of specific software components for tackling the different challenges of the autonomous navigation. The *Navigation Control* module manages the mission requests and the overall navigation behavior. *Deliberative Path Planning* generates local driving trajectories for the *Guidance* module to follow, while avoiding the tunnel walls and obstacles. *Command Executor* maintains a queue of consistent and smooth commands to guarantee short-term operation, while simultaneously maintaining system safety. Finally, global and local localization allows maintaining an estimation of the pose of the LHD inside the mine.

When comparing the automation system with a seasoned human operator, it shows a slightly slower performance (about 10% in terms of average instant speed), which is not that serious when taking into consideration all the safety and operational benefits of the system. Besides being faster, the human operator showed smoother driving and more control of the LHD, but this did not necessarily reflect on the performance of the system, or at least it was not noticeable when supervising the operation. It needs to be considered that the tunnel was very narrow and the system needed to be tuned to drive very near to the walls, at a distance of about 10 [cm], in order to be able to drive through some parts of the tunnel (the LHD manufacturer recommends a minimum distance of 50 [cm] to each side of the tunnel).

One of the major problems during testing on site was the lack of a wireless communication infrastructure with the capabilities of high speed roaming. This caused preemptive stops and/or speed reductions while going through the tunnel, hindering the optimizing process of the system and hurting the overall performance. A video showing the operator's graphic interface while the system is driving the LHD autonomously through the tunnel can be found at https://youtu.be/4Q34N25XjpA (accessed on 14 July 2021).

The system is now being installed and tested in a room and pillar mine in Germany, where a more robust, and better performing, network infrastructure will be used.

Author Contributions: Conceptualization, M.M., I.P.-T., C.T. and J.R.-d.-S.; methodology, M.M., I.P.-T., C.T. and J.R.-d.-S.; software, I.P.-T. and C.T.; validation, M.M., I.P-T. and C.T.; resources, J.R.-d.-S.; data curation, M.M.; writing—original draft preparation, M.M., C.T. and J.R.-d.-S.; writing—review and editing, M.M., I.P.-T., C.T. and J.R.-d.-S.; funding acquisition, J.R.-d.-S. All authors have read and agreed to the published version of the manuscript.

Funding: This research was funded by the Chilean National Research Agency ANID under project grant Basal AFB180004 and FONDECYT 1201170.

Institutional Review Board Statement: Not applicable.

Informed Consent Statement: Not applicable.

Data Availability Statement: The data presented in this study is contained in the article itself.

Acknowledgments: We thank Paul Vallejos for the valuable discussions and support for on-site execution of the experiments, and Felipe Inostroza and Daniel Cárdenas for their valuable help in computing metrics for the results section. We acknowledge Compañía Minera San Gerónimo for providing the mine infrastructure for testing the system, and GHH Chile for supplying the LHD machine needed for this work.

Conflicts of Interest: The authors declare no conflict of interest.

References

1. Salvador, C.; Mascaró, M.; Ruiz-del-Solar, J. Automation of unit and auxiliary operations in block/panel caving: Challenges and opportunities. In Proceedings of the MassMin2020—The 8th International Conference on Mass Mining, Santiago, Chile, 9–11 December 2020.
2. Scheding, S.; Dissanayake, G.; Nebot, E.; Durrant-Whyte, H. Slip modelling and aided inertial navigation of an LHD. In Proceedings of the International Conference on Robotics and Automation, Albuquerque, NM, USA, 25 April 1997; Volume 3 pp. 1904–1909.
3. Madhavan, R.; Dissanayake, M.W.M.G.; Durrant-Whyte, H.F. Autonomous underground navigation of an LHD using a combined ICP-EKF approach. In Proceedings of the IEEE International Conference on Robotics and Automation, Leuven, Belgium 16–20 May 1998; Volume 4, pp. 3703–3708.
4. Scheding, S.; Dissanayake, G.; Nebot, E.M.; Durrant-Whyte, H. An experiment in autonomous navigation of an underground mining vehicle. *IEEE Trans. Robot. Autom.* **1999**, *15*, 85–95. [CrossRef]
5. Poole, R.A.; Golde, P.V.; Baiden, G.R.; Scoble, M. A review of INCO's mining automation efforts in the Sudbury Basin. *CIM Bull* **1998**, *91*, 68–74.
6. Mäkelä, H. Overview of LHD navigation without artificial beacons. *J. Robot. Auton. Syst.* **2001**, *36*, 21–35. [CrossRef]
7. Roberts, J.M.; Duff, E.S.; Corke, P.I.; Sikka, P.; Winstanley, G.J.; Cunningham, J. Autonomous control of underground mining vehicles using reactive navigation. In Proceedings of the 2000 ICRA: Millennium Conference—IEEE International Conference on Robotics and Automation, San Francisco, CA, USA, 24–28 April 2000; Volume 4, pp. 3790–3795.
8. Ridley, P.; Corke, P. Autonomous control of an underground mining vehicle. In Proceedings of the Australian Conference on Robotics and Automation, Sydney, Australia, 14–15 November 2001.
9. Duff, E.S.; Roberts, J.M.; Corke, P.I. Automation of an underground mining vehicle using reactive navigation and opportunistic localization. In Proceedings of the 2003 IEEE/RSJ International Conference on Intelligent Robots and Systems, Las Vegas, NV USA, 27–31 October 2003; Volume 3, pp. 3775–3780.
10. Larsson, J. Reactive Navigation of an Autonomous Vehicle in Underground Mines. Ph.D. Thesis, Örebro University, Örebro Sweden, 2007.
11. Marshall, J.; Barfoot, T.; Larsson, J. Autonomous underground tramming for center-articulated vehicles. *J. Field Robot.* **2008**, *25*, 400–421. [CrossRef]
12. Larsson, J.; Appelgren, J.; Marshall, J. Next generation system for unmanned lhd operation in underground mines. In Proceedings of the Annual Meeting and Exhibition of the Society for Mining, Metallurgy and Exploration—SME, Phoenix, AZ, USA, 28 February–3 March 2010.
13. Gleeson, D. Sandvik to Automate New LHD fleet at Codelco's El Teniente Copper Mine. 2016. Available online: https://im-mining.com/2021/02/16/sandvik-to-automate-new-lhd-fleet-at-codelcos-el-teniente-copper-mine (accessed on 14 July 2021).
14. Schunnesson, H.; Gustafson, A.; Kumar, U. Performance of automated LHD machines: A review. In Proceedings of the International Symposium on Mine Planning and Equipment Selection, Banff, AB, Canada, 16–19 November 2009. Available online: http://urn.kb.se/resolve?urn=urn:nbn:se:ltu:diva-38494 (accessed on 14 July 2021).
15. Walker, S. The drivers of autonomy. *Eng. Min. J.* **2012**, *213*, 52.
16. Paraszczak, J.; Gustafson, A.; Schunnesson, H. Technical and operational aspects of autonomous LHD application in metal mines. *Int. J. Min. Reclam. Environ.* **2015**, *29*, 391–403.
17. Dekker, L.G.; Marshall, J.A.; Larsson, J. Experiments in feedback linearized iterative learning-based path following for center-articulated industrial vehicles. *J. Field Robot.* **2019**, *36*, 955–972. [CrossRef]
18. Jang, H.; Topal, E. Transformation of the Australian mining industry and future prospects. *Min. Technol.* **2020**, *129*, 120–134. [CrossRef]
19. Li, J.G.; Zhan, K. Intelligent mining technology for an underground metal mine based on unmanned equipment. *Engineering* **2018**, *4*, 381–391. [CrossRef]
20. Gao, L.; Ma, F.; Jin, C. A model-based method for estimating the attitude of underground articulated vehicles. *Sensors* **2019**, *19*, 5245. [CrossRef] [PubMed]
21. Bai, G.; Liu, L.; Meng, Y.; Luo, W.; Gu, Q.; Ma, B. Path tracking of mining vehicles based on nonlinear model predictive control. *Appl. Sci.* **2019**, *9*, 1372. [CrossRef]
22. Silver, D.; Ferguson, D.; Morris, A.; Thayer, S. Feature extraction for topological mine maps. In Proceedings of the IEEE/RSJ International Conference on Intelligent Robots and Systems, Sendai, Japan, 28 September–2 October 2004; Volume 1, pp. 773–779.
23. Konolige, K.; Marder-Eppstein, E.; Marthi, B. Navigation in hybrid metric-topological maps. In Proceedings of the IEEE International Conference on Robotics and Automation, Shanghai, China, 9–13 May 2011; pp. 3041–3047.
24. Yoon, Y.; Shin, J.; Kim, H.J.; Park, Y.; Sastry, S. Model-predictive active steering and obstacle avoidance for autonomous ground vehicles. *Control Eng. Pract.* **2009**, *17*, 741–750. [CrossRef]
25. Backman, J.; Oksanen, T.; Visala, A. Navigation system for agricultural machines: Nonlinear model predictive path tracking. *Comput. Electron. Agric.* **2012**, *82*, 32–43. [CrossRef]
26. Ji, J.; Khajepour, A.; Melek, W.W.; Huang, Y. Path planning and tracking for vehicle collision avoidance based on model predictive control with multiconstraints. *IEEE Trans. Veh. Technol.* **2016**, *66*, 952–964. [CrossRef]

27. Ruiz-del-Solar, J.; Vallejos, P.; Correa, M. Robust autonomous navigation for underground vehicles. In Proceedings of the Automining—5th International Congress on Automation in Mining, Antofagasta, Chile, 30 November–2 December 2016.
28. Dijkstra, E. A note on two problems in connexion with graphs. *Numer. Math.* **1959**, *1*, 269–271. [CrossRef]
29. Koenig, N.; Howard, A. Design and use paradigms for gazebo, an open-source multi-robot simulator. In Proceedings of the 2004 IEEE/RSJ International Conference on Intelligent Robots and System, Sendai, Japan, 28 September–2 October 2004; Volume 3, pp. 2149–2154.
30. Quigley, M.; Conley, K.; Gerkey, B.; Faust, J.; Foote, T.; Leibs, J.; Wheeler, R.; Ng, A. ROS: An open-source robot operating system. In Proceedings of the ICRA Workshop on Open Source Software, Kobe, Japan, 12–17 May 2009.

applied sciences

MDPI

Article

Autonomous Loading System for Load-Haul-Dump (LHD) Machines Used in Underground Mining

Carlos Tampier, Mauricio Mascaró and Javier Ruiz-del-Solar *

Advanced Mining Technology Center, Department of Electrical Engineering, Universidad de Chile, Santiago 8370451, Chile; carlos.tampier@amtc.cl (C.T.); mauricio.mascaro@amtc.cl (M.M.)
* Correspondence: jruizd@ing.uchile.cl

Abstract: This paper describes an autonomous loading system for load-haul-dump (LHD) machines used in underground mining. The loading of fragmented rocks from draw points is a complex task due to many factors including: bucket-rock interaction forces that are difficult to model, humidity that increases cohesion forces, and the possible presence of boulders. The proposed system is designed to integrate all the relevant tasks required for ore loading: rock pile identification, LHD positioning in front of the ore pile, charging and excavating into the ore pile, pull back and payload weighing. The system follows the shared autonomy paradigm: given that the loading process may not be completed autonomously in some cases, it takes into account that the machine/agent can detect this situation and ask a human operator for assistance. The most novel component of the proposed autonomous loading system is the excavation algorithm, and the disclosure of the results obtained from its application in a real underground production environment. The excavation method is based on the way that human operators excavate: while excavating, the bucket is tilted intermittently in order to penetrate the material, and the boom of the LHD is lifted on demand to prevent or correct wheel skidding. Wheel skidding is detected with a patented method that uses LIDAR-based odometry and internal measurements of the LHD. While a complete loading system was designed, the validation had to be divided in two stages. One stage included the rock pile identification and positioning, and the other included the charging, excavation, pull back, and weighting processes. The stage concerning the excavation algorithm was validated using full-scale experiments with a real-size LHD in an underground copper mine in the north of Chile, while the stage concerning the rock pile identification was later validated using real data. The tests showed that the excavation algorithm is able to load the material with an average of 90% bucket fill factor using between three and four attempts (professional human operators required between two and three loading attempts in this mine).

Keywords: autonomous loading system; field robotics; mining automation

Citation: Tampier, C.; Mascaró, M.; Ruiz-del-Solar, J. Autonomous Loading System for Load-Haul-Dump (LHD) Machines Used in Underground Mining. *Appl. Sci.* **2021**, *11*, 8718. https://doi.org/10.3390/app11188718

Academic Editors: Luis Gracia and Carlos Perez-Vidal

Received: 12 August 2021
Accepted: 3 September 2021
Published: 18 September 2021

1. Introduction

The automation of mining equipment is an important requirement in the mining industry. This is because mining operations need to increase the safety of the workers, as well as to augment the productivity, efficiency, and predictability of the processes. Safety is, without doubt, a key factor, and has been the top priority of mining companies during the past decades. This is particularly true for underground operations, with hazardous environments where workers are exposed to constant risks of rock falls, rock bursts, and mud rushes, and where the presence of dust in the air can result in a number of associated occupational diseases in workers [1]. All of these hazards have been steadily increasing as mine operations have gone deeper, and geomechanical conditions therefore become more extreme. As a consequence, great effort has been invested in increasing the automation level of underground mining machines, especially those that operate in high-risk areas [1], one of them being load-haul-dump (LHD) machines (also known as scoop trams).

An LHD is a four-wheeled, center-articulated vehicle with a frontal bucket used to load and transport ore on the production levels of an underground mine (see Figure 1). The automation of the transport phase has been covered in many previous publications [2–6] and is currently commercially available. However, although the loading part has also been a subject of scientific research, [7–17], its application in mining environments is not as widely used as other types of automation solutions such as autonomous navigation, assisted tele-operation, or collision avoidance. The slower-paced progress in this area could be due to the complexity of the addressed problem: during excavation, the interaction between the bucket and the material is difficult to model, because the bucket-material interaction forces may vary greatly depending on the properties of the material (e.g., humidity, hardness, fragmentation), the rock pile geometry, and the LHD dynamics (mass, speed, etc.). For this reason, the development of autonomous loading solutions requires full-scale experiments under conditions that can be found only in real mining environments, since physical interactions between the machine and the fragmented rock cannot be easily reproduced elsewhere. This adds to the fact that loading maneuvers and techniques are mostly developed and perfected through the experiences of human operators.

Figure 1. Full-scale LHD machine used during the test of the excavation algorithm.

In this paper, a complete system for the automation of the loading process using LHDs is proposed. This system is designed to integrate all the relevant tasks required for ore loading: rock pile identification, the LHD's positioning in front of the ore pile, charging, excavation, pull back, and payload weighing (see Figure 2). Assuming that in some cases the loading may not be completed autonomously, the system can detect this situation and request the help of a human operator (by tele-operation). Thus, the proposed system falls under the shared autonomy paradigm. In the mining exploitation context, fleets of LHDs are normally supervised by humans, so providing assistance to an autonomous equipment is not an unusual requirement, especially considering that the most common practice is for LHD automation systems to rely on tele-operators to handle the loading task. It must be noticed that providing assistance on demand, i.e., only in the few cases that the autonomous system is not able to load, is much more efficient than the current semi-autonomous operation used in most commercial systems, where navigation is executed autonomously but loading is teleoperated.

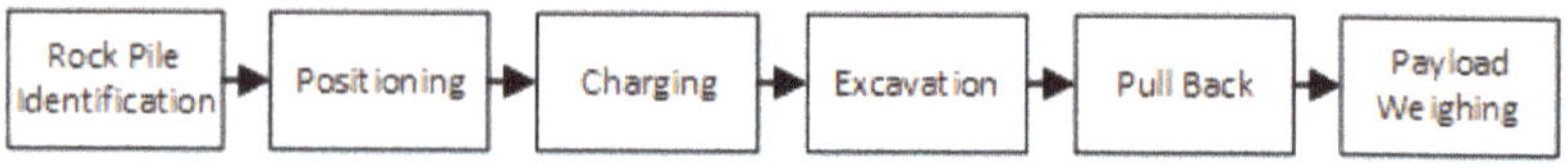

Figure 2. Steps of the proposed autonomous loading system.

The core of the autonomous loading system is the excavation algorithm, which is based on the way that human operators excavate: the bucket is tilted intermittently while excavating in order to penetrate the material, and the boom of the LHD is lifted on demand to prevent or correct wheel skidding. Wheel skidding is detected with a patented method that uses LIDAR-based odometry and internal measurements of the LHD's actuators [18]. The excavation algorithm was validated in an underground mine (a sublevel stoping copper mine located in Chile), using full-scale excavation experiments with a real LHD and a typical production rock pile. The validation process in a productive area of a real mine and the lessons learned are fully disclosed here. The 2D-pile modeling was validated afterwards with data from these tests.

It is also important to mention that commercial solutions to the problems of autonomous loading need to be compatible with the 24 × 7 operation of mines, where production throughput is one of the main requirements. That means that the LHD needs to be able to load without stopping during the transition from navigation to loading. The proposed automation framework considers this requirement, and it is able to characterize the rock pile, without stopping the LHD, while it approaches the pile.

The main contributions of this paper are:

- A system that models and implements the whole loading process required in underground mining operations, from rock pile identification to payload weighing. The system is based on the shared autonomy paradigm, which allows the system to obtain assistance on demand from a human operator.
- An excavation algorithm that is based on the way that human operators excavate, and that uses a patented method for detecting wheel skidding, which is required for successful excavation of fragmented material in draw points.
- Full-scale rock excavation experiments using a commercial LHD in a production level of a sublevel stoping mine.

This paper is organized as follows: Section 2 presents the background and related work on loading automation for LHDs. Section 3 describes the proposed autonomous loading system. In Section 4, results of the full-scale experiments are shown and discussed. Finally, in Section 5, the main conclusions of this work are drawn.

2. Background and Related Work

2.1. Problem Description

Autonomous loading of fragmented rock involves a sequence of steps: As the LHD's operator approaches the draw point, a swift assessment of the condition of the rock pile is made. The bucket of the LHD is tilted downwards until the tip is pressed against the ground. Next, the machine is commanded to charge at the rock pile. After the LHD makes contact, it buries the bucket in the rock pile, and bucket tilt and lift movements are issued as it advances through the fragmented rock. When sufficient penetration is achieved, the bucket is fully retracted, and the vehicle is withdrawn with a reverse motion. Then, the operator shakes the bucket briefly and assesses whether or not enough ore has been loaded. If necessary, a new loading maneuver is carried out.

The step in which the machine is being controlled to dig into the rock pile will herein be referred to as the excavation process. During excavation, bucket-rock interaction forces affect the bucket motion through the material, and ultimately define the amount of ore loaded. Granular material inter-particle forces, as well as bucket-rock interactions, have been studied previously, but their complexity has led only to stochastic modeling [19]. In consequence, an analytical approach to bucket motion control is not possible, and accurately replicating these interactions to develop a system in simulation is highly unlikely. Hence, only a real environment offers suitable conditions for development of, and experimentation with, an excavation procedure. Furthermore, fragmented rock, unlike piled sand or gravel, is much more difficult to load. This is aggravated due to the environmental conditions in underground mining. Depending on the mining method, the column of blasted rock might exert large compression forces on the open face of the draw point (see Figure 3). Humidity can also increase cohesion forces, making the overall process more difficult [20]. Moreover, large rocks can be present in the draw point. For this reason, the operator needs to detect the presence of large rocks while approaching the draw point, and, depending on their size, could be forced to change the goal of the maneuver from filling the bucket to just loading one large rock. If the rock is too big to be loaded, the task is aborted. Furthermore, other conditions, such as narrow tunnels, uneven or unprepared terrain, and/or significant amounts of water and mud near the draw point, can make the loading process very challenging. An autonomous system should be prepared to handle all these situations in addition to performing certain processing and steps that a human operator deals with while driving the LHD, such as detecting an accurate enough position of the rock pile before charging at it, making sure not to hit the tunnel while pulling back from the rock, and estimating if the bucket is full enough.

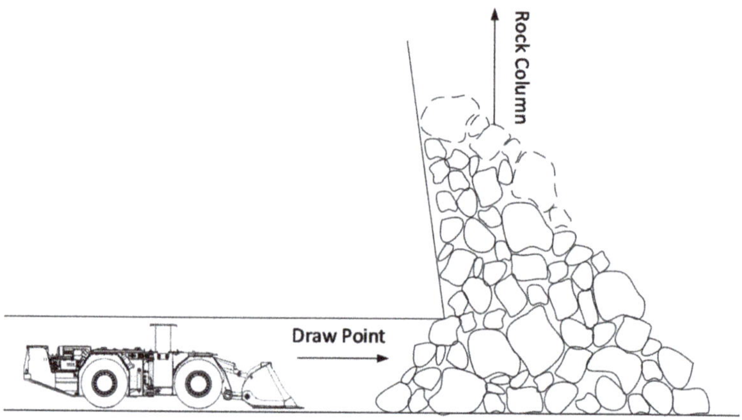

Figure 3. Sublevel stoping draw point.

2.2. How Human Operators Excavate

The excavation process is the core of the loading maneuver; it is the crucial step and, since bucket-rock interactions cannot be predicted in advance, there is no optimal method for performing this operation. Nevertheless, human operators have had to deal with this situation by learning the loading technique from experience and shared good practices. It is natural, therefore, to analyze which ideas can be borrowed from them. Through several interviews with LHD operators of Chilean block caving and sublevel stoping mining companies, and discussions and analysis of data from mining operation manuals, some common points in their practices were identified:

- Before engaging with the rock pile, the bucket must be fully extended.
- Forward motion is selected for maximum traction (first gear).
- Intermittent tilt commands are the basis of the excavation process.
- Lift commands are mainly issued to prevent or correct wheel skidding.

Figure 4 shows a graphical representation of an excavation maneuver. First, the LHD engages with the rock pile, then the bucket is tilted in order to excavate on the muck pile. When the wheels lose traction, the boom is lifted to regain traction. After filling the bucket, the LHD pulls back. Figure 5 shows a graphic of some of the relevant variables during a line-of-sight remote-controlled loading maneuver. As mentioned, bucket tilting ("tilt command") is activated intermittently. Its duration and shape vary among operators and loading attempts. Techniques also using negative tilt commands (downwards tilting of the bucket) were also encountered in combination with positive lift commands ("lift command"). The common factor in all of them was the intermittency of the commands. As for the lift command, its use was seen to be highly variable, which correlates with the notion of using it mainly to prevent wheel skidding. In the example of Figure 5, the lift command was not used at all. The "pedal command" is the one used for machine acceleration and thus, in turn, engine revolutions per minute (RPM). "Tilt angle" and "lift angle" are the bucket tilt and boom lift angle values, estimated through the encoders installed in the machine. "Transmission pressure" is the hydrostatic transmission pressure of the machine, a variable that can be associated with the engine's power output. "Skidding factor" is an estimated skidding factor that is explained in Section 3.3.

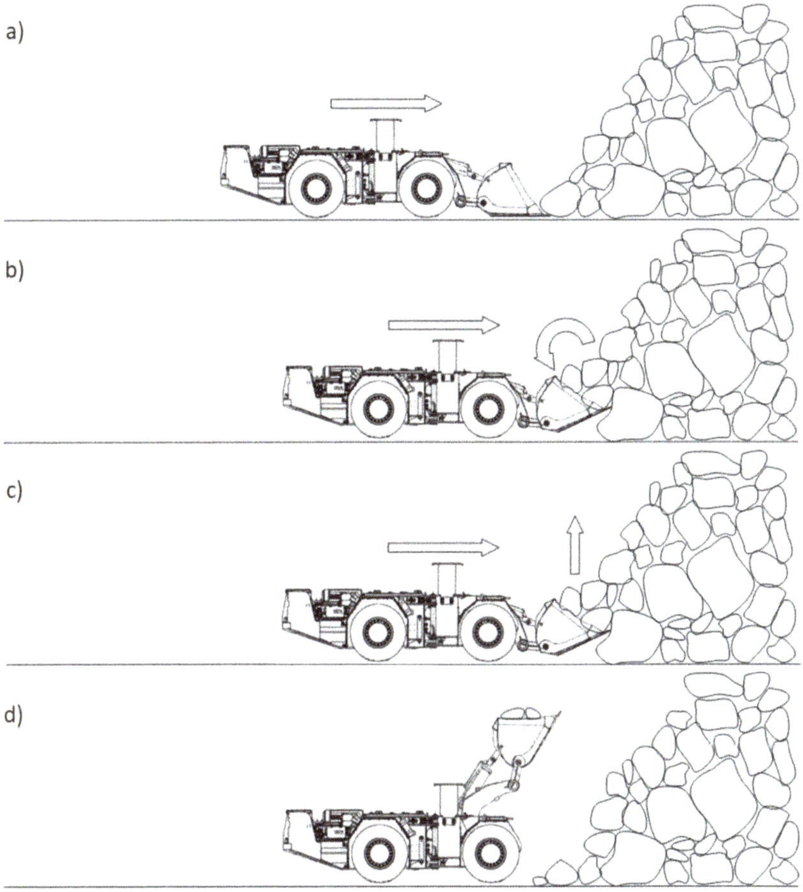

Figure 4. Excavation method diagram. (**a**) LHD engages with the rock pile. (**b**) Bucket tilt used to excavate. (**c**) Boom lift used to control wheel traction. (**d**) End of the excavation phase.

2.3. Related Work

Mikrived analyzed the interaction forces in excavation processes and proposed that measurements of the resistive forces could be used for controlling them [21]. Hemami studied the LHD loading process and dealt with theoretical formulations on bucket trajectory and resistive forces during a scooping motion [7,8,10]. Other work has addressed the loading problem using laboratory-scale prototypes, under less realistic conditions [9,11,12,22]. Research on autonomous loading showed scarce progress up until the past decade. In [13], full-scale experiments with a real LHD were performed using a proposed admittance-based controller for bucket commands. This work was further extended in [15–17] with support from the LHD original equipment manufacturer (OEM) Atlas Copco (now Epiroc), a study that led to a patented method [23]. It should be noted that this method addresses only the problem of ore excavating once the bucket is inside the rock pile. On the other hand, our approach comprehends the complete loading process, while also benefiting from human experience.

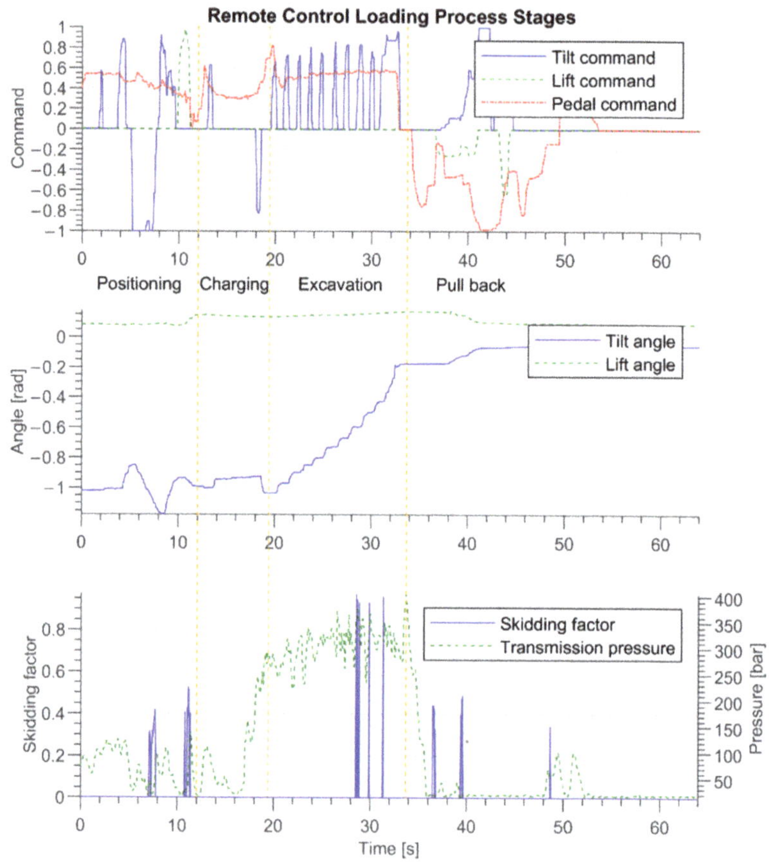

Figure 5. Relevant variables during the loading maneuver: Lift/Tilt command: manual bucket lifting/tilting command. Pedal command: manual machine acceleration command. Transmission Pressure: hydrostatic transmission pressure of the machine. Tilt Angle: tilt angle of the bucket. Lift Angle: lift angle of the boom. Skidding factor: estimated skidding factor. See the definitions of these variables in Sections 3.3 and 3.4.

Caterpillar, another LHD OEM, is known to have been working on technical solutions for the autonomous loading problem, as demonstrated by patents [24–26]. While their patents show that they have tackled different issues of the autonomous loading problem, a general system has not been described, and their research has not been made available to the scientific community. However, it is uncertain if either of these solutions are able to handle the loading problem in block/panel caving draw points, which is significantly harder than loading from an ore pile or from draw points originated by other mining methods. As mentioned, loading from draw points is difficult due to many factors including: bucket-rock interaction forces that are difficult to model, humidity that increases cohesion forces, and the possible presence of boulders.

According to the public information shared on the web sites of LHD OEMs, only Epiroc, Sandvik, and RCT Global have commercially available products that focus on the autonomous loading task. However, specific features, limitations, and applicability for block/panel caving or sublevel stoping underground mining are not publicly disclosed, and therefore no comments can be made on them. At least in Chile, where all major mining companies operate and where one third of the world's copper is produced, no autonomous loading system for LHD is in use.

Work associated with standard front-end loaders was not considered in this analysis, because they operate on the surface, where the difficulties associated with underground draw points are not present. In addition, these types of vehicles usually deal with material that has already been processed by milling equipment.

Our previous work on the topic of autonomous loading with LHDs was focused on laboratory-scale trials [27]. While useful for an initial approach to an experimental setup, laboratory experiments are limited due to the large deviation between real-world operating conditions and data from a small-scale and controlled environment. Other laboratory work is presented by Shi [22], which has some similarities with the here-presented method, as excavation is modeled after the experience of human operators. However, since their experiments are carried out using a robotic arm, it is difficult to apply those findings to a four-wheeled machine such as an LHD.

As can be seen, little work has been published on the particular case of autonomous loading with LHD machines in real underground mining environments, and, as far as we know, no previous method has validated an excavation technique using human operator experiential knowledge under real underground mining conditions.

3. Proposed Autonomous Loading System

3.1. Methodology Overview

The hardware used in this work included a full-scale LHD machine (a GHH LF11H with 11-ton tramming capacity and hydraulic powertrain) retrofitted with two 2D laser scanners (SICK LMS511), encoders in all joints (IFM RN7012), pressure sensors along the bucket's tilt and the boom's lift hydraulic lines (manufacturer specific, 600 bar), and an onboard industrial computer (Advantech ARK-2151-59AIE) (see Figure 1). All components are of industrial grade, suitable for the rigorous environment of underground mining, and readily available as off-the-shelf components used widely in a broad range of commercial applications. In order to better integrate the system with commercial autonomous navigation solutions for LHDs, which only use 2D scanners, 3D laser scanners were not used in the current version of the proposed system. The software implementation was performed using ROS as the central middleware [28], CABSL library [29] for state-machine codification, and ROS actions to interface with the main processing function in each state. The system was initially developed and tested in a simulated environment using Gazebo [30].

An autonomous loading routine can be thought of as a sequence of several steps that perform specific actions, as presented in the diagram of Figure 2. There, rock pile identification is the process of finding the draw point's location and status (whether it is suited for loading or not). Positioning refers to orienting the LHD machine to ensure that a forward thrust will end in a collision with the rock pile. Charging is the step that considers lowering the bucket and accelerating towards the pile. Excavation is the machine control that performs the digging action. Pull back comprises the backward motion of the vehicle and bucket shake. Finally, Payload weighing estimates the amount of ore loaded. Most of the mentioned steps could fail due to the highly variable environment in underground mining. It is possible that the onboard sensors fail to find the rock pile accurately or that the machine positioning cannot avoid colliding with a tunnel wall. The collision between the machine and the rock pile might not be detected, and the LHD's bucket could become stuck during excavation. In any of these cases, human assistance would be required to overcome the problematic situation. In fact, human interaction with automated mobile machines for mining operations is needed in the state-of-the-art systems, at the very least to supervise a fleet of machines. Since large-scale operations have a fleet of vehicles, it is common for one or more operators to supervise and assist multiple semi-autonomous machinery, as each machine does not require constant attention. For these reasons, in our view, it is imperative that this human interaction is considered at the design stage of the autonomous loading system and, thus, included in the formulation of the agent's behavior. Figure 6 shows the proposed state machine for the autonomous loading process with the human in the loop. There are two assistance states: positioning assistance allows the operator to take control of the machine to position it in front of the rock pile, while loading assistance provides the operator with full control of the LHD to perform the remaining parts of the loading maneuver up until payload weighing, which is needed to record the performance of the task. The positioning assistance is called either when the rock pile identification cannot characterize the rock or when positioning is not able to place the LHD to start the charging procedure, while loading assistance is called when the charging, the excavation, or the pull back procedures cannot be successfully completed.

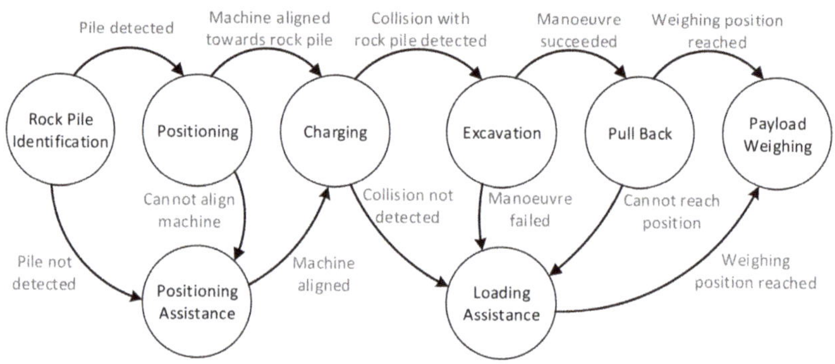

Figure 6. Finite state machine formulation for the autonomous loading with human operators in-the-loop.

3.2. Environment Modeling and Rock Pile Identification

Of the six steps of the autonomous loading routine, the first one, rock pile identification, is the one that most strongly relies on the processing of data acquired from the environment. The aim of this step is to find the location and status of the draw point accurately, in real time, and without the need of extra steps or delays. This means that the LHD is required to continue moving while scanning the pile. The characterization of the rock pile is performed through integration of consecutive laser scans from the frontal 2D LIDAR sensor mounted on the LHD. The integration of the laser scans relies on a LIDAR-based odometry method [31], and the output of the machine's inclinometer data, which is filtered and integrated in the local coordinate frame in order to benefit from the natural tilting of the LHD while moving across uneven terrain. The aforementioned odometry method consists of an adaptive frame rate modification of the range-flow based odometry (RF2O) [32]. While RF2O computes displacement by comparing two scans, using the "range flow constraint equation", the adaptive frame rate implements a control over the processing frequency, increasing it at higher machine speeds.

A block diagram for the pile identification step is presented in Figure 7. The LIDAR odometry enables the integration of 2D laser scans into a 3D point cloud of the environment. Then, the rock pile is segmented, and its features are computed (center, limits, and slope).

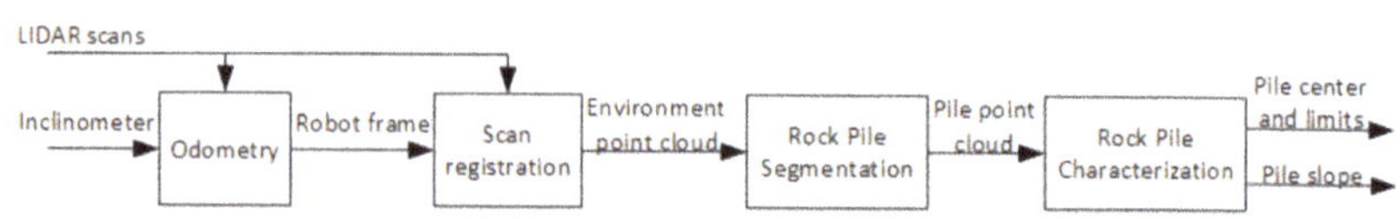

Figure 7. Rock pile identification block diagram. LIDAR odometry and machine's inclinometer data are used to register 2D laser scans into a 3D model of the environment. A segmentation of the rock pile is then carried out, and its center, limits, and slope are extracted.

A conceptual diagram of the complete pile identification process is shown in Figure 8. The succession of images (a, b, and c) illustrates how the integration is carried out. For each new scan (colored circles), the data is registered in an accumulated point cloud, using the computed odometry. It is important to note that the whole process must be executed in real time without stopping, or even slowing down, the LHD. Moreover, the fact that the LIDAR reaches the bottom of the pile from a long distance away, because it is tilted down 2.5 degrees, results in the point cloud having very low density, all of which increases the complexity of the problem.

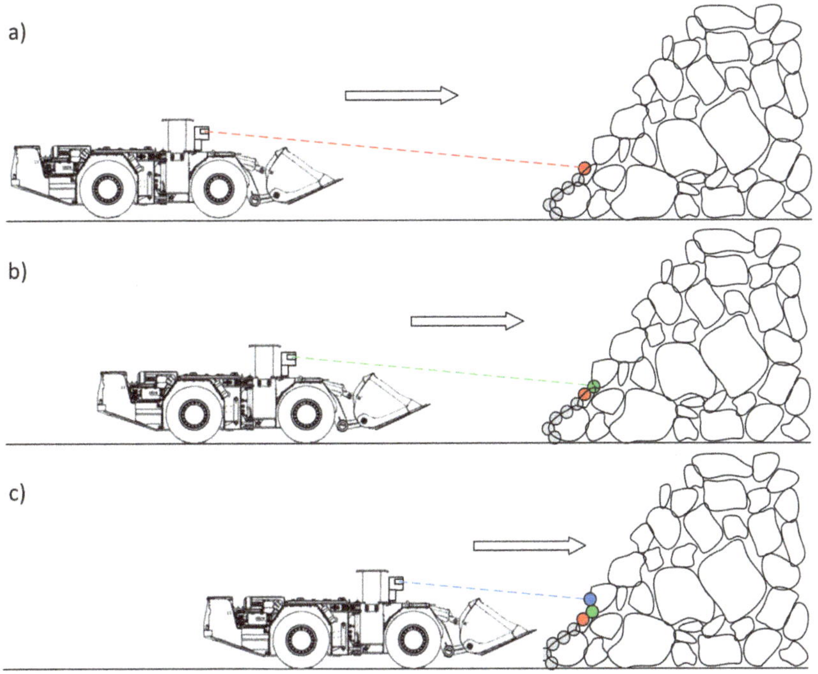

Figure 8. Rock pile reconstruction process through laser-scan integration. As the machine approaches the pile (images (**a**–**c**)), the new laser scans are merged with the previous ones using the computed odometry. A slightly tilted LIDAR is used for this purpose.

Once enough scans have been integrated, the rock pile is segmented using the normal vectors of the surface, a curvature filter, and neighbors count in a small radius. The normal vectors provide information about the points belonging to the walls or to the pile (normal ny is negative on the left wall, near zero on the pile, and positive on the right wall). The curvature provides a rotation-invariant description of the points, which can be used for further segmenting the pile. Finally, neighbor counts provide information valuable for rejecting outliers (points that are isolated from all others). Once a pile is segmented successfully, its mean slope is estimated and is used to classify the status of the pile: a too shallow slope indicates there is not enough ore to perform a loading, and a too steep slope signals the presence of a tunnel wall instead of a rock pile, or even a rock pile in a "hang-up" state, a condition that can arise in certain underground mining methods, such as block caving, in which the fragmented rock stops flowing, generating an unstable and hazardous condition. The determination of the status of the pile is performed by a simple thresholding of its slope.

For clarity purposes, the described algorithm for segmenting and characterizing the pile is detailed in Figure 9.

```
Start with an empty point cloud

while distance to (cx, cy, cz) >15 and pile point cloud has less than 100 points do
Read new scan
Apply a self-filter to remove scan points that correspond to the LHD body
Transform the captured laser scan into the local_map reference frame
Aggregate the points to the integrated ones
Apply a voxel grid (with cell size 0.2 m in XY, 0.1 m in Z)
if the resulting point cloud is not empty,
Compute normals (nx, ny, nz) per point, using a radius 0.75 meters
Remove points with normal values ny < −0.3 or ny > 0.3
Remove points with curvature < 0.05 or curvature > 0.4
Remove points which have less than 80 neighbors inside a radius of 0.75 m
Compute the gravity center of the pile: (cx, cy, cz)
Compute left/right bounds for the pile: min_y, max_y
end if
end while

Compute the final normals (nx, ny, nz) using points within a radius of 3 m
Compute the pile slope from the average of the normals
Set the pile characterization procedure as succeeded
```

Figure 9. Pseudo code of pile segmentation and characterization algorithm.

Besides the pile status, the locations of the horizontal bounds are needed for the next step of the loading routine, the machine positioning. The left (E_L) and right (E_R) limits of the rock pile are required to be able to direct the machine without colliding with the tunnel walls. Using the information available from the pile segmentation algorithm, the limits are estimated as $E_L = max\_y$, and $E_R = min\_y$ (see definitions in Figure 9).

The normal vectors of the points provide valuable information for segmenting the pile, as shown in Figure 10. In this figure, low values of the normal component *ny* are colored in blue, while high values are colored in red.

Oversized rock detection can also be performed in this identification step; however, the description of the functionality is outside the scope of this paper (see, for example, the detection system described in [33]).

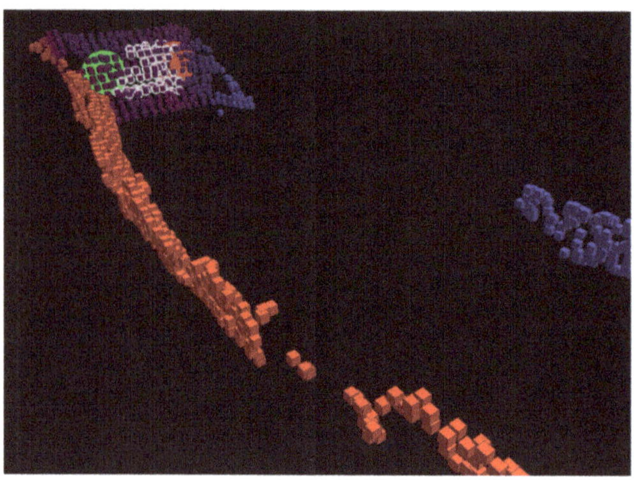

Figure 10. Point cloud showing pile reconstruction (point color corresponds to the *ny* component of the normal).

If the identification of the rock pile has not finished by the time the LHD is closer than a certain predetermined distance to the pile (in the experiments, a threshold of 15 m was used), a transition to the positioning assistance state is forced in the behavior state machine from Figure 6. In the experiments, a threshold of 15 m was used as the predetermined distance. This value was deemed sufficient for the LHD to successfully achieve the following positioning in front of the rock pile. The overall loading routine is, nevertheless, not sensitive to small variations of the selected parameter, and its value should be selected based on the particular conditions of the operating environment.

3.3. Positioning and Charging

After the draw point location and status have been determined, the loading routine continues with the positioning of the machine. Here, the LHD is continued to be driven forward using a "guidance" algorithm [34] until the machine's reference point (in this case its central articulation joint) is at a set distance from the center of the pile (computed as the center of gravity of the segmented points). Once this distance is met, small corrections to the steering angle are issued to ensure that the projected path lands within the limits of the rock pile. Finally, the bucket is lowered until a tight contact with the ground is achieved. A block diagram of this process is shown in Figure 11.

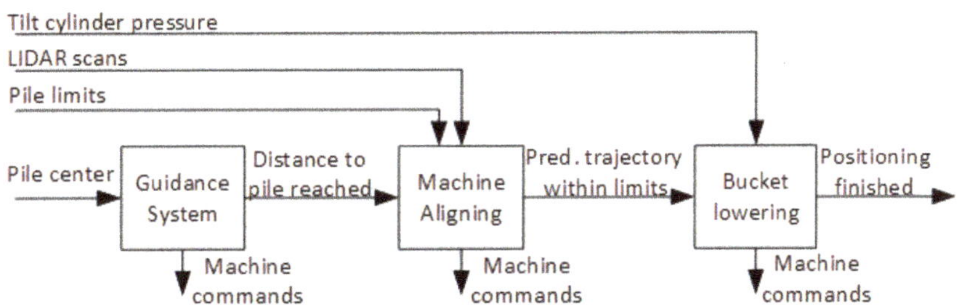

Figure 11. Positioning block diagram. The machine is driven by a guidance system until a predefined distance to the rock pile is achieved. The steering angle of the LHD is then adjusted to ensure collision with the pile, and the bucket is lowered.

The projected path for the machine aligning procedure (see Figure 12) is computed using the LHD's kinematic equations, reported in several previous publications, such as [35]:

$$\partial[x, y, \theta]/\partial t = \left[v\ \cos(\theta), v\ \sin(\theta), (v\ \sin(\gamma) + L_r\omega)/\left(L_f \cos(\gamma) + L_r\right)\right] \tag{1}$$

$$\partial\gamma/\partial t = \omega \tag{2}$$

where $[x, y, \theta]$ is the pose of the LHD, v is the linear speed, γ is the steering angle of the LHD's pivot, $\omega = \dot{\gamma}$ is the steering speed of the LHD's pivot, L_r is the length from the LHD's pivot to the rear wheel axis, and L_f is the length from the LHD's pivot to the front wheel axis.

Since the next step of the autonomous loading routine, the charging against the rock pile, does not involve steering commands, a fixed steering angle (constant γ, $\omega = 0$) can be considered, which simplifies the model:

$$\partial[x, y, \theta]/\partial t = [v\ \cos(\theta), v\ \sin(\theta), v \cdot K] \tag{3}$$

where K is a constant value that depends on the LHD dimensions and the fixed steering angle. Equation (3) shows that an LHD charging against a rock pile will follow a circular trajectory, characterized by a center (C) and a curvature radius (R). Because the *rock pile identification* step computed the left (E_L) and right (E_R) pile edges, the problem of determining if the LHD will collide with the rock pile simplifies to checking if the line between E_L and E_R intersects with both the circular path of the left and right sides of the LHD. Figure 12 shows a diagram of the described scenario.

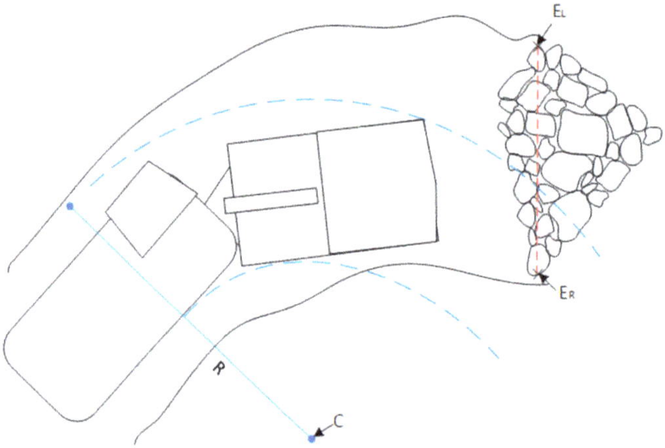

Figure 12. Intersection of the predicted trajectory with the detected pile limits. Left and right pile limit points denoted as E_L and E_R, respectively (with the face of the pile between them in a red segmented line). LHD's left side circular trajectory center and curvature radius denoted with C and R, respectively. In a blue dotted line both sides of the predicted LHD's trajectory is presented.

The solution for the intersection between a circle and a line segment is well known in the common literature and it involves computing the following discriminant value:

$$\Delta = \left(2\,\vec{E}\cdot\vec{EC}\right)^2 - 4\left|\vec{E}\right|^2\left(\left|\vec{EC}\right|^2 - R^2\right) \tag{4}$$

where $\vec{E}$ is the vector going from E_R to E_L and $\vec{EC}$ is the vector going from E_R to C. If Δ results in a negative value, then the infinite line defined by the points E_R and E_L never intersects the given circular path. On the other hand, if Δ is positive, a solution exists. Let S_1 and S_2 be solutions to the intersection between the infinite line defined by E_R and E_L and the circle of center C and curvature radius R:

$$\begin{aligned} S_1 &= \frac{-2\,\vec{E}\cdot\vec{EC} - \sqrt{\Delta}}{2\left|\vec{E}\right|} \\ S_2 &= \frac{-2\,\vec{E}\cdot\vec{EC} + \sqrt{\Delta}}{2\left|\vec{E}\right|} \end{aligned} \tag{5}$$

then, if $0 < S_1 < 1$ or $0 < S_2 < 1$, the circular path intersects the segment $\overline{E_L E_R}$. Otherwise, if $\min(|S_1|, |S_2|) < \min(|S_1 - 1|, |S_2 - 1|)$, the circular path passes to the right of the segment $\overline{E_L E_R}$. If none of the previous conditions is met, then the path passes to the left of segment $\overline{E_L E_R}$.

The above process is executed for the circular paths defined by the left and right side of the LHD. If one of them does not intersect the pile, the steering angle is adjusted in the appropriate direction until both paths land within the rock pile limits. During this steering adjustment, a simple collision check is computed periodically over a bounding box for each body of the LHD. Figure 13 shows a diagram of the collision check for the front body of the LHD. A bounding box is defined for the front body. Then, vectors going from the bottom left corner to the top left corner ($\vec{S_f}$) and to the bottom right corner ($\vec{B_f}$) are defined.

In order to check if point P of the detected tunnel wall is inside the bounding box (and hence causing a collision), the following condition is evaluated:

$$0 < \vec{S_f} \cdot \vec{d_f} < \left|S_f\right|^2 \wedge 0 < \vec{B_f} \cdot \vec{d_f} < \left|B_f\right|^2 \quad (6)$$

where $\vec{d_f}$ is the vector going from the bottom left corner of the bounding box to point P. If the condition is true, then point P is inside the collision bounding box.

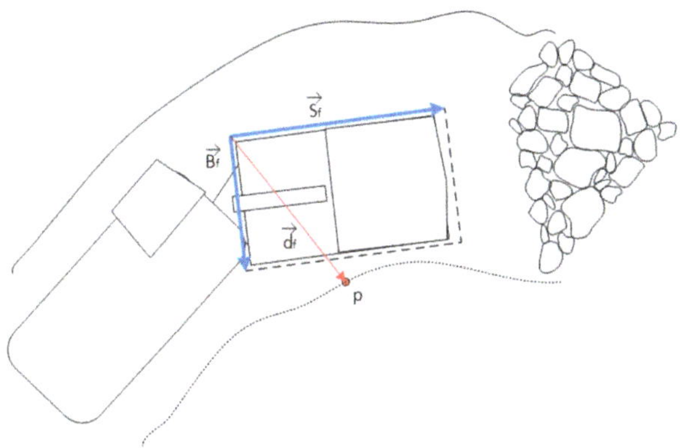

Figure 13. Simple collision check between the front body of the LHD and point P of the detected tunnel wall. Vectors Sf and Bf (in blue) of the collision bounding box are used alongside vector df (in red) to determine if point P is inside the bounding box.

Once the machine's steering angle has been adjusted so that its projected path lands within the rock pile limits, the bucket of the LHD is lowered using the tilt hydraulic cylinder. The tilt cylinder is commanded to extend until the pressure in its hydraulic line rises over a predefined threshold, signaling a tight contact between the bucket tip and the ground. An example from full-scale experimental data is shown in Figure 14, where a threshold of 23 bars was selected.

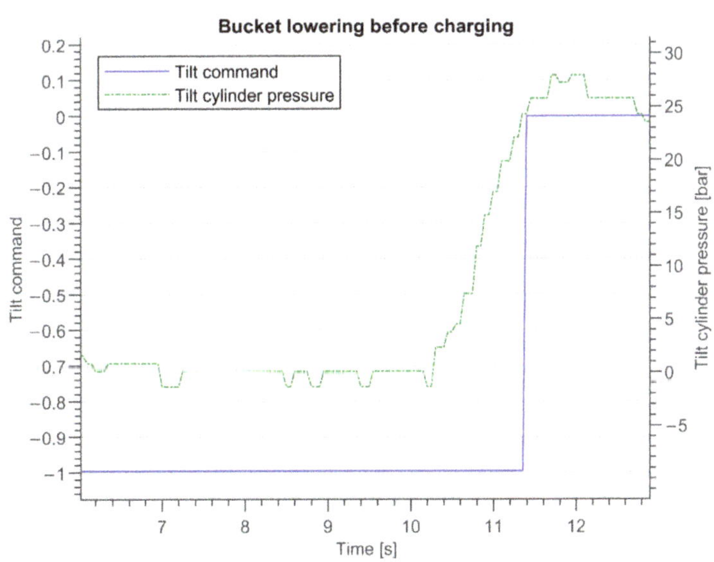

Figure 14. Cylinder pressure increase as the bucket is tightly pressed against the ground in preparation for the charging maneuver.

If the above positioning process is not completed within a predefined time limit (in the experiments, a threshold of 30 s was used), an operator assistance request is generated, effectively putting the system in the positioning assistance state from Figure 6. Otherwise, once the vehicle is in position and with the bucket down, it is commanded to drive forward in first gear. The transition to the excavation phase occurs as the collision with the rock pile is detected by a patented method that analyzes the machine's transmission pressure (engine load) and a skidding factor, estimated from the LIDAR odometry and the machine's tachometer data [18]. A block diagram is shown in Figure 15.

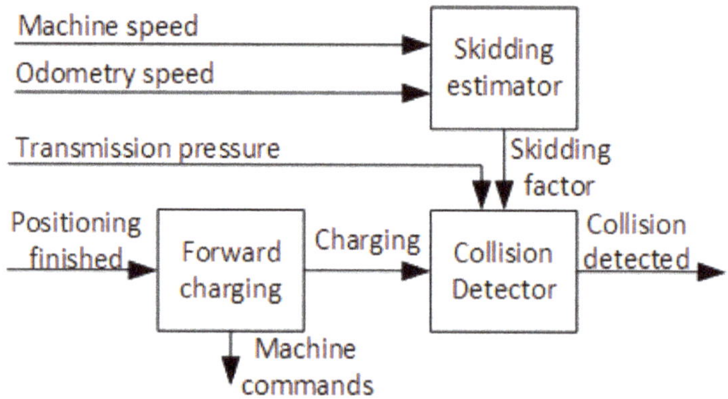

Figure 15. Charging block diagram. The machine is driven forward while a collision detection routine analyzes the machine's transmission pressure and an estimated skidding factor.

The skidding estimator from Figure 15 computes the skidding factor with the following formula:

$$skid = (v_T - v_L)/v_{Mm} \tag{7}$$

where *skid* is the skidding factor; v_T is the speed measured from the machine's tachometer; v_L is the speed estimated with the LIDAR-based odometry method mentioned in Section 3.2., and v_{Mm} is a normalizing factor, computed as the maximum between v_T and v_L in a fixed time window.

Simultaneously, the collision detector module monitors the pressure of the hydrostatic transmission system (for the used LF11H LHD, this variable correlates with the engine's power output). If either the skidding factor or the transmission pressure rises above a predefined threshold, a collision detection event is flagged. Figure 16 shows a plot of the skid factor and the transmission pressure signals, alongside the tilt command, for a teleoperated loading maneuver. The collision with the rock pile happens just before the human operator activates the tilt command. As mentioned above, a significant rise in the transmission pressure, and of the skidding factor, can be seen at that point. Extensive analysis of experimental data, such as shown in Figure 16, led to transmission pressure threshold to be set at 300 bar, and the skidding factor threshold at 0.5.

If a collision detection is not generated within a predefined time limit (in the experiments, a threshold of 10 s was used), an operator assistance request is issued, forcing the system into loading assistance state from Figure 6.

3.4. Excavation Algorithm

The excavation algorithm is based on the techniques used by experienced human operators from block caving, panel caving and sublevel stoping operations (which are the most common type of underground operations in Chile). From a control theory point of view, the proposed method acts as a traction controller during the excavation process. A diagram is shown in Figure 17. The pregenerated commands is simply a module in charge of generating precomputed commands for the pedal and tilt signals. These commands are similar to the commands observed during manual operation. Specifically, the acceleration pedal is maintained at a constant output value and the bucket tilt is intermittently activated in the form of a ramp function. A ramp function was only selected in order to have an extra parameter, although a parameter sensibility analysis could not be performed in the full-scale experiments due to time constrains. The traction controller acts on the lift command in order to suppress wheel skidding, a variable that is estimated from Equation (7), as explained in Section 3.3. In practice, a simple "on/off" controller was implemented, where the lift command was set at the maximum value if the estimated skidding factor raised over a predefined threshold. This choice was made to keep the methods simple in favor of achieving a validation of the complete system.

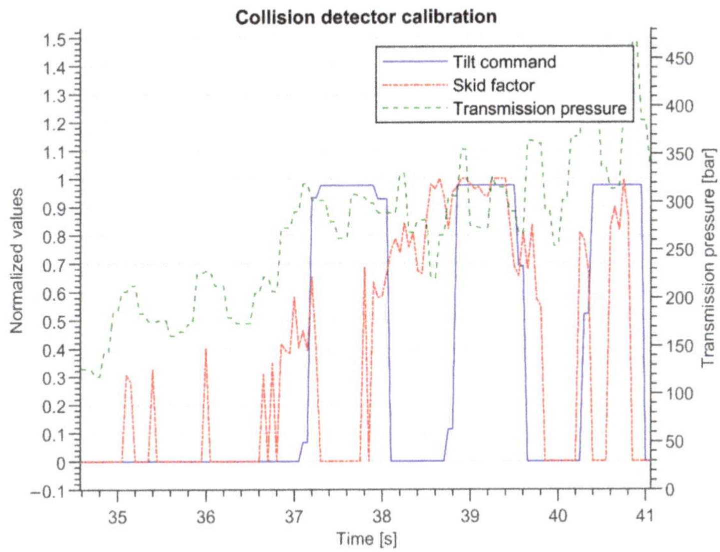

Figure 16. Detection of the collision with the rock pile. Through analysis of the manual operation data, the transmission pressure signal and the estimated skidding factor were selected and used in a threshold detector.

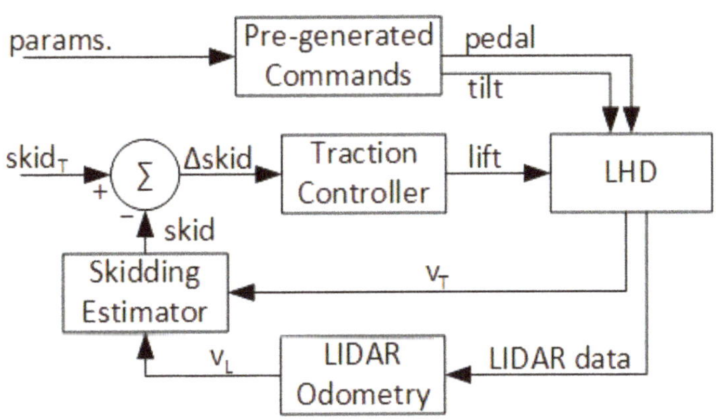

Figure 17. Control diagram of the excavation method. Tilt and pedal commands are predetermined from the human operator control method. The lift command is adjusted depending on the detected machine skidding, which is in turn estimated from the LIDAR odometry and machine tachometer data.

The implementation of the excavation method was carried out using a two-state machine. In the first state, called S_{tilt}, a fixed acceleration and a ramp-shaped bucket tilt command are selected. In the second state, called S_{push}, the fixed acceleration command persists, and wheel skidding is analyzed. If the wheel-skidding factor rises above a predefined threshold, the boom lift command of the machine is activated in order to compensate for it. A machine-specific option, named "full RPM", is also activated during this state. This option forces the engine to maximize its power output, providing additional penetration force. Formally, the following equations apply for the machine commands:

$$\begin{aligned} pedal &= P_0 \\ tilt &= \begin{cases} R_0(t - t_{tilt}) & if\ state = S_{tilt} \wedge b(t - t_{tilt}) \leq 1 \\ 1 & if\ state = S_{tilt} \wedge b(t - t_{tilt}) > 1 \\ 0 & if\ state = S_{push} \end{cases} \\ lift &= \begin{cases} 0 & if\ state = S_{tilt} \vee skid < skid_{max} \\ 1 & if\ state = S_{push} \wedge skid \geq skid_{max} \end{cases} \\ fullRpm &= \begin{cases} 0 & if\ state = S_{tilt} \\ 1 & if\ state = S_{push} \end{cases} \end{aligned} \tag{8}$$

where P_0, R_0, and $skid_{max}$ are predefined constant values; t_{tilt} is the start time of state S_{tilt}, and *skid* is the skidding factor. As for the output commands, *pedal* is the signal sent to the machine's internal control unit that regulates acceleration (similar to the signal sent when pressing the physical pedal in the machine's cabin), *tilt* is the signal sent to retract (positive values) or extend (negative values) the cylinder that controls the movement of the bucket of the LHD machine, and $lift$ is the signal that commands the "boom" cylinder of the machine, that is, the arm that can lift (positive values) or lower (negative values) the bucket. Finally, the $fullRpm$ signal is, as mentioned, a machine-specific command that forces the engine's RPM to the maximum, temporarily increasing the force of the hydraulic mechanism. It should be noted that analysis and correction of wheel skidding is only performed during the S_{push} state. This decision was made based on the empirical observation that wheel skidding is much less likely to occur during bucket retraction and the fact that activating the lift cylinder effectively reduces the power available for the tilt hydraulics. In Figure 18, a flow chart of the excavation process is presented, where EP represents engine power (a variable observed, in this case, through the transmission pressure of the machine).

Both states, S_{tilt} (tilt state) and S_{push} (push state), have upper and lower bounds to their execution time, so it is guaranteed that their execution alternates during the excavation. Besides the time limit of S_{push}, the system can transition back to S_{tilt} if the hydrostatic transmission pressure (or, more generally, the engine power output) climbs over a threshold value, meaning that the resistive force from the rock pile is too large, and the bucket is not able to penetrate further without a scooping motion. This alternation between states continues until any of the following stop criteria are met: the bucket is fully retracted, the bucket is lifted over a maximum angle, or the machine has advanced further than a maximum displacement value. This method was patented as reported in [36]. For clarity purposes, a pseudo code of the described algorithm is shown in Figure 19. Table 1 shows the value of the parameters used for the full-scale validation experiments reported in Section 4. The numerical values were selected based on operation data.

In the case that the described excavation process is not completed within a predefined time limit (in the experiments, a threshold of 30 s was used), an operator assistance request is generated, thus putting the system in the loading assistance state from Figure 6.

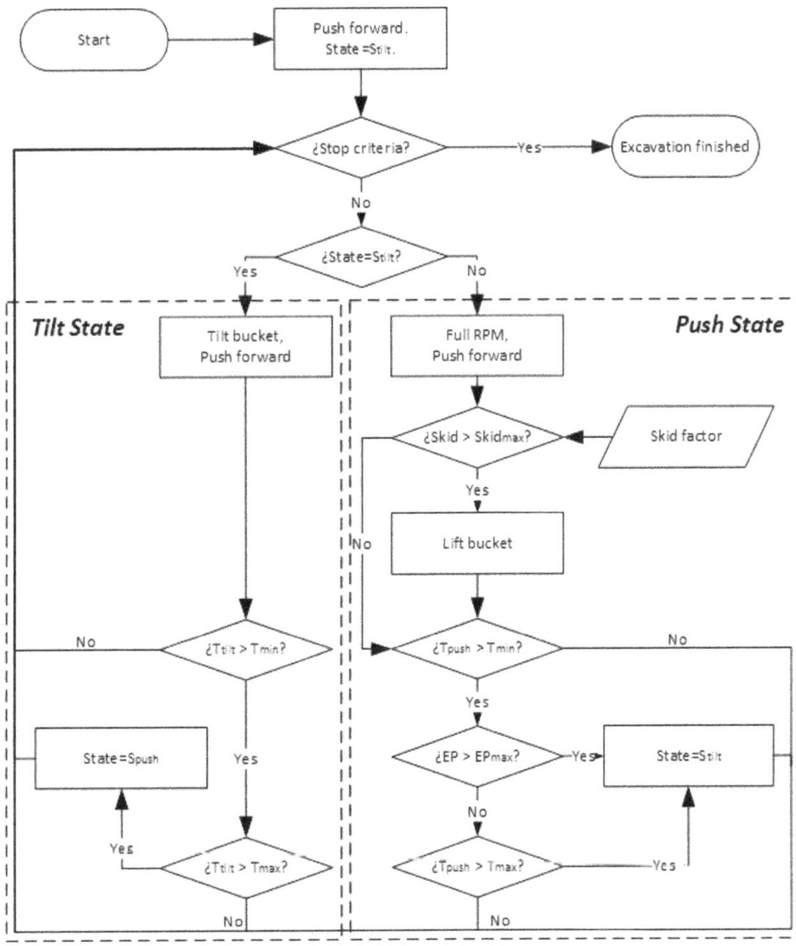

Figure 18. Flow chart of the excavation algorithm.

3.5. Pull Back and Payload Weighing

Once the excavation is finished, the bucket gets fully retracted and the vehicle is driven backwards, using the same "guidance" navigation module from the positioning step. A predefined time limit is assigned to this step (in the experiments, 60 s were used). If a timeout occurs, the system is put in the loading assistance state from Figure 6.

After a predefined distance has been covered within the aforementioned time limit, the bucket is commanded to perform short downward and upward tilting movements, known as the "bucket shake". Then, the bucket is lifted up to a fixed angle (β_{weigh}), and the pressure sensors in the boom hydraulic line are used to estimate the bucket fill factor. A fill factor is used instead of the mass weight, since it simplifies calibration, while providing the same information for the evaluation process. The relationship between the lift cylinder pressure (P) and the fill factor was found experimentally and corresponds to the following affine model:

$$f_{fill} = mP - n \tag{9}$$

where f_{fill} is the relative fill factor (0 for an empty bucket and 1 for a full bucket), and m and n are the constants to be calibrated through experimentation. Table 2 shows the parameters used for the particular conditions of the field trials presented in this work (machine and rock characteristics).

```
Require: initial state S = S_tilt, acceleration pedal value P_0, max. lift value β_max, max.
tilt value α_max, max. displacement value d_max, max. duration Δt_max, min. duration
Δt_min, max. skidding value Skid_max, max. engine power EP_max

read: start time t_s, lift angle β, tilt angle α, odometry odo, engine power EP

while β < β_max and α < α_max and odo < d_max do
read: current time t, skidding factor skid, β, α, odo, EP
                Δt ← t − t_s

if S = S_tilt do
pedal ← P_0, lift ← 0, tilt ← ramp(Δt), fullRpm ← 0
if Δt > Δt_max do
read: t_s
                S ← S_push
end if
end if

if S = S_push do
pedal ← P_0, tilt ← 0, fullRpm ← 1
if skid > skid_max do
lift ← 1
else do
                lift ← 0
end if
if EP > EP_max and Δt > Δt_min or Δt > Δt_max do
read: t_s
S←S_tilt
end if
end if

end while
```

Figure 19. Pseudo code of the excavation algorithm.

Table 1. Excavation method parameters used for the full-scale validation.

Parameter	Value
P_0	0.5
R_0	1.2
$skid_{max}$	0.2
β_{max}	0.5 [rad]
α_{max}	0.0 [rad]
d_{max}	2.0 [m]
Δt_{max}	1.0 [s]
Δt_{min}	0.5 [s]
EP_{max}	300 [bar]

Table 2. Weight model parameters used for the full-scale validation.

Parameter	Value
β_{weigh}	0.2
m	0.0207
n	1.0774

4. Experiments and Analysis

Full-scale experiments with a GHH LF11H model LHD (see picture in Figure 1) were carried out in a production tunnel of a sublevel stoping mine located in the Coquimbo Region (Chile), facilitated by the Chilean mining company, CMSG. Pictures of the mine site and the LHD inside its tunnel are shown in Figure 20. Figure 20a shows the mechanical workshop just outside the mine's main entrance, and Figure 20b shows the LHD during surface tests. Figure 20c shows the LHD outside the test tunnel with an almost fully loaded bucket.

Figure 21 shows the layout of the tunnel where experiments were carried out. The LHD traveled from the "start–end point" shown in the figure to the end of the red line, where the stope was originally located. A stope is a large open underground space that is produced by the sublevel stoping mining method. The LHDs load ore inside the stope from a large pile of material, also sometimes referred to as the loading point or draw point. Once the LHD has a fully loaded bucket, it has to travel back to the "start–end point" to dump the ore.

(a) (b) (c)

Figure 20. LHD machine and mining facility. (**a**) Mechanical workshop just outside the mine's main entrance; (**b**) LHD during surface tests; (**c**) LHD outside the test tunnel with an almost fully loaded bucket.

Figure 21. Layout of the mine's tunnel where experiments were carried out.

The experiments reported here were carried out in a fully functional tunnel of the mine, i.e., a tunnel used for extracting mineral. In fact, the experiments took place during day hours, while the mine's night shift used the LHD in manual mode for production. As a result of this, the shape and location of the stope and the draw point changed every day (as the ore runs out, the inner wall of the stope is blasted as part of the mining process).

Because of this, and since in situ experimentation was time-limited, experiments were separated in two parts. First, the field trial experiments regarding charging and collision detection of the LHD and the pile, and the LHD's excavation and payload weighing, are reported in Section 4.1. Then, pile identification experiments from an offline processing of the collected field data are presented in Section 4.2.

The mining tunnel was narrower than recommended by the LHD manufacturer, so the LHD's positioning was more complex than initially expected. This also contributed to dividing the experiments into the two groups described above. Figure 22 shows a picture of the draw point as seen from the teleoperation system cameras.

Figure 22. Operator's point of view of the loading point used for onsite experiments.

4.1. Full-Scale Rock Excavation Experiments

Since the pile identification method was not fully optimized at the time of the onsite tests, and time for executing the experiments in the mine was limited, the development team divided the loading procedure in two stages in order to maximize the total number of full-scale rock excavation experiments and evaluate each stage separately. One stage included rock pile identification and positioning, and the other included the charging, excavation, pull back, and weighting processes. To perform this, loading experiments required a human operator to position the machine five to ten meters in front of the draw point, effectively bypassing the rock pile identification and positioning stages, and starting the system's state machine in "positioning assistance", as described in Section 3.

A total of 30 complete loading experiments were carried out. Similar to in manual operation, multiple loading attempts were required in order to fill the bucket during each experiment. Manual operation of LHD in sublevel stoping mines usually needs two or three attempts to achieve a proper bucket fill factor, since the procedure is performed using line-of-sight radio control with limited visibility behind the LHD (safety regulations forbid operators to enter the stope area). In some attempts, the tip of the bucket was lowered too much, causing high resistive forces from the ground during charging, and thus triggering the collision detection too early. A total of 81 loading attempts were performed in order to fulfill the 30 complete loads. Of these, 7 failed (early collision trigger) and 74 were successful, resulting in the 91% success rate of the charging and collision detection method.

For each experiment, a fixed value of 0.5 was used for the pedal command. An operator monitoring the system's behavior during the complete loading process from a remote-control station decided when the bucket was full enough, and hence if a new attempt was needed. The operator relied only on the visual input of the machine's forward camera to make this decision.

Figure 23 shows an example of the relevant variables in an autonomous loading attempt, which has some resemblance to the operator-controlled loading shown in Figure 5.

Table 3 shows the number of experiments carried out, classified by the number of attempts to achieve a full bucket, and the average fill factor for each category. It can be seen that most experiments required three loading attempts, and that the operator continued to reattempt loading until the fill factor was about 90%. This can be seen more easily on Table 4, where the average fill factor and average number of attempts for all experiments are shown. It must be noted that while the fill factor at the end of each experiment met the criteria for manual operation (since it was the same operator deciding when the bucket was full), the average number of attempts was found to be higher than that in manual operation, which, according to manual operators was between two and three attempts.

Table 5 and Figures 24 and 25 show more detail about progressive filling of the bucket through loading attempts for the 30 loading experiments. Table 5 shows the average durations of the excavation step and the resulting fill factor, classified according to the progressive sequence of attempts, as well as the total number of experiments in each case. On average, the first attempt managed to fill 62% of the bucket capacity; then, the second attempt achieved 77%; the third attempt achieved 82%; and for the six experiments that needed a fourth attempt, the average that resulted was a 90% fill factor. The average excavation duration was about 10–12 s, which comprised the time between the collision of the LHD with the rock pile and the end of the excavation algorithm. The required time for positioning, charging, and weighing was not taken into account in this measurement. It can be seen that the average duration of the excavation in the first attempt is slightly longer than the others.

Figure 24 shows a box plot of the excavation step duration for all loading attempts. The time taken for the excavation ranges from 6 to almost 20 s, depending on the conditions of the rock pile and the specific interaction between the bucket and the rock pile. Figure 25 shows a box plot of the fill factor that was obtained progressively through loading attempts. It can be seen that the biggest jump in performance happened in the second loading attempt.

Some outliers with performance above 100% are also depicted for the second and third attempts and appear as a consequence of loading a large boulder.

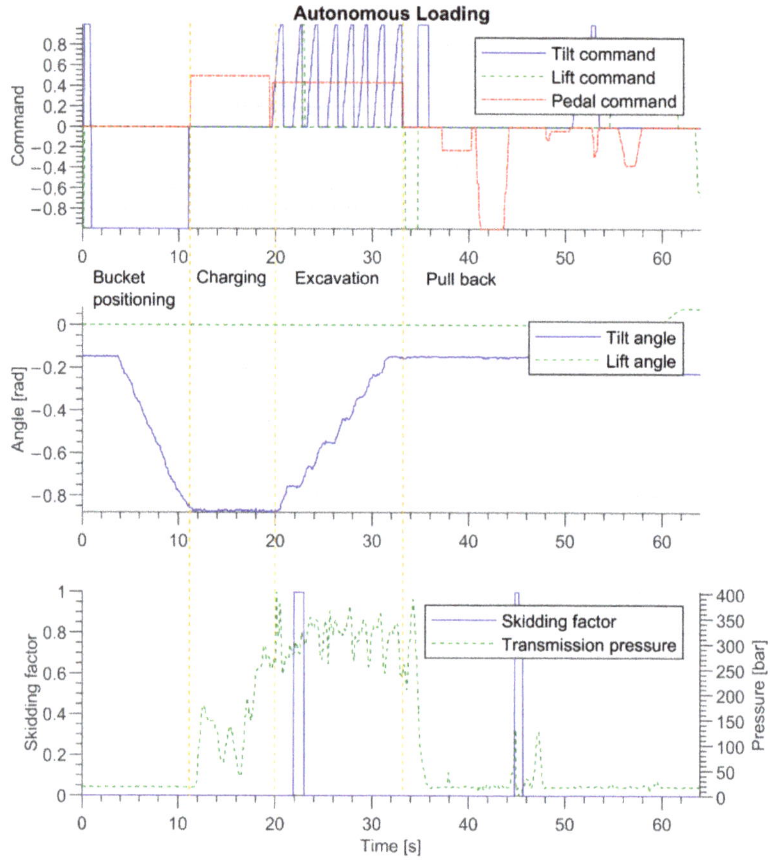

Figure 23. Relevant variables in an autonomous loading attempt.

Table 3. Loading experiments and average fill factor per number of loading attempts. N-Attempts: number of attempts required to fill bucket. N-Exp: number of experiments. Total-Attempts: total number of performed attempts. %-Experiments: % of the experiments. Fill-Factor: average fill factor.

N-Attempts	N-Exp	Total-Attempts	%-Experiments	Fill-Factor
1 attempt	9	9	30%	88%
2 attempts	4	8	13%	107%
3 attempts	11	33	37%	87%
4 attempts	6	24	20%	90%
TOTAL	30	74	100%	-

Table 4. Total average number of attempts and fill factors.

Total Average Number of Attempts	Total Average Fill Factor
3.6	90%

Table 5. Duration and fill factor in consecutive attempts. P-Attempts: progressive sequence of attempts. N-Exp: number of experiments. Duration: average duration. Fill-Factor: average fill factor.

P-Attempts	N-Exp	Percentage	Duration	Fill-Factor
First attempt	30	100%	12.7 s	62%
Second attempt	21	70%	10.2 s	77%
Third attempt	17	57%	10.8 s	82%
Fourth attempt	6	20%	10.7 s	90%

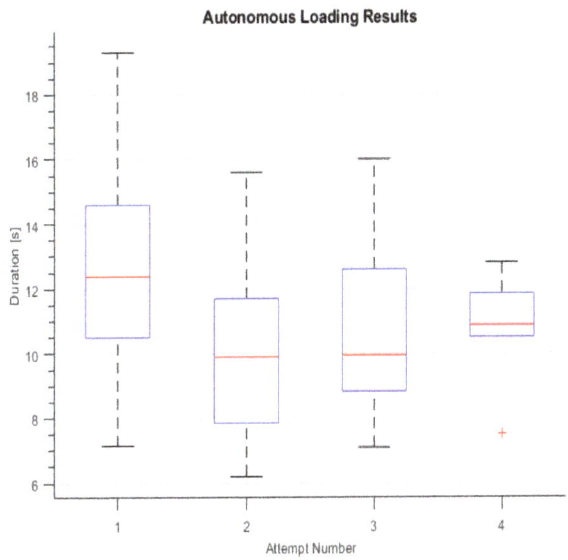

Figure 24. Excavation step duration per loading attempt. Red "+" signs represent outliers.

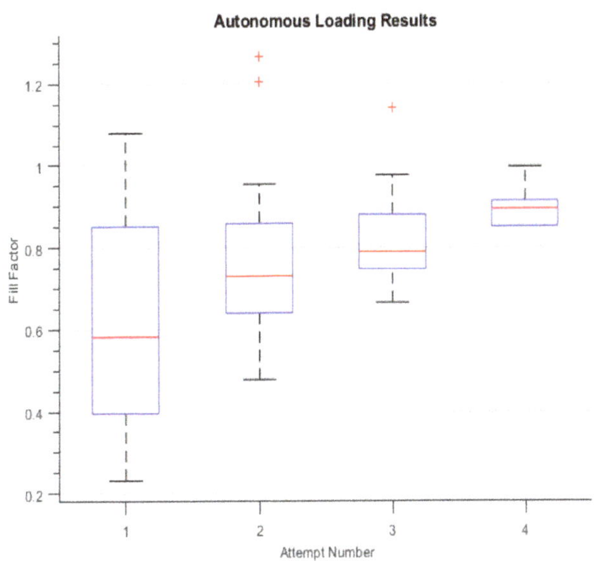

Figure 25. Progressive fill factor across loading attempts. Red "+" signs represent outliers.

During experimentation, the value of accurately placing the bucket of the machine against the ground, prior to the charging against the rock pile, was recognized. A bucket tilted too far down would cause an early trigger of a collision detection, while a bucket tilted just slightly upwards has a relevant impact on how far the machine can penetrate into the rock pile. An even more relevant factor that affects the performance of the loading maneuver is the status of the rock pile. Newly blasted ore is far easier to load than draw points that have had more time to settle and compress due to the weight of the rock. Draw points with large amounts of ore are also easier to load than those that have little material left. Despite these relevant factors, the proposed method was able to extract full buckets each time, at the expense of the operator having to perform more than the usual number of reattempts required under manual operation.

It is important to note that in a real mining operation, the efficiency of an LHD is measured in the amount of ore that it is able to haul from the extraction point to the dumping point (usually in tons/hour). Therefore, the percentage of bucket filling is not the only relevant factor when evaluating the system, but also the amount of time it uses to load the bucket. This is especially important when deciding if the system should make another attempt at the excavation procedure to achieve a fuller bucket. This criterion varies among different operations, as the dumping and haulage time is different.

4.2. Offline Results Using Field Data: 2.5D Modeling of the Extraction Point

Personnel cannot be inside a stope as it is forbidden for security reasons (i.e., rock can fall from the ceiling of the stope), so it was not possible to obtain accurate measurements of the ore pile characteristics. However, an accurate modeling of the ore pile is not really necessary, as it only needs to be detected, as well as having its width and inclination roughly estimated in order for the excavation algorithm to work.

To obtain the datasets required to characterize our 2.5D modeling algorithm, the LHD was driven from a fixed distance of about 50 m from the loading point to the entrance of the stope, while all sensor and machine data was recorded. These datasets were captured at different times so the drawing point was not located in the same place, nor did the ore pile have the same shape. First, in each case the point cloud was computed and then the pile's width and inclination estimated. Figure 26 shows an example of the point cloud obtained while the LHD is approaching an extraction point, after registering the position of the points using the LIDAR-based odometry and the inclinometer data (see details in Section 3.2). The color of the points represents the distance to the LHD (in the x-axis) at the time of registration. In the case of the rock pile at the end of the tunnel, red indicates that the machine was further away when that reading was taken.

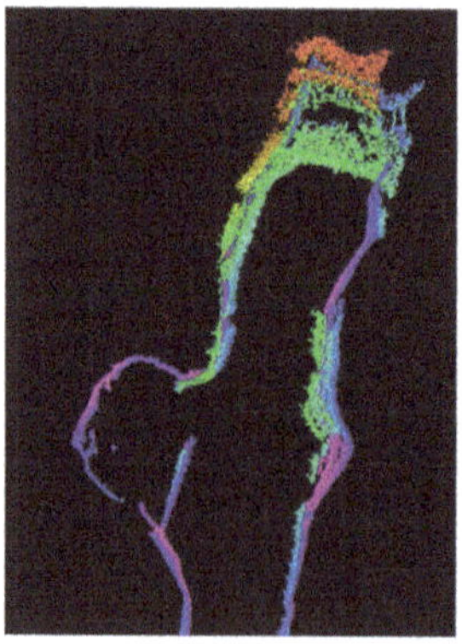
(a)

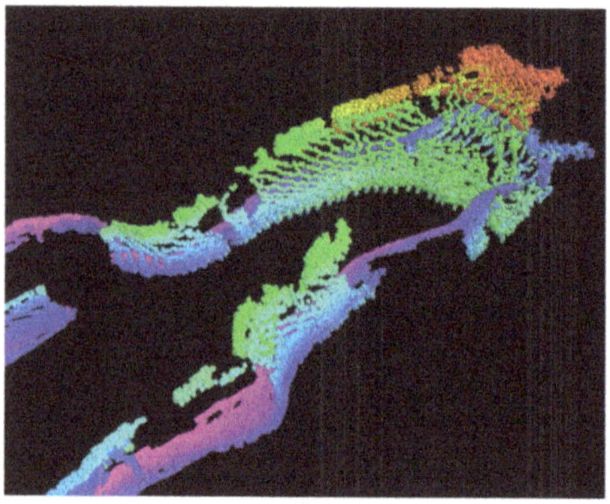
(b)

Figure 26. Example of the scanning obtained while the LHD is approaching an extraction point. (**a**) Top view of the pile reconstruction. (**b**) Isometric view of the pile reconstruction.

The results of several modeling attempts where the pile's width and inclination were estimated are summarized in Table 6. They consider ground truth width, predicted width, ground truth inclination, and predicted inclination. Both ground truth width and ground truth inclination were determined manually from the integrated point clouds. The average absolute error in the predicted width is 0.58 m, while the average absolute error in the inclination is 1.7 degrees. The precision achieved by the characterization procedure is sufficient for the task of autonomous loading.

Table 6. Results of modeling extraction points.

Modeling Attempt	Ground Truth Width (m)	Predicted Width (m)	Ground Truth Inclination (°)	Predicted Inclination (°)
1	3.25	2.89	−53.0	−52.6
2	3.10	5.27	−50.1	−48.7
3	3.10	3.16	−52.4	−50.5
4	3.44	3.29	−50.6	−50.5
5	3.20	3.72	−50.0	−53.4
6	3.03	3.25	−49.5	−46.5

5. Conclusions

A complete autonomous loading system for LHD machines for underground mining was presented. The loading system considers identification of the rock pile, positioning of the machine in front of it, charging against the rock pile, excavating, moving away from the draw point, and estimating the bucket weight or fill factor. Despite that the proposed system was fully implemented (including the necessity for a human operator to be involved in the process in order to complete the task when the system fails), it could not be tested as a whole. The proposed system which implements the whole loading process, the excavation algorithm, and the tests in a real production environment are the most important contributions of this work.

The experimentation phase was divided between onsite execution and validation of the excavation algorithm, and offline data processing for the rock pile identification method. In the onsite experiments, identification and positioning were bypassed in favor of teleoperation by a human operator. It was also the operator's choice whether or not to perform multiple loading attempts in order to fill the bucket. A total number of 30 excavation experiments were carried out, most of them requiring multiple attempts to achieve a full bucket. An average of 3.6 attempts per experiment was needed in order to obtain a bucket fill factor of 90%. By comparison, manual operation usually needs two or three attempts to achieve this bucket fill factor. This performance may seem inferior to previous published work: however, there are other factors involved that prevent direct comparison of the results. Equipment, type of ore, and mining method should be considered. Thus, comparison to the human operators of the specific mining site should be preferred. It is important to mention that the main driver for automating the loading process is to increase the safety of workers rather than obtain higher efficiency. In this regard, the proposed excavation algorithm fulfills our expectations.

Since these experiments are time-consuming in an industry where time is an expensive resource, only enough experiments needed to validate the excavation algorithm were able to be executed. Part of our future work will be to carry out new experiments in an underground mine in order to validate the complete autonomous loading system, and to measure its performance more accurately. Despite this, the problem remains relevant and mining companies are looking forward to integrating autonomous loading to LHDs in their operations, as it would enable them to close the loop for a fully autonomous production cycle.

A video showing the operator's graphic interface while the system is autonomously performing an excavation procedure can be found in https://youtu.be/Oa11kTBJf2Y (accessed on 7 September 2021).

The system is now being installed and tested in a room and pillar mine in Germany, where it will be tested as a whole system.

Author Contributions: Conceptualization, C.T., M.M. and J.R.-d.-S.; methodology, M.M., C.T. and J.R.-d.-S.; software, C.T.; validation, M.M. and C.T.; resources, J.R.-d.-S.; data curation, C.T.; writing—original draft preparation, C.T., M.M. and J.R.-d.-S.; writing—review and editing, C.T., M.M. and J.R.-d.-S.; funding acquisition, J.R.-d.-S. All authors have read and agreed to the published version of the manuscript.

Funding: This research was funded by the Chilean National Research Agency ANID under project grant Basal AFB180004 and FONDECYT 1201170.

Institutional Review Board Statement: Not applicable.

Informed Consent Statement: Not applicable.

Data Availability Statement: The data presented in this study is contained in the article itself.

Acknowledgments: We thank Paul Vallejos for the valuable discussions and support for onsite execution of the experiments; we thank David Leottau for his valuable work in the early implementations of the 2.5D modeling of the ore pile. We also thank Patricio Loncomilla and Martin Calvo for their valuable work enhancing the 2.5D modeling algorithm and its associated results. We also acknowledge Compañía Minera San Gerónimo for providing the mine infrastructure for testing the system, and GHH Chile for supplying the LHD machine needed for this work.

Conflicts of Interest: The authors declare no conflict of interest.

References

1. Salvador, C.; Mascaró, M.; Ruiz del Solar, J. *Automation of Unit and Auxiliary Operations in Block/Panel Caving: Challenges and Opportunities*; Massmin: Santiago, Chile, 2020.
2. Scheding, S.; Nebot, E.M.; Stevens, M.; Durrant-Whyte, H.; Roberts, J.; Corke, P.; Cunningham, J.; Cook, B. Experiments in Autonomous Underground Guidance. In Proceedings of the 1997 IEEE International Conference on Robotics and Automation, Albuquerque, NM, USA, 25–25 April 1997; pp. 1898–1903.
3. Cunningham, J.; Roberts, J.M.; Corke, P.; Durrant-Whyte, H. Automation of Underground LHD and Truck Haulage. In Proceedings of the AusIMM Annual Conference—"The Mining Cycle" (AUSIMM 1998), Mount Isa, Australia, 19–23 April 1998.
4. Mäkelä, H. Overview of LHD Navigation without Artificial Beacons. *Robot. Auton. Syst.* **2001**, *36*, 21–35. [CrossRef]
5. Duff, E.S.; Roberts, J.M.; Corke, P.I. Automation of an underground mining vehicle using reactive navigation and opportunistic localization. In Proceedings of the 2003 IEEE/RSJ International Conference on Intelligent Robots and Systems, Las Vegas, NV, USA, 27–31 October 2003; Volume 3, pp. 3775–3780.
6. Larsson, J. Reactive Navigation of an Autonomous Vehicle in Underground Mines. Ph.D. Thesis, Örebro University, Örebro, Sweden, 2007.
7. Hemami, A. Modelling, analysis and preliminary studies for automatic scooping/loading in a mechanical loader, International Journal of Surface Mining. *Reclam. Environ.* **1992**, *6*, 151–159.
8. Hemami, A. Study of bucket trajectory in automatic scooping with load-haul-dump loaders. *Trans. Inst. Min. Metall.* **1993**, *102*, 37–42.
9. Ji, Q.; Sanford, R.L. Autonomous excavation of fragmented rock using machine vision. In *Emerging Computer Techniques for the Minerals Industry*; Society for Mining Metallurgy: Englewood, CO, USA, 1993; pp. 221–228.
10. Hemami, A. Study of forces in the scooping operation of a mechanical loader. *Trans. Can. Soc. Mech. Eng.* **1994**, *18*, 191–205.
11. Petty, M.K.; Billingsley, J.; Tran-Cong, T. Autonomous LHD loading. In Proceedings of the Fourth Annual Conference on Mechatronics and Machine Vision in Practice, Toowoomba, QLD, Australia, 23–25 September 1997.
12. Takahashi, H.; Hasegawa, M.; Nakano, E. Analysis on the resistive forces acting on the bucket of a load-haul-dump machine and a wheel loader in the scooping task. *Adv. Robot.* **1999**, *13*, 97–114. [CrossRef]
13. Marshall, J.A.; Murphy, P.F.; Daneshmend, L.K. Toward autonomous excavation of fragmented rock: Full-scale experiments. *IEEE Trans. Autom. Sci. Eng.* **2008**, *5*, 562–566. [CrossRef]
14. Hemami, A.; Hassani, F. An overview of autonomous loading of bulk material. In Proceedings of the 26th International Symposium on Automation and Robotics in Construction, Austin, TX, USA, 24–27 June 2009.
15. Dobson, A.; Marshall, J.; Larsson, J. Admittance control for robotic loading: Design and experiments with a 1-tonne loader and a 14-tonne load-haul-dump machine. *J. Field Robot.* **2017**, *34*, 123–150. [CrossRef]

16. Heshan, F.; Marshall, J.A.; Almqvist, H.; Larsson, J. Towards Controlling Bucket Fill Factor in Robotic Excavation by Learning Admittance Control Setpoints. In *Field and Service Robotics*; Springer: Cham, Switzerland, 2018.
17. Heshan, F.; Marshall, J.A.; Larsson, J. Iterative learning-based admittance control for autonomous excavation. *J. Intell. Robot. Syst.* **2019**, *96*, 493–500.
18. Ruiz del Solar, J.; Vallejos, P.; Tampier, C.; Leottau, D. Un Método de Estimación de Derrape y Detección de Colisión con una Pila de Mineral. Chilean Patent 201803481, 8 July 2020.
19. Erikson, J.M.; Mueggenburg, N.W.; Jaeger, H.M.; Nagel, S.R. Force distributions in three-dimensional compressible granular packs. *Phys. Rev. E* **2002**, *66*, 040301. [CrossRef] [PubMed]
20. Herminghaus, S. Dynamics of wet granular matter. *Adv. Phys.* **2005**, *54*, 221–261. [CrossRef]
21. Mikhirev, P.A. Design of automated loading buckets. *Sov. Min.* **1986**, *22*, 292–298. [CrossRef]
22. Shi, X.; Lever, P.; Wang, F. Experimental robotic excavation with fuzzy logic and neural networks. In Proceedings of the 1996 IEEE International Conference on Robotics and Automation, Minneapolis, MN, USA, 22–28 April 1996; Volume 1, pp. 957–962.
23. Dobson, A.; Marshall, J. Autonomous Loading Vehicle Controller. WO Patent 2015/109392, 30 July 2015.
24. Rocke, D.J. Control System for Automatically Controlling a Work Implement of an Earth Working Machine to Capture Material. U.S. Patent Application No. 5,528,843, 25 June 1996.
25. Alshaer, B.J.; Ingram, R.G.; Krone, J.J.; Berry, J.K.; Harris, J.J. Automatic Digging and Loading System for a Work Machine. U.S. Patent Application No. 7,555,855, 7 July 2009.
26. Fletcher, J.G.; Jones, D.A.; Hewavisenthi, R.D.; Chow, R.K. Excavation System Having Adaptive Dig Control. U.S. Patent Application No. 09,587,369, 7 March 2017.
27. Asenjo, R.; Boulard, A.; Mascaró, M.; Tampier, C.; Ruiz del Solar, J. Autonomous Loading for an LHD. In Proceedings of the Automining 2016, Antofagasta, Chile, 30 November–2 December 2016.
28. Quigley, M.; Conley, K.; Gerkey, B.; Faust, J.; Foote, T.; Leibs, J.; Wheeler, R.; Ng, A. ROS: An Open-Source Robot Operating System; ICRA Workshop on Open Source Software. 2009. Available online: http://www.cim.mcgill.ca/~{}dudek/417/Papers/quigley-icra2009-ros.pdf (accessed on 7 September 2021).
29. Röfer, T. CABSL—C-Based Agent Behavior Specification Language. In *RoboCup 2017: Robot World Cup XXI*; Springer: Cham, Switzerland, 2017; pp. 135–142.
30. Koenig, N.; Howard, A. Design and use paradigms for gazebo, an open-source multi-robot simulator. In Proceedings of the 2004 IEEE/RSJ International Conference on Intelligent Robots and System, Sendai, Japan, 28 September–2 October 2004; Volume 3, pp. 2149–2154.
31. Leottau, D.; Loncomilla, P.; Mascaró, M.; Tampier, C.; Ruiz del Solar, J. LIDAR-based odometry estimation for autonomous loading of material with LHD. In Proceedings of the Massmin 2020, Santiago, Chile, 9–11 December 2020.
32. Jaimez, M.; Monroy, G.; Gonzalez, J. Planar Odometry from a Radial Laser Scanner. A Range Flow-based Approach. In Proceedings of the IEEE International Conference on Robotics and Automation, Stockholm, Sweden, 16–21 May 2016.
33. Lobos, K.; Loncomilla, P.; Ruiz-del-solar, J. Rocky-YOLO: Detección de rocas usando deep learning para aplicaciones en minería. In Proceedings of the SIMIN 2019, Santiago, Chile, 21–23 August 2019. (In Spanish).
34. Mascaró, M.; Tampier, C.; Ruiz del Solar, J. Topological Navigation in Tunnels: LHD Navigation as a Case Study. In Proceedings of the Automining 2020, Santiago, Chile, 30 November–4 December 2020.
35. Marshall, J.; Barfoot, T.; Larsson, J. Autonomous underground tramming for center-articulated vehicles. *J. Field Robot.* **2008**, *25*, 400–421. [CrossRef]
36. Ruiz del Solar, J.; Vallejos, P.; Tampier, C. Método de Carguío Autónomo Para Cargadores Frontales. Chilean Patent 201803004, 26 December 2019.

applied sciences

MDPI

Article

Estimating the Soundscape Structure and Dynamics of Forest Bird Vocalizations in an Azimuth-Elevation Space Using a Microphone Array

Reiji Suzuki [1,*], Koichiro Hayashi [2], Hideki Osaka [3], Shiho Matsubayashi [4], Takaya Arita [1], Kazuhiro Nakadai [5] and Hiroshi G. Okuno [6,7]

1 Graduate School of Informatics, Nagoya University, Nagoya 464-8601, Japan
2 School of Informatics and Sciences, Nagoya University, Nagoya 464-8601, Japan
3 toriR Lab., Echizen 915-0242, Japan
4 Graduate School of Engineering Science, Osaka University, Toyonaka 560-8531, Japan
5 Department of Systems and Control Engineering, Tokyo Institute of Technology, Meguro-ku, Tokyo 152-8552, Japan
6 Graduate School of Inform, Kyoto University, Kyoto 606-8501, Japan
7 Future Robotics Organization, Waseda University, Shinjuku, Tokyo 169-0072, Japan
* Correspondence: reiji@nagoya-u.jp; Tel.: +81-52-789-4258

Abstract: Songbirds are one of the study targets for both bioacoustic and ecoacoustic research. In this paper, we discuss the applicability of robot audition techniques to understand the dynamics of forest bird vocalizations in a soundscape measured in azimuth and elevation angles with a single 16-channel microphone array, using HARK and HARKBird. First, we evaluated the accuracy in estimating the azimuth and elevation angles of bird vocalizations replayed from a loudspeaker on a tree, 6.55 m above the height of the array, from different horizontal distances in a forest. The results showed that the localization error of azimuth and elevation angle was equal to or less than 5 degrees and 15 degrees, respectively, in most of cases when the horizontal distance from the array was equal to or less than 35 m. We then conducted a field observation of vocalizations to monitor birds in a forest. The results showed that the system can successfully detect how birds use the soundscape horizontally and vertically. This can contribute to bioacoustic and ecoacoustic research, including behavioral observations and study of biodiversity.

Keywords: bird song; soundscape; ecoacoustics; sound source localization; robot audition; HARK

Citation: Suzuki, R.; Hayashi, K.; Osaka, H.; Matsubayashi, S.; Arita, T.; Nakadai, K.; Okuno, H.G. Estimating the Soundscape Structure and Dynamics of Forest Bird Vocalizations in an Azimuth-Elevation Space Using a Microphone Array. *Appl. Sci.* **2023**, *13*, 3607. https://doi.org/10.3390/app13063607

Academic Editors: Luis Gracia and Carlos Perez-Vidal

Received: 26 December 2022
Revised: 27 February 2023
Accepted: 4 March 2023
Published: 11 March 2023

1. Introduction

Songbirds are one of the study targets for the purpose of ecoacoustic research [1,2]: an interdisciplinary science that investigates natural and anthropogenic sounds and their relationship to the environment across multiple scales of time and space [3], as well as bioacoustic research. This is because their vocalizations (1) can tell us a suite of useful information about the environment for monitoring, (2) have rich and complex variety of structures [4], which are used as benchmark problems for classification tasks (e.g., BirdCLEF [5]), and (3) enable bird individuals to interact in complex ways, behaving as complex systems [6].

There are several approaches for using microphone arrays to localize bird vocalizations. Rhinehart et al. recently surveyed applications of acoustic localization using autonomous recording units in terrestrial environments [7], and pointed out that ecologists will make better use of acoustic localization; it can collect large-scale animal position data with minimizing the influence on the environment if recording hardware and automated localization and classification software are more available, and their algorithms are improved for outdoor measurement.

Some of these studies focused on both azimuth and elevation estimation [8–10]. Hedley et al. developed a 4-channel microphone array unit by combining two stereo microphones and evaluated the accuracy to estimate azimuth and elevation angles of replayed songs of a few species [9]. The results showed that most of sounds were estimated within 12 degrees of the true direction of arrivals (DOA) in the azimuth angle and 9 degrees in the elevation angle within a range of at least 30 m. It was also discussed that the DOA estimation may improve the ability to assess abundance in biodiversity surveys. However, the experiment was conducted in an open space, and the elevation angle was limited to −10 to 15 degrees from horizontal. As a different approach, Gayk et al. constructed a microphone array system to estimate 3D position of flying songbirds with a wireless microphone array. The system was consisted of four 7-m poles arranged in a 25 m square, and each pole had two microphone channels that are placed on top and bottom of the pole. They adopted a triangulation method based on time-of-arrival differences of a sound recorded at these microphones to cross-correlate and estimate sound position. They showed that both broadcasted bird calls and calls of natural migratory birds were successfully triangulated with the accuracy and estimated accuracy of less than 3 m. In addition, there is increasing interest and development for sound event localization and detection (SELD) of various environmental sounds using microphone arrays and ambisonic microphones [11]. We expect that practical experimental analyses of natural sounds such as bird vocalizations in forests can further contribute to better use of such microphone array-based techniques in natural fields.

We have been proposing that robot audition techniques [12], especially an open source software for robot audition HARK (Honda Research Institute Japan Audition for Robots with Kyoto University), can contribute to bioacoustics and ecoacoustics. It not only provides the DOA estimation of sounds, but also allows us to separate them and perform further signal processing on them, even in real time. We developed HARKBird, a collection of Python scripts for localizing bird songs in fields using HARK [13]. Previously, we confirmed the effects of playback of conspecific song on song or call responses by measuring the changes in their localized direction [14] and changes in their 2D position using a set of microphone arrays [15].

It is recognized that data characterizing the vertical structure of vegetation are becoming increasingly useful for biodiversity applications as remote sensing techniques such as radar and lidar become more readily available [16]. We believe that direct observation of the vertical and horizontal soundscape of vocalizations among birds would also contribute to this field, as well as to bioacoustic analysis of bird behavior. There is initial work on 3D localization of bird songs using multiple microphone array units [10] and observation of nocturnal birds with a single microphone array unit based on the azimuth-elevation estimation [17]. However, we still need to investigate how HARK or HARKBird can estimate both azimuth and elevation angles of bird songs to capture the dynamics and structures of the soundscape of bird songs. In particular, a systematic evaluation of the localization accuracy of elevation angles and an estimation of the structure of soundscape in a realistic situation where multiple bird species are vocalizing are important for the practical use of the system.

This paper aims to demonstrate a systematic evaluation of the localization accuracy of azimuth-elevation angles of replayed bird vocalizations in a practical forest environment, and show an example field observation of the structure and dynamics of birdsong soundscape. For this purpose, we use a self-developed 16-channel microphone array, called DACHO, using HARK and HARKBird. Suzuki et al. [18] used the same microphone array to conduct spatiotemporal analysis of acoustic interactions between great reed warblers (*Acrocephalus arundinaceus*). They conducted a 2D localization of their vocalizations using multiple arrays and estimated the location of two individuals' song posts with mean error distance of 5.5 ± 4.5 m from the location of observed song posts. They then evaluated the temporal localization accuracy of the songs by comparing the duration of localized songs around the song posts with those annotated by human observers, with an accuracy score of average 0.89 for one bird that stayed at one song post. However, the localization accuracy of songs in the elevation angle was not evaluated, and thus a systematic analysis of the accuracy of elevation angle estimation in field conditions would supplement and strengthen our knowledge about the application of robot audition techniques to ecoacoustic research.

We used a single microphone array unit because it is a minimal system and its cost is low for field deployment. We think that sound source localization is useful to passively monitor auditory behaviors of rare or nocturnal birds. Localized results can be used to estimate the abundance and the distribution of those birds. The high portability and low deployment cost are both essential in such a case.

First, we evaluated the accuracy in estimating the azimuth and elevation angles of bird vocalizations replayed from a loudspeaker on a tree, 6.55 m above the height of the array, from different horizontal distances in a forest. The results showed that the localization error of azimuth and elevation angle was equal to or less than 5 degrees and 15 degrees, respectively, in most of cases when the horizontal distance from the array was equal to or less than 35 m. We then conducted field observation of vocalizations to monitor birds in a forest. The results showed that the system can successfully capture how birds use the soundscape horizontally and vertically. This can contribute to bioacoustic and ecoacoustic research, including behavioral observations and study of biodiversity.

The organization of the paper is as follows: We firstly introduce two cases of experimental trials: a speaker test and field observation of soundscape dyamics of bird vocalizations, and introduce the sound source localization method based on HARK and HARKBird in Section 2. Then, we show experimental results of the two trials in Section 3, and finally summarize and discusses the significance of the findings and their implications for further contribution to ecoacoustics and related fields in Section 4.

2. Methods

2.1. Speaker Test

We conducted a speaker test to investigate whether and how bird vocalizations can be localized in a forest environment using azimuth and elevation angles. The experiment was conducted at Nagano park, Kawachinagano, Osaka, Japan on 3 December 2018 (Figure 1). Figure 2 shows a schematic diagram of two experimental setups. In Experiment 1, we placed a loudspeaker on a tripod (height = 1.3 m). A 16-ch microphone array DACHO (WILD-BIRD-SONG-RECORDER; SYSTEM IN FRONTIER Inc., Tokyo, Japan) was also placed on a tripod. The array was specifically developed for bird observations in the field. It consists of 16 microphones, arranged within an egg-shaped frame, which is 17 cm in height and 13 cm in width. It records using a 16-channel, 16 bit, 16 kHz format. Recorded raw data are stored in SD cards and can be exported in wave format for further analysis. One can schedule a recording by preparing the time settings in a micro-SD card. See [18] for more detail and an example of using this microphone array in open fields. We changed the distance between the loudspeaker and the microphone array from 0 to 65 m, with an interval of 5 m, by moving the microphone array along a straight path. This is because the maximum length of the ridge that could be considered straight was 65 m around the loudspeaker. Within this distance, a spacing of 5 m was chosen as it was sufficient to measure the effect of the difference in loudspeaker height and the horizontal difference between the array and the loudspeaker.

We replayed a sound file containing four vocalizations of Scaly Thrush (*Zoothera dauma*) at each location as shown in Figure 3. The distance between the loudspeaker and the microphone was 30 m (Experiment 1). In this figure, four vocalizations of the replayed songs were localized successfully, and at the same time, other sound sources were localized around 1 and 8 s. This species is known to sing this type of songs mainly at night. In this experiment, we adopted this vocalization as the playback sound, to simulate observations of such nocturnal vocalizations, which are not easily observed by other methods such as video recordings.

In Experiment 2, we attached the loudspeaker on a tree, 6.55 m above the height of the microphone array. This is because it was the maximum height at which we could safely place the loudspeaker and at which we could study the effect of the height of the loudspeaker on the localization accuracy of the replayed sound. We performed the same speaker experiment as in Experiment 1.

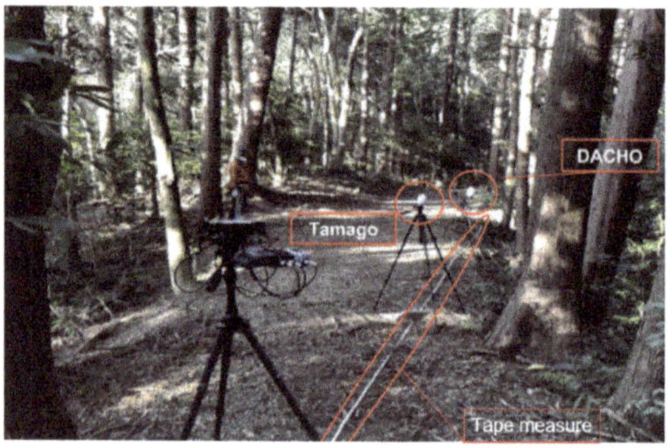

Figure 1. A snapshot of the experiment. We used DACHO, a 16-ch microphone array, for recording replayed songs from a loudspeaker.

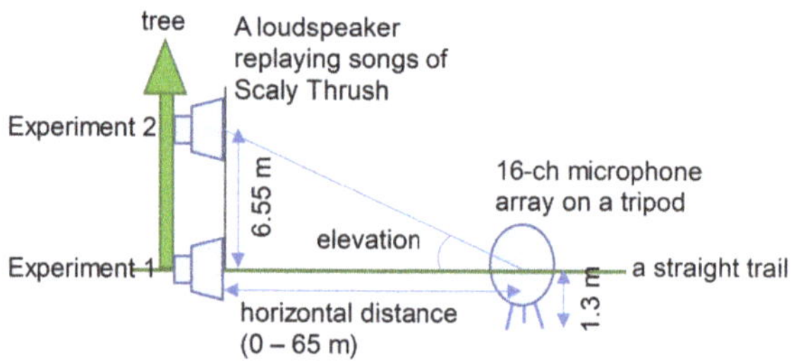

Figure 2. A schematic image of the experimental condition.

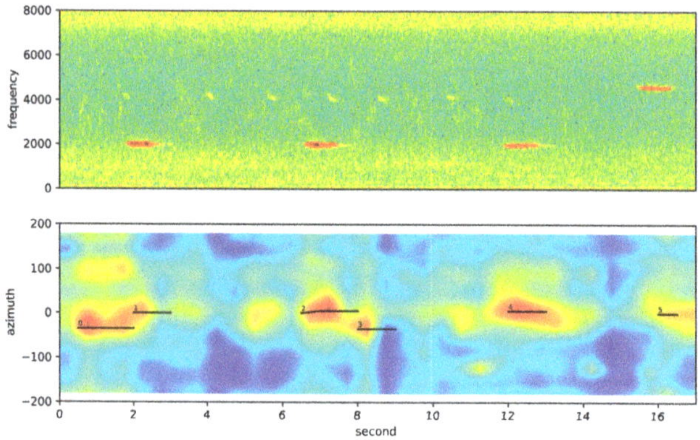

Figure 3. An example of the recording of a replayed sound (**top**) and the localization results (**bottom**). We used the latter part of a replayed sound that include four vocalizations of Scaly Thrush, which is shown in the top figure. The bottom figure shows a heat map of the MUSIC spectrum, whose value represents the strength of sound existence in the corresponding direction. Each black line represents the duration and direction of a localized sound.

2.2. Field Observation of Soundscape Dynamics of Bird Vocalizations

We also conducted a field observation with the same microphone array set up in the Inabu field, the experimental forest of the Field Science Center, Graduate School of Bioagricultural Sciences, Nagoya University, Japan. The forest is mainly a conifer plantation with small patches of broad leaf trees. We placed a DACHO on a path around a patch of broad leaf trees as shown in Figure 4.

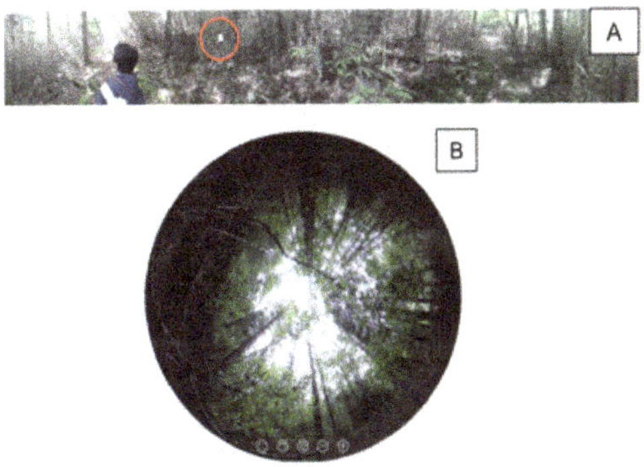

Figure 4. A panorama (**A**) and 360 degree (**B**) photos around the microphone array.

Recording was conducted from 4 April to 7 May 2018. Common bird species actively vocalized during the breeding season here. In particular, Blue-and-white Flycatchers (*Cyanoptila cyanomelana*) tend to sing, advertising their territories, on top of tall trees in their territories. We focused on a 1000-s recording from 8:00 AM on 3 May (Figure 5), where such a typical pattern of bird vocalizations was observed.

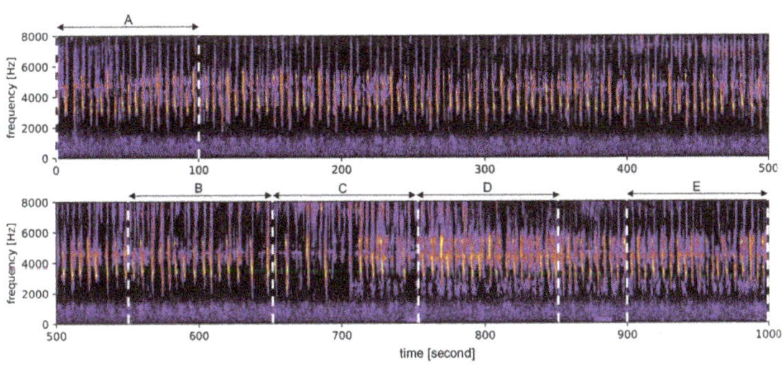

Figure 5. A spectrogram of the whole recording and 5 time slots on which we focus in the analysis.

2.3. Bird Song Localization Using HARKBird

We used HARKBird 2.0 [13], a collection of Python scripts for bird song localization, to estimate the DOA of sound sources in recordings, using sound source localization and separation functions in HARK.

The employed sound source localization algorithm is based on the multiple signal classification (MUSIC) [19] using multiple spectrograms obtained by short-time Fourier transformation (STFT). The MUSIC method is a widely used high-resolution algorithm, and is based on the eigenvalue decomposition of the correlation matrix of multiple signals from a microphone array. We adopted the standard eigenvalue decomposition (SEVD) MUSIC method implemented as one of the sound source localization methods in HARK. All localized sounds are separated the sounds as wave files (16 bit, 16 kHz) using geometric high-order decorrelation-based source separation (GHDSS) method [20], which is also implemented in HARK. For more details on HARKBird (http://www.alife.cs.i.nagoya-u.ac.jp/~reiji/HARKBird/), accessed on 25 December 2022, see [13,21] and on HARK, see Nakadai et al. [12]. In order to optimize localization performance, we can adjust some parameters of HARKBird, such as the source tracking and the lower bound frequency for MUSIC, to reduce noise, etc.

We used a transfer function of the microphone array created from a numerical simulation based on the geometry of the channels of the microphone array using HARKTool5, assuming that there no effects of the body of the array unit on sound transmission. The resolution of DOA for azimuth angle was 5 degrees. The resolutions for elevation angles were 5 and 15 degrees for a speaker test and a bird observation, respectively.

For the DOA estimation of replayed vocalizations in the speaker test, we used the limited frequency range (1800–5000 Hz) for sound source localization, which included the replayed songs. We found that the amplitude of the replayed vocalizations became weaker the farther as the microphone array was from the speaker. Therefore, we gradually decreased the threshold parameter (THRESH), which determines the minimum value of the MUSIC spectrum to detect a sound source, from 29 to 20 with increasing distance. we determined these threshold values empirically according to the acoustic condition around the microphone array. However, this resulted in HARKBird localize vocalizations of other bird species more frequently. Therefore, we manually selected the localized sounds that were detected as replayed vocalizations, and excluded other localized sounds from the analyses. We also lowered the threshold in degree to distinguish multiple sound sources in different directions when there were other sound sources in closer directions to the replayed songs. This is to avoid recognizing them as a part of the replayed vocalizations. We used default values for the other parameters of HARKBird.

For the field observation, we focused on five 100-s time slots (A–E) during which a Blue-and-white Flycatcher (Figure 5) sang on top of tall trees, along with other species such as Varied Tit (*Sittiparus varius*) and Coal Tit (*Periparus ater*). We focused on the behavioral changes in the individual of Blue-and-white Flycatcher, and chose the durations that well illustrated his different behavioral patterns.

We adjusted parameters of HARKBird to localize their songs and exclude other sound sources. We plotted the distribution of songs in the polar-coordinate system representing the azimuth-elevation space for each slot. We then calculated the elevation and azimuth variations of the localized sounds to see if such statistical metrics could reflect the soundscape structures of bird songs.

3. Results

3.1. Speaker Test

Figure 6 shows the estimated azimuth (left) and elevation (right) of the replayed vocalizations in Experiment 1 (top) and 2 (bottom). A red line represents the expected value. Each box plot represents the distribution of localized values when the microphone array was placed x m from the loudspeaker. In Experiment 1, the expected azimuth and elevation angles were 0 degrees. The errors of observed azimuth were equal to or less than 5 degrees when the distance was equal to or less than 50 m. The errors of observed elevation were equal to or less than 10 degrees except in the case of 15 m distance. The slightly larger error when the distance = 15 m was expected to be due to the vocalization of another species (Brown-eared Bulbul (*Hypsipetes amaurotis*)). The large error when the distance = 0 m is expected to be due to that the DOA substantially became different among localized sounds because they can change drastically even with small noise if the microphone is right under the loudspeaker. Thus, both the elevation and azimuth angles of replayed vocalizations were successfully estimated in this experiment.

In Experiment 2, the expected azimuth was 0 degrees, while the expected elevation decreased inversely proportional to the distance of 90 degrees, as shown in Figure 6 (bottom right). The errors of the observed azimuth were equal to or less than 5 degrees until the horizontal distance was equal to or less than 35 m, while it became larger than Experiment 1. This result was expected because the net distance between the microphone and the loudspeaker was larger. This was also expected because other species were vocalizing in the same direction as the speaker, causing the localization of replayed sounds to deviate from the expected value, which was sometimes observed in Experiment 2.

The observed elevation also reflected well the expected value; it decreased inversely proportional to the distance and errors were less than 15 degrees, except in some cases (e.g., 15 or 40 m away). We expect that the errors can be reduced if we adopt a transfer function with a higher resolution of elevation angles.

Overall, we were able to correctly estimate both the elevation and azimuth of bird vocalizations even when the songs were far away from the microphone array, if there were no other vocalizations or sounds in the similar direction as the target sounds.

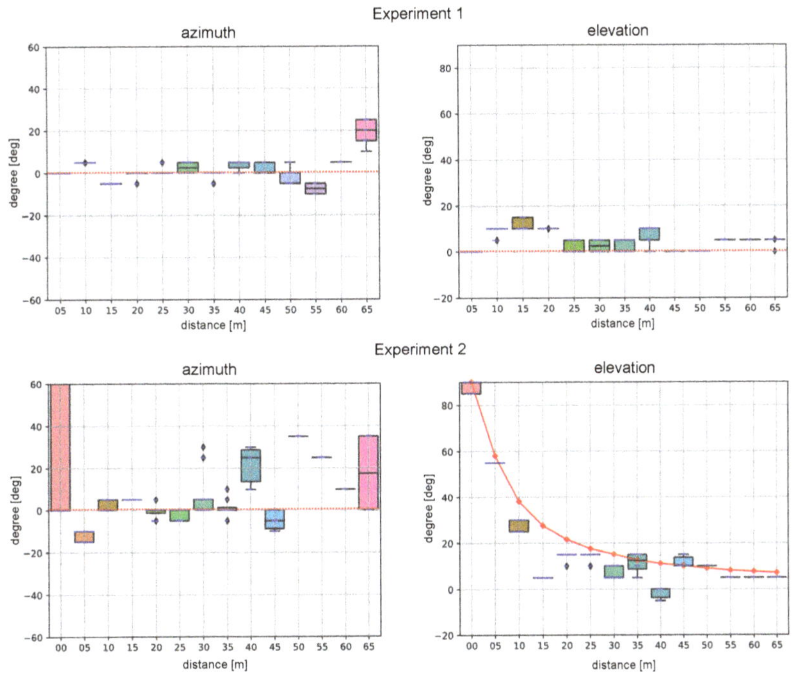

Figure 6. An estimated azimuth (**left**) and elevation (**right**) of replayed songs in Experiment 1 (**top**) and 2 (**bottom**).

3.2. Field Observation of Soundscapes of Bird Vocalizations

Figure 7 shows the song distribution on polar coordinate of azimuth (angle) and elevation (radius) in each time slot (A–E) in Figure 5. Each plot represents the direction of localized bird song (with annotations for species names). We mainly observed songs of Blue-and-white Flycatcher (*Cyanoptila cyanomelana*), Varied Tit (*Sittiparus varius*) and Coal Tit (*Periparus ater*). When we focus on a Blue-and-white Flycatcher, the individual tended to sing at much higher positions than other species, repeatedly moving to other high positions and singing a few times over B to C and returning to the starting position (A) in D. This reflects the fact that this species tends to sing on high trees along streams in his territory.

In contrast, the songs of the other two species tended to be localized at lower elevation angles, suggesting that they tend to sing at lower positions around the microphone. We also see that the localized positions formed multiple clusters, indicating they tended to move slightly. Thus, we could quantitatively observe the spatial structure of the soundscape in which one species tended to occupy the high elevation range, while the other species occupied the lower range.

Table 1 shows some indices on the localized sounds in each time slot. We observe see several changes in the soundscape structure of bird songs. The number of localized songs gradually increased over the time slots, indicating that actively singing individuals (i.e., Varied Tit) entered this acoustic scene. Elevation angle variation became smallest at

C, indicating that the Blue-and-white Flycatcher was probably at a relatively distant tree, considering this species tends to sing on top of a tree. The high values of azimuth variation in B and C reflect that the Blue-and-white Flycatcher moved during the period and other species sang in the opposite direction. Thus, we can grasp the dynamics of the soundscape structures around the microphone array by looking at the changes in these types of indices.

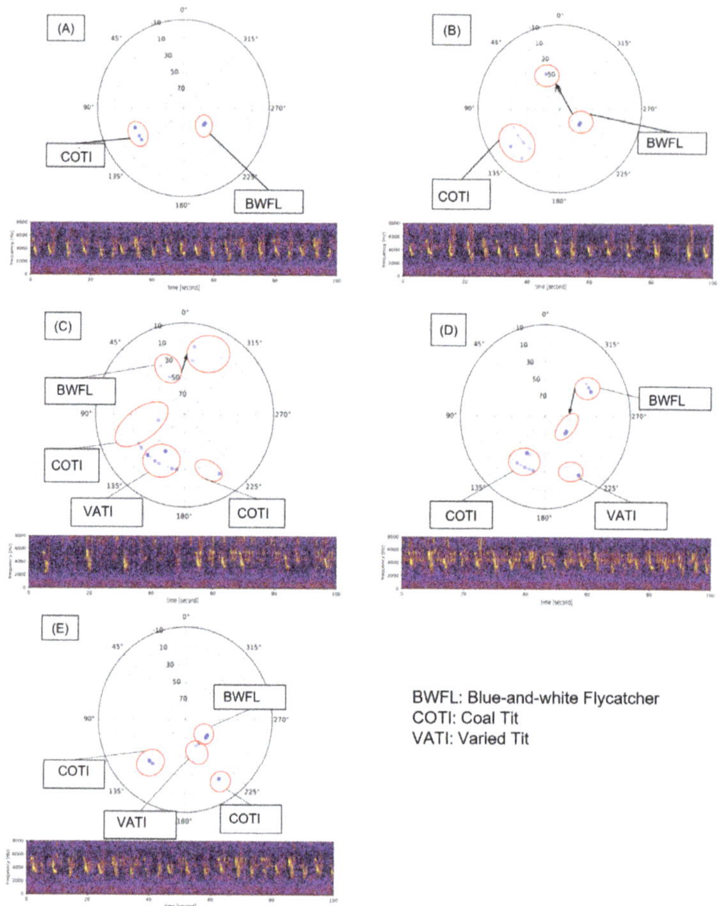

Figure 7. A song distribution on the polar coordinate of azimuth (angle) and elevation (radius) in each time slot (A–E) in Figure 5.

Table 1. The variations in azimuth and elevation of localized sounds in each time slot.

Time Slot (s)	# of Localized Songs	Azimuth Variation (rad^2)	Elevation Variation (rad^2)
A (0–100)	37	0.87	0.07
B (550–650)	31	1.15	0.10
C (650–750)	46	0.81	0.04
D (750–850)	57	0.76	0.06
E (900–1000)	45	0.48	0.09

4. Discussion and Conclusions

We discussed the applicability of robot audition techniques to understand the soundscape dynamics of bird vocalizations in forests. We focused on the elevation information of localized bird songs in addition to azimuth information. A speaker test for DOA estimation of replayed vocalizations showed that the observed DOA of a distant target sound matched well with expected values in both azimuth and elevation angle when no other bird vocalizations were present near the target. A field observation of several individuals reflected well the ecologically plausible structures of the soundscape of bird species in the experimental forest, showing vertical species structures of bird vocalizations. Several statistical indices of localized songs can also summarize the detailed changes in the structures of the soundscape.

The localization of bird vocalization is based on various components including both hardware (i.e., microphone arrays) and software (i.e., HARK). Finer resolution of the estimated DOA would be an important factor for this purpose because many sound sources other than those of the target species or individuals always coexist in fields. Improving the resolution of the MUSIC spectrum by increasing the number of steering vectors (i.e., candidate directions for DOA estimation) would be useful, but requires a great deal of computational cost, especially for azimuth-elevation estimation. The interpolation of the MUSIC spectrum used for finer 2D localization of bird vocalizations with two microphone arrays [22] would be efficient in this case. The balanced settings of DOA resolution along with interpolation would be beneficial for long-term analyses for biodiversity surveys. Further consideration of the effects of microphone channel geometry on the localization accuracy of bird vocalizations is part of our future work.

A systematic comparison of other sound source localization and separation techniques, including adaptive filtering, is important for more practical applications of robot audition techniques to bird behavioral observations. In this study, we employed the simplest and most standard methods (SEVD-MUSIC and GHDSS) employed in HARK, expecting that it will provide a baseline result because the method has been shown to be applicable to field observations of birds in previous studies as introduced in Introduction. We also expected that using such a simple method would be appropriate to examine the basic effects of acoustic noise in the natural environment, and advanced methods can improve the results (e.g., the MUSIC based on generalized singular value decomposition (GSVD-MUSIC) for better speech recognition [23], and the MUSIC method based on incremental generalized eigenvalue decomposition (iGEVD-MUSIC) for drone audition [24]).

Also, this research has an experimental rather than a theoretical aspect. Still, we believe it is important for considering the trends and challenges in robotic applications to show an example of the application of robotics to field observations of natural sounds. At the same time, we believe that a report on sound source localization in both elevation and azimuth angles is particularly important for birds that can fly. The report will contribute to the practical application of related techniques to ecoacoustics, as microphone arrays are expected to be used more frequently in this field.

The spatial localization of bird songs using multiple microphone arrays (i.e., an array of arrays) is a promising approach to determine the precise location of vocalizations. A system with three microphone array units estimated the location of two color banded Great-Reed Warbler's song posts in a reed marsh with a mean error distance of 5.5 m from the location of the observed song posts [18]. Also, various types of animal vocalization systems based on many microphone units deployed over fields have been proposed recently [25]. Gayk et al. successfully 3D triangulated, using a time difference of arrival (TDOA) approach, calls of warblers using a large microphone array unit system in which channels were far apart from each other [26]. However, it could be costly to deploy and calibrate multiple units in field observations. Our approach based on a single but multi-channel array unit, showing good accuracy of azimuth-elevation angles of bird vocalizations, suggest another possibility to better capture bird vocalizations while keeping deployment costs low.

Our results show how our observation method could be used to noninvasively monitor rare birds in the field. For example, Matsubayashi et al. evaluated the practical effectiveness of localization technology for auditory monitoring of endangered Eurasian bittern (*Botaurus stellaris*) which inhabits wetlands in remote areas with thick vegetation, using a 8-ch microphone array unit [27]. They successfully localized booming calls of at least two males in a reverberant wetland, surrounded by thick vegetation and riparian trees. In addition to the non-invasiveness to the ecosystem where the target birds inhabit, our recording system has lower deployment cost for field observers. We believe that our monitoring system, given advantages and limitations presented in this study, offers a practical tool for field ecologists, e.g., to estimate abundance and distribution of rare species.

However, estimating the distance of sound sources from microphones and two-dimensional (spatial) localization of them are important or essential for more detailed ecological surveys. We think that extracting any complementary information about their distance from separated sounds (e.g., relative amplitude [28]) would be a novel direction to better capture the structure of the soundscape with a single microphone array unit.

From another perspective, there is increasing interest and development of sound event localization and detection for various environmental sounds using microphone arrays. For example, the workshop on detection and classification of acoustic scenes and events (DCASE) provided a dataset (STARSS22) for sound source localization and classification of domestic sounds in indoor environments [29]. A competition of sound source localization and classification has been conducted, and participants discuss issues arising from the task (e.g., [11]). Experimental reports on the sound source localization of distant and elevated calls in a forest environment where many species of birds coexist, which was investigated in this study, could contribute to further progress in these fields because it may provide different insights into sound localization in harsher conditions unique to natural acoustic environments.

Although camera trap-based animal monitoring combined with object detection algorithms is widely used [30], it is challenging to capture small animals, such as songbirds, because they are basically far distant from the device, and there exists the problem of back-lighting. This experiment shows that it is possible to quantitatively extract the dynamics of the use of niches among species, which could only be described verbally or roughly before, even when the method is based only on azimuth and elevation angle information.

The increasing interest and popularity of 3D audio in public has made portable 3D recording equipment more accessible (e.g., Zoom H3-VR; Zoom Inc., GoPro MAX; GoPro Inc.). It is worth mentioning that these microphone units (or cameras with multiple microphones) are inexpensive, portable and easily affordable, even with the significant disadvantage of poor sound source localization performance due to their small size. This study also suggests the possibility of using this type of portable and easily available microphone array in ecoacoustics, which can contribute to citizen science of ornithology [31] in addition to the recent development of bird song extraction apps based on deep learning techniques (e.g., BirdNet [32]). One of the problems in the application of these approaches is the low accuracy in detection of vocalizations overlapped with each other or with other environmental sound sources. The robot audition techniques can resolve this problem by separating sound sources by making use of spectogram information from multiple channels, as discussed in this paper.

The future work includes practical comparisons of the efficiency of bird song localization and separation between the microphone arrays adopted in this study and such commercially available ones in order to further explore the applicability of robot audition techniques to ecoacoustic research.

Author Contributions: Conceptualization, K.H., H.O. and R.S.; field experiment K.H. and H.O.; formal analysis and writing, K.H. and R.S., supervision, S.M., T.A., K.N. and H.G.O., funding: R.S. and K.N. All authors have read and agreed to the published version of the manuscript.

Funding: This research was funded by JSPS/MEXT KAKENHI: JP21K12058, JP20H00475, JP19KK0260.

Institutional Review Board Statement: The experimental procedures have been approved by the planning and evaluation committee in the Graduate School of Information Science, Nagoya University (GSI-H30-1).

Data Availability Statement: The experimental data are available upon request.

Acknowledgments: We thank Noriyoshi Kaneko and Shiro Murahama for designing and conducting a speaker test. We also thank Naoki Takabe (Nagoya Univ.) for conducting a field recording in the experimental forest.

Conflicts of Interest: The authors declare no conflict of interest.

References

1. Gasc, A.; Francomano, D.; Dunning, J.B.; Pijanowski, B.C. Future directions for soundscape ecology: The importance of ornithological contributions. *Auk* **2016**, *134*, 215–228. [CrossRef]
2. Stowell, D. Computational Bioacoustic Scene Analysis. In *Computational Analysis of Sound Scenes and Events*; Virtanen, T., Plumbley, M.D., Ellis, D., Eds.; Springer: Berlin/Heidelberg, Germany, 2018; Chapter 11, pp. 303–333.
3. Farina, A.; Gage, S.H. *Ecoacoustics: The Ecological Role of Sounds*; John Wiley and Sons: Hoboken, NJ, USA, 2017.
4. Catchpole, C.K.; Slater, P.J.B. *Bird Song: Biological Themes and Variations*; Cambridge University Press: Cambridge, UK, 2008.
5. Goëau, H.; Gloti, H.; Vellinga, W.P.; Planqué, R.; Joly, A. LifeCLEF Bird Identification Task 2016: The arrival of Deep learning. In Proceedings of the CLEF: Conference and Labs of the Evaluation Forum, Évora, Portugal, 5–8 September 2016.
6. Suzuki, R.; Cody, M.L. Complex systems approaches to temporal soundspace partitioning in bird communities as a self-organizing phenomenon based on behavioral plasticity. *Artif. Life Robot.* **2019**, *24*, 439–444. [CrossRef]
7. Rhinehart, T.A.; Chronister, L.M.; Devlin, T.; Kitzes, J. Acoustic localization of terrestrial wildlife: Current practices and future opportunities. *Ecol. Evol.* **2020**, *10*, 6794–6818. [CrossRef] [PubMed]
8. Harlow, Z.; Collier, T.; Burkholder, V.; Taylor, C.E. Acoustic 3D localization of a tropical songbird. In Proceedings of the IEEE China Summit and International Conference on Signal and Information Processing (ChinaSIP), Beijing, China, 6–10 July 2013; pp. 220–224.
9. Hedley, R.W.; Huang, Y.; Yao, K. Direction-of-arrival estimation of animal vocalizations for monitoring animal behavior and improving estimates of abundance. *Avian Conserv. Ecol.* **2017**, *12*, 6. [CrossRef]
10. Gabriel, D.; Kojima, R.; Hoshiba, K.; Itoyama, K.; Nishida, K.; Nakadai, K. Case study of bird localization via sound in 3D space. In Proceedings of the 36th Annual Conference of the Robotics Society of Japan, Tokyo, Japan, 21 August 2018; p. RSJ2018AC1I2–06.
11. Nguyen, T.N.T.; Watcharasupat, K.N.; Lee, Z.J.; Nguyen, N.K.; Jones, D.L.; Gan, W.S. What makes sound event localization and detection difficult? Insights from error analysis. In Proceedings of the 6th Workshop on Detection and Classification of Acoustic Scenes and Events (DCASE 2021), online, 15–19 November 2021; pp. 120–124.
12. Nakadai, K.; Okuno, H.G.; Mizumoto, T. Development, Deployment and Applications of Robot Audition Open Source Software HARK. *J. Robot. Mechatronics* **2017**, *27*, 16–25. [CrossRef]
13. Sumitani, S.; Suzuki, R.; Matsubayashi, S.; Arita, T.; Nakadai, K.; Okuno, H.G. An integrated framework for field recording, localization, classification and annotation of birdsongs using robot audition techniques - HARKBird 2.0. In Proceedings of the IEEE International Conference on Acoustics, Speech, and Signal Processing, Brighton, UK, 12–17 May 2019; pp. 8246–8250.
14. Suzuki, R.; Sumitani, S.; Naren, N.; Matsubayashi, S.; Arita, T.; Nakadai, K.; Okuno, H.G. Field observations of ecoacoustic dynamics of a Japanese bush warbler using an open-source software for robot audition HARK. *J. Ecoacoustics* **2018**, *2*, EYAJ46. [CrossRef]
15. Sumitani, S.; Suzuki, R.; Matsubayashi, S.; Arita, T.; Nakadai, K.; Okuno, H.G. Fine-scale observations of spatio-spectro-temporal dynamics of bird vocalizations using robot audition techniques. *Remote Sens. Ecol. Conserv.* **2020**, *7*, 18–35. [CrossRef]
16. Huang, Q.; Swatantran, A.; Dubayah, R.; Goetz, S.J. The Influence of Vegetation Height Heterogeneity on Forest and Woodland Bird Species Richness across the United States. *PLoS ONE* **2014**, *9*, 103236. [CrossRef]
17. Matsubayashi, S.; Saito, F.; Suzuki, R.; Matsubayashi, S.; Arita, T.; Nakadai, K.; Okuno, H.G. Observing Nocturnal Birds Using Localization Techniques. In Proceedings of the 2021 IEEE/SICE International Symposium on System Integrations (SII), Virtual, 11–14 January 2021; pp. 493–498.
18. Suzuki, R.; Matsubayashi, S.; Saito, F.; Murate, T.; Masuda, T.; Yamamoto, Y.; Kojima, R.; Nakadai, K.; Okuno, H.G. A Spatiotemporal Analysis of Acoustic Interactions between Great Reed Warblers (Acrocephalus arundinaceus) Using Microphone Arrays and Robot Audition Software HARK. *Ecol. Evol.* **2018**, *8*, 812–825. [CrossRef]
19. Schmidt, R. Bayesian Nonparametrics for Microphone Array Processing. *IEEE Trans. Antennas Propag. (TAP)* **1986**, *34*, 276–280. [CrossRef]
20. Nakajima, H.; Nakadai, K.; Hasegawa, Y.; Tsujino, H. Blind source separation with parameter-free adaptive step-size method for robot audition. *IEEE Trans. Audio Speech Lang. Process.* **2010**, *18*, 1476–1484. [CrossRef]
21. Suzuki, R.; Matsubayashi, S.; Nakadai, K.; Okuno, H.G. HARKBird: Exploring acoustic interactions in bird communities using a microphone array. *J. Robot. Mechatronics* **2017**, *27*, 213–223. [CrossRef]
22. Sumitani, S.; Suzuki, R.; Matsubayashi, S.; Arita, T.; Nakadai, K.; Okuno, H.G. Extracting the relationship between the spatial distribution and types of bird vocalizations using robot audition system HARK. In Proceedings of the 2018 IEEE/RSJ International Conference on Intelligent Robots and Systems, Madrid, Spain, 1–5 October 2018; pp. 2485–2490.
23. Nakamura, K.; Nakadai, K.; Okuno, H.G. A real-time super-resolution robot audition system that improves the robustness of simultaneous speech recognition. *Adv. Robot.* **2013**, *27*, 933–945. [CrossRef]

24. Okutani, K.; Yoshida, T.; Nakamura, K.; Nakadai, K. Outdoor Auditory Scene Analysis Using a Moving Microphone Array Embedded in a Quadrocopter. In Proceedings of the 2012 IEEE/RSJ International Conference on Intelligent Robots and Systems (IROS 2012), Algarve, Portugal, 7–12 October 2012; pp. 3288–3293.
25. Verreycken, E.; Simon, R.; Quirk-Royal, B.; Daems, W.; Barber, J.; Steckel, J. Bio-acoustic tracking and localization using heterogeneous, scalable microphone arrays. *Commun. Biol.* **2021**, *4*, 1275. [CrossRef] [PubMed]
26. Gayk, Z.; Mennill, D.J. Pinpointing the position of flying songbirds with a wireless microphone array: Three-dimensional triangulation of warblers on the wing. *Bioacoustics* **2019**, *29*, 375–386. [CrossRef]
27. Matsubayashi, S.; Nakadai, K.; Suzuki, R.; Ura, T.; Hasebe, M.; Okuno, H.G. Auditory Survey of Endangered Eurasian Bittern Using Microphone Arrays and Robot Audition. *Front. Robot. AI* **2022**, *9*, 854572. [CrossRef] [PubMed]
28. Hedley, R.W.; Wilson, S.J.; Yip, D.A.; Li, K.; Bayne, E.M. Distance truncation via sound level for bioacoustic surveys in patchy habitat. *Bioacoustics* **2021**, *30*, 303–323. [CrossRef]
29. Politis, A.; Shimada, K.; Sudarsanam, P.; Adavanne, S.; Krause, D.; Koyama, Y.; Takahashi, N.; Takahashi, S.; Mitsufuji, Y.; Virtanen, T. STARSS22: A dataset of spatial recordings of real scenes with spatiotemporal annotations of sound events. *arXiv* **2022**, arXiv:2206.01948.
30. Tan, M.; Chao, W.; Cheng, J.K.; Zhou, M.; Ma, Y.; Jiang, X.; Ge, J.; Yu, L.; Feng, L. Animal Detection and Classification from Camera Trap Images Using Different Mainstream Object Detection Architectures. *Anim. Open Access J. MDPI* **2022**, *12*, 1976. [CrossRef] [PubMed]
31. Tulloch, A.I.; Possingham, H.P.; Joseph, L.N.; Szabo, J.; Martin, T.G. Realising the full potential of citizen science monitoring programs. *Biol. Conserv.* **2013**, *165*, 128–138. [CrossRef]
32. Wood, C.M.; Kahl, S.; Rahaman, A.; Klinck, H. The machine learning–powered BirdNET App reduces barriers to global bird research by enabling citizen science participation. *PLoS Biol.* **2022**, *20*, e3001670. [CrossRef] [PubMed]

applied sciences

MDPI

Article

An Efficient Stochastic Constrained Path Planner for Redundant Manipulators

Arturo Gil Aparicio [1] and Jaime Valls Miro [2,*]

1 System Engineering Department, Miguel Hernandez University, 03202 Elche, Alicante, Spain; arturo.gil@umh.es
2 Robotics Institute, University of Technology Sydney, Sydney, NSW 2007, Australia
* Correspondence: jaime.vallsmiro@uts.edu.au

Abstract: This brief proposes a novel stochastic method that exploits the particular kinematics of mechanisms with redundant actuation and a well-known manipulability measure to track the desired end-effector task-space motion in an efficient manner. Whilst closed-form optimal solutions to maximise manipulability along a desired trajectory have been proposed in the literature, the solvers become unfeasible in the presence of obstacles. A manageable alternative to functional motion planning is thus proposed that exploits the inherent characteristics of null-space configurations to construct a generic solution able to improve manipulability along a task-space trajectory in the presence of obstacles. The proposed Stochastic Constrained Optimization (SCO) solution remains close to optimal whilst exhibiting computational tractability, being an attractive proposition for implementation on real robots, as shown with results in challenging simulation scenarios, as well as with a real 7R Sawyer manipulator, during surface conditioning tasks.

Keywords: manipulator motion planning; manipulability; stochastic planner

Citation: Gil Aparicio, A.; Valls Miro, J. An Efficient Stochastic Constrained Path Planner for Redundant Manipulators. *Appl. Sci.* **2021**, *11*, 10636. https://doi.org/10.3390/app112210636

Academic Editors: Luis Gracia and Carlos Perez-Vidal

Received: 30 September 2021
Accepted: 4 November 2021
Published: 11 November 2021

1. Introduction

Mapping the path of an end-effector onto a configuration trajectory for the robot to accomplish a desired collision-free task is a well-known problem in robotics [1]. The consideration of redundancy, where the actuated degrees of freedom of the manipulator exceed the end-effector variables defining its functionality in the task space, adds an interesting dimension to the planning problem. It effectively facilitates a scheme where additional objectives can also be incorporated along the way. Beyond obstacle avoidance, constraints such as minimal energy, jerk-free paths, anthropomorphism and so forth can thus be considered. A particularly attractive scenario in motion planning is the avoidance of undesirable singularities in joint space [2], which limits the ability to move in certain task space directions. Increasing the manipulability of the robotic system at each time step is regarded as an effective means of moving away from the neigborhood of such configurations [3], thus reducing the hazardous condition whereby small task space movements may translate to large joint velocities. Avoiding near-singular regions is also a particularly concerning situation when the manipulator might be operating in close proximity to human operators. Moreover, operating away from singularity regions also relaxes the effect of undesirable dynamics that otherwise impose additional perturbances to the robot controllers, hence permitting superior end-effector precision whilst executing the desired task.

This brief proposes a stochastic method that exploits the particular kinematics of closed-chain mechanisms with redundant actuation and a well-known manipulability measure [4] to track the desired end-effector task-space motion in an efficient manner. The approach departs from global solutions with high computational costs, or optimal formulations that can only be solved numerically, without any guarantee of success except for simple obstacle-free problems. The approach has been tested through simulation on a number of redundant multibody topologies, and via experimental deployment on the Rethink Robotics 7R Sawyer arm.

The rest of the paper is organised as follows: Section 2 provides broad coverage of techniques in relation with this motion planning for redundant manipulators in the presence of obstacles. Next, Section 3 describes the kinematics of redundant manipulators and the exploitation of the null space. The stochastic algorithm to generate collision-free trajectories is described in Section 4. Section 5 presents a set of experiments carried out both in simulation and with a real platform. Finally, the main conclusions are described in Section 6.

2. Related Work

A robotic manipulator is considered to be redundant when it exhibits more degrees of freedom than those needed to perform the task. Typical examples of these redundant robots include serial manipulators with seven or eight degrees of freedom, mobile robots equipped with serial arms, humanoid robots and many others.

Several authors have formulated strategies to exploit the redundant degrees of freedom to improve the quality of the task being carried out. In this manner, a main task can be accomplished by the robot while the other redundant degrees of freedom are used to solve other sub-tasks [3,5] that may include: avoiding joint limits, eluding kinematic singularities and preventing the collision with obstacles in the workspace [6].

Kinematic singularities are often avoided by trying to maximize the volume of the end-effector's velocity ellipsoid [4]. When the robot is placed at a kinematic singularity, the volume of this ellipsoid is zero and the robotic manipulator loses one or more degrees of freedom. Maximizing the volume of this ellipsoid is regarded as an effective means to avoid singularities and expand on the robot's motion capabilities at a given configuration. Moreover, the optimization of the manipulability allows for faster end effector velocities (linear and angular speed), which in turn benefits the applicability of the chosen control strategies and, given the reciprocity between manipulabity and force ellipsoids, it also procures access to more precise forces and contacts. On the other hand, sub-optimal reduced manipulability often means that the contact of the end effector with the surface cannot be assured.

A global optimal control of redundancy is formulated in [7] based on Pontryagin's maximum principle. The method employs the redundant degrees of freedom to optimize manipulability or joint smoothness while attaining the same position and orientation along a trajectory. However, though reliable and fast, the method cannot be applied if the gradient of the cost function with respect to the obstacles cannot be computed. Another off-line technique along a given path is presented in [8],where a novel combination with a smoothing interpolation based on Bezier curves is proven able to avoid sharp edges and high accelerations. The method also confers the robot with the ability to avoid collisions, by accounting for the velocity of dynamic obstacles and previewing its next position in order to plan the optimal correction of the trajectory. This method was further extended to the case where redundant manipulators operate in a spatial workspace, with the added capability of avoiding sphere-like obstacles [9]. Finally, [10] deals with an optimization problem in the case where the robot is redundant along the end effector's tool axis. The technique allows the finding of a sequence via points in order to minimize the time between target points whilst avoiding obstacles.

A number of authors have proposed planning algorithms based on the discretization of the workspace. This discretization usually implies that kinematic constraints cannot be exactly satisfied and often lack in the quality of the paths produced. In [6], a deterministic approach to path planning is presented. The solution is based on the discretization of the Jacobian null-space and a backtracking strategy to prevent the incursion into kinematic singularities.

A technique based on gradient descent is presented in [11]. The method relies on computing a gradient for a cost function based on smoothness and obstacles. The trajectory of the robot inside the workspace is free and the planned position and orientation is based on a potential-based cost function.

Stochastic optimization is presented in [12]. The planner is based on a stochastic method that iteratively optimizes non-smooth cost functions, most distinctly when they cannot be easily represented by closed functions. The algorithm defines a cost function as the sum of obstacle and torque costs, plus errors in the position and orientation of the end effector. Being a sum of factors, however, entails that the optimization of the cost function cannot ensure that the end effector achieves a given position and orientation exactly.

A significant branch of stochastic planners is based on Rapidly-exploring Random Trees (RRTs) [13]. These efficient algorithms usually work by building two trees rooted at the beginning and end configurations of the trajectory. A simple greedy heuristic is presented in [14,15] to grow the trees and explore the high dimensional space while trying to connect both trees. The planner has been successfully applied to a variety of path planning problems for the computation of collision-free grasping and manipulation tasks. The growing phase of the tree considers a random generation of new samples in order to explore new solutions in the joint space, but does not include any feature that enables the optimization of the generated trajectories that are, essentially, random. A problem that arises in some of these RRT-based planning algorithms is that, often, a continuous cyclic path in task space does not correspond to a closed path in joint space. As a result, the behavior is not predictable and constitutes a risk if, for instance, a human agent is operating in the vicinity of the robot. In [16,17], a variation of an RRT-based planning algorithm is proposed that satisfies the constrains of the path and, additionally, ensures the attainment of joint trajectories that are cyclic.

The method presented here presents the following characteristics:

- Handles kinematic restrictions on the end effector exactly and does not rely on a discretization of the task space or configuration space;
- Exploits the null-space at each configuration along the path to maximize the manipulability of the robot while avoiding obstacles;
- Produces smooth trajectories that can be directly commanded to the robot without the need for a posterior smoothing phase;
- Delivers a fitting experimental performance for challenging motion problems.

3. Kinematics of Redundant Manipulators

A task generally requires a given number of degrees of freedom to be described and solved. In this sense, a manipulator is considered redundant when it possesses more degrees of freedom than those required to complete the task. For example, placing the robot's end effector at a given position and orientation inside the robot's workspace typically require six degrees of freedom. This is a main requirement, for example, during polishing applications, where the tool must keep close contact with the surface being treated [18]. As a result, two different spaces can be defined:

- The task space $\mathbb{R}^{\mathrm{m}}$: In our case we have six Degrees of Freedom (DoF) in our polishing application;
- The join space $\mathbb{R}^{\mathrm{n}}$, as the number of DoF of the robotic manipulator. In this case we have 7DoF.

The degrees of redundancy of the manipulator is $n - m$, which means that infinite solutions $\mathbf{q}$ allow reaching the same position/orientation in space. Thus, in a polishing application, 6DoF are needed to place the end effector at a particular position and orientation, while controlling the pressure on the surface. The usage of a robot with additional DoF (in our case is $n - m = 1$) may be justified by the need to avoid obstacles in the workspace when performing the task or attaining specific poses.

The inverse kinematic problem in a redundant manipulator possesses, in general, infinite solutions. This means that, if we require the end-effector to be placed at a given position and orientation inside its workspace, a self-motion of the kinematic chain can be performed. This implies that the arm joints can be reconfigured while the end effector maintains the same position and orientation in space. This fact gives these kinds of manipulators the ability to find different poses that attain dissimilar characteristics, while still complying with the task requirements. Well-known examples include, for instance, the Rethink Robotics Sawyer robot or the Willow Garage PR2 arms, with 7DoF available through a set of seven rotational joints. Additionally, combinations of robotic arms with mobile platforms can be considered within this category of redundant manipulators.

The relationship between the position and orientation of the end effector can be expressed as:

$$\mathbf{x} = \mathbf{f}(\mathbf{q}), \tag{1}$$

where $\mathbf{x}$ is the $(m \times 1)$ vector of task variables defining the position and orientation of the end effector, $\mathbf{f}$ represents a known nonlinear transformation vector and $\mathbf{q}$ is the $(n \times 1)$ vector of joint variables. The robot must then reach a set of goals defined as $G = \{\mathbf{x}_1, \mathbf{x}_2, \dots \mathbf{x}_N\}$ sampled from the desired surface to be tracked defined in the task space.

The above Equation (1) can be differentiated with respect to time as:

$$\dot{\mathbf{x}} = J(\mathbf{q})\dot{\mathbf{q}}, \tag{2}$$

where J represents the $(m \times n)$ manipulator Jacobian matrix (also defined as $\partial \mathbf{f}/\partial \mathbf{q}$). The upper dot denotes time derivative. For simplicity $J(\mathbf{q})$ is written as J. As a result, Equation (2) defines the direct kinematics of the manipulator in terms of the end effector's velocity.

Our particular path planning problem can be posed as follows. Considering a given trajectory $\mathbf{x}(t)$ as known in the task space, find a joint space trajectory $\mathbf{q}(t)$ that satisfies $\mathbf{f}(\mathbf{q}(t)) = \mathbf{x}(t)$ for any t. In our case, we try to find a set of joint vectors $Q = \{\mathbf{q}_1, \mathbf{q}_2, \dots \mathbf{q}_N\}$ such that $\mathbf{f}(\mathbf{q}_i) = \mathbf{x}_i$ for any of the points of the trajectory G. In this sense, finding an optimal trajectory with respect to the manipulability index can be stated as in Algorithm 1, which essentially proposes the optimization of the manipulability along the whole trajectory, considering that the manipulability can be expressed in a differential form with respect to the joint coordinates. Solutions to this problem have been proposed ([7,19]) at significant computational expense, becoming intractable in the presence of complex obstacle settings.

Algorithm 1 Global Manipulability Optimization.

1: Optimize: $\sum_{i=1}^{N} \omega_i = \sum_{i=1}^{N} det(J_i J_i^T)$
2: s.o. $\mathbf{q}_{i+1} = \mathbf{q}_i + \dot{\mathbf{q}}_i * \Delta t$.
3: s.o. $\dot{\mathbf{q}}_i = J_i^{\dagger}\dot{\mathbf{x}} + k_i * (I - J_i^{\dagger} J_i)\dot{\mathbf{q}}_0$
4: RETURN $Q = \{\mathbf{q}_1, \mathbf{q}_2, \dots, \mathbf{q}_N\}$ the joint path

A well-known solution to invert the kinematic Equation (2) is:

$$\dot{\mathbf{q}} = J^{\dagger}\dot{\mathbf{x}}, \tag{3}$$

where $J^{\dagger}$ is the Moore–Penrose pseudo-inverse [20], defined as $J^{\dagger} = J^T(JJ^T)^{-1}$. This pseudo-inverse has nice properties, since it minimizes the norm $|\dot{\mathbf{q}}^T\dot{\mathbf{q}}|$. Thus, given an initial joint position $\mathbf{q}_r$, the length of the computed joint trajectory is, by nature, minimal. This equation allows us to compute the joint positions required to reach a set of positions and orientations $G = \{\mathbf{x}_1, \mathbf{x}_2, \dots, \mathbf{x}_N\}$ of the end effector. This minimum square solution is represented in Algorithm 2. The algorithm, for each time step, generates a joint speed that brings the solution $\mathbf{x}$ closer to the desired goal $\mathbf{x}_i$. The final solution of the algorithm depends on the initial seed used $\mathbf{q}_r$, thus, infinite solutions can be obtained by initializing $\mathbf{q}_r$ randomly. However, this solution may produce a path that includes kinematic singularities [3] or collides with obstacles in the environment. This simple planner makes use of Algorithm 3, computing at each time step the inverse kinematics solution from a given random joint position $\mathbf{q}_r$ along the initial trajectory.

Algorithm 2 Simple Planner (G, $\mathbf{q}_r$)

INPUT: $G = \{\mathbf{x}_1, \mathbf{x}_2, \ldots \mathbf{x}_N\}$, set of positions and orientations
$\mathbf{q}_r$, an initial random seed for joint positions
OUTPUT: $\{\mathbf{q}_1, \mathbf{q}_2, \ldots, \mathbf{q}_N\}$, the joints configuration path

function SIMPLE_PLANNER(G)
  $\mathbf{q} = \mathbf{q}_r$
  for $i = 1, 2, \ldots, N$ **do**
    $\mathbf{x}_i = G\{i\}$
    $\mathbf{q}_i$ = Inverse_Kinematics($\mathbf{x}_i, \mathbf{q}$)
    $\mathbf{q} = \mathbf{q}_i$
  end for
  return $Q = \{\mathbf{q}_1, \mathbf{q}_2, \ldots, \mathbf{q}_N\}$ (the joint path)
end function

Algorithm 3 Inverse Kinematics($\mathbf{x}_i$, $\mathbf{q}_r$)

INPUT:
$\mathbf{x}_i$, the position and orientation of the end effector
$\mathbf{q}_r$, the initial seed of the algorithm
OUTPUT:
$\mathbf{q}$, joint positions

function INVERSE_KINEMATICS($\mathbf{x}_i$, $\mathbf{q}_r$)
  $\mathbf{q_i} = \mathbf{q_r}$
  while $\mathbf{x} \neq \mathbf{x}_i$ **do**
    $\dot{\mathbf{x}} = Compute\_VW(\mathbf{x}_i, \mathbf{x})$
    J_i = Jacobian($\mathbf{q}_i$)
    $\dot{\mathbf{q}}_i = J_i^{\dagger}\dot{\mathbf{x}}$
    $\mathbf{q} = \mathbf{q} + \dot{\mathbf{q}}_i * \Delta t$
    $\mathbf{x} = f(\mathbf{q})$
  end while
return q
end function
function COMPUTE_VW($\mathbf{x}_i$, $\mathbf{x}$)
  //Computes linear and angular speed
  //to reduce the error in $\mathbf{x} - \mathbf{x}_i$
  $\dot{\mathbf{x}} = (\mathbf{x} - \mathbf{x}_i)/\Delta t$ **return** $\dot{\mathbf{x}}$
end function

A more general solution to (3) can be written as:

$$\dot{\mathbf{q}} = J^{\dagger}\dot{\mathbf{x}} + (I - J^{\dagger}J)\dot{\mathbf{q}}_0, \tag{4}$$

where I is an $(n \times n)$ identity matrix and $\dot{\mathbf{q}}_0$ is an arbitrary $(n \times 1)$ joint velocity vector. This solution includes the projector operator $(I - J^{\dagger}J)$, which allows us to project a vector $\dot{\mathbf{q}}_0$ on the null space of the initial solution provided by $J^{\dagger}\dot{\mathbf{x}}$. Gradient projection methods exploit this property and compute a joint speed vector $\dot{\mathbf{q}}$ as:

$$\dot{\mathbf{q}} = J^{\dagger}\dot{\mathbf{x}} + (I - J^{\dagger}J)\frac{\partial p_0}{\partial \mathbf{q}} \tag{5}$$

p_0 being an arbitrary cost function that needs to be optimized and $\frac{\partial p_0}{\partial q}$ its gradient. Equation (5) indicates that the redundant degrees of freedom can be used to attain additional constraints, such as obtaining greater manipulability along the trajectory or avoiding collisions with the environment. Usually, manipulability is measured with the index introduced by Yoshikawa [4]:

$$\omega = \sqrt{det(JJ^T)} \tag{6}$$

Additionally, using the manipulability has a strong impact when trying to avoid kinematic singularities. In particular, that situation happens whenever the matrix J, at some joint position $\mathbf{q}$, has a rank less than m. This situation can be easily detected, since the manipulability index ω becomes null and the manipulator loses, at least, one degree of freedom, which jeopardizes its ability to complete the task.

4. Trajectory Tracking Optimisation

The manipulator task studied in this work corresponds to a tracking problem, whereby an end-effector task space trajectory needs to be mapped to a corresponding joint space trajectory. Given $\mathbf{x}_i$, representing the position and orientation of the end effector with respect to a base reference system, the proposed methodology can be regarded as a trajectory optimisation problem for the set of N waypoint goals $G = \{\mathbf{x}_1, \mathbf{x}_2, \dots \mathbf{x}_N\}$ defining the contact surface that the robot must follow with precision. Maximising the manipulability index along the trajectory whilst avoiding obstacles allows us to move away from singularities whilst facilitating the desired motion along the desired surface for the contact/visiting task being pursued.

The method starts by generating a set of K hypotheses on the path:

$$Q_k = \{\mathbf{q}_1, \mathbf{q}_2 \cdots \mathbf{q}_N\}. \tag{7}$$

Each hypothesis K of the path ensures that the robot's end effector reaches the goal $\mathbf{x}_i$ for $i = 1, \dots N$, and utilises an inverse kinematic method based on the Moore–Penrose pseudo inverse as described in Algorithm 2. The initial joint positions $\mathbf{q}_r$ of the manipulator are initialized randomly and, as a result, the resulting K paths $Q_k = \{\mathbf{q}_1, \mathbf{q}_2 \cdots \mathbf{q}_N\}$ are completely random with the property of $|\dot{\mathbf{q}}^T\dot{\mathbf{q}}|$ being minimal for each different trajectory at each time step.

This set of K joint paths form a set of hypotheses that start from different and arbitrary initial $\mathbf{q}_1$ joint configurations. Each of the generated trajectories is optimized at each time step $i = 1, \dots N$ in order to increase its manipulability index indicated in Equation (6) and, at the same time, avoid obstacles. For each of the K hypotheses over the trajectory, our method performs a sampling on the null space at each step i of the trajectory. The sampling is achieved by generating H samples around each joint configuration $\mathbf{q}_i$. To generate each new sample, we compute a vector belonging to the null space as:

$$\dot{\mathbf{q}}_h = (I - J^{\dagger}J)\dot{\mathbf{q}}_0, \tag{8}$$

where $\dot{\mathbf{q}}_0$ represents an arbitrary vector that generates a sample of the null space. Next, H random movements are generated as:

$$\mathbf{q}_h = \mathbf{q}_i + \alpha \cdot \dot{\mathbf{q}}_h \Delta t. \tag{9}$$

Any of these new samples $\mathbf{q}_h$ around each joint configuration $\mathbf{q}_i$ does not alter the position and orientation of the end effector since $\dot{\mathbf{q}}_h$ is only a self-motion and thus $J\dot{\mathbf{q}}_h = \mathbf{0}$. The variable α is chosen from a normal distribution and Δt is an integration time. The time integration step Δt and α are parameters of the algorithm.

At each time step, each of the new $\mathbf{q}_h$ (for $h = 1, \dots H$) is a self motion around the joint positions $\mathbf{q}_i$ that can potentially improve its manipulability index while avoiding obstacles. For each $\mathbf{q}_h$, we then compute a weight that accounts for the manipulability:

$$\omega = \sqrt{JJ^T} \tag{10}$$

$$c = 1/(\omega + \delta) \tag{11}$$

$$W_\omega = e^{\frac{-c}{\omega_{max}}}, \tag{12}$$

where ω_{max} is the maximum observed manipulability of the mechanism. The parameter δ avoids the division by zero if the mechanism incurs in a singularity.

In order to consider the presence of obstacles, for each sample $\mathbf{q}_h$, we compute the closest distance d of all the points of the robot arm to obstacles and with itself. The distance d is negative whenever any point of the manipulator lies inside an obstacle. We then compute:

$$x = \max(\epsilon - d, 0) \tag{13}$$

$$W_o = exp(-x/\lambda). \tag{14}$$

ϵ is the minimum required distance to the obstacles and λ is a parameter that smooths the computed weights. During the experiments we have used $\epsilon = 0.2$ m and $\lambda = 0.5$ m.

A weight is computed that accounts for the manipulability index and the distance to obstacles as:

$$W_{q_h} = W_\omega \cdot W_o. \tag{15}$$

Finally, a sample $\mathbf{q}_h$ from the null-space is selected that maximizes the weight W_{q_h}. The final path of the robot is obtained by selecting the path with the higher weight W_{q_h}. The complete algorithm is described in Algorithm 4.

Algorithm 4 Stochastic Planner(G, O)

INPUT:
$G = \mathbf{x}_1, \mathbf{x}_2, \ldots \mathbf{x}_N$, set of positions and orientations.
OUTPUT:
$Q = \mathbf{q}_1, \mathbf{q}_2, \ldots \mathbf{q}_N$, path of the robot.

function STOCHASTICPLANNER(G, O)
  //Build K initial random paths.
  for $k = 1, 2, \ldots, K$ **do**
    $\mathbf{q}_r = uniform(1, m)$
    for $i = 1, 2, \ldots, N$ **do**
      $\mathbf{q}_{i+1}$ = InverseKinematics($\mathbf{x}_i$, $\mathbf{q}_i$)
    end for
    //Store a random path for the K particle.
    $Q_k = \{\mathbf{q}_1, \mathbf{q}_2, \ldots, \mathbf{q}_N\}$
  end for
  while convergence of $\sum W_k$ **do**
    for $k = 1, 2, \ldots, K$ **do**
      $p_k = G_k$
      for $i = 1, 2, \ldots, N$ **do**
        q_i = SampleFromNullSpace($G_{k,i}$)
        w_h = ComputeWeights(q_h)
        //Find $\mathbf{q}$ that maximizes $W = \{w_1, \ldots w_H\}$
        $\hat{\mathbf{q}} = argmax_{q_h}(W_h)$
        //Add $\hat{\mathbf{q}}$ to the path
      end for
    end for
  end while
return $Q = \{\mathbf{q}_1, \mathbf{q}_2, \ldots, \mathbf{q}_N\}$, the joint path
end function
function SAMPLEFROMNULLSPACE($G_{k,i}$) **INPUT:** Current joint position i at path k.
OUTPUT: H samples q_h from null space.
  for $h = 1, 2, \ldots, H$ **do**
    $\dot{\mathbf{q}}_h = (I - J^{\dagger} J)\dot{\mathbf{q}}_0$
    $\mathbf{q}_h = \mathbf{q}_i + \epsilon \cdot \dot{\mathbf{q}}_h \Delta t$
  end for
  return $q_h = \{q_1, q_2, \ldots, q_H\}$, samples from null space.
end function

5. Results

The proposed approach is hereby demonstrated through both simulation with a 4DoF planar manipulator, and experimentally on a real robot—the 7R Rethink Robotics Sawyer manipulator. The simulations have been performed in Matlab using the *ARTE* toolbox (freely available at https://arvc.umh.es/arte, last accessed on 11 November 2021).

5.1. Preliminaries

During all the experiments, the robot's end effector must be able to achieve a given position and orientation inside the workspace thus ensuring a precise contact with the surface that allows us to complete a hypothetical surface treatment task (polishing, deburring etc.). The details of the two different robots are given next.

- A simulated 4DoF planar robot (shown in Figure 1 in the three simulated workspaces first considered): This robot is composed of four rotational joints and four links of lengths $l_1 = l_2 = l_3 = 1$ m and $l_4 = 0.3$ m. The robot is used to track a line defined in the workspace. The robot end effector must thus reach a set of goal points in the workspace $\mathbf{x}_i = (x_i, y_i, \phi_i)$ with $i = 1, \ldots, N$ representing the different task goals (x_i, y_i), and a given orientation ϕ_i set to remain always perpendicular to the 1D surfaces (lines) to be traced. Since the robot has 4DoF, there exists a free degree of freedom. At each step i, our proposed algorithm will compute the robot's path q exploiting the null space of the previous task point as described in Section 4.
- A real 7DoF manipulator: The Sawyer robot, by Rethink Robotics, is composed of seven rotational joints. The manipulator is made to follow a virtual straight path that lies exactly on a surface. Thus, the robot end effector must be capable of visiting a set of positions defined by $\mathbf{x}_i$ with $i = 1, \ldots N$ that sit on the surface, and do so with an orientation of the end effector perpendicular to the surface. Since the robot has 7DoF, there exists a free degree of freedom. Again, at each time step i, our proposed algorithm will compute the robot's path $\mathbf{q}_i$, exploiting the null space of the previous task point just visited.

During the experiments, our proposed algorithm will be used to find a trajectory that is able to complete the required task while, at the same time, optimising the manipulability at each time step and avoiding collisions.

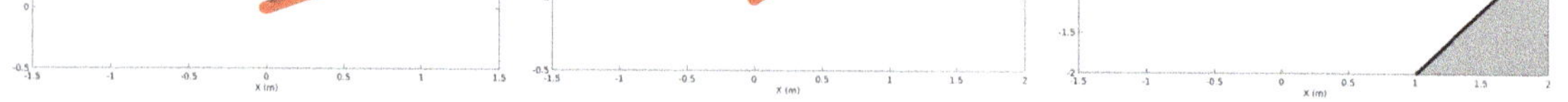

(a) (b) (c)

Figure 1. Workspaces for simulation experiments. Experiment I (**a**), Experiment II (**b**) and Experiment III (**c**). The manipulator fixed base is located at the $(0, 0)$ coordinates. The dark grey sections define the obstacle walls, with its surface identifying the contact line to be tracked by the manipulator end effector.

5.2. Simulation Study

Three different case studies have been carried out in simulation with the same 4DoF planar manipulator. The workspace has been made to be increasingly challenging to test the effectiveness of the scheme under controlled but demanding conditions:

- Case study I: In order to complete the task, the robot must follow a trajectory perpendicular to the given surface shown in Figure 1a. During the trajectory the robot must not collide with the surface or with itself. This restriction imposes that any point belonging to links 1, 2 and 3 must be at a distance higher than $\delta = 0.2$ m to the surface. In addition, any joint in the robot must be at a distance higher than $\delta = 0.2$ m from any other joint center;
- Case study II: In this simulation, the same restrictions considered in case study I apply. However, the surface path to be tracked is made more demanding, as can be seen in Figure 1b. The changes in the normal to the surfaces are accomplished by a set of smooth changes;
- Case study III: As above, but a workspace with a near-full overhang around the robot makes tracking particularly challenging for collision checking during the surface following task, almost encroaching the robot fully by its surroundings and severely limiting its mobility, as depicted in Figure 1c.

An extensive set of runs has been carried out in order to compare the outcome of our proposed approach. For each trajectory produced by the algorithm, the mean manipulability value is computed as $\hat{\omega} = (1/N)\sum w_i$, considering that the trajectory has N different waypoints. In addition, the whole trajectory is computed to check for collisions with the environment and self-collisions. The success rate accounts for the number of generated trajectories that do not collide with any obstacle.

Figure 2 presents the results in terms of manipulability at each time step along the single line environment (case studies I, II and III). For all graphs, the blue line represents the joint trajectory that possesses the higher mean manipulability that the algorithm can achieve. The trajectory with minimum mean manipulability is also shown in red, whilst green presents a trajectory that lies in the mean value of all the manipulabilities obtained. We compare the results of the proposed Stochastic Constrained Optimization (SCO) method on the right (Figure 2b,d,f) with the output of Algorithm 2 on the left (Figure 2a,c,e), representing the path solution recovered from an inverse kinematic solution based on the Moore–Penrose pseudo inverse. The initial position $\mathbf{q}_r$ used for this algorithm is selected randomly. It is clearly apparent how SCO is able to consistently produce trajectories that attain a higher degree of manipulability along the trajectory, while, at the same time, accomplishing the task at hand. Please, note also the low dispersion achieved in the manipulabilities represented in Figure 2b,d,f, indicative of the dependable ability of SCO to produce high manipulability regardless.

Table 1 presents the results of case studies I, II and III. The first row of each simulation, named "Simple Planner", represents the results obtained using Algorithm 2 when applied to each of the workspaces represented in Figure 1. The following rows, represent the results of the SCO algorithm with varying number of samples K. In order to evaluate the success rate, we check that the robot does not collide with obstacles or derives in self-collision during the resulting trajectory. As expected, the success rate of this algorithm is low during the three simulations. In addition, it also produces trajectories that exhibit low manipulability overall along the path. The first row of simulation IB uses the proposed SCO Algorithm 1 using a single hypothesis on the path ($K = 1$). In this case, the planner is able to produce a significantly higher success rate (76%) compared to Algorithm 2 (simple planner). In addition, the trajectory is optimized, both avoiding obstacles and increasing the mean manipulability. As the number of hypotheses K increases, we observe a higher success rate, while, at the same time, increasing the manipulability along the whole trajectory. It is worth noting that, for $K = 10$ hypotheses, the proposed algorithm achieves a success rate of 100%. Further increasing $K = 20, 50$ hypotheses maintains the same success rate and allows us to obtain an even higher manipulability index along the trajectories.

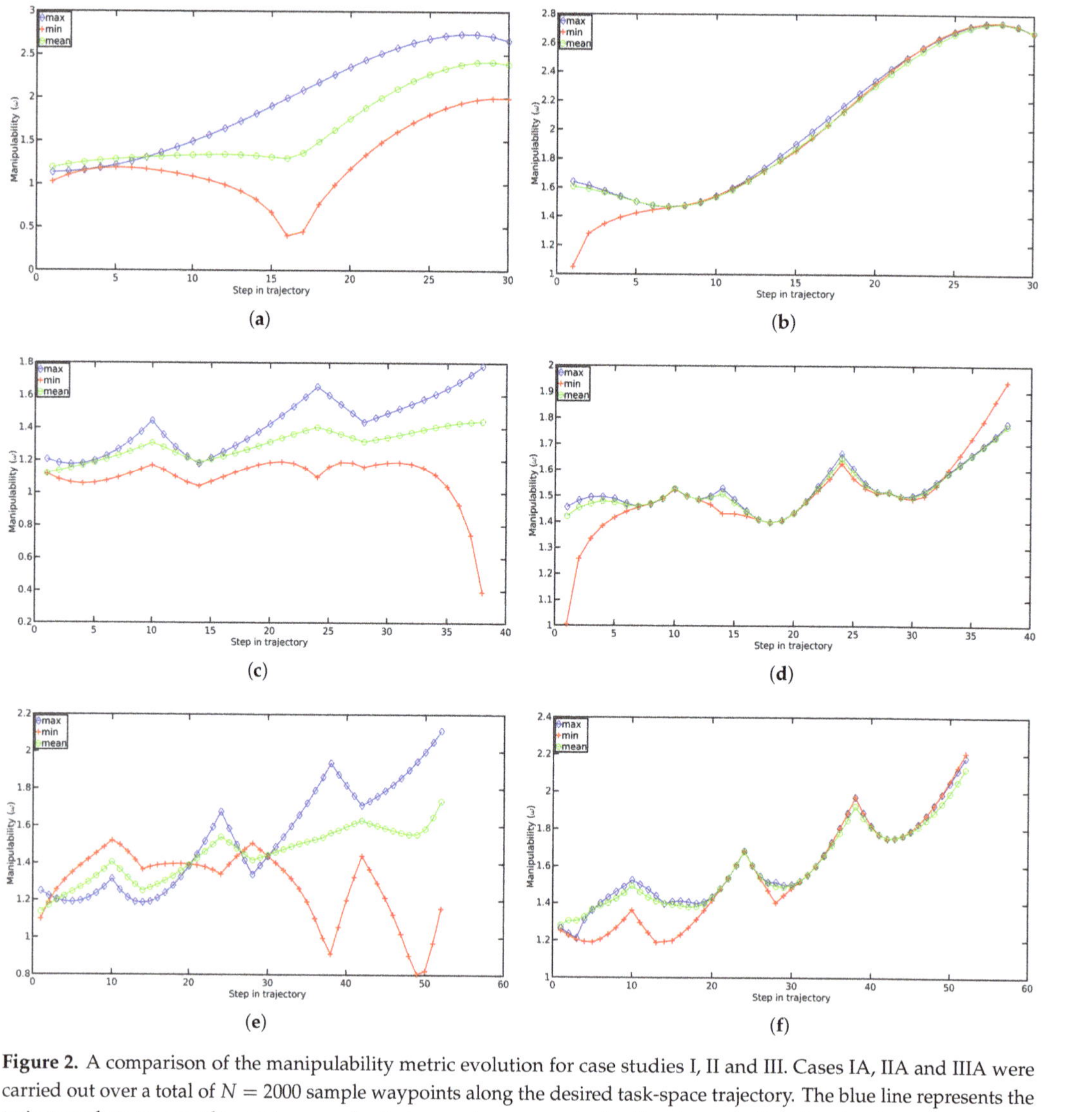

Figure 2. A comparison of the manipulability metric evolution for case studies I, II and III. Cases IA, IIA and IIIA were carried out over a total of $N = 2000$ sample waypoints along the desired task-space trajectory. The blue line represents the trajectory that presents the maximum value in terms of mean manipulability along the line that the algorithm is able to achieve, red is the minimum and green the mean value across all the runs. The corresponding simulations IB, IIB and IIIB were obtained by using SCO with K=20 samples. (**a**) Case study I A, (**b**) Case study I B, (**c**) Case study II A, (**d**) Case study II B, (**e**) Case study III A, (**f**) Case study III B.

Table 1. Results of case studies I, II and III.

Case	K	N	Success Rate (%)	Manip. Mean $\hat{\omega}$	Manip. Max.	Comp. Time (Sec.)
I A (Simple_Planner)	-	2000	44	1.64 ± 0.48	1.95	0.99 ± 0.01
I B	1	30	76	1.96 ± 0.16	2.04	4.32 ± 0.11
I B	5	30	90	1.98 ± 0.11	2.04	13.38 ± 0.17
I B	10	30	96	2.01 ± 0.05	2.04	24.27 ± 0.62
I B	20	30	100	2.02 ± 0.03	2.04	45.92 ± 2.13
I B	50	30	100	2.03 ± 0.02	2.28	114.18 ± 2.25
II A (Simple_Planner)	-	2000	26	1.29 ± 0.21	1.41	0.56 ± 0.01
II B	1	30	56	1.47 ± 0.12	1.52	4.99 ± 0.12
II B	5	30	80	1.49 ± 0.13	1.52	17.18 ± 0.19
II B	10	30	100	1.50 ± 0.06	1.52	34.66 ± 0.74
II B	20	30	100	1.52 ± 0.02	1.53	64.74 ± 3.33
II B	50	30	100	1.53 ± 0.006	1.53	156.64 ± 4.62
III A (Simple_Planner)	-	2000	12	1.20 ± 0.43	1.61	0.77 ± 0.01
III B	1	30	33	1.58 ± 0.05	1.60	6.37 ± 0.18
III B	5	30	80	1.58 ± 0.06	1.61	23.63 ± 0.66
III B	10	30	97	1.60 ± 0.05	1.61	44.85 ± 0.77
III B	20	30	100	1.60 ± 0.03	1.61	88.99 ± 3.15
III B	50	30	100	1.61 ± 0.009	1.61	219.88 ± 4.68

Table 1 also presents the results of case study II. As before, the "Simple Planner" in the first row refers to the results of the path obtained with Algorithm 2, where the simulation samples the path at a total of $N = 2000$ and, for a successful run, collisions with itself and the environment for a given trajectory must be avoided. The success rate of this algorithm is, again, very low. Given the increased complexity of the task when compared to run IA, the success rate is even lower. The results with SCO are collected under case study IIB, and as below different cases are investigated with varying initial hypotheses along the path: $K = \{1, 5, 10, 20, 30\}$. It can be observed how the proposed SCO is successful in producing a 56% of feasible solutions for $K = 1$ and 80% for $K = 5$, a marked improvement over the "Simple Planner" even with these low number of hypotheses. Moreover, as the number of hypotheses K increases, so does the success rate, while, at the same time, increasing the manipulability along the length of the trajectory. It is worth noting that, for $K = 10$, the proposed algorithm is able to achieve full success (rate of 100%). Further increasing K to 20 or 50 maintains flawless success and allows us to obtain even marginally higher manipulability along the trajectories. Table 1 also collects the results of case study III. As before, "Simple Planner" in the first row refers to the results of the path obtained with Algorithm 2. The success rate of this algorithm is, again, very low. The task in case study III is more demanding, which explains the low success rate, compared to case studies IA and IIA. Again, the results with SCO are collected under simulation IIIB with $K = \{1, 5, 10, 20, 30\}$. Following the same trends observed before, the proposed SCO produces a 33% of successful solutions for $K = 1$ and 80% for $K = 5$. The proposed algorithm also succeeds in this more demanding scenario, and as the number of hypotheses K increases, the success rate and the manipulability are increased. It is worth noting that for $K = 20$ the proposed algorithm is able to achieve full success (rate of 100%). As observed earlier, further increasing K beyond this point (to the maximum of 50 hypotheses in this case) only slightly increases the optimized manipulability while maintaining the same success rate.

Table 1 also presents computation times of each simulation carried out on an Intel™ Core i7 CPU @ 2.90GHz × 16 running Ubuntu 20.04 and Matlab™ R2018a. Mean computation times (seconds) and 2σ intervals are collected. As expected, a linear trend is observed in the computational effort.

5.3. Real Experiments

An experiment has been conducted on a physical manipulator to verify the performance of the proposed method under realistic settings. Figure 3 presents the real experimental setup with the 7R Rethink Robotics Sawyer cobot. The experiment consists of tracking a reference linear path on a flat surface whist keeping the tool in an orientation perpendicular to the surface throughout the motion. A force controller developed to maintain contact with the surface was implemented [18] to aid with the task at hand of simulating a surface conditioning assignment such as polishing, whilst the pose adopted along the path is set by the proposed SCO path planning strategy. The experimental setup is depicted in Figure 3a, consisting of a 7R Rethink Robotics Sawyer cobot, a force sensor Nano25 SI-25-25 attached to the robot end-effector, a small polishing tool proxy attached to the sensor (a cylinder of $43 \times 43 \times 10$ mm), and a target flat surface to polish.

As per the test methodology adopted in the simulation cases to show the capabilities of the proposed method, the routine undertaken includes a comparison with the standard "Simple Planner" derived from the the Moore–Penrose solution (Algorithm 2). The initial conditions are set to be the same in both cases: an obstacle-free initial pose with average manipulability established prior. This was the initial condition fed to both trajectory planners. A video is supplied (https://arvc.umh.es/arte/AppliedSciences21.mp4, accessed on 11 November 2021) to visualise the full experimental setting and resulting motion, with several stills depicting the starting, mid- and end-points of the test trajectories also being collected in Figure 3. It can be observed how the optimality in manipulability being sought out by the SCO proposed planners derives configurations that keep the elbow link down, whereas in the "Simple Planner" case that is not the case, ultimately even compromising stability by traversing near-singular regions. The reader is referred to the video linked in the manuscript where the undesirable dynamic disturbances induced in the controller are clearly apparent with vibrations at compromised locations along the path, most notably at the mid-point (Figure 3e).

The final manipulability attained in both instances, shown in Figure 4, corroborates the ability of the algorithm to seek areas with higher manipulability, hence permitting operations with manipulator configurations away from singularity regions, ultimately leading to superior end-effector precision, less energy expenditure and overall safer trajectories whilst seeking to abide by the desired end effector path during execution.

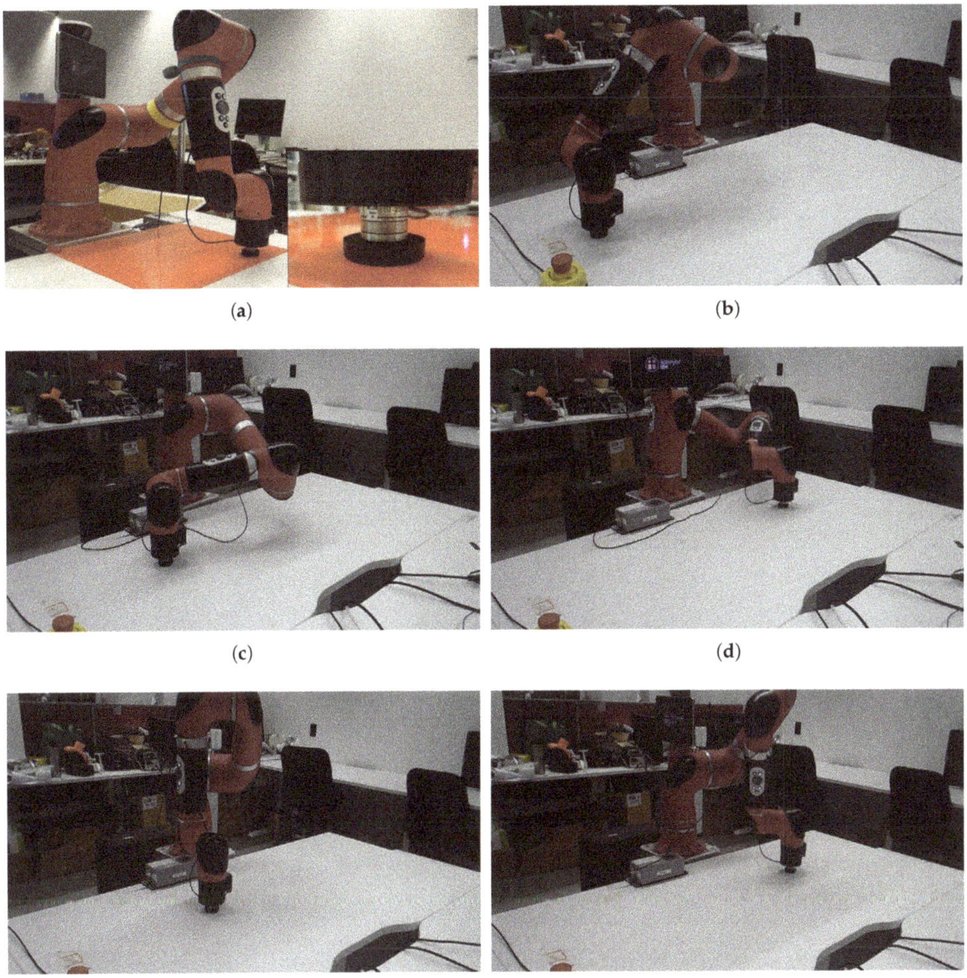

(a) (b) (c) (d) (e) (f)

Figure 3. Sawyer polishing task experiment: (**a**) setup: 7R manipulator with a force sensor attached to the robot end-effector, a tool consisting of a cylinder of 43 × 43 × 10 mm attached to the sensor and a flat surface target, (**b**) initial pose with minimal manipulability for both tests, (**c**,**d**) proposed SCO algorithm, (**e**,**f**) "Simple Planner" case.

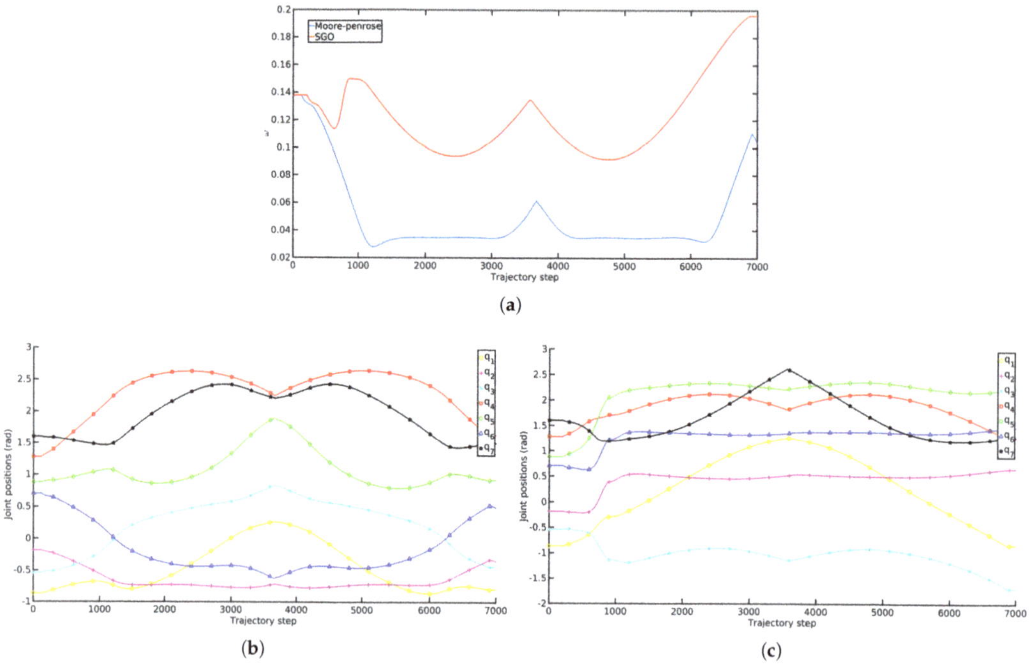

Figure 4. Real polishing experiment with the Sawyer robot: (**a**) Manipulability comparison: attained manipulability evolution of the traversed path illustrated in Figure 3, with the proposed SCO scheme shown in red (top line), and the "Simple Case" Moore–Penrose solution depicted in blue (bottom line). Corresponding (**b**) "Simple Planner" trajectory evolution, and (**c**) Evolution of joint trajectories using the SCO algorithm.

6. Conclusions

An efficient stochastic algorithm able to produce obstacle-free configuration trajectories for a given workspace has been proposed. The random process exploits the particular kinematics of closed-chain mechanisms with redundant actuation to increase manipulability along a desired end-effector task-space motion in an iterative process. The stochastic solution remains close to optimal whilst affording computational tractability, being an attractive proposition for implementation on real robots. Results from tests in challenging simulation scenarios, as well as with a 7R manipulator constrained to undertake surface treatment tasks, have been presented to show the suitability of the proposed Stochastic Constrained Optimization (SCO) trajectory planner for redundant manipulators to be able to track arbitrary task-space paths. The challenging aspect of planning trajectories, where the robot must remain in close contact with non-smooth irregular surfaces whilst optimizing the manipulability index at each time step, represents an on-going research effort continuing with the line of work presented in this manuscript.

Author Contributions: Conceptualization, A.G.A. and J.V.M.; methodology, A.G.A. and J.V.M.; software, A.G.A.; validation, A.G.A. and J.V.M.; formal analysis, A.G.A.; investigation, A.G.A.; resources, J.V.M.; writing—original draft preparation, A.G.A. and J.V.M.; writing—review and editing, A.G.A. and J.V.M.; visualization, A.G.A. and J.V.M.; supervision, J.V.M. All authors have read and agreed to the published version of the manuscript.

Funding: This research received no external funding.

Institutional Review Board Statement: Not applicable.

Informed Consent Statement: Not applicable.

Conflicts of Interest: The authors declare no conflict of interest.

References

1. Sciavicco, L.; Siciliano, B. *Modelling and Control of Robot Manipulators*, 2nd ed.; Springer: London, UK, 2000.
2. Zlatanov, D.; Fenton, R.G.; Benhabib, B. Singularity analysis of mechanisms and robots via a motion-space model of the instantaneous kinematics. In Proceedings of the 1994 IEEE International Conference on Robotics and Automation, San Diego, CA, USA, 8–13 May 1994; pp. 980–985. [CrossRef]
3. Siciliano, B.; Slotine, J. A general framework for managing multiple tasks in highly redundant robotic systems. In Proceedings of the Fifth International Conference on Advanced Robotics (ICAR'91), Pisa, Italy, 19–22 June 1991; pp. 1211–1216.
4. Yoshikawa, T. Manipulability of Robotic Mechanisms. *Int. J. Robot. Res.* **1985**, *4*, 3–9. [CrossRef]
5. Solanes, J.E.; Gracia, L.; Muñoz-Benavent, P.; Valls Miro, J.; Girbés, V.; Tornero, J. Human-robot cooperation for robust surface treatment using non-conventional sliding mode control. *ISA Trans.* **2018**, *80*, 528–541. [CrossRef] [PubMed]
6. Quispe, A.H.; Stilman, M. Deterministic Motion Planning for redundant robots along End-Effector Paths. In Proceedings of the 2012 IEEE-RAS International Conference on Humanoid Robots, Osaka, Japan, 29 November–1 December 2012; pp. 785–790. [CrossRef]
7. Nakamura, Y.; Hanafusa, H. Optimal Redundancy Control of Robot Manipulators. *Int. J. Robot. Res.* **1987**, *6*, 32–42. [CrossRef]
8. Scoccia, C.; Palmieri, G.; Palpacelli, M.C.; Callegari, M. A Collision Avoidance Strategy for Redundant Manipulators in Dynamically Variable Environments: On-Line Perturbations of Off-Line Generated Trajectories. *Machines* **2021**, *9*, 30. [CrossRef]
9. Palmieri, G.; Scoccia, C. Motion Planning and Control of Redundant Manipulators for Dynamical Obstacle Avoidance. *Machines* **2021**, *9*, 121. [CrossRef]
10. Bottin, M.; Rosati, G. Trajectory Optimization of a Redundant Serial Robot Using Cartesian via Points and Kinematic Decoupling. *Robotics* **2019**, *8*, 101. [CrossRef]
11. Ratliff, N.; Zucker, M.; Bagnell, J.A.; Srinivasa, S. CHOMP: Gradient optimization techniques for efficient motion planning. In Proceedings of the 2009 IEEE International Conference on Robotics and Automation, Kobe, Japan, 12–17 May 2009; pp. 489–494. [CrossRef]
12. Kalakrishnan, M.; Chitta, S.; Theodorou, E.; Pastor, P.; Schaal, S. STOMP: Stochastic trajectory optimization for motion planning. In Proceedings of the 2011 IEEE International Conference on Robotics and Automation, Shanghai, China, 9–13 May 2011; pp. 4569–4574. [CrossRef]
13. LaValle, S.M.; Kuffner, J.J. Randomized kinodynamic planning. In Proceedings of the IEEE International Conference on Robotics and Automation, Detroit, MI, USA, 10–15 May 1999; pp. 473–479.
14. Kuffner, J.; LaValle, S. RRT-connect: An efficient approach to single-query path planning. In Proceedings of the 2000 ICRA—Millennium Conference—IEEE International Conference on Robotics and Automation, Symposia Proceedings (Cat. No.00CH37065), San Francisco, CA, USA, 24–28 April 2000; Volume 2, pp. 995–1001. [CrossRef]
15. Berenson, D.; Srinivasa, S.; Kuffner, J. Task Space Regions: A framework for pose-constrained manipulation planning. *Int. J. Robot. Res.* **2011**, *30*, 1435–1460. [CrossRef]
16. Cefalo, M.; Oriolo, G.; Vendittelli, M. Planning safe cyclic motions under repetitive task constraints. In Proceedings of the 2013 IEEE International Conference on Robotics and Automation, Karlsruhe, Germany, 6–10 May 2013; pp. 3807–3812. [CrossRef]
17. Oriolo, G.; Cefalo, M.; Vendittelli, M. Repeatable Motion Planning for Redundant Robots Over Cyclic Tasks. *IEEE Trans. Robot.* **2017**, *33*, 1170–1183. [CrossRef]
18. Solanes, J.E.; Gracia, L.; Muñoz-Benavent, P.; Esparza, A.; Valls Miro, J.; Tornero, J. Adaptive robust control and admittance control for contact-driven robotic surface conditioning. *Robot.-Comput.-Integr. Manuf.* **2018**, *54*, 115–132. [CrossRef]
19. Mayorga, R.V.; Wong, A.K.C. On the Optimal Path Planning for Robot Manipulators. In *Kinematic and Dynamic Issues in Sensor Based Control*; Taylor, G.E., Ed.; Springer Berlin Heidelberg: Berlin/Heidelberg, Germany, 1990; pp. 179–200.
20. Moore, E.H. On the Reciprocal of the General Algebraic Matrix. *Bull. Am. Math. Soc.* **1920**, *26*, 394–395. [CrossRef]

applied sciences

MDPI

Article

A Novel Inverse Kinematic Solution of a Six-DOF Robot Using Neural Networks Based on the Taguchi Optimization Technique

Teodoro Ibarra-Pérez [1,*], José Manuel Ortiz-Rodríguez [2], Fernando Olivera-Domingo [1], Héctor A. Guerrero-Osuna [3], Hamurabi Gamboa-Rosales [3] and Ma. del Rosario Martínez-Blanco [2,*]

1 Instituto Politécnico Nacional, Unidad Profesional Interdisciplinaria de Ingeniería Campus Zacatecas (UPIIZ), Zacatecas 98160, Mexico
2 Laboratorio de Innovación y Desarrollo Tecnológico en Inteligencia Artificial (LIDTIA), Universidad Autónoma de Zacatecas, Zacatecas 98000, Mexico
3 Unidad Académica de Ingeniería Eléctrica, Universidad Autónoma de Zacatecas, Zacatecas 98000, Mexico
* Correspondence: tibarrap@ipn.mx (T.I.-P.); mrosariomb@uaz.edu.mx (M.d.R.M.-B)

Abstract: The choice of structural parameters in the design of artificial neural networks is generally based on trial-and-error procedures. They are regularly estimated based on the previous experience of the researcher, investing large amounts of time and processing resources during network training, which are usually limited and do not guarantee the optimal selection of parameters. This paper presents a procedure for the optimization of the training dataset and the optimization of the structural parameters of a neural network through the application of a robust neural network design methodology based on the design philosophy proposed by Genichi Taguchi, applied to the solution of inverse kinematics in an open source, six-degrees-of-freedom robotic manipulator. The results obtained during the optimization process of the structural parameters of the network show an improvement in the accuracy of the results, reaching a high prediction percentage and maintaining a margin of error of less than 5%.

Keywords: backpropagation; optimization methods; inverse kinematics; robotics

Citation: Ibarra-Pérez, T.; Ortiz-Rodríguez, J.M.; Olivera-Domingo, F.; Guerrero-Osuna, H.A.; Gamboa-Rosales, H.; Martínez-Blanco, M.d.R. A Novel Inverse Kinematic Solution of a Six-DOF Robot Using Neural Networks Based on the Taguchi Optimization Technique. *Appl. Sci.* **2022**, *12*, 9512. https://doi.org/10.3390/app12199512

Academic Editor: Luis Gracia

Received: 24 August 2022
Accepted: 17 September 2022
Published: 22 September 2022

1. Introduction

One of the main problems in the design of neural networks is the selection of the structural parameters of the network and their corresponding values before performing the training. In this work, the robust design artificial neural network (RDANN) methodology is used. The main focus of this methodology is based on reducing the number of experiments that can be carried out using the factorial fractional method, a statistical procedure based on the robust design philosophy proposed by Genichi Taguchi. This technique allows one to set the optimal settings on the control factors to make the process insensitive to noise factors [1,2].

Currently, the selection of the structural parameters in the design of artificial neural networks (ANNs) remains a complex task. The design of neural networks implies the optimal selection of a set of structural parameters in order to obtain greater convergence during the training process and high precision in the results. In [1], the feasibility of this type of approach for the optimization of structural parameters in the design of a backpropagation artificial neural network (BPANN) for the determination of operational policies in a manufacturing system is demonstrated, where it is shown that the Taguchi method allows designers to improve the performance in the learning speed of the network and the precision in the obtained results.

Most designers select an architecture type and determine the various structural parameters of the chosen network. However, there are no clear rules on how to choose those parameters in the selected network architecture, although these parameters determine the success of the network training. The selection of the structural parameters of the network is generally carried out through the implementation of conventional procedures based on trial and error, as shown in Figure 1, where a significant number of ANN models are generally implemented in comparison with other unconventional procedures [3–6].

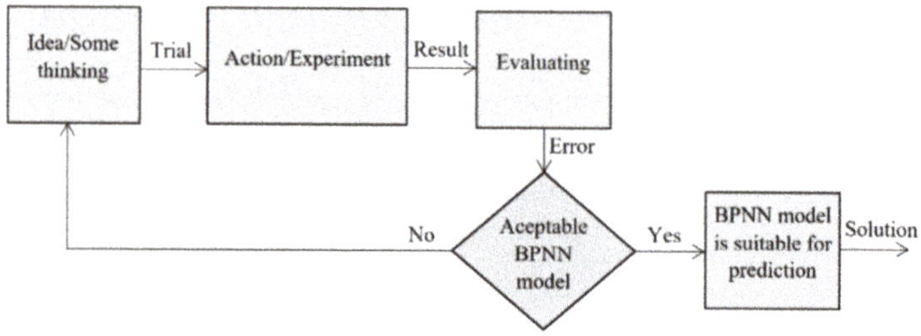

Figure 1. Trial-and-error procedure.

In this case, if a desired level of performance is not maintained, the levels in the previously established design parameters are changed until the desired performance is obtained. In each experiment, the responses are observed in order to determine the appropriate levels in the design of the structural parameters of the network [7].

A drawback in the use of this type of procedure is that one parameter is evaluated, while the others are kept at a single level, so the level selected in a variable may not necessarily be the best at the end of the experiment, since it is very likely that most of the layout variables involved will change their value. A possible solution could be that all the possible combinations in the parameters are evaluated, that is, to carry out a complete factorial design. However, the number of combinations can be very large due to the number of levels and previously established design parameters, so this method could be computationally expensive and time-consuming.

Due to all these limitations, the scientific community has shown special interest in the implementation of new approaches and procedures applied to the optimization of structural parameters in the search to generate better performance in ANNs [8–13].

Currently, ANNs can be trained to solve problems that can be complex for a human or a conventional computer, since they allow obtaining results with a high degree of precision and a significant reduction in error in real-time applications. In recent decades, the use of ANNs has been successfully applied in different fields, including pattern recognition, classification, identification, voice, vision, control systems, and robotics, the latter of which has raised special interest among researchers in the field, particularly the solution of the inverse kinematics in manipulators with six or more degrees of freedom, due to the great flexibility of control that they present for the execution of very complex tasks [14–17].

In [18], a BPNN algorithm is proposed, optimized by Fruit Fly Optimization Algorithm (FOA), to find the solution of the inverse kinematics in a four-DOF robot, obtaining an output error range −0.04686–0.1271 smaller than that obtained by a BPNN. In [19], a BPNN algorithm, optimized by means of particle swarm optimization (PSO), is studied to solve the inverse kinematic problem in a six-DOF UR3 robot applied in puncture surgery, where convergence in the precision of the results, as well as the speed and generalization capacity of the proposed network, is improved. In [20], a deep learning approach is proposed to solve the inverse kinematics in a seven-DOF manipulator. The approach used allows it to be fast, easy to implement, and more stable, allowing less sensitivity in hyperparameters. In [21], a combination of swarm intelligence (SI) and the product of exponentials (PoEs) is used to solve the inverse kinematics in a seven-DOF manipulator, where they are compared with the conventional inverse kinematics and standard PSO algorithms. In [22], the main approach is based on a redundant manipulator inverse kinematic problem that is formulated as a quadratic programming optimization problem solved by different types of recurrent neural networks. In [23], an approach is proposed to address the complexity of solving the inverse kinematics in a seven-DOF serial manipulator through an algorithm based on the Artificial Bee Colony (ABC) optimization algorithm, where two control parameters are used in order to adjust the search to optimize the distribution of the sources. In [24], an optimization approach is shown in the planning of the trajectories applied in a five-bar parallel robot for real-time control, minimizing the trajectory time and avoiding singularities in the parallel manipulator, achieving an error of less than 0.7° at the joints.

Factorial experimental design is a statistical technique used to identify and measure the effect that one variable has on another variable of interest. In 1920, R. A. Fisher studied multiple factors in the agricultural field to determine the effect of each factor on the response variable, as well as the effect of the interactions between factors on this variable. This method is known as the factorial design of experiments. Factors are variables that determine the functionality of a product or process and significantly influence system performance and can usually be controlled. To evaluate the impact of each variable, the factors must establish at least two levels; therefore, given k factors with l levels, a complete factorial design that includes all the possible combinations between these factors and levels will produce a total of l^k experimental runs. Obviously, as k or l increases, the number of experiments may become unfeasible to carry out, since a significant number of factors would imply a large number of experiments. For this reason, fractional factorial designs have been introduced, which require only a fraction of a run, unlike a complete factorial design, and which allow estimating a sufficient number of effects [25,26].

Genichi Taguchi is considered to be the author of robust parameter design through a procedure focused on reducing variation and/or sensitivity to noise in the design of products or processes, which is based on the concept of fractional factorial design. Through the implementation of orthogonal arrays (OA) and fractional factorial design, it is possible to analyze a wide range of parameters through a reduced number of experiments, ensuring a balanced comparison between the factors involved and the interaction with their different levels [2,27,28].

The Taguchi method is applied in four stages:

1. *Selection of design and noise variables*. In this stage, the most important parameters for the product/process are considered, taking into account the quality characteristics. Generally, there are variables that can be controlled by the user and others that cannot. These types of variables are known as design and noise factors, respectively, which have an important influence on the operation of the product/process. They can be determined mainly by answering the following questions: What is the optimal design condition? What factors contribute to the results and to what extent? What will be the expected result?
2. *Design and experimentation*. An OA is established, which contains the organization of the experiment taking into account the levels established for each of the factors in order to minimize the effects produced by noise factors. In other words, the adjustments made to the factors must be determined in such a way that there is the least variation in the response of the product/process, and the mean is established as close as possible to the desired objective. The OA allows the implementation of a balanced design in the weighting of the pre-established levels for each factor involved since it is possible to evaluate various factors with a minimum number of tests, obtaining a considerable amount of information through the application of few tests. The mean and variance of the response obtained in the OA configuration are combined into a single performance measure known as the signal-to-noise ratio (S/N).
3. *Analysis of results*. The S/N ratio is a quality indicator by which the effect produced on a particular parameter can be evaluated. The variation in the response obtained in dynamic characteristics, the S/N ratio, is shown below in the following equation:

$$S/N = 10 \cdot log_{10}\left(\frac{\beta_i}{MSE_i}\right), \tag{1}$$

where β_i is the square of the largest value of the signal, and MSE_i represents the root mean square deviation in the performance of the neural network, or in other words, the mean square of the distance between the measured response and the best fit line.

A valid robustness measure is related to obtaining the highest values in the S/N ratio, because the configurations of control factors that minimize the effects on noise factors can be identified.

4. *Execution and confirmation of tests in optimal conditions.* In this stage, a confirmation experiment is carried out by performing training with optimal design conditions in order to calculate the performance robustness measure and verify if this value is close to the predicted value.

Inverse Kinematics with ANNs

During the last decade, robotics had an outstanding development in the industry, particularly in aerospace, military, and medical areas, among others, especially in manipulators with a large number of degrees of freedom (DOF), due to their high flexibility and control to perform complex tasks [17,29,30].

Modern manipulators, usually kinematically redundant, allow complex tasks to be solved with high precision in the results. These types of manipulators have at least six DOF, allowing greater flexibility and mobility to perform complex tasks. The complexity in manipulator control design based on an inverse kinematic solution approach can be computationally complex, due to the nonlinear differential equation systems that are usually present. Traditional methods with geometric, iterative, and algebraic approaches have certain disadvantages and can often be generically inappropriate or computationally expensive [16,31].

The ANNs present major advantages related to nonlinearity, parallel distribution, high learning capacity, and great generalization capacity, and they can maintain a high calculation speed, thus fulfilling the real-time control requirements. Consequently, various approaches have been proposed by the scientific community in the use of intelligent algorithms applied to the control of robotic manipulators such as the use of ANNs [19,20,32,33], genetic algorithms [31,33–38], recurrent neural networks (RNNs) [37], [38], optimization algorithms [18,23,39,40], and the use of neural networks and optimization methods for parallel robots [24,41].

The organization of this work is as follows: In Section 2.1, the kinematic model of the Quetzal manipulator is established. Section 2.2 describes the procedure for generating the training and testing dataset. Section 2.3 describes the implementation of the RDANN methodology for the optimization of structural parameters in the BPNN. In Section 3, the results obtained are subjected to a reliability test stage through the use of a cross-validation method to verify that the dataset is statistically consistent. The results of training in the optimized BPNN show a significant improvement in the accuracy of the results obtained compared with the use of conventional procedures based on trial-and-error tests.

2. Materials and Methods

In this paper, a robust design model is presented through a methodological and systematic approach based on the design philosophy proposed by Genichi Taguchi. The integration of optimization processes and ANN design are methodological tools that allow the performance and generalization capacity in ANN models to be improved. In this study, an RDANN methodology was used, which was initially proposed for the reconstruction of spectra in the field of neutron dosimetry by means of ANNs [7].

The RDANN methodology was used to estimate the optimal structural parameters in a BPNN to calculate the inverse kinematics in a six-DOF robot, where the main objective was the development of accurate and robust ANN models. In other words, it was sought that the selection of the structural parameters of the proposed model allows us to obtain the best possible performance in the network.

2.1. Kinematic Analysis

The robot called Quetzal is based on an open source, 3D-printable, and low-cost manipulator [42]. The modeling and representation were carried out using the Denavit–

Hartemberg (D–H) parameters to obtain a kinematic model through four basic transformations that are determined based on the geometric characteristics of the Quetzal manipulator to be analyzed [43]. The D–H parameters are shown in Table 1.

Table 1. D–H parameters of the Quetzal manipulator.

i	Link Offset d_i (cm)	Joint Angle θ_i (rad)	Link Length a_{i-1} (cm)	Twist Angle α_{i-1} (rad)
1	20.2	θ_1	0	$\pi/2$
2	0	θ_2	16	0
3	0	$\theta_3+\pi/2$	0	$\pi/2$
4	19.5	θ_4	0	$-\pi/2$
5	0	θ_5	0	$\pi/2$
6	6.715	θ_6	0	0

The basic transformations represent a sequence of rotations and translations, where the reference system of element i is related to the system of element $i-1$. The transformation matrix is given by Equation (2).

$$ {}^{i-1}_{i}T = R_Z(\theta_i)D_Z(d_i)D_{X,}(a_{i-1})R_X(\alpha_{i-1}) \tag{2} $$

Carrying out the multiplication of the four matrices, Equation (3) is obtained:

$$ {}^{i-1}_{i}T = \begin{bmatrix} c\theta_i & -s\theta_i c\alpha_{i-1} & s\theta_i s\alpha_{i-1} & a_{i-1}c\theta_i \\ s\theta_i & c\theta_i c\alpha_{i-1} & -c\theta_i s\alpha_{i-1} & a_{i-1}s_{\theta i} \\ 0 & s\alpha_{i-1} & c\alpha_{i-1} & d_i \\ 0 & 0 & 0 & 1 \end{bmatrix} \tag{3} $$

where ${}^{i-1}_{i}T$ is the D–H transformation matrix from coordinate system i to $i-1$. $R_Z(\theta_i)$ is the rotation matrix representing a rotation θi around the Z axis, $D_Z(d_i)$ is the translation matrix representing a translation of d_i on the Z axis, $D_{X,}(a_{i-1})$ is the translation matrix representing a translation of a_{i-1} on the X axis, $R_X(\alpha_{i-1})$ is the rotation matrix representing a rotation α_{i-1} around the X axis, $c\theta_i$ is shorthand for cos $cos(\theta_i)$, $s\theta_i$ is shorthand for $sin(\theta_i)$, etc. The transformation matrices of each of the joints are multiplied to obtain the initial position of the end effector in the base reference system, as shown in Equation (4).

$$ {}^{0}_{6}T = {}^{0}_{1}A \cdot {}^{1}_{2}A \cdot {}^{2}_{3}A \cdot {}^{3}_{4}A \cdot {}^{4}_{5}A \cdot {}^{5}_{6}A = \begin{bmatrix} {}^{0}_{6}R_{3\times 3} & {}^{0}_{6}P_{3\times 1} \\ 0 & 1 \end{bmatrix} = \begin{bmatrix} n_x & o_x & a_x & p_x \\ n_y & o_y & a_y & p_y \\ n_z & o_z & a_z & p_z \\ 0 & 0 & 0 & 1 \end{bmatrix} \tag{4} $$

Therefore, the equation of the forward kinematics of the Quetzal manipulator can be expressed as shown in Equation (5).

$$ F_{forward\_k}(\theta_1,\theta_2,\theta_3,\theta_4,\theta_5,\theta_6) = (p_x,p_y,p_z,n_x,n_y,n_z,o_x,o_y,o_z,a_x,a_y,a_z) \tag{5} $$

As shown in Equation (5), the position of the end effector of the manipulator can be obtained from the angular values of the six joints of the manipulator. However, in practice, it is necessary to obtain the angles at each of the joints through a given position, so it is necessary to calculate the inverse kinematics, which can be expressed as shown in Equation (6).

$$ F_{inverse\_k}(p_x,p_y,p_z,n_x,n_y,n_z,o_x,o_y,o_z,a_x,a_y,a_z) = (\theta_1,\theta_2,\theta_3,\theta_4,\theta_5,\theta_6) \tag{6} $$

Solving (4) gives the orientation and position of the final effector in regard to the reference system, as shown in Equation (7), where the position vector $[p] = \{42.215, 0, 20.2\}$ and the orientation vector $[n\ o\ a] = \{0, 0, 1, 0, -1, 0, 1, 0, 0\}$.

$$T_0^6 = \begin{bmatrix} 0 & 0 & 1 & 42.215 \\ 0 & -1 & 0 & 0 \\ 1 & 0 & 0 & 20.2 \\ 0 & 0 & 0 & 1 \end{bmatrix} \tag{7}$$

The graphic representation of the initial position of the Quetzal robotic manipulator is shown in Figure 2 through a simulation carried out with the Robotics Toolbox for MATLAB software [44].

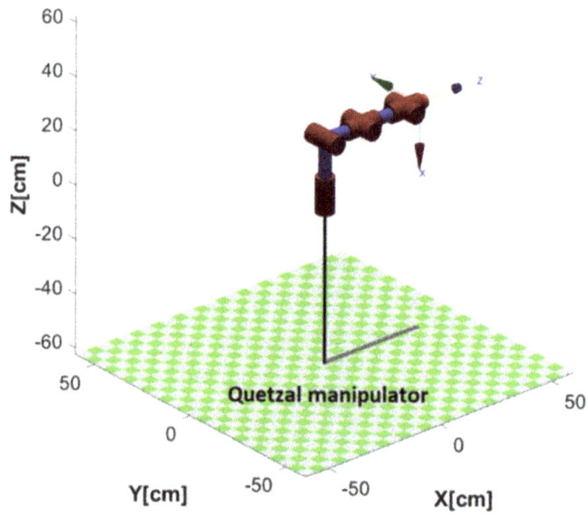

Figure 2. Representation of the initial position of the Quetzal manipulator.

2.2. Training and Testing Datasets

The dataset was generated from the equations obtained in the forward kinematics of the Quetzal manipulator. The variables involved in the proposed dataset were the orientation vector $[n\ o\ a] = \{n_x, n_y, n_z, o_x, o_y, o_z, a_x, a_y, a_z\}$, the position vector $[p] = \{p_x, p_y, p_z\}$, and the vector of joint angles $[\theta] = \{\theta_1, \theta_2, \theta_3, \theta_4, \theta_5, \theta_6\}$, for a total of 18 variables.

Table 2 shows the ranges of movement established for each of the joints in the workspace of the manipulator. Dataset X was generated with a spatial resolution of $25 \times 25 \times 25 \times 25 \times 25 \times 25 \times 25$ in a six-dimensional matrix with 18 variables involved. The total size of the dataset was 4,394,531,250 data, occupying an approximate physical memory space of 32.74 Gb due to the data class of type $Single - precision$ used in the study, where the data are stored as 4-byte (32-bit) floating point values.

Table 2. Range of motion and step size at each of the joints.

Description	θ_1	θ_2	θ_3	θ_4	θ_5	θ_6
Min	0	0	2π	0	2π	0
Max	2π	π	$\pi/2$	2π	$\pi/2$	2π
Step	$\pi/12$	$\pi/24$	$\pi/24$	$\pi/12$	$\pi/24$	$\pi/12$
Total steps	25	25	25	25	25	25

Figure 3 shows a graphical representation of the workspace using a 3D data scatter plot corresponding to the position vector $[p]$ based only on the joints $\theta_1, \theta_2, \theta_4, \theta_5$ and θ_6, where it is possible to appreciate the workspace of the robotic manipulator without taking into account joint θ_3. The illustrated workspace was generated from the forward kinematic equations as a function of the six joints.

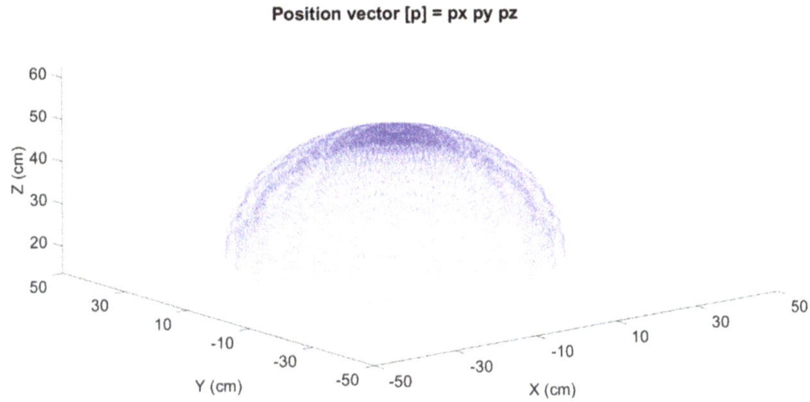

Figure 3. Position vector $[p] = \{p_x, p_y, p_z\}$ in function of joints $\theta_1, \theta_2, \theta_4, \theta_5$ and θ_6.

In order to process the enormous volume of data in a conventional processor, a data reduction filter (DRF) based on linear systematic sampling (LSS) was applied to reduce the set to a size of 190.7 Kb in memory [45]. The data were processed on an 8-core AMD Ryzen 7 5000 series processor with a base clock of 1.8 GHz, 16 GB of RAM, and integrated Radeon 16-thread graphics with a maximum clock of 4.3 GHz [46]. The dataset and code are available in Supplementary Materials.

Figure 4 shows the scatter matrices corresponding to the position dataset before and after applying the FRD filter, where a reduction of 99.99% of the data was obtained with a size of 24,410 data. In Figure 4b, it is shown that the data maintain a constant and uniform distribution with respect to the dataset of Figure 4a.

The data were normalized with a mean of zero in the range from −1 to 1 using Equation (8), where *data* is the data to normalize, *min* is the minimum value of the dataset, *range* is a value established by the difference between the maximum value and the minimum value of the dataset, *h* and *l* are the maximum and the minimum desired values for normalization.

$$DataNorm = \left(\frac{data - min}{range} \times (h - l)\right) + l \quad (8)$$

2.3. Robust Design Methodology

Figure 5 shows the RDANN methodology based on the Taguchi philosophy that consists of four stages. The designer must know the problem and choose an appropriate network model, as well as the parameters involved in the design of the network for its optimization (planning stage). By implementing an OA and systematically training a small number of ANN models (experimentation stage), the response to be analyzed is determined using the S/N relationship of the Taguchi method (analysis stage). Finally, through a confirmation process, the best performance values of the model are obtained (confirmation stage).

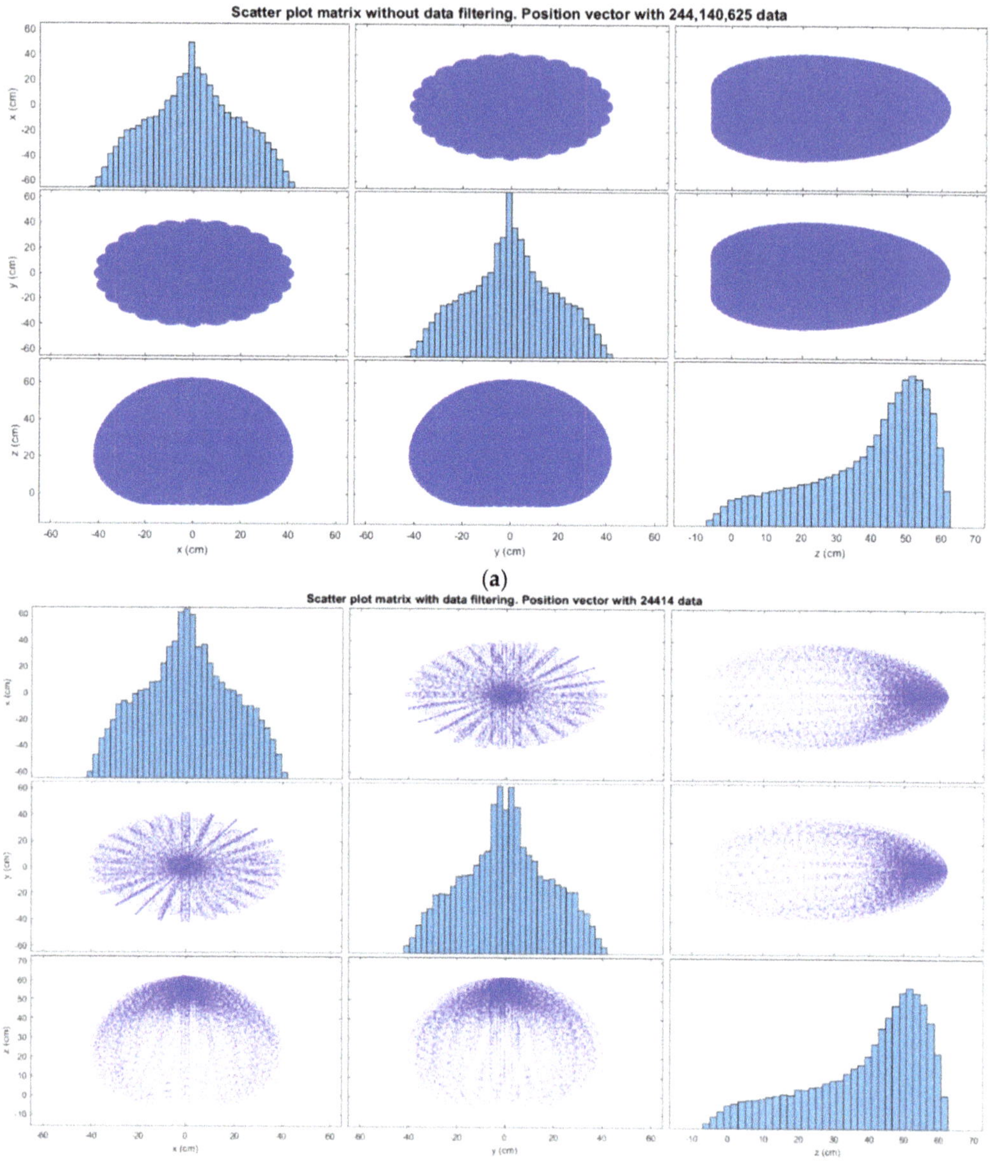

Figure 4. Scatter matrix of the position dataset: (**a**) before applying the reduction filter; (**b**) after applying the data reduction filter.

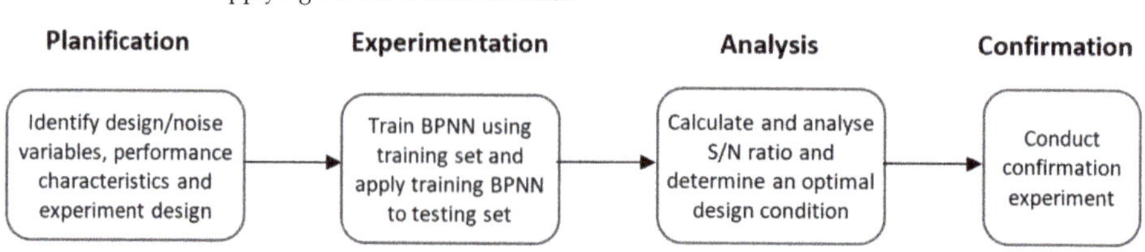

Figure 5. Robust design methodology for the optimization of structural parameters.

The graphic representation of the BPNN used in this work is shown in Figure 6, with 12 input variables and 6 output variables that correspond to the position vector $[p] = \{p_x, p_y, p_z\}$ and orientation vector $[n\ o\ a] = \{n_x, n_y, n_z, o_x, o_y, o_z, a_x, a_y, a_z\}$ as input and the vector of joint angles $[\theta] = \{\theta_1, \theta_2, \theta_3, \theta_4, \theta_5, \theta_6\}$ as output.

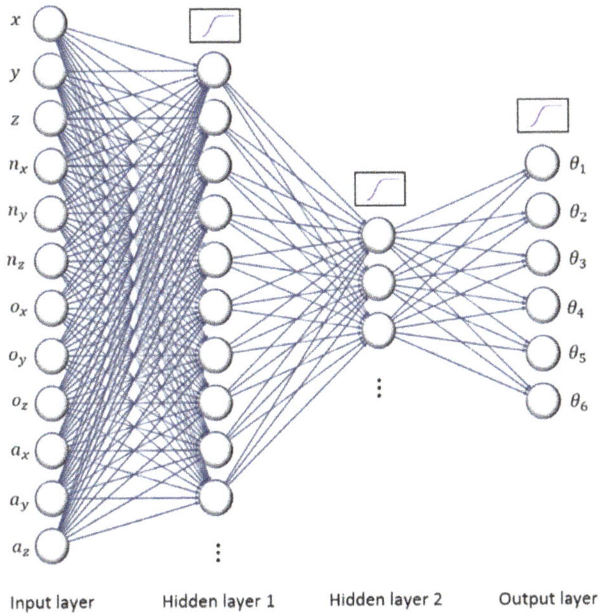

Figure 6. BPNN network topology used in this study.

2.3.1. Planning Stage

In this stage, the design variables, noise, and the objective function are identified. The objective function is defined according to the purpose and requirements of the system. In this work, the objective function is related to the prediction or classification errors between the calculator data and the data predicted by the ANN model during the testing stage. The performance at the output of the ANN or the mean square error (MSE) is used as the objective function and is described in the following equation:

$$MSE = \sqrt{\frac{1}{N}\sum_{i=1}^{N}(\theta_i^{PREDICTED} - \theta_i^{ORIGINAL})^2} \quad (9)$$

In this case, N represents the number of attempts, $\theta_i^{PREDICTED}$ represents the set of joint values that are predicted by the BPANN, and $\theta_i^{ORIGINAL}$ represents the set of joint values.

The design variables correspond to those that can be controlled by the user, such as the number of neurons in the first layer, the number of neurons in the second layer, the momentum constant, and the learning rate. By contrast, the noise variables are commonly not directly controlled by the user in most cases, such as the initialization of synaptic weights that are generally assigned randomly, the size of the training sets versus test sets, and the random selection of training and test sets. According to the requirements of the problem, the user can choose the factors related to variation in the system during the optimization process. Four design variables and three noise variables were selected because they were directly involved with the performance of the ANN, as described below in Table 3 with their respective configuration levels.

In terms of variables, *A* is the number of neurons in the first hidden layer; *B* is the number of neurons in the second hidden layer; *C* is the momentum constant, which allows the stabilization of the updating of each of the synaptic weights taking into account the sign of the gradient; *D* is the learning rate, which allows us to define the cost that the gradient has in updating a weight because the increase or decrease in the synaptic weight is related to the magnitude of the proposed value, so it may or may not affect the convergence of the MSE, causing instability and divergence [47]. *X* is the initial set of weights, *Y* is the size in proportions of the dataset, and *Z* is the random selection from the training and testing set

Table 3. Design and noise variables.

Variables	Level 1	Level 2	Level 3
A	L1	L2	L3
B	L1	L2	L3
C	L1	L2	L3
D	L1	L2	L3
X	L1	L2	L3
Y	L1	L2	L3
Z	L1	L2	L3

Once the variables and their respective levels were chosen, a suitable OA was chosen to carry out the training sessions. An OA is described as $L_r(S^c)$, where r represents the number of rows, c represents the number of columns, and s represents the number of levels in each of the columns. In this experiment, the columns of the OA represent the parameters to be optimized, and the rows represent the tests carried out by combining the three proposed levels.

2.3.2. Experimentation Stage

The success in this stage depends on an adequate choice of the OA because, in this process, a series of calculations are carried out in order to evaluate the interaction and the effects produced between the variables involved through a reduced number of experiments For the implementation of a robust design, Taguchi suggests the use of a configuration in two crossed OAs with $L_9(3^4)$ and $L_4(3^2)$, as shown below in Table 4.

Table 4. Recording of responses in trials using crossed OAs with $L_9(3^4)$ and $L_4(3^2)$ configurations.

Trial No.	A	B	C	D	G1	G2	G3	G4	Average	S/N
1	1	1	1	1	$Resp_{1,1}$	$Resp_{2,1}$	$Resp_{3,1}$	$Resp_{4,1}$	Avg1	SN1
2	1	2	2	2	$Resp_{1,2}$	$Resp_{2,2}$	$Resp_{3,2}$	$Resp_{4,2}$	Avg2	SN2
3	1	3	3	3	$Resp_{1,3}$	$Resp_{2,3}$	$Resp_{3,3}$	$Resp_{4,3}$	Avg3	SN3
4	2	1	2	3	$Resp_{1,4}$	$Resp_{2,4}$	$Resp_{3,4}$	$Resp_{4,4}$	Avg4	SN4
5	2	2	3	1	$Resp_{1,5}$	$Resp_{2,5}$	$Resp_{3,5}$	$Resp_{4,5}$	Avg5	SN5
6	2	3	1	2	$Resp_{1,6}$	$Resp_{2,6}$	$Resp_{3,6}$	$Resp_{4,6}$	Avg6	SN6
7	3	1	3	2	$Resp_{1,7}$	$Resp_{2,7}$	$Resp_{3,7}$	$Resp_{4,7}$	Avg7	SN7
8	3	2	1	3	$Resp_{1,8}$	$Resp_{2,8}$	$Resp_{3,8}$	$Resp_{4,8}$	Avg8	SN8
9	3	3	2	1	$Resp_{1,9}$	$Resp_{2,9}$	$Resp_{3,9}$	$Resp_{4,9}$	Avg9	SN9

2.3.3. Analysis Stage

Through the S/N ratio, a quantitative evaluation is carried out, where the mean and the variation in the responses measured by the ANN with different design parameters are considered. The unit of measure is the decibel, and the formula is described as follows:

$$S/N = 10 \cdot log_{10}(MSD) \tag{10}$$

In this case, MSD is the root mean square deviation in the ANN performance. The best topology is considered when more signals and less noise are obtained; therefore, a high S/N ratio at this stage allows us to identify the best design values in the BPANN with the help of statistical analysis with the JMP software.

2.3.4. Confirmation Stage

In this stage, the value of the robustness measure is obtained based on the specifications and optimal conditions of the design. A confirmation experiment is carried out using the optimal design conditions that were previously chosen, in order to verify if the calculated value is close to the value predicted by the BPANN.

3. Results

In this work, the RDANN methodology was used for the optimal selection of the structural parameters in a feed-forward backpropagation network, known as BPNN, to find the solution to the inverse kinematics in a Quetzal robot. For the BPNN training, the "resilient backpropagation" training algorithm and $mse = 1E^{-4}$ were selected. In accordance with the RDANN methodology, an OA corresponding to the design and noise variables, respectively, was implemented in configurations $L_9(3^4)$ and $L_4(3^2)$ to determine the response to the tests during the 36 training sessions carried out.

The results obtained after applying the RDANN methodology are presented in the next sections.

3.1. Planning Stage

Table 5 shows the design and noise variables with their respective assigned values for the different levels proposed during the experiment.

Table 5. Design and noise variables with their assigned levels.

Variables	Level 1	Level 2	Level 3
A	80	100	120
B	30	60	90
C	0.1	0.2	0.3
D	0.01	0.1	0.2
X	Set1	Set2	Set3
Y	7:3	8:2	9:1
Z	Tr1/Tst1	Tr2/Tst2	Tr3/Tst3

The values for the three levels established in each of the tests regarding the number of neurons for the first hidden layer were $A = 80$, 100, and 120, respectively; for the number of neurons in the second hidden layer, they were $B = 30$, 60, and 90, respectively; for the constant momentum, they were $C = 0.1$, 0.2, and 0.3, respectively; and for the learning rate, they were $D = 0.01$, 0.1, and 0.2, respectively.

The values for the initial sets of weights X were randomly determined at all three levels. The values set in the proportions of the dataset for level 1 were $Y = 70\%$ training and 30% testing; for level 2, they were 80% and 20%, and for level 3, they were 90% and 30%, respectively; finally, the random selection of the training and testing set for level

1 was $Z = Training1/Test$; for level 2, it was $Training2/Test2$, and for level 3, it was $Training3/Test3$.

3.2. Experimentation Stage

A total of 36 training sessions were carried out by implementing the OA in $L_9(3^4)$ and $L_4(3^2)$ configurations, where the network architectures were trained and tested, obtaining the results shown in Table 6.

Table 6. Responses measured during the implementation of the crossed OA.

Trial No.	G1	G2	G3	G4	Average	S/N
1	0.0825789	0.08150994	0.08200742	0.08102935	0.0817814	−41.797957
2	0.0651544	0.06734919	0.06683558	0.0666657	0.0665012	−36.960574
3	0.0568985	0.05618239	0.05984281	0.05778079	0.0576761	−31.219083
4	0.074502	0.06984874	0.07645081	0.07592863	0.0741825	−27.856466
5	0.0618532	0.06118982	0.06071765	0.05706847	0.0602073	−28.973660
6	0.0525082	0.05237245	0.05415778	0.05406061	0.0532747	−34.832469
7	0.070358	0.06704642	0.06866684	0.07036554	0.0691092	−32.761969
8	0.0549823	0.04965534	0.05378926	0.05494074	0.0533419	−26.523630
9	0.0465695	0.04604935	0.04942250	0.05026433	0.0480764	−27.302767

For the analysis of the S/N ratio, an analysis of variance (ANOVA) was performed using the statistical software program JMP. The S/N ratio and the mean value of the MSE are two of the specific criteria for determining the appropriate levels in the variables involved in network design, and their choice is determined through a series of validation procedures carried out in the next stage, as described below.

3.3. Analysis Stage

Figure 7a shows the best network topology obtained through the normal profile; Figure 7b describes the best topology through the desirable profile, and Figure 7c describes the best network topology using the maximized desirable profile. The three network profiles were obtained through statistical analysis in the JMP software to identify the optimal values in each of the proposed profiles. After performing the analysis of the S/N ratio, the values in which the levels for each of the variables involved were nearest to the average and S/N ratio red lines on the X axis were chosen, which are described in Table 7.

Table 7. Best design values with normal desirable and maximized profiles.

Profile	A	B	C	D
Normal	80	30	0.01	0.1
Desirable	80	30	0.01	0.1
Maximized	100	30	0.2	0.2

For the choice of the best network profile obtained, three training sessions were carried out for each of the three profiles in order to contrast them based on the size of the training and test data and their generalization capacity, estimating the percentage of correct answers in the prediction of the data, obtaining the results shown in Table 8.

The best topology corresponds to the maximized desirable profile, with the percentage of obtained hits being 87.71% with a margin of error of less than 5% in the tests. Once the best topology was chosen, the statistical tests of correlation and chi-square were performed, showing the best and worst prediction of the network, as shown in Figures 8 and 9, respectively.

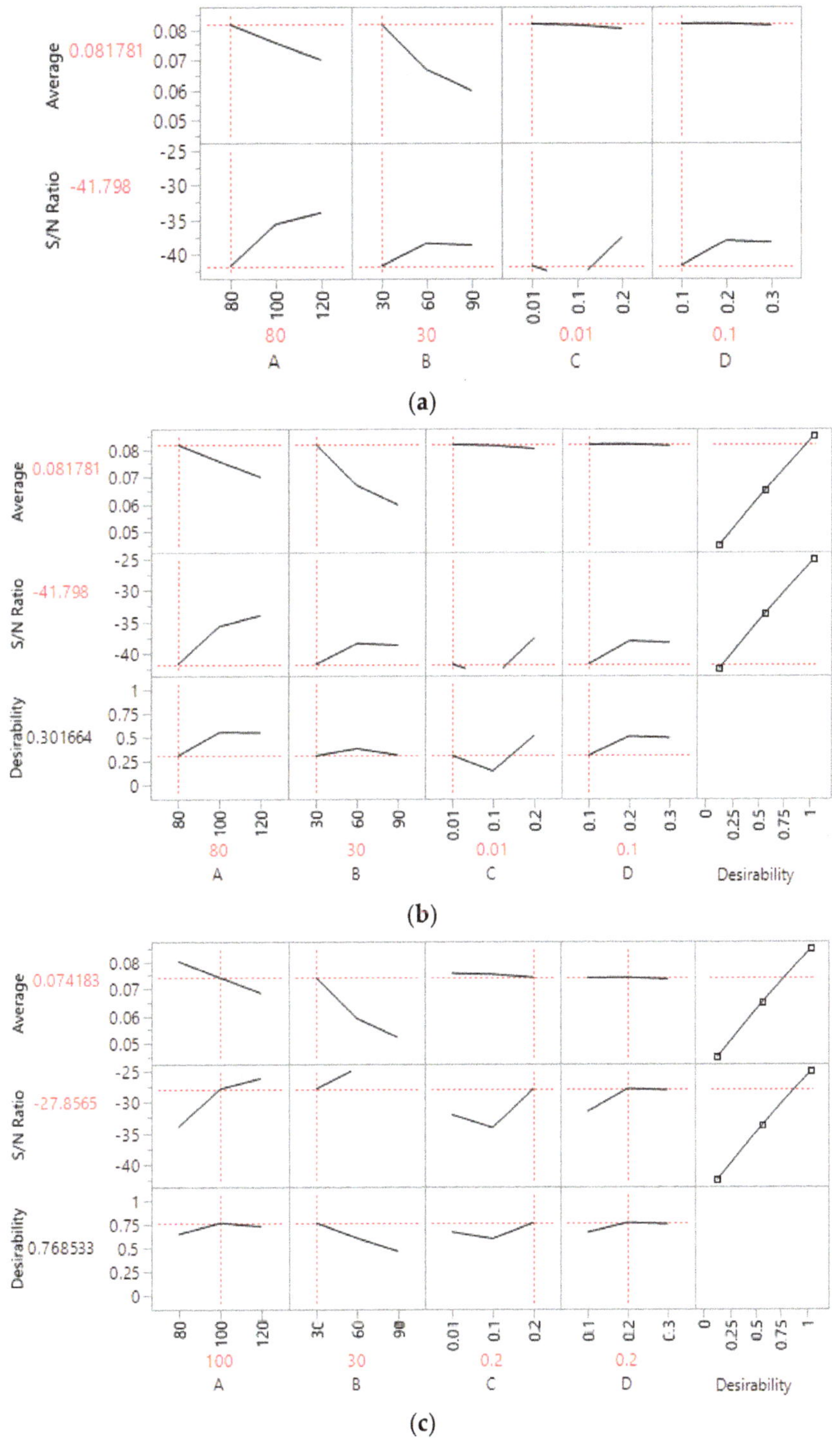

Figure 7. S/N analysis for the determination of the optimal parameters of the network: (**a**) normal profile; (**b**) desirability profile; (**c**) maximized desirability profile.

Table 8. Comparison of density and percentage of hits in the three best profiles.

Training	Profile	Density	Training time (m)	% of Hits $\chi^2<5\%$
1	Normal	70:30	13.2649	85.72
2	Normal	80:20	16.1627	85.83
3	Normal	90:10	18.3473	86.40
4	Desirable	70:30	13.4989	87.39
5	Desirable	80:20	16.1765	85.32
6	Desirable	90:10	18.5385	85.50
7	Maximized	70:30	16.3393	87.71
8	Maximized	80:20	17.7275	86.30
9	Maximized	90:10	19.5463	86.86

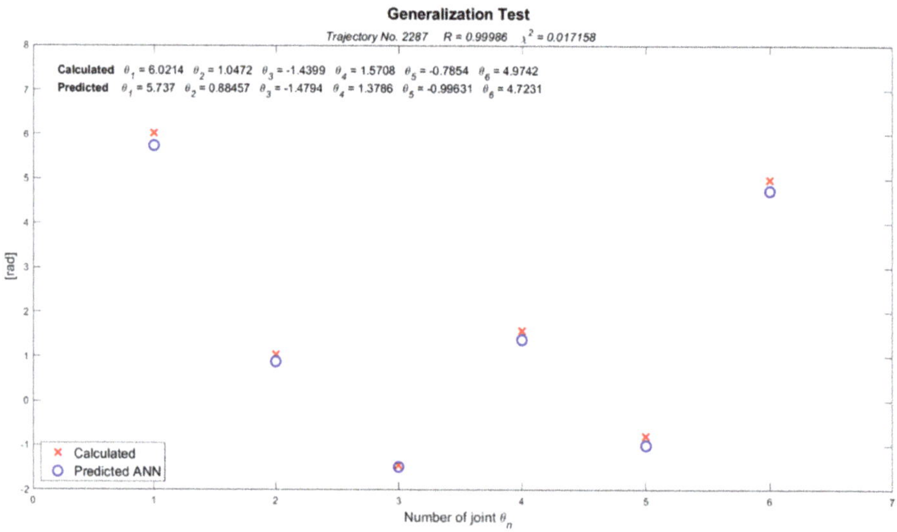

(**a**)

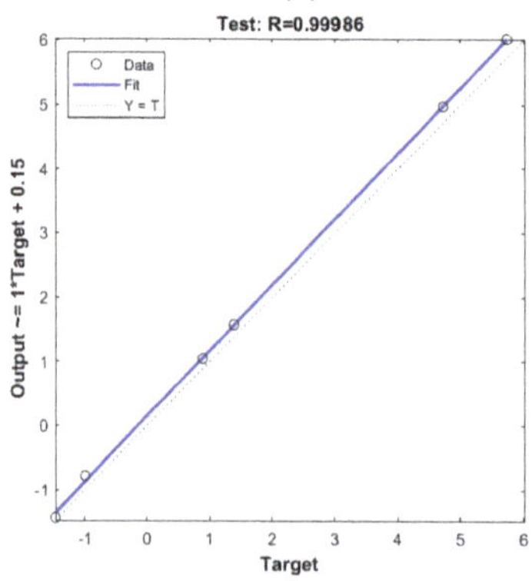

(**b**)

Figure 8. Trajectory of the manipulator with the best prediction and correlation test: (**a**) best predicted values; (**b**) correlation test.

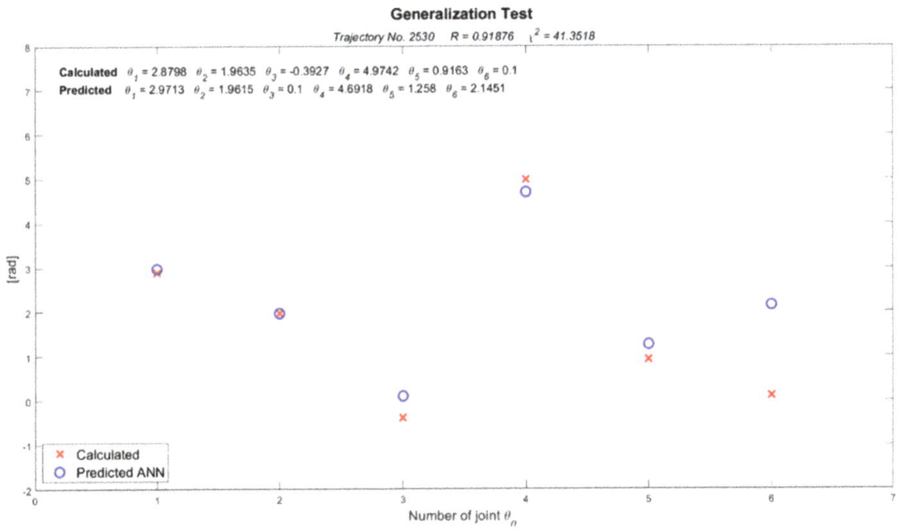

(a)

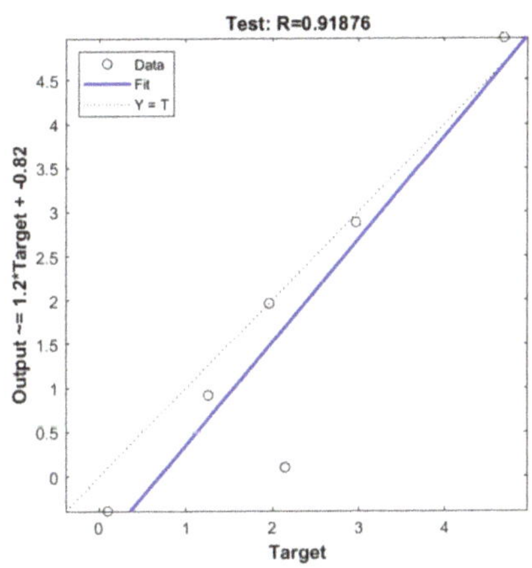

(b)

Figure 9. Trajectory of the manipulator with the worst prediction and correlation test: (**a**) worst predicted values; (**b**) correlation test.

To determine if the predicted data are statistically reliable, the cross-validation method was used by splitting the training and testing datasets. The set was split into five subsets of the same size, as shown in Figure 10. The validation subset in each training session was used to measure the generalization error, in other words, the misclassification rate of the model with data dissimilar from those previously applied during the training procedure. The cross-validation procedure was implemented on the training and testing datasets, and the average value of MSE and the standard deviation obtained were very close to those obtained in the confirmation stage [48].

Table 9 shows the results obtained in the cross-validation process, where it is observed that the average training value was equal to 17.5099, the average percentage of hits considering an error of less than 5% was equal to 87.86%, and the average value of MSE was equal to 17.5099, with standard deviations of 1.5848, 1.8974, and 0.0059, respectively.

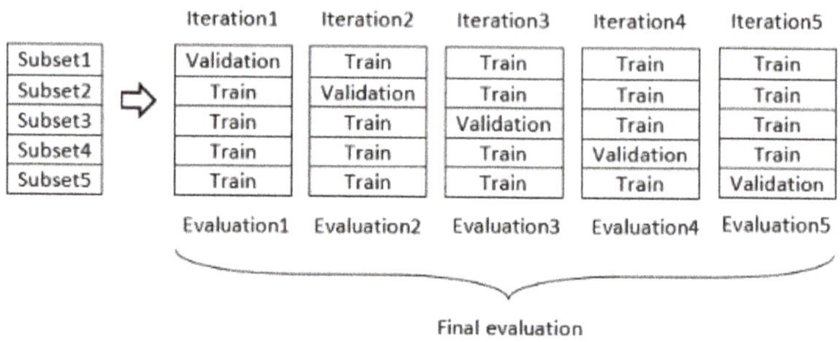

Figure 10. Cross-validation model.

Table 9. Cross-validation results.

Profile	Training Time (m)	% of Hits χ^2<5%	MSE
1	14.6750	85.17	0.0757
2	18.2350	86.95	0.0690
3	18.2165	87.93	0.0669
4	18.2044	89.75	0.0633
5	18.2189	89.51	0.0603
Average	17.5099	87.86	0.0670
Standard deviation	1.5848	1.8974	0.0059

In relation to the three profiles analyzed, the choice of the appropriate levels for the structural parameters of the best network topology were those corresponding to the maximized desirable profile with 100 and 30 neurons, respectively, a momentum of 0.2, and a learning rate of 0.2.

Figure 11 shows the layered surface diagram of the neural network used in this work. The training was performed using MATLAB software. The ANN was composed of an input layer with 100 neurons, a hidden layer with 30 neurons, and an output layer with 6 neurons. All three layers used the activation function. The training algorithm used to adjust the weighting of the synaptic weights was *resilient backpropagation*.

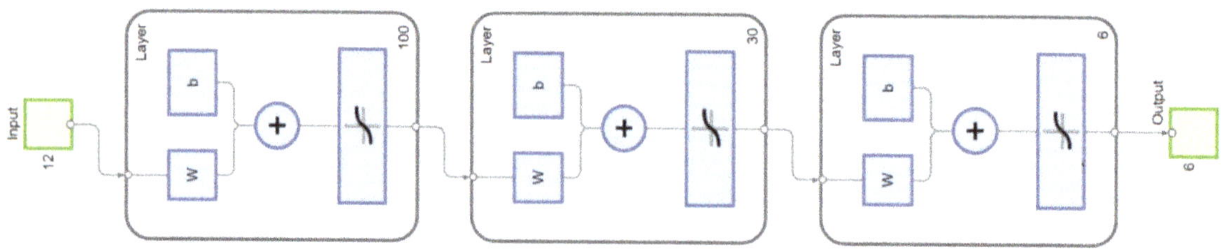

Figure 11. Best maximized desirable topology used in this study.

3.4. Implementation Results Compared with Simulation Results

Table 10 shows the measurement of the 10 trajectories predicted by the Quetzal manipulator and the error generated in comparison with the calculated trajectory. To analyze the data, 10 trajectories were chosen from the training dataset, and the simulation of each of them was carried out in order to obtain the distance traveled from the initial position to the final point.

Table 10. Trajectory comparison.

Trajectory	Calculated Final Position (cm)	Calculated Measure	Predicted Position (cm)	Measured Distance (cm)	Error %
6	[−0.93, −3.81, 59.34]	4.9862	[5.51, 4.22, 58.91]	5.2	4.2
13	[−4.56, −19.98, 52.47]	22.78	[−8.78, −17.90, 52.97]	22.5	1.2
161	[−11.32, −35.76, 22]	55.13	[−27.17, −25.88, 18.96]	54.5	1.1
216	[−22.27, −26.95, 28]	48.97	[−27.18, −23.14, 28]	51.5	5.1
236	[−14.97, −25.94, 25.77]	47.33	[−28.22, −13.01, 20]	51	7.7
317	[14.32, −27.56, 8.46]	62.25	[14.58, −24.46, 4.94]	63.5	2
663	[4.36, −18.87, 46.35]	25.16	[5.12, −19.25, 46.33]	26	3.3
988	[30.75, −12.49, 33.98]	43.70	[26.63, −20.93, 29.13]	45.5	4.1
1025	[−12.41, 1.87, 57.09]	13.63	[−11.11, 4.18, 58.39]	14	2.7
1216	[23.41, 24.49, 37.90]	41.82	[19.59, 28.94, 34.78]	43.5	4
				Average	3.5

The greatest error observed was in trajectory number 236, with a value of 7.7% compared with the calculated one, while for trajectory number 6, the error was 1.1% compared with the calculated one. A mean error of 3.5% was obtained for the implementation of the 10 physically realized trajectories using the low-cost (approximately USD 1500) 3D-printed Quetzal manipulator.

3.5. Comparative Analysis

Table 11 shows the values obtained in the design of the optimized BPNN in comparison with the BPNN based on trial and error and other methods used in the optimization of the structural parameters in ANN. As can be seen, the conventional BPNN method based on trial and error shows a greater difficulty in determining the optimal parameters, whereas the optimized BPNN results in a shorter time in the training process than the other methods; in addition, it involves noise parameters that are necessary to generate greater robustness in the network design.

Table 11. Comparison of results with other methods.

Method	Iterations	Training Time	Total Tested Networks	Prediction Error
BPNN trial and error	Often millions	Several hours	100 in several hours	Undetermined
PSO conventional [49]	5000	Not specified	10	0.00913
PSO [11]	100	03:51:38	10 in 39 h	0.0456
This study	10,000	00:17:30	36 in 178 min	0.0171

4. Conclusions and Discussion

Various approaches and powerful learning algorithms of great value have been introduced in recent decades; however, the integration of the various approaches in ANN optimization has allowed researchers to improve performance and generalizability in ANNs. The results of this work revealed that the proposed systematic and experimental approach is a useful alternative for the robust design of ANNs since it allows simultaneously considering the design and the noise variables, incorporating the concept of robustness in the design of ANNs. The RDANN methodology used in this work was initially proposed in the field of neutron dosimetry, so it was adapted for implementation in the field of robotics, allowing us to improve the performance and generalization capacity in an ANN to find the solution to the inverse kinematics in the Quetzal manipulator.

The time spent during the network design process was significantly reduced comparec with the conventional methods based on trial and error. In the methods that are generally proposed by the previous experience of the researcher, the design and modeling of the network can take from several days to a few weeks or even months to test the differen ANN architectures, which can lead to a relatively poor design. The use of the RDANN methodology in this study allowed the network design to be carried out in less time, with approximately 13 h of training, due to the orthogonal arrangement corresponding to the 36 training sessions performed using a conventional AMD Ryzen 7 5700 series processor with an integrated graphics card.

Depending on the complexity of the problem, the use of this methodology allows handling times ranging from minutes to hours to determine the best robustness parameters in the network architecture. Therefore, it is possible to streamline the process and reduce efforts, with a high degree of precision in network performance. The use of the RDANN methodology allowed the analysis of the interaction between the values in the design variables that were involved, in order to consider their effects on network performance thus allowing a reduction in the time and effort spent in the modeling stage and speeding up the selection and interpretation of the optimal values in the structural parameters of the network. The quality of the data in the training sets, without a doubt, can significantly help to increase the performance, generalization capacity, and precision of the results obtained Although the proposed method was implemented and tested in a low-cost manipulator in this study, in future work, we plan to implement it in an industrial-type robot controller The implementation of the proposed method in parallel robotic manipulators, where the solution of the kinematics is more complex, is also considered.

Supplementary Materials: The following are available online at https://www.mdpi.com/article/10.3390/app12199512/s1, dataset: dataSetQuetzalRobot.mat, code: workspaceQuetzal.m.

Author Contributions: Conceptualization, T.I.-P. and M.d.R.M.-B.; methodology, J.M.O.-R. and T.I.-P.; software, H.G.-R. and H.A.G.-O.; validation, H.G.-R. and T.I.-P.; formal analysis, J.M.O.-R and M.d.R.M.-B.; investigation, T.I.-P.; resources, F.O.-D. and H.A.G.-O.; data curation, H.G.-R. and H.A.G.-O.; writing—original draft preparation, T.I.-P.; writing—review and editing, T.I.-P. and F.O.-D. visualization, H.G.-R. and H.A.G.-O.; supervision, M.d.R.M.-B.; project administration, M.d.R.M.-B. funding acquisition, J.M.O.-R. and T.I.-P. All authors have read and agreed to the published version of the manuscript.

Funding: This research was supported by the Instituto Politécnico Nacional (IPN) under grants number CPE/COTEBAL/14/2021, the Consejo Nacional de Ciencia y Tecnología (CONACYT-BECAS) under grants number 431101/498318 and the Consejo Zacatecano de Ciencia, Tecnología e Innovación (COZCyT).

Institutional Review Board Statement: Not applicable.

Informed Consent Statement: Not applicable.

Data Availability Statement: Not applicable.

Acknowledgments: This research was supported by the Instituto Politécnico Nacional (IPN) under grants number CPE/COTEBAL/14/2021, the Consejo Nacional de Ciencia y Tecnología (CONACYT-BECAS) under grants number 431101/498318 and the Consejo Zacatecano de Ciencia, Tecnología e Innovación (COZCyT). The authors gratefully acknowledge these support agencies. We sincerely thank the people who provided support and advice for this paper, as well as the reviewers for their comments and suggestions.

Conflicts of Interest: The authors declare no conflict of interest.

References

1. Khaw, J.F.C.; Lim, B.S.; Lim, L.E.N. Optimal Design of Neural Networks Using the Taguchi Method. *Neurocomputing* **1995**, *7*, 225–245. [CrossRef]
2. Rankovic, N.; Rankovic, D.; Ivanovic, M.; Lazic, L. A New Approach to Software Effort Estimation Using Different Artificial Neural Network Architectures and Taguchi Orthogonal Arrays. *IEEE Access* **2021**, *9*, 26926–26936. [CrossRef]
3. Snoek, J.; Larochelle, H.; Adams, R.P. Practical Bayesian Optimization of Machine Learning Algorithms. In *Advances in Neural Information Processing Systems 25 (NIPS 2012)*; Pereira, F., Burges, C.J., Bottou, L., Weinberger, K.Q., Eds.; Curran Associates, Inc.: Red Hook, NY, USA, 2012; Volume 25.
4. Bergstra, J.; Bengio, Y. Random Search for Hyper-Parameter Optimization. *J. Mach. Learn. Res.* **2012**, *13*, 281–305.
5. Kosarac, A.; Mladjenovic, C.; Zeljkovic, M.; Tabakovic, S.; Knezev, M. Neural-Network-Based Approaches for Optimization of Machining Parameters Using Small Dataset. *Materials* **2022**, *15*, 700. [CrossRef] [PubMed]
6. Teslyuk, V.; Kazarian, A.; Kryvinska, N.; Tsmots, I. Optimal Artificial Neural Network Type Selection Method for Usage in Smart House Systems. *Sensors* **2021**, *21*, 47. [CrossRef]
7. Ortiz-Rodríguez, J.M.; Martínez-Blanco, M.d.R.; Viramontes, J.M.C.; Vega-Carrillo, H.R. Robust Design of Artificial Neural Networks Methodology in Neutron Spectrometry. In *Artificial Neural Networks*; Suzuki, K., Ed.; IntechOpen: Rijeka, Croatia, 2013.
8. Zajmi, L.; Ahmed, F.Y.H.; Jaharadak, A.A. Concepts, Methods, and Performances of Particle Swarm Optimization, Backpropagation, and Neural Networks. *Appl. Comput. Intell. Soft Comput.* **2018**, *2018*, 9547212. [CrossRef]
9. Sun, W.; Huang, C. A Carbon Price Prediction Model Based on Secondary Decomposition Algorithm and Optimized Back Propagation Neural Network. *J. Clean. Prod.* **2020**, *243*, 118671. [CrossRef]
10. Huang, D.; Wu, Z. Forecasting Outpatient Visits Using Empirical Mode Decomposition Coupled with Back-Propagation Artificial Neural Networks Optimized by Particle Swarm Optimization. *PLoS ONE* **2017**, *12*, e0172539. [CrossRef]
11. Gaxiola, F.; Melin, P.; Valdez, F.; Castro, J.R.; Castillo, O. Optimization of Type-2 Fuzzy Weights in Backpropagation Learning for Neural Networks Using GAs and PSO. *Appl. Soft Comput.* **2016**, *38*, 860–871. [CrossRef]
12. Huang, H.-X.; Li, J.-C.; Xiao, C.-L. A Proposed Iteration Optimization Approach Integrating Backpropagation Neural Network with Genetic Algorithm. *Expert Syst. Appl.* **2015**, *42*, 146–155. [CrossRef]
13. Amirsadri, S.; Mousavirad, S.J.; Ebrahimpour-Komleh, H. A Levy Flight-Based Grey Wolf Optimizer Combined with Back-Propagation Algorithm for Neural Network Training. *Neural Comput. Appl.* **2018**, *30*, 3707–3720. [CrossRef]
14. Li, S.; Zhang, Y.; Jin, L. Kinematic Control of Redundant Manipulators Using Neural Networks. *IEEE Trans. Neural Netw. Learn. Syst.* **2017**, *28*, 2243–2254. [CrossRef]
15. Gao, R. Inverse Kinematics Solution of Robotics Based on Neural Network Algorithms. *J. Ambient Intell. Humaniz. Comput.* **2020**, *11*, 6199–6209. [CrossRef]
16. Kramar, V.; Kramar, O.; Kabanov, A. An Artificial Neural Network Approach for Solving Inverse Kinematics Problem for an Anthropomorphic Manipulator of Robot SAR-401. *Machines* **2022**, *10*, 241. [CrossRef]
17. Šegota, S.B.; Anđelić, N.; Mrzljak, V.; Lorencin, I.; Kuric, I.; Car, Z. Utilization of Multilayer Perceptron for Determining the Inverse Kinematics of an Industrial Robotic Manipulator. *Int. J. Adv. Robot. Syst.* **2021**, *18*, 1729881420925283. [CrossRef]
18. Bai, Y.; Luo, M.; Pang, F. An Algorithm for Solving Robot Inverse Kinematics Based on FOA Optimized BP Neural Network. *Appl. Sci.* **2021**, *11*, 7129. [CrossRef]
19. Jiang, G.; Luo, M.; Bai, K.; Chen, S. A Precise Positioning Method for a Puncture Robot Based on a PSO-Optimized BP Neural Network Algorithm. *Appl. Sci.* **2017**, *7*, 969. [CrossRef]
20. Malik, A.; Lischuk, Y.; Henderson, T.; Prazenica, R. A Deep Reinforcement-Learning Approach for Inverse Kinematics Solution of a High Degree of Freedom Robotic Manipulator. *Robotics* **2022**, *11*, 44. [CrossRef]
21. Malik, A.; Henderson, T.; Prazenica, R. Multi-Objective Swarm Intelligence Trajectory Generation for a 7 Degree of Freedom Robotic Manipulator. *Robotics* **2021**, *10*, 127. [CrossRef]
22. Hassan, A.A.; El-Habrouk, M.; Deghedie, S. Inverse Kinematics of Redundant Manipulators Formulated as Quadratic Programming Optimization Problem Solved Using Recurrent Neural Networks: A Review. *Robotica* **2020**, *38*, 1495–1512. [CrossRef]
23. Zhang, L.; Xiao, N. A Novel Artificial Bee Colony Algorithm for Inverse Kinematics Calculation of 7-DOF Serial Manipulators. *Soft Comput.* **2019**, *23*, 3269–3277. [CrossRef]
24. Bourbonnais, F.; Bigras, P.; Bonev, I.A. Minimum-Time Trajectory Planning and Control of a Pick-and-Place Five-Bar Parallel Robot. *IEEEASME Trans. Mechatron.* **2015**, *20*, 740–749. [CrossRef]
25. Arboretti, R.; Ceccato, R.; Pegoraro, L.; Salmaso, L. Design of Experiments and Machine Learning for Product Innovation: A Systematic Literature Review. *Qual. Reliab. Eng. Int.* **2022**, *38*, 1131–1156. [CrossRef]
26. Ou, Z.; Zhang, M.; Qin, H. Tripling of Fractional Factorial Designs. *J. Stat. Plan. Inference* **2019**, *199*, 151–159. [CrossRef]
27. Rahman, M.A.; Muniyandi, R.c.; Albashish, D.; Rahman, M.M.; Usman, O.L. Artificial Neural Network with Taguchi Method for Robust Classification Model to Improve Classification Accuracy of Breast Cancer. *PeerJ Comput. Sci.* **2021**, *7*, e344. [CrossRef] [PubMed]
28. Manni, A.; Saviano, G.; Bonelli, M.G. Optimization of the ANNs Predictive Capability Using the Taguchi Approach: A Case Study. *Mathematics* **2021**, *9*, 766. [CrossRef]

29. Zhang, Y.; Guo, D.; Li, Z. Common Nature of Learning Between Back-Propagation and Hopfield-Type Neural Networks for Generalized Matrix Inversion With Simplified Models. *IEEE Trans. Neural Netw. Learn. Syst.* **2013**, *24*, 579–592. [CrossRef] [PubMed]
30. Alebooyeh, M.; Urbanic, R.J. Neural Network Model for Identifying Workspace, Forward and Inverse Kinematics of the 7-DOF YuMi 14000 ABB Collaborative Robot. *IFAC PapersOnLine* **2019**, *52*, 176–181. [CrossRef]
31. Köker, R. A Genetic Algorithm Approach to a Neural-Network-Based Inverse Kinematics Solution of Robotic Manipulators Based on Error Minimization. *Inf. Sci.* **2013**, *222*, 528–543. [CrossRef]
32. Volinski, A.; Zaidel, Y.; Shalumov, A.; DeWolf, T.; Supic, L.; Tsur, E.E. Data-Driven Artificial and Spiking Neural Networks for Inverse Kinematics in Neurorobotics. *Patterns* **2022**, *3*, 100391. [CrossRef]
33. Fang, G.; Tian, Y.; Yang, Z.-X.; Geraedts, J.M.P.; Wang, C.C.L. Efficient Jacobian-Based Inverse Kinematics With Sim-to-Real Transfer of Soft Robots by Learning. *IEEEASME Trans. Mechatron.* **2022**, 1–11, *in press*. [CrossRef]
34. Phuoc, L.M.; Martinet, P.; Lee, S.; Kim, H. Damped Least Square Based Genetic Algorithm with Ggaussian Distribution of Damping Factor for Singularity-Robust Inverse Kinematics. *J. Mech. Sci. Technol.* **2008**, *22*, 1330–1338. [CrossRef]
35. Köker, R.; Çakar, T. A Neuro-Genetic-Simulated Annealing Approach to the Inverse Kinematics Solution of Robots: A Simulation Based Study. *Eng. Comput.* **2016**, *32*, 553–565. [CrossRef]
36. Qie, X.; Kang, C.; Zong, G.; Chen, S. Trajectory Planning and Simulation Study of Redundant Robotic Arm for Upper Limb Rehabilitation Based on Back Propagation Neural Network and Genetic Algorithm. *Sensors* **2022**, *22*, 4071. [CrossRef]
37. Liu, R.; Liu, C. Human Motion Prediction Using Adaptable Recurrent Neural Networks and Inverse Kinematics. *IEEE Control Syst. Lett.* **2021**, *5*, 1651–1656. [CrossRef]
38. Reinhart, R.F.; Steil, J.J. Reaching Movement Generation with a Recurrent Neural Network Based on Learning Inverse Kinematics for the Humanoid Robot ICub. In Proceedings of the 2009 9th IEEE-RAS International Conference on Humanoid Robots, Paris, France, 7–10 December 2009; pp. 323–330.
39. Yiyang, L.; Xi, J.; Hongfei, B.; Zhining, W.; Liangliang, S. A General Robot Inverse Kinematics Solution Method Based on Improved PSO Algorithm. *IEEE Access* **2021**, *9*, 32341–32350. [CrossRef]
40. Ghosh, A.; Singh, O.; Ray, A.K. Inverse Kinematic Solution of a 7 DOF Robotic Manipulator Using Boundary Restricted Particle Swarm Optimization. *IFAC PapersOnLine* **2022**, *55*, 101–105. [CrossRef]
41. Wang, D.; Wu, J.; Wang, L.; Liu, Y. A Postprocessing Strategy of a 3-DOF Parallel Tool Head Based on Velocity Control and Coarse Interpolation. *IEEE Trans. Ind. Electron.* **2017**, *65*, 6333–6342.
42. Larrañaga, A. 3D Printable Robotic Arm. Available online: https://github.com/AngelLM (accessed on 5 July 2022).
43. Denavit, J.; Hartenberg, R.S. A Kinematic Notation for Lower-Pair Mechanisms Based on Matrices. *J. Appl. Mech.* **1955**, *22*, 215–221. [CrossRef]
44. Corke, P.I.; Khatib, O. *Robotics, Vision and Control: Fundamental Algorithms in MATLAB.*; Springer: Berlin/Heidelberg, Germany, 2011; Volume 73, ISBN 3-642-20144-X.
45. Mostafa, S.A.; Ahmad, I.A. Remainder Linear Systematic Sampling with Multiple Random Starts. *J. Stat. Theory Pract.* **2016**, *10*, 824–851. [CrossRef]
46. Martínez-Blanco, M.d.R.; Ibarra-Pérez, T.; Olivera-Domingo, F.; Ortiz-Rodríguez, J.M. Optimization of Training Data Set Based on Linear Systematic Sampling to Solve the Inverse Kinematics of 6 DOF Robotic Arm with Artificial Neural Networks. In *Frontiers of Data and Knowledge Management for Convergence of ICT, Healthcare, and Telecommunication Services*; Paul, S., Paiva, S., Fu, B., Eds.; Springer International Publishing: Cham, Switzerland, 2022; pp. 85–112. ISBN 978-3-030-77558-2.
47. Gurney, K. *An Introduction to Neural Networks*, 1st ed.; CRC Press: Boca Raton, FL, USA, 1997; ISBN 1-315-27357-8.
48. Paneiro, G.; Rafael, M. Artificial Neural Network with a Cross-Validation Approach to Blast-Induced Ground Vibration Propagation Modeling. *Undergr. Space* **2021**, *6*, 281–289. [CrossRef]
49. Dereli, S.; Köker, R. IW-PSO Approach to the Inverse Kinematics Problem Solution of a 7-DOF Serial Robot Manipulator. *Sigma J. Eng. Nat. Sci.* **2018**, *36*, 77–85.

MDPI
St. Alban-Anlage 66
4052 Basel
Switzerland
www.mdpi.com

Applied Sciences Editorial Office
E-mail: applsci@mdpi.com
www.mdpi.com/journal/applsci

www.ingramcontent.com/pod-product-compliance
Lightning Source LLC
Chambersburg PA
CBHW061452220926
51850CB00004B/73
9783036586342